Energy Transfer Processes in Condensed Matter

NATO ASI Series

Advanced Science Institutes Series

A series presenting the results of activities sponsored by the NATO Science Committee, which aims at the dissemination of advanced scientific and technological knowledge, with a view to strengthening links between scientific communities.

The series is published by an international board of publishers in conjunction with the NATO Scientific Affairs Division

A Life Sciences **B Physics**	Plenum Publishing Corporation New York and London
C Mathematical and Physical Sciences	D. Reidel Publishing Company Dordrecht, Boston, and Lancaster
D Behavioral and Social Sciences **E Engineering and Materials Sciences**	Martinus Nijhoff Publishers The Hague, Boston, and Lancaster
F Computer and Systems Sciences **G Ecological Sciences**	Springer-Verlag Berlin, Heidelberg, New York, and Tokyo

Recent Volumes in this Series

Volume 111—Monopole '83
edited by James L. Stone

Volume 112—Condensed Matter Research Using Neutrons: Today and Tomorrow
edited by Stephen W. Lovesey and Reinhard Scherm

Volume 113—The Electronic Structure of Complex Systems
edited by P. Phariseau and W. M. Temmerman

Volume 114—Energy Transfer Processes in Condensed Matter
edited by Baldassare Di Bartolo

Volume 115—Progress in Gauge Field Theory
edited by G.'t Hooft, A. Jaffe, H. Lehmann, P. K. Mitter, I. Singer, and R. Stora

Volume 116—Nonequilibrium Cooperative Phenomena in Physics and Related Fields
edited by Manuel G. Velarde

Volume 117—Moment Formation in Solids
edited by W. J. L. Buyers

Series B: Physics

Energy Transfer Processes in Condensed Matter

Edited by

Baldassare Di Bartolo

Boston College
Chestnut Hill, Massachusetts

Assistant Editor

Aliki Karipidou

Boston College
Chestnut Hill, Massachusetts

Plenum Press
New York and London
Published in cooperation with NATO Scientific Affairs Division

Proceedings at a NATO Advanced Study Institute on Energy Transfer Processes in Condensed Matter, held June 16–30, 1983, in Erice, Sicily, Italy

Library of Congress Cataloging in Publication Data

NATO Advanced Study Institute on Energy Transfer Processes in Condensed Matter (1983: Erice, Italy)
Energy transfer processes in condensed matter.

(NATO ASI series. Series B, Physics; v. 114)
"Proceedings of a NATO Advanced Study Institute on Energy Transfer Processes in Condensed Matter, held June 16–30, 1983, in Erice, Sicily, Italy"—T.p. verso.
"Published in cooperation with NATO Scientific Affairs Division."
Includes bibliographical references and index.
1. Condensed matter—Congresses. 2. Energy transfer—Congresses. I. Di Bartolo, Baldassare. II. Karipidou, Aliki. III. Title. IV. Series.
QC173.4.C65N376 1983 530.4 84-17830

ISBN-13: 978-1-4612-9467-2 e-ISBN-13: 978-1-4613-2407-2
DOI: 10.1007/978-1-4613-2407-2

Softcover reprint of the hardcover 1st edition 1984
A Division of Plenum Publishing Corporation
233 Spring Street, New York, N.Y. 10013

FERD WILLIAMS

This book is dedicated to the memory of
Professor Ferd Williams

Quels livres valent la peine d'être écrits,
hormis les Mémoires?

André Malraux.

PREFACE

This book presents an account of the NATO Advanced Study Institute on "Energy Transfer Processes in Condensed Matter", held in Erice, Italy, from June 16 to June 30, 1983. This meeting was organized by the International School of Atomic and Molecular Spectroscopy of the "Ettore Majorana" Centre for Scientific Culture.

The objective of the Institute was to present a comprehensive treatment of the basic mechanisms by which electronic excitation energy, initially localized in a particular constituent or region of a condensed material, transfers itself to the other parts of the system. Energy transfer processes are important to such varied fields as spectroscopy, lasers, phosphor technology, artificial solar energy conversion, and photobiology. This meeting was the first encounter of this sort entirely dedicated to this important topic.

A total of 65 participants came from 47 laboratories and 16 nations (Belgium, Czechoslovakia, F.R. of Germany, France, Greece, India, Ireland, Israel, Italy, The Netherlands, Poland, Portugal, Switzerland, Turkey, United Kingdom, and the United States of A America). The secretaries of the course were: Ms. Aliki Karipidou for the scientific aspects and Mr. Massimo Minella for the administrative aspects of the meeting.

The Institute opened with an overview of energy transfer processes in various condensed matter systems. This task was the responsibility of the first lecturer. The next two lecturers presented the framework necessary to describe the physical processes resulting in energy deposition and transfer in condensed matter and the methods used to observe transfer phenomena. Once the basic principles were established, specific treatments of energy transfer in various systems were presented. A lecturer covered the case of energy transfer in nonconducting materials, another lecturer the case of energy transfer in solid rare gases etc. The subjects of energy transfer in ruby crystals, energy localization, energy transfer in glasses, laser systems, and organic crystals and solar energy conversion were also treated. Specific aspects of the theory were presented in seminar form together with several applications to various areas of current interest. This book includes also the contributions sent by Professors Ern, Duval and Powell who,unfortunately were not able to come.

I would like to thank for their help Dr. A. Gabriele, Ms. P. Savalli, and all the personnel of the "Ettore Majorana" organization in Erice, Prof. R. Uritam, Chairman of the Department of Physics at Boston College, Prof. V. Adragna, Dr. G. Denaro, Rag. M. Strazzera, and Avv. G. Luppino.

I would like to thank the members of the organizing committee (Professors Williams, Knox and Scharmann) and Prof. Kenkre for their valuable help and advice.

I am grateful to the secretaries of the course Ms. Karipidou and Mr. Minella whose collaboration during the course was especially valuable. I wish to acknowledge also Ms. Karipidou's work as patient and intelligent assistant editor of this book.

It is difficult to describe the spirit of true collaboration and the interest with which everybody participated in this Institute. It was a privilege to direct this meeting and to meet so many fine people who came from so different and various places and worked together in the congenial atmosphere of the town of Erice. I am already looking forward to the next 1985 meeting of the International School of Spectroscopy; I am sure I will see again many of you, my friends. Arrivederci a presto!

B. Di Bartolo
Editor and Director of
the Institute

Erice, June 1983

CONTENTS

LONG SEMINARS

SHORT SEMINARS

INTRODUCTION TO ENERGY TRANSFER AND RELEVANT SOLID-STATE CONCEPTS

J.E. Bernard, D.E. Berry and F. Williams

Physics Department
University of Delaware
Newark, Delaware 19711, U.S.A.

ABSTRACT

In order to describe energy transfer involving electronic states of condensed matter the many-body problem is first separated into electronic and nuclear problems, and then the many-electron problem is approximated by one-electron band structure. These analyses are reviewed. Because several mechanisms for electronic energy transfer depend on coupling with nuclear motion, lattice dynamics and electron-phonon interaction are also reviewed. Energy transfer is then considered rather broadly, and resonant and non-resonant transfer, excitonic transfer, Auger transitions, and charge transport by thermal or hot electrons leading to energy transfer, respectively, by capture or collisional excitation are introduced. A glossary of terms relevant to electronic energy transfer completes this introductory chapter.

I. INTRODUCTION

A well-known mechanism for the transfer of energy from one site to another in crystals, amorphous materials, solutions and biological systems is resonant energy transfer. This mechanism was explained quantum mechanically by Förster [1] in the dipole approximation for transfer between organic molecules. Dexter [2] generalized the theory to higher order interactions, including exchange interaction, and applied the theory to energy transfer between dopant ions in inorganic solids. In resonant transfer the energy donor in its excited electronic state equilibrates with the lattice modes before transfer occurs to the energy acceptor. The theory was developed for donor and acceptor which are weakly coupled to each other but each, separately, strongly coupled to the lattice. As a consequence of this strong electron-phonon interaction, energy can be conserved during the transfer. The transfer probability is found to depend on the overlap of the emission spectrum of the energy donor and the absorption spectrum of the energy acceptor. The basic interaction which drives the transfer is the electromagnetic coupling between the donor and acceptor. In the dipole approximation it is essentially the same as the Van der Waals interaction between gas molecules.

For dopants which are weakly coupled to the lattice the energy mismatch between donor and acceptor obviates transfer of electronic energy alone. Phonon creation and/or annihilation can take care of energy conservation during transfer. Single phonon, two phonon and multiphonon processes occur. Even for transfer between identical dopants weakly coupled to the lattice, thus with small or no energy mismatch, phonon assistance can be effective in transfer, as shown by Orbach [3] for quadrupole-quadrupole transfer in cases for which that multipole transfer is forbidden.

A well-known mechanism for electronic energy transfer in crystals is the exciton. This elementary excitation is an eigenstate of a periodic Hamiltonian, and thus its motion is distinguishable in most cases from resonant energy transfer. The exciton consists of a coupled electron and positive hole and therefore transfers no charge, only energy. Excitons divide into two types depending on the strength of the electron-hole coupling: Wannier [4] which are weakly coupled; Frenkel [5] which are strongly coupled. In principle their motion can be wave-like (coherent), however, diffusive motion is commonly observed for excitons. Even the tightly coupled Frenkel excitons move from site to site fast enough so that lattice relaxation around the excitation does not destroy the lattice periodicity. Otherwise the exciton is said to be self trapped, and pure excitonic transfer is less probable, though a hopping mechanism involving coordinated lattice and electronic dynamics can continue. It should be noted that in general the concept of energy transfer implies a subdivision of the total system, which is only clearly valid for weakly coupled donor and acceptor, though for strongly coupled components, as for some excitonic transport, the concept retains some validity.

In some cases the separation of resonant energy transfer and excitonic transport becomes difficult. If we have an exciton which moves carrying lattice polarization with it, then the exciton together with the polarization can be considered as a quasi-particle whose eigenstates are those of a periodic Hamiltonian. Transport of energy to dopants or defects can thus occur by motion of this quasi-particle but also by resonant transfer from the relaxed excitonic state. The range of the resonant transfer interaction can of course be included in the cross section for capture of excitons.

Energy transfer, interpreted broadly, includes many other mechanisms as well. The conduction electron and the valence band positive hole separately carry useful energy equal to the energy of the eigenstate in which each moves, measured from the Fermi level of the material. If these electronic particles are thermalized with the lattice at their respective band edges then they carry potential energy which can be transferred to dopants on their capture; if these electronic particles are not thermalized with the lattice, that is, are hot carriers, then they carry kinetic energy which can be transferred to dopants or deep centers by collisional excitation. Interchanges between kinetic and potential energy occur spatially at semiconductor heterojunctions, and even at semiconductor/liquid electrolyte junctions. Mobile electrons and positive holes which carry lattice polarization with them are polarons.

The mechanisms of energy transfer discussed so far are for transfer in position coordinate space. There are other mechanisms which involve energy transfer in momentum space. These include Auger mechanisms, as noted by Landsberg [6]. Transfer of hot electrons into the minima of higher conduction bands involves transfer of kinetic energy into potential energy in which there is transfer in wave vector (quasi-momentum) space.

The mechanisms can be subdivided in several ways: (1) by the continuous or discrete nature of the initial and final electronic states for the transfer; (2) by the vehicle for transfer being a charged particle or an uncharged elementary excitation; or (3) by the intrinsic or extrinsic character of the states and transitions, that is, characteristic of the perfect crystal or of dopants. In Table 1 we categorize various mechanisms of energy transfer based on (1) and (2).

Radiative and radiationless de-excitation are the dominant processes which compete with energy transfer. These were the subjects of two previous courses in this school [7], [8]and will not be discussed further.

Applications of energy transfer in condensed matter are technologically important. Fluorescent lamps utilize resonant

Table 1. Energy Transfer Mechanisms in Condensed Matter

Electronic States	Charged Particle Transfer	Excitation Transfer Only	Charged Particle plus Excitation
continuous ↔ continuous	electrons and positive holes-thermalized and hot	intraband (subband) transitions; excitons	heterojunction tunneling
discrete ↔ discrete	hopping; polaron	resonant energy transfer; phonon-assisted transfer	charge transfer
continuous ↔ discrete	tunneling	Auger excitation	collisional

transfer between dopants in phosphors. Cathode ray tubes and radiation detectors operate with excitonic intermediate states. Sensitization of the photographic process is basically energy transfer. Dyes and pigments are stabilized by de-excitation channels which depend on energy transfer. Photovoltaic and photoconversion solar devices utilize electronic charge or excitonic transport. Photosynthesis itself depends on energy transfer. Phenomena relevant to applications will be discussed in subsequent chapters.

In Section II we present analyses which provide some basic concepts prerequisite for developing these mechanisms. The separation of the many-body problem of condensed matter into a many-electron problem and a lattice or molecular dynamics problem is basic to crystals, amorphous materials, solutions and biological systems. The one-electron approximation to the many-electron problem is equally basic. Electronic band structure and lattice dynamics have largely developed for crystals utilizing the translational invariance of ordered lattices. Electron-phonon interaction plays a key role in several energy transfer mechanisms and is analyzed in the final subsection of Section II. In Section III the mechanisms just introduced and some others noted in Table 1 are treated in some detail. Finally, a glossary of terms used in analyses of energy transfer completes the chapter.

II. BASIC CONCEPTS UNDERLYING ENERGY TRANSFER IN SOLIDS

II. A. Separation of Electronic and Nuclear Motion

It is useful, when dealing with many-body problems in which the particles fall into two or more subgroups in which the motion occurs on widely different time scales, to try to separate the problem of determining the motion (or states) of one subgroup from that of the other(s). For example, in molecules and solids it is common for the electrons to be moving at much higher speeds than the nuclei, due to the large difference in their masses. It is reasonable to assume then that the electrons are to a large extent able to follow smoothly the motion of the nuclei without undergoing abrupt transitions between states.

Indeed, the electronic sub-system can be viewed as having a Hamiltonian which is slowly varying in time, and the adiabatic theorem of quantum mechanics assures that in the limit of infinitely slow time dependence, transitions among non-degenerate states do not occur. A discussion of this theorem can be found in any quantum mechanics text; the original proof, under the assumptions of discreteness of the spectrum and non-degeneracy except at crossing points, is due to Born and Fock [9], and a more general proof avoiding these assumptions has been given by Kato [10]. It is this theorem which underlies the adiabatic approximation, a commonly used technique for separating electronic and nuclear motions. A closely related method known as the Born-Oppenheimer [11],[12] approximation can be used to accomplish the same thing via a perturbation expansion without explicit reference to the notion of adiabaticity. Variations on these can be used to obtain more accurate (or less accurate) solutions based on separation of the electronic and nuclear motions. We should also like to point out that these techniques are by no means limited to separation into two sub-systems. It is possible to effect separation into more sub-systems, such as is sometimes done in the case of solids with local modes due to impurities, which may have characteristic time constants intermediate between those of the electrons and the lattice modes

The derivation we present here is due to Born [13], and is readily accessible in a number of other works [12],[14].

The Hamiltonian for a molecule or solid can be written as the sum of contributions due to the electrons and the nuclei (or, more practically, ions consisting of the nuclei plus core electrons)

$$H = T_N + V_N + T_{el} + V_{el} + V_{N-el}, \qquad (1)$$

where T_N is the nuclear (ionic) kinetic energy operator, V_N the nuclear potential energy, T_{el} and V_{el} the corresponding electronic quantities, and V_{N-el} the interaction energy between the electrons and the nuclei. It is common to group the electronic kinetic energy and all the potential energy into the electronic Hamiltionian

$$H_{el} = T_{el} + V_T, \tag{2}$$

where

$$V_T = V_{el} + V_{N-el} + V_N \tag{3}$$

is the total potential energy of the system. This defines the problem of determining the states of the electrons in the field of static nuclei, where the nuclear positions serve as parameters

$$H_{el}(r;R)\psi_\alpha(r;R) = E_\alpha(R)\psi_\alpha(r;R) \tag{4}$$

where $r = (r_1,\ldots,r_n)$ denotes the 3n electron coordinates, $R = (R_1,\ldots,R_N)$ the 3N nuclear coordinates, and α the complete set of quantum numbers for the electronic problem. We should point out that spin-dependent interactions, such as the spin-orbit interaction are sometimes included in H_{el} when they are expected to be significant. It is assumed that the solutions of (4) are known for each set of values of the nuclear coordinates. If in fact the motion of the nuclei were infinitely slow, the adiabatic theorem would guarantee that (neglecting perturbations due to external influences and approximations made in solving the electronic problem) there would be no transitions between the electronic states characterized by different values of the quantum number α, and the motion of the nuclei could be described classically.

It is, however, important to assess the degree to which the finite speed of the nuclei affects the electronic state (causing transitions), and to determine the actual state of the nuclei, which move in a potential in part determined by the electronic state. To do this, we write down an expression for the wave function of the entire system in the form of an expansion in terms of the electronic wave functions taken as a complete set where the expansion coefficients $\chi_\alpha(R)$ are (in the adiabatic limit) the nuclear wave functions.

$$\Psi(r,R) = \sum_\alpha \chi_\alpha(R)\psi_\alpha(r;R). \tag{5}$$

Then the Schrödinger equation for the complete system is

$$[H-E_{total}]\Psi(r,R) = [H_{el}(r,R) + T_N(R) - E_{total}] \sum_\alpha \chi_\alpha(R)\psi_\alpha(r;R)$$

$$= 0. \qquad (6)$$

Taking account of equation (4), multiplying by $\psi_\beta^*(r;R)$, and integrating over the electronic coordinates results in a set of coupled equations for the coefficients χ_α (R).

$$[T_N + E_\beta - E_{total}]\chi_\beta(R) + \sum_\alpha C_{\beta\alpha}\chi_\alpha(R) = 0 \qquad (7)$$

where

$$C_{\beta\alpha} := A_{\beta\alpha} + B_{\beta\alpha}$$

$$A_{\beta\alpha} := - \sum_{i=1}^{N} \frac{\hbar^2}{M_i} \int \psi_\beta^*(r;R)\nabla_{R_i}\psi_\alpha(r;R)\,dr\cdot\nabla_{R_i}$$

$$B_{\beta\alpha} := \int \psi_\beta^*(r;R)T_N\psi_\alpha(r;R)\,dr. \qquad (8)$$

Now let us consider the $C_{\beta\alpha}$ in more detail. Differentiation of the orthonormality condition for electronic wave functions gives

$$\nabla_{R_i} \int \psi_\beta^*(r;R)\psi_\alpha(r;R)\,dr$$

$$= \int \psi_\beta^*(r;R)\nabla_{R_i}\psi_\alpha(r;R)\,dr + \int \psi_\alpha(r;R)\nabla_{R_i}\psi_\beta^*(r;R)\,dr$$

$$= \nabla_{R_i}\delta_{\alpha\beta} = 0, \text{ for all } i \text{ and all } \beta,\alpha. \qquad (9)$$

This is a statement of the conservation of particle number for small variations in the R_i. In particular, for $\beta = \alpha$

$$\mathrm{Re}[\int \psi_\beta^*(r;R)\nabla_{R_i}\psi_\alpha(r;R)\,dr] = 0, \quad \text{for all } i. \qquad (10)$$

In the absence of a magnetic field, it is always possible to write the ψ_α as real functions, so that the diagonal elements of A vanish in that case. The off-diagonal elements of C can be taken as a measure of the electron-phonon coupling and can be neglected in a first approximation if they are small compared to the separation between the electronic energy levels. This is a relatively good approximation in the case of many molecules and insulating and non-degenerate semiconducting crystals, but not in the case of metals and degenerate semiconductors, where there are continuous bands of accessible levels. In Section IIE below we consider various means of dealing with these neglected terms in the cases where they are not negligible.

It is common to keep the diagonal elements of B as part of the effective potential in which the nuclei move. It is, in fact, possible to retain the off-diagonal elements of B in the electronic Hamiltonian as well, since they, unlike the matrix elements of A, are not operators. Then, only the off-diagonal elements of A serve to effect transitions among the electronic states. It is, however, more usual in the literature to find the off-diagonal elements of B neglected entirely, with those of A being treated as the electron-phonon coupling. The term "adiabatic approximation" is most often applied to the approximation in which the diagonal elements of B are included in the electronic Hamiltonian, and all other elements of C are neglected rather than the one in which all matrix elements of C are neglected. Either form of the approximation can be obtained by starting with the assumption that the total wave function is a simple product of an electronic function and a nuclear function, so it is common simply to begin any treatment involving an adiabatic approximation by writing the wavefunction as a simple product

$$\Psi(r,R) = \chi(R)\psi(r;R). \qquad (11)$$

It is perhaps worth stressing that there is considerable variation in the literature in the meaning of the term "adiabatic approximation". We have tried to point out several of the variations, but there are surely more.

II.B. One-electron Approximation

In this section we outline the approximations and assumptions made in going from the full many-electron problem to the one-electron approximation to this problem, the latter of which is the starting point in the calculation of the electronic energies and wavefunctions for condensed materials. We assume that the adiabatic approximation has been made and we are now interested in the solutions of the electronic problem of the adiabatic approximation.

We consider the problem in which the ions are located at their equilibrium positions and in which there are N ions and n electrons. Then the electronic problem is the many-electron problem defined by the following Schrödinger equation:

$$H_{el}\psi_\alpha(r;R) = E_\alpha(R)\psi_\alpha(r;R) \tag{12}$$

where

$$H_{el} = \sum_j \frac{p_j^2}{2m} + \sum_{i<j} \frac{e^2}{|r_i - r_j|} - \sum_{i,J} \frac{e^2}{|r_i - R_J|}$$

where $r = (r_1,\ldots,r_j,\ldots,r_n)$; r_j is a three vector locating the jth electron, $R = (R_1,\ldots,R_J,\ldots,R_N)$; R_J is a three vector locating the Jth ion, $\alpha = (\alpha_1,\ldots,\alpha_\nu,\ldots,\alpha_{n'})$; α is the complete set of quantum numbers indexing the different eigenstates of the many-electron problem which has been subdivided into sets of quantum numbers α_ν where ν indexes the different subsets of quantum numbers, and p_j is the momentum of the jth electron.

The formal problem described by equation (12) for the case of a crystal or other condensed phase of matter, including large molecules, is not in a tractable form and, therefore, we must make some approximations to the many-electron problem. We begin by using the Hartree-Fock approximation, in which $\psi_\alpha(r;R)$ is represented by a Slater determinant wavefunction composed of the one-electron wave functions

$$\phi_{\alpha_\nu}(r_j;R)S_{\sigma_\nu}(\zeta_j).$$

Here, $\phi_{\alpha_\nu}(r_j;R)$ represents the space portion of the wavefunction of the jth electron which is in an eigenstate indexed by the set of quantum numbers α_ν. The spin portion of the wavefunction of this electron is $S_{\sigma_\nu}(\zeta_j)$ which is indexed by the quantum number σ_ν. In what follows we will assume, as we have implictly done in equation (12), that there are no spin-spin nor spin-orbit interactions. Therefore, the spin portion of the wavefunction will not enter into the final equations of the one-electron approximation and the quantum number sets α_ν for the space portion of the electronic states will be doubly degenerate. Note that the particular one-electron wavefunctions to be used in the Slater determinant wavefunction depend on the state of electronic excitation of the material being described.

We begin by subdividing the electrons into two subgroups designated as the core electrons for which $j=n'+1,\ldots,n$ and non-core electrons for which $j=1,\ldots,n'$. The conditions met by the core and non-core electrons which define this division will be specified later. If, for the many-electron wavefunction in equation (12), we substitute the approximate wavefunction given by the Slater determinant and if we keep track of the core and non-core electrons, we then get:

For the non-core one-electron states:

$$\frac{p_j^2}{2m}\phi_{\alpha_\nu}(r_j) - \sum_{J=1}^{N} \frac{e^2}{|r_j-R_J|}\phi_{\alpha_\nu}(r_j) + \sum_{\mu=n'+1}^{n} e^2\left[\int d^3r_o \frac{\phi^*_{\alpha_\mu}(r_o)\phi_{\alpha_\mu}(r_o)}{|r_j-r_o|}\phi_{\alpha_\nu}(r_j) - \int d^3r_o \frac{\phi^*_{\alpha_\mu}(r_o)\phi_{\alpha_\nu}(r_o)}{|r_j-r_o|}\phi_{\alpha_\mu}(r_j)\right] \tag{13}$$

$$+ \sum_{\substack{\mu=1\\ \mu\neq\nu}}^{n'} e^2\left[\int d^3r_o \frac{\phi^*_{\alpha_\mu}(r_o)\phi_{\alpha_\mu}(r_o)}{|r_j-r_o|}\phi_{\alpha_\nu}(r_j) - \int d^3r_o \frac{\phi^*_{\alpha_\mu}(r_o)\phi_{\alpha_\nu}(r_o)}{|r_j-r_o|}\phi_{\alpha_\mu}(r_j)\right] = \varepsilon_{\alpha_\nu}\phi_{\alpha_\nu}(r_j)$$

for $1 \leq \nu \leq n'$.

For the core one-electron states:

$$\frac{p_j^2}{2m}\phi_{\alpha_\nu}(r_j) - \sum_{J=1}^{N} \frac{e^2}{|r_j-R_J|}\phi_{\alpha_\nu}(r_j)$$

$$+ \sum_{\substack{\mu=n'+1\\ \mu\neq\nu}}^{n} e^2\left[\int d^3r_o \frac{\phi^*_{\alpha_\mu}(r_o)\phi_{\alpha_\mu}(r_o)}{|r_j-r_o|}\phi_{\alpha_\nu}(r_j) - \int d^3r_o \frac{\phi^*_{\alpha_\mu}(r_o)\phi_{\alpha_\nu}(r_o)}{|r_j-r_o|}\phi_{\alpha_\mu}(r_j)\right] +$$

$$+ \sum_{\mu=1}^{n'} e^2 \left[\int d^3 r_o \frac{\phi^*_{\alpha_\mu}(r_o)\phi_{\alpha_\mu}(r_o)}{|r_j - r_o|} \phi_{\alpha_\nu}(r_j) \right. \quad (14)$$

$$\left. - \int d^3 r_o \frac{\phi^*_{\alpha_\mu}(r_o)\phi_{\alpha_\nu}(r_o)}{|r_j - r_o|} \phi_{\alpha_\mu}(r_j) \right] = \varepsilon_{\alpha_\nu} \phi_{\alpha_\nu}(r_j)$$

for $n' + 1 \leq \nu \leq n$.

Here we have suppressed the parametric dependence of the wavefunctions and eigenvalues on the nuclear coordinates R and $\varepsilon_{\alpha_\nu} = \varepsilon_{\alpha_\nu}(R)$ are the one-electron energy eigenvalues. We can define the following quantity:

$$\varepsilon_{\nu core} = \sum_{\mu=1}^{n'} e^2 \int d^3 r_j \int d^3 r_o \phi^*_{\alpha_\nu}(r_j)\phi^*_{\alpha_\mu}(r_o)$$

$$\times \left[\frac{\phi_{\alpha_\mu}(r_o)\phi_{\alpha_\nu}(r_j) - \phi_{\alpha_\nu}(r_o)\phi_{\alpha_\mu}(r_j)}{|r_j - r_o|} \right] \quad (15)$$

for $n' + 1 \leq \nu \leq n$.

This is just the energy of interaction between the νth core electron and all the non-core electrons. We then define the division between core and non-core one-electron states by the condition that

$$\varepsilon_{\nu core} << \varepsilon_{\alpha_\nu}$$

for $n' + 1 \leq \nu \leq n$.

This condition is just the condition that the interaction of the non-core electrons with a core electron is small compared to the energy of the given core electron. We usually make a single division into core/non-core one-electron states by using the subdivision made for the ground state problem for the excited states as well.

We can decouple the problem of finding the core one-electron states from the problem of finding the non-core states by neglecting the terms corresponding to $\varepsilon_{\nu core}$ in equation (14).

We then can assume in what follows that we have found a solution to equation (14). Usually it is assumed that a good approximation for one of these one-electron states is just a linear combination of the corresponding one-electron atomic orbitals. These linear combinations are chosen so that the translational symmetry is satisfied. For further details on this point see the discussion on linear combinations of atomic orbitals done in connection with the tight binding approximation discussed later in this subsection. In connection with this approximation, the core electrons are then assumed to be the non-valence electrons of the atoms. Since these electrons are so localized they do not overlap with one another very much in the condensed phase and, therefore, do not perturb one another in this phase. We can now rewrite equation (13) for the non-core one-electron states as:

$$\frac{p_j^2}{2m}\phi_{\alpha_\nu}(r_j) + V_o(r_j)\phi_{\alpha_\nu}(r_j) - \sum_{\mu=n'+1}^{n} e^2 \int d^3r_o \frac{\phi^*_{\alpha_\mu}(r_o)\phi_{\alpha_\nu}(r_o)}{|r_j-r_o|}\phi_{\alpha_\mu}(r_j) \tag{16}$$

$$+ \sum_{\substack{\mu=1\\ \mu\neq\nu}}^{n'} e^2\left[\int d^3r_o \frac{\phi^*_{\alpha_\mu}(r_o)\phi_{\alpha_\mu}(r_o)}{|r_j-r_o|}\phi_{\alpha_\nu}(r_j) + \int d^3r_o \frac{\phi^*_{\alpha_\mu}(r_o)\phi_{\alpha_\nu}(r_o)}{|r_j-r_o|}\phi_{\alpha_\mu}(r_j)\right] = \varepsilon_{\alpha_\nu}\phi_{\alpha_\nu}(r_j)$$

for $1 \le \nu \le n'$,

where

$$V_o(r_j) = -\sum_{J=1}^{N}\frac{e^2}{|r_j-R_J|} + \sum_{\mu=n'+1}^{n} e^2\int d^3r_o \frac{\phi^*_{\alpha_\mu}(r_o)\phi_{\alpha_\mu}(r_o)}{|r_j-r_o|} .$$

We then make the additional approximation that we can neglect the exchange interaction between the core electrons and the non-core electrons, that is, that we can neglect the third term on the left hand side of equation (16). Finally, we also neglect, in the one-electron approximation, all correlation between the non-core electrons. That is, the last two terms on the left hand side of equation (16) are neglected. With these assumptions we then get to the one-electron approximation for equation (12) which is

$$\frac{p_j^2}{2m}\phi_{\alpha_\nu}(r_j) + V_o(r_j)\phi_{\alpha_\nu}(r_j) = \varepsilon_{\alpha_\nu}\phi_{\alpha_\nu}(r_j) \qquad (17)$$

for $1 \leq \nu \leq n'$.

Here $V_o(r_j)$, the one-electron potential, can be approximated by the sum of the potentials of the ions forming the solid. That is, by:

$$V_o(r_j) = \sum_J V_{ionJ}(r_j - R_J)$$

where $V_{ionJ}(r_o)$ is the electrostatic potential of the Jth ion formed from the Jth atom by removal of its valence electrons. Equation (17) is a set of simple decoupled potential equations to solve for the one-electron states in the one-electron approximation. Note that the one-electron states are introduced into the calculation by the Hartree-Fock approximation but the one-electron approximation makes additional assumptions

We can improve on the one-electron approximation by making a slightly different set of approximations in equation (16). First we replace the two exhange terms, that is, the third and last terms on the left hand side of equation (16), by an approximation having the form of a self-consistent potential, instead of neglecting these terms as we did in the one-electron approximation. Slater [15] has shown that we can approximate equation (16).

$$\frac{p_j^2}{2m}\phi_{\alpha_\nu}(r_j) + V_o(r_j)\phi_{\alpha_\nu}(r_j)$$

$$- \beta_{ex}\left[\sum_{\mu=n'+1}^{n} \phi^*_{\alpha_\mu}(r_j)\phi_{\alpha_\mu}(r_j)\right]^{1/3}\phi_{\alpha_\nu}(r_j)$$

$$+ \sum_{\substack{\mu=1 \\ \mu\neq\nu}}^{n'} e^2 \int d^3r_o \frac{\phi^*_{\alpha_\mu}(r_o)\phi_{\alpha_\mu}(r_o)}{|r_j - r_o|}\phi_{\alpha_\nu}(r_j) + \qquad (18)$$

$$-\beta_{ex}\left[\sum_{\substack{\mu=1\\ \mu\neq\nu}}^{n'} \phi^*_{\alpha_\mu}(r_j)\phi_{\alpha_\mu}(r_j)\right]^{1/3}\phi_{\alpha_\nu}(r_j) = \varepsilon_{\alpha_\nu}\phi_{\alpha_\nu}(r_j)$$

for $1 \leq \nu \leq n'$

where Slater [15] has determined that $\beta_{ex} = e^2(81/8\pi)^{\frac{1}{2}}$. Kohn and Sham [16] have disagreed on the value of β_{ex}, contending that it should depend on the details of which states are occupied. In equation (18) we have also kept the self-consistent coulomb correlation term from the non-core electrons. These are coupled one-electron equations and are, therefore, not as easily solved as equation (17). A first approximation to the solution of equation (18) can be gotten from equation (17).

Finally, we note that the total energy of the non-core electrons in the one-electron approximation is given by

$$E = \sum_{\nu=1}^{n'} \varepsilon_{\alpha_\nu} - \sum_{\substack{\nu,\mu\\ \nu<\mu}}^{n'} e^2 \int d^3r_0 \int d^3r_1 \times \frac{\phi^*_{\alpha_\mu}(r_0)\phi_{\alpha_\mu}(r_0)\phi^*_{\alpha_\nu}(r_1)\phi_{\alpha_\nu}(r_1)}{|r_1-r_0|} \tag{19}$$

$$+ \beta_{ex}\sum_{\nu=1}^{n'}\int d^3r_0\left[\sum_{\substack{\mu=1\\ \mu\neq\nu}}^{n}\phi^*_{\alpha_\mu}(r_0)\phi_{\alpha_\mu}(r_0)\right]^{1/3} \times \phi^*_{\alpha_\nu}(r_0)\phi_{\alpha_\nu}(r_0).$$

II.C. Electronic Band Structure

In this subsection we consider properties of the solutions to the equations of the one-electron approximation for a three dimensional periodic array of ions. In particular, we want to determine some of the general properties of the one-electron eigenvalues and eigenfunctions for the set of equations

$$H_{el}\phi_{\alpha_\nu}(r_o) = \frac{p_o^2}{2m}\phi_{\alpha_\nu}(r_o) + V(r_o)\phi_{\alpha_\nu}(r_o) = \varepsilon_{\alpha_\nu}\phi_{\alpha_\nu}(r_o) \tag{20}$$

for $1 \leq \nu \leq n$ = number of valence electrons,

where $V(r_o)$ is the algebraic sum of the electrostatic potentials due to the ions forming the three dimensional array.

In order to investigate the properties of ε_{α_ν} and $\phi_{\alpha_\nu}(r_o)$, we must first develop some notation to describe the array and the space of eigenvalues $\{\alpha_\nu\}_{\nu=1}^{n}$. The properties of the array can be described in terms of the primitive basis vectors $\hat{a}_i$, i = 1,2,3 and the concept of a unit cell of an array. The unit cell must satisfy the condition that it is a subdivision of the array such that the space occupied by the array can be completely filled by identical, non-overlapping unit cells. By identical unit cells, is meant that if we did the gedanken experiment of cutting the array into the non-overlapping unit cells then, by translations only, the different unit cells could be moved to a common place and that all the ions and/or parts of ions in each of the unit cells would be at the same place after the translation. We define the unit cell of an array as the smallest unit of the array which meets the above condition. We define the three primitive basis vectors as follows: We choose a point in one of the unit cells and then we consider all the vectors connecting this point to all the equivalent points in all of the other unit cells composing the array. From this set of vectors is chosen a set of vectors which are the shortest non-pairwise collinear vectors.

We then define the direct Bravais lattice as the lattice composed of the points given by

$$X_n = n_1\hat{a}_1 + n_2\hat{a}_2 + n_3\hat{a}_3, \quad -\frac{N_i}{2} \leq n_i \leq \frac{N_i}{2} . \tag{21}$$

Here n_i and N_i are integers. The number of unit cells, N, in the array is given in terms of the N_i as $N = N_1N_2N_3$ and the number of ions in the array is N times the number of ions per unit cell. The X_n designate the lattice vectors. Note there is one point in the direct Bravais lattice for each unit cell in the array; however, since the unit cells can contain more than one ion, the number of points in the direct Bravais lattice is not in general equal to the number of ions in the array. In addition, every vector X_n corresponds to the directed distance between equivalent points for a given pair of unit cells of the array. And the directed distance between equivalent points of any two unit cells is equal to an X_n.

If we define a set of vectors Y_γ; $\gamma = 1,2,\ldots,\Gamma$ = number of ions in a unit cell; which locate the ions in the unit cell relative to one of the ions of the unit cell, then the potential $V(r_o)$ of equation (20) is defined in terms of the ionic potentials $V_\gamma(r_o)$, as

$$V(r_o) = \sum_{n_1,n_2,n_3} V_{cell}(r_o - X_n) , \qquad (22)$$

where

$$V_{cell}(r_o) = \sum_{\gamma}^{\Gamma} V_\gamma(r_o - Y_\gamma). \qquad (23)$$

This definition of $V(r_0)$ is in accordance with the general philosophy of the division into core and non-core states of the one-electron approximation, in that we assumed that the core states, which in this case we take to be the states of the electrons of each ion, are not affected significantly by the formation of the crystal.

In order to make this study of the properties of the wavefunctions and eigenvalues a tractable problem, we must have some convenient way of handling the boundary conditions met by the wave functions at the crystal surfaces. We will assume that the array is large enough so that only a few of the one-electron states are affected to any significant degree by the boundaries and that these states are localized near the surfaces of the array, and therefore, for these few states a perturbation-like calculation can be used to determine their properties. The perturbation is the boundary conditions at the surfaces of the array. We therefore idealize the array either as: (A) an array which is infinite in all directions so that there are no surfaces for the array or (B) we identify each pair of surfaces perpendicular to each of the directions defined by the three primitive basis vectors, $\hat{a}_i$. This latter idealization forms a cyclic finite array containing N unit cells such that a translation by $X_{N_i} = N_i\hat{a}_i$, $i = 1,2,3$, returns us to the same point in the array. This way of dealing with the problem of the boundary conditions is known as the Born-von Karman boundary conditions. We will use both idealizations in our discussion of the solutions to equation (20)

Using either of the given idealizations and the form of $V(r_0)$ given in equations (22) and (23), we see that

$$V(r_o) = V(r_o + X_n) \qquad (24)$$

where X_n is given by equation (21).

We can define a second Bravais lattice called the reciprocal Bravais lattice as follows: Its primitive basis vectors are defined as

$$b_i = 2\pi\, \varepsilon_{ijk} \frac{\hat{a}_j x \hat{a}_k}{\hat{a}_1 \cdot \hat{a}_2 x \hat{a}_3} \tag{25}$$

from which it is found that $\hat{a}_i \cdot \hat{b}_j = 2\pi\delta_{ij}$. Note that within the definition of the primitive basis vectors of the reciprocal lattice given here there is an additional factor of 2π as compared to the definition used in scattering theory and that which we will use in connection with our discussion of phonons. The reciprocal Bravais lattice is composed of the points given by the vectors

$$K_\nu = \nu_1 \hat{b}_1 + \nu_2 \hat{b}_2 + \nu_3 \hat{b}_3, \; -\frac{N_i}{2} \leq \nu_i \leq \frac{N_i}{2} \tag{26}$$

where the ν's are integers. We define the Brillouin zone boundary in the reciprocal lattice space as the planes defined by the condition

$$k_o \cdot K_\nu + \tfrac{1}{2} K_\nu^{\,2} = 0.$$

There is a Brillouin zone boundary defined for each K_ν given above. The first Brillouin zone is the portion of the reciprocal lattice space such that the line joining the origin and any point in the first Brillouin zone does not intersect any of the zone boundaries.

We begin the considerations of the solutions to equation (20) with the translational symmetry of the Hamiltonian. Given any lattice vector X_n, then

$$H_{el}(r_o) = H_{el}(r_o + X_n) \equiv T_{X_n} H_{el}(r_o) T^+_{X_n} \tag{27}$$

where we have defined the unitary operators, T_{X_n}, as the translation operators corresponding to the translations through distances X_n. First note that the operators corresponding to any two translations commute with one another. Since the operators T_{X_n} commute with each other and with the Hamiltonian of equation (20) then there exists a set of solutions which are simultaneous solutions to equation (20) and the following set of equations:

$$T_{X_n} \phi_{\alpha_\nu}(r_o) \equiv \phi_{\alpha_\nu}(r_o + X_n) = \tau_{\alpha_\nu}(X_n)\phi_{\alpha_\nu}(r_o). \tag{28}$$

Note there is one equation for each X_n. We begin by finding the form of all the solutions to the equation set (28). By using the normalization condition and equation set (28) plus the unitarity of the T_{X_n} operators we have

$$\int d^3r_o |\phi_{\alpha_\nu}(r_o)|^2 = \int d^3r_o \phi^*_{\alpha_\nu}(r_o) T^+_{X_n} T_{X_n} \phi_{\alpha_\nu}(r_o) =$$

$$= \int d^3r_o |\phi_{\alpha_\nu}(r_o + X_n)|^2$$

$$= |\tau_{\alpha_\nu}(X_n)|^2 \int d^3r_o |\phi_{\alpha_\nu}(r_o)|^2.$$

This gives

$$|\tau_{\alpha_\nu}(X_n)|^2 = 1. \tag{29}$$

From the group properties of the T_{X_n}, that is, that

$$T_{X_n} T_{X_m} = T_{X_n+X_m},$$

it follows that

$$\tau_{\alpha_\nu}(X_n)\tau_{\alpha_\nu}(X_m) = \tau_{\alpha_\nu}(X_n + X_m). \tag{30}$$

Then from equation (29), it is found that $\tau_{\alpha_\nu}(X_n) = \exp[it_{\alpha_\nu}(X_n)]$ where $t_{\alpha_\nu}(X_n)$ is a real function of the X_n, and from equation (30) that

$$t_{\alpha_\nu}(X_n) + t_{\alpha_\nu}(X_m) = t_{\alpha_\nu}(X_n + X_m).$$

This implies that $t_{\alpha_\nu}(X_n)$ has the form

$$t_{\alpha_\nu}(X_n) = k_\nu \cdot X_n$$

for some k_ν.

Therefore, we have that

$$\phi_{\alpha_\nu}(r_o + X_n) = e^{ik_\nu \cdot X_n} \phi_{\alpha_\nu}(r_o). \tag{31}$$

This can be satisfied for all X_n if $\phi_{\alpha\nu}(r_o)$ is chosen to have the form

$$\phi_{k_\nu \beta_\nu}(r_o) = e^{ik_\nu \cdot r_o} u_{k_\nu \beta_\nu}(r_o) \tag{32}$$

where $u_{k}\beta_\nu$ (r_o) has the periodicity of the lattice. We have taken the set of quantum numbers, $\{\alpha_\nu\}$, for the νth electronic state to be $\{k_\nu, \beta_\nu\}$. Equations (31) or (32) are just two ways of stating Bloch's Theorem for the solutions of equation (20) when $V(r_o)$ is periodic. We can choose the k_ν's to be contained in the reciporcal lattice space. If we do so, then the k_ν's have the form

$$k_\nu = k_{\nu_1} \hat{b}_1 + k_{\nu_2} \hat{b}_2 + k_{\nu_3} \hat{b}_3$$

where the k_{ν_i} are not necessarily integers.

All of the representations of the translation group are one dimensional and given by $\exp[ik_\nu \cdot X_n]$. Therefore, two solutions of the set of equations (28) are the same if exp $[ik_\nu \cdot X_n]$ has the same value for all X_n's. Consider two solutions indexed by k_ν and $k_{\bar{\nu}}$, if

$$k_\nu \cdot X_n = k_{\bar{\nu}} \cdot X_n + 2\pi \eta_n \tag{33}$$

for all X_n where η_n are integers, then

$$e^{ik_\nu \cdot X_n} = e^{ik_{\bar{\nu}} \cdot X_n}$$

for all X_n.

Now if

$$k_{\bar{\nu}} = k_\nu + K_L$$

where K_L is a reciprocal lattice vector, then equation (33) is satisfied. If we choose the set of k_ν's that index the independent representations of the translation group to be those k_ν's nearest the origin of the reciprocal space, then all of the independent representations will be indexed by k_ν's in the first Brillouin zone. And every k_ν in the first Brillouin zone indexes an independent representation. Finally, it is noted that any solution of equation (20) must be a solution of the equation set (28).

If the Born-von Karman boundary conditions are used, then we must satisfy the additional condition,

$$e^{ik_\nu \cdot X_{N_i}} = 1, \quad X_{N_i} = N_i a_i$$

which implies that

$$k_\nu \cdot X_{N_i} = 2\pi \nu_i$$

where the ν_i are integers, which gives

$$k_\nu \cdot X_{N_i} = 2\pi k_{\nu_i} N_i = 2\pi \nu_i .$$

From which it is found that the k_ν's for the independent representations are

$$k_\nu = \frac{\nu_1}{N_1} \hat{b}_1 + \frac{\nu_2}{N_2} \hat{b}_2 + \frac{\nu_3}{N_3} \hat{b}_3, \; -\frac{N_i}{2} \leq \nu_i \leq \frac{N_i}{2} . \qquad (34)$$

We now turn to the properties of the solutions of equation (20). We will consider two limiting cases: case I for nearly free electrons and case II for tightly bound electrons. For more details on each of these cases see, for example, the book by J. D. Patterson [17].

1. Case I: Nearly Free Electrons. In this case, we assume that the non-core electrons are loosely bound to the array. The equations to be solved are

$$[-\frac{\hbar^2}{2m} \nabla_o^2 + V(r_o)]\phi_{k_\nu \beta_\nu}(r_o) = \varepsilon_{\beta_\nu}(k_\nu)\phi_{k_\nu \beta_\nu}(r_o) \qquad (35)$$

where

$$\phi_{\mathbf{k}_\nu \beta_\nu}(r_o) = \exp[ik_\nu \cdot r_o] u_{k_\nu \beta_\nu}(r_o) . \qquad (36)$$

Recall that

$$u_{k_\nu \beta_\nu}(r_o + X_n) = u_{k_\nu \beta_\nu}(r_o).$$

Kittel [18] has given a proof of the theorem which states that any function f(r) which has the periodicity of the array can be expanded in a Fourier series of the reciprocal lattice vectors, K_ν. That is, if

$$f(r_o + X_n) = f(r_o)$$

then

$$f(r_o) = \sum_m f(K_m)\exp[iK_m \cdot r_o].$$

Since both $u_{k_\nu \beta_\nu}(r_o)$ and $V(r_o)$ have the periodicity of the array, we can expand both in Fourier series as follows:

$$u_{k_\nu \beta_\nu}(r_o) = \sum_m u(K_m)\exp[iK_m \cdot r_o] \quad (37)$$

$$V(r_o) = \sum_m V(K_m)\exp[iK_m \cdot r_o] , \quad (38)$$

where

$$K_m = m_1\hat{b}_1 + m_2\hat{b}_2 + m_3\hat{b}_3$$

The m_i's are integers. Substitution of equations (36),(37) and (38) into equation (35) gives the following set of coupled equations

$$\left[\frac{\hbar^2}{2m}|k_\nu + K_n|^2 - \varepsilon(k_\nu) + V(0)\right]u(K_n) = -\sum_{m\neq 0} V(K_m)u(K_n - K_m) \quad (39)$$

where $K_o = 0$.

By nearly-free electrons is meant electrons in a potential meeting the condition that

$$V(K_m) << V(0) \quad \text{when } K_m \neq K_o.$$

We begin solving equation (39) by considering those k_ν such that $u(K_m) << u(0)$, then equation (39) for $K_n = 0$ is

$$\left[\frac{\hbar^2}{2m}k_\nu^2 - \varepsilon(k\ ) + V(0)\right]u(0) = -\sum_{m\neq 0} V(K_m)u(-K_m). \quad (40)$$

The right hand side of this equation is small, since both $V(K_m)$ and $u(-K_m)$ for $K_m \neq K_o$ are assumed to be small. Since, at least one of the u's must be large, we assume that u(0) is not small. Therefore, we must have

$$\varepsilon(k_\nu) = V(0) + \frac{\hbar^2 k_\nu^2}{2m}. \qquad (41)$$

Now looking at equation (39) for $K_n \neq 0$, then we have, using equation (41)

$$\left[\frac{\hbar^2}{2m}(2k_\nu \cdot K_n + K_n^2)\right]u(K_n) + V(K_n)u(0)$$

$$= - \sum_{m\neq 0,n} V(K_m)u(K_n - K_m). \qquad (42)$$

By assumption, the right hand side of this equation is second order in small quantities, whereas the left hand side is first order in small quantities if $2k_\nu \cdot K_n + K_n^2$ is not a small quantity. Since the points of the Brillouin zone boundaries are defined by the condition that $2k_B \cdot K_n + K_n^2 = 0$, then the above condition means that we are far from the Brillouin zone boundaries. In that case, the right hand side of equation (42) can be neglected. We then have

$$u(K_n) = - \frac{V(K_n)u(0)}{\frac{\hbar^2}{2m}[2k_\nu \cdot K_n + K_n^2]} \qquad (43)$$

which satisfies the condition that the $u(K_n) \ll u(0)$ as was assumed. On the other hand, if $2k_\nu \cdot K_n + K_n^2$ is a small quantity for a given k_ν and for some K_n, that is, we are considering a k_ν near the Brillouin zone boundary defined by the reciprocal lattice vector K_n, then the corresponding $u(K_n)$ for the given k_ν is not small. If this is the case, we cannot neglect all the terms involving $u(K_n)$ on the right hand side of equation (39) in comparison to those on the left hand side. We then must assume that $u(0)$ and $u(K_n)$ for the given K_n are of the same order of magnitude and that $u(K_L) \ll u(0)$ for $K_L \neq K_o, K_n$. Then equation (39) gives

$$\left[\frac{\hbar^2}{2m}k_\nu^2 - \varepsilon(k_\nu) + V(0)\right]u(0) + V(-K_n)u(K_n)$$

$$= - \sum_{m\neq 0,n} V(-K_m)u(K_m) \qquad (44)$$

$$\left[\frac{\hbar^2}{2m}k_\nu^2 + \frac{\hbar^2}{2m}(2k_\nu \cdot K_n + K_n^2) - \varepsilon(k_\nu) + V(0)\right]u(K_n)$$

$$+ V(K_n)u(0) = - \sum_{m\neq 0,n} V(K_m)u(K_n - K_m), \qquad (45)$$

where the right hand sides of these last two equations do not involve either u(0) or $u(K_n)$, and therefore are of second order in small quantities, whereas the left hand sides are first order in small quantities. Therefore, the right hand sides of these equations can be neglected. In order for there to be solutions for which u(0) and $u(K_n)$ are not both zero, we must satisfy the following secular equation:

$$\begin{vmatrix} \left[\frac{\hbar^2 k_\nu^2}{2m} + V(0) - \varepsilon(k_\nu)\right] & V(-K_n) \\ V(K_n) & \left[\frac{\hbar^2 k_\nu^2}{2m} + \frac{\hbar^2}{2m}(2k_\nu \cdot K_n + K_n^2) + V(0) - \varepsilon(k_\nu)\right] \end{vmatrix} = 0$$

If V(r) is a real potential, then $V(-K_n) = V^*(K_n)$. Solving for $\varepsilon(k_\nu)$ gives

$$\varepsilon\left(\frac{-K_n}{2} + \delta k_\nu\right) = V(0) + \frac{\hbar^2}{2m}\left[\frac{K_n^2}{4} + (\delta k_\nu)^2\right] \pm \left[|V(K_n)|^2 + \frac{\hbar^4}{4m^2}\left[(\delta k_\nu)\cdot K_n\right]^2\right]^{\frac{1}{2}} \quad (46)$$

where we have written $k_\nu = -K_n/2 + \delta k_\nu$ with δk_ν a small quantity.

Turning to the problem of determining the form of the allowed energy values, it is seen that equation (41) gives the value of $\varepsilon(k_\nu)$ far from the Brillouin zone boundaries. If k_ν is replaced by $-K_n/2 + \delta k_\nu$ in this equation then it takes on the following form:

$$\varepsilon\left(\frac{-K_n}{2} + \delta k_\nu\right) = V(0) + \frac{\hbar^2}{2m}\left[\frac{K_n^2}{4} + (\delta k_\nu)^2\right] - \frac{\hbar^2(\delta k_\nu)K_n}{2m} \quad (47)$$

and if the square root in equation (46) is expanded then its form is:

$$\varepsilon\left(\frac{-K_n}{2} + \delta k_\nu\right) = V(0) + \frac{\hbar^2}{2m}\left[\frac{K_n^2}{4} + (\delta k_\nu)^2\right] \pm \frac{\hbar^2}{2m}\left[|(\delta k_\nu)\cdot K_n| - \frac{2m^2|V(K_n)|^2}{\hbar^4|(\delta k_\nu)\cdot K_n|}\right] \quad (48)$$

so that when $\hbar^2 \delta k_\nu \cdot K_n/m >> V(K_n)$, then equation (46) is to a good approximation the same as equation (41). The plus sign in equation (46) corresponds to negative values of $\delta k_\nu \cdot K_n$ in equation (47) and the minus sign to positive $\delta k_\nu \cdot K_n$. In order to picture the results given in equations (47) and (48) we consider one direction in the reciprocal lattice space defined by $k = \kappa \hat{b}_1$.

In Figure 1 the darkened circles on the k-axis represent the reciprocal Bravais lattice points. The vertical dashed lines represent the Brillouin zone boundaries. The dashed curve represents the allowed energies versus k when $V(K_n) = 0$ for all $K_n \neq 0$. The solid curves represent the allowed energies versus k as given by equations (47) and (48).

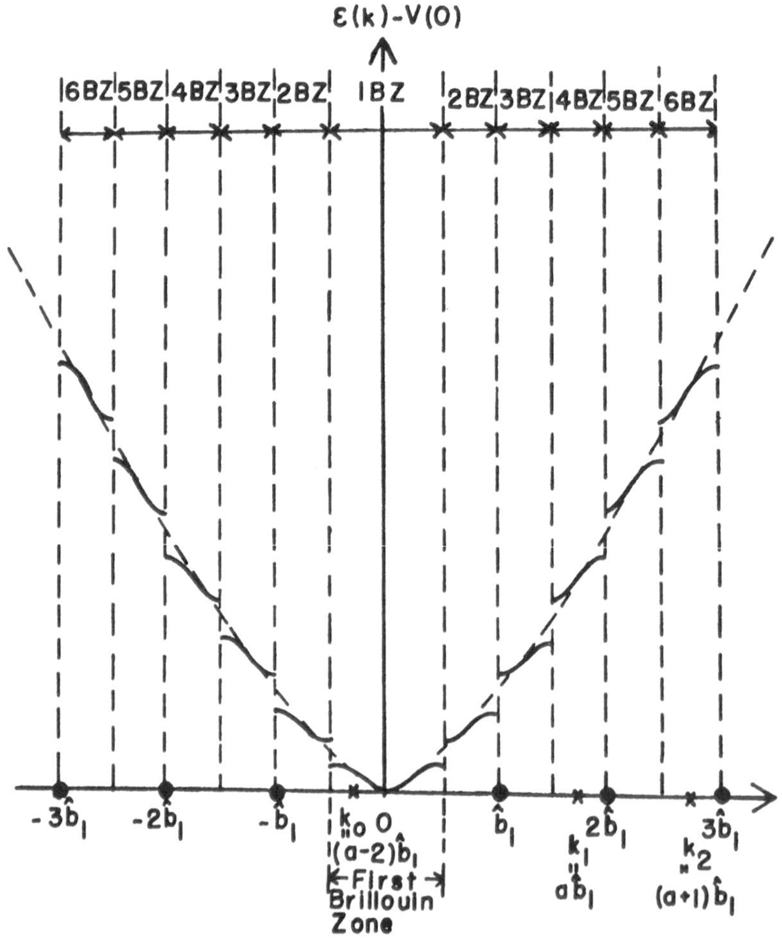

Fig. 1. Extended zone diagram of allowed energies vs k-values for nearly free-electron approximation.

The vectors $k=k_1$ and $k=k_2$ correspond to identical representations of the translation group and therefore are indistinguishable reciprocal lattice vectors. However, since the energy is different for these two k-vectors, the energy must depend on an additional quantum number which can be taken to be the Brillouin zone number indicated across the top of Figure 1. Therefore, the set of quantum numbers β_ν, indexing the one-electron states and energies in equation (34) can be subdivided into $\{n_\nu, k_\nu\}$ where n_ν corresponds to the number of the Brillouin zone to which the k_ν vector of the electron would belong when all of the $V(K_n)$ are zero. The values of n_ν for the energy states corresponding to k_1 and k_2 are 5 and 6. The independent k_ν are all in the first Brillouin zone as indicated by equation (34). In Figure 1, k_0 is the corresponding k-vector for both k_1 and k_2 in the first Brillouin Zone.

In Figure 2, we have enlarged the first two Brillouin zones. Since the onl independent k-values are contained in the first Brillouin zone, we have redrawn Figure 1 for the allowed energy values versus k in the reduced zone picture. The solid curve gives the allowed energies versus k for $V(K_n) \neq 0$ and the dashed curves for $V(K_n) = 0$. Note that the allowed energies are periodic with the periodicity of the reciprocal Bravais lattice. See Kittel's Theorem 8 pages 185 [18]. Also indicated in Figure 2 are the location of the energy eigenvalues corresponding to the k_1 and the k_2 vectors on Figure 1. On the right hand side of Figure 2 is indicated the ranges of allowed energies, the shaded regions usually designated as the allowed bands, and the unallowed energies designated as the forbidden bands.

So far, we have surpressed any consideration of the spin degree of freedom of the electrons. If the spin quantum number is to be explicitly indicated, then the notation used for the one-electron energies will be $\varepsilon_{n_\nu s_\nu \gamma_\nu}(k_\nu)$, where n_ν indexes the allowed energy bands, $s_\nu \in \{\uparrow, \downarrow\}$ indexes the spin state of the electron, $k_\nu \in$ {first Brillouin zone} indexes the translational representation of the energy state and γ_ν are any additional quantum numbers needed to uniquely specify the state of the electron, for example, quantum numbers connected with the rotational symmetry of the array.

If there are no external magnetic fields present then it can be shown [18].

1. From time reversal symmetry that

$$\varepsilon_{n_\nu \uparrow \gamma_\nu}(k_\nu) = \varepsilon_{n_\nu \downarrow \gamma_\nu}(k_\nu). \tag{49}$$

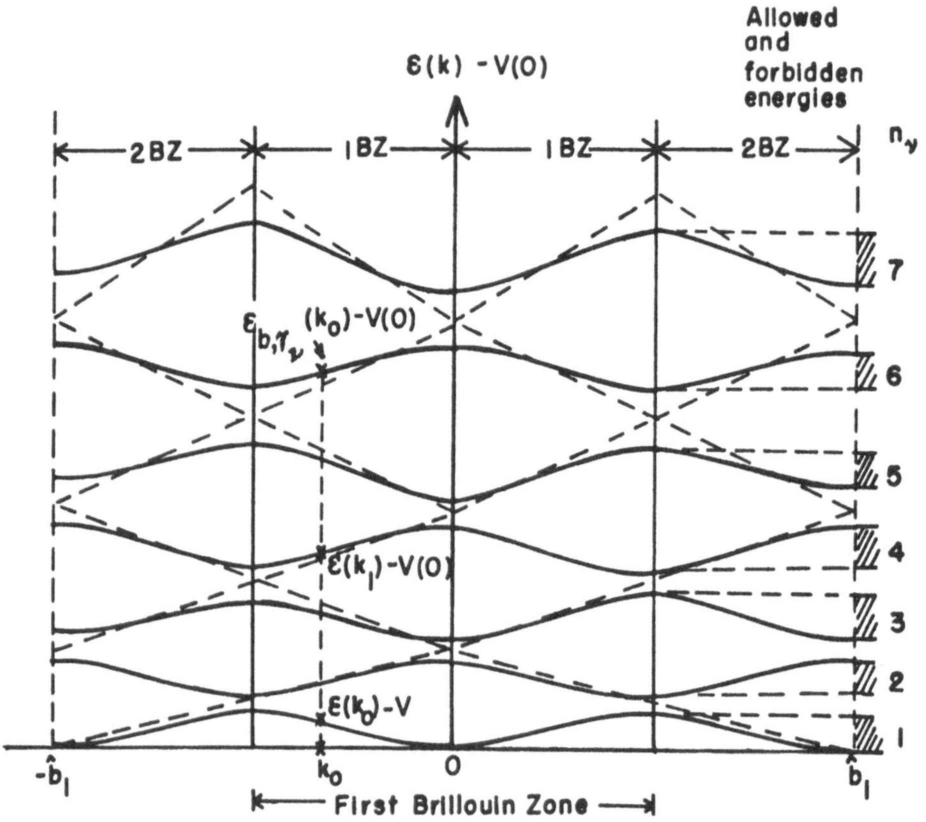

Fig. 2. Allowed energies vs. k-values for first two zones in nearly free-electron approximation.

2. From space inversion symmetry; that is, there is a center of symmetry; that

$$\varepsilon_{n_\nu \uparrow \gamma_\nu}(k_\nu) = \varepsilon_{n_\nu \uparrow \gamma_\nu}(-k_\nu). \tag{50}$$

3. From translational symmetry that

$$\varepsilon_{n_\nu \uparrow \gamma_\nu}(k_\nu) = \varepsilon_{n_\nu \uparrow \gamma_\nu}(k_\nu + K_n). \tag{51}$$

2. Case II: Tightly Bound Electrons. We have just consider-

ed the case of a weak potential and found that there are allowed and forbidden energy bands. This particular result is, however, not a consequency of the potential being weak. In order to show this, we will consider the opposite limiting case, that of a tightly bound electron. Here it is assumed that the non-core electrons are tightly bound to the ions, although not as tightly bound as the core electrons. The energy eigenvalues and states can be found by using perturbative techniques on equation (20) where the zero order approximate wavefunctions are linear combinations of atomic orbitals. For simplicity, we will assume that there is but one ion per unit cell. The ionic potentials of all the ions of the array are the same and are designated by $V_o(r_o)$ as was done in equation (23) The atomic orbitals for the non-core electrons for a single ion satisfy the equation

$$H_{at}(r_o)\tilde{\phi}_\ell(r_o) = E_\ell\tilde{\phi}_\ell(r_o) , \qquad (52)$$

where

$$H_{at}(r_o) = \frac{p_o^2}{2m} + V_o(r_o).$$

Here ℓ is a set of quantum numbers indexing the one-electron states of the ion.

The zero order approximate wavefunctions $\phi_{\alpha_\nu}(r_o)$ are given in terms of the $\tilde{\phi}_\ell(r_o)$ by

$$\phi_{\ell k_\nu \beta_\nu}(r_o) = \sum_n \exp[ik_\nu \cdot X_n]\tilde{\phi}_\ell(r_o - X_n) , \qquad (53)$$

where there exists one $\phi_{\ell k_\nu \beta_\nu}(r_o)$ for each k_ν in the first Brillouin zone and one for each atomic orbital of the non-core electrons of the ions. Note that this choice of the $\phi_{\ell k_\nu \beta_\nu}(r_o)$ satisfies Bloch's theorem as given by equation (31) or (32), for all X_n. Substitution of equation (53) into equation (20) gives

$$H\phi_{\ell k_\nu \gamma_\nu}(r_o) = \sum_n [H_{at}(r_o - X_n)\tilde{\phi}_\ell(r_o - X_n) + \upsilon_{X_n}(r_o)\tilde{\phi}_\ell(r_o - X_n)]e^{ik_\nu \cdot X_n}$$

$$= \sum_n \varepsilon_{\ell\gamma_\nu}(k_\nu)\tilde{\phi}_\ell(r_o - X_n)e^{ik_\nu \cdot X_n} \qquad (54)$$

where $\upsilon_{X_n*}(r_o) \equiv V(r_o) - V_o(r_o - X_n)$. Multiplication of equation (54) by $\tilde{\phi}_\ell(r - X_n)$ and integration over the r_o - space yields (when equation (52) is used)

$$[\varepsilon_{\ell\gamma_\nu}(k_\nu)-E_\ell]\left[\sum_{\substack{n\\n\neq p}}\int d^3r_o\tilde{\phi}_{\ell'}(r_o-X_p)\tilde{\phi}_\ell(r_o-X_n)e^{ik_\nu\cdot X_n}\right.$$

$$\left.+\ \delta_{\ell\ell'}e^{ik_\nu\cdot X_p}\right] = \sum_n e^{ik_\nu\cdot X_n}\int d^3r_o\tilde{\phi}_{\ell'}(r_o-X_p)\upsilon_{X_n}(r_o)\tilde{\phi}_\ell(r_o-X_n).$$

(55)

By tightly bound electrons is meant that the electrons on one ion do not overlap the electrons on any of the other ions in the array to a good approximation. Therefore, we can assume that

$$\sum_{\substack{n\\n\neq p}}\int d^3r_o\tilde{\phi}^*_{\ell'}(r_o-X_p)\tilde{\phi}_\ell(r_o-X_n)\ \tilde{\approx}\ 0\quad\text{for all } \ell,\ell',$$

and then equation (IIC-36) gives

$$\varepsilon_{\ell\gamma_\nu}(k_\nu) = E_\ell+\sum_{\ell'}\sum_m \exp[ik_\nu\cdot R_m]\int d^3r_o\tilde{\phi}^*_{\ell'}(r_o+X_m)\upsilon_{X_o}(r_o)\tilde{\phi}_\ell(r_o).$$

(56)

In addition, in the tight binding approximation being considered here it is assumed that the integrals in equation (56) are all small and decrease rapidly as the magnitude of X_m increases. This means that only the first few terms in the sum over "m" need to be kept to get a good approximation to $\varepsilon_{\ell\gamma_\nu}(k_\nu)$. Let $\Theta_\ell(X_n)$ be defined as

$$\Theta_\ell(X_m) = \frac{1}{\eta}\sum_{\ell'}\int d^3r_o\tilde{\phi}_{\ell'}(r_o+X_m)\upsilon_{X_o}(r_o)\tilde{\phi}_\ell(r_o), \qquad (57)$$

where $\Theta_\ell(X_n)$ will be designated as the generalized overlap integral, and

$$\varepsilon_{\ell\gamma_\nu}(k_\nu) = E_\ell + \sum_m \exp[ik_\nu\cdot X_m]\Theta_\ell(X_m). \qquad (58)$$

Here η is the number of non-core electrons per ion. Note that sums over ℓ' in equations (56) and (57) are over the occupied non-core electron states. Equation (58) can be written out explicitly for some simple arrays. See, for example, Patterson [17].

1. Simple cubic array

$$\varepsilon_{\ell\gamma_\nu}(k_\nu) = E_\ell + \theta_\ell(0) + 2\theta_\ell(X_1)[\cos k_x a + \cos k_y a + \cos k_z a]$$

where $X_1 = a(1,0,0)$.

2. Body-centered cubic array

$$\varepsilon_{\ell\gamma_\nu}(k_\nu) = E_\ell + \theta_\ell(0) + 8\theta_\ell(X_1)\cos\left(\frac{k_x a}{2}\right)\cos\left(\frac{k_y a}{2}\right)\cos\left(\frac{k_z a}{2}\right)$$

where $X_1 = \frac{a}{2}(1,1,1)$.

Here, we have used $k_\nu = (k_x, k_y, k_z)$ and $\hat{a}_1 = (a,0,0)$, $\hat{a}_2 = (0,a,0)$ and $\hat{a}_3 = (0,0,a)$.

We turn to the interpretation of the form of the allowed energies in the tight binding approximation as derived here. The perturbation expansion can be viewed as being in terms of the overlap of the electronic wavefunctions for the non-core electrons centered on different ionic sites where the expansion is in terms of nearest neighbors, second nearest neighbors, etc. with the zero order approximation being the situation when there is no overlap For infinitely large lattice spacing, we have N atoms each with the same discrete energy levels. Therefore, in the zero order approximation of the electronic energy eigenvalues for the array are discrete and degenerate. The degeneracy of the ith level is N, the number of ions forming the array, times ν_i, the degeneracy of ith atomic level. When the energy eigenvalues for the array are considered using degenerate perturbation theory as done in equations (52) through (58), then the degeneracy is broken and we get bands of allowed energy values separated by forbidden energy values. The width of the allowed energy bands increases with decreasing lattice spacing, since the perturbation increases with decreasing lattice spacing. The centroid of each band is equal to the non-interacting energy value of the degenerate states from which the bands are formed. Finally, each band contains $\nu_i N$ discrete levels. See Figure 3. For a macroscopic crystal and taking account of the zero point motion of the ions, we have broadening, so that each band becomes a continuum. However, each band can only have $\nu_i N$ electrons in it. In addition, the width of the allowed bands is independent of the number of ions forming the array, and, therefore, when an array is formed containing of the order of 100 ions, the states remain discrete. When an array has more than one ion per unit cell, then the N used above is taken to be the number of unit cells in the array, and the zero order approximate wavefunctions could be taken as linear combinations of molecular orbitials of the molecule formed from the ions contained in a unit cell.

Both the tight binding and the nearly free-electron models have given rise to more realistic methods of calculating the energy levels of real materials. We will mention here but one arising from each simple model.

From the nearly free-electron model arises the orthogonalized plane wave method in which the zero order approximate wavefunctions for the non-core electrons are taken as Schmidt's orthogonalized plane waves where the orthogonalization is to the core electron states. These orthogonalized plane waves replace the plane waves used in the simple model and are given by

$$\phi_{k_\nu \beta_\nu}(r_o) \equiv \eta_{k_\nu \beta_\nu} e^{ik_\nu \cdot r_o} - \frac{N}{V} \sum_{\mu=n'+1}^{n} \int d^3 r_o e^{ik_\nu \cdot r_o} \phi^*_{\alpha_n}(r_o)\}$$

for $1 \leq \nu \leq n'$, (59)

where V is the volume of the array. By the use of these as starting wavefunctions for the band calculation inclusion of the matching of the non-core electron states to the core electron states in each unit has been partially accomplished, and therefore, convergence of the perturbation calculation is helped. See, for example, Callaway [19].

From the tight binding model, the concept of the Wannier functions arises. The problem with the simple tight binding approximation is that the atomic orbitals used are non-orthogonal for orbitals located on different ions in the array. In principle this problem can be circumvented by using the Wannier functions which are defined as

$$A_{\beta_\nu}(r_o - X_n) = \frac{1}{\sqrt{N}} \sum_{k_\nu} e^{-ik_\nu \cdot X_n} \phi_{k_\nu \beta_\nu}(r_o), \tag{60}$$

where the summation is over the first Brillouin zone. If we define the following matrix elements

$$H_{\beta_\nu}(X_n) = \frac{1}{N} \sum_{k_\nu} e^{ik_\nu \cdot X_n} \varepsilon_{\beta_\nu}(k_\nu) = \int d^3 r_o A^*_{\beta_\nu}(r_o + X_n - X_m) H A_{\beta_\nu}(r_o - X_m). \tag{61}$$

then equation (20) takes the form

$$HA_{\beta_\nu}(r_o - X_n) = \sum_m H_{\beta_\nu}(X_m - X_n) A_{\beta_\nu}(r_o - X_m). \tag{62}$$

H is the one-electron Hamiltonian used in equation (20). From either equation (61) or (62) it follows that

$$\varepsilon_{\beta_\nu}(k_\nu) = \sum_n H_{\beta_\nu}(X_n) e^{-ik\cdot X_n}.$$

When we are interested in differences between electronic eigenvalues for a given ionic configuration of the lattice and not the energy changes due to changes in the ionic configuration, we can neglect the last two terms in equation (19) as a first approximation. These energy changes can be calculated by keeping track of the one-electron states whose occupancies change with excitation. To do this it is sufficient to keep track of the occupied electron states above the Fermi level and the unoccupied states below the Fermi level, the latter designated as being occupied by positive holes. The energy difference between two many-electron states in a first approximation is taken equal to the energy differences of the sum of the energies of all of the electrons above and all the holes below the Fermi level. This approximation is a good one, if there is a low density of uncorrelated electrons and holes. If this condition is not met, then account must be taken of the interaction between the holes and electrons.

The question of the mass of these electrons and holes when they are in an allowed energy band state must be considered, so that we can describe their reaction to external forces. If particles are strongly interacting it is usually incorrect to assign individual masses to each of them independently because their reaction to an applied force is as a unit not as individual particles. For example, in the case of an electron in a crystal lattice, if you pull on it you also pull on the lattice indirectly, and therefore the mass assigned to this electron must describe the reaction of this unit of electron plus lattice to an applied force. From the relationship between momentum, kinetic energy and mass for a free particle, the following generalized definition of the mass, designated as the effective mass for the ijth component of motion of an electron in the nth band can be made

$$m^*_{nij}(k_\nu) = \hbar^2 \left[\frac{\partial^2 \varepsilon_{n\gamma_\nu}(k_\nu)}{\partial k_{\nu_i}\, \partial k_{\nu_j}}\right]^{-1} \tag{63}$$

where k_ν is the ith component of the three vector k_ν. For positive holes the sign of (63) changes because hole motion is equivalent to electron motion in the opposite direction.

The effective mass at the Brillouin zone boundary defined by $(\delta k_\nu)\cdot K_n = 0$ for the nearly-free electron model is

$$\frac{1}{m^*_{\pm i}} = \frac{1}{m} \pm \frac{K_{n_i}^2 \hbar^2}{4m^2 |V(K_n)|^3} \tag{64}$$

where "m" is the free electron mass.

The effective mass at $k_\nu = 0$ for the tight binding case for the simple cubic and the body-centered cubic arrays is

$$m^* = \frac{-\hbar^2}{2a^2 \theta_\ell (X_1)} \quad . \tag{65}$$

II.D. Lattice Dynamics and Phonons

As shown in equation (7) the ions of a many-ions system move, to a good approximation in most cases, in the adiabatic potential which is the electronic energy eigenvalue with its parametric dependence on the ionic coordinates. With the ions at equilibrium with each other the force on any ion is zero. The dynamics of small displacements from the equilibrium position can therefore be treated by a Taylor expansion around the equilibrium coordinate positions, retaining only the harmonic terms, choosing propagating wave solutions for the equations of motion, to obtain the dispersion relations between the energy and the wave vector; and finally identifying the phonons after quantization. We consider a linear chain of ions, all of the same type, then a linear chain with links of two types of ions of different masses, and finally a three dimensional lattice with arbitrary number of ions of different types in the unit cell.

For a linear chain of ions with equilibrium positions, X, the expansion of the adiabatic potential for small displacements, $u = R-X$ is

$$V(R) = V(X) + \frac{\partial V(X)}{\partial X}\cdot u + \tfrac{1}{2}\frac{\partial^2 V(X)}{\partial X \partial X}\cdot\cdot uu + -- \; . \tag{66}$$

In general, at the equilibrium position X, we have

$$\frac{\partial V(X)}{\partial X} \cdot u = 0,$$

The following notation has been used above:

$$\frac{\partial V(X)}{\partial X} = \frac{\partial V(X+u)}{\partial u}\bigg|_{u=0} .$$

In the harmonic approximation, the force on the Lth-ion from interaction with all the other ions is

$$F_L = \sum_s \frac{\partial^2 V(X)}{\partial(X_{s+L}-X_L)\partial X_o} \cdot (u_{s+L}-u_L). \quad (67)$$

We will define the force constant κ_s as

$$\kappa_s \equiv \frac{\partial^2 V(X)}{\partial(X_{s+L}-X_L)\partial X_o} .$$

For the case of identical ions and in the approximation of only nearest neighbor interactions, the equations of motion are:

$$M \frac{d^2 u_L}{dt^2} = F_L = \kappa_1[u_{L+1} + u_{L-1} - 2u_L] \quad (68)$$

where M is the mass of the ions. Note, from inversion symmetry at the Lth lattice site, it can be shown that $\kappa_s = \kappa_{-s}$. The propagating wave solutions have the form

$$u_L = \mu \ \exp[\,i(qLa-\omega t)\,], \quad (69)$$

where q is the wave vector, "a" the unit cell distance and ω the orbital frequency of the motion. Substitution of equation (69) into (68) yields the dispersion relation:

$$\omega(q) = 2\sqrt{\frac{\kappa_1}{M}} \ \sin(\frac{qa}{2}). \quad (70)$$

For additional details see Kittel [20].

For the linear chain with alternate ions with one type of ion located at the positions indexed by the odd integers and the other type of ion located at the positions indexed by even integers having masses M_1 and M_2 we can again consider the motion of small oscillations of the ions about their equilibrium positions. Let $\kappa_{2s}(00)$ be the force constant between two ions of type one separated by an equilibrium distance sa; $\kappa_{2s}(EE)$ be the force constant between two ions of type two separated by an equilibrium distance sa and $\kappa_{2s+1}(E0)$ be the force constant between unlike ions separated by an equilibrium distance $(s+\frac{1}{2})a$. If there is inversion symmetry, as there must be in a linear chain, at each ionic equilibrium position, then it can be shown that

$$\begin{aligned}\kappa_{2s}(EE) &= \kappa_{-2s}(EE)\\ \kappa_{2s}(00) &= \kappa_{-2s}(00)\\ \kappa_{2s+1}(E0) &= \kappa_{-2s-1}(E0)\ .\end{aligned} \tag{71}$$

In the harmonic approximation:

The force on the $2L^{th}$ ion is

$$F_{2L} = \sum_s \{\kappa_{2s}(EE)[u_{2L+2s}-u_{2L}]+\kappa_{2s+1}(E0)[u_{2L+2s+1}-u_{2L}]\} \tag{72}$$

and the force on the $2L+1^{th}$ ion is

$$F_{2L+1} = \sum_s \{\kappa_{2s+1}(E0)[u_{2L+1+2s+1}-u_{2L+1}] + \kappa_{2s}(00)[u_{2L+1+2s}-u_{2L+1}]\}. \tag{73}$$

The equations of motion for the ions are:

$$\frac{d^2u_{2L+1}}{dt^2} = \frac{1}{M_1}\sum_s \{\kappa_{2s+1}(E0)[u_{2L+2s+2}-u_{2L+1}] + \kappa_{2s}(00)[u_{2L+2s+1}-u_{2L+1}]\} \tag{74}$$

and

$$\frac{d^2u_{2L}}{dt^2} = \frac{1}{M_2}\sum_s \{\kappa_{2s}(EE)[u_{2L+2s}-u_{2L}] + \kappa_{2s+1}(E0)[u_{2L+2s+1}-u_{2L}]\}. \tag{75}$$

Again, the propagating solutions are assumed of the form:

$$u_{2L+1} = \mu_0 \exp\{i[(L+\tfrac{1}{2})qa-\omega t]\} \tag{76}$$

and

$$u_{2L} = \mu_E \exp\{i[Lqa-\omega t]\}. \tag{77}$$

When equations (76) and (77) are substituted into equations (74) and (75) and equation (71) is used, then it is found that μ_0 and μ_E must satisfy

$$\frac{2}{M_1}\sum_{s\geq 0}\{\kappa_{2s+1}(E0)\cos[(s+\tfrac{1}{2})qa]\mu_E + [\kappa_{2s}(00)(\cos sqa-1) - \kappa_{2s+1}(E0)]\mu_0\} + \omega^2\mu_0 = 0$$

$$\frac{2}{M_2}\sum_{s\geq 0}[\kappa_{2s}(EE)(\cos sqa-1) - \kappa_{2s+1}(E0)]\mu_E + \kappa_{2s+1}(E0)\cos[(s+\tfrac{1}{2})qa]\mu_0\} + \omega^2\mu_E = 0. \tag{78}$$

From the secular equation for these coupled equations we can determine the dispersion relations between ω^2 and q. The secular equation is

$$\begin{vmatrix} A-\omega^2 & B_1 \\ B_2 & C-\omega^2 \end{vmatrix} = 0, \tag{79}$$

where A, B_i and C are real functions of q and are given by

$$A \equiv \frac{2}{M_1}\sum_{s\geq 0}\{\kappa_{2s}(00)[1-\cos sqa] + \kappa_{2s+1}(E0)\}$$

$$B_i \equiv -\frac{2}{M_i}\sum_{s\geq 0}\{\kappa_{2s+1}(E0)\cos[(s+\tfrac{1}{2})qa]\} \tag{80}$$

$$C \equiv \frac{2}{M_2}\sum_{s\geq 0}\{\kappa_{2s}(EE)[1-\cos sqa] + \kappa_{2s+1}(E0)\}.$$

We now investigate some of the properties of the solutions of equations (78), which can be written in matrix form as

$$\begin{pmatrix} A & B_1 \\ B_2 & C \end{pmatrix}\begin{pmatrix} \mu_0 \\ \mu_E \end{pmatrix} = \omega^2\begin{pmatrix} \mu_0 \\ \mu_E \end{pmatrix}. \tag{81}$$

Note that each unit cell contains an ion of mass M_1 and one of mass M_2. We will choose the Lth unit cell so that the $2L^{th}$ and $2L+1^{th}$ ions are in it. For polar materials the two different ions would have opposite charge. If the ions within a unit cell move in the same direction, then the net change in the local polarization would be very small. On the other hand, if the ions within a cell move in the opposite direction, then the net change in the local polarization would be large in comparison to the first case. That is, for the first case the two sublattices, where the two sublattices are defined as the lattices composed of ions of a single type, move in the same direction and for the second case they move in opposite directions. It is of interest to keep track of whether or not there is a large varying polarization created for a given motion. We therefore define a new set of basis vectors μ_p and μ_{np} where μ_p corresponds to a large polarization and μ_{np} to a small polarization. The transformation from μ_O, μ_E to μ_{np}, μ_p is defined by

$$U = \frac{1}{\sqrt{2}} \begin{pmatrix} \exp(\frac{iqa}{2}) & 1 \\ \exp(\frac{iqa}{2}) & -1 \end{pmatrix},$$

where

$$\begin{pmatrix} \mu_{np} \\ \mu_p \end{pmatrix} = U \begin{pmatrix} \mu_O \\ \mu_E \end{pmatrix}.$$

We can transform equation (81) to

$$\begin{pmatrix} \Gamma_{11} & \Gamma_{12} \\ \Gamma_{21} & \Gamma_{22} \end{pmatrix} \begin{pmatrix} \mu_{np} \\ \mu_p \end{pmatrix} = 2\omega^2 \begin{pmatrix} \mu_{np} \\ \mu_p \end{pmatrix}, \tag{82}$$

where

$$\Gamma_{11} \equiv A + C + B_1 e^{iqa/2} + B_2 e^{-iqa/2}$$

$$\Gamma_{22} \equiv A + C - B_1 e^{iqa/2} - B_2 e^{-iqa/2}$$

$$\Gamma_{12} \equiv A - C - B_1 e^{iqa/2} + B_2 e^{-iqa/2}$$

$$\Gamma_{21} \equiv A - C + B_1 e^{iqa/2} - B_2 e^{-iqa/2} .$$

The coupled equations given in equation (78), (81) or (82) define two modes of vibration for each q-value in the first Brillouin zone, where the first Brillouin zone is defined as those q-values such that $-\pi/a \leq q \leq \pi/a$. The q's are analogous to the reciprocal lattice vectors $k_{\alpha\nu}$ for the electronic dynamics. Note that the q's have been defined slightly differently from the k's in that the 2π factor has not been incorporated into the q's. Since the $\omega(q)$'s vary slowly with q it is natural to group the vibrational modes for various q's into sets designated as branches. Each branch contains one mode from each set of modes for a given q-value.

The conditions that the Γ_{ij}'s must satisfy in order that the modes corresponding to the two branches separate into ones which are primarily nonpolarizing and ones which are primarily polarizing is that the off-diagonal elements of equation (88) must be small in comparison to the diagonal elements. If only the Γ_{12} element meets this requirement for a given q-value, then one mode for that q-value is primarily of type μ_p, that is, a polarizing mode; while the other is of mixed type, μ_p and μ_{np}. On the other hand, if only the Γ_{21} element meets this requirement then one mode is primarily of type μ_{np} and the other is of mixed type.

We now consider the solution to equation (81) and/or (82) for some special cases. The following notation is used

$$
\begin{aligned}
\langle\kappa_n(EE)\rangle &\equiv \sum_{s\geq 0} s^n \kappa_{2s}(EE) \\
\langle\kappa_n(00)\rangle &\equiv \sum_{s\geq 0} s^n \kappa_{2s}(00) \qquad (83) \\
\langle\kappa_n(E0)\rangle &\equiv \sum_{s\geq 0} s^n \kappa_{2s+1}(E0).
\end{aligned}
$$

1. Case 1: The case of Small q-Values. First it is noted that $\Gamma 12$'s leading term in q is $4\langle K_0(EO)\rangle\times(1/M_1 - 1/M_2)$ and that Γ_{21}'s leading term is $-i\langle K_0(EO)\rangle\times(1/M_1 + 1/M_2)qa$. Therefore, for small enough q-values we have modes of type μ_{np}, designated as the acoustical branch, and the other set of modes is of mixed type, μ_p and μ_{np} which are designated as the optical branch. The corresponding dispersion relations are:

Optical Branch

$$
\omega_{op}^2 = 2\left(\frac{1}{M_1} + \frac{1}{M_2}\right)\langle\kappa_0(EO)\rangle + \left\{M_2\left[\frac{\langle\kappa_2(00)\rangle}{M_1} - \frac{\langle\kappa_2(EE)\rangle}{M_2}\right] - 2\left[\langle\kappa_2(EO)\rangle + \langle\kappa_1(EO)\rangle + \frac{1}{4}\langle\kappa_0(EO)\rangle\right]\right\}\frac{q^2a^2}{M_1+M_2} \qquad (84)
$$

Acoustical Branch

$$\omega_{ac}^2 = \{M_1[\frac{\langle\kappa_2(OO)\rangle}{M_1} - \frac{\langle\kappa_2(EE)\rangle}{M_2}]$$

$$+ 2[\langle\kappa_2(EO)\rangle + \langle\kappa_1(EO)\rangle + \frac{1}{4}\langle\kappa_0(EO)\rangle]\} \frac{q^2a^2}{M_1+M_2}. \quad (85)$$

Note from the definitions of the $\langle K_n\rangle$ given in equation (83) (that the larger the value of n the more important are the long range interaction terms in determining the values of the $\langle K_n\rangle$.

2. Case II: The Case of $q = \pi/a$. In this case, we are considering the q-value corresponding to the Brillouin zone boundary. It is easily seen that $B_1 = B_2 = 0$ and

$$A = \frac{2}{M_1}\langle\kappa_0(EO)\rangle + \frac{4}{M_1}\sum_{s\geq 0}\kappa_{4s+2}(OO) \equiv \omega_o^2$$

$$C = \frac{2}{M_2}\langle\kappa_0(EO)\rangle + \frac{4}{M_2}\sum_{s\geq 0}\kappa_{4s+2}(EE) \equiv \omega_E^2 \quad ,$$

where the mode corresponding to ω_o^2 is given by the solution $\mu_O = \mu$ and $\mu_E = 0$ which implies that $\mu_{np} = \mu_p = i\mu$, while the mode corresponding to ω_E^2 is given by $\mu_O = 0$ and $\mu_E = \mu$ which implies that $\mu_{np} = -\mu_p = \mu$. Therefore, both branches at the zone boundary are equal mixtures of polarizing and non-polarizing vibrational motions. In between the small q-values and the zone boundary, the mixture of polarizing and non-polarizing components for the acoustical branch varies from 0-100 to 50-50. The larger of the two ω^2's above corresponds to the optical branch.

The analysis of the general three dimensional lattice is analogous to the analysis of the one dimensional alternating ions situation. Equation (66) becomes in this general situation

$$V(R) = V(X) + \sum_{J,I}\frac{\partial V(X)}{\partial X_{JI}}\cdot u_{JI}$$

$$+ \frac{1}{2}\sum_{J,I}\sum_{J',I'}\frac{\partial^2 V(X)}{\partial X_{JI}\, X_{J'I'}} u_{JI}u_{J'I'} + \dots ,$$

(86)

where X_{JI} and u_{JI} are three dimensional vectors with J indexing the unit cell in which the corresponding ion is located and I indexing the location within the cell. The three degrees of freedom for each ion lead to one longitudinal and two transverse modes per ion in the unit cell. From equation (86) the equations of motion are found to be

$$\frac{d^2u_{JI}}{dt^2} = -\frac{1}{M_I}\sum_{J',I'} \kappa(JI,J'I')\cdot u_{J'I'}, \qquad (87)$$

where

$$\kappa(JI, J'I') \equiv \frac{\partial^2 V(X)}{\partial X_{JI}\partial X_{J'I'}}$$

is the general force tensor. In equation (87), $I,I' = 1,2,\ldots,$ Γ = number of ions per unit cell. Substitution of the propagating plane wave solutions of the form

$$u_{JI} = \mu e_I \exp\{i[X_{JI}\cdot q-\omega t]\}$$

into equation (87)

$$\sum_{J'I'} \{ \frac{\kappa(JI,J'I')}{M_I} \cdot\exp[i(X_{J'I'}-X_{JI})\cdot q]-\omega^2\delta_{JJ'}\delta_{II'}\}e_{I'} = 0. \qquad (88)$$

Note that equation (88) is a set of 3Γ coupled equations and that q is a three dimensional vector in this case. The solutions to equation (88) give the 3Γ dispersion relations $\omega\gamma$ (q); $\gamma = 1,2,\ldots,3\Gamma$; of the 3Γ branches and the corresponding $e_I(\gamma,q)$. In equation (88) $\mu = \mu(\gamma,q)$ is the amplitude of the γ, q mode of vibration and the $e_I(\gamma,q)$'s are normalized three dimensional polarization vectors.

For the case of alternating ions in one dimension for $q=\pi/a$, the $e_I(\gamma,q)$'s and $\omega^2(\gamma,q)$ are

$$\omega^2(1,\pi/a) = \omega_0^2 \qquad \omega^2(2,\pi/a) = \omega_E^2$$

$$e_0(1,\pi/a) = 1 \qquad e_0(2,\pi/a) = 0$$

$$e_E(1,\pi/a) = 0 \qquad e_E(2,\pi/a) = 1 .$$

In general, the $e_I(\gamma,q)$ satisfy

$$\sum_I e_I^*(\gamma,q)\cdot e_I(\gamma',q) = \delta_{\gamma\gamma'} \qquad \text{orthornormality} \tag{89}$$

and

$$\sum_\gamma e_{Ij}^*(\gamma,q)e_{I'j'}(\gamma,q) = \delta_{II'}\delta_{jj'} \qquad \text{completeness}, \tag{90}$$

where j,j' index the vector components.

For each mode specified by $\{\gamma,q\}$ quantization leads to harmonic oscillator states for the $\mu(\gamma,q)$ with energies given by

$$E_n(\gamma,q) = \hbar\omega_\gamma(q)(n+\tfrac{1}{2}), \tag{91}$$

where n specifies the phonon occupancy.

It will be convenient in what follows to use a second quantization notation for the phonon which is defined in terms of the creation and annihilation operators $b_{q\gamma}^+$ and $b_{q\gamma}$ for the phonon eigenstate having $\omega_\gamma(q)$ and vibrational form

$$u_{JI} = e_I(\gamma,q)\exp\{i[X_{JI}\cdot q-\omega_\gamma(q)t]\}. \tag{92}$$

In particular the operator $b_{q\gamma}^+$ operating on a state of the lattice vibration creates a state for which the lattice vibration is increased by one phonon belonging to the γth branch having wave vector q. These operators satisfy the commutation relations

$$[b_{q\gamma},b_{q'\gamma'}^+]=\delta_{qq'}\delta_{\gamma\gamma'}\ . \tag{93}$$

The displacement and momentum operators for the Ith ion in the Jth unit cell can be given in terms of the creation and annihilation operators as

$$u_{JI} = \sum_q\sum_\gamma \left[\frac{\hbar}{2NM_I\omega_\gamma(q)}\right]^{\frac{1}{2}} e_I(\gamma,q)\{b_{q\gamma}^+\exp[iq\cdot X_{JI}]$$

$$+ b_{q\gamma}\exp[-iq\cdot X_{JI}]\} \tag{94}$$

and

$$P_{JI} = i\sum_q\sum_\gamma \left[\frac{\hbar M_I\omega_\gamma(q)}{2N}\right]^{\frac{1}{2}} e_I(\gamma,q)\{b_{q\gamma}^+\exp[iq\cdot X_{JI}] +$$

$$- b_{q\gamma} \exp[-iq \cdot X_{JI}]\} , \quad (95)$$

where

$$P_{JI} = -i\hbar \frac{\partial}{\partial u_{JI}} = -i\hbar \frac{\partial}{\partial R_{JI}} .$$

For further details see Jones and March[21].

II.E. The Electron-Phonon Interaction

As shown in subsection IIA the electron-phonon interaction is given by the off-diagonal elements of the A matrix given in equation (8) which has the form

$$A_{k\alpha;\, k'\beta} = - \sum_J \sum_I \frac{\hbar^2}{M_I} \int d^{3n} r \psi^*_{k\alpha}(r;R) \nabla_{R_{JI}} \psi_{k'\beta}(r;R) \cdot \nabla_{R_{JI}} , \quad (96)$$

where $\psi_{k\alpha}(r;R)$ is the many-electron electronic wavefunction of the adiabatic approximation. The Bloch theorem applies to the many-electron wavefunction as well as to the one-electron wavefunction and has the form

$$\psi_{k\alpha}(r+X_n;R) = e^{ik \cdot X_n} \psi_{k\alpha}(r;R) , \quad (97)$$

where

$$k = \sum_\nu k_\nu ,$$

with k_ν's corresponding to the wavevectors of the occupied one-electron states when that approximation is used. Since $P_{JI} = -i\hbar\nabla_{R_{JI}}$ where P_{JI} is the momentum operator for the JI^{th} ion, equation (96) can be written as
equation

$$A_{k\alpha;k'\beta} = \hbar \sum_\gamma \sum_q \sum_J \sum_I \left[\frac{\hbar\omega_\gamma(q)}{2NM_I} \right]^{\frac{1}{2}} e_I(\gamma,q) \cdot \int d^{3n} r \psi^*_{k\alpha}(r;R) \quad (98)$$

$$\times \nabla_{R_{JI}} \psi_{k'\beta}(r;R) \quad [b^+_{q\gamma} e^{iq \cdot X_{JI}} - b_{q\gamma} e^{-iq \cdot X_{JI}}],$$

where equation (95) has been used to represent P_{JI}.

We consider the formal evaluation of the integral in equation

(97) when the magnitude of the oscillations of each ion is small. Equation (4) specifies the wavefunction $\psi_{k\alpha}(r;R)$. $V_T(r,R)$ can be approximated using the expansion given in equation (66) keeping only the first two terms. In this approximation equation (4) has the form

$$[T_{el}+V_T(r,X) + \sum_{J,I} u_{JI}\cdot\nabla_{X_{JI}} V_T(r,X_{JI})]\psi_{k\alpha}(r;X+u) \qquad (99)$$

$$= E_{k\alpha}(X+u)\psi_{k\alpha}(r;X+u) \ .$$

Let the last term on the left hand side of this equation be handled perturbatively since it is assumed that the u_{JI}'s are small, then $\psi_{k\alpha}(r;X+u)$ is to first order terms in the u_{JI}'s

$$\psi_{k\alpha}(r;X+u) = \psi_{k\alpha}(r;X) + \sum_{\substack{k'\beta \\ k'\neq k \\ \beta\neq\alpha}} \int d^{3n}r'\psi^*_{k'\beta}(r';X)[\sum_{J,I} u_{JI}$$

$$\cdot\nabla_{X_{JI}} V_T(r',X)]\ \psi_{k\alpha}(r';X)\ \frac{\psi_{k'\alpha}(r;X)}{[E_{k\alpha}(X)-E_{k'\beta}(X)]} \ . \qquad (100)$$

Since the $\{\psi_{k\alpha}(r;X)\}$ are an orthonormal set, then from equation (100) we have, to the approximation used in the equation,

$$\int d^{3n}r\psi^*_{k\alpha}(r;R)\nabla_{R_{JI}}\psi_{k'\beta}(r;R) = \int d^{3n}r\psi^*_{k\alpha}(r;R)\nabla_{u_{JI}}\psi_{k'\beta}(r;R)$$

$$= \frac{\int d^{3n}r\psi^*_{k\alpha}(r;X)[\nabla_{X_{JI}} V_T(r,X)]\psi_{k'\beta}(r;X)}{[E_{k'\beta}(X) - E_{k\alpha}(X)]}$$

$$\equiv \frac{\vec{W}_{JI}(k\alpha;k'\beta)}{[E_{k'\beta}(X)-E_{k\alpha}(X)]} \ . \qquad (101)$$

The conservation of energy, which must hold because of the time translational symmetry of the Hamiltonian in equation (1). implies that for real transitions

$$E_{k'\beta}(X) - E_{k\alpha}(X) = \pm\hbar\omega_\gamma(q) \qquad (102)$$

for some γ and q. The plus sign corresponds to the creation of a phonon of energy $h\omega_\gamma(q)$ and the minus sign to the annihilation of a phonon. In addition, by space translational symmetry of the Hamiltonian for translation through a distance of a lattice vector, X_n, conservation of quasi-momentum must hold, that is

$$k'-k = \pm q + K_\nu , \tag{103}$$

where K_ν is a reciprocal lattice vector. From equations (98) (101),(102) and (103) it is found that

$$A_{k\alpha;k'\beta} = \sum_\gamma \sum_q \sum_J \sum_I \left[\frac{\hbar}{2NM_I\omega_\gamma(q)} \right]^{\frac{1}{2}} \vec{W}_{JI}(k\alpha;k'\beta)\cdot e_I(\gamma,q)$$

$$\Big\{ b^+_{q\gamma} e^{iq\cdot X_{JI}} \delta(E_{k'\beta}(X)-E_{k\alpha}(X)-h\omega_\gamma(q))\,\delta(k'-k-q-K_\nu)$$

$$+ b_{q\gamma} e^{-iq\cdot X_{JI}} \delta(E_{k'\beta}(X)-E_{k\alpha}(X)+h\omega_\gamma(q))\,\delta(k'-k+q-K_\nu)\Big\}. \tag{104}$$

We again look at equation (4) for an arbitrary u, but this time write it as

$$\{T_{el} + V_T(r,X)+[V_T(r,X+u)-V_T(r,X)]\}\psi_{k\alpha}(r;X+u)$$

$$= E_{k\alpha}(X + u)\psi_{k\alpha}(r;X+u). \tag{105}$$

The term in square brackets is treated perturbatively. In the zero order approximation the ionic motion is separated from the electronic motion completely. This zero order approximation is known as the Crude Adiabatic Approximation. See for example, the review of Williams, Berry and Bernard [22]. The states, in this approximation, have well defined numbers of phonons and electronic energy, and the perturbation causes transitions between these states. The transition matrix elements, which will be designated by $A_{k\alpha;k'\beta}$, are

$$A_{k\alpha;k'\beta} \equiv \int d^{3n}r\psi^*_{k\alpha}(r;X)[V_T(r,X+u)-V_T(r,X)]\psi_{k'\beta}(r;X). \tag{106}$$

From equations (66) and (94) to first order in u, we have

$$V_T(r,X+u)-V_T(r,X) = \sum_\gamma \sum_q \sum_J \sum_I \left[\frac{\hbar}{2NM_I\omega_\gamma(q)} \right]^{\frac{1}{2}} e_I(\gamma,q)\cdot\nabla_{X_{JI}} V_T(r,X_{JI})$$

$$\times \{b^+_{q\gamma} e^{iq\cdot X_{JI}} + b_{q\gamma} e^{-iq\cdot X_{JI}}\}. \qquad 107)$$

Equations (106) and (107) together with conservation of energy and momentum give

$$\mathcal{A}_{k\alpha;k'\beta} = A_{k\alpha;k'\beta}$$

within the approximation used, Bloch [23], [24] first used $A_{k\alpha;k'\beta}$ as the electron-phonon coupling matrix. For additional details of the connection between the Bloch matrix and the electron-phonon adiabatic interaction matrix, see Sham and Ziman [25].

Note that going from the adiabatic electron-phonon interaction to the Bloch form depends critically on the use of perturbation theory. Since the correct form of the electron-phonon interaction is that given by the adiabatic approximation, whenever the Bloch form is used in a situation where the perturbation calculation is invalid the results could be in error. Sham and Ziman [25] point out that calculations involving virtual transitions would be in error if the Bloch form is used. In what follows we will use the Bloch formulation, as is usually done.

When the array is deformed there are two possible effects on the electrons due to the deformation. First, for all arrays there is a local change in the potential near each lattice site of a displaced ion, but the combined displacements of all ions are such that there is no long range electric field established in the array. Second, if there are ions in the array having different charges, then there can be a long range interaction between the phonons and the electrons which arises due to the phonons inducing changes in the dipole moments within the unit cell. Usually, this second interaction is the larger, and thus, the first can be neglected. The short range interaction is treated by the theory of the deformation potential, and long range interaction is included in the Fröhlich Hamiltonian formulation.

1. Deformation Potential Theory. Bardeen and Shockley [26] introduced deformation potential theory, in which the changes in the energy due to a uniform strain are related to the Block form of the electron-phonon interaction. Here, we will follow the presentation of Jones and March [27].

Consider a uniform strain ϵ_{ij} defined as follows: Given any distortion of a periodic array where the Jth ion is displaced by an amount u_J where we are considering an array having one ion per unit cell, we define a vector function $u(r_o)$, r_o being the vector coordinate of a single electron, such that it is the smoothest function having the property

$$u(X_J) = u_J. \tag{108}$$

Then ϵ_{ij} is

$$\epsilon_{ij} = \frac{\partial u_i(r_o)}{\partial r_{oj}} ,$$

where i and j index the vector components of the vector function $u(r_o)$ and r_o, respectively.

Begin by finding the change in $\varepsilon_\beta(k)$, the one-electron energy, given by equation (20) due to a uniform strain ε_{ij}. Note that for a uniform strain the periodic array is distorted into another periodic array with changed lattice constant. Therefore, there exist Bloch states for both arrays, but they are not the same states. Since the distortion can be represented mathematically by the transformation of the r_o space of equation (20) where the transformation is given by

$$\zeta_o = r_o - u(r_o) \approx r_o - \epsilon \cdot r_o ,$$

we can find the energies of the distorted Bloch waves in terms of the energies of the undistorted Bloch waves by solving the transform of equation (20) perturbatively. The result, to the first order in the strain, is

$$\varepsilon_\beta(k;\text{strained}) = \varepsilon_\beta(k;\text{unstrained}) + \sum_{i,j} \epsilon_{ij} D_{ij}(k,\beta) , \tag{109}$$

where

$$D_{ij}(k,\beta) = \int d^3 r_o \psi^*_{k\beta}(r_o) \mathcal{D}_{ij} \psi_{k\beta}(r_o) ,$$

with

$$\mathcal{D}_{ij} = -\frac{1}{m} P_{oi} P_{oj} + \sum_n (X_{ni} - r_{oi}) \frac{\partial V(r_o, X)}{\partial X_{nj}} + \frac{\hbar k_i}{m} P_{oj} .$$

Here i,j index the components of vectors and tensors, and the $\psi_{k\beta}(r_o)$ are the unstrained wavefunctions.

On the other hand, for non-uniform strains where $u(r_o)$ is given by

$$u(r_o) = \mu e(\gamma,q)\exp[iq\cdot r_o], \tag{110}$$

then

$$\epsilon_{ij} = \frac{\partial u_i(r_o)}{\partial r_{oj}} = iq_j u_i(r_o) \ . \tag{111}$$

If equation (110) is substituted into the Bloch form of the electron-phonon interaction matrix given by equation (106) and calculated in the same approximation used to get equation (109) then

$$A_{k\alpha;k'\beta} = \sum_{ij} \int d^3r_o \psi^*_{k\alpha}(r_o)\epsilon_{ij}(r_o)h_{ij}(r_o)\psi_{k'\beta}(r_o), \tag{112}$$

where

$$h_{ij}(r_o) = -\frac{1}{m} P_{oi}P_{oj} + \sum_n (X_{ni}-r_{oi}) \frac{\partial V(r_o,X)}{\partial X_{nj}} \pm \frac{i\hbar\omega_\gamma(q)q_j P_{oi}}{q^2} . \tag{113}$$

The upper sign is for creation of a phonon and the lower for annihilation. The deformation potential is defined as

$$\delta V(r_o) \equiv \sum_{i,j} \epsilon_{ij}(r_o)h_{ij}(r_o). \tag{114}$$

The last term in $h_{ij}(r_o)$ can be approximated by $\pm i\hbar C_j(\gamma,q)P_{oi}$ for longitudinal acoustical modes with $\omega_\gamma(q)$ approximated by

$$\sum_j C_j(\gamma,q)q_j ,$$

where $C(\gamma,q)$ is the phase velocity of the phonon mode indexed by $\{\gamma,q\}$. By comparing equations (109) and (119) it is seen that in this approximation the deformation potential matrix elements are

$$A_{k\beta;k'\beta} = \sum_{i,j} \int d^3r_o \psi^*_{k\beta}(r_o)\epsilon_{ij}(r_o)\Big[D_{ij}(k\beta) - \frac{\hbar k_i}{m} P_{oj} \pm i\hbar C_j(\gamma,q)P_{oi}\Big] \psi_{k'\beta}(r_o). \tag{115}$$

The last two terms in the integrand are zero at the band extrema. Therefore, the electron-phonon interaction is

$$A_{k\beta,k'\beta} = \sum_{i,j} D_{ij}(k,\beta) \int d^3r_o \psi^*_{k\beta}(r_o)\varepsilon_{ij}(r_o)\psi_{k'\beta}(r_o), \quad (116)$$

where $D_{ij}(k,\beta)$ is determined from static uniform strain measurements. Note first that we are using a diagonal matrix element of $\mathcal{D}_{ij}$ to determine the above off-diagonal matrix elements. If $D_{ij}(k,\beta)$ depends strongly on k then some type of average of $D_{ij}(k,\beta)$ and $D_{ij}(k',\beta)$ would have to be used. Second, if we take the limit of the ionic potential going to zero, the "empty lattice test", then the matrix elements for the deformation potential are not zero. On the other hand the matrix elements for the electron-phonon interaction are zero, as are those in equation (112)

2. Fröhlich Hamiltonian. Fröhlich developed the polaron interaction given below [28]. In this case, we are interested in an electron interacting with phonons which cause lattice polarization. For these phonons, there will be both a deformation interaction and an additional interaction due to the electric field produced by the polarization. The Frohlich Hamiltonian is usually developed for longitudinal optical phonons and has the form

$$H_{F\gamma}(r_o) = \sum_q \{V_{q\gamma} b^+_{q\gamma} \exp[iq\cdot r_o] - V_{q\gamma} b_{q\gamma} \exp[-iq\cdot r_o]\}, \quad (117)$$

where

$$V_{q\gamma} = \frac{4\pi ie}{q} \left[\frac{\omega_\gamma(q)}{8\pi} \left(\frac{1}{\varepsilon_\infty} - \frac{1}{\varepsilon_0} \right)\right]^{1/2} .$$

Here ε_0 and ε_∞ are the low and high frequency dielectric constants, and the values of γ correspond to any longitudinal polarizing mode. The essential steps in deriving equation (117) are:

(a) to use equation (94) for u_{JI} to define the relative displacement in each unit cell as

$$u_J^{(rel)}(X) = \sum_I u_{JI} . \quad (118)$$

(b) to use the long wavelength limit for the phonons, that is, assume small q-values.

(c) to assume that a continuous approximation is valid, so that X in equation (118) is replaced by r_o.

(d) to write the polarization in terms of the relative displacment

$$u_J^{(rel)}(r_o) = \sum_I u_{JI} \qquad (119)$$

Note that the motion of the polarizing phonons can be described either in terms of the motion of the ions or in terms of the dynamical polarization produced by them.

(e) to use the dynamical equations for the ions and the connection between the ions and the polarization to develop the dynamical equations of the polarization in terms of microscopic quantities.

(f) to note that ε_0 includes the polarization induced by ionic motion and due to electronic motion whereas ε_∞ includes the polarization due to the electronic motion only. Since we are interested in the polarization induced by the ionic motion only, the appropriate polarization to be related to the phonons is the difference between the low and high frequency polarizations.

(g) once having the above inter-relation between the macroscopic polarization and the ionic motion, then a connection between the microscopic quantities and macroscopic quantities can be made which gives rise to equation (119) as the interaction of the long wavelength longitudinal polarizing phonons and the electrons. For further details of this derivation see Kubo and Nagamiya [29].

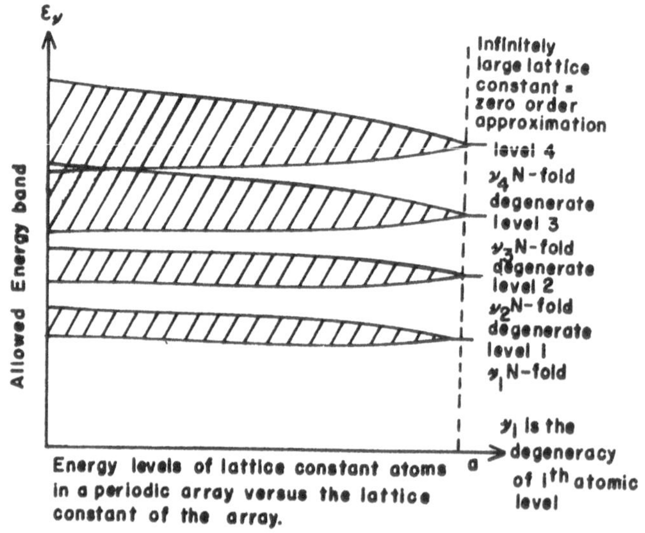

Fig. 3. Allowed energies vs inter-atomic distance for tightly bound electron case.

III. GENERAL METHODS OF ENERGY TRANSFER

III.A. Resonant Energy Transfer

This process is important for energy transfer between dopants or defects in both semiconductors and insulators. It arises from a weak electromagnetic coupling between the electrons on the energy donors and those on the acceptors, which are not to be confused with charged donors and acceptors in semiconductors; these energy donors and acceptors undergo only changes in their state of excitation not in their state of ionization. A key concept in the analysis of resonant energy transfer is that the transfer occurs between the excited donor, which has relaxed with respect to the nuclear coordinates, and the ground state of the acceptor, which is also relaxed with respect to the nuclear coordinates. When this is not the case, as for nearest neighbors, departures from the Förster predictions are not unexpected. Each relaxed nuclear configuration in a unique way is characteristic of its specific electronic state. The transfer takes place in a time of the same order as radiative transitions, and therefore the nuclear coordinates cannot change during the transition. In addition, no real photons or phonons are involved in this process. It is common to distinguish between radiative transfer, exciton motion, non-resonant transfer and resonant transfer as follows: In radiative transfer a real photon is created by the energy donor and is then absorbed by the energy acceptor; For pure excitonic motion the nuclear configuration remains unrelaxed, however, motion of the excitonic polaron involves at least partial relaxation; For non-resonant transfer phonons are created or annihilated as part of the transfer process, whereas for resonant transfer phonons are neither created nor destroyed.

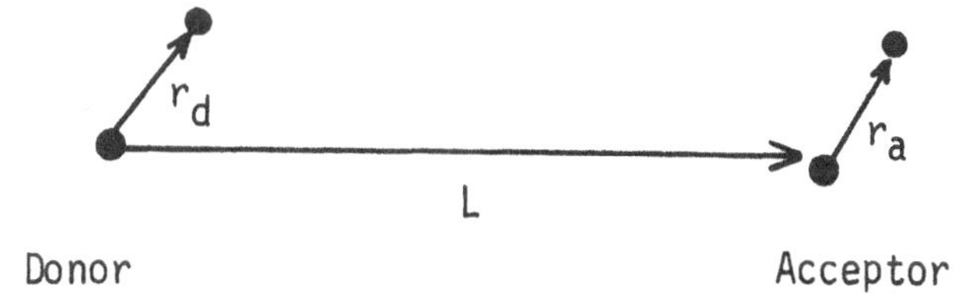

Fig. 4. Relative locations of the electrons and ion cores of the energy donor and acceptor.

The essentials of the process of resonant energy transfer are evident from an analysis of the transfer between an energy donor and acceptor each having a single electron changing its state, as illustrated in Figure 4. Here we assume that all electronic states involved are singlets. The adiabatic approximation and a simple Hartree product wavefunction are assumed for the two-electron system of the donor and acceptor. The second assumption is possible if we neglect exchange, which is a good approximation when the distance between the donor and acceptor is large. The initial and final states for the transfer are thus:

$$\phi_i(r_d,r_a;R_d,R_a) = \phi_{D^*}(r_d;R_d)\phi_A(r_a;R_a) \tag{120}$$

and

$$\phi_f(r_d,r_a;R_d,R_a) = \phi_D(r_d;R_d)\phi_{A^*}(r_a;R_a) , \tag{121}$$

where $\phi_D{}^*$ and ϕ_D are the excited and ground states of the donor, having energies $\varepsilon_D{}^*$ and ε_D, and $\phi_A{}^*$ and ϕ_A are the excited and ground states of the acceptor, having energies $\varepsilon_A{}^*$ and ε_A. The transfer probability per unit time between given initial and final states, ϕ_i and ϕ_f, is given by Fermi's golden rule in terms of the matrix element of the electromagnetic interaction, H_{DA}, as

$$P_{fi} = \frac{2\pi}{\hbar} \left|\langle\phi_f|H_{DA}|\phi_i\rangle\right|^2 \rho(E_f)\delta(E_f-E_i), \tag{122}$$

where $\rho(E_f)$ is the density of final states, and the Dirac brackets denote integration over all electronic coordinates. Due to H_{DA} being small the density of final states takes the form of a product, $\rho(E_f)=\rho_D(\varepsilon_D)\rho_A(\varepsilon_A{}^*)$, and the initial and final state energies are $E_i=\varepsilon_A+\varepsilon_D{}^*$ and $E_f=\varepsilon_A{}^*+\varepsilon_D$. The transition energy for the donor is $\Delta E=\varepsilon_D{}^*-\varepsilon_D$, and the delta function in equation (122) implies that $\varepsilon_{A^*}-\varepsilon_A=\Delta E$ as well.

In order to compute the transfer rate it is, of course, necessary to take a weighted average over all initial states and sum over all final states which contribute to the process. With the assumption of no electronic degeneracy, the only remaining summation needed over initial and final states is then due to the phonon broadening of the levels. It is convenient, in the weighted average over initial states, to employ probability distributions $W_D{}^*(\varepsilon_D{}^*)$ and $W_A(\varepsilon_A)$ which are normalized to unity. Then the transfer rate is given by

$$P = \int d\varepsilon_{D^*} d\varepsilon_A d\varepsilon_D d\varepsilon_{A^*} P_{fi} W_{D^*}(\varepsilon_{D^*}) W_A(\varepsilon_A). \qquad (123)$$

Substitution of equation (122) into this expression gives

$$P = \frac{2\pi}{\hbar} \int d(\Delta E)\, d\varepsilon_{D^*} d\varepsilon_A \left| \langle \phi_f; \varepsilon_A + \Delta E, \varepsilon_{D^*} - \Delta E | H_{DA} | \phi_i; \varepsilon_A, \varepsilon_{D^*} \rangle \right|^2$$

$$\rho_D(\varepsilon_{D^*} - \Delta E)\, \rho_A(\varepsilon_A + \Delta E)\, W_{D^*}(\varepsilon_{D^*})\, W_A(\varepsilon_A), \qquad (124)$$

where $|\phi;\varepsilon_A,\varepsilon_D\rangle$ is the state ϕ in which the acceptor has energy ε_A and the donor has energy ε_D. The interaction Hamiltonian, H_{DA}, is

$$H_{DA} = \frac{e^2}{\epsilon} \left\{ \frac{1}{|L + r_a - r_d|} - \frac{1}{|L + r_a|} - \frac{1}{|L + r_d|} + \frac{1}{L} \right\} \qquad (125)$$

which can be seen from Figure 4. Here ε is the appropriate dielectric constant for the material. Note the last three terms in equation (125) cannot couple initial and final states in which both the donor and acceptor have made a transition, and therefore, do not contribute to the transfer rate. When only the first term is retained, and the result is expanded in powers of L

$$H_{DA} = \frac{e^2}{\epsilon L^5} [r_d \cdot r_a L^2 - 3(r_d \cdot L)(r_a \cdot L)]. \qquad (126)$$

From (126) and (124), we find

$$P = \frac{4\pi e^4}{3\hbar \epsilon^2 L^6} \int d(\Delta E) \left\{ \int d\varepsilon_A [|\langle \varepsilon_A + \Delta E | r_a | \varepsilon_A \rangle|^2 \rho_A(\varepsilon_A + \Delta E) W_A(\varepsilon_A)] \right. \qquad (127)$$

$$\left. \int d\varepsilon_{D^*} [|\langle \varepsilon_{D^*} - \Delta E | r_d | \varepsilon_{D^*} \rangle|^2 \rho_D(\varepsilon_{D^*} - \Delta E) W_{D^*}(\varepsilon_{D^*})] \right\},$$

where averaging over all possible orientations of L has been employed.

We now write equation (127) in terms of the transition probabilities for the emission and absorption of a real photon in the dipole approximation. Note that since the interaction involved in (127) is the coulomb interaction, the coupling involves virtual longitudinal photons, whereas real photon processes involve transverse photons. However, in the non-relativistic dipole approximation these two interactions are proportional to one another.

The dipole transition rate for emission at an energy donor is

$$P_{D^*}(\varepsilon_{D^*}) = \frac{4e^2(\Delta E)^3}{3\hbar^4 c^3} \left|\langle\varepsilon_{D^*}-\Delta E|r_d|\varepsilon_{D^*}\rangle\right|^2 \rho_D(\varepsilon_{D^*}-\Delta E)$$

and for absorption (per incident photon) at an energy acceptor is

$$P_A(\varepsilon_A) = \frac{4\pi^2 e^2(\Delta E)}{3\hbar c} \left|\langle\varepsilon_A+\Delta E|r_a|\varepsilon_A\rangle\right|^2 \rho_A(\varepsilon_A+\Delta E),$$

where ΔE is the energy of the photon emitted or absorbed. We can define the normalized line shape functions for the emission and absorption as

$$f_{D^*}(\Delta E) \equiv \eta_{em} \int d\varepsilon_{D^*} W_D(\varepsilon_{D^*}) P_{D^*}(\varepsilon_D) \tag{128}$$

and

$$f_A(\Delta E) \equiv \eta_{ab} \int d\varepsilon_A W_A(\varepsilon_A) P_A(\varepsilon_A) \tag{129}$$

where η_{em} and η_{ab} are normalization constants. They are related to the mean radiative life time, τ_{D^*}, of the donor state ϕ_{D^*} and the total absorption, σ_A, of the acceptor from the state ϕ_A to the state ϕ_{A^*} as follows: $\eta_{em}=\tau_{D^*}$ and $\sigma_A=1/\eta_{ab}$. We can now use equations (128) and (129) in equation (127) to find that

$$P = \frac{3c^4\hbar^4\sigma_A}{4\pi\epsilon^2 L^6 \tau_{D^*}} \int d(\Delta E)\, \frac{f_{D^*}(\Delta E) f_A(\Delta E)}{(\Delta E)^4}\,. \tag{130}$$

This equation gives the probability of resonant energy transfer between dopants and defects, with broad band optical spectra where the broadening is due to electron-phonon interaction, in terms of the overlap of the emission spectrum of the donor and the absorption spectrum of the acceptor, as illustrated in Figure 5.

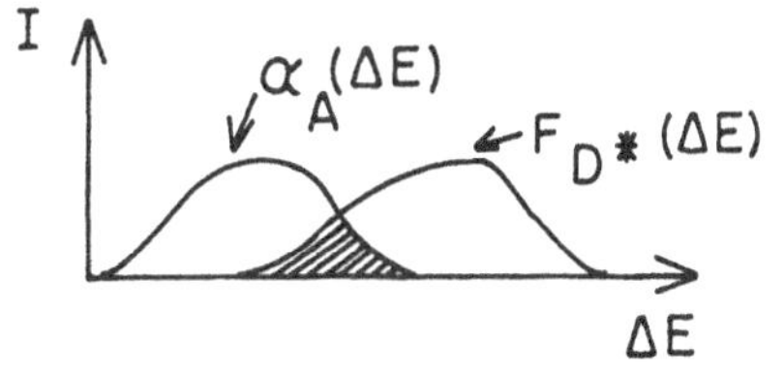

Fig. 5. Emission and absorption spectra of the energy donor and acceptor, respectively, used in the calculation of the resonant energy transfer process.

If the leading non-zero coupling term is a dipole-quadrupole or quadrupole-quadrupole term in equation (126) the the transfer probability is proportional to L^{-8} or L^{-10}, respectively. These are important when the L^{-6} dipole-dipole term vanishes due to symmetry. If determinantal Hartree-Fock two-electron functions are used in place of equation (120) and (121) exchange terms result giving an exponential dependence on L. The exchange terms are important for energy donor and acceptor systems for which the separation distance is small.

Resonant energy transfer provides an additional decay mechanism for the excited donor, and therefore, the lifetime of the donor as measured by the decay of its characteristic fluorescence is decreased. This clearly distinguishes this process from radiative transfer by sequential donor emission and acceptor absorption, which would not change the lifetime of the donor excited state. Note that the dependencies on donor and acceptor concentrations of the transfer probability can be used to extract the multipole character of the transfer. The dependence on the overlap of the donor emission and acceptor absorption contains evidence of the lattice relaxation before transfer, in contrast to exciton motion. Conservation of electronic energy during the transfer distinguishes resonant transfer from non-resonant transfer.

If we consider the theory outlined above from the point of view of electromagnetic field theory, the essentials of the calculation connecting the energy transfer probability to the emission spectrum of the energy donor and the absorption spectrum of the acceptor are: (1) that the electromagnetic interaction is a vector interaction, and therefore, there are really three parts to the interaction; (2) that in free space, these three parts correspond to transfer of either a transverse photon, with two independent polarization components, or a longitudinal photon; (3) that the longitudinal photon interaction is a pure coulomb interaction and that the transverse photon interaction involves emission or absorption of radiation; (4) that the energy transfer calculation involves the coulomb interaction whereas the radiative transitions involve the transverse photon interaction; and (5) that in the calculation considered above, we expressed the longitudinal photon interaction in terms of the transverse photon interaction. This latter is done by expanding each interaction as a power series of the coordinates of the electronic charges and then eliminating these coordinates.

Included in both the energy transfer process and the emission and absorption processes are the variations of the electronic charge distributions due to the motion of the ions of the lattice. The important changes in the charge distributions for the energy donor and acceptor for this problem are summed up in the probability functions $W_{D^*}(\varepsilon_{D^*})$ and $W_A(\varepsilon_A)$. The explicit forms of these functions are not needed, since they can be eliminated from the energy transfer calculation by using the measured results for the emission spectrum of the donor and the absorption spectrum of the acceptor, in which they are included. This is only possible if the initial state of the lattice plus electronic particles of interest is the same for the energy transfer process and for the radiative transition process. That is, that the lattice motion is the same in both cases. In particular, that for both processes the initial state is the relaxed lattice state, which is only possible for the transfer process if the transfer rates are slow compared to the relaxation process.

III.B. Nonresonant Energy Transfer

For dopants with weak electron-phonon interactions, for example, trivalent rare earth dopants (RE^{3+}), the optical spectra consist of zero-phonon (ZP) lines and their phonon replicas. Energy transfer between these types of dopants may not be possible by the resonant transfer mechanism due to a lack of overlap of the spectra of the donor and acceptor. However, phonon-assisted transfer is possible. This may involve, for moderately small energy mismatch, creation and/or annihilation of a phonon at the sites of either, or both, the donor and the acceptor, and for large energy mismatch, many-phonon processes. For very small mismatch, the creation of one phonon and annihilation of another of a slightly different energy may be required. The latter, two-phonon process can have a higher probability than a single-phonon process involving a low energy phonon because of the low density of states of these latter phonons. In addition to phonon-assisted processes which create or annihilate real phonons, there are also phonon-assisted transfer processes involving virtual phonon exchange, in which a phonon is first annihilated/created at one site and then created/annihilated at the second site. Orbach [3] first showed the importance of the phonon-assisted processes in connection with the transfer between isolated Cr^{3+} ions in ruby, which leads to capture of the excitation at Cr^{3+} pairs. Imbusch [30] had suggested earlier that phonon-assisted transfer could be responsible for this since the calculated resonant transfer process was too small to account for the observed transfer rate. In these situations the calculated resonant transfer rate is small because of the large separation of the emission and absorption bands of the energy donor and acceptor in comparison to the homogeneous line widths.

The simplest non-resonant transfer is the single-phonon creation or annihilation process. The corresponding energy transferred to the acceptor is $E_{ZPD} \mp \hbar\omega_q$, where E_{ZPD} is the zero phonon transition energy of the donor, $\hbar\omega_q$ is the phonon energy and the plus and minus signs go with the annihilation and creation of the phonon, respectively. The Stokes process, in which a phonon is created, is the more probable. The phonon involved may correspond to either a local or normal mode depending on the dopant and the crystal. For the situation in which there are many phonons involved, that is, $|E_{ZPD}-E_{ZPA}|$ is greater than any of the phonon energies available for energy conservation, where E_{ZPA} is the zero phonon transition energy of the acceptor; the probability of a sequential phonon creation or annihilation of n phonons all having the same energy, $\hbar\omega_q$, is proportional to $\exp[-\beta\Delta\varepsilon]$, where $\hbar\omega_q=\Delta\varepsilon/n$ and $\Delta\varepsilon = |E_{ZPD}-E_{ZPA}|$. This process is observed for RE^{3+} doped crystals and glasses [31].

The actual calculation of the phonon-assisted energy transfer is more complex than for resonant transfer; however, the basic ideas are the same. In a lattice at any temperature there is a given magnitude and form of the lattice vibration. If we view this lattice vibration in terms of phonons, then there exists a corresponding equilibrium distribution of phonons. It is this equilibrium distribution and its effects on the energies and charge distributions of the energy donors and acceptors which are represented in the resonant energy transfer calculation by $W_{D^*}(\varepsilon_{D^*})$ and $W_A(\varepsilon_A)$, and which are also present in the calculation of the non-resonant energy transfer process probability. In the non-resonant energy transfer calculation, we use the radiative transition spectra of the donor and acceptor to evaluate the integrals of these functions times the electronic transition matrix elements involved in the transfer process,as was done in the resonant energy transfer calculation. However, the difference between the resonant and the non-resonant transfer calculations is that the latter also involves the creation or annihilation of a real phonon, that is, an extra or missing phonon with respect to the equilibrium phonon distribution. The perturbation used in the non-resonant transfer calculation is the sum of the electromagnetic interaction term used in the treatment of the resonant transfer problem and the electron-phonon interaction terms discussed in subsection IIE. In first order perturbation theory, the electromagnetic interaction corresponds to the resonant transfer process, radiative transitions, etc., and the electron-phonon interaction to phonon creation and annihilation plus the interaction between the virtual phonons and the energy donors and acceptors, which causes the broadening of their levels. In higher order perturbation theory, there are terms which cause changes in the number of phonons only, terms which cause pure electronic transitions that can give rise to additional effects in the resonant energy transfer and in the radiative transitions, and finally terms which involve both the electron-phonon interaction and the electromagnetic interaction that describe, in part, the non-resonant energy transfer process. For a single-phonon creation or annihilation transfer process we can use the mixed terms of the second order perturbation calculation.

In this case the matrix element to be used in Fermi's golden rule for the calculation of the transfer probability per unit time between the initial and final states is

$$\begin{aligned}\langle\phi_F|H_{DA}^{eff}|\phi_I\rangle = \sum \int dq' \Big\{ &\mp\langle\phi_D,\phi_{A^*},\{n_q\pm\delta_{qq'}\}|H_{DA}|\phi_{D^*},\phi_A,\{n_q\pm\delta_{qq'}\}\rangle \\ &\cdot \langle\phi_{D^*},\phi_A,\{n_q\pm\delta_{qq'}\}|H_{el-ph}|\phi_{D^*},\phi_A,\{n_q\}\rangle/\hbar\omega_{q'} \\ &\mp \langle\phi_D,\phi_{A^*},\{n_q\}|H_{DA}|\phi_{D^*},\phi_A,\{n_q\}\rangle \\ &\cdot \langle\phi_D,\phi_{A^*},\{n_q\pm\delta_{qq'}\}|H_{el-ph}|\phi_D,\phi_{A^*},\{n_q\}\rangle/\hbar\omega_{q'} \Big\}, \end{aligned} \quad (131)$$

where the summation is over terms with the upper or lower signs. The state $|\phi_D, \phi_A, \{n_q\}>$ is a state of the donor, acceptor and phonons where ϕ_D and ϕ_A indicate the states of the donor and acceptor, in accordance with the notation of equations (120) and (121) and $\{n_q\}$indicates the state of the phonons, where the n_q's are the phonon occupation numbers. Note that we must keep track of whether the phonon was created or destroyed at the donor or the acceptor. Note also that involved in equation (131) is the same matrix element of H_{DA} as was used in equations (122) and (124), except that the energy of the donor or acceptor has been shifted by $\hbar\omega_q$. The matrix element given in equation (131) is used with Fermi's golden rule which is then multiplied by $W_D{}^*W_A$, and integrated over $\varepsilon_D{}^*$, ε_A, $\varepsilon_A{}^*$ and ε_D to give the non-resonant transfer probability. The result of this calculation can be related to the radiative spectra of the donor and acceptor, from which it is seen that the non-resonant transfer rate contains terms which are proportional to

$$\rho_{ph}(\hbar\omega_q) \int d(\Delta E) \frac{f_{D^*}(\Delta E) f_A(\Delta E \pm \hbar\omega_q)}{(\Delta E)^3 (\Delta E \pm \hbar\omega_q)}$$

where $\rho_{ph}(\hbar\omega_q)$ is the density of states of the phonons and $\Delta E = \varepsilon_D{}^* - \varepsilon_D$ which is equal to E_{ZPD} for non-resonant zero phonon transitions. For details of this type of calculation for the non-resonant transfer rate see Orbach [3]. Note that due to the factor $\rho_{ph}(\hbar\omega_q)$ the transfer rate will increase as the energy mismatch increases so long as $\Delta\varepsilon$ is less than the maximum energy of any phonon that is available for the non-resonant process.

Finally, we note the effect designated as energy localization which is due to the spectra of the donor and/or the acceptor being inhomogeneously broadened. The inhomogeneous contribution to band widths can be distinguished and arises because dopants are located at sites with different crystal or ligand fields. The results for resonant and non-resonant transfer rates calculated above use the homogeneous line shape functions. It is found that the transfer rate, in general, decreases rapidly with increasing average separation between the donors and acceptors, the latter of which can be determined from the concentrations. Inhomogeneous broadening reduces the number of donors that can participate in transfer with a particular set of acceptors and vice versa, so that the rate will be smaller than that suggested by a calculation of the average separation which is based solely on the concentrations. A system in which this occurs is referred to as having energy localization. This effect is particularly important in disordered materials, such as glasses, but also exists for randomly distributed dopants in crystals, especially for higher dopant concentrations. This effect is significant for resonant energy transfer if the inhomogeneous broadening is large compared to the homogeneous broadening due to the electron-phonon interaction.

III.C. Electronic Charge Transport and Energy Transfer

In this subsection the transfer of energy associated with charge transport is discussed. We begin with some observations and discussion of terminology to be used. The energy associated with a charge is equal to the difference between its energy and the Fermi energy. In most systems either the mobile electrons or the mobile holes predominate in the equilibrium state. Since the characteristics of the charge transport depend more on whether or not it is the predominant carriers that are involved in the transport than the sign of the carriers, we differentiate the carriers as majority or minority. The term majority carriers designates those carriers predominant in number in the equilibrium state and the term minority carriers those which are not. When the electrons are the majority carriers the Fermi level is closer to the conduction band whereas the Fermi level would be closer to the valence band when the holes are the majority carriers. Although all charges which are transported through a material and have non-zero energy relative to the Fermi level can transfer energy, the only useable energy is that which the carriers as a whole have in excess of what they would have in the equilibrium state. This excess energy can exist either because there are excess electrons and/or holes as compared to the number in the equilibrium situation or because the carriers have additional kinetic energy. Note that here we are using the terms electron and hole to indicate an electron in a state above the Fermi level and an electron state below the Fermi level which is empty, respectively. It is only the excess energy that is useable since at equilibrium the lattice and charges have the same temperature. If energy could be extracted from the charges when they are in the equilibrium state, then their temperature would be decreased below that of the lattice, and this is only possible thermodynamically if there is a reservoir colder than the lattice available to the electrons.

Because of the electron-phonon interaction, carriers that are out of equilibrium with the lattice because they have excess kinetic energy usually return to equilibrium rapidly by creating phonons. Non-equilibrium can be maintained, however, if energy is supplied to the carriers at a rate equal to or greater than this rate of loss. On the other hand, when the non-equilibrium in a semiconductor or insulator is due to an excess of electrons and/or holes equilibrium can only be re-established by the recombination of carriers of opposite type or by charge transport. Recombination can be with either free or bound carriers and can be done either by multi-phonon or radiative recombination. The rate of recombination of electrons and holes from different bands by multi-phonon emission can be small if there are few states in the middle of the forbidden gap, which is the case for many semiconducting and insulating materials. If there are few charged donors and acceptors which can bind holes and electrons, and the radiative recombination rate is small, then the excess carriers can exist for a long enough time so that they can transfer energy while being transported through the material.

The question of supplying energy to the carriers in order to maintain them in a state in which they have an excess kinetic energy must be considered. Energy can be supplied by an electric field, F, to an electronic particle at the rate $e^2F^2\tau/m^*$, where e is the electronic charge, m^* the effective mass and τ the momentum relaxation time. In metals the fields that can be generated by external means are usually small, and therefore, the kinetic energy gained per unit time is usually smaller than the energy lost to the phonons. For this reason the carriers in metals are rarely far out of equilibrium with the lattice for any significant length of time, and therefore, the useable energy transferred when charges are transported is small.

On the other hand, for semiconductors and insulators there can be significant amounts of energy transferred via charge transport. There are two reasons for this. First, if there are excess minority carriers, then their associated energies are large, since these carriers are far from the Fermi level. Their energies are at least equal to half the width of the forbidden gap. This is the case whether or not they have large kinetic energies. Second, any of the carriers can have large excess energy because large electric fields can be developed in these materials, which can maintain the carriers in a state having excess kinetic energy. The large fields are possible because there are fewer mobile carriers in these materials as compared to metals. However, for this same reason there is less transport of charge in these materials even though there is large transport of energy per charge. It must also be noted that although the fields that can be maintained in these materials can be quite large, the distribution of the carriers which can be maintained or reached with these fields cannot have an arbitrarily large excess kinetic energy since most of the electron-phonon interaction matrix elements increase with the momentum of the phonon created, which in turn increases with the initial energy of the carriers being scattered. The one exception is the electron-optical phonon interaction in polar materials. Note that the minority carriers have a large potential energy in addition to their kinetic energy, whereas the majority carriers' energy is primarily kinetic. It is usual to distinguish carrier distributions having excess energy by whether or not they have large excess kinetic energy. The usual designation for carrier distributions with a large excess kinetic energy is to call them "hot" carriers. However, this terminology has some problems associated with it. First, it implies that there is only one way in which the carrier distribution can be out of equilibrium when there is an excess kinetic energy, which is not true. Second, it implies that all non-equilibrium distributions can be characterized by a temperature, that is, that they are either Maxwellian- or Fermi-Dirac-like distributions which are defined by a temperature, and this temperature is greater than the lattice temperature. However, there are distributions for which there is no characteristic temperature that can be defined. For example, there are distributions which contain ballistic and/or streaming electrons, which are not characterized by a temperature. Ballistic electrons are electrons for which the electron-phonon interaction is negligible, and therefore, do not lose any significant amount of energy to the lattice during transport. Streaming electrons are electrons for which there is only a weak coupling to the optical phonons, which can happen when they have a substantial amount of kinetic energy. This is possible in polar materials, since for large kinetic energy the electron-optical phonon interaction decreases with the energy of the electrons. If either ballistic or streaming electrons are present in a given situation, the distribution would have peaks narrow in comparison to the width of the Fermi-Dirac or Maxwellian peak for the same situation.

To study the transport properties we use either the Boltzmann equation or the master equation and/or its generalization. In the rest of this subsection we discuss the Boltzmann equation and the master equation. The generalized master equation will be discussed in the subsection on excitons. For a detailed discussion of the Boltzmann equation see the proceedings of the conference on the "Physics of Nonlinear Transport in Semiconductors" [32], and for the master equation see the works of Dresden [33] and Chester [34].

The Boltzmann equation has the form

$$\frac{\partial f}{\partial t} + \frac{p_o}{m^*} \cdot \nabla_{r_o} f(r_o,p_o,t) + eF\cdot\nabla_{p_o} F(r_o,p_o,t) = \left(\frac{\partial f(r_o,p_o,t)}{\partial t}\right)_{coll}, \qquad (132)$$

where $f=f(r_o,p_o,t)$ is the carrier distribution function and F the electric field. r_o and p_o are the position and momentum of a single carrier and $(\partial f/\partial t)_{coll}$ gives the change in the distribution due to collisions. It is this latter term which is the most difficult to determine, and it is this term in which the properties of the particular system are contained. We can rewrite the collision term as follows:

$$\left(\frac{\partial f}{\partial t}\right)_{coll} = \int d^3p_o'[f(p_o')w(p_o'\to p_o) - f(p_o)w(p_o\to p_o') \qquad (133)$$

where $w(p_1\to p_2)$ is the probability per unit time of a carrier being scattered from the state in which it has momentum p_1 to the state with momentum p_2. In the case of scattering by phonons, $w(p_1\to p_2)$ is given by

$$\sum_{\alpha\ \beta} A_{p_2\alpha;p_1\beta} \ .$$

$A_{p_2\alpha;p_1\beta}$ is given by equation (96). It should be noted that the terms in equation (132) containing $\nabla_{ro}f$ and $_{po}f$ are usually designated as the streaming terms.

The master equation has the form

$$\frac{\partial P_\alpha(t)}{\partial t} = \sum_{\substack{\beta \\ \beta\neq\alpha}} [P_\beta(t)\Omega_{\alpha\beta} - P_\alpha(t)\Omega_{\beta\alpha}], \qquad (134)$$

where $P_\alpha(t)$ is the probability at time t that the quantum mechanical system is in a group of states indexed by α where $\alpha=\{\alpha_1,\alpha_2,\ldots,\alpha_\nu,\ldots\}$. The $P_\alpha(t)$ are designated as the many-particle probability functions. The α_ν's are subsets of the quantum number set α indexing the states of the individual particles of the many-particle system. Note that the grouping of the states into groups indexed by the α's is usually done by having some of the particles' quantum numbers specified, with the rest of the particles having their quantum numbers varying over all possible values. For example, the electronic particles might belong to the first set of particles and the phonons to the second set. Note that the quantum numbers for the phonons would be their occupation numbers. The matrix elements $\Omega_{\alpha\beta}$ of the operator Ω are the transition probabilities per unit time from the group of states indexed by β to those indexed by α. The properties of the particular system being described are incorporated in the $\Omega_{\alpha\beta}$. Note that one of the assumptions made in deriving the master equation from the dynamical equation for the quantum mechanical density function is that the states indexed by the α's are weakly interacting and therefore the matrix elements of the operator Ω should be small.

We can define the one-, two-, etc. particle probability functions by summing the many-particle probability functions over all but one, two, etc. of the particle indexes. Dresden [33] has shown that under appropriate assumptions and with the use of the one-particle probability functions that the Boltzmann equation can be derived from the master equation. However, the streaming terms are missing because there are no corresponding terms in the master equation from which he started.

Note that the Boltzmann equation is a semiclassical equation in that the collision term is usually found quantum mechanically whereas the streaming terms imply a classical specification. In addition we should note that it is implicit in the Boltzmann equation that only infinitesimal changes are described. On the other hand, the master equation is more suited to systems in which discrete quantum jumps are important. Also note that once we have either f or $P_\alpha(t)$ then any prediction of experimental results that can be made can be found by taking weighted averages of the appropriate quantity weighted by f or $P_\alpha(t)$.

When solving the Boltzmann equation for real systems, since the collision term is usually complicated, it is usual to look for approximate solutions. We assume that the carriers are characterized by a temperature, which will be the case if the electron-electron scattering rate is large in comparison to the rate of change of the energy of the carriers as a whole. If this is the situation, a commonly used approximation is the displaced Maxwellian given by

$$f(P_o) = A \exp\left\{ -\left(\frac{P_o^2}{2m^*} - v_d \cdot P_o \right) / k_B T_e \right\}, \qquad (135)$$

where v_d and T_e are parameters to be determined, k_B the Boltzmann constant and A a normalization constant. The parameter T_e gives the elevated temperature of the carriers caused by the applied electric field. Note the analogy between the assumption made in this approximation for the Boltzmann equation and the assumption of the adiabatic approximation. In both, because of an internal fast rate of adjustment in a subsystem as compared to the energy supplied to or lost from the subsystem, a partial separation can be made between the internal subsystem, in this case the electrons, and the external subsystem, in this case the lattice and electric field, by including the effects of the external system on the internal system parametrically. Although it seems reasonable to assume this form as an approximation to the solution of the Boltzmann equation when the above conditions are satisfied, it is probably better to think of this solution as a particular two-parameter fit to the solution. The displaced Maxwellian is appropriate for situations in which the carriers are non-degenerate. If the carriers were degenerate then a displaced Fermi-Dirac distribution would be more appropriate.

The values of v_d and T_e can be determined in terms of the field, F, and the parameters describing the scattering mechanisms by making use of the balance equations for the energy and momentum. These equations are derived from the Boltzmann equation by first multiplying the Boltzmann equation by either the energy or the momentum and then integrating over the momentum space. The results are:

Momentum Balance Equation

$$eF = \int d^3p_o \int d^3p'_o(p'_o - p_o)w(p'_o \rightarrow p_o)f(p'_o) \qquad (136)$$

Energy Balance Equation

$$ev_d \cdot F = \int d^3p_o \int d^3p'_o \left(\frac{p'^2_o}{2m^*} - \frac{p_o^2}{2m^*} \right) w(p'_o \rightarrow p_o)f(p'_o). \qquad (137)$$

Equations (136) and (137) can be solved for v_d and T_e with which $f(p_o)$ is then given. Note that the displaced Maxwellian solution is appropriate for situations in which there is spatial homogeneity. If an improved parametric solution is wanted, a ten parameter solution could be used since there are ten independent conserved quantities, in general, for any three dimensional problem to which there correspond ten balance equations. These ten balance equations could be used therefore to specify the ten parameters. In essence we have chosen arbitrarily six of these parameters when we use the displaced Maxwellian approximation. The reason for doing this is that in most situations it is difficult to specify the form of the additional balance equations. If we have found the $f(p_0)$ for the displaced Maxwellian distribution then we can determined the momentum and energy relaxation times, which are defined as follows:

$$\tau_m = \frac{d|\langle p_o \rangle|}{d(e|F|)} \tag{138}$$

and

$$\tau_E = \frac{d\langle p_o^2/2m^* \rangle}{d(eF\cdot v_d)} , \tag{139}$$

where $\langle\phi\rangle \equiv \int dp_0\, f(p_0)\phi(p_0)$, which is the average value of $\phi(p_0)$. The two parameters τ_m and τ_E can be used to define an approximate parametric solution of equation (132) which is more general than the displaced Maxwellian approximation. In most cases τ_E is much longer than τ_m, which allows for some interesting effects. For example, the approach to equilibrium does not have to be monotonic. That is, there can be an overshoot for a short length of time [35]. In the above discussion of the displaced Maxwellian distribution it was implicitly assumed that the energy bands were parabolic. We can take care of the explicit band shape by using the actual energy-momentum relationship in the displaced Maxwellian and in the energy balance equation. The simplest way to do this is by assuming that the effective mass is momentum dependent.

If we consider a transport problem for which there is a fundamental quantum discreteness then it would be more appropriate to use either the master equation or a generalization of it. An example of this type of problem is the transport of carriers across an interface where the quantum discreteness involved is that the carriers are either on one side of the interface or the other. In this situation a possible subdivision of the states is into states for which the particle is on one or the other side of the interface. Equation (134) then takes the form

$$\frac{\partial P(i,\varepsilon,t)}{\partial t} = \int d\varepsilon'[P(j,\varepsilon',t)\Omega_{ij}(\varepsilon,\varepsilon')-P(i,\varepsilon,t)\Omega_{ji}(\varepsilon',\varepsilon)], \tag{140}$$

where $i\neq j$, $i,j=1,2$ index the states for the two sides of the interface and ε is the energy of the particle. Once the Ω's and $P(i,\varepsilon,0)$ are given, then we have a set of coupled linear integro-differential equations to solve for $P(i,\varepsilon,t)$. Note that the linearity of the master equation is in contradistinction to the Boltzmann equation which is nonlinear in f.

When there is transport of charge the energy of the carriers can be transferred to charged donors or acceptors by capture of cool carriers. The kinetic energy of the "hot" carriers can be transferred to localized centers by collision excitation or by Auger processes. The latter two processes will be discussed later in this section.

III.D. Energy Transfer by Excitons

1. Exciton Structure. As introduced in Section I, excited states of crystals in which the electron and positive hole are correlated and propagate as a neutral particle, the exciton, constitute and important mechanism of energy transfer in semiconductors and insulators. The exciton is an elementary excitation, a boson, and an eigenstate of a periodic Hamiltonian, however, its description goes beyond the one-electron approximation of subsection II.B and the electronic band structure of subsection II.C. Nevertheless, the analyses of the properties of excitons depend on the concepts developed in these subsections. As noted in Section I, excitons can be divided into two classes: Wannier (effective mass) and excitons and Frenkel (tight binding) excitons. In the following we shall outline the formulation of these, then consider generalized excitons as elementary excitations, and then analyze their energy transport properties.

A simplified treatment of the Wannier exciton can be done using effective mass theory. A derivation of the effective mass approximation is presented in the appendix. Here we use a generalization of two-particle states. The zero-order Hamiltonian is now

$$H_o = T_e + T_h + V_{e-L} + V_{h-L}, \tag{141}$$

where the potentials are the interactions between the particles of interest and the lattice, and T_e and T_h are the kinetic energy operators. In the approximation of a continuous dielectric background the interaction between the electron and hole is

$$H'(r_e, r_h) = -\frac{e^2}{\varepsilon |r_e - r_h|}, \tag{142}$$

where r_e is the electron coordinate, r_h is the hole coordinate and ε is the dielectric constant. H' is assumed weak compared to the crystal potential; equivalently the average separation is assumed large compared to unit cell dimensions in order for the continuum approximation of equation (142) to be valid. The zero-order wavefunctions are simple products of Bloch functions since the particles are distinguishable. The exciton wavefunction is then represented as an expansion in these functions

$$\phi_{ex} = \sum_{n,m} \int F_{nm}(k_e, k_h)\psi_n(k_e, r_e)\psi_m(k_h, r_h) d^3k_e d^3k_h, \tag{143}$$

where the sum is over bands, and the integrations are over a Brillouin zone. The Schrödinger equation for the exciton wavefunction is then

$$[H_o + H']\phi_{ex} = E\phi_{ex}. \tag{144}$$

From here, the procedure is essentially the same as that shown in the appendix, except that the basis states are the two-particle products used in equation (143). The resulting equation is

$$\begin{aligned}
&[E_{nm}(k_e, k_h) - E]F_{nm}(k_e, k_h) \\
&+ \sum_{n',m'} \sum_{\alpha,\beta} \int F_{n'm'}(k_e', k_h') C_{nn'}(k_e, k_e', K_\alpha) C_{mm'}(k_h, k_h', K_\beta) \\
&\times H'(q) d^3k_e' d^3k_h' = 0.
\end{aligned} \tag{145}$$

Here $q = k_e - k_e' - K_\alpha + k_h + k_h' + K_\beta$, H'(q) is the Fourier transform of the H' given in equation (142), and E is the eigenvalue in equation (144). The approximations made here will be the same as those made in the appendix, namely that a single band is important for each particle (a conduction band for the electron and a valence band for the hole); the bands are isotropic, with extrema at k=0; the F functions are localized about k=0; and all the C's are approximately equal to 1. Then the following momentum-space Schrödinger equation results:

$$\left[\frac{\hbar^2 k_e^2}{2m_e^*} + \frac{\hbar^2 k_h^2}{2m_h^*} + H'(q) + E_g - E\right] F(k_e,k_h) = 0, \tag{146}$$

where E_g, the energy gap, merely sets the position of the energy scale. In order to transform this equation to real space, we define the double Fourier transform of $F(k_e,k_h)$ by

$$F(k_e,k_h) = \int F(r_e,r_h)\exp[-ik_e\cdot r_e - ik_h\cdot r_h]d^3k_e d^3k_h. \tag{147}$$

Then the effective mass equation is

$$\left[-\frac{\hbar^2}{2m_e^*}\nabla_e^2 - \frac{\hbar^2}{2m_h^*}\nabla_h^2 + H'(r_e,r_h) + E_g - E\right]F(r_e,r_h) = 0. \tag{148}$$

This is, of course, just the Schrödinger equation for the hydrogen atom. The internal and center of mass motion can be separated by making the coordinate definitions

$$r = r_e - r_h,$$

$$R = \frac{m_e^* r_e + m_h^* r_h}{m_e^* + m_h^*}. \tag{149}$$

Then the effective mass function F can be written as a product

$$F(r_e,r_h) = f(r)G(R), \tag{150}$$

and f and G satisfy

$$\left[-\frac{\hbar^2}{2\mu^*}\nabla_r^2 - \frac{e^2}{\varepsilon r}\right]f(r) = (E - E_g - \mathcal{E})f(r), \tag{151}$$

and

$$-\frac{\hbar^2}{2(m_e^*+m_h^*)}\nabla_R^2 G(r) = \mathcal{E}G(r), \tag{152}$$

respectively, where

$$\mu^* = \frac{m_e^* m_h^*}{m_e^*+m_h^*}. \tag{153}$$

The eigenvalues of equation (151) are just those of the hydrogen atom in the center of mass frame, and the eigenvalue of equation (152) is just the total kinetic energy of the exciton. Note that contrary to the usual practive in solving the hydrogen atom problem, we do not assume that the total momentum of the electron and hole system is zero. Thus, the energy of the exciton (E in equation (144) is

$$E = E_g + \frac{\hbar^2 K^2}{2(m_e^* + m_h^*)} - \frac{\mu^* e^4}{\varepsilon^2 2\hbar^2 n^2}. \qquad (154)$$

This approach is rather limited in its applicability to real systems but is of interest nevertheless due to the clear physical picture it affords. Unfortunately, it is somewhat difficult to generalize this calculation, so for accurate calculations it is best to use an approach which is more general from the start. We refer the reader to the review by Knox [36] for a more general treatment.

The Frenkel exciton includes excitons which can be approximated in terms of intramolecular excitation and those approximated by charge transfer transitions. Excitonic states of alkali halides were early investigated by von Hippel [37] and their energies approximated from charge transfer between nearest-neighbor anion and cation

$$E_{ex} = \chi - I + \frac{(2A - 1)}{R_0} + E_{pol}, \qquad (155)$$

where χ is the electron affinity of the anion, I is the first ionization energy of the cation, A is Madelung's constant, R_0 is nearest-neighbor anion-cation distance in the crystal and E_{pol} is the polarization energy.

Intramolecular excitons have been investigated more recently by more sophisticated methods. For these a tight binding approach using approximations to the Wannier functions is useful. For crystals, such as rare gas solids and some organic crystals, the interactions between the molecules are weak, i.e. van der Waals interaction only, and excitonic wavefunctions can be described as follows:

$$\phi_{ex}(K) = \sum_j A_*(r-R_j) \prod_{i \neq j} A(r-R_i) e^{iK\cdot(r-R_j)}, \qquad (156)$$

where $A_*(r - R_j)$ describes an excited molecule at the jth site; $A(r - R_i)$, ground state molecules at the ith site; and the sum over sites including the exponential factor makes $\phi_{ex}(K)$ an eigenstate of the periodic Hamiltonian of the crystal. Frenkel excitons in aromatic organic crystals such as anthracene have been extensively investigated by Pope and collaborators [38]; excitons in rare gas crystals by Zimmerer [39] and Hahn and Schwentner [40]. Intracore excitons have been investigated for some materials, including alkali halides [41].

Excitons can be generalized as elementary excitations and formulated as follows:

$$\phi_{ex}(K) = \sum_{i,j} f(R_i - R_j) e^{iK \cdot R_j} c_{jc}^{+} c_{iv} \phi_g, \qquad (157)$$

where ϕ_g is the electronic wave function and is usually taken to be of the Hartree-Fock form, c_{iv} is the annihilition operator for a valence electron at the ith site, c_{jc}^{+} is the creation operator for a conduction electron at the jth site, and $f(R_i - R_j)$ is a correlation function for the two operators. Some limiting cases are:

$f(R_i - R_j) = \delta_{ij}$	intrasite exciton, e.g. core exciton
$f(R_i - R_j) = \delta_{i,i\pm 1}$	charge transfer exciton, e.g. alkali halide
$f(R_i - R_j) = F(r_e, r_h)$	effective mass exciton, e.g. III-V semiconductors.

2. Exciton Transport. Now we shall briefly examine the problem of exciton transport. It has been pointed out [42] that the study of exciton transport can be viewed an encompassing all forms of spatial energy transfer in which mass and charge are not transported. Here the point of view is that any quantum of electronic excitation whether localized at a lattice site or impurity, or delocalized, regardless of the state of relaxation of the lattice is an exciton. In this view, the resonant and non-resonant energy transfer problems can be seen as a part of the somewhat more general topic of exciton motion. Thus, the study of the motion of excitons is of considerable importance in the process of sensitized luminescence, whether in organic or inorganic crystals. It is also fundamental to many other processes. For example, it is known that in photosynthesis the initial excitation must migrate to particular sites where trapping occurs and the chemistry of sugar synthesis takes place.

The microscopic interactions responsible for the motion of excitons have been discussed earlier in the sections on resonant and non-resonant energy transfer, so we will deal in this section only with the macroscopic, or statistical, treatment of the transport problem. It is instructive to consider first the two limiting cases for the type of motion undergone by excitons as this is one of the principal issues involved in modern theories of exciton transport. In one limit the motion can be characterized as wave like, that is, an exciton wave characterized by a wave vector k_0 propagates through the crystal with little scattering by phonons. This type of motion is generally termed "coherent", and it is well described by the microscopic equation governing the system, the Schrodinger equation. In the opposite limit, the motion is that of a random walk through the lattice, that is, a diffusion process. This type of motion is called "incoherent" or "diffusive" and is best described by a master equation. The latter type of motion is clearly an irreversible process, whereas the former is reversible. The source of the irreversibility is the interaction between the excitons and phonons.

In general, the situation in real crystals lies somewhere between these two limits, and in order to relate the microscopic interactions responsible for exciton motion to macroscopic observables it is necessary to develop statistical procedures which include aspects of both. Previous techniques, such as the diffusion equation approach and the master equation approach, dealt only with incoherent motion. A great deal of work on this problem has appeared recently, principally dealing with the motion of Frenkel excitons in molecular crystals. Since the classic review of Knox [36], several new approaches have been developed to treat both the coherent and incoherent aspects of exciton motion. Among these are: 1) a polaron-like treatment, due to Grover and Silbey [43], [44], 2) a stochastic Liouville equation approach, initiated by Haken and Strobl [45], and 3) a generalized master equation approach, originated by Kenkre and Knox [42], [46]. Reviews of the latter two, together with extensive lists of references, have been published recently by Kenkre [47] and Reineker [48]. Our discussion is based primarily on these reviews.

The application of generalized master equations (GME's) to exciton transport is the most recent of the methods to appear. It was developed in order to unify the treatment of coherent and incoherent exciton motion in a very general way within a single framework. The appropriate GME is

$$\frac{dP_m(t)}{dt} = \int_0^t \sum_n [W_{mn}(t-t')P_n(t') - W_{nm}(t-t')P_m(t')]dt', \quad (158)$$

where $P_m(t)$ is the probability of occupancy of site m, and the $W_{mn}(t - t')$ are "memory functions", that is, functions which serve to include an appropriate amount of dependence on "past history" in the equation. This non-locality in time means that the GME is non-Markovian. In the limit of infinitely rapidly decaying memory, $W_{mn}(t - t') = \Omega_{mn}\delta(t - t')$, the GME becomes the ordinary Markovian master equation and yields

$$\frac{dP_m(t)}{dt} = \sum_n [\Omega_{mn}P_n(t) - \Omega_{nm}P_m(t)], \quad (159)$$

and is appropriate to the description of completely incoherent motion. The cause of the decay in the memory functions is the interaction of the excitons with the reservoir of phonons. In the limit of no interaction with the phonon reservoir (infinitely long memory), the GME becomes the microscopic equation for the diagonal elements of the density matrix, and is then a description of completely coherent motion. Thus, the fundamental problem of the GME approach is now that of determining the memory functions from the microscopic dynamics. In some cases this can be done exactly. It is quite interesting that in some cases in which an exact calculation is not possible, the memory functions can be related directly to measurements of spectra. For more details on the calculation of memory functions in specific cases, the reader is referred to the review of Kenkre [47].

The actual memory functions appropriate for a given system are, in general, somewhere between the limits of infinitely short and infinitely long memory, and thus, their effects may or may not be apparent on comparison with particular experiments. The solutions of the GME show coherent behavior only for times short in comparison to a parameter called the "coherence time". On this basis, one would expect the effects of the coherence to be observable only in experiments with temporal resolution finer than the coherence time for the system being observed. However, even in steady-state sensitized luminescence experiments coherence can be observed if the radiative lifetime of the excitons is comparable to, or shorter than the coherence time. In this case, the excitons move coherently during most of their existence, and transfer rates are affected. At high temperatures, the increased phonon density causes the motion of excitons to be less coherent than at low temperatures. Therefore, in this regime, as well as in those steady-state experiments where the exciton radiative lifetime is larger than the coherence time, the memory functions would be close to those appropriate for incoherent motion, and an ordinary Markovian master equation serves to describe the transport properties adequately.

Whereas the GME approach unifies the treatment of coherent and incoherent motion in a very general way through the use of memory functions, the stochastic Liouville equation (SLE) approach treats the two types of motion in a more specific way as the sum of two contributions. The method was developed by Haken and Strobl [45], and has undergone considerable development since then by Haken, Reineker, Strobl and others [49], [50]. Kenkre [47], [51] has pointed out that the method developed by Grover and Silbey [44] results in a transport equation which is formally identical to the SLE. For this reason, we will not discuss the Grover and Silbey method separately. The form of the SLE is

$$\frac{d\rho_{mn}(t)}{dt} = -i\,[H_0,\rho]_{mn} - 2\delta_{mn}\sum_k \gamma_{|m-n|}(\rho_{mm} - \rho_{nn}) \qquad (160)$$

$$- 2(1 - \delta_{mn})\Gamma\rho_{mn} + 2(1 - \delta_{mn})\bar{\gamma}_{|m-n|}\rho_{nm},$$

where the ρ_{mn} are the density matrix elements, $[\cdot,\cdot]$ denotes the commutator, H_0 is the non-core electronic Hamiltonian,

$$\Gamma = \sum_n \gamma_{|m-n|}$$

and the various γ's are rates which we discuss briefly below. The first term on the right-hand side represents the coherent motion of excitons. Note that it, together with the left-hand side, is formally identical to the von Neumann equation for the density matrix. The second term on the right-hand side represents incoherent motion with rates given by $2\gamma_{|m-n|}$ for sites $|m - n|$ apart. The diagonal matrix elements present in this term represent the probability of occupancy of particular lattice sites by the exciton. Note that if only this term is kept on the right-hand side, the master equation results. The remaining terms, involving only off-diagonal elements, represent interactions with the phonons. The first of these describes the decay of phase relations between the lattice sites, the decay constant 2Γ being independent of site.

Kenkre [47] has shown that the SLE with its last term neglected can be put into the form of a GME with memory functions given by

$$W_{mn}(t) = e^{-2\Gamma t}\, W^c_{mn}(t) + 2\gamma_{mn}\delta(t), \qquad (161)$$

where $W^c_{mn}(t)$ is the memory function appropriate to completely coherent motion, i.e. with no interaction with the phonons. The second term clearly represents incoherent motion, due to its infinitely rapid decay in time. We can see from this comparison that the SLE method appears to contain somewhat less generality in the type of coherence it can describe. However, this is offset by the fact that it is an equation which gives off-diagonal as well as diagonal elements of the density matrix, whereas, the GME only gives the diagonal matrix elements.

III.E. Auger Processes as Energy Transfer

Auger processes involve the transfer of energy between electronic particles in such a way that one of the particles in the final state lies in a continuum. In solids this can be viewed as a three-particle process in which two carriers collide, causing one of them to recombine with a carrier of opposite type, the binding energy being given up to the remaining collision partner. The recombining particles may be intrinsic carriers, or one or both may be bound to dopants or defects, or they may be bound together as an exciton, either free or trapped. The concept of Auger transitions in solids has been applied to metals as well as semiconductors, but it is of greatest utility in the latter, and the bulk of research has been concentrated there.

While Auger transitions do not themselves involve spatial energy transfer over significant distances, they do involve transfer of energy from one particle to another, and the final state always involves having at least one particle in a continuum state, usually with non-zero kinetic energy, so that energy is then available to be transferred by particle transport. For this reason, we take the view that Auger processes constitute a form of energy transfer, but in themselves, only in momentum space (the latter remark is due to Landsberg [6]), that is, there is transfer of energy from one state or group of states localized in momentum space to another. It is worth emphasizing that Auger decay of an exciton or excited impurity is a competitive process to both spatial energy transfer (i.e. resonant or non-resonant transfer, or exciton transport) and radiative decay.

It is possible to envision a wide variety of different types of Auger processes in semiconductors. Generally it is convenient to classify them into four distinct groups according to whether they are intrinsic (occur in a pure material) or extrinsic (involve impurity or defect states), and whether they are phonon-less or phonon assisted. Distinctions are also often made between processes occurring in direct-gap materials (direct processes) and those occurring in indirect-gap materials (indirect processes). We shall follow the convention of denoting those processes involving collision between two electrons with one of them receiving the energy from the recombination of the other with a hole as eeh processes, and the corresponding collisions of two holes as ehh processes.

The simplest of the Auger processes is the phononless band-band Auger process in direct-gap materials, which was investigated in detail by Beattie and Landsberg [52]. The eeh version of this process is shown in Figure 6. Here the lack of phonon assistance means that energy and momentum conservation must be satisfied by the two colliding particles (electrons or holes). As a result, it is not possible for either of the particles to undergo a band edge-to-band edge transition ($\Delta E=E_g$, $\Delta k=0$). Thus, there must be an activation energy associated with the process, and it is related to the size of the energy gap. For direct-gap materials with large gaps, the activation energy is large, so that the probability of this type of transition is expected to be small.

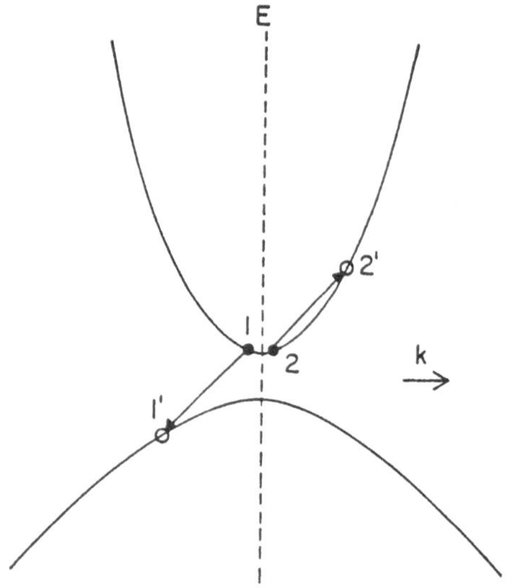

Fig. 6. Direct, Phononless band-band Auger process (eeh), showing conservation of energy and momentum.

In their treatment of this process, Beattie and Landsberg used perturbation theory, starting from zero-order states which were taken to be Bloch states, with the perturbation being the screened electron-electron interaction

$$H' = \frac{e^2}{\varepsilon r} e^{-\lambda r}, \tag{162}$$

where ε is the dielectric constant and $1/\lambda$ is a screening length. The dielectric constant takes account of screening by inner electrons, and $1/\lambda$ takes account of screening by conduction electrons.

In order to obtain the net recombination probability, it is necessary to take the difference between the probabilities for the forward (Auger) process and the reverse (impact ionization) process. If energy is conserved, the net probability is zero unless the electrons in the bands of interest are not in equilibrium. Beattie and Landsberg considered the electrons within each band to be in equilibrium among themselves but not with those of the other band. They defined quasi-Fermi levels for each band, and it was the difference in these quasi-Fermi levels that was responsible for the existence of a net non-zero Auger transition probability. They calculated carrier lifetimes (under the assumption that the Auger process was the only available recombination process) and compared them with experimental results in InSb. Above about 250K they found reasonable (order of magnitude) agreement, suggesting that Auger transitions form the dominant recombination process in InSb at high temperatures.

It can be shown that the dependence of the band-band Auger transition rate on carrier density depends on n^2p (eeh process) or np^2 (ehh process), where, as usual, n denotes the negative carrier concentration and p the positive carrier concentration. Following Haug [53] we write the Auger recombination probability for the eeh process in the form

$$W = A \int |M|^2 P \delta(E_1'+E_2'-E_1-E_2)\delta(k_1'+k_2'-k_1-k_2)d^3k_1 d^3k_2 d^3k_1' d^3k_2', \tag{163}$$

where A is a constant factor, M is a matrix element of H', and the particle indices are shown in Figure 6. P consists of the appropriate occupation probabilities and contains two terms, the first for the forward process and the second for the reverse process:

$$P = f_e(k_1)f_e(k_2)f_h(k_1')[1-f_e(k_2')] \tag{164}$$

$$- [1-f_e(k_1)][1-f_e(k_2)][1-f_h(k_1')]f_e(k_2').$$

Here f_e and f_h refer to Fermi-Dirac distributions involving appropriate quasi-Fermi levels F_e and F_h for the electrons and holes, respectively:

$$f_e = \frac{1}{1 + \exp[(E-F_e)/k_BT]}$$

$$f_h = \frac{1}{1 + \exp[(F_h-E)/k_BT]}, \qquad (165)$$

where k_B is Boltzmann's constant. Taking energy conservation into account we can rewrite P in the form

$$P = \{1 - \exp[(F_h-F_e)/k_BT]\}f_e(k_1)f_e(k_2)f_h(k_1')[1-f_e(k_2')]. \qquad (166)$$

Now if the Fermi-Dirac distributions are approximated by Boltzmann distributions

$$f_e \approx e^{(F_e-E)/k_BT}$$

$$f_h \approx e^{(E-F_h)/k_BT}, \qquad (167)$$

where $1-f_e \approx 1$, $1-f_h \approx 1$. f_e and f_h are proportional to n and p, respectively [54]. Thus, the forward rate goes as n^2p, and the inverse rate as nn_op_o, where the subscript indicates the equilibrium densities. Thus, the net Auger recombination rate depends on n^2p. This is a faster increase with carrier concentration than that for radiative rates by a factor of n (or p in the case of the ehh processes). Note that the analysis on which this is based breaks down for degenerate semiconductors, since it requires the approximation of the Fermi-Dirac distribution function by a Boltzmann distribution.

While in direct-gap semiconductors phononless Auger transitions require an activation energy, it is theoretically possible that a phononless Auger transition could occur in an indirect-gap material with no threshold energy. Such a material would have to have a rather special band structure. However, it turns out that the band structure of germanium is a reasonably close approximation to that required for phononless ehh processes without activation energy. The band structure of silicon, on the other hand, does not favor this process. Both materials have band structures that would seem to favor the corresponding eeh process, but Pilkuhn [55] has pointed out that it appears to be improbable because of a small matrix element and a small region of k space for the transition.

Phonon-assisted Auger recombination was studied by Eagles [52] for direct-gap materials, and by Huldt [53] for indirect-gap materials. These are second-order processes, and as such, would generally be expected to be less probable than the phononless processes. However, phonon assistance makes possible the satisfaction of the momentum conservation condition more easily, and this may compensate for the decreased probability due to the higher order. Indeed it is often observed experimentally that Auger transition rates have only weak temperature dependence, which is consistent with phonon participation because the activation energy is thus obviated. This suggests that the phonon assisted process may be of considerable importance in many common semiconductors.

Phononless Auger processes in indirect-gap materials were investigated by Huldt [54] and by Hill and Landsberg [55], with the result that the recombination coefficients were lower than those determined experimentally.

The case of degenerate semiconductors has been considered by Haug [56], [57] for both direct and indirect-gap semiconductors. He showed that phononless Auger recombination is highly improbable in degenerate materials, so both papers concentrate on phonon-assisted processes. In the later paper he applies the theory to the phenomenon of electron-hole drops and finds fairly good agreement with experiment.

Other types of Auger processes include those involving localized centers, or traps. These are known as band-trap Auger transitions. Here a conduction electron recombines with a trapped hole, or a valence hole with a trapped electron, the recombination energy being given to a free carrier. Auger transitions involving two trapped particles are also possible, as in electron transfer between a donor and an acceptor with the energy again given up to a free carrier. Annihilation of free or bound excitons in which an electronic particle receives the recombination energy are yet another type of Auger process. Pilkuhn [58] has remarked that Auger transitions involving bound excitons are important in silicon.

III.F. Inelastic Collisions: Hot Electron Excitation

Impact excitation in solids involves the transfer of energy from an energetic charge carrier to a localized state, which may be an impurity ion (localized in real space), or some quantum of excitation of the crystal, such as an exciton or plasmon (localized in either real or momentum space). The projectile may be quite energetic, as in cathode ray bombardment, or it may have only a few eV of energy, as in electroluminescence. This inelastic scattering process is just the inverse of a particular type of Auger process in which the recombination occurs between bound charges on a single site. As such, it can be considered to be a form of energy transfer.

We shall be concerned here only with the field of electroluminescence, where the excitation energy is of the order of two or three electron volts, much greater than thermal energies. As a consequence, the carriers which participate must be more energetic than those which are equilibrated with the lattice, and so they are dubbed "hot carriers", even though it may not be correct to assume that they can be characterized by a temperature.

There remains some controversy in the literature as to whether the mechanism of excitation is direct impact excitation or impact ionization of the lattice followed by energy transfer to the luminescent center [59, 60]. However, it appears that the bulk of experimental work supports the impact excitation mode at least in the case of ZnS:Mn and ZnS doped with various rare earths [61], [62], [63], [64], [60].

Tanaka et al. [60] have performed experiments involving ZnS:Mn in both photoluminescence (PL) and electroluminescence (EL). In the former case, time resolved spectroscopy showed recombination luminescence followed by a build-up of emission from the manganese, clearly revealing lattice ionization followed by energy transfer as the mechanism of excitation. However, in the case of EL, no recombination luminescence was observed, suggesting that energy transfer was not the excitation mechanism. This suggests further that the carrier energy distribution tails off quite sharply in the region between the manganese excitation energy (about 2.2 eV) and the band gap (about 3.7 eV). This last conclusion is also supported by a result of Zhong and Bryant [65] which can be used to estimate the temperature of the hot carriers as about 1800K, assuming that it is reasonable to characterize the distribution by a temperature. For a Maxwellian distribution at that temperature less than one percent of the carriers have sufficient energy to excite the manganese.

Evidence against direct impact excitation has been published by Walentynowicz et al. [59]. Their experiment, also using ZnS:Mn as the active layer, showed the Mn^{2+} emission peaking about 200 μs after excitation in both PL and EL. Also, only one peak was observed in both cases, with no change in the intensity distribution. Thus, they claim that non-radiative transfer is the excitation mechanism. They also measured the delay in the emission peak vs concentration of Mn^{2+} and found a substantial decrease above one percent by weight. They claim that this is consistent with the dependence of non-radiative transfer on activator-sensitizer separation. It should be pointed out, however, that it is also consistent with the dependence of resonant transfer among Mn^{2+} centers on the separation between the Mn^{2+} ions. This would cause concentration quenching, but no mention was made of this phenomenon. Having noted this controversy, we shall be concerned below only with the process of direct impact excitation of the luminescent centers.

Allen [66], [67], [68] has presented a simple model describing the impact excitation process. In the case considered, the device is a reverse-biased Schottky diode, in which the acceleration of the energetic carriers takes place in the same region in which the excitation occurs. If N_L and N_L^* represent the concentration of luminescent centers and the concentration of excited centers, respectively, in the region of interest, then

$$\frac{J}{q}\sigma W(N_L - N_L^*) = \frac{N_L^*}{\tau}W, \tag{168}$$

where W is the width of the region, J the current density, q the charge, τ the lifetime of the excited state, and

$$\sigma = \frac{1}{\langle v\rangle}\int_0^\infty v(E)\sigma(E)f(E)dE \tag{169}$$

is a sort of average cross section. $\sigma(E)$ is the cross section for the excitation process, f(E) is the carrier energy distribution, v is the carrier velocity, and $\langle v\rangle$ is the velocity averaged over the distribution function. From (168) the fraction of excited centers is obtained:

$$\frac{N_L^*}{N_L} = \frac{J}{J + \frac{q}{\sigma\tau}}. \tag{170}$$

Now, the number of centers excited by an electron in crossing the region is given by $\sigma(N_L - N_L^*)W$, so the quantum efficiency is

$$\begin{aligned}\eta_q &= \sigma(N_L - N_L^*)W\eta_{rad} \\ &= \sigma N_L W\eta_{rad}\frac{q/\tau\sigma}{J + q/\tau\sigma},\end{aligned} \tag{171}$$

where η_{rad} is the radiative quantum efficiency. Unfortunately, there is no clear physical interpretation available for σ as defined in (169). In most situations of interest there will be significant variations in carrier velocity across the region containing the luminescent centers, so that the variation of σ across the region must be taken into account. In Allen's case, there are high fields within the active region which are responsible for acceleration of the carriers. The variation of the field within the region and of σ with the strength of the field are thus important. Allen has taken these into account by simply assuming particular functional forms. He is then able to determine an expression for the dependence of the quantum efficiency on applied voltage. The expression agrees reasonably well with experimental values when the voltage is not too high. We do not present the expression or its derivation here, as it may not be directly applicable to more general situations.

The problem of calculating the cross sections for the excitation process has not been extensively investigated. There are probably a number of reasons for this. Among them are the difficulty of the problem. At the relatively low energies involved the Born approximation is known to represent the cross sections poorly, and more precise methods, such as the close-coupling methods, are quite unwieldy when dealing with the large numbers of electrons present in the typical luminescent center (transition metal ions and rare earth ions are common). In addition, there are unknown effects due to the presence of the crystal lattice, the states involved in the transitions arising from degenerate states split by both the crystal field and spin-orbit interaction. As a result it is difficult to obtain reliable wave functions for the states involved.

For these reasons Bernard, Martens and Williams [69], approached the problem from a semi-classical point of view, using a very simple calculation. It was based on an adiabatic picture of the scattering process, in which the incoming hot electron was viewed as perturbing the energy levels of the center via the interaction between itself and the induced dipole moment of the center. The levels have different polarizabilities (with the excited states having the greater polarizabilities), so that the separation of the levels varies with the distance between the projectile and the target. In particular, the level separation decreases as the distance decreases. It was assumed that the two levels of interest could be considered in isolation and that at some distance the two levels would become degenerate (cross), the probability of occurrence of the transition being greatest at that distance. This separation was then considered to be the effective geometric radius for the cross section for that transition. In addition, it was assumed that the interaction could be described as that between a point charge and a polarizable center at large distance. Under these assumptions, the interaction energy is

$$\Delta E_n = -\frac{\alpha_n e^2}{2r^4}, \tag{172}$$

where α_n is the polarizability of the nth level and r is the distance between the incoming electron and the target. If r_0 is the separation distance at which crossing occurs, the cross section is then given by

$$\sigma = \pi r_0^2 = \pi e\left[\frac{(\alpha_e - \alpha_g)}{2(E_e - E_g)}\right]^{1/2}, \tag{173}$$

where E_g and E_e are the unperturbed ground and excited state energies, respectively.

Some of the effects of the host lattice were taken into account by using the energy levels (with respect to the vacuum) as measured in the crystal in the determination of the polarizability difference between the levels, as well as in the denominator of equation (173). The final expression for the cross section was from (174), [69], [70]

$$\sigma = \frac{\pi e}{|E_g|}\left\{\frac{-Kk}{1 + \frac{\Delta E}{E_g}}\left[b + \frac{k}{E_g}\left(\frac{1 + \frac{1}{2}\frac{\Delta E}{E_g}}{1 + \frac{\Delta E}{E_g}}\right)\right]\right\}, \tag{174}$$

where k, K, and b are constants for a given center and are determined by the mean square radii of the electron orbits, the polarizability of the ground state, and the binding energy (in the free ion) of one of the electrons occupying the level in which the transition is said to occur. ΔE is the separation of the energy levels when the projectile is an infinite distance from the target. In the usual case $\Delta E/E_g$ is small compared to unity and k/E_g is small compared to b, so that σ is approximately inversely proportional to $|E_g|$. Thus, deeper centers should be expected to have smaller cross sections.

Comparisons of the results obtained from (174) [69],[70] showed reasonably good agreement with experimental information currently available in the case of ZnS:Mn [71] and ZnF_2:Mn [72], [73], [74]. In all these cases the cross section was estimated to be about the same as the geometric cross section of the free ion, about 10^{-16} cm^2. Support was also provided in the case of ZnS:Mn by Törnqvist [75] who indicated that an early estimate by

Allen [67] was too low, and by a more recent comment by Allen [68], who stated that the cross section for ZnS:Mn may be as high as the geometric cross section. However, there was relatively poor agreement with a cross section inferred [70] from results of Suyama et al. [76] in the case of $Y_2O_3:Eu^{3+}$. In the latter case, however, the inferred experimental cross section is in considerable doubt. We are aware of no other available experimental information on cross sections for rare earth dopants.

It was pointed out in [70] that the theoretical calculation described above is a semi-classical approximation to a physical process which can only be properly treated by a completely quantum mechanical calculation. This is particularly true for the manganese and the trivalent rare earth ions, in which the transition involves a spin flip of an electron within a particular configuration, $3d^5$ in the case of Mn^{2+} and $4f^n$ in the case of the trivalent rare earths. This can occur by an electron exchange process which can be fully described only via the formalism of quantum mechanics.

IV. CLOSING REMARKS

Condensed matter is a many-body system, and the concept of electronic energy transfer is based on two separate subdivisions of the system. On the one hand, into the energy donor and energy acceptor; on the other hand, into the electronic energy and the energy associated with lattice dynamics. Both separations are approximations: the separation into electronic and vibrational energies is based on the quantum mechanical adiabatic approximation; the separation into energy donor and energy acceptor is handled differently depending on the strength of the coupling between these two subsystems.

The many-electron subsystem of condensed matter is usually further approximated in terms of one-electron states. These allow determination of the electronic band structure, which has been well developed for crystals, that is, condensed matter whose structure has translational invariance. From the lattice dynamics for these structures phonon dispersion has been similarly developed. The electronic bands and phonon states of crystals, plus the effects of electron-phonon interaction provide the background for theoretical analyses of the mechanisms of electronic energy transfer in condensed matter. However, translational periodicity is not essential for treatment of some similar mechanisms which occur in amorphous solids and in liquids.

The conventional definition of electronic energy transfer encompasses transport of energy in position-coordinate space. The classic mechanism is resonant transfer between inequivalent and weakly-coupled dopants which are separately equilibrated with the lattice in their respective electronic states before transfer. Non-resonant and excitonic transfer are also included within the scope of this definition. These mechanisms do not depend on translational invariance of the structure.

Excitons were conceived originally as electronic eigenstates of a periodic potential with correlation of an electron and a positive hole so that the coupled electron and hole moved together, transporting no charge. The concept is now applied more widely, including any excited, non-conducting states of condensed matter, amorphous and liquid as well as crystalline, and thus includes many extrinsic as well as intrinsic mechanisms for the transfer of electronic energy in position-coordinate space.

Electronic transfer in which net charge is transferred may also transport energy. This is most evident in semiconductors for which the minority charges carry potential energy almost equal to the band gap. Hot electrons in insulators, semiconductors or metals carry kinetic energy. In general, electrons or positive holes may transport energy equal to their eigenenergies as measured from the Fermi energy of the material.

A general definition of electronic energy transfer includes the transport of electronic energy in momentum-coordinate, as well as in position-coordinate, space. That is, transfer between sets of states having their probability densities localized in momentum space or between a set localized in position space and another localized in momentum space is now included. Thus, Auger transitions and charge transfer between subbands now fit within the category of energy transfer processes. The possibility is suggested, of course, of applying the techniques of energy transfer in position space to transfer in momentum space, and vice versa. It will be of some interest to see whether, in the future, attempts to do this prove fruitful.

APPENDIX: EFFECTIVE MASS APPROXIMATION FOR DOPANTS WITH COULOMB FIELDS

Our derivation follows closely that given by Pantelides [77]. We consider a system consisting of an electron in a crystal, under the influence of the lattice potential and an impurity potential, the latter assumed to be hydrogenic. The Hamiltonian is

$$H = H_0 + H', \qquad \text{(A-1)}$$

where

$$H_0 = T + V, \qquad \text{(A-2)}$$

and

$$H'(r) = -\frac{e^2}{\epsilon r}. \qquad \text{(A-3)}$$

Here T is the electron kinetic energy, V the interaction with the lattice, and ϵ the dielectric constant. The wavefunction for the electron is expanded in terms of eigenfunctions of H_0 (Bloch functions):

$$\phi = \sum_n \int F_n(k)\psi_n(k,r)d^3k, \qquad \text{(A-4)}$$

where the integration is over a Brillouin zoné, and the sum is over bands. Now take the inner product

$$\int \psi_n^*(k,r)H\phi(r)d^3r = F_n(k)E_n(k) + \sum_{n'} \int F_{n'}(k')H'_{nn'}(k,k')d^3k'$$

$$= EF_n(k), \qquad \text{(A-5)}$$

where

$$H'_{nn'}(k,k') = \int \psi_n^*(k,r)H'(r)\psi_{n'}(k',r)d^3r. \qquad \text{(A-6)}$$

Here we have made use of the fact that

$$H_0\psi_n(k,r) = E_n(k)\psi_n(k,r). \qquad \text{(A-7)}$$

The Bloch functions have the form

$$\psi_n(k,r) = e^{ik\cdot r}u_n(k,r), \qquad \text{(A-8)}$$

where u_n is a function having the periodicity of the lattice (see the glossary entry for Bloch's theorem), so equation (A-6) can be rewritten as

$$H'_{nn'}(k,k') = \sum_{\alpha} e^{-ik\cdot r} u_n^*(k,r)H'(r)e^{ik'\cdot r} u_{n'}(k',r)d^3r. \quad \text{(A-9)}$$

The product $u_n^* u_{n'}$ can be expanded in terms of plane waves

$$u_n^*(k,r)u_{n'}(k',r) = \sum_{\alpha} C_{nn'}(k,k',K_\alpha)e^{iK_\alpha\cdot r}, \quad \text{(A-10)}$$

where the sum is taken over reciprocal lattice vectors. Thus, (A-9) becomes

$$H'_{nn'}(k,k') = \sum_{\alpha} C_{nn'}(k,k',K_\alpha)\tilde{H}'(k-k'-K_\alpha), \quad \text{(A-11)}$$

where $\tilde{H}'(q)$ is the Fourier transform of $H'(r)$

$$\tilde{H}'(q) = \int H'(r)e^{-iq\cdot r}d^3r = -4\pi\frac{e^2}{\epsilon q^2}. \quad \text{(A-12)}$$

Thus, equation (A-5) can now be put into the form

$$[E-E_n(k)]F_n(k) + \sum_{n'}\sum_{\alpha}\int F_{n'}(k')C_{nn'}(k,k',K_\alpha)\tilde{H}'(k-k'-K_\alpha)d^3k'$$
$$= 0. \quad \text{(A-13)}$$

Now we introduce a number of approximations. First, it is assumed that the sum over bands can be reduced to a single band. This is expected to be valid if the state of the electron is not too different from that of an electron in a particular band (a shallow state). Also, it is assumed that the band is isotropic and has a non-degenerate extremum at $k=0$. In this case, it is expected that $F(k)$ will be fairly well localized about the extremum, so that the contributions to the integral will only come from the region with k and k' small compared to the non-zero K_α's. From equation (A-12) we can see that $\tilde{H}'(q)$ falls off fairly rapidly with q. This, together with the fact that we need only consider small values of k and k' (and, hence, $|k-k'|$), motivates the omission of all terms in the sum on α for which $K_\alpha \neq 0$.

The constants $C(k,k')$ are given by

$$C(k,k',K_\alpha) = \int u^*(k,r)u(k',r)e^{-iK_\alpha\cdot r}d^3r. \quad \text{(A-14)}$$

Note that the band index has been dropped. The orthonormality of the Bloch functions then gives $C(k,k,0)=1$, and we assume this is approximately true for $k \neq k'$ as well. The energy E(k) for an isotropic band with extremum at k=0 can be expanded to give

$$E(k) = E_o + \frac{\hbar^2 k^2}{2m^*}, \qquad (A-15)$$

where m^* is the effective mass.

Then, after making these approximations, the integrations on k are extended to all of k space, and equation (A-13) becomes

$$\left[E_o - E + \frac{\hbar^2 k^2}{2m^*}\right]F(k) + \int \bar{H}'(k-k')F(k')d^3k' = 0, \qquad (A-16)$$

which is just a momentum-space Schrödinger equation. Using the Fourier transform

$$F(r) = \int F(k)e^{ik\cdot r}d^3k, \qquad (A-17)$$

we obtain the effective-mass equation:

$$\left[-\frac{\hbar^2}{2m^*}\nabla^2 + H'(r)\right]F(r) = (E_o-E)F(r). \qquad (A-18)$$

GLOSSARY

Adiabatic approximation: For a system composed of two interacting subsystems which have greatly differing inertia an adiabatic approximation to the dynamics of the system can be made. In this approximation the wavefunction is assumed to be a product of coupled wavefunctions for the individual subsystems. The wavefunction for the small inertia subsystem (usually referred to as the fast subsystem) depends only on the static properties of the large inertia subsystem (usually referred to as the slow subsystem), not on the rate change of the latter. This assumption can be made because we assume that the motion of the large inertia subsystem is slow compared to that of the small inertia subsystem, so that the latter, to a good approximation, never lags behind in its adjustment to the former. On the other hand, the small inertia subsystem is viewed as producing a force on the large inertia subsystem via an average potential. With these assumptions about the interactions between the parts of the system, the product wavefunction can be taken to have the form $\phi_s(r;R)\chi_{1s}(R)$, where $\phi_s(r;R)$ is the wavefunction for the small inertia subsystem, whose coordinates are r, and $\chi_{1s}(R)$ is the wavefunction for the large inertia subsystem, whose coordinates are R. Note that χ depends on the state of the small inertia subsystem, whereas ϕ depends on the configuration of the large inertia subsystem. Here we have differentiated the subsystems in terms of their inertia, since it is their relative ability to adjust to each other's changes that is important.

<u>Adiabatic theorem of quantum mechanics</u>: In general, there are no stationary solutions to the time-dependent Schrödinger equation for a time-dependent Hamiltonian H(t). However, for any fixed value of t stationary solutions do exist. The theorem asserts that in the limit of infinitely slow variation of H(t) with time the stationary states for a particular value of t change smoothly into those for adjacent values of t.

<u>Allowed and forbidden energy bands</u>: The energy eigenvalues of the one-electron states for crystalline materials group into sets of closely spaced energy levels, that is, sets of quasi-continuous levels. The range of energies corresponding to each set is designated as an allowed energy band. The intervening energy ranges in which there are no closely spaced energy eigenvalues are designated forbidden bands.

<u>Auger process</u>: In atoms, an autoionization, in which an atom in an excited state of energy greater than the first ionization energy decays into an ion and a free electron. In solids, a collision process between two electronic particles in which one recombines with a third electronic particle of opposite type, with the energy released by the recombination being given to the other particle participating in the collision. The latter always lies in a continuum of energy levels in the final state, though it may not in the initial state.

<u>Bloch's theorem</u>: states that the solutions to Schrödinger's equation for a periodic potential have the form

$$\phi_{k\beta}(r) = e^{ik\cdot r} u_{k\beta}(r),$$

where $u_{k\beta}(r)$ has the same periodicity as the potential. Here k and r are 3n-vectors.

<u>Born-von Karman boundary condition</u>: The Born-von Karman boundary

condition is defined by the identification of the surfaces of an array in pairs which are perpendicular to any given primitive basis vector. This identification removes the effects of the surfaces on the consideration of the form of the wavefunction for the particles of the system. With this identification the array is considered to be periodic, with periods $N_i\hat{a}_i$, i=1,2,3.

<u>Brillouin zone/Brillouin zone boundaries</u>: The Brillouin zone boundaries are those planes in the reciprocal Bravais lattice space whose points correspond to those wave vectors k which satisfy the equations $2k \cdot K_\nu + K_\nu{}^2 = 0$, where K_ν is any reciprocal Bravais lattice vector. These planes are the perpendicular bisectors of the reciprocal lattice vectors. The Brillouin zones are defined as the connected subregions of the reciprocal lattice space when this space is subdivided along the Brillouin zone broundaries.

<u>Charge transport energy transfer</u>: When charge is transported through a lattice, energy that can be extracted from the charge distribution is also transferred through the system. The energy transferred is the energy of the charge-carrier distribution in excess of that of the equilibrium distribution, in other words, the energy of excess carriers with reference to the Fermi level.

<u>Core electron and non-core electron</u>: The electrons in any system can be subdivided into two groups designated as the core and non-core electrons. This subdivision is made in the following way: For a core electron the one-electron energy is large compared to its total interaction energy with all the non-core electrons, whereas for a non-core electron the one-electron energy is small or of the same order as its interaction energy with all of the core electrons. It is usual to assume that in a condensed system the non-core electrons are the outer shell electrons of the atoms, which can be easily removed by interaction with other atoms, and the core electrons are the electrons left on the positive ion of the atom. Both conduction and valence band electrons are non-core electrons.

<u>Crude adiabatic approximation</u>: is the approximation within the framework of the adiabatic approximation in which the wavefunctions of the small inertia subsystem are assumed to be independent of the coordinates of the large inertia subsystem as well as independent of the rate of change of the configuration of the large inertia subsystem. See entry for the Adiabatic approximation.

Deformation potential: The deformation potential is a short range interaction between phonons and electronic particles which arises because of the local deformation of the lattice. This local deformation alters the potential seen by the electronic particles in the region and causes scattering. The potential describing this interaction can be approximated by that for a static uniform deformation. This is designated as the deformation potential. A further approximation can be made wherein the deformation potential is approximated as a constant, E_n, the deformation potential parameter, times the local strain. The deformation potential parameter for the electronic particles in the nth band is given by $E_n = \delta\varepsilon_n(0)/\Delta$, where $\Delta = \delta V$ is the static uniform volume dilatation.

Direct Bravais lattice/lattice vectors: The direct Bravais lattice is the lattice composed of the points given by the direct Bravais lattice vectors, which are defined by $X_n=n_1\hat{a}_1+n_2\hat{a}_2+n_3\hat{a}_3$, where n_i is an integer between $\pm N_i/2$. Here N_i is the number of unit cells in the direction defined by $\hat{a}_i$ in the array.

Effective mass: Electronic particles in crystals react to an applied force as if they had an inertial mass which is different from the free-electron inertial mass. This mass is designated as the effective mass. The difference between the effective mass and the free-electron mass is due to the interaction between the electronic particles and the periodic potential. The effective mass, $m^*_{ni}(k)$, for the nth allowed band in the ith direction is defined as

$$m^*_{n_i}(k) = \pm\hbar^2\left[\frac{\partial^2\varepsilon_n(k)}{\partial k_i^2}\right]^{-1},$$

where the plus sign is for electrons and the minus sign is for holes.

Electronic energy transfer: A restrictive definition of electronic energy transfer can be inferred from the original energy transfer processes studied by Förster [1], Dexter [2] and Orbach [3]. These authors considered the transfer of electronic energy localized in one region of the lattice to a second localized region which did not overlap the first, for example, the transfer of the electronic excitation from an isolated Cr^{3+} ion to a pair of Cr^{3+} ions or another isolated Cr^{3+} ion. More generally, we consider any transfer of energy from one set of electronic states to a second set of states, in which the two sets of electronic states differ from one another in a specifible way, to be energy transfer.

In the case of the more restrictive definition the characteristic which differentiates the two sets of states is that their probability densities are localized to different regions of position space. In the more general case energy transfer takes place between sets of states which are distinguished by having their probability densities localized to different regions of either position or momentum space. In addition, we can consider energy transfer between sets of states which are differentiated by one set having its probability densty localized in momentum space, while the other is localized in position space.

Energy acceptor: The subsystem responsible for the set of states to which electronic energy is transferred.

Energy band structure: is the designation for the relationship between the one-electron energy eigenvalues and the quasi-momentum, k, that is, the relationship $\varepsilon_n(k)$. The term also refers to the energy ranges of the allowed and forbidden energy bands.

Energy donor: The subsystem responsible for the set of states from which electronic energy is transferred.

Exciton: Any excited electronic state of a lattice, involving correlation of the electron and positive hole. Thus, it may be viewed as a bound electron-hole pair. If the separation between the electron and hole is small enough that they can be considered to occupy the same unit cell, the name Frenkel exciton is applied. The term Wannier exciton is used if the separation is large compared to the size of a unit cell. Note that motion of an exciton involves no transport of charge or real mass. The term "exciton" is also often applied to electronic excitations in disordered solids and other systems.

Excitonic energy transfer: Energy is transferred during exciton motion. In this energy transfer process the sets of states are localized in position space, but not so localized as not to be localized in quasi-momentum space as well. The initial configuration distribution for excitonic energy transfer is an unrelaxed distribution.

Fermi energy: The Fermi energy is that energy for which the occupation probability is one-half. The Fermi energy is also referred to as the Fermi level.

Fermi's golden rule: gives the transition probability per unit time, P_{fi}, from initial state ϕ_i to all final states ϕ_f having energy E_f as

$$P_{fi} = \frac{2\pi}{\hbar}\int \phi_f^*(r_o)H_I\phi_i(r_o)\rho_f(E_f)\delta(E_f-E_i)d^3r_o,$$

where $\rho_f(E_f)$ is the density of the final states having energy E_f and H_I is the interaction causing the transition.

Fröhlich Hamiltonian: describes the long-range interaction of the polaron theory. This arises in polar materials because of the interaction between electronic particles and the electric field produced by the polarization of the lattice by those phonons which are associated with large polarization. The form of the Fröhlich Hamiltonian is given by equation (IIE-22).

Hartree-Fock approximation: In the Hartree-Fock approximation the many-electron wavefunction, $\psi_\alpha(r;R)$, for the n-electron system with electronic coordinates $r=(r_1,r_2,\ldots,r_n)$, where r_i is the three-vector coordinate of the ith electron, is approximated by

$$\psi_\alpha(r;R) = \begin{vmatrix} \phi_{\alpha_1}(r_1) & \phi_{\alpha_2}(r_1) & \cdots & \phi_{\alpha_n}(r_1) \\ \phi_{\alpha_1}(r_2) & & & \cdot \\ \cdot & & & \cdot \\ \cdot & & & \cdot \\ \cdot & & & \cdot \\ \phi_{\alpha_1}(r_n) & \cdot & \cdot \quad \cdot & \phi_{\alpha_n}(r_n) \end{vmatrix}$$

Here the $\phi_{\alpha_i}(r_j)$'s are one-electron wavefunctions, and the α_i's are the sets of quantum number indexing these one-electron states. The right-hand side of this equation is known as a Slater determinant wavefunction.

Intraband energy transfer: Intraband energy transfer takes place between subbands of a given band. When the subbands' minima have different energy values the process involves conversion of kinetic energy to potential energy or vice versa. The sets of states between which this transfer process takes place are localized in momentum space, and the configuration distribution is the relaxed distribution.

Markov process: Any dynamical process in which transitions among regions of space or states or groups of states are completely chaotic, that is, have no dependence on past history. A classic example is Brownian motion. The transition probabilities for a Markov process satisfy the Chapman-Kolmogorov equation [78]:

$$\Omega(x,t;x',t') = \int_{t}^{t'}\int \Omega(x,t;x'',t'')\Omega(x'',t'';x',t')dx''dt'', \quad (1)$$

where $\Omega(x,t;x',t')$ is the probability that the system originally at x at time t will be at x' at time t'. In terms of transitions among states or groups of states, equation (1) is written

$$\Omega_{\alpha\beta} = \sum_{\gamma} \Omega_{\alpha\gamma}\Omega_{\alpha\beta} \quad . \quad (2)$$

Nearly free electron approximation: In the nearly free electron approximation we assume that all Fourier coefficients for non-zero K-values of the potential of the non-core electrons are small compared to the zero K-value coefficient.

Non-resonant energy transfer: Non-resonant energy transfer, like resonant transfer, takes place between two sets of multi-electronic-particle states whose charge distributions are localized in position space. The initial and final states of the system are products of two multi-electronic-particle wavefunctions, one from each of the sets of states between which the energy is transferred. The initial configuration distribution is that corresponding to the relaxed initial state. That is, the ions about the energy donor are in the vibrational state that corresponds to the excited electronic state of the donor and the temperature of the lattice, while the ions about the energy acceptor are in the vibrational state that corresponds to the ground electronc state of the acceptor and the temperature of the lattice. In the non-resonant transfer process there are real phonons created and/or annihilated.

One-electron approximation: In the one-electron approximation for the non-core electrons, we approximate the many-electron wavefunction by a Slater determinant of one-electron functions and assume that the interaction among the non-core electrons is zero and that there is no exchange interaction between the core and non-core electrons. The equations determining the one-electron functions for the non-core electrons are:

$$\frac{p_o^2}{2m}\phi_{\alpha_o}(r_o) + V(r_o)\phi_{\alpha_o}(r_o) = \varepsilon_{\alpha_o}\phi_{\alpha_o}(r_o),$$

where $V(r_o)$ is the sum of the ionic potentials.

Phonon branch: The energy function of the phonons, $\hbar\omega(q)$, is a quasi-continuous, multi-valued function of q, where q indexes the different allowed propagating wave solutions of the array. Here the function $\omega(q)$ is defined for q between $\pm\pi/a$. The multi-valued function $\omega(q)$ is equivalent to a set of single-valued quasi-continuous functions of q designated by $\omega_\gamma(q)$. Each $\omega_\gamma(q)$ defines a phonon branch. Note that each $\{q,\gamma\}$ set of values corresponds to a normal mode of the lattice vibration.

Phonons: are the quanta of the vibrations of a crystal. For each normal mode of vibration of an ordered array there is a corresponding phonon. The term is sometimes applied to quantized vibrations of disordered solids and even to those of molecules.

Polaron: Longitudinal optical (LO) vibrational modes in ionic crystals cause relatively large polarization of a crystal lattice. The resulting dipole fields couple strongly with mobile electrons. Thus, moving charges tend to carry along with them a localized lattice polarization, which can be thought of as a "cloud" of phonons. The coupled system of an electron and LO phonons is called a polaron.

Primitive basis vectors: We defined the primitive basis vectors of any periodic array as the set of vectors that are the shortest non-pairwise collinear vectors that can be chosen from the set of all vectors which connect the origin of one unit cell with the origins of all the other unit cells in the periodic array. In a three dimensional lattice the three primitive basis vectors are denoted by $\hat{a}_i$, i=1,2,3.

Quasi-momentum: The quasi-momenta are the eigenvalues of the translation symmetry group of a lattice.

Reciprocal Bravais lattice/lattice vector/primitive basis vectors: The reciprocal Bravais lattice is the set of points given by the vectors $K_\nu=\nu_1\hat{b}_1+\nu_2\hat{b}_2+\nu_3\hat{b}_3$, where ν_i is an integer between

$\pm N_i/2$. The primitive basis vectors, $\hat{b}_i$, are defined by equation (25). An important property is that each of the primitive reciprocal lattice vectors is orthogonal to two of the primitive lattice vectors of the direct lattice.

Resonant energy transfer: Resonant energy transfer, like non-resonant transfer, takes place between two sets of multi-electronic-particle states whose charge distributions are localized in position space. The initial and final states of the system are products of two multi-electronic-particle wavefunctions, one from each of the sets of states between which the energy is transferred. The initial configuration distribution is that corresponding to the relaxed initial state. That is, the ions about the energy donor are in the vibrational state that corresponds to the excited electronic state of the donor and the temperature of the lattice, while the ions about the energy acceptor are in the vibrational state that corresponds to the ground electronc state of the acceptor and the temperature of the lattice. In the resonant transfer process no real phonons are created or annihilated.

Tight binding approximation: In the tight binding approximation we assume that all the overlaps between Wannier functions for the non-core electrons centered in different unit cells are, to a good approximation, zero.

Unit cell: is a subdivision of a periodic array such that the complete space occupied by the array can be filled by identical, non-overlapping unit cells. By identical unit cells is meant cells which can, by translation only, be located at a common place in space such that each ion in each unit cell lies at the same place as the corresponding ions in all the other unit cells of the array. The unit cell is the smallest repeatable unit of a periodic array.

Wannier functions: are defined as a complete set of localized states which are the Fourier transforms of the Bloch functions. In terms of the Bloch functions the Wannier functions are defined as

$$A_\beta(r_o - X_n) = \frac{1}{\sqrt{N}} \sum_k e^{-ik \cdot X_n} \phi_{k\beta}(r_o),$$

where the summation is over the first Brillouin zone and X_n is a direct Bravais lattice vector. The Wannier functions are localized about a unit cell of the periodic array located by the vector X_n.

REFERENCES

1. Th. Förster, Ann. Phys. (Leipzig) 2, 55 (1948).

2. D. L. Dexter, "A theory of sensitized luminescence in solids," J. Chem. Phys. 21(5), 836-850 (May 1953).

3. R. Orbach, "Relaxation and energy transfer," in Optical Properties of Ions in Solids, edited by Baldassare DiBartolo (Plenum, New York, 1975), pp. 355-399.

4. G. H. Wannier, "The structure of electronic excitation levels in insulating crystal," Phys. Rev. 52(3), 191-197 (1 August 1937).

5. J. Frenkel, "On the transformation of light into heat in solids. I," Phys. Rev. 37, 17-44 (1931).

6. P. T. Landsberg, personal communication.

7. Optical Properties of Ions in Solids, edited by Baldassare DiBartolo (Plenum, New York, 1975).

8. Radiationless Processes, edited by B. Di Bartolo, (Plenum, New York, 1980).

9. M. Born and V. Fock, "Beweis des Adiabatensatzes," Z. Phys. 51, 165-180 (1928).

10. T. Kato, "On the adiabatic theorem of quantum mechanics," J. Phys. Soc. Jpn. 5, 435-439 (1950).

11. M. Born and R. Oppenheimer, "Zur Quantentheorie der Molekeln," Ann. Phys. (Leipzig) 84(20), 457-484 (2 November 1927).

12. M. Born and K. Huang, Dynamical Theory of Crystal Lattices (Oxford, London, 1954).

13. M. Born, Gött. Nachr. math. phys. Kl. 1, (1951).

14. G. V. Chester and A. Houghton, "Electron-phonon interaction in metals I: The harmonic approximation," Proc. Phys. Soc. London 73(472), 609-621 (1 April 1959).

15. J. C. Slater, "A simplification of the Hartree-Fock method," Phys. Rev. 81, 385-390 (1951).

16. W. Kohn and L. J. Sham, "Self-consistent equations including exchange and correlation effects," Phys. Rev. A140, A1133-A1138 (1965).

17. J. D. Patterson, Introduction to the Theory of Solid State Physics (Addison-Wesley, Reading, Mass., 1971).

18. C. Kittel, Quantum Theory of Solids, (Wiley, New York, 1963).

19. J. Callaway, Energy Band Theory (Academic, New York, 1964).

20. C. Kittel, "General Introduction," in Phonons in Perfect Lattices with Point Imperfections, edited by R. W. H. Stevenson (Plenum, New York, 1966), p. 1.

21. W. Jones and N. H. March, Theoretical Solid State Physics (Wiley, London, 1973), Vol. 1: Perfect Lattices in Equilibrium.

22. F. Williams, D. E. Berry and J. E. Bernard, "Present Trends in the Theory of Radiationless Processes," in Radiationless Processes, edited by Baldassare DiBartolo (Plenum, New York, 1980), pp. 1-37.

23. F. Bloch, "Uber die Quantenmechanik der Elektronen in Kristallgittern," Z. Phys. 52, 555-600 (1928).

24. F. Bloch, "Zum elektrochen Widerstandsgesetz bei tiefen Temperaturen," Z. Phys. 59, 208-214 (1930).

25. L. J. Sham and J. M. Ziman, "The Electron-Phonon Interaction," in Solid State Physics, edited by F. Seitz and D. Turnbull (Academic, New York, 1963), Vol. 15, pp. 221-298.

26. J. Bardeen and W. Shockley, "Deformation potentials and mobilities in non-polar crystals," Phys. Rev. 80, 72-80 (1950).

27. W. Jones and N. H. March. Theoretical Solid State Physics (Wiley, London, 1973), Vol. 2: Non-equilibrium and Disorder.

28. H. Fröhlich, "Electrons in Lattice Fields," Adv. Phys. 3, 325-361 (July 1954).

29. Solid State Physics, edited by R. Kubo and T. Nagamiya (McGraw-Hill, New York, 1969), p. 136.

30. R. Orbach, "Phonon Sidebands and Energy Transfer," in Optical Properties of Ions and Crystals, edited by H. M. Crosswhite and H. W. Moos (Interscience, New York, 1967), pp. 445-455.

31. R. Reisfeld, "Multiphonon Relaxation in Glasses," in Radiationless Processes, edited by D. Di Bartolo (Plenum, New York, 1980), pp. 489-498.

32. Physics of Nonlinear Transport in Semiconductors, edited by D. K. Ferry, J. R. Barker, and C. Jacobi (Plenum, New York, 1979).

33. M. Dresden, "A Study of Models in Non-equilibrium Statistical Mechanics," in Studies in Statistical Mechanics, edited by J. DeBoer and G. E. Uhlenbeck (North-Holland, Amsterdam, 1962), Vol. 1, pp. 303-343.

34. G. V. Chester, "The Theory of Irreversible Processes," Rep. Prog. Phys. 26, 411-472 (1963).

35. J. R. Barker, "Hot electron phenomena and electroluminescence," J. Lumin. 23(1,2), 101-126 (July-August 1981). (Proceedings of the Liège Workshop on the Physics of Electroluminescence).

36. R. S. Know, "Theory of Excitons," in Solid State Physics, Supplement 5, edited by F. Seitz and D. Turnbull (academic, New York, 1963).

37. A. von Hippel, "Einige prinzipielle Gesichtspunkte zur Spektroskpie Ionenkristalle und ihre Anwendung auf die Alkalihalogenide," Z. Phys. 101(11-12), 680-720 (11 August 1936).

38. M. Pope and C. E. Swenberg, Electronic Processes in Organic Crystals (Oxford, New York, 1982).

39. G. Zimmerer, "Luminescence properties of rare gas solids," J. Lumin. 18/19, 875-891 (1979). (Proceedings of the 1978 International Conference on Luminescence in Paris).

40. U. Hahn and N. Schwentner, "Electronic relaxation cascades in Wannier and Frenkel type exciton states," J. Lumin. 18/19, 23-26 (1979). (Proceedings of the 1978 International Conference on Luminescence in Paris).

41. W. Gudat, C. Kunz, and H. Petersen, "Core exciton and band structure in LiF," Phys. Rev. Lett. 32(24), 1370-1373 (17 June 1974).

42. V. M. Kenkre and R. S. Knox, "Generalized-master-equation theory of excitation transfer," Phys. Rev. B 9(12), 5279-5290 (15 June 1974).

43. M. K. Grover and R. Silbey, "Exciton-phonon interactions in molecular crystals," J. Chem. Phys. 52(4), 2099-2108 (15 February 1970).

44. M. Grover and R. Silbey, "Exciton migration in molecular crystals," J. Chem. Phys. 54(11), 4843-4851 (1 June 1971).

45. H. Haken and G. Strobl, "Exact treatment of coherent and incoherent triplet exciton migration," in The Triplet State, edited by A. Zahlan (Cambridge University, London, 1967), pp. 311-314.

46. V. M. Kenkre and R. S. Knox, "Theory of fast and slow exciton transfer rates," Phys. Rev. Lett. 33(14), 803-806 (30 September 1974).

47. V. M. Kenkre, "The master equation approach: Coherence, energy transfer, annihilation, and relaxation," in Exciton Dynamics in Molecular Crystals and Aggregates, Springer Tracts in Modern Physics, Vol. 94 (Springer-Verlag, New York, 1982), pp. 1-109.

48. P. Reineker, "Stochastic Liouville equation approach: Coupled coherent and incoherent motion, optical line shapes, magnetic resonance phenomena," in Exciton Dynamics in Molecular Crystals and Aggregates, Springer Tracts in Modern Physics, Vol. 94 (Springer-Verlag, New York, 1982), pp. 111-226.

49. H. Haken and P. Reineker, "The coupled coherent and incoherent motion of excitons and its influence on the line shape of optical absorption," Z. Phys. 249(3), 253-268 (1972).

50. H. Haken and G. Strobl, "An exactly solvable model for coherent and incoherent exciton motion," Z. Phys. 262(2), 135-148 (1973).

51. V. M. Kenkre, "Relations among theories of excitation transfer," Phys. Rev. B 11(4), 1741-1745 (15 February 1975).

52. D. M. Eagles, "The phonon-assisted Auger effect in semiconductors," Proc. Phys. Soc. London 78(2), 204-216 (August 1961).

53. L. Huldt, "Phonon-assisted Auger recombination in germanium," Phys. Status Solidi A 33(2), 607-614 (16 February 1976).

54. L. Huldt, "Band-to-band Auger recombination in indirect gap semiconductors," Phys. Status Solidi A 8, 173-187 (1971).

55. D. Hill and P. T. Landsberg, "A formalism for the indirect Auger effect. I," Proc. R. Soc. London, Ser. A 347, 547-564 (1976).

56. A. Haug, "Phonon-assisted Auger recombination in degenerate semiconductors," Solid State Commun. 22(8), 537-539 (May 1977).

57. A. Haug, "Auger recombination of electron-hole drops," Solid State Commun. 25, 477-479 (1978).

58. M. H. Pilkuhn, "Non-radiative recombination and luminescence in silicon," J. Lumin. 18/19, 81-87 (1979).

59. E. Walentynowicz and S. Szuba, "The excitation mechanism of Mn luminescence centres in ZnS:Mn thin films," Phys. Status Solidi A 49(2), K201-K203 (October 1978).

60. S. Tanaka, H. Kobayashi, H. Sasakura, and Y. Hamakawa, "Evidnece for the direct impact excitation of Mn centers in electroluminescent ZmS:Mn films," J. Appl. Phys. 47(12), 5391-5393 (December 1976).

61. D. Khang, "Electroluminescence of rare-earth and transition metal molecules in II-VI compounds via impact excitation," Appl. Phys. Lett. 13(6), 210-212 (15 September 1968).

62. D. C. Krupka, "Hot electron impact excitation of Tb^{3+} luminescence in ZnS:Tb^{3+} thin films," J. Appl. Phys. 43(2), 476-481 (February 1972).

63. H. Kobayashi, S. Tanaka, and H. Sasakura, "Excitation mechanism of electroluminescent ZnS:Er^{3+} thin films," Jpn. J. Appl. Phys. 12(10), 1637-1638 (October 1973).

64. H. Kobayashi, S. Tanaka, H. Sasakura and Y. Hamakawa, "Excitation mechanism of electroluminescent ZnS this films doped with rare-earth ions," Jpn. J. Appl. Phys. 13(7), 1110-1114 (July 1974).

65. G. Z. Zhong and F. J. Bryant, "Direct current electroluminescence and spectral intensity in erbium- doped zinc sulphide thin films," J. Lumin. 24/25, 909-912 (November 1981). (Proceedings of the 1981 International Conference on Luminescence in Berlin).

66. J. W. Allen, A. W. Livingstone, and K. Turvey, "Electroluminescence in reverse-biassed zinc selenide Schottky diodes," Solid State Electron. 15(12), 1363-1369 (December 1972).

67. J. W. Allen, "Electroluminescence in reverse-biassed Schottky diodes," J. Lumin. 7, 228-240 (1973).

68. J. W. Allen, "Impact excitation and ionization," J. Lumin. 23(1,2), 127-139 (July-August 1980). (Proceedings of the Liège Workshop on the Physics of Electroluminescence).

69. J. E. Bernard, M. F. Martens and F. Williams, "Collision excitation cross-sections and energy levels of deep and very deep centers in electroluminescence," J. Lumin. 24/25, 893-896 (November 1981). (Proceedings of the 1981 International Conference on Luninescence in Berlin).

70. J. E. Bernard, M. F. Martens, D. C. Morton, and F. Williams, "Mechanisms of thin-film electroluminescence," IEEE Trans. Elect. Devices ED-30(5), 448-452 (May 1983).

71. D. H. Smith, "Modeling a.c. thin-film electroluminescent devices," J. Lumin. 23(1,2), 209-235 (July-August 1980). (Proceedings of the Liège Workshop on the Physics of Electroluminescence).

72. D. C. Morton and F. Williams, "A multilayer thin film electroluminescent display," in SID International Symposium Digest of Technical Papers (Society for Information Display, Los Angeles, April 1981), pp. 30-31.

73. F. Williams and D. C. Morton, "Some recent advances in thin film electroluminescence," Condensed Mat. Phys. (Proc. Solid State Div. European Phys. Soc.) 1, 429 (1981).

74. D. C. Morton and F. E. Williams, "A new thin-film electroluminescent material--ZnF_2:Mn," Appl. Phys. Lett. 35(9), 671-672 (1 November 1979).

75. R. O. Törnqvist and T. O. Tuomi, "DC Electroluminescence in $InSn_xO_y$-Ta_2O_5-ZnS:Mn- Ta_2O_5-Al thin film structures," J. Lumin. 24/25, 901-904 (November 1981). (Proceedings of the 1981 International Conference on Luminescence in Berlin).

76. T. Suyama, K. Okamoto and Y. Hamakawa, "New type of thin-film electroluminescent device having a multilayer structure," Appl. Phys. Lett. 41(5), 462-464 (1 September 1982).

77. S. T. Pantelides, "The electronic structure of impurities and other point defects in semiconductors," Rev. Mod. Phys. 50(4), 797-858 (October 1978).

78. K. Yosida, Functional Analysis, 4th ed. (Springer-Verlag, New York, 1974), p. 379.

ENERGY TRANSFER AMONG IONS IN SOLIDS

B. Di Bartolo

Department of Physics
Boston College
Chestnut Hill, Massachusetts 02167

ABSTRACT

The subject of interactions among atoms is introduced by considering first the static and then the dynamic effects of these interactions in a two-atom system and in a linear chain of atoms. Subsequently the different types of interactions (multipolar, exchange and electromagnetic) are examined. After reviewing the different modes of excitation of a solid containing both donors and acceptors, a "statistical" treatment of energy transfer is presented by considering first the case of energy transfer without migration among donors, and then the case of energy transfer when such migration occurs. Finally the concept of collective excitations in solids is introduced and a general theory of these excitations is presented.

I. INTERACTION AMONG ATOMS

I.A. Two-Atom System

Let us consider for simplicity two atoms of hydrogen with the two nuclei located at positions a and b and the two electrons at positions (x_1, y_1, z_1) and (x_2, y_2, z_2) as in Figure 1. Let the internuclear distance R be such that

$$R \gg x_1, y_1, z_1, x_2, y_2, z_2 \qquad (1)$$

so that the overlap between the wavefunctions of the two atoms may

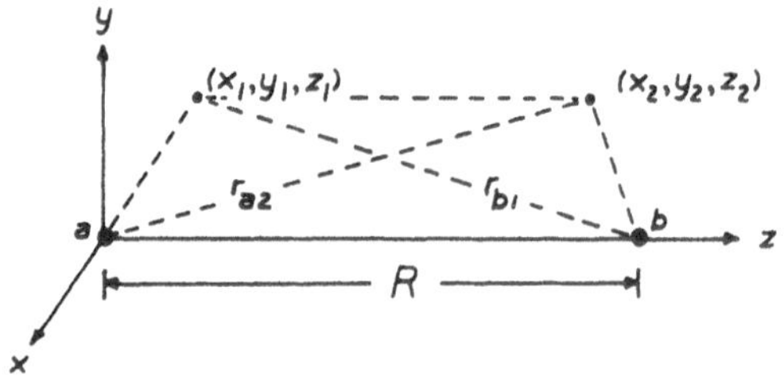

Fig. 1. System consisting of two hydrogen atoms.

be neglected. The wavefunction of the system, neglecting the interatomic interaction is given by [1]

$$\Psi(1,2) = \psi_a(1)\ \psi_b(2) \tag{2}$$

where $\psi_a(1)$ and $\psi_b(2)$ are the wavefunctions of the individual atoms; 1 and 2 stand for the coordinates of the two electrons. The interaction Hamiltonian is given by

$$H' = e^2\left(\frac{1}{r_{ab}} + \frac{1}{r_{12}} - \frac{1}{r_{a2}} - \frac{1}{r_{b1}} \right) \tag{3}$$

where

$$\begin{cases} r_{ab} = R \\ r_{12} = \sqrt{(x_1 - x_2)^2 + (y_1 - y_2)^2 + (z_1 - z_2 - R)^2} \\ r_{a2} = \sqrt{x_2^2 + y_2^2 + (R + z_2)^2} \\ r_{b1} = \sqrt{x_1^2 + y_1^2 + (z_1 + R)^2} \end{cases} \tag{4}$$

The correction to the energy eigenvalue to first order in H' is

$$E' = \langle \psi_a(1)\ \psi_b(2) | H' | \psi_a(1)\ \psi_b(2) \rangle \tag{5}$$

We use the expansion

$$\frac{1}{\sqrt{1 + \varepsilon}} = 1 - \frac{\varepsilon}{2} + \frac{3}{8}\varepsilon^2 - \frac{15}{48}\varepsilon^3 + \ldots \tag{6}$$

and retain only the first two powers of the electrons' coordinates. The result is

$$H' = \frac{e^2}{R^3}(x_1x_2 + y_1y_2 - 2z_1z_2) \tag{7}$$

and

$$E' = \langle \psi_a(1)\ \psi_b(2) | H' | \psi_a(1)\ \psi_b(2) \rangle$$

$$= \frac{e^2}{R^3} \langle \psi_a(1)\ \psi_b(2) | (x_1x_2 + y_1y_2 - 2z_1z_2) | \psi_a(1)\ \psi_b(2) \rangle = 0 \tag{8}$$

because H' is an odd operator and ψ_a and ψ_b have definite parity. The correction to the ground state energy to first order in H' is zero. But to second order

$$E'' = \sum_{k \neq 0} \frac{\langle 0 | H' | k \rangle \langle k | H' | 0 \rangle}{E_o - E_k} \tag{9}$$

where $|0\rangle$ and $|k\rangle$ are ground and excited states of the two-atom system, respectively. The energy levels of the hydrogen atom are given by

$$E_n = -\frac{e^2}{2n^2 a_o} \tag{10}$$

where $a_o = \hbar^2/me^2$. The ground state energy E_o is equal to $-(e^2/a_o)$. The states $|k\rangle$ that contribute to the sum in (9) are made up of products of odd wavefunctions of the two atoms; the energy of the lowest such state is $-(e^2/4a_o)$ and the energy of the highest such state is zero. Therefore $(E_o - E_k)$ lies in the interval

$$\left(-\frac{3}{4}\frac{e^2}{a_o}\ ,\ -\frac{e^2}{a_o}\right) \quad ;$$

we shall approximate this quantity with the value $-(e^2/a_o)$ for all k's.

Going back to (9) we find

$$E'' = -\frac{1}{e^2/a_o} \sum_{k\neq 0} \langle 0|H'|k\rangle\langle k|H'|0\rangle$$

$$= -\frac{1}{e^2/a_o} \left[\sum_{k\neq 0} \langle 0|H'|k\rangle\langle k|H'|0\rangle + \langle 0|H'|0\rangle\langle 0|H'|0\rangle \right]$$

$$= -\frac{1}{e^2/a_o} \sum_{k} \langle 0|H'|k\rangle\langle k|H'|0\rangle = -\frac{1}{e^2/a_o} \langle 0|(H')^2|0\rangle \qquad (10)$$

where we have used the relation (8) reexpressed as

$$\langle 0|H'|0\rangle = 0 \qquad (12)$$

and also

$$\sum_{k} \langle 0|H'|k\rangle\langle k|H'|0\rangle = \langle 0|(H')^2|0\rangle \qquad (13)$$

We have

$$\langle 0|(H')^2|0\rangle = \frac{e^4}{R^6} \langle \psi_a(1)\ \psi_b(2)|x_1^2x_2^2 + y_1^2y_2^2 + 4z_1^2z_2^2|\psi_a(1)\ \psi_b(2)\rangle$$

$$= \frac{e^4}{R^6} (\overline{x_1^2}\,\overline{x_2^2} + \overline{y_1^2}\,\overline{y_2^2} + 4\overline{z_1^2}\,\overline{z_2^2}) \qquad (14)$$

In the ground state of the hydrogen atom

$$\overline{x^2} = \overline{y^2} = \overline{z^2} = \frac{1}{3}\overline{r^2} \qquad (15)$$

and

$$\overline{r^2} = 3a_o^2 \qquad (16)$$

Then

$$<0|(H')^2|0> = \frac{e^4}{R^6}\left(\frac{1}{9}\overline{r_1^2}\,\overline{r_2^2} + \frac{1}{9}\overline{r_1^2}\,\overline{r_2^2} + \frac{4}{9}\overline{r_1^2}\,\overline{r_2^2}\right)$$

$$= \frac{2}{3}\frac{e^4}{R^6}\overline{r_1^2}\,\overline{r_2^2} = \frac{6e^4}{R^6}a_0^4 \tag{17}$$

, and

$$E'' = -\frac{1}{e^2/a_0}<0|(H')^2|0> = -\frac{6e^2 a_0^5}{R^6} \tag{18}$$

If we extend the expansion of H' and include higher power of the electrons' coordinates the result is [2]

$$E'' = -\frac{6e^2 a_0^5}{R^6} - \frac{135e^2 a_0^7}{R^8} - \frac{1416e^2 a_0^9}{R^{10}} \tag{19}$$

This energy, intrinsically negative, is called the *van der Waals' energy*. Note that this energy goes to zero as $h \to 0$.

For many electron atoms the perturbation Hamiltonian is represented by a sum of terms of type (7) with one term for each pair of electrons [3].

The van der Waals Hamiltonian is responsible for lowering the *ground state energy* of the two-atom system. The van der Waals bonding is an effect of this interaction. We shall see later that the van der Waals interaction is also responsible for the time evolution of a two-atom *excited* system, i.e., it leads to energy transfer from one excited atom to the other atom. This is one of several examples of interactions that produce both static and dynamical effects.

I.B. Dynamical Effects of the Interaction

Consider a system with a time-independent Hamiltonian H_0. The time-dependent Schrödinger equation is

$$H_0\Psi = i\hbar\frac{\partial\Psi}{\partial t} \tag{20}$$

If the system is in a stationary state labeled i

$$\Psi(t) = \Psi_i(t) = e^{-i(E_i/\hbar)t} \Psi_i(0) \tag{21}$$

where the energy values are given by

$$H_o \Psi_i(0) = E_i \Psi_i(0) \tag{22}$$

We shall assume that the wavefunctions $\Psi_i(t)$ are orthonormal.

Let us now suppose that the system is subjected to a time-dependent perturbation represented by $H'(t)$. The system will be represented by a wavefunction $\Psi(t)$ such that

$$H\Psi(t) = (H_o + H')\Psi(t) = i\hbar \frac{\partial \Psi(t)}{\partial t} \tag{23}$$

We can expand $\Psi(t)$ in terms of the complete set $\Psi_i(t)$

$$\Psi(t) = \sum_i c_i(t) \Psi_i(t) \tag{24}$$

If $H' = 0$, the coefficients c_i's are time-independent. Replacing Eq. (24) in Eq. (23)

$$(H_o + H') \sum_i c_i(t) \Psi_i(t) = i\hbar \left[\sum_i c_i(t) \frac{\partial \Psi_i(t)}{\partial t} + \sum_i \frac{\partial c_i(t)}{\partial t} \Psi_i(t) \right] \tag{25}$$

Then

$$\sum_i c_i(t) H' \Psi_i(t) = i\hbar \sum_i \frac{\partial c_i(t)}{\partial t} \Psi_i(t) \tag{26}$$

where we have taken advantage of Eqs. (21) and (22). Multiplying by $\Psi_k^*(t)$ and integrating over all space we obtain

$$\begin{aligned} i\hbar \frac{\partial c_k(t)}{\partial t} &= \sum_i c_i(t) \langle \Psi_k(t) | H' | \Psi_i(t) \rangle \\ &= \sum_i c_i(t) \langle \Psi_k(0) | H' | \Psi_i(0) \rangle e^{i\omega_{ki} t} \end{aligned} \tag{27}$$

where

$$\omega_{ki} = \frac{E_k - E_i}{\hbar} \tag{28}$$

1. Coherent Energy Transfer in a Two-Atom System. We shall now make the following assumptions:

(a) The system has only two energy states, say 1 and 2; and

(b) the perturbation H' is constant, but is turned on at time t = 0.

In a two-atom A and B system, state 1(2) consists of atom A(B) excited and B(A) deexcited.

The coupled equations (27) become

$$\begin{cases} i\hbar\, \dot{c}_1(t) = c_1(t)\, \langle\Psi_1(0)|H'|\Psi_1(0)\rangle \\ \qquad + c_2(t)\, \langle\Psi_1(0)|H'|\Psi_2(0)\rangle\, e^{i\omega_{12}t} \\ i\hbar\, \dot{c}_2(t) = c_1(t)\, \langle\Psi_2(0)|H'|\Psi_1(0)\rangle\, e^{i\omega_{21}t} \\ \qquad + c_2(t)\, \langle\Psi_2(0)|H'|\Psi_2(0)\rangle \end{cases} \tag{29}$$

Set

$$\begin{cases} \langle\Psi_i(0)|H'|\Psi_i(0)\rangle = V_i \\ \langle\Psi_i(0)|H'|\Psi_k(0)\rangle = M_{ik} \end{cases} \tag{30}$$

Then

$$i\hbar\, \dot{c}_1(t) = c_1(t)\, V_1 + c_2(t)\, M_{21}^{*}\, e^{-i\omega_{21}t}$$

$$i\hbar\, \dot{c}_2(t) = c_2(t)\, M_{21}\, e^{i\omega_{21}t} + c_2(t)\, V_2 \tag{31}$$

If at time t = 0, the system is in state 1:

$$c_1(0) = 1 \quad , \qquad c_2(0) = 0 \tag{32}$$

the time evolution of the system is given by the equations

$$\begin{cases} c_1(t) = \cos at + \dfrac{i}{2a\hbar}\,[(E_2 + V_2) - (E_1 + V_1)]\,\sin at \\ \qquad \cdot\, e^{-i\left(\frac{V_1 + V_2}{2\hbar}\right)t}\, e^{-i\frac{\omega_{21}}{2}t} \\ c_2(t) = \dfrac{M_{21}}{i\hbar a}\,\sin at\; e^{-i\left(\frac{V_1 + V_2}{2\hbar}\right)t}\, e^{i\frac{\omega_{21}}{2}t} \end{cases} \tag{33}$$

where

$$a = \left\{\frac{|M_{21}|^2}{\hbar^2} + \frac{[(E_2 + V_2) - (E_1 + V_1)]^2}{4}\right\}^{1/2} \sec^{-1} \tag{34}$$

It is

$$|c_1(t)|^2 + |c_2(t)|^2 = 1 \tag{35}$$

at all times. Assume now

$$E_1 = E_2 \quad , \quad V_1 = V_2 \tag{36}$$

In this case

$$\begin{cases} |c_1(t)|^2 = \cos^2 at \\ |c_2(t)|^2 = \sin^2 at \end{cases} \tag{37}$$

where

$$a = \frac{|M_{12}|}{\hbar} \tag{38}$$

and the time evolution of the system is represented in Figure 2.

At time $t = 0$

$$|c_1(0)|^2 = 1 \quad , \quad |c_2(0)|^2 = 0 \tag{39}$$

At time $t = \dfrac{\pi}{2a} = \dfrac{\pi\hbar}{2|M_{12}|}$

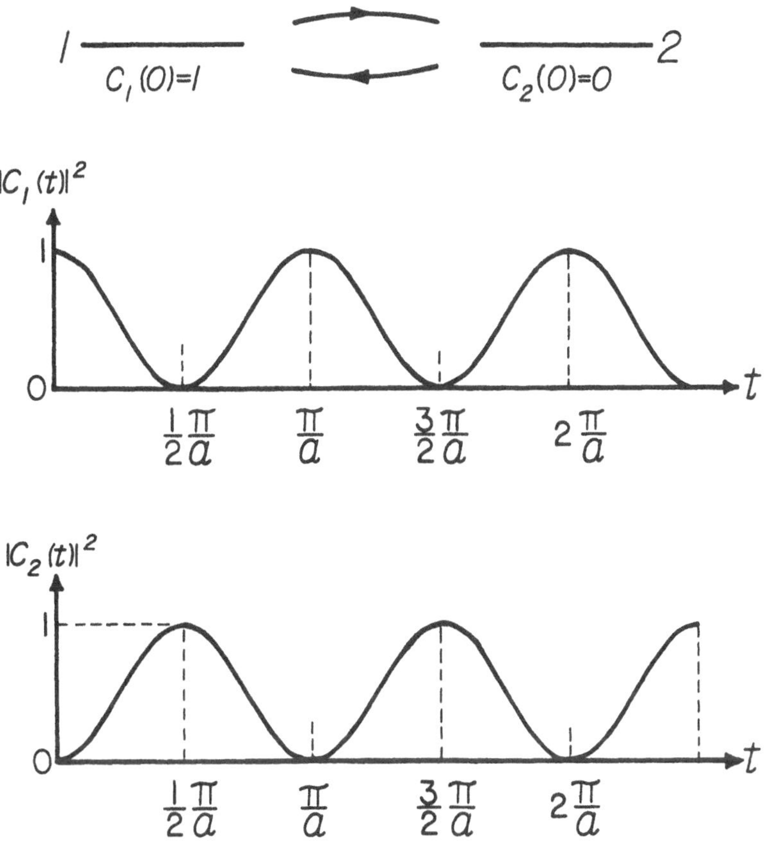

Fig. 2. Coherent time evolution of a two-atom system.

$$|c_1(\frac{\pi}{2a})|^2 = 0 \quad , \quad |c_2(\frac{\pi}{2a})|^2 = 1 \tag{40}$$

This time evolution is called *coherent*.

2. Incoherent Energy Transfer in a Two-Atom System. If at time t = T (small) the phase of the wavefunction is interrupted, the change in $|c_2(t)|^2$ in time T is

$$\Delta|c_2(t)|^2 \simeq a^2 T^2 \tag{41}$$

and in time t >> T

$$|c_2^2(t)|^2 \simeq \frac{t}{T}\Delta|c_2(t)|^2 = \frac{t}{T} a^2 T^2 = a^2 Tt \tag{42}$$

The probability $|c_2(t)|^2$ is now proportional to t and a probability per unit time can be defined

$$w_{12} = \frac{|c_2(t)|^2}{t} = a^2\ T = \frac{|M_{21}|^2}{\hbar^2}\ T \tag{43}$$

But

$$T = \frac{h}{\Delta E} \tag{44}$$

where ΔE = width of the transition. Then

$$w_{12} = \frac{|M_{21}|^2}{h^2}\ \frac{h}{\Delta E} = \frac{2\pi}{\hbar}\ |\langle\Psi_2(0)|H'|\Psi_1(0)\rangle|^2\ g(E) \tag{45}$$

where $g(E)$ = density of final states. If both state 1 and state 2 are smeared

$g_1(E_1)\ dE_1$ = probability that state 1 has energy in $(E_1, E_1 + dE_1)$

$g_2(E_2)\ dE_2$ = probability that state 2 has energy in $(E_2, E_2 + dE_2)$

and $g(E)$ in (45) is replaced by

$$\int g_1(E_1)[\int g_2(E_2)\ \delta(E_2 - E_1)\ dE_2]\ dE_1 = \int g_1(E_1)\ g_2(E_1)\ dE_1$$

$$= \int g_1(E)\ g_2(E)\ dE \tag{46}$$

The question now is: If at time $t = 0$, the system is in state 1, how will it evolve? And the answer is: Since probabilities per unit time are appropriate, a *master equation* treatment must be used. Let

$$P_1(t) = |c_1(t)|^2$$

$$P_2(t) = |c_2(t)|^2 \tag{47}$$

Then, since $w_{12} = w_{21} = w$

$$\begin{cases} \dfrac{dP_1(t)}{dt} = w[P_2(t) - P_1(t)] \\ \\ \dfrac{dP_2(t)}{dt} = w[P_1(t) - P_2(t)] \end{cases} \qquad (48)$$

Given the initial conditions

$$P_1(0) = 1 \quad , \quad P_2(0) = 0 \qquad (49)$$

the solutions of (48) are

$$\begin{cases} P_1(t) = \frac{1}{2}(1 + e^{-2wt}) \\ \\ P_2(t) = \frac{1}{2}(1 - e^{-2wt}) \end{cases} \qquad (50)$$

and are represented in Figure 3.

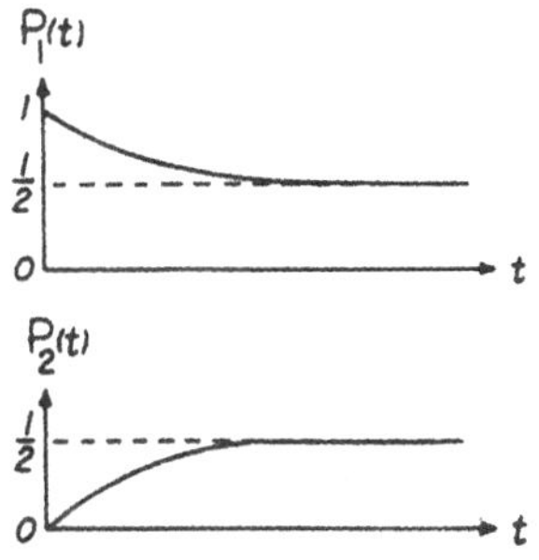

Fig. 3. Incoherent time evolution of a two-atom system.

This time evolution is called *incoherent*.

We want at this point to associate some key words and expressions to coherent and incoherent evolutions.

Coherent evolution

Reversibility
Oscillations
Characteristic time $\propto \frac{1}{M}$
Transfer rate $\propto M$

Incoherent evolution

Irreversibility
Approach to equilibrium
Characteristic time $\propto \frac{1}{M^2}$
Transfer rate $\propto M^2$

3. Coherent Energy Transfer in a Linear Chain. Consider a linear chain of equal interacting atoms. Let state Ψ_m be that corresponding to atom m being excited. From (27) we have

$$i\hbar \dot{c}_m(t) = \sum_i c_i(t) M_{mi} e^{i\omega_{mi}t} \tag{51}$$

where

$$M_{mi} = \langle\Psi_m(0)|H'|\Psi_i(0)\rangle \tag{52}$$

Since the atoms are all equal $\omega_{mi} = 0$. If we assume only nearest neighbor interaction and take the matrix element M as real for simplicity, Eqs. (51) become

$$i\hbar \dot{c}_m(t) = (c_{m-1} + c_{m+1}) M \tag{53}$$

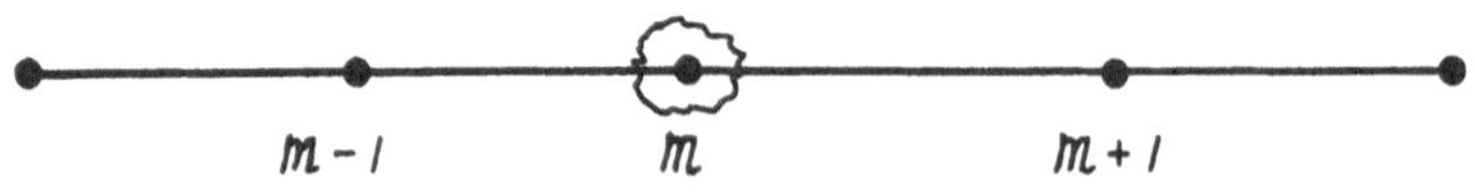

Fig. 4. A linear chain of interacting atoms.

If the excitation is localized on atom m at time t = 0:

$$c_m(0) = 1 \quad , \qquad c_{n\neq m}(0) = 0 \tag{54}$$

Let

$$c_m(t) = i^{-m} f_m\left(\frac{2M}{\hbar} t\right) = i^{-m} f_m(x) \tag{55}$$

where

$$x = \frac{2M}{\hbar} t$$

Then

$$\dot{c}_m(t) = i^{-m} \frac{df_m(x)}{dx} \frac{dx}{dt} = i^{-m} \frac{2M}{\hbar} \frac{df_m(x)}{dx}$$

$$c_{m-1}(t) = i^{-(m-1)} f_{m-1}(x) = i^{-m+1} f_{m-1}(x)$$

$$c_{m+1}(t) = i^{-m-1}$$

It is

$$i\hbar\, \dot{c}_m(t) = i\hbar\, i^{-m} \frac{2M}{\hbar} \frac{df_m(x)}{dx} = i^{-m+1}\, 2M \frac{df_m(x)}{dx}$$

Therefore

$$i^{-m+1}\, 2M \frac{df_m(x)}{dx} = Mi^{-m+1} f_{m-1}(x) + Mi^{-m-1} f_{m+1}(x)$$

$$2 \frac{df_m(x)}{dx} = f_{m-1}(x) - f_{m+1}(x) \tag{56}$$

The solutions of the above equation are Bessel functions:

$$f_m(x) = J_m(x) \tag{57}$$

Then

$$c_m(t) = i^{-m} J_m\left(\frac{2M}{\hbar} t\right) \tag{58}$$

and the probability of finding atom m excited at time t is

$$P_m(t) = |c_m(t)|^2 = J_m^2\left(\frac{2M}{\hbar} t\right) \tag{59}$$

4. <u>Incoherent Energy Transfer in a Linear Chain</u>. Let us consider again a linear chain of interacting equal atoms and let us again assume only nearest neighbor interactions. This time, however, we assume an incoherent time evolution with w being the transition probability per unit time that the excitation moves from, say, atom m to atom (m - 1) or (m + 1). The equation to solve is a master equation

$$\dot{P}_m(t) = w[P_{m+1}(t) - P_{m-1}(t) - 2P_m(t)] \tag{60}$$

where $P_m(t) = |c_m(t)|^2$. Let, as before, the initial conditions be

$$P_m(0) = 1 \quad , \quad P_{n\neq m}(0) = 0 \tag{61}$$

Set

$$P_m(t) = e^{-2wt} g_m(2wt) = e^{-x} g_m(x) \tag{62}$$

where

$$x = 2wt$$

Then

$$\dot{P}_m(t) = 2w \frac{\partial}{\partial x} [e^{-x} g_m(x)] = 2w[-e^{-x} g_m(x) + e^{-x} \frac{dg_m(x)}{dx}]$$

$$= -2w P_m(t) + 2w e^{-x} \frac{dg_m(x)}{dx}$$

The master equation then gives us

$$2 \frac{dg_m(x)}{dx} = g_{m+1}(x) + g_{m-1}(x) \tag{63}$$

$g_m(x)$ are modified Bessel functions:

$$g_m(x) = I_m(x) \quad , \quad g_m(2wt) = I_m(2wt)$$

Then

$$P_m(t) = e^{-2wt} I_m(2wt) \tag{64}$$

where

$$I_m(x) = \text{modified Bessel functions}$$

I.C. The Relevant Energy Transfer Hamiltonian

Let us now consider again a two-atom system as in Figure 5. Let H_A and H_B be the Hamiltonian of atom A and B, respectively. The Hamiltonian of the two-atom system is then given by

$$H = H_A + H_B + H_{AB} \tag{65}$$

where

$$\begin{aligned} H_{AB} &= \frac{Z_a Z_b e^2}{R} + \frac{e^2}{r_{12}} - \frac{Z_a e^2}{r_{a2}} - \frac{Z_b e^2}{r_{b1}} \\ &= \frac{Z_a Z_b e^2}{R} + \frac{e^2}{r_{12}} - \frac{Z_a e^2}{|\vec{R} + \vec{r}_{b2}|} - \frac{Z_b e^2}{|-\vec{R} + \vec{r}_{a1}|} \end{aligned} \tag{66}$$

Let $|a\rangle$ and $|a'\rangle$ be the ground state and excited state wavefunctions of atom A, respectively and $|b\rangle$ and $|b'\rangle$ the ground state and excited state wavefunctions of atom B, respectively. We shall consider a transition from an initial state $|1\rangle$ to a final state $|2\rangle$ of the two-atom system, where

(67)

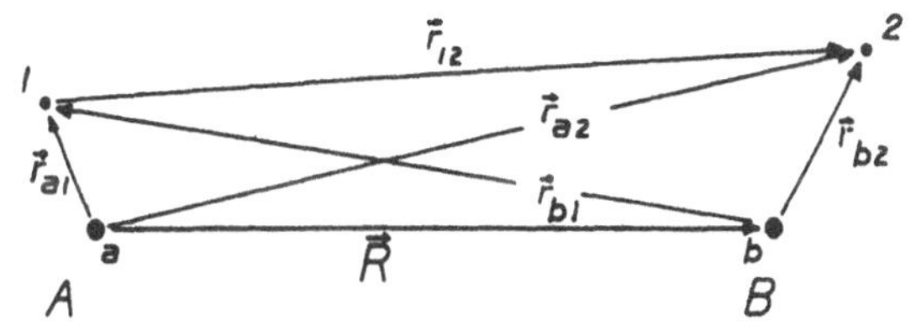

Fig. 5. Two-atom system.

$$|1\rangle = |a'b\rangle \quad , \quad |2\rangle = |ab'\rangle \tag{67}$$

The relevant matrix element is given by

$$\begin{aligned}
&\langle a'(1)\ b(2)|H_{AB}|a(1)\ b'(2)\rangle \\
&= Z_a Z_b \frac{e^2}{R} \langle a'(1)|a(1)\rangle\langle b(2)|b'(2)\rangle \\
&+ \langle a'(1)\ b(2)\left|\frac{e^2}{r_{12}}\right|a(1)\ b'(2)\rangle \\
&- Z_a e^2 \langle a'(1)|a(1)\rangle\langle b(2)\left|\frac{1}{|\vec{R}+\vec{r}_{b2}|}\right|b'(2)\rangle \\
&- Z_b e^2 \langle a'(1)\left|\frac{1}{|-\vec{R}+\vec{r}_{a1}|}\right|a(1)\rangle\langle b(2)|b'(2)\rangle \\
&= \langle a'(1)\ b(2)\left|\frac{e^2}{r_{12}}\right|a(1)\ b'(2)\rangle
\end{aligned} \tag{68}$$

Taking into account the overlap of the wavefunctions we replace the relevant product wavefunctions as follows:

$$\begin{cases}
|a'(1)\ b(2)\rangle \to \frac{1}{\sqrt{2}}|a'(1)\ b(2) - b(1)\ a'(2)\rangle \\
|a(1)\ b'(2)\rangle \to \frac{1}{\sqrt{2}}|a(1)\ b'(2) - b'(1)\ a(2)\rangle
\end{cases} \tag{69}$$

Then

$$\begin{aligned}
&\langle a'(1)\ b(2)|H_{AB}|a(1)\ b'(2)\rangle \\
&\to \frac{1}{2} \langle a'(1)\ b(2)|H_{AB}|a(1)\ b'(2)\rangle \\
&+ \frac{1}{2} \langle b(1)\ a'(2)|H_{AB}|b'(1)\ a(2)\rangle \\
&- \frac{1}{2} \langle b(1)\ a'(2)|H_{AB}|a(1)\ b'(2)\rangle
\end{aligned}$$

$$- \frac{1}{2} \langle a'(1)\ b(2) | H_{AB} | b'(1)\ a(2) \rangle$$

$$= \langle a'(1)\ b(2) | H_{AB} | a(1)\ b'(2) \rangle$$

$$- \langle a'(1)\ b(2) | H_{AB} | b'(1)\ a(2) \rangle \qquad (70)$$

and

$$\langle a'(1)\ b(2) | H_{AB} | b'(1)\ a(2) \rangle$$

$$= Z_a Z_b \frac{e^2}{R} \langle a'(1) | b'(1) \rangle \langle b(2) | a(2) \rangle$$

$$+ \langle a'(1)\ b(2) \left| \frac{e^2}{r_{12}} \right| b'(1)\ a(2) \rangle$$

$$- Z_a e^2 \langle a'(1) | b'(1) \rangle \langle b(2) \left| \frac{1}{|\vec{R} + \vec{r}_{b2}|} \right| a(2) \rangle$$

$$- Z_b e^2 \langle b(2) | a(2) \rangle \langle a'(1) \left| \frac{1}{|-\vec{R} + \vec{r}_{a1}|} \right| b'(1) \rangle$$

$$\simeq \langle a'(1)\ b(2) \left| \frac{e^2}{r_{12}} \right| b'(1)\ a(2) \rangle \qquad (71)$$

The relevant matrix element is then given by

$$\langle | H_{AB} | \rangle = \langle a'(1)\ b(2) \left| \frac{e^2}{r_{12}} \right| a(1)\ b'(2) \rangle$$

$$- \langle a'(1)\ b(2) \left| \frac{e^2}{r_{12}} \right| b'(1)\ a(2) \rangle \qquad (72)$$

where the first term is called the *direct term* and the second term the *exchange term*.

I.D. Interaction Between Two Atoms in Solids

Let A and B be two atoms in fixed position in a solid; Figure 6 reports the relevant coordinates. It is

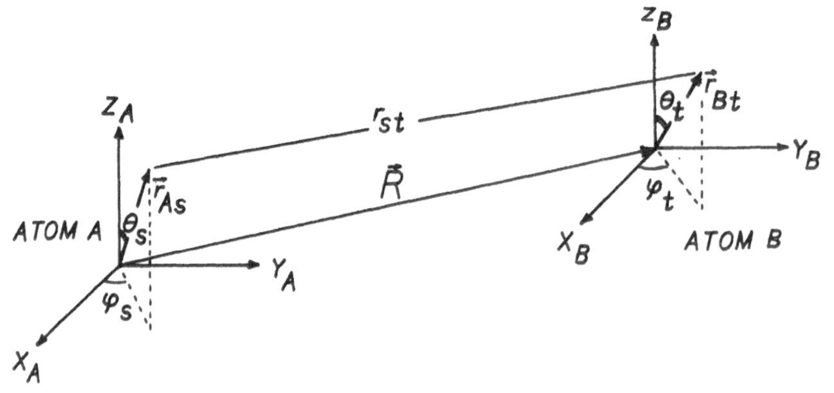

Fig. 6. Two-atom system.

$$\vec{R} \equiv (R, \Theta, \phi) \quad , \qquad R \gg r_{As}, r_{Bt}$$

$$\vec{r}_{st} = \vec{R} + \vec{r}_{Bt} - \vec{r}_{As}$$

The relevant energy transfer Hamiltonian is

$$H_{AB} = \sum_{s,t} \frac{e^2}{r_{st}} = \sum_{s,t} \frac{e^2}{|\vec{R} + \vec{r}_{Bt} - \vec{r}_{As}|} \tag{73}$$

where the sum is over the electrons of the two atoms.

Carlson and Rushbrooke [4] have carried out the expansion of H_{AB} in spherical harmonics taking into account the fact that $R \gg r_{As}, r_{Bt}$. The result of their calculation is

$$H_{AB} = e^2 \sum_{\ell_1=0}^{\infty} \sum_{\ell_2=0}^{\infty} \sum_{m_1=-\ell}^{\ell} \sum_{m_2=-\ell}^{\ell} \frac{1}{R^{\ell_1+\ell_2+1}}$$

$$\cdot\; G_{12}\, C_{m_1+m_2}^{\ell_1+\ell_2}(\Theta, \phi)^*\, D_{m_1}^{\ell_1}(A)\, D_{m_2}^{\ell_2}(B) \tag{74}$$

where

$$G_{12} = (-1)^{\ell_1} \sqrt{\frac{(2\ell_1 + 2\ell_2 + 1)!}{(2\ell_1)!(2\ell_2)!}} \begin{pmatrix} \ell_1 & \ell_2 & \ell_1+\ell_2 \\ m_1 & m_2 & -m_1-m_2 \end{pmatrix} \tag{75}$$

$$D_m^{\ell}(A) = \sum_s r_s^{\ell} \, C_m^{\ell}(\theta_s, \phi_s) \tag{76}$$

$$C_m^{\ell}(\Omega) = \sqrt{\frac{4\pi}{2\ell + 1}} \, Y_{\ell m}(\Omega) \tag{77}$$

$$\underbrace{\begin{pmatrix} j_1 & j_2 & j_3 \\ m_1 & m_2 & m_3 \end{pmatrix}}_{\text{3j symbols}} = (-1)^{j_1-j_2-m_3} \sqrt{2j_3 + 1}$$

$$\cdot \underbrace{\langle j_1 \, m_1 \, j_2 \, m_2 | j_1 \, j_2 \, j_3 - m_3 \rangle}_{\text{Clebsch-Gordan coefficients}}$$

3j symbols = 0 unless

$$\Delta(j_1 \, j_2 \, j_3) \quad ; \quad m_1 + m_2 + m_3 = 0$$

<u>Example</u>

$$Y_{10} = \sqrt{\frac{3}{4\pi}} \cos\theta = \sqrt{\frac{3}{4\pi}} \frac{z}{r}$$

$$Y_{1,\pm 1} = \mp \sqrt{\frac{3}{8\pi}} e^{\pm i\phi} \sin\theta = \mp \sqrt{\frac{3}{8\pi}} \frac{x \pm iy}{r}$$

Then

$$C_0^1 = \sqrt{\frac{4\pi}{3}} Y_{10} = \sqrt{\frac{4\pi}{3}} \sqrt{\frac{3}{4\pi}} \frac{z}{r} = \frac{z}{r}$$

$$C_{\pm 1}^1 = \sqrt{\frac{4\pi}{3}} Y_{1\pm 1} = \sqrt{\frac{4\pi}{3}} \left[\mp \sqrt{\frac{3}{8\pi}} \frac{x \pm iy}{r} \right] = \mp \frac{1}{\sqrt{2}} \frac{x \pm iy}{r}$$

and

$$D^1_0 = rC^1_0 = z$$

$$D^1_1 = rC^1_1 = -\frac{1}{\sqrt{2}}(x + iy)$$

$$D^1_{-1} = rC^1_{-1} = \frac{1}{\sqrt{2}}(x - iy)$$

Then the direct term of $<|H_{AB}|>$ is

$$<a'b|H_{AB}|ab'> = \sum_{\substack{\ell_1\ell_2\\ m_1m_2}} \frac{e^2}{R^{\ell_1+\ell_2+1}} G_{12}\, C^{\ell_1+\ell_2 *}_{m_1+m_2} \cdot <a'|D^{\ell_1}_{m_1}|a> \cdot <b|D^{\ell_2}_{m_2}|b'> \tag{78}$$

The relevant quantity that enters the transition probability is

$$|<a'b|H_{AB}|ab'>|^2 = \sum_{\substack{\ell_1\ell_2\\ m_1m_2}} \sum_{\substack{\ell_3\ell_4\\ m_3m_4}} \frac{e^4}{R^{\ell_1+\ell_2+\ell_3+\ell_4+2}} G_{12}\, G_{34} \cdot C^{\ell_1+\ell_2 *}_{m_1+m_2}\, C^{\ell_3+\ell_4}_{m_3+m_4} \cdot <a'|D^{\ell_1}_{m_1}|a><a'|D^{\ell_3}_{m_3}|a>^* \cdot <b|D^{\ell_2}_{m_2}|b'><b|D^{\ell_4}_{m_4}|b'>^* \tag{79}$$

We average over Θ, ϕ:

$$\frac{1}{4\pi} \iint \sin\Theta \, d\Theta \, d\phi \, C_{m_1+m_2}^{\ell_1+\ell_2 \, *} \, C_{m_3+m_4}^{\ell_3+\ell_4}$$

$$= \frac{1}{\sqrt{(2\ell_1 + 2\ell_2 + 1)(2\ell_3 + 2\ell_4 + 1)}}$$

$$\cdot \, \delta_{\ell_1+\ell_2, \ell_3+\ell_4} \, \delta_{m_1+m_2, m_3+m_4}$$

and use the following approximations

(1) Neglect cross terms: $\ell_1 \neq \ell_3$, $\ell_2 \neq \ell_4$, $m_1 \neq m_3$, $m_2 \neq m_4$

$$|\langle|H_{AB}|\rangle|^2 \simeq \sum_{\substack{\ell_1\ell_2 \\ m_1m_2}} \left[\frac{e^2}{R^{\ell_1+\ell_2+1}}\right]^2 G_{12}^2 \frac{1}{2\ell_1 + 2\ell_2 + 1}$$

$$\cdot \; |\langle a'|D_{m_1}^{\ell_1}|a\rangle|^2 |\langle b|D_{m_2}^{\ell_2}|b'\rangle|^2$$

where

$$G_{12}^2 = \frac{(2\ell_1 + 2\ell_2 + 1)!}{(2\ell_1)!(2\ell_2)!} \begin{pmatrix} \ell_1 & \ell_2 & \ell_1+\ell_2 \\ m_1 & m_2 & -m_1-m_2 \end{pmatrix}^2$$

(2) Take average of G_{12}^2:

$$\langle G_{12}^2 \rangle = \frac{1}{(2\ell_1 + 1)(2\ell_2 + 1)} \sum_{m_1, m_2} G_{12}^2$$

$$= \frac{1}{(2\ell_1 + 1)(2\ell_2 + 1)} \sum_{m_1, m_2} \frac{(2\ell_1 + 2\ell_2 + 1)!}{(2\ell_1)!(2\ell_2)!}$$

$$\cdot \begin{pmatrix} \ell_1 & \ell_2 & \ell_1+\ell_2 \\ m_1 & m_2 & -m_1-m_2 \end{pmatrix}^2$$

$$= \frac{(2\ell_1 + 2\ell_2 + 1)!}{(2\ell_1 + 1)!(2\ell_2 + 1)!} \underbrace{\sum_{m_1, m_2} \begin{pmatrix} \ell_1 & \ell_2 & \ell_1+\ell_2 \\ m_1 & m_2 & -m_1-m_2 \end{pmatrix}^2}_{1}$$

$$= \frac{(2\ell_1 + 2\ell_2 + 1)!}{(2\ell_1 + 1)!(2\ell_2 + 1)!}$$

Therefore

$$|\langle|H_{AB}|\rangle|^2 = \sum_{\substack{\ell_1 \ell_2 \\ m_1 m_2}} \left(\frac{e^2}{R^{\ell_1+\ell_2+1}} \right)^2$$

$$\cdot \frac{(2\ell_1 + 2\ell_2 + 1)!}{(2\ell_1 + 1)!(2\ell_2 + 1)!} \frac{1}{2\ell_1 + 2\ell_2 + 1}$$

$$\cdot |\langle a'|D_{m_1}^{\ell_1}|a\rangle|^2 |\langle b|D_{m_2}^{\ell_2}|b'\rangle|^2$$

$$\cdot \sum_{\ell_1 \ell_2} \left(\frac{e^2}{R^{\ell_1+\ell_2+1}} \right)^2 \frac{(2\ell_1 + 2\ell_2)!}{(2\ell_1 + 1)!(2\ell_2 + 1)!}$$

$$\cdot \left[\sum_{m_1} |\langle a'|D_{m_1}^{\ell_1}|a\rangle|^2 \right] \left[\sum_{m_2} |\langle b|D_{m_2}^{\ell_2}|b'\rangle|^2 \right] \tag{80}$$

II. DIFFERENT TYPES OF INTERACTIONS

II.A. Multipolar Electric Interactions

An electric multipole of a charge distribution $\rho(\vec{x})$ with $\vec{x} \equiv (r, \theta, \phi)$ is defined as follows:

$$D_{\ell,m} = \sqrt{\frac{4\pi}{2\ell + 1}} \int d\tau \; \rho(\vec{x}) \; r^{\ell} \; Y_{\ell m}(\theta, \phi) \tag{81}$$

If we set

$$\rho(\vec{x}) = \sum_s e\delta(\vec{x} - \vec{x}_s) \tag{82}$$

we obtain

$$D_{\ell,m} = \sqrt{\frac{4\pi}{2\ell + 1}} \int d\tau \sum_s e\delta(\vec{x} - \vec{x}_s)\, r^\ell\, Y_{\ell m}(\theta, \phi)$$

$$= e \sum_s \sqrt{\frac{4\pi}{2\ell + 1}}\, r_s^\ell\, Y_{\ell m}(\theta_s, \phi_s) \tag{83}$$

These multipoles are, apart e, the same quantities D_m^ℓ defined in (76).

The square of the matrix element of the multipolar interaction can be written as

$$|\langle|H_{AB}|\rangle|^2 = \frac{C^{(6)}}{R^6} + \frac{C^{(8)}}{R^8} + \frac{C^{(10)}}{R^{10}} + \ldots \tag{84}$$

where

$$C^{(6)} = \text{dipole-dipole term}$$

$$= \frac{e^4}{R^6}\frac{4!}{3!3!}\left[\sum_{m=-1}^{1} |\langle a'|D_m^1|a\rangle|^2\right]\left[\sum_{m=-1}^{1} |\langle b|D_m^1|b'\rangle|^2\right] \tag{85}$$

$$C^{(8)} = \text{dipole-quadrupole term} = \frac{e^4}{R^8}\frac{6!}{3!5!}$$

$$\cdot \left\{\left[\sum_{m=-1}^{1} |\langle a'|D_m^1|a\rangle|^2\right]\left[\sum_{m=-2}^{2} |\langle b|D_m^2|b'\rangle|^2\right]\right.$$

$$\left. + \left[\sum_{m=-2}^{2} |\langle a'|D_m^2|a\rangle|^2\right]\left[\sum_{m=-1}^{1} |\langle b|D_m^1|b'\rangle|^2\right]\right\} \tag{86}$$

$$C^{(10)} = \text{quadrupole-quadrupole term} = \frac{e^4}{R^{10}} \frac{8!}{5!5!}$$

$$\cdot \left[\sum_{m=-2}^{2} |\langle a'|D_m^2|a\rangle|^2 \right] \left[\sum_{m=-2}^{2} |\langle b|D_m^2|b'\rangle|^2 \right] \qquad (87)$$

The quantities $|\langle a'|D_m^\ell|a\rangle|^2$, $|\langle b'|D_m^\ell|b\rangle|^2$ can in principle be derived from spectroscopic data.

<u>Example</u>: dipole-dipole term

In classical electrodynamics the power irradiated by an oscillating electric dipole M cos ωt is given by

$$P = \frac{\omega^4}{3c^3} M^2 \qquad (88)$$

A classical oscillator of amplitude M has two Fourier components

$$M \cos \omega t = \frac{1}{2} Me^{i\omega t} + \frac{1}{2} Me^{-i\omega t} \qquad (89)$$

with frequency ω and -ω. Classically we do not distinguish frequency ω from frequency -ω, namely photons absorbed from photons emitted. Quantum mechanics, however, allows only one of the two components to enter the relevant matrix element

$$M_{classical} \rightarrow 2|M_{QM}| \qquad (90)$$

and

$$P_{QM} = \frac{\omega^4}{3c^3} |2M|^2 = \frac{4\omega^4}{3c^3} |M|^2 \qquad (91)$$

Therefore

$$A = \text{rate of decay} = \frac{\text{energy emitted per unit time}}{\text{energy of a photon}}$$

$$= \frac{1}{\tau_o} = \frac{4\omega^4 |M|^2}{3c^3 \hbar\omega} = \frac{8\pi\omega^3}{3hc^3} |M|^2 \qquad (92)$$

where τ_o = radiative lifetime. Then

$$e^2 \sum_m |\langle a'|D_m^1|a\rangle|^2 = |M|^2 = \frac{3hc^3}{8\pi\omega^3}\frac{1}{\tau_o} \tag{93}$$

The *f number*, a quantity usually derived from absorption data, is defined as follows:

$$f = \frac{2m\omega}{3\hbar e^2}|M|^2 \tag{94}$$

We can then write

$$e^2 \sum_m |\langle b|D'_m|b'\rangle|^2 = |M|^2 = \frac{3\hbar e^2}{2m\omega} f \tag{95}$$

If we use (93) and (95) in (85) we obtain

$$C^{(6)} = |\langle|H_{AB}|\rangle|dd = \frac{1}{R^6}\frac{4!}{3!3!}\left(\frac{3hc^3}{8\pi\omega^3}\frac{1}{\tau_{oA}}\right)\left(\frac{3\hbar e^2}{2m\omega}f_B\right)$$

$$= \frac{1}{R^6}\frac{3e^2c^3\hbar^6}{4mE^4}\frac{f_B}{\tau_A}\varepsilon \tag{96}$$

where $E = \hbar\omega$

$$\varepsilon = \frac{\text{probability of radiative decay}}{\text{probability of radiative decay + probability of nonradiative decay}}$$

$$= \frac{1/\tau_{oA}}{1/\tau_A} = \text{quantum efficiency} \tag{97}$$

where τ_{oA} = radiative lifetime of atom A and τ_A = effective lifetime of atom A.

The transfer rate is then given by

$$w_{AB} = \frac{2\pi}{\hbar}|\langle|H_{AB}|\rangle|^2 \int g_A(E)\, g_B(E)\, dE$$

$$= \frac{2\pi}{\hbar}\frac{1}{R^6}\frac{3e^2c^3\hbar^6}{4m}\frac{f_B}{\tau_A}\varepsilon \int \frac{g_A(E)\, g_B(E)}{E^4}\, dE$$

$$= \frac{3e^2c^3\hbar^5\pi}{2mR^6}\frac{f_B}{\tau_A}\varepsilon\int\frac{g_A(E)\ g_B(E)}{E^4}dE = \frac{1}{\tau_A}\left(\frac{R_o}{R}\right)^6 \tag{98}$$

$$R_o = \varepsilon f_B\left(\frac{3e^2c^3\hbar^5\pi}{2m}\right)\int\frac{g_A(E)\ g_B(E)}{E^4}dE$$

= radius at which the transfer rate is equal to the decay rate (99)

Let w_{AB}^{dd}, w_{AB}^{dq} and w_{AB}^{qq} be the energy transfer rates by dipole-dipole, dipole-quadrupole and quadrupole-quadrupole mechanisms, respectively. We can compare the magnitudes of w_{AB}^{dd} and w_{AB}^{dq}:

$$w_{AB}^{dd} = \frac{e^4}{R^6}\frac{2}{3}|\langle D^1\rangle|^2|\langle D^1\rangle|^2 \tag{100}$$

$$w_{AB}^{dq} = \frac{e^4}{R^8}|\langle D^1\rangle|^2|\langle D^2\rangle|^2 \tag{101}$$

$$\frac{w_{AB}^{dq}}{w_{AB}^{dd}} = \frac{1}{R^2}\frac{3}{2}\frac{|\langle D^2\rangle|^2}{|\langle D^1\rangle|^2} \simeq \frac{1}{R^2}\frac{a_o^4}{a_o^2} = \left(\frac{a_o}{R}\right)^2 \tag{102}$$

If the electric dipole transition in atom B is not allowed, then $|\langle D^1\rangle|^2 < a_o$ and it is possible that

$$w_{AB}^{dq} > w_{AB}^{dd} \tag{103}$$

Note that

$$\frac{w_{AB}^{qq}}{w_{AB}^{dd}} \simeq \frac{1}{R^4}\frac{|\langle D^2\rangle|^2|\langle D^2\rangle|^2}{|\langle D^1\rangle|^2|\langle D^1\rangle|^2} \simeq \left(\frac{a_o}{R}\right)^4 \tag{104}$$

II.B. Exchange Interactions

We shall now examine more closely the matrix element (72). The direct term can be written as follows:

$$\langle a'(1)\, b(2) \left| \frac{e^2}{r_{12}} \right| a(1)\, b'(2) \rangle = \iint a'(\vec{r}_1)^* \, a(\vec{r}_1) \, \frac{e^2}{r_{12}} \cdot b(\vec{r}_2)^* \, b'(\vec{r}_2) \, d\tau_1 \, d\tau_2 \tag{105}$$

and represents the Coulomb interaction between the charge distributions $ea'(\vec{r}_1)^* \, a(\vec{r}_1)$ and $eb(\vec{r}_2)^* \, b'(\vec{r}_2)$ at distance $\vec{R}$ from each other. The exchange term can be written

$$-\langle a'(1)\, b(2) \left| \frac{e^2}{r_{12}} \right| b'(1)\, a(2) \rangle = \iint a'(\vec{r}_1)^* b'(\vec{r}_1) \, \frac{e^2}{r_{12}} \cdot a(\vec{r}_2) \, b(\vec{r}_2)^* \, d\tau_1 \, d\tau_2 \tag{106}$$

and represents the Coulomb interaction between the charge distributions $ea'(\vec{r}_1)^* \, b'(\vec{r}_1)$ and $ea(\vec{r}_2) \, b(\vec{r}_2)^*$ at distance $\vec{R}$ from each other; these two charge distributions are very small if $\vec{R}$ is large. Therefore the exchange term is small if $\vec{R}$ is not small.

The exchange term can also be written

$$\langle a'(1)\, b(2) \left| \frac{-e^2}{r_{12}} \right| b'(1)\, a(2) \rangle = -\langle a'(1)\, b(2) \left| \frac{e^2}{r_{12}} P_{12} \right| a(1)\, b'(2) \rangle \tag{107}$$

where P_{12} is an operator that interchanges the two electron coordinates. For many electron atoms

$$\langle a'b | H_{AB} | ab' \rangle \text{ (exchange term)} = \langle a'b \left| -\sum_{i,j} \frac{e^2}{r_{ij}} P_{ij} \right| ab' \rangle \tag{108}$$

The overlap of the electron charges makes the condition $R > r_{As}, r_{Bt}$ invalid. However, if the overlap is small the multipolar part can be treated as previously. As for the exchange part,

$$-\sum_{i,j} \frac{e^2}{r_{ij}} P_{ij}$$

it can be replaced by an equivalent operator [5]:

$$H_{AB}^{exch} = -\sum_{s,t}\sum_{\ell,\ell'}\sum_{m,m'} j_{\ell\ell'}^{mm'}(\Theta, \phi)$$

$$\cdot\, C_m^{\ell}(\theta_s, \phi_s)\, C_{m'}^{\ell'}(\theta_t, \phi_t)\left(\frac{1}{2} + 2\vec{s}_s \cdot \vec{s}_t\right)$$

$$= -\sum_{s,t} J_{st}\left(\frac{1}{2} + 2\vec{s}_s \cdot \vec{s}_t\right) \tag{109}$$

where

$$J_{st} = \sum_{\ell,\ell'}\sum_{m,m'} j_{\ell\ell'}^{mm'}(\Theta, \phi)\, C_m^{\ell}(\theta_s, \phi_s)\, C_{m'}^{\ell'}(\theta_t, \phi_t) \tag{110}$$

This equivalent operator operates only on the angular and spin part of the wavefunctions. The quantities $j_{\ell\ell'}^{mm'}$ contain the radial integrals.

We can make the following observations:

(1) The sum $\sum_{s,t}$ is over all the electrons of the unfilled 3d (transition metal ions) or 4f (rare earth ions) shells. For each couple of electrons we have a parameter

$\langle J_{st}\rangle$ = matrix element of J_{st} taken between angular parts of the orbital wavefunctions

(2) The number of independent $\langle J_{st}\rangle$ parameters is reduced if the symmetry is high.

(3) $\langle J_{st}\rangle$ depends exponentially on R, for large R.

(4) If all the arbitals have the same asymptotic radial dependence e^{-r/r_0} (r_0 = effective Bohr radius), then $\langle H^{exch}\rangle \propto e^{-2R/r_0}$ in the limit of large R.

(5) For rare earth ions $r_0 \lesssim 0.3$ Å [6] and the separation of even the nearest ions may be large compared to r_0, so that the exchange interaction is small. For transition metal ions $r_0 \lesssim 0.6$ Å [7]. For both types of ions *superexchange* is important; the form of the equivalent operator in this case is the same as for the simple exchange case. However, the values of the parameters $\langle J_{st}\rangle$ are more difficult to estimate.

(6) Magnetic order in solids is a "static" effect of the exchange interactions.

II.C. Electro-Magnetic Interactions

Let us consider two atoms A and B at distance $\vec{R}$ from each other, as in Figure 7. Let $\vec{H}$ be the magnetic field at B, due to currents in A. It is

$$\vec{H} = \frac{e}{c} \frac{\vec{R} \times \vec{v}_A}{R^3} = -i \frac{e\hbar}{mc} \frac{\vec{R} \times \vec{\nabla}_A}{R^3} \tag{111}$$

since

$$\vec{v}_A = - \frac{i\hbar\vec{\nabla}_A}{m} \tag{112}$$

On the other hand

$$\langle\psi_f|[H,x]|\psi_i\rangle = \langle\psi_f|Hx - xH|\psi_i\rangle = (E_f - E_i)\langle\psi_f|x|\psi_i\rangle$$

$$= \hbar\omega\langle\psi_f|x|\psi_i\rangle$$

where $\omega = \dfrac{E_f - E_i}{h}$. It is

$$\dot{x} = \frac{i}{\hbar} [H,x]$$

$$\dot{p}_x = m\dot{x} = \frac{im}{\hbar} [H,x]$$

Then

$$\langle p\rangle_{fi} = \frac{im}{\hbar}\hbar\omega\langle\psi_f|x|\psi_i\rangle = im\omega\langle x\rangle_{fi}$$

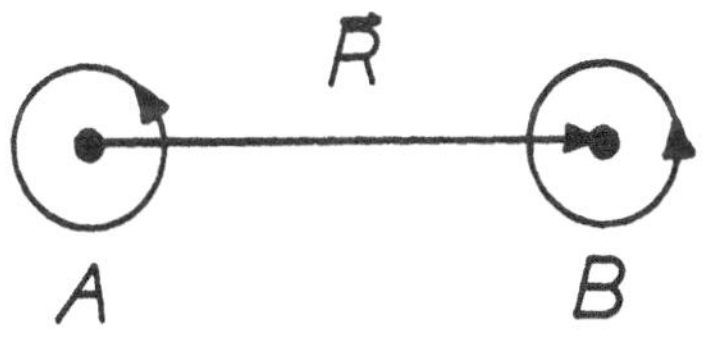

Fig. 7. A system consisting of two atoms.

$$<-i\hbar\vec{\nabla}> = im\omega<\vec{r}>$$

$$<\vec{\nabla}> = -\frac{mE}{\hbar^2}<\vec{r}> \tag{113}$$

and

$$<H> \simeq \frac{e\hbar}{mc}\frac{1}{R^2}<\nabla_A> = \frac{e\hbar}{mcR^2}\frac{mE}{\hbar^2}<r_A> = \frac{eE}{c\hbar R^2}<r> \tag{114}$$

This magnetic field will interact with the magnetic dipole moment of B, giving an interaction energy

$$H^{dd}(em) \simeq \mu_B<H> = \frac{e\hbar}{2mc}\frac{eE}{c\hbar R^2}<r> = \frac{eE<er>}{2mc^2R^2} \tag{115}$$

In order to compare the magnitude of this interaction with the dipole-dipole electrostatic interaction we evaluate the ratio

$$\frac{<H^{dd}(em)>}{<H^{dd}(el)>} = \frac{eE<er>/(2mc^2R^2)}{<er><er>/R^3} = \frac{ER}{2mc^2<r>} = \frac{ER}{2mc^2(\hbar^2/me^2)}$$

$$= \frac{ERe^2}{2c^2\hbar} \simeq \frac{E(e^2/\hbar c)^2}{e^2/R} \tag{116}$$

Note that

$$\frac{e^2}{\hbar c} = \text{fine structure constant} = 7.3 \times 10^{-3}$$

We have

$$E \simeq \frac{hc}{\lambda} = \frac{6.625 \times 10^{-27} \times 3 \times 10^{10}}{5000 \times 10^{-8}} \simeq 4 \times 10^{-12}$$

$$\frac{e^2}{R} \simeq \frac{(4.8 \times 10^{-10})^2}{5 \times 10^{-8}} \simeq 4.6 \times 10^{-12}$$

$$\left(\frac{e^2}{\hbar c}\right)^2 = (7.3 \times 10^{-3})^2 = 5.3 \times 10^{-5}$$

and

$$\frac{\langle H^{dd}(em)\rangle}{\langle H^{dd}(el)\rangle} \simeq \frac{4 \times 10^{-12} \times 5.3 \times 10^{-5}}{4.6 \times 10^{-12}} = 4.6 \times 10^{-5}$$

$$\frac{|\langle H^{dd}(em)\rangle|^2}{|\langle H^{dd}(el)\rangle|^2} \simeq 2 \times 10^{-9} \tag{117}$$

The electromagnetic interactions are then negligible.

It can also be shown that the effects of the interactions among magnetic multipoles are negligible.

II.D. Phonon-Assisted Energy Transfer

The probability for energy transfer between two ions in solids is proportional to the overlap integral

$$\int g_A(E)\, g_B(E)\, dE = \hbar \int g_A(\omega)\, g_B(\omega)\, d\omega \tag{118}$$

where $g_A(\omega)$ and $g_B(\omega)$ are the line shape functions for ions A and B, respectively. If we consider the case of two Lorentzian lines of width $\Delta\omega_A$ and $\Delta\omega_B$, centered at ω_A and ω_B, respectively, we find

$$\int g_A(\omega)\, g_B(\omega)\, d\omega = \frac{1}{\pi} \frac{\Delta\omega}{(\Delta\omega)^2 + (\omega_A - \omega_B)^2} \tag{119}$$

where $\Delta\omega = \Delta\omega_A + \Delta\omega_B$. For sharp and well-separated lines the value of the integral (119) and the probability for energy transfer become negligible. At low temperatures, where the lines in solids tend generally to be Gaussian, the value of the integral may be even smaller. In these circumstances the energy transfer process may be favored by the emission or absorption of a phonon whose energy compensates for the energy mismatch between the two transitions and ensures the conservation of energy in the process. The transition probability per unit time of the energy transfer process accompanied by the production of a phonon, if $\Delta E = E_A - E_B > 0$ as in Figure 8, is given by [8]

$$w_{AB} = \frac{2\pi}{\hbar} |\langle a'b|H_{AB}|ab'\rangle|^2\, S[n(\omega) + 1] \int g_A(E)\, g_B(E - \hbar\omega)\, dE \tag{120}$$

where S = ion-vibrations coupling parameter and $\hbar\omega = \Delta E$. If $\Delta E < 0$ then the energy transfer process is accompanied by the annihilation

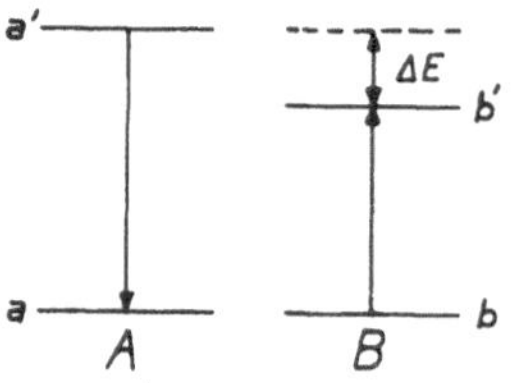

Fig. 8. Two-atom system requiring phonon assistance for energy transfer.

of a phonon and

$$w_{AB} = \frac{2\pi}{\hbar} |\langle a'b|H_{AB}|ab'\rangle|^2 \; S[n(\omega)] \int g_A(E) \; g_B(E + \hbar\omega) \; dE \tag{121}$$

If the transfer rates in both A → B and B → A directions are much greater than the intrinsic decay rates of A and B, phonon-assisted energy transfer processes may establish a Boltzmann distribution of populations between the excited states of A and B [9].

If $\Delta E \gg \hbar\omega_m$ where ω_m is the maximum phonon frequency, energy transfer is assisted by the creation of many phonons and the probability per unit time of the process is given by [10, 11]:

$$w_{AB}(\Delta E) = w_{AB}(0) \; e^{-\beta\Delta} \tag{122}$$

where β = temperature-dependent parameter related to similar parameter for multiphonon decay rate.

III. STATISTICAL TREATMENT OF ENERGY TRANSFER. MODES OF EXCITATION

III.A. Introduction

A typical sequence of events that includes the transfer of energy from one atom or ion called *sensitizer* or *energy donor (D)*, to an atom or ion called *activator* or *energy acceptor (A)* consists of:

i) absorption of a photon by D,

ii) energy transfer from D to A, and

iii) emission of a photon by A.

In the present treatment we shall assume that D and A are weakly interacting, so that the energy level shifts due to the interaction are smaller than the width of the D and A levels. This means that the absorption bands of D and A are identifiable.

Also, the present treatment will not consider the process of *radiative transfer* which consists of the emission of a photon by D and the absorption of the same photon by A. When such a process occurs the lifetime of D is in general not affected by the presence of A. If only one ion, say D, is present in the sample and if its concentration is high, the D → D radiative transfer may lead to trapping of the radiation and to an increase of the measured lifetime. In such case this lifetime may depend on the size and shape of the sample.

The process we will be considering consists of the *nonradiative transfer* of energy from D to A (step ii above).

III.B. Pulsed Excitation

Assume that we have a number N_D of donors and N_A of acceptors and call w_{DA} the probability of D → A energy transfer per unit time. Assume also that w_{DD} the probability of D → D energy transfer is negligible.

The question we shall try to answer is the following: If we excite a number of donors with a light pulse, how will the system respond? Let the pulse of light begin at time t = -T and end at time t = 0, and let T be much smaller than w_{DA}^{-1}:

$$T \ll w_{DA}^{-1} \tag{123}$$

Let

$N_{d'}(0)$ = number of excited donors at time t = 0

$N_{a'}(0)$ = number of excited acceptors at time t = 0

We shall put

$$N_{a'}(0) = 0 \tag{124}$$

because during the short interval of time (-T, 0) no relevant $D \rightarrow A$ transfer takes place. If

$N_d(0)$ = number of donors in the ground state at time $t = 0$

$N_a(0)$ = number of acceptors in the ground state at time $t = 0$

it will be

$$N_d(0) = N_D - N_{d'}(0) \tag{125}$$

$$N_a(0) = N_A \tag{126}$$

We shall also call τ the lifetime of the donor, in the absence of the activator:

$$\tau^{-1} = P + w_{nr} \tag{127}$$

where

P = probability of spontaneous emission per unit time

w_{nr} = probability of nonradiative decay per unit time

We shall define

$\rho_i(t)$ = probability that the donor at position $\vec{R}_i$ is excited at time t

and

$\bar{\rho}(t)$ = statistical average of $\rho_i(t)$

The number of excited donors at time t is given by $N_{d'}(0)\,\bar{\rho}(t)$ and the probability of finding a donor excited at time t by

$$\bar{\rho}(t)\,\frac{N_{d'}(0)}{N_D} \xrightarrow[t\to 0]{} \frac{N_{d'}(0)}{N_D} \tag{128}$$

The number of quanta emitted as luminescence by the donors per unit time is

$$P\,N_{d'}(0)\,\bar{\rho}(t) \tag{129}$$

The total number of quanta emitted as luminescence by the donors is

$$\mathcal{N} = P\, N_{d'}(0) \int_0^\infty \bar{\rho}(t)\, dt \tag{130}$$

The total number of quanta emitted by the donors, in the absence of activators, is

$$\mathcal{N}_0 = P\, N_{d'}(0) \int_0^\infty e^{-(t/\tau)}\, dt = P\, N_{d'}(0)\, \tau \tag{131}$$

The *quantum yield of luminescence* is

$$\frac{\mathcal{N}}{\mathcal{N}_0} = \frac{1}{\tau} \int_0^\infty \bar{\rho}(t)\, dt \tag{132}$$

III.C. Continuous Excitation

If we excite the donors with a light pulse beginning at $t_0 - T$ and ending at t_0, $\bar{\rho}(t - t_0)$ is the response of the donor system. The response of this system to N short pulses of equal amplitude will be

$$\phi(t) = \sum_{k=1}^{N} \bar{\rho}(t - t_k) \tag{133}$$

The luminescence signal due to this excitation is

$$S(t) = \text{const} \sum_{k=1}^{N} \bar{\rho}(t - t_k) \xrightarrow[(t_{k+1}-t_k)\to 0]{} \text{const} \int_{-\infty}^{+\infty} \bar{\rho}(t - t')\, dt' \tag{134}$$

where the constant depends on the intensity of light and the donor absorption transition probability. By making the interval between pulses go to zero we are essentially exciting the donor system continuously. If we replace $+\infty$ with zero in the integral we get

$$S(t) = \text{const} \int_{-\infty}^{0} \bar{\rho}(t - t')\, dt' = \text{const} \int_t^\infty \bar{\rho}(t')\, dt' \tag{135}$$

and

$$S(0) = \text{const} \int_0^{\infty} \overline{\rho}(t)\, dt \qquad (136)$$

The quantity S(0) is proportional to the number of excited donors at the time of observation under continuous excitation.

In order to find the value of the constant let us consider the donor system with the same type of excitation, in the absence of acceptors, as in Figure 9. w is the absorption transition probability per unit time; it depends on the intensity of light and on the $d \to d'$ absorption cross section. In steady state

$$w\, N_d - \frac{1}{\tau} N_{d'} = 0 \qquad (137)$$

But

$$N_{d'} = N_D - N_d \qquad (138)$$

Then

$$N_{d'} = \frac{w\, N_d}{w + \frac{1}{\tau}} = \frac{w\, \tau\, N_d}{1 + w\tau} \xrightarrow[w \ll \tau^{-1}]{} w\, \tau\, N_D \qquad (139)$$

On the other hand, using the result (136)

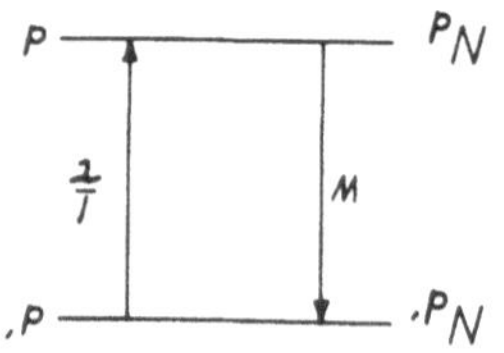

Fig. 9. Donor system in the absence of activators.

$$N_{d'} = \text{const} \int_0^\infty \bar{\rho}(t)\,dt = \text{const} \int_0^\infty e^{-(t/\tau)}\,dt = \text{const}\ \tau \tag{140}$$

Therefore the value of the constant is wN_D and in general

$$N_{d'} = w\,N_D \int_0^\infty \bar{\rho}(t)\,dt \tag{141}$$

The quantum yield of luminescence is given by

$$\frac{N_{d'} \text{ in presence of activators}}{N_{d'} \text{ in absence of activators}} = \frac{wN_D \int_0^\infty \bar{\rho}(t)\,dt}{wN_D\,\tau}$$

$$= \frac{1}{\tau} \int_0^\infty \bar{\rho}(t)\,dt \tag{142}$$

Comparing this result with (132) we can conclude that quantum yields can be measured by either the pulsed excitation method or the continuous excitation method.

IV. STATISTICAL TREATMENT OF ENERGY TRANSFER. CASE WITH NO MIGRATION AMONG DONORS

IV.A. Basic Equation

Following the pulsed excitation of the donor system, the probability that the donor at position $\vec{R}_i$ is excited at time t is given by the function $\rho_i(t)$ whose time evolution is described by the equation [12]

$$\frac{d}{dt}\rho_i(t) = \left[-\frac{1}{\tau} - \sum_{j=1}^{N_A} w_{DA}(|\vec{R}_i - \vec{R}_j|)\right]\rho_i(t) \tag{143}$$

where $\vec{R}_j$ = position of acceptor j. In the equation above we assume that only a small number of acceptors is excited at all times, so that it is always $N_a(t) \simeq N_A$.

The solution of the equation above, if the initial condition is

$$\rho_i(0) = 1 \tag{144}$$

is given by

$$\rho_i(t) = \exp\left[-\frac{t}{\tau} - t\sum_{j=1}^{N_A} w_{DA}(|\vec{R}_i - \vec{R}_j|)\right]$$

$$= e^{-(t/\tau)} \prod_{j=1}^{N_A} \exp\left[-t\, w_{DA}(|\vec{R}_i - \vec{R}_j|)\right] \tag{145}$$

The mode of decay described by $\rho_i(t)$ depends on the particular environment of donor i.

Let $\bar{\rho}(t)$ be the average of $\rho_i(t)$ over a large number of donors. Let also w(R) be the probability distribution of D - A distances R in the volume of the solid V; if this probability is uniform within the solid, as we assume, w will be given by 1/V:

$$\bar{\rho}(t) = e^{-(t/\tau)} \lim_{\substack{N_A\to\infty \\ V\to\infty}} \left[\int_V e^{-tw_{DA}(R)}\, w(R)\, d^3\vec{R}\right]^{N_A}$$

$$= e^{-(t/\rho)} \lim_{\substack{N_A\to\infty \\ V\to\infty}} \left[\frac{4\pi}{V}\int_0^{R_V} e^{-tw_{DA}(R)}\, R^2\, dR\right]^{N_A}$$

$$= e^{-(t/\tau)} \lim_{\substack{N_A\to\infty \\ V\to\infty}} [I(t)]^{N_A} \tag{146}$$

where $V = (4/3)\pi R_V^3$ and R_V = radius of the largest spherical volume, and

$$I(t) = \frac{4\pi}{V}\int_0^{R_V} e^{-tw_{DA}(R)}\, R^2\, dR \tag{147}$$

Note that the limit in (146) is taken for a large solid, but in such a way that the concentration of activators N_A/V remains constant.

In order to evaluate I(t) and then $\bar{\rho}(t)$ we need to know the function $w_{DA}(R)$.

IV.B. Simple Models

1. Perrin Model. In this model [13]

$$w_{DA}(R) = \begin{cases} \infty & R < R_o \\ 0 & R > R_o \end{cases} \tag{148}$$

Then

$$I(t) = \frac{4\pi}{V}\int_0^{R_V} e^{-tw_{DA}(R)} R^2\, dR = \frac{4}{3}\frac{R_V^3 - R_o^3}{V} = 1 - \frac{R_o^3}{R_V^3}$$

$$= 1 - \frac{c_A}{c_o N_A} \tag{149}$$

where

$$c_A = \text{concentration of acceptors} = \frac{N_A}{V} = \frac{N_A}{4\pi R_V^3/3} \tag{150}$$

$$c_o^{-1} = \frac{4\pi}{3} R_o^3 = \text{volume of donor's "sphere of influence"} \tag{151}$$

$$\frac{c_A}{c_o} = \text{number of acceptors in the sphere of influence of donor} \tag{152}$$

We have then

$$\bar{\rho}(t) = e^{-(t/\tau)} \lim_{\substack{N_A\to\infty \\ V\to\infty}} [I(t)]^{N_A} = \lim_{\substack{N_A\to\infty \\ V\to\infty}} e^{-(t/\tau)} \left(1 - \frac{c_A}{c_o N_A}\right)^{N_A}$$

$$= e^{-(t/\tau)}\, e^{-(c_A/c_o)} \tag{153}$$

Note that if $R_o = \infty$, $c_o = 0$ and $\bar{\rho}(t) = 0$ (immediate transfer). If $R_o = 0$, $c_o = \infty$ and $\bar{\rho}(t) = e^{-(t/\tau)}$ (no transfer).

The quantum yield of the donor luminescence is given by

$$\frac{N}{N_o} = \frac{1}{\tau}\int_0^\infty \bar{\rho}(t)\,dt = \frac{1}{\tau}\int_0^\infty e^{-(t/\tau)-(c_A/c_o)}\,dt = e^{-(c_A/c_o)} \tag{154}$$

The *transfer yield* is

$$1 - \frac{N}{N_o} = 1 - e^{-(c_A/c_o)} \tag{155}$$

2. Stern-Volmer Model. In this model [14] w_{DA} has the form

$$w_{DA}(R) = w = \text{const} \tag{156}$$

Then

$$I(t) = \frac{4\pi}{V}\int_0^{R_V} e^{-tw_{DA}(R)}\,R^2\,dR = \frac{4\pi}{V}\,e^{-tw}\,\frac{R_V^3}{3} = e^{-tw} \tag{157}$$

and

$$\bar{\rho}(t) = e^{-(t/\tau)}\,e^{-N_A wt} = e^{-(1+N_A w\tau)(t/\tau)} \tag{158}$$

where purposely we have not taken the limit $N_A \to \infty$.

The quantum yield of the donor luminescence is

$$\frac{N}{N_o} = \frac{1}{\tau}\int_0^\infty \bar{\rho}(t)\,dt = \frac{1}{1 + N_A w} \tag{159}$$

The transfer yield is

$$1 - \frac{N}{N_o} = \frac{N_A w\tau}{1 + N_A w\tau} \tag{160}$$

IV.C. Multipolar Interactions

For multipolar interactions

$$w_{DA}(R) = \frac{C^{(6)}}{R^6} + \frac{C^{(8)}}{R^8} + \frac{C^{(10)}}{R^{10}} + \ldots \tag{161}$$

Assume a prevalent multipolar interaction

$$w_{DA}(R) = \frac{C^{(n)}}{R^n} \tag{162}$$

We define a radius R_o as follows

$$w_{DA}(R) = \frac{C^{(n)}}{R^n} = \frac{1}{\tau} \left(\frac{R_o}{R} \right)^n \tag{163}$$

R_o is the distance at which the energy transfer rate is equal to the decay rate of the donor.

We set

$$I_n(t) = \frac{4\pi}{V} \int_{R_m}^{R_V} R^2 \, e^{-(t/\tau)(R_o/R)^n} \, dR \tag{164}$$

where

R_m = smallest possible D - A distance

R_V = radius of largest spherical volume

Also

$$x = \left(\frac{R_o}{R} \right)^n \frac{t}{\tau} \tag{165}$$

$$x_m = \left(\frac{R_o}{R_m} \right)^n \frac{t}{\tau} \tag{166}$$

$$x_V = \left(\frac{R_o}{R_V} \right)^n \frac{t}{\tau} \tag{167}$$

Then

$$I_n(t) = \frac{3}{n} x_V^{3/n} \int_{x_V}^{x_m} x^{-1-(3/n)} \, e^{-x} \, dx \tag{168}$$

$$\int_{x_V}^{x_m} x^{-1-(3/n)}\, e^{-x}\, dx = \int_{x_V}^{x_m} e^{-x}\, d(-\frac{n}{3}\, x^{-3/n})$$

$$\overset{IP}{=} [e^{-x}(-\frac{n}{3}\, x^{-3/n})]_{x_V}^{x_m} - \frac{n}{3} \int_{x_V}^{x_m} (x^{-3/n}\, e^{-x})\, dx$$

$$= -\frac{n}{3}\, x_m^{-3/n}\, e^{-x_m} + \frac{n}{3}\, x_V^{-3/n}\, e^{-x_V} - \frac{n}{3} \int_o^\infty x^{-3/n}\, e^{-x}\, dx$$

$$+ \frac{n}{3} \int_o^{x_V} x^{-3/n}\, e^{-x}\, dx + \frac{n}{3} \int_o^\infty x^{-3/n}\, e^{-x}\, dx \qquad (169)$$

where IP indicates integration by parts. Note that

$$\int_o^\infty x^{3/n}\, e^{-x}\, dx = \Gamma(1 - \frac{3}{n}) \qquad (170)$$

We shall consider the limiting situation where

$$x_m = (\frac{R_o}{R_m})^n \frac{t}{\tau} \xrightarrow[t\to\infty]{} \infty \qquad (171)$$

$$x_V = (\frac{R_o}{R_V})^n \frac{t}{\tau} \xrightarrow[V\to\infty]{} 0 \qquad (172)$$

Note the following asymptotic expansions of incomplete Γ functions

$$\int_{x_m}^\infty x^{-3/n}\, e^{-x}\, dx \xrightarrow[x_m\to\infty]{} x_m^{-3/n}\, e^{-x_m} (1 - \frac{3}{n}\, x_m^{-1} + \ldots)$$

$$= x_m^{-3/n}\, e^{-x_m} - \frac{3}{n}\, x_m^{-(3/n)-1}\, e^{-x_m} + \ldots \qquad (173)$$

$$\int_o^{x_V} x^{-3/n}\, e^{-x}\, dx \xrightarrow[x_V\to 0]{} x_V^{1-(3/n)} [1 - (\frac{n}{2n-3})\, x_V + \ldots]$$

$$= x_V^{1-(3/n)} - \frac{n}{2n-3}\, x_V^{2-(3/n)} + \ldots \qquad (174)$$

We have

$$I_n(t) = \frac{3}{n} x_V^{3/n} \int_{x_V}^{x_m} x^{-1-(3/n)} e^{-x} dx$$

$$= \frac{3}{n} x_V^{3/n} \Big[- \frac{n}{3} e^{-x_m} x_m^{-(3/n)} + \frac{n}{3} x_V^{-(3/n)} e^{-x_V} - \frac{n}{3} \Gamma(1 - \frac{3}{n})$$

$$+ \frac{n}{3} x_m^{-(3/n)} e^{x_m} - x_m^{-(3/n)-1} e^{-x_m} + \frac{n}{3} x_V^{1-(3/n)}$$

$$- \frac{n}{3} \frac{n}{2n - 3} x_V^{2-(3/n)} \Big]$$

$$= e^{-x_V} - x_V^{3/n} \Gamma(1 - \frac{3}{n}) - \frac{3}{n} x_V^{3/n} x_m^{-(3/n)-1} e^{-x_m}$$

$$+ x_V - \frac{n}{2n - 3} x_V^2 \xrightarrow[x_m \to \infty]{x_V \to 0} 1 - x_V^{3/n} \Gamma(1 - \frac{3}{n}) \tag{175}$$

Then

$$I_n(t) = 1 - x_V^{3/n} \Gamma(1 - \frac{3}{n}) = 1 - \left[(\frac{R_o}{R_V})^n \frac{t}{\tau} \right]^{3/n} \Gamma(1 - \frac{3}{n})$$

$$= 1 - \frac{R_o^3}{R_V^3} (\frac{t}{\tau})^{3/n} \Gamma(1 - \frac{3}{n}) \tag{176}$$

But

$$c_A = \frac{N_A}{V} = \frac{N_A}{(4\pi/3) R_V^3} \tag{177}$$

$$c_o^{-1} = \frac{4\pi R_o^3}{3} \tag{178}$$

$$\frac{R_o^3}{R_V^3} = \frac{c_A}{c_o N_A} \tag{179}$$

Then

$$I_n(t) = 1 - \frac{c_A}{c_o N_A} \left(\frac{t}{\tau} \right)^{3/n} \Gamma\left(1 - \frac{3}{n}\right) \tag{180}$$

Also

$$\lim_{\substack{N_A \to \infty \\ V \to \infty}} I_n(t) = \exp\left[- \frac{c_A}{c_o N_A} \left(\frac{t}{\tau} \right)^{3/n} \Gamma\left(1 - \frac{3}{n}\right) \right] \tag{181}$$

.

and finally

$$\bar{\rho}(t) = e^{-(t/\tau)} [I_n(t)]^{N_A} = \exp\left[- \frac{t}{\tau} - \frac{c_A}{c_o} \Gamma\left(1 - \frac{3}{n}\right) \left(\frac{t}{\tau} \right)^{3/n} \right] \tag{182}$$

$$(x_m \to \infty \; ; \; x_V \to 0)$$

This result was first derived by Förster [15]. We want to discuss the conditions under which this formula has been derived. $x_m >> 1$ means

$$\left(\frac{R_o}{R_m} \right)^n \frac{t}{\tau} >> 1$$

and

$$\frac{t}{\tau} >> \left(\frac{R_m}{R_o} \right)^n$$

Take $R_m = 3A$, $R_o = 10A$ and $n = 6$

$$\frac{t}{\tau} >> \left(\frac{3}{10} \right)^6 = 7 \times 10^{-4}$$

$$t >> 7 \times 10^{-4} \tau$$

This shows that the approximation $x_m >> 1$ may be good even for very short times.

$x_V << 1$ means

$$\left(\frac{R_o}{R_V} \right)^n \frac{t}{\tau} << 1$$

$$\frac{t}{\tau} << \left(\frac{R_V}{R_o} \right)^n$$

Take R_V = 1cm, R_o = 10A and n = 6

$$\frac{t}{\tau} << \left(\frac{10^8}{10} \right)^6 = 10^{42}$$

$$t << 10^{42}\tau$$

which is indeed always the case.

Figure 10 reports a plot of the function $\bar{\rho}(t)$ given in (182) for the case n = 6 (dipole-dipole interaction).

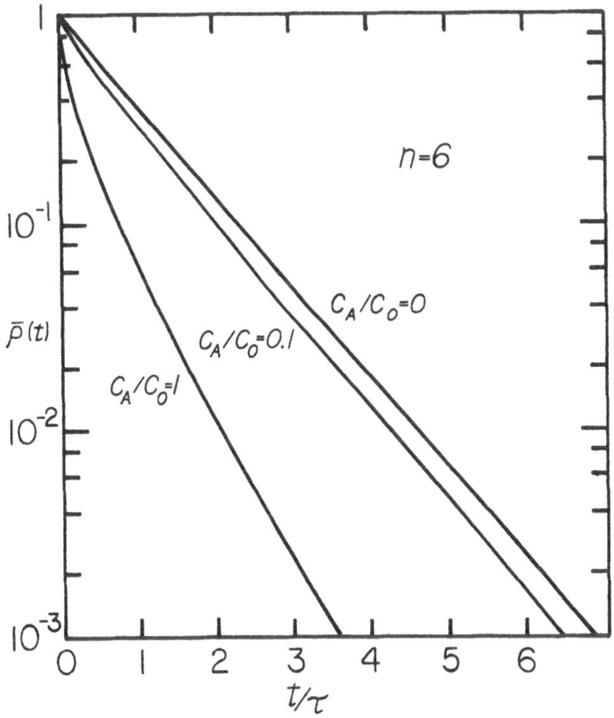

Fig. 10. The average probability of donor excitation in the case of dipole-dipole interaction.

The following observations can be made regarding Figure 10:

(1) At the beginning the decay is faster, because the donors close to the acceptors decay first.

(2) After a while other donors, which are farther away from acceptors, and remained excited, start transferring energy.

(3) At very long times donors that are very far from acceptors will finally decay with their own lifetime.

(4) The greater is c_A/c_0 = number of acceptors in donor's sphere of influence, the longer one has to wait for the $\bar{\rho}(t)$ curve to become parallel to the one for $c_A/c_0 = 0$.

In Figure 11 a comparison is made of the $\bar{\rho}(t)$ curves for the various processes: n = 6 (dipole-dipole), n = 8 (dipole-quadrupole) and n = 10 (quadrupole-quadrupole).

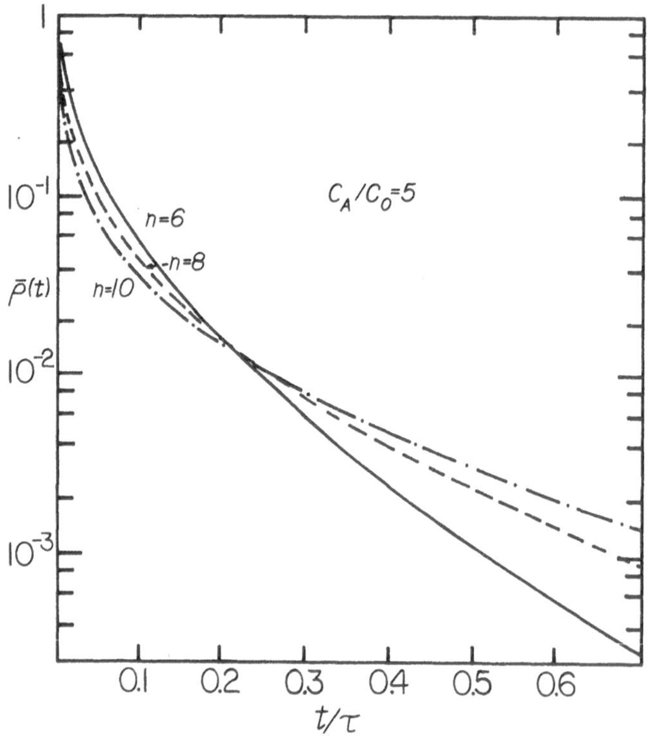

Fig. 11. The average probability of donor excitation for n = 6 (dipole-dipole interaction), n = 8 (dipole-quadrupole interaction), and n = 10 (quadrupole-quadrupole interaction). The ratio c_A/c_0 is equal to 5. (Reproduced from [12] by courtesy of R. K. Watts and permission from Plenum Press.)

We have seen that the conditions $x_V \ll 1$ and $x_m \gg 1$ are in accord with the practical cases of interest. However, we want to consider the limits of very short times by making

$$x_V \to 0 \quad \text{or} \quad V \to \infty \quad \text{(as before)} \tag{183}$$

$$x_m \to 0 \tag{184}$$

One should expect $\bar{\rho}(t)$ to decay exponentially, since at very short times only donors with acceptors as nearest neighbors transfer their energy.

The relevant integral in this case is still

$$\int_{x_V}^{x_m} x^{-1-(3/n)} e^{-x} dx \simeq \int_{x_V}^{x_m} x^{-1-(3/n)} (1 - x) dx$$

$$= \frac{n}{3} x_V^{-(3/n)} - \frac{n}{3} x_m^{-(3/n)} - \frac{n}{n-3} x_m^{1-(3/n)} + \frac{n}{n-3} x_V^{1-(3/n)} \tag{185}$$

Then

$$I_n(t) = \frac{3}{n} x_V^{3/n} \int_{x_V}^{x_m} x^{-1-(3/n)} e^{-x} dx$$

$$= 1 - \left(\frac{x_V}{x_m} \right)^{3/n} - \frac{3}{n-3} x_V^{3/n} x_m^{1-(3/n)} + \frac{3}{n-3} x_V \tag{186}$$

But

$$x_V = \left(\frac{R_o}{R_V} \right)^n \frac{t}{\tau} \tag{187}$$

$$x_m = \left(\frac{R_o}{R_m} \right)^n \frac{t}{\tau} \tag{188}$$

$$\left(\frac{x_V}{x_m} \right)^{3/n} = \left(\frac{R_m}{R_V} \right)^3 \tag{189}$$

Then

$$I_n(t) \simeq 1 - \left(\frac{R_m}{R_V}\right)^3 - \frac{3}{n-3}\left(\frac{R_o}{R_V}\right)^3\left(\frac{R_o}{R_m}\right)^{n-3}\frac{t}{\tau} + \frac{3}{n-3}\left(\frac{R_o}{R_V}\right)^n\frac{t}{\tau}$$

$$= 1 - \frac{1}{N_A}\left[\frac{c_A}{c_m} + \frac{3}{n-3}\left(\frac{c_A}{c_o}\right)\left(\frac{c_m}{c_o}\right)^{(n/3)-1}\frac{t}{\tau} + \frac{3}{n-3}\left(\frac{c_A}{c_o}\right)^{n/3}\frac{t}{\tau}\frac{1}{N_A^{(n/3)-1}}\right] \tag{190}$$

$(c_m^{-1} = \frac{4}{3}\pi R_m^3)$

For very large N_A

$$I_n(t) \simeq 1 - \frac{1}{N_A}\left[\frac{c_A}{c_m} + \frac{3}{n-3}\left(\frac{c_A}{c_o}\right)\left(\frac{c_m}{c_o}\right)^{(n/3)-1}\frac{t}{\tau}\right] \tag{191}$$

and

$$[I_n(t)]^{N_A} \simeq \left\{1 - \frac{1}{N_A}\left[\frac{c_A}{c_m} + \frac{3}{n-3}\left(\frac{c_A}{c_o}\right)\left(\frac{c_m}{c_o}\right)^{(n/3)-1}\frac{t}{\tau}\right]\right\}^{N_A}$$

$$\xrightarrow[N_A\to\infty]{} \exp\left[-\frac{c_A}{c_m} - \frac{3}{n-3}\left(\frac{c_A}{c_o}\right)\left(\frac{c_m}{c_o}\right)^{(n/3)-1}\frac{t}{\tau}\right] \tag{192}$$

Finally

$$\bar{\rho}(t) = e^{-(t/\tau)}[I_n(t)]^{N_A} = e^{-(c_A/c_m)}\, e^{-(t/\tau')} \tag{193}$$

where

$$\frac{1}{\tau'} = \frac{1}{\tau}\left[1 + \frac{3}{n-3}\frac{c_A}{c_o}\left(\frac{c_m}{c_o}\right)^{(n/3)-1}\right] \tag{194}$$

Therefore at very short time the function $\bar{\rho}(t)$ decays exponentially as expected.

We wish to consider now the quantum yield of donor luminescence.

If n = 6 (dipole-dipole interaction)

$$\bar{\rho}(t) = \exp\left[-\frac{t}{\tau} - \frac{c_A}{c_o}\Gamma\left(\frac{1}{2}\right)\left(\frac{t}{\tau}\right)^{1/2}\right] = \exp\left[-\frac{t}{\tau} - \frac{c_A}{c_o}\sqrt{\pi}\left(\frac{t}{\tau}\right)^{1/2}\right] \tag{195}$$

Then

$$\frac{N}{N_o} = \frac{1}{\tau}\int_o^\infty \exp\left[-\frac{t}{\tau} - \frac{c_A}{c_o}\sqrt{\pi}\left(\frac{t}{\tau}\right)^{1/2}\right]dt = \int_o^\infty e^{-t-2qt^{1/2}}\, dt$$

$$= e^{q^2}\int_o^\infty e^{-(q+t^{1/2})^2}\, dt = 1 - 2qe^{q^2}\int_q^\infty e^{-x^2}\, dx \tag{196}$$

where

$$2q = \sqrt{\pi}\,\frac{c_A}{c_o}$$

But

$$\text{erf } q = 1 - \frac{2}{\sqrt{\pi}}\int_q^\infty e^{-x^2}\, dx = \frac{2}{\sqrt{\pi}}\int_o^q e^{-x^2}\, dx \tag{197}$$

Then [15]-[17]

$$\frac{N}{N_o} = 1 - q\, e^{q^2}\sqrt{\pi}(1 - \text{erf } q) \tag{198}$$

We shall examine the following two limiting cases:

$q \to 0$: $\quad c_A \to 0 \quad , \quad \text{erf } q \to 0$

$$\frac{N}{N_o} \to 1 - \frac{\pi}{2}\frac{c_A}{c_o} \to 1$$

$q \to \infty$: $\quad c_A \to \infty$ or $c_o^{-1} \to \infty \quad , \quad \text{erf } q \to 1$

$$\frac{N}{N_o} \to 0$$

IV.D. Exchange Interactions

We refer back to the result (72) and in particular to the second term, called the exchange term. The transition probability due to this term in the interaction Hamiltonian is given by

$$w_{DA} = \frac{2\pi}{\hbar} \left|\langle d'(1)\ a(2) \left| \sum \frac{e^2}{r_{12}} \right| a'(1)\ d(2)\rangle\right|^2 \int g_D(E)\ g_A(E)\ dE \tag{199}$$

where the symbol $\sum$ indicates the sum over the electrons. We can make the following observations:

i) The exchange term is strongly dependent on the D - A distance.

ii) In contrast to the case of the multipolar interaction, the matrix element of the interaction is not proportional to the oscillator strengths of D or A, and cannot be related to any spectroscopic characteristic of D or A.

We shall assume that [18]

$$\left|\langle d'(1)\ a(2) \left| \frac{e^2}{r_{12}} \right| a'(1)\ d(2)\rangle\right|^2 = Z^2 = K^2\ e^{-(2R/L)} \tag{200}$$

where

K = const with dimension of energy

L = const, called "effective Bohr radius"

Then

$$w_{DA}(R) = \frac{2\pi}{\hbar} K^2\ e^{-(2R/L)} \int g_D(E)\ g_A(E)\ dE \tag{201}$$

Set

$$\frac{2\pi}{\hbar} K^2 \int g_D(E)\ g_A(E)\ dE = \frac{e^{2R_o/L}}{\tau} \tag{202}$$

$$\frac{2R_o}{L} = \gamma \tag{203}$$

Then

$$w_{DA}(R) = \frac{1}{\tau} e^{2R_o/L} e^{-(2R/L)} = \frac{1}{\tau} e^{2(R_o/L)(1-R/R_o)}$$

$$= \frac{1}{\tau} e^{\gamma[1-(R/R_o)]} \tag{204}$$

Note that if $R = R_o$, $w_{DA} = \tau^{-1}$. We have from (147) and (204)

$$I(t) = \frac{4\pi}{V} \int_0^{R_V} e^{-tw_{DA}(R)} R^2 \, dR = \frac{4\pi}{V} \int_0^{R_V} e^{-(t/\tau)e^{\gamma[1-(R/R_o)]}} R^2 \, dR \tag{205}$$

Let

$$z = e^{\gamma} \frac{t}{\tau} = e^{2R_o/L} \frac{t}{\tau}$$

$$y = e^{-\gamma(R/R_o)}$$

Then

$$e^{-zy} = e^{-(t/\tau)} e^{\gamma[1-(R/R_o)]}$$

It is

$$dy = -\frac{\gamma}{R_o} e^{-\gamma(R/R_o)} dR = -\frac{\gamma}{R_o} y \, dR$$

$$dR = -\frac{R_o}{\gamma} \frac{dy}{y}$$

$$\ln y = -\gamma \frac{R}{R_o}$$

$$R = -\frac{R_o}{\gamma} \ln y$$

and

$$R^2 \, dR = \frac{R_o^2}{\gamma^2} (\ln y)^2 \left(-\frac{R_o}{\gamma} \frac{dy}{y}\right) = -\frac{R_o^3}{\gamma^3} \frac{(\ln y)^2}{y} dy$$

Also

$$\frac{4\pi}{V} = \frac{4\pi}{(4\pi\ R_V^3)/3} = \frac{3}{R_V^3}$$

Then

$$\frac{4\pi}{V} R^2\ dR = -\frac{3}{\gamma^3}\left(\frac{R_o}{R_V}\right)^3 \frac{(\ln y)^2}{y}\ dy$$

and

$$I(t) = \frac{4\pi}{V}\int_0^{R_V} e^{-(t/\tau)}\ e^{\gamma[1-(R/R_o)]}\ R^2\ dR$$

$$= \int_1^{y_V} -\frac{3}{\gamma^3}\left(\frac{R_o}{R_V}\right)^3 \frac{(\ln y)^2}{y}\ e^{-zy}\ dy$$

$$= \frac{3}{\gamma^3}\left(\frac{R_o}{R_V}\right)^3 \int_{y_V}^1 \frac{e^{-zy}(\ln y)^2}{y}\ dy \qquad (206)$$

where $y_V = e^{-\gamma(R_V/R_o)}$. For small y_V $(R_V \to \infty)$

$$\int_{y_V}^1 \frac{e^{-zy}(\ln y)^2}{y}\ dy = \int_{y_V}^1 \frac{1}{3}\left[\frac{d}{dy}(\ln y)^3\right] e^{-zy}\ dy$$

$$= \left[\frac{1}{3}(\ln y)^3\ e^{-zy}\right]_{y_V}^1 + \frac{1}{3}\int_{y_V}^1 (\ln y)^3\ z\ e^{-zy}\ dy$$

$$= -\frac{1}{3}\ e^{-zy_V}(\ln y_V)^3 + \frac{1}{3}\ z\left[\int_0^1 e^{-zy}(\ln y)^3\ dy\right.$$

$$\left. - \int_0^{y_V} e^{-zy}(\ln y)^3\ dy\right]$$

$$\to -\frac{1}{3}(\ln y_V)^3 - \frac{1}{3}\left[-z\int_0^1 e^{-zy}(\ln y)^3\ dy\right]$$

$$- \frac{1}{3} z \int_0^{y_V} e^{-zy} (\ln y)^3 \, dy$$

$$\simeq - \frac{1}{3} (\ln y_V)^3 - \frac{1}{3} g(z) = \frac{1}{3} \gamma^3 \left(\frac{R_V}{R_o} \right)^3 - \frac{1}{3} g(z) \qquad (207)$$

where

$$g(z) = -z \int_0^1 e^{-zy} (\ln y)^3 \, dy \qquad (208)$$

Then

$$\bar{\rho}(t) = \lim_{\substack{N_A \to \infty \\ V \to \infty}} e^{-(t/\tau)} \left[\frac{3}{\gamma^3} \left(\frac{R_o}{R_V} \right)^3 \int_{y_V}^1 \frac{e^{-zy} (\ell n\, y)^2}{y} \, dy \right]^{N_A}$$

$$= \lim_{\substack{N_A \to \infty \\ V \to \infty}} e^{-(t/\tau)} \left\{ \frac{3}{\gamma^3} \left(\frac{R_o}{R_V} \right)^3 \left[\frac{\gamma^3}{3} \left(\frac{R_V}{R_o} \right)^3 - \frac{1}{3} g(z) \right] \right\}^{N_A}$$

$$= \lim_{\substack{N_A \to \infty \\ V \to \infty}} e^{-(t/\tau)} \left[1 - \frac{1}{\gamma^3} \left(\frac{R_o}{R_V} \right)^3 g(z) \right]^{N_A} \qquad (209)$$

Taking (179) into account we obtain

$$\bar{\rho}(t) = \lim_{\substack{N_A \to \infty \\ V \to \infty}} e^{-(t/\tau)} \left[1 - \frac{1}{\gamma^3} \frac{c_A}{c_o N_A} g(z) \right]^{N_A}$$

$$= \exp - \frac{t}{\tau} - \frac{1}{\gamma^3} \frac{c_A}{c_o} g(e^{\gamma t/\tau}) \Big] \qquad (210)$$

where $\gamma = \frac{2R_o}{L}$.

We note, in regard to g(z) [18]:

(1) $g(0) = 0$.

(2) $g(z)$ is positive and monotonically increasing for $z > 0$.

(3) Expansion of the exponential e^{-zy} and integration term by term gives the series

$$g(z) = 6z \sum_{m=0}^{\infty} \frac{(-z)^m}{m!(m+1)^4} \tag{211}$$

For small values of z this series converges rapidly.

(4) For z > 10 g(z) can be approximated by the expression

$$g(z) = (\ln z)^3 + 1.7316 (\ln z)^2 + 5.934 \ln z + 5.445 \tag{212}$$

V. STATISTICAL TREATMENT OF ENERGY TRANSFER. CASE WITH MIGRATION AMONG DONORS

V.A. Migration

The case treated in the previous part IV deals with direct transfer from donor to acceptor, with no migration among donors and is exemplified in Figure 12(a). We shall now consider the case in which $w_{DD} \neq 0$, and energy transfer processes can take place among donors, so that the energy of excitation may reach an acceptor after hopping resonantly among donors as in Figure 12(b).

Since migration among donors can be viewed as a diffusion

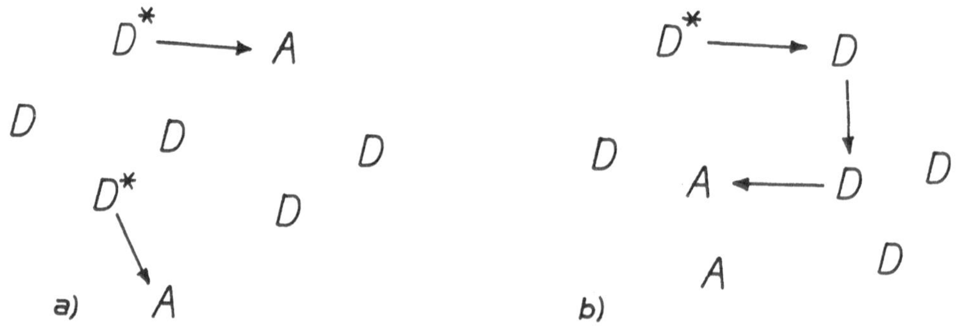

Fig. 12(a). Direct transfer; (b) Transfer with migration.

process we shall review some elementary considerations regarding the phenomenon of diffusion.

V.B. Diffusion

Consider a system consisting of N similar molecules in a volume V, and let n be the density N/V. Let n_1 be the density of especially labelled molecules, for example, of radioactive molecules. Assume $n_1 = n_1(x)$, i.e. a non-equilibrium situation, but assume also n = const, so that no net motion of whole substance occurs during diffusion.

The mean number of molecules of type 1 crossing the unit area of a plane perpendicular to the x direction in the x direction in the unit time is given by

$$J_x = -D \frac{\partial n_1}{\partial x} \qquad [\mathrm{cm}^{-2}\ \mathrm{sec}^{-1}] \tag{213}$$

where

$$D = \text{coefficient of self diffusion} \quad [\mathrm{cm}^2\ \mathrm{sec}^{-1}]$$

The above formula (213) is valid for gases, liquids and isotropic solids [19]. If $\partial n_1/\partial x > 0$, $J_x < 0$ and the flow takes place in the -x direction.

Let us consider now the crossing by molecules of type 1 of an area A perpendicular to the x direction, as in Figure 13. It is

$$\frac{\partial}{\partial t}(n_1\ A\ dx) = A\ J_x(x) - A\ J_x(x + dx) \tag{214}$$

and

$$\frac{\partial n_1}{\partial t} dx = J_x(x) - [J_x(x) + \frac{\partial J_x(x)}{\partial x} dx] = - \frac{\partial J_x(x)}{\partial x}$$

or, because of (213)

$$\frac{\partial n_1(x,t)}{\partial t} = D \frac{\partial^2 n_1(x,t)}{\partial x^2} \tag{215}$$

The above equation is called *diffusion equation.*

We shall consider the initial condition of N_1 molecules

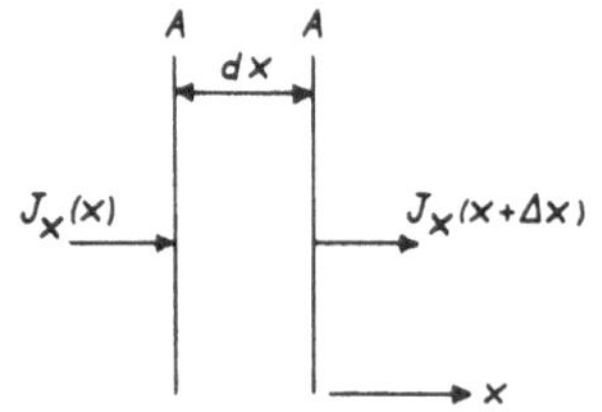

Fig. 13. Crossing of plane by especially labelled molecules.

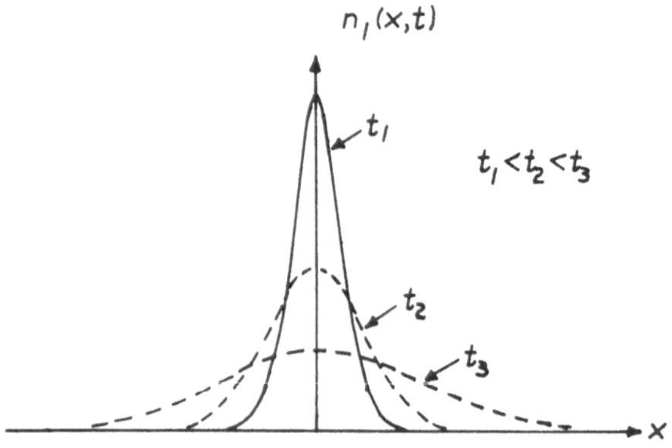

Fig. 14. Evolution of the density function of especially labelled molecules.

introduced at the time t = 0 near the plane x = 0:

$$n_1(x,0) = N_1 \ \delta(x) \tag{216}$$

The solution of the diffusion equation under this condition is

$$n_1(x,t) = \frac{N_1}{\sigma\sqrt{2\pi}} e^{-(x^2/2\sigma^2)} = \frac{N_1}{\sqrt{4\pi Dt}} e^{-(x^2/4Dt)} \tag{217}$$

The shape of the $n_1(x,t)$ curve is always gaussian (see Figure 14) with the standard deviation

$$\sigma = \sqrt{2Dt} \tag{218}$$

It is also

$$\langle x^2\rangle = \frac{1}{N_1} \int_{-\infty}^{+\infty} x^2 \ n_1(x,t) \ dx = \sigma^2 = 2Dt \tag{219}$$

In three dimensions the diffusion equation is

$$\frac{\partial}{\partial t} n_1(\vec{r},t) = D \ \nabla^2 \ n_1(\vec{r},t) \tag{220}$$

If the initial condition is

$$n_1(\vec{r},t) = N_1 \ \delta(\vec{r}) \tag{221}$$

the solution is given by

$$n_1(\vec{r},t) = \frac{N_1}{(4\pi \ Dt)^{3/2}} e^{-(r^2/4Dt)} \tag{222}$$

Also

$$\langle r^2\rangle = \frac{1}{N_1} \int r^2 \ n_1(\vec{r},t) \ d^3\vec{r} = 6Dt \tag{223}$$

V.C. Migration as Diffusion Process

We shall be concerned with a function $\rho(\vec{R},t)$ which we define by stating that $\rho(\vec{R},t) \ d^3\vec{R}$ is the probability that the donor with coordinate in $(\vec{R}, \vec{R} + d\vec{R})$ is excited at time t. We shall now consider the following cases.

1. <u>Diffusion Only</u>. In this simplest case the diffusion equation gives us

$$\frac{\partial}{\partial t}\rho(\vec{R},t) = D\,\nabla^2\,\rho(\vec{R},t) \tag{224}$$

If the excitation is initially localized at $\vec{R} = 0$, then

$$\rho(\vec{R},t) = \frac{1}{\sqrt{4\pi\ Dt}}\,e^{-(R^2/4Dt)} \tag{225}$$

with

$$\langle R^2\rangle = 6Dt \tag{226}$$

2. <u>Diffusion and Relaxation</u>. In this case a term has to be added to the relevant equation

$$\frac{\partial}{\partial t}\rho(\vec{R},t) = D\,\nabla^2\,\rho(\vec{R},t) - \frac{1}{\tau}\rho(\vec{R},t) \tag{227}$$

If we set

$$\rho(\vec{R},t) = \psi(\vec{R},t)\;e^{-(t/\tau)} \tag{228}$$

we obtain for $\psi(\vec{R},t)$

$$\frac{\partial}{\partial t}\psi(\vec{R},t) = D\,\nabla^2\,\psi(\vec{R},t) \tag{229}$$

If, again, the excitation is initially localized at $\vec{R} = 0$, we obtain

$$\vec{\rho}(\vec{R},t) = e^{-(t/\tau)}\,\frac{1}{(4\pi\ Dt)^{3/2}}\,e^{-(R^2/4Dt)} \tag{230}$$

It is

$$\int \vec{\rho}(\vec{R},t)\;d^3\vec{R} = e^{-(t/\tau)} \tag{231}$$

3. <u>Diffusion, Relaxation and Transfer</u>. In this case the relevant equation is

$$\frac{\partial}{\partial t}\rho(\vec{R},t) = \left[D\,\nabla^2 - \frac{1}{\tau} - \sum_{j=1}^{N_A} w_{DA}(|\vec{R} - \vec{R}_j|)\right]\rho(\vec{R},t) \tag{232}$$

where the first term in the [ ] parentheses deals with diffusion among donor, the second with the self-decay of donors, and the third with donor-acceptor energy transfer.

Equation (232) is analogous to (143), but differs from it in two respects. First, the diffusion term is included in (232) and, also, the equivalent of $\rho_j(t)$ is $\rho(\vec{R},t)\ d^3\vec{R}$. The form of $\rho_j(t)$, given the initial condition $\rho_j(0) = 1$, is given by (145); the average of $\rho_j(t)$ over a large number of donors is given by (146) and finally by

$$\bar{\rho}(t) = \begin{cases} \exp\left[-\frac{t}{\tau} - \frac{c_A}{c_o}\Gamma(1-\frac{3}{n})(\frac{t}{\tau})^{3/n}\right] & \text{(multipolar interactions)} \\ \exp\left[-\frac{t}{\tau} - \frac{1}{\gamma^3}\frac{c_A}{c_o}g(e^{\gamma t/\tau})\right] & \text{(exchange interactions)} \end{cases} \tag{233}$$

where

$$\gamma = \frac{2R_o}{L}$$

$$g(z) = -z\int_0^1 e^{-zy}(\ln y)^3\ dy$$

In the present case the analogous of $\bar{\rho}(t)$ is the function

$$\phi(t) = \frac{1}{N_D}\int \rho(\vec{R},t)\ d^3\vec{R} \tag{234}$$

$\phi(t)$ reduces to the $\bar{\rho}(t)$ in (233) in the limit $D \to 0$.

No general expression for $\phi(t)$ has been found. However, Yokota and Tanimoto [20] have obtained an expression for $\phi(t)$ in the case of dipole-dipole interaction when

$$w_{DA} = \frac{C_{DA}}{R^6} = \frac{1}{\tau}\left(\frac{R_o}{R}\right)^6 \tag{235}$$

This expression was obtained by using the method of the Padé approximants and is now reported

$$\phi(t) = \exp[-\frac{t}{\tau} - \frac{c_A}{c_o}\Gamma(\frac{1}{2})(\frac{t}{\tau})^{1/2}(\frac{1+10.87x+15.5x^2}{1+8.743x})^{3/4} \tag{236}$$

where

$$x = \frac{D\tau}{R_o^2}(\frac{t}{\tau})^{2/3} = DC_{DA}^{-1/3}\ t^{2/3} \tag{237}$$

For $x << 1$ $\phi(t)$ reduces to

$$\phi(t) = \exp\left[-\frac{t}{\tau} - \frac{c_A}{c_o}\Gamma(\frac{1}{2})(\frac{t}{\tau})^{1/2}\right] \tag{238}$$

At early times diffusion is not important. Only donors with nearby acceptors are decaying; the time is not sufficient for the excitation to diffuse among the donors before being transferred to the acceptors.

$$DC_{DA}^{-1/3}\ t^{2/3} << 1$$

or

$$t << t^* = \frac{C_{DA}^{1/2}}{D^{3/2}} \tag{239}$$

Note that if C_{DA} is large (D - A interaction strong) t^* is large, and if D is large (fast diffusion among donors) t^* is small. Before the characteristic time t^* diffusion is negligible. During the time t^* the excitation, if initially localized in one place, would diffuse a distance R^* given by

$$R^{*2} = 6Dt^*$$

It is

$$R^* = \sqrt{6Dt^*} = \sqrt{6D\frac{C_{DA}^{1/2}}{D^{3/2}}} \simeq (\frac{C_{DA}}{D})^{1/4} \tag{240}$$

The asymptotic behavior of $\phi(t)$ can be found by letting $t \to \infty$, namely $x \to \infty$. In this case.

$$\phi(t) \xrightarrow[x\to\infty]{} \exp[-\frac{t}{\tau} - \frac{c_A}{c_o}\Gamma(\frac{1}{2})(\frac{t}{\tau})^{1/2}(\frac{15.5x}{8.743})^{3/4}]$$

$$= \exp[-\frac{t}{\tau} - \frac{c_A}{c_o}\sqrt{\pi}\ (\frac{t}{\tau})^{1/2}(1.733x)^{3/4}] \tag{241}$$

Now

$$(1.733x)^{3/4} = 1.536[\frac{D\tau}{R_o^2}(\frac{t}{\tau})^{2/3}]^{3/4} = 1.536\frac{D^{3/4}\ \tau^{3/4}}{R_o^{3/2}}(\frac{t}{\tau})^{1/2} \tag{242}$$

and

$$\frac{c_A}{c_o}\sqrt{\pi}\ (\frac{t}{\tau})^{1/2}(1.733x)^{3/4} = 4\pi\ c_A \frac{D^{3/4}\ R_o^{3/2}}{\tau^{1/4}}\ 0.907t$$

$$= 4\pi\ c_A \frac{D^{3/4}}{\tau^{1/4}}\tau^{1/4}\ C_{DA}^{1/4}\ 0.907t$$

$$\simeq 4\pi\ Dc_A\ 0.91(\frac{C_{DA}}{D})^{1/4}t = 4\pi\ Dc_A\ R_D\ t$$

where

$$R_D = 0.91(\frac{C_{DA}}{D})^{1/4} \simeq R^* \tag{242}$$

Then

$$\phi(t) \xrightarrow[t\to\infty]{} \exp(-\frac{t}{\tau} - 4\pi\ D\ c_A\ R_D\ t)$$

$$= \exp(-\frac{t}{\tau} - K_D\ t) = e^{-(t/\tau_{eff})} \tag{243}$$

where

$$K_D = 4\pi\ D\ c_A\ R_D \qquad [R_D \simeq (\frac{C_{DA}}{D})^{1/4}] \tag{244}$$

$$\tau_{eff}^{-1} = \tau^{-1} + K_D \tag{245}$$

At these "later" times the only donors that are still excited are those far away from any acceptor. They now transfer their energy by first diffusing it, i.e., by sending it to a donor near an acceptor.

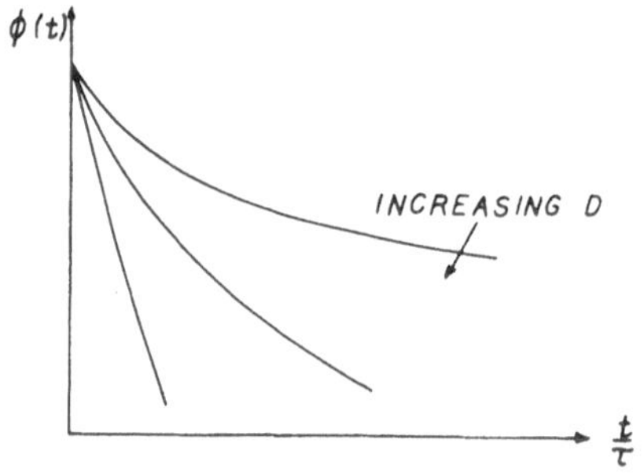

Fig. 15. Behavior of the average probability of donor excitation in the case of migration among donors.

The behavior of the function $\phi(t)$ is sketchily presented in Figure 15.

Summarizing the results just obtained:

(1) $\phi(t)$ is non-exponential for t small enough that migration is not important. In this limit $\phi(t)$ has the previous form found for the no-migration case.

(2) $\phi(t)$ is exponential for large t; it decays at a rate determined by migration.

(3) As migration becomes more rapid, the boundary between these two regions shifts.

$$t^* \simeq \frac{C_{DA}^{1/2}}{D^{3/2}}$$

shifts to shorter times, until, for sufficiently fast migration, the decay appears to be entirely exponential.

V.D. Migration as Random Walk

Consider a donor 1, excited at $t = 0$. The probability that donor 1 is still excited at time t is

$$\psi_1(t) = \exp\left[-\frac{t}{\tau} - t \sum_{i=1}^{N_A} w_{DA}(\vec{R}_1 - \vec{R}_i)\right] \quad (246)$$

$$(0 \leq t < t_1)$$

At time $t = t_1$, the excitation hops to donor 2. The probability that donor 2 is still excited at time $t > t_1$ is given by

$$\psi_2(t) = \psi_1(t_1) \exp\left[-\frac{t - t_1}{\tau} - (t - t_1) \sum_i w_{DA}(\vec{R}_2 - \vec{R}_i)\right]$$

If ψ_i is the probability of finding donor i excited at time t, then [21]

Hop #

1 $\psi_1(t_1)$

2 $\psi_2(t_2) = \psi_1(t_1)\, \psi_2(t_2 - t_1)$

3 $\psi_3(t_3) = \psi_2(t_2)\, \psi_3(t_3 - t_2) = \psi_1(t_1)\, \psi_2(t_2 - t_1)\, \psi_3(t_3 - t_2)$

....

k $\psi_k(t_k)\, \psi_{k-1}(t_k)\, \psi_k(\tau - t_k) = \psi_1(t_1)\, \psi_2(t_2 - t_1)\, \psi_3(t_3 - t_2)$

$\ldots\, \psi_k(\tau - t_k)$

The actual path of the excitation is affected by two random quantities (see Figure 16)

i) $t_j - t_{j-1}$ = interval of time between jumps, and

ii) $\sum_i w_{DA}(\vec{R} - \vec{R}_i)$ = energy transfer probability per unit time

We shall consider the two random quantities independent of each other. In addition, we shall call s the generic time interval among jumps and shall assume that the probability that s is in (s, s + ds) is given by

$$P(s)\ ds = \frac{e^{-(s/\tau_o)}}{\tau_o}\ ds \tag{247}$$

where

$$\tau_o = \langle s \rangle = \int_0^\infty s\ P(s)\ ds \tag{248}$$

The probability of a sequence of intervals

$$0 - t_1 - t_2 - t_3 - \ldots\ t_k - \tau$$

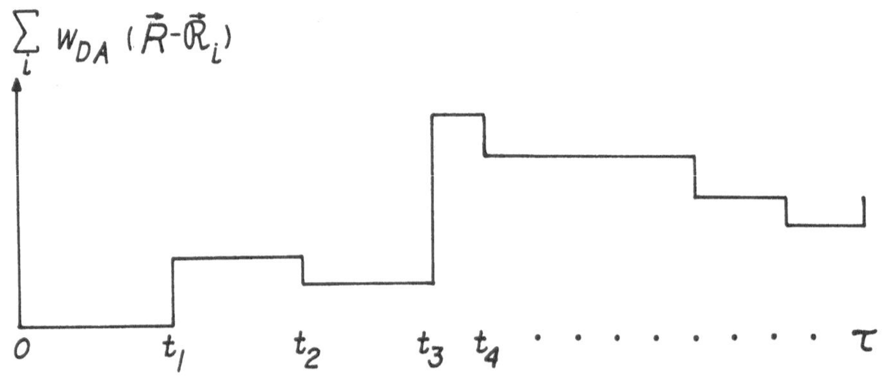

Fig. 16. Energy transfer probability per unit time as affected by the hopping of excitation among donors.

is given by

$$e^{-(t_1/\tau_0)} \frac{dt_1}{\tau_0} \cdot e^{-(t_2-t_1)/\tau_0} \frac{dt_2}{\tau_0} \cdot e^{-(t_3-t_2)/\tau_0} \frac{dt_3}{\tau_0} \cdot$$

$$\cdots\cdots e^{-(\tau-t_k)/\tau_0} \frac{dt_k}{\tau_0}$$

$$= \frac{e^{-[(t_1/\tau_0)+(t_2-t_1)/\tau_0+(t_3-t_2)/\tau_0+\ldots+(\tau-t_k)/\tau_0]}}{\tau_0^k}$$

$$\cdot\, dt_1\, dt_2 \ldots dt_k = e^{-(\tau/\tau_0)} \prod_{i=1}^{k} \frac{dt_i}{\tau_0} \tag{249}$$

The meaningful function $\phi(\tau)$ expressing the probability of donor excitation, has to be obtained by averaging over the two random processes. In evaluating the function $\phi(\tau)$ we have to sum over the different sequences of events that can take place in the interval $(0, \tau)$. In particular, one term will correspond to "no-jump", the following terms to "one-jump", "two-jumps", etc.

For no jumps:

$$\phi_0(\tau) = e^{-(\tau/\tau)} \bar{\rho}(\tau) \tag{250}$$

For one jump:

$$\phi_1(\tau) = \frac{e^{-(\tau/\tau_0)}}{\tau_0} \int_0^{\tau} dt_1\, \bar{\rho}(t_1)\, \rho(\tau - t_1) \tag{251}$$

For two jumps:

$$\phi_2(\tau) = \frac{e^{-(\tau/\tau_0)}}{\tau_0^2} \int_0^{\tau} dt_2\, \bar{\rho}(\tau - t_2) \int_0^{t_2} dt_1\, \bar{\rho}(t_2 - t_1)\, \bar{\rho}(t_1)$$

. . . . (252)

Summing over all the terms $\phi_i(\tau)$:

$$\phi(\tau) = e^{-(\tau/\tau_o)}\,\bar{\rho}(\tau) + \frac{e^{-(\tau/\tau_o)}}{\tau_o}\int_o^\tau dt_1\,\bar{\rho}(\tau - t_1)\,\bar{\rho}(t_1)$$

$$+ \frac{e^{-(\tau/\tau_o)}}{\tau_o^2}\int_o^\tau dt_2\,\bar{\rho}(\tau - t_2)\int_o^{t_2} dt_1\,\bar{\rho}(t_2 - t_1)\,\bar{\rho}(t_1)$$

$$+ \ldots\ldots\ldots$$

$$= e^{-(\tau/\tau_o)}\,[\bar{\rho}(\tau) + S_1(\tau)] \qquad (253)$$

where

$$S_1(\tau) = \sum_{\ell=1}^{\infty}\left(\frac{1}{\tau_o}\right)^{\ell}\int_o^\tau dt_\ell\,\bar{\rho}(\tau - t_\ell)\int_o^{t_\ell} dt_{\ell-1}\,\bar{\rho}(t_\ell - t_{\ell-1})$$

$$\ldots\ldots\ldots\ldots \int_o^{t_2} dt_1\,\bar{\rho}(t_2 - t_1)\,\bar{\rho}(t_1)$$

$$= \frac{1}{\tau_o}\int_o^\tau dt_1\,\bar{\rho}(\tau - t_1)\,\bar{\rho}(t_1)$$

$$+ \sum_{\ell=2}^{\infty}\left(\frac{1}{\tau_o}\right)^{\ell}\int_o^\tau dt_\ell\,\bar{\rho}(\tau - t_\ell)\int_o^{t_\ell} dt_{\ell-1}\,\bar{\rho}(t_\ell - t_{\ell-1})$$

$$\ldots\ldots\ldots\ldots \int_o^{t_2} dt_1\,\bar{\rho}(t_2 - t_1)\,\bar{\rho}(t_1)$$

$$= e^{\tau/\tau_o}\,\phi(\tau) - \bar{\rho}(\tau) \qquad (254)$$

Multiplying (253) by $e^{\tau/\tau_o}\,\bar{\rho}(T - \tau)\,\frac{d\tau}{\tau_o}$ and integrating over τ

$$\frac{1}{\tau_o} \int_o^T e^{\tau/\tau_o} \phi(\tau)\, \bar{\rho}(T - \tau)\, d\tau$$

$$= \frac{1}{\tau_o} \int_o^T \bar{\rho}(\tau)\, \bar{\rho}(T - \tau)\, d\tau + S_2(T) \qquad (255)$$

where

$$S_2(T) = \frac{1}{\tau_o} \int_o^T \bar{\rho}(T - \tau)\, S_1(\tau)\, d\tau$$

$$= \frac{1}{\tau_o^2} \int_o^T \bar{\rho}(T - \tau) \int_o^\tau dt_1\, \bar{\rho}(\tau - t_1)\, \bar{\rho}(t_1)$$

$$+ \sum_{\ell=2}^{\infty} \left(\frac{1}{\tau_o} \right)^{\ell+1} \int_o^T d\tau\, \bar{\rho}(T - \tau) \int_o^\tau dt_\ell\, \bar{\rho}(\tau - t_\ell)$$

$$\cdot \int_o^{t_\ell} dt_{\ell-1}\, \bar{\rho}(t_\ell - t_{\ell-1}) \ldots \int_o^{t_2} dt_1\, \bar{\rho}(t_2 - t_1)\, \bar{\rho}(t_1) \qquad (256)$$

On the other hand

$$S_1(T) = \frac{1}{\tau_o} \int_o^T dt_1\, \bar{\rho}(T - t_1)\, \bar{\rho}(t_1)$$

$$+ \left\{ \frac{1}{\tau_o^2} \int_o^T d\tau\, \bar{\rho}(T - \tau) \int_o^\tau dt_1\, \bar{\rho}(\tau - t_1)\, \bar{\rho}(t_1) \right.$$

$$+ \frac{1}{\tau_o^3} \int_o^T d\tau\, \bar{\rho}(T - \tau) \int_o^\tau dt_2\, \bar{\rho}(\tau - t_2) \int_o^{t_2} dt_1$$

$$\left. \cdot\, \bar{\rho}(t_2 - t_1)\, \bar{\rho}(t_1) + \ldots\ldots\ldots\ldots \right\}$$

$$= \frac{1}{\tau_o} \int_o^T dt_1\, \bar{\rho}(T - t_1)\, \bar{\rho}(t_1) + S_2(T) \qquad (257)$$

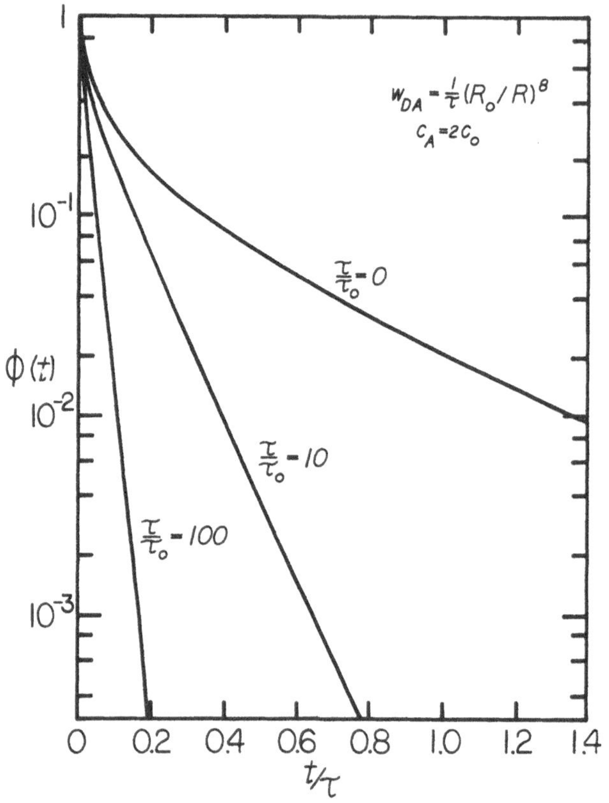

Fig. 17. Decay of donor luminescence in the energy hopping model; $c_A = 2c_0$ and $n = 8$ (dipole-quadrupole interaction). (Reproduced from [12] by courtesy of R. K. Watts and permission from Plenum Press.)

But

$$S_2(T) = S_1(T) - \frac{1}{\tau_0}\int_0^T dt_1\, \bar{\rho}(T - t_1)\, \bar{\rho}(t_1)$$

$$= e^{T/\tau_0}\phi(T) - \bar{\rho}(T) - \frac{1}{\tau_0}\int_0^T dt_1\, \bar{\rho}(T - t_1)\, \bar{\rho}(t_1) \qquad (258)$$

Then (255) becomes

$$\frac{1}{\tau_0}\int_0^T e^{\tau/\tau_0}\phi(\tau)\, \bar{\rho}(T - \tau) = e^{T/\tau_0}\phi(T) - \bar{\rho}(T)$$

or

$$\phi(T) = \bar{\rho}(T)\, e^{-(T/\tau_o)} + \frac{1}{\tau_o}\int_o^T \phi(\tau)\,\bar{\rho}(T-\tau)\, e^{-(T-\tau/\tau_o)}\, d\tau \tag{259}$$

which can be written [22]

$$\phi(t) = \bar{\rho}(t)\, e^{-(t/\tau_o)} + \frac{1}{\tau_o}\int_o^t \phi(t')\,\bar{\rho}(t-t')\, e^{-(t-t'/\tau_o)}\, dt'$$

$$\cdot = \bar{\rho}(t)\, e^{-(t/\tau_o)} + \frac{1}{\tau_o}\int_o^t \phi(t-t')\,\bar{\rho}(t')\, e^{-(t'/\tau_o)}\, dt' \tag{260}$$

The following observations can be made on Eq. (260):

(1) This equation is a renewal equation, the value of ϕ at time t depending on the value of ϕ at time t - t'.

(2) The solution of this equation is generally found by numerical methods.

(3) $\bar{\rho}(t)$ in Eq. (260) takes a form appropriate to the type of D - A interaction: multipolar or exchange.

Figure 17 reports the decay of the donor luminescence following an exciting pulse, for the case $c_A/c_o = 2$ and n = 8 (dipole-quadrupole interaction)

Figure 18 reports the donor luminescence yield for n = 6, 8 and 10 obtained by integrating the solution of Eq. (260) [12]. Note that the smaller is τ_o (the greater the hopping rate), the weaker is the luminescence yield of the donors.

A couple of more things ought to be said regarding Eq. (260). First, let us cite a theorem that applies to Laplace transforms: "Given two functions $f_1(\theta)$ and $f_2(\theta)$ with Laplace transforms $F_1(s)$ and $F_2(s)$, respectively

$$F_1(s)\, F_2(s) = \int_0^\theta f_1(\theta-\theta')\, f_2(\theta')\, d\theta' \text{"} \tag{261}$$

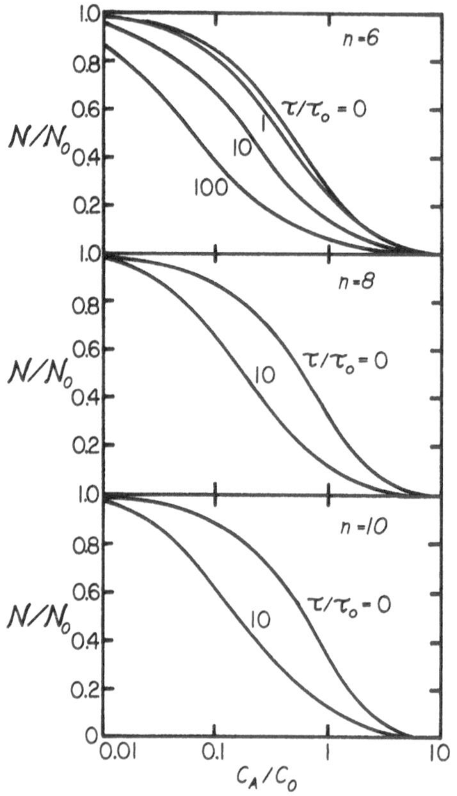

Fig. 18. Yield of donor luminescence obtained by integrating the solution of Eq. (260). (Reproduced from [12] by courtesy of R.K. Watts and permission from Plenum Press).

Let us now use this theorem to deal with Eq. (260). If we set

$$\begin{cases} \tau = a\,\tau_o \\ \dfrac{t}{\tau} = \theta \quad , \quad \dfrac{t}{\tau_o} = a\,\dfrac{t}{\tau} = a\theta \end{cases} \tag{262}$$

Eq. (260) is written

$$\phi(\theta) = \overline{\rho}(\theta)\, e^{-a\theta} + a\int_o^\theta e^{-a(\theta-\theta')}\, \overline{\rho}(\theta - \theta')\, \phi(\theta')\, d\theta' \tag{263}$$

Assume now that

$$f_1(\theta) = e^{-a\theta}\, \bar{\rho}(\theta) \quad , \quad f_2(\theta) = \phi(\theta)$$

It is

$$F_1(s) = \bar{\rho}(s + a) \quad , \quad F_2(s) = \phi(s)$$

Then because of (261)

$$\int_0^\theta e^{-a(\theta-\theta')}\, \bar{\rho}(\theta - \theta')\, \phi(\theta')\, d\theta' = \bar{\rho}(s + a)\, \phi(s)$$

and (263) gives

$$\phi(s) = \bar{\rho}(s + a) + a\, \bar{\rho}(s + a)\, \phi(s)$$

or

$$\phi(s) = \frac{\bar{\rho}(s + a)}{1 - a\, \bar{\rho}(s + a)} \tag{264}$$

$\phi(\theta)$ is the inverse Laplace transform of (264) above.

Second, we want to check the general validity of (260). We note that

$$\phi(t) \xrightarrow[\tau_0 \to \infty]{} \bar{\rho}(t) \tag{265}$$

Let us now assume that no acceptors are present or, that the w_{DA} = const as in the Stern-Volmer model. In this case we expect that the hopping of excitation should have no effect.

Let

$$\bar{\rho}(t) = e^{-(t/\tau)}$$

$$\tau = a\, \tau_0$$

Then

$$\bar{\rho}(t)\, e^{-(t/\tau_0)} = e^{-(t/\tau)(1+a)}$$

$$\bar{\rho}(t-t')\, e^{-(t-t')/\tau_o} = e^{-(1+a)(t/\tau)}\, e^{(1+a)(t'/\tau)}$$

and (260) becomes

$$\phi(t) = e^{-(1+a)(t/\tau)} + \frac{a}{\tau}\, e^{-(1+a)(t/\tau)} \int_o^t e^{(1+a)(t'/\tau)} \phi(t')\, dt' \tag{267}$$

which gives

$$- \frac{1+a}{\tau}\, \phi(t) + \frac{1+a}{\tau}\, e^{-(1+a)(t/\tau)}$$

$$= - \frac{a(1+a)}{\tau^2}\, e^{-(1+a)(t/\tau)} \int_o^t e^{(1+a)(t'/\tau)}\, \phi(t')\, dt' \tag{268}$$

On the other hand

$$\frac{d\phi(t)}{dt} = - \frac{1+a}{\tau}\, e^{-(1+a)(t/\tau)} - \frac{a(1+a)}{\tau^2}\, e^{-(1+a)(t/\tau)}$$

$$\cdot \int_o^t e^{(1+a)(t'/\tau)}\, \phi(t')\, dt' + \frac{a}{\tau}\, \phi(t)$$

$$= - \frac{1+a}{\tau}\, e^{-(1+a)(t/\tau)} - \frac{1+a}{\tau}\, \phi(t)$$

$$+ \frac{1+a}{\tau}\, e^{-(1+a)(t/\tau)} + \frac{a}{\tau}\, \phi(t) = - \frac{1}{\tau}\, \phi(t)$$

or

$$\phi(t) = e^{-(t/\tau)} = \bar{\rho}(t) \tag{269}$$

V.E. <u>Comparison of Two Models</u>

We summarize now the results found for the two models.

1. Diffusion Model

$$\phi(t) = \exp\left\{-\frac{t}{\tau} - \frac{c_A}{c_o}\Gamma\left(\frac{1}{2}\right)\left(\frac{t}{\tau}\right)^{1/2}\left[\frac{1 + 10.87x + 15.5x^2}{1 + 8.743x}\right]^{3/4}\right\} \qquad (270)$$

where

$$x = D\tau R_o^{-2}\left(\frac{t}{\tau}\right)^{2/3} = \frac{D}{C_{DA}^{1/3}} t^{2/3} \qquad (n = 6 \text{ only}) \qquad (271)$$

2. Random Walk (Hopping) Model

$$\phi(t) = \bar{\rho}(t)\, e^{-(t/\tau_o)} + \frac{1}{\tau_o}\int_o^t \phi(t')\,\bar{\rho}(t - t')\, e^{-(t-t'/\tau_o)}\, d\tau' \qquad (272)$$

for any $\bar{\rho}(t)$.

Watts has compared the two equations (270) and (272) and found agreement, as shown in Figure 19, for the case $n = 6$, $c_A = 2c_o$ and

$$x = 0 \longrightarrow \tau_o = \infty$$

$$x = 0.14\left(\frac{t}{\tau}\right)^{2/3} \longrightarrow \tau_o = \tau$$

$$x = 1.33\left(\frac{t}{\tau}\right)^{2/3} \longrightarrow \tau_o = \frac{\tau}{10}$$

V.F. Calculation of Transfer Rates

1. Diffusion Model. We call c_D the concentration of donors:

$$c_D = \frac{N_D}{V} \qquad (273)$$

The average D - D distance is given by

$$L = \left(\frac{4\pi\, c_D}{3}\right)^{-1/3} \qquad (274)$$

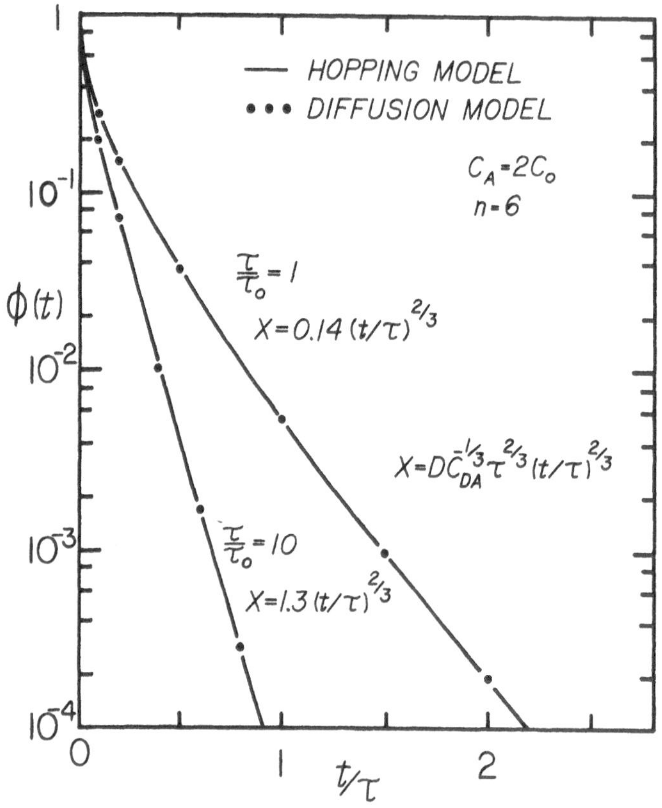

Fig. 19. Decay of donor luminescence according to the diffusion model and to the hopping model. (Reproduced from [12] by courtesy of R. K. Watts and permission from Plenum Press.)

Considering the general relation $D = R^2/6t$, we have [23]

$$dD = \frac{R^2}{6t} c_D R^2 \sin\theta \, dR \, d\theta \, d\phi$$

$$= \frac{R^2}{6} \frac{C_{DD}}{R^6} c_D R^2 \sin\theta \, dR \, d\theta \, d\phi = \frac{1}{6} C_{DD} c_D \frac{1}{R^2} \sin\theta \, dR \, d\theta \, d\phi \tag{275}$$

and

$$D = \frac{4\pi}{6} C_{DD} c_D \int_L^\infty \frac{1}{R^2} dR = \frac{4\pi}{6} C_{DD} c_D L^{-1}$$

$$= \frac{1}{2} \left(\frac{4\pi}{3} \right)^{4/3} c_D^{4/3} \, C_{DD} = 3.375 \, c_D^{4/3} \, C_{DD} \tag{276}$$

Measured value for D for rare earth ions are 10^{-11} - 10^{-9} cm^2 sec^{-1} [24].

We shall now use the expression (244) to find

$$K_D = 4\pi \, D \, c_A \, R_D = 4\pi(3.375 \, c_D^{4/3} \, C_{DD}) \, c_A \left(\frac{C_{DA}^{1/4}}{D^{1/4}} \right)$$

$$\simeq 42 \, c_D^{4/3} \, c_A \, C_{DD} \frac{C_{DA}^{1/4}}{(3.375)^{1/4} \, c_D^{1/3} \, C_{DD}^{1/4}} \simeq 30 \, c_D \, c_A \, C_{DD}^{3/4} \, C_{DA}^{1/4} \tag{277}$$

2. <u>Hopping Model</u>. If n = 6 (dipole-dipole interaction) Eq. (182) gives us

$$\bar{\rho}(t) = \exp\left[- \frac{t}{\tau} - \frac{c_A}{c_o} \Gamma\left(\frac{1}{2} \right) \left(\frac{t}{\tau} \right)^{1/2} \right] \tag{278}$$

But

$$\frac{c_A}{c_o} \frac{\Gamma\left(\frac{1}{2} \right)}{\tau^{1/2}} = c_A \frac{4\pi \, R_o^3}{3} \frac{\sqrt{\pi}}{\tau^{1/2}} = \frac{4\pi}{3} \sqrt{\pi} \, c_A \frac{R_o^3}{\tau^{1/2}} = \frac{4\pi}{3} \sqrt{\pi} \, c_A \, C_{DA}^{1/2}$$

Therefore we can write [22]

$$\bar{\rho}(t) = \exp\left[- \frac{t}{\tau} - (At)^{1/2} \right]$$

where

$$A = \frac{16\pi^3}{9} \, c_A^2 \, C_{DA} \tag{279}$$

If $A \gg \frac{1}{\tau}$

$$\bar{\rho}(t) \simeq e^{-(At)^{1/2}} \tag{280}$$

Consider the following distribution

$$\phi(W) = \left(\frac{A}{4\pi} \right)^{1/2} W^{-3/2} e^{-(A/4W)} \tag{281}$$

We shall show that this distribution is normalized. If we set

$$W^{-1/2} = z \quad , \qquad -\frac{1}{2} W^{-3/2} \, dW = dz$$

$$\frac{1}{W^{3/2}} \, dW = -2dz$$

We obtain

$$\int_0^\infty \phi(W) \, dW = \left(\frac{A}{4\pi} \right)^{1/2} \int_0^\infty W^{-3/2} e^{-(A/4W)} \, dW$$

$$= 2\left(\frac{A}{4\pi} \right)^{1/2} \int_0^\infty e^{-(A/4)z^2} \, dz$$

$$= 2\left(\frac{A}{4\pi} \right)^{1/2} \frac{\sqrt{\pi}}{2\sqrt{A/4}} = 1$$

Let us now calculate the following integral

$$\int_0^\infty e^{-Wt} \phi(W) \, dW = \left(\frac{A}{4\pi} \right)^{1/2} \int_0^\infty W^{-3/2} e^{-Wt-(A/4W)} \, dW$$

$$= \left(\frac{A}{4\pi} \right)^{1/2} 2 \int_0^\infty e^{-(A/4)z^2-(t/z^2)} \, dz \tag{282}$$

Set

$$\sqrt{\frac{A}{4}} \, z = x$$

$$\frac{A}{4} z^2 = x^2$$

$$dz = \frac{2}{\sqrt{A}} \, dx$$

Then

$$\int_0^\infty e^{-Wt}\,\phi(W)\,dW = \sqrt{\frac{A}{4\pi}}\;2\int_0^\infty \frac{2}{\sqrt{A}}\,dx\left[e^{-x^2-(At/4)(1/x^2)}\right]$$

$$= \frac{2}{\sqrt{\pi}}\int_0^\infty e^{-x^2-(At/4)(1/x^2)}\,dx = e^{-(At)^{1/2}} = \bar{\rho}(t) \qquad (283)$$

We may then say that $\phi(W)\,dW$ is the probability that the energy transfer rate is in $(W, W + dW)$. The most probable migration rate is obtained from (281) and is equal to $A/6$; we then set

$$\frac{1}{\tau_o} = \frac{A}{6} = \left(\frac{2}{3}\pi\right)^3 c_D^2\, C_{DD} \qquad (284)$$

where we have replaced C_{DA} with C_{DD} because we are dealing with $D \rightarrow D$ transfer. In this derivation it is assumed that $A >> 1/\tau$, namely that the hopping rate is much larger than the intrinsic donor decay rate.

For very high rare earth donor concentrations $\tau_o < 10^{-9}$ sec [25].

We define a critical radius R_c for the donor such that

$$w_{DA} = \frac{C_{DA}}{R^6} = \frac{1}{\tau}\frac{R_o^6}{R^6} = \frac{1}{\tau_o}\frac{R_c^6}{R^6} \qquad (285)$$

R_c is the distance at which the $D \rightarrow A$ transfer rate is equal to the $D \rightarrow D$ transfer rate. R_o, as before, is the distance at which the $D \rightarrow A$ transfer rate is equal to the D decay rate. The number of acceptors inside the sphere of radius R_c is

$$\frac{4\pi}{3}R_c^3\, c_A = \frac{4\pi}{3}\pi\, c_A\left(\frac{R_o^6}{\tau}\tau_o\right)^{1/2} = \frac{4}{3}\pi\, c_A\left(C_{DA}\,\tau_o\right)^{1/2} \qquad (286)$$

The $D \rightarrow A$ energy transfer probability is greater than $1/\tau_o$ for all the acceptors inside this sphere.

The rate of $D \rightarrow A$ transfer is equal to the number of acceptors that fall into the strong interaction zone around the migrating donor excitation per unit time:

$$K_H = \frac{4}{3}\pi R_c^3 c_A \frac{1}{\tau_o} = \frac{4}{3}\pi c_A (C_{DA}\tau_o)^{1/2}\frac{1}{\tau_o}$$

$$= \frac{4}{3}\pi c_A C_{DA}^{1/2}\frac{1}{\tau_o^{1/2}} = \frac{4}{3}\pi c_A C_{DA}^{1/2}\frac{c_D C_{DD}^{1/2}}{\left(\frac{27}{8\pi^3}\right)^{1/2}}$$

$$\simeq 40\, c_A c_D C_{DD}^{1/2} C_{DA}^{1/2} \qquad (287)$$

V.G. Regimes of Donor Decay

There are three regimes of donor decay:

1. No Diffusion. c_D is small. The decay of donor excitation is given by $\overline{\rho}(t)$. $\overline{\rho}(t)$ is the average of exponentials corresponding to the various transfer rates $w_{DA}(R)$. An extreme case is that of a donor which is far away from all acceptors; the donor decay function is

$$\rho(t) = e^{-(t/\tau)} \qquad (288)$$

where τ = intrinsic lifetime of donor. The other extreme case is that of a donor which has an acceptor as nearest neighbor; in this case

$$\rho(t) = \exp\left[-\frac{t}{\tau} - t\, w_{DA}(R_m)\right] \qquad (289)$$

where R_m = minimum D - A distance.

2. Diffusion-Limited Decay. c_D is higher. For $t \ll C_{DA}^{1/2}/D^{3/2}$, $\overline{\rho}(t)$ is the same as for the case of no migration. For $t \gg C_{DA}^{1/2}/D^{3/2}$

$$\overline{\rho}(t) \sim e^{-(t/\tau)-K_D t} \qquad (290)$$

$t^* = C_{DA}^{1/2}/D^{3/2}$ represents a boundary between the "transfer without migration" region and the "transfer with migration" region.

3. Fast Diffusion. c_D is still higher. An increase in the concentration of donors produces a faster migration of energy, i.e. a larger D, and a larger K_D. The characteristic time t^* becomes shorter and shorter until it reaches the value $w_{DA}^{-1}(R_m)$, R_m being

the shortest D - A distance. Any additional increase of c_D has effect on D, but no effect on K_D which takes a "saturated" value [26]. In these conditions the decay of the donor luminescence is purely exponential, since the fast diffusion averages out all the different donor environments.

In the present case we set

$$K_D = K\, c_A \tag{291}$$

where K = const, independent of c_D, or, taking into account the possibility that a large fraction of acceptors are excited

$$K_D = K\, n_A \tag{292}$$

where n_A = concentration of ground state acceptors. In this case the evolution of the excited donor population follows the *rate equation*

$$\frac{d}{dt} n_{d'}(t) = -\frac{n_{d'}(t)}{\tau} - K\, n_A(t)\, n_{d'}(t) \tag{293}$$

Rate equations have been used for explaining energy transfer in rare-earth doped systems [27].

V.H. Migration in the Case of Inhomogeneous Broadening of Donors Levels

Inhomogeneous broadening results when the sites occupied by the optically active ions are not exactly equal. In glasses the disorder produces an inhomogeneous line broadening of $\sim$100 cm^{-1} for rare earths.

Even in a "perfect" crystal there may be environments which are different for the various donors. The difference of the environments can be due to slight changes in the local crystalline field. Also, a donor may have an acceptor nearby, another donor may be far away from all acceptors, and this may also affect slightly the levels of the donor.

Consider now the donor system where donors are in different environments exemplified in Figure 20. Donor-to donor energy transfer in this case takes place among nonresonant levels.

If $g_i(E)$ is the homogeneous line shape function of level i, the transfer rate between level i and level k is proportional to the overlap integral of the line shape functions corresponding to the two levels:

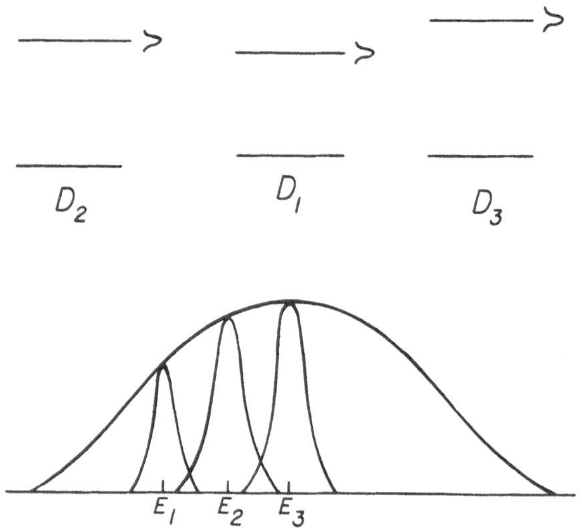

Fig. 20. Inhomogeneous broadening of donor levels.

$$w_{ik} \propto \int g_i(E - E_i)\, g_k(E - E_k)\, dE \tag{294}$$

This means that the energy transfer rate between two donors i and k is now a function not only of their relative distance $|\vec{R}_i - \vec{R}_k|$, but also of the difference of their excited state energies $(E_i - E_k)$:

$$w_{ik} = w_{ik}(|\vec{R}_i - \vec{R}_k|, E_i - E_k) \tag{295}$$

Consider now the case in which $|\vec{R}_i - \vec{R}_k|$ is large. Because of the large distance energy transfer is improbable, even if $E_i = E_k$; the inhomogeneous broadening has then no effect in this case.

If $|\vec{R}_i - \vec{R}_k|$ is small, the inhomogeneous broadening may reduce the overlap integral (294) and consequently prevent the

$i \rightarrow k$ energy transfer.

If $\rho_i(t)$ is the probability that ion i is excited at time t:

$$\frac{d}{dt}\rho_i(t) = -\frac{1}{\tau}\rho_i(t) - \rho_i(t)\sum_k w_{ik} + \sum_k w_{ki}\,\rho_k(t) \tag{296}$$

This relation gives us a system of equations which are coupled through the last term in the right member.

Inhomogeneous broadening may have relevant effects on the spectral characteristics of an emitting system, even if this system consists of ions of one type. Consider Figure 21, where for simplicity only two different environments are represented. In general the distribution of the ions in the available sites may be random, but, by the use of monochromatic (laser) light it is possible to excite selectively the ions residing in a particular environment, say D_1. Ions D_1 are then brought to level E_1' from which they decay nonradiatively to level E_1. Assume the exciting light intensity to be constant in time and let τ and τ' be the intrinsic lifetimes of level E_1 and E_1', respectively. If $\tau' >> (\sum_k w'_{1k})^{-1}$ *or* $\tau >> (\sum_k w_{1k})^{-1}$, then it takes the excited D_1 ions much less time to transfer their energy to the other D_k ions than to relax. Under these circumstances the selective excitation has no effect, since the energy spreads rapidly among the ions in different sites; the luminescence spectrum will be similar to the one obtained with wideband excitation.

If, on the other hand, $\tau' << (\sum_k w'_{1k})^{-1}$ *and* $\tau << (\sum_k w_{1k})^{-1}$ then it takes the excited D_1 ions much less time to relax than to transfer their energy to the other D_k ions. Under these circumstances only the D_1 ions will luminesce with a definite photon energy E_1; the luminescence spectrum will be much narrower than the one obtained with wide band excitation. This effect is called *fluorescence line narrowing* (FLN) [28].

VI. COLLECTIVE EXCITATIONS

VI.A. Introduction

We have considered the energy transfer processes taking place among donor centers and between donor and acceptor centers. In the systems examined donors and acceptors are distributed over the

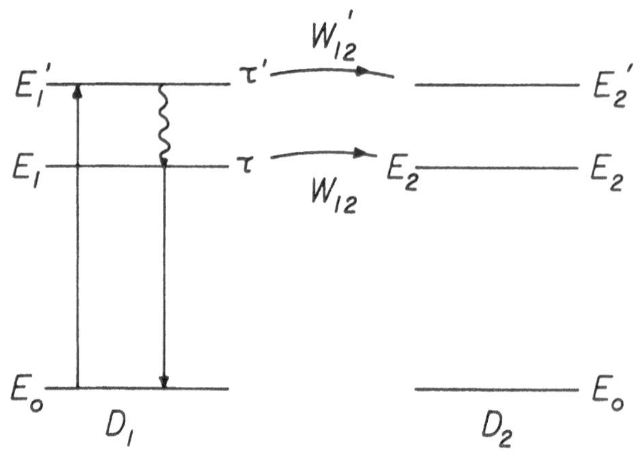

Fig. 21. Energy levels of ions in different environments.

available sites and replace substitutionally an intrinsic constituent of the solid. We wish to consider now the case of an ordered solid consisting entirely of identical atoms and to study how the interactions among these ions lead to excitations of *collective* type. We shall begin our treatment with a linear system and extend later this treatment to three dimensions.

Consider a linear crystal consisting of a chain of N identical atoms with a unit cell of length a.

Assume that the Hamiltonian of the system is the following

$$H = H_o + H' \tag{297}$$

where

$$H_o = \sum_{s=1}^{N} H_s \tag{298}$$

$$H' = \frac{1}{2} \sum_{\substack{s,s' \\ s \neq s'}} V_{ss'} \tag{299}$$

The H_s are one-atom Hamiltonians and the $H_{ss'}$ are terms representing interactions between different atoms. The ground state wavefunction of H_0 is given by

$$\psi_g = u_1 \, u_2 \, \cdots \, u_{N-1} \, u_N \tag{300}$$

The first excited state of H_0 corresponds to the following N degenerate wavefunctions

$$\begin{cases} \psi_1 = v_1 \, u_2 \, \cdots \, u_N \\ \psi_2 = u_1 \, v_2 \, \cdots \, u_N \\ \cdots\cdots\cdots\cdots\cdots \\ \psi_N = u_1 \, u_2 \, \cdots \, v_N \end{cases} \tag{301}$$

In these wavefunctions u_i represents a localized one-atom ground state and v_i a localized one-atom excited state. Both u_i and v_i are assumed to be nondegenerate.

The ground state of the Hamiltonian H_0 is nondegenerate; the first excited state of H_0 is N-fold degenerate, because there are N distinct ways to excite only one atom. For all these wavefunctions the energy is E_0:

$$H_o \, \psi_\ell = E_o \, \psi_\ell \tag{302}$$

The eigenfunctions of H are given by

$$H \, \psi'_\nu = (H_o + H') \, \psi'_\nu = E'_\nu \, \psi'_\nu \tag{303}$$

The excited-state eigenfunctions of H can be expressed as follows:

$$\psi'_\nu = \sum_{\ell=1}^{N} a_{\nu\ell} \, \psi_\ell \tag{304}$$

Replacing this in Eq. (303) we find

$$\sum_\ell a_{\nu\ell}(E_o - E'_\nu) \, \psi_\ell + \sum_\ell a_{\nu\ell} \, H'\psi_\ell = 0 \tag{305}$$

Multiplying by ψ_m^* and integrating

$$a_{\nu m}(E'_\nu - E_o - H'_{mm}) - \sum_{\ell \neq m} a_{\nu\ell} H'_{m\ell} = 0 \tag{306}$$

We note the following:

E'_ν = total energy of the system

$E_o + H'_{mm}$ = energy of the chain if the excitation cannot move

$E'_\nu - (E_o + H'_{mm}) = \varepsilon_\nu$ = energy associated with the motion of the excitation

We see from Eq. (306) that the motion of the excitation is related to the off-diagonal terms of the Hamiltonian H!

VI.B. Eigenfunctions

In order to look for functions that approximate well the wavefunctions ψ'_ν that satisfy the eigenvalue equation (303) we want to take advantage of the symmetry of the system. We shall assume periodic boundary conditions to take care of "surface" effects and translational symmetry for the chain

$$H(x + \ell a) = H(x) \tag{307}$$

Later in this treatment we shall generalize our results to three-dimensional crystals. It seems appropriate at this time to digress from the present sequence of derivations and express a general result for the three-dimensional case.

Let us assume that an electron is in a periodic field of force and its Hamiltonian is such that

$$H(\vec{r} + \vec{R}_n) = H(\vec{r}) \tag{308}$$

where $\vec{R}_n$ = lattice vector. The eigenfunctions of H are given by the equation

$$H\psi(\vec{r}) = E\psi(\vec{r}) \tag{309}$$

We shall assume these functions to be orthonormal. It is always possible [29], even if E is degenerate, to express $\psi(\vec{r})$ in such a way that

$$\psi(\vec{r} + \vec{R}_n) = e^{i\vec{k}\cdot\vec{R}_n}\, \psi(\vec{r}) \tag{310}$$

The vector $\vec{k}$ is defined, apart from any reciprocal lattice vector $\vec{K}_s$. For any vector of the latter type

$$e^{i\vec{K}_s\cdot\vec{R}_n} = 1 \tag{311}$$

for any $\vec{R}_n$. This gives us the possibility of keeping $\vec{k}$ within the *first Brillouin zone.*

The values of $\vec{k}$ are determined by the periodic boundary conditions that we impose on the wavefunctions.

We apply now these considerations to our linear crystal. As we noted in Eq. (304), the desired wavefunctions have the form

$$\psi'_k = \sum_{\ell=1}^{N} a_{k\ell}\, \psi_\ell \tag{312}$$

where we have changed the wavefunction index from ν to k. In the expressions above k runs over N possible values and we note that the wavefunctions ψ'_k may involve some degeneracy.

As we established in Eqs. (301), the first excited state of H_0 is related to the wavefunctions

$$\begin{cases} \psi_1(x) = v_1(x)\ u_2(x - a)\ u_3(x - 2a)\ \ldots\ u_N[x - (N - 1)\,a] \\ \psi_2(x) = u_1(x)\ v_2(x - a)\ u_3(x - 2a)\ \ldots\ u_N[x - (N - 1)\,a] \\ \cdots\cdots\cdots\cdots\cdots\cdots\cdots\cdots\cdots\cdots \\ \psi_N(x) = u_1(x)\ u_2(x - a)\ u_3(x - 2a)\ \ldots\ v_N[x - (N - 1)\,a] \end{cases} \tag{313}$$

Let us consider the effect of a shift of the origin in the +x direction by an amount a:

$$\psi_i(x) \rightarrow \psi_i(x - a) \tag{314}$$

It is

$$\psi_1(x - a) = v_1(x - a)\ u_2(x - 2a)\ u_3(x - 3a)\ \dots\ u_N(x - Na)$$
$$= v_1(x - a)\ u_2(x - 2a)\ u_3(x - 3a)\ \dots\ u_N(x)$$
$$= \psi_2(x) \qquad (315)$$

We have made use of the periodic boundary conditions in the intermediate step. In general it may be seen that

$$\psi_i(x - a) = \psi_{i+1}(x) \qquad (316)$$

Let us now examine the effect of this shift of origin on the wavefunctions ψ_k'. First, according to the result (310),

$$\psi_k'(x - a) = e^{-ika}\ \psi_k'(x)$$
$$= e^{-ika}(a_{k1}\ \psi_1 + a_{k2}\ \psi_2 + \dots\ a_{kN}\ \psi_N) \qquad (317)$$

On the other hand, we may also write

$$\psi_k(x - a) = a_{k1}\ \psi_2 + a_{k2}\ \psi_3 + \dots\ a_{kN}\ \psi_1 \qquad (318)$$

The comparison of the coefficients in Eqs. (317) and (318) gives us

$$a_{k\ell} = e^{i(\ell-1)ka}\ a_{k1} \qquad (319)$$

The desired wavefunctions can now be written

$$\psi_k = a_{k1}(\psi_1 + e^{ika}\ \psi_2 + e^{i2ka}\ \psi_3 + \dots) \qquad (320)$$

The normalization of ψ_k' yields

$$|a_{k1}|^2 = \frac{1}{N} \qquad (321)$$

and we may choose the phase of a_{k1} so that

$$a_{k1} = \frac{1}{\sqrt{N}}\ e^{ika} \qquad (322)$$

resulting in

$$a_{k\ell} = \frac{1}{\sqrt{N}} e^{i\ell ka} \tag{323}$$

Therefore the approximate wavefunctions of H are

<u>ground state</u>

$$\psi_g = |u_1 u_2 \dots u_N> \tag{324}$$

<u>first excited state</u>

$$\psi'_k = \frac{1}{\sqrt{N}} \sum_{\ell=1}^{N} e^{i\ell ka} |u_1 u_2 \dots v_\ell \dots u_N> \tag{325}$$

The above wavefunctions are approximate for the following reasons:

(1) We treated the localized electronic wavefunctions u_j as non-overlapping. In fact, any functions used to approximate them *do* overlap.

(2) We did not take explicitly into account the interaction terms V_{ss} in generating the ψ_g and ψ'_k.

Finally, we consider the allowed values of k which are, as we said, determined by the boundary conditions that we impose. If we choose periodic boundary conditions, $\psi'_k(x + Na) = \psi'_k(x)$ and

$$a_{k,N+1} = a_{k1} \tag{326}$$

or

$$e^{ik(N+1)a} = e^{ika} \tag{327}$$

This implies

$$e^{ikNa} = 1 \tag{328}$$

with the solutions

$$k = \frac{2\pi n}{Na} \tag{329}$$

If we take

$$-\frac{N}{2} < n \leq \frac{N}{2}$$

the range for n is compatible with k being in the first Brillouin zone.

VI.C. Dispersion Relations

We have already seen in the previous section that

$$a_{k\ell} = \frac{1}{\sqrt{N}} e^{i\ell ka} \tag{330}$$

These coefficients appear in Eq. (306) as follows

$$a_{km}\, \varepsilon_k - \sum_{\ell \neq m} a_{k\ell}\, H'_{m\ell} = 0 \tag{331}$$

where

$$\varepsilon_k = E'_k - E_o - H'_{mm} \tag{332}$$

Substituting (330) in Eq. (331), we obtain a relationship between ε_k and k

$$\frac{1}{\sqrt{N}} e^{imka}\, \varepsilon_k - \sum_{\ell \neq m} \frac{1}{\sqrt{N}} e^{i\ell ka}\, H'_{m\ell} = 0$$

and

$$\varepsilon_k = \sum_{\ell \neq m} e^{i(\ell - m)ka}\, H'_{m\ell} \tag{333}$$

This *dispersion relation* is independent of m; in addition, the N different values of k generate N values for ε_k. We may recall that

$$H'_{m\ell} = \langle \psi_m | H' | \psi_\ell \rangle \tag{334}$$

which in the present case reduces to

$$H'_{m\ell} = \langle v_m\, u_\ell | H' | u_m\, v_\ell \rangle \tag{335}$$

In many cases only nearest neighbor interactions are non-negligible; for such systems the matrix elements may be written

$$H'_{m\ell} = M\,\delta_{\ell,m\pm 1} \tag{336}$$

where M is the strength of the interaction. The dispersion relation represented by Eq. (333) then becomes

$$\varepsilon_k = Me^{ika} + Me^{-ika} = 2M\cos ka \tag{337}$$

The total energy of the system is then of the form

$$E'_k = E_o + H'_{mm} + \varepsilon_k = E_o + H'_{mm} + 2M\cos ka \tag{338}$$

where we have taken into account the expression (332). We observe that, as a result of the perturbation H', the N-fold degenerate state has become a band of N states (recall the N allowed values of k).

The dispersion relation is sketched in Figure 22.

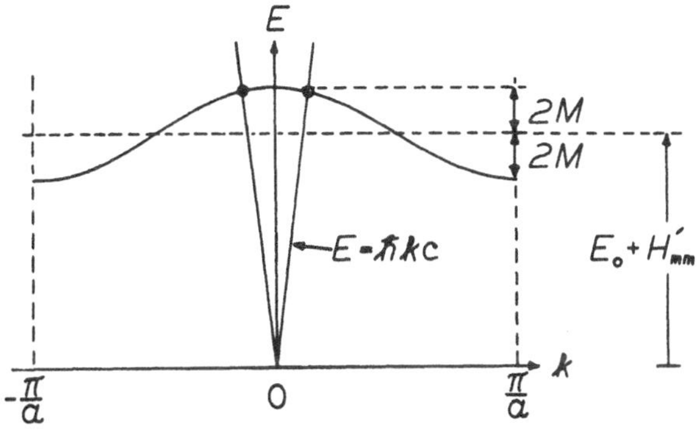

Fig. 22. Dispersion relation for collective excitations of a linear chain.

VI.D. Effective Mass

The collective excitations can be treated as quasi-particles with an effective mass. By forming wave packets with a spread $\Delta\vec{k}$ about $\vec{k}$ we can define a particle velocity as the group velocity

$$v_k = \frac{d\omega}{dk} = \frac{1}{\hbar}\frac{\partial\varepsilon_k}{\partial k} \tag{339}$$

We note that this formula is also valid for electromagnetic waves.

Since, for a free particle,

$$\varepsilon_k = \frac{\hbar^2 k^2}{m} \tag{340}$$

we can associate to our quasi-particle the mass

$$m^* = \frac{\hbar^2}{\left(\frac{\partial^2\varepsilon}{\partial k^2}\right)} \tag{341}$$

In the present case

$$\varepsilon_k = 2M \cos ka \tag{342}$$

and

$$\frac{\partial\varepsilon_k}{\partial k} = -2Ma \sin ka \tag{343}$$

$$\frac{\partial^2\varepsilon_k}{\partial k^2} = -2Ma^2 \cos ka \tag{344}$$

Therefore

$$v_k = \frac{1}{\hbar}\frac{\partial\varepsilon_k}{\partial k} = -\frac{2Ma \sin ka}{\hbar} \tag{345}$$

and

$$m^* = \frac{\hbar^2}{\left(\frac{\partial^2 \varepsilon_k}{\partial k^2}\right)} = -\frac{\hbar^2}{2Ma^2 \cos ka} \tag{346}$$

ε_k, v_k and m^* are represented in Figure 23. We note that for small k

$$v_k = -\frac{2Mka^2}{\hbar} \tag{347}$$

$$m^* = -\frac{\hbar^2}{2Ma^2} \tag{348}$$

In order to establish some qualitative features of the effective mass approach, let us treat only the absolute values of v and m^*. We note:

(1) The velocity is linearly dependent on both M and k. As the strength of the interaction increases, the speed of excitation propagation also increases (all other factors being equal). The dependence of v on the interatomic distance a is more subtle, since a portion of this dependence is "hidden" in Mk. As a typical example, let us consider the electric dipole-electric dipole interaction between nearest neighbors. This interaction goes as a^{-3}. In this case, we have $v \propto a^{-2}$, since $k \propto a^{-1}$. We therefore see that the velocity decreases (possibly dramatically) as a increases.

(2) The effective mass is inversely proportional to M. A strong interaction between nearest neighbors would then result in a relatively small effective mass (all other factors being equal). This reinforces our general notion that a strong interaction enhances the delocalization of excitation energy in the system. As for the velocity, the dependence of m^* on a requires some specification of the interaction M. For the electric dipole-electric dipole case given above, it is easy to verify that $m^* \propto a^1$. This results in the physically reasonable behavior that an increased separation distance hampers the movement of excitation energy.

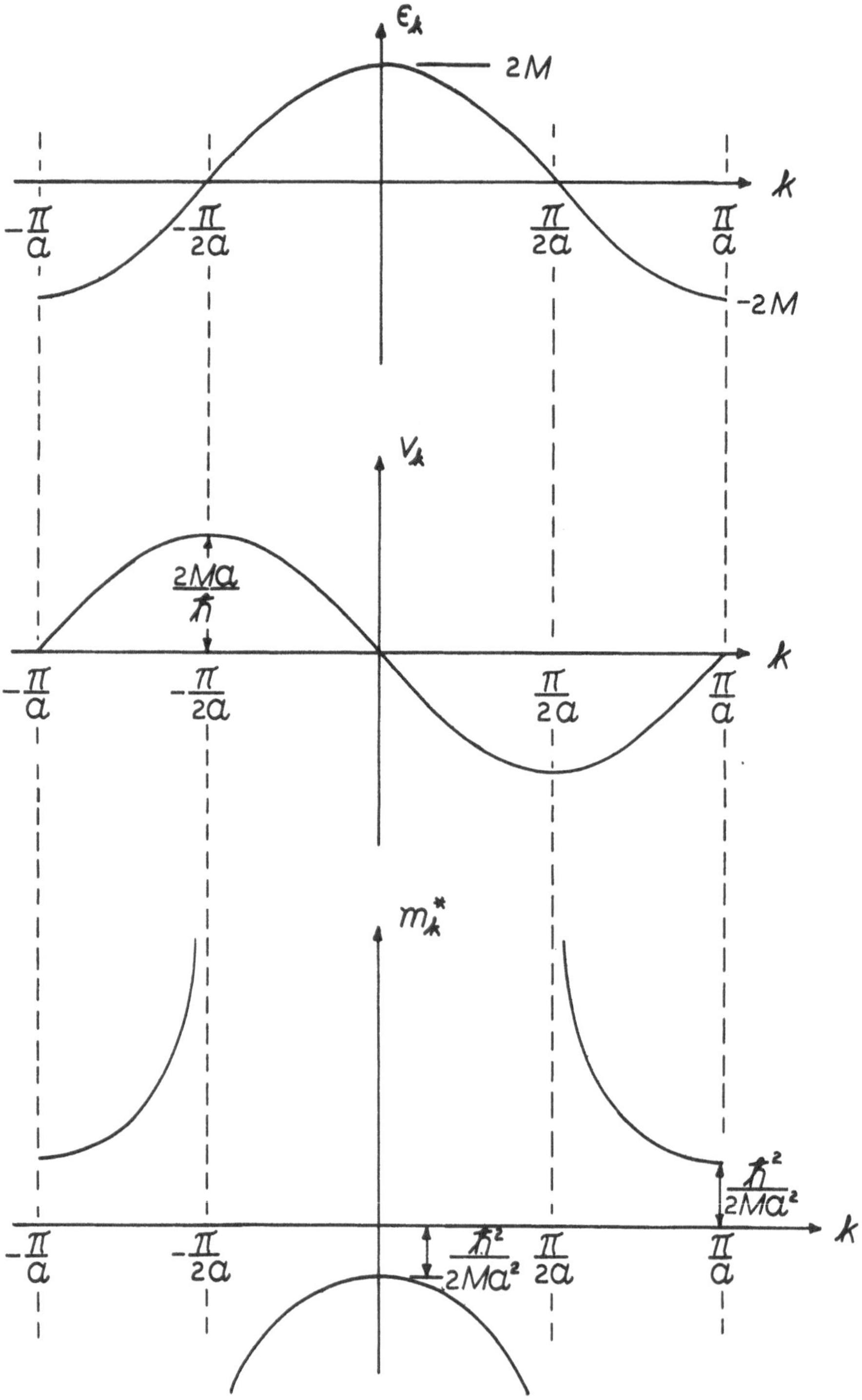

Fig. 23. Energy, velocity and equivalent mass of collective excitations.

VI.E. Generalization to Three Dimensions

We generalize our treatment to an ordered solid in three dimensions. Let $\vec{a}_1$, $\vec{a}_2$ and $\vec{a}_3$ be the primitive *basic lattice vectors* and $N = N_1 \cdot N_2 \cdot N_3$ the total number of atoms. We introduce the notion of *reciprocal lattice*, a geometrical construction which consists of an array of points. The primitive basic vectors of the reciprocal lattice are given by

$$\begin{cases} \vec{b}_1 = 2\pi \dfrac{\vec{a}_2 \times \vec{a}_3}{\vec{a}_1 \cdot \vec{a}_2 \times \vec{a}_3} \\[2ex] \vec{b}_2 = 2\pi \dfrac{\vec{a}_3 \times \vec{a}_1}{\vec{a}_1 \cdot \vec{a}_2 \times \vec{a}_3} \qquad (\vec{b}_i \cdot \vec{a}_j = 2\pi\, \delta_{ij}) \\[2ex] \vec{b}_3 = 2\pi \dfrac{\vec{a}_1 \times \vec{a}_2}{\vec{a}_1 \cdot \vec{a}_2 \times \vec{a}_3} \end{cases} \tag{349}$$

In the reciprocal lattice we carve out the *Brillouin Zone*, another geometrical construction in $\vec{k}$ space. It is constructed by starting from a lattice point of the reciprocal lattice, drawing lines to the nearest neighbor points and cutting these lines halfway with perpendicular planes; the smallest volume enclosed by these planes is the (first) Brillouin Zone.

The wavefunction of the crystal corresponding to the first excited state is now designated as follows:

$$\psi_{\vec{k}} = \frac{1}{\sqrt{N}} \sum_{\ell} e^{i\vec{k}\cdot\vec{R}_\ell} |u_1\, u_2 \ldots v_\ell \ldots u_N> \tag{350}$$

The allowed values of $\vec{k}$ are determined by the periodic boundary conditions.

The dispersion relation is given by

$$\varepsilon_{\vec{k}} = \sum_{\substack{\ell \\ \ell \neq m}} e^{i\vec{k}\cdot(\vec{R}_\ell - \vec{R}_m)} H'_{m\ell} \tag{351}$$

The group velocity is given by

$$\vec{v}_{\vec{k}} = \frac{1}{\hbar} \vec{\nabla}_{\vec{k}} \varepsilon_{\vec{k}} \tag{352}$$

and the *mass tensor* by

$$m^* = \frac{\hbar^2}{\vec{\nabla}_{\vec{k}} \vec{\nabla}_{\vec{k}} \varepsilon_{\vec{k}}} \tag{353}$$

VI.F. Periodic Boundary Conditions and Density of States

Consider a crystal with

N_1 cells in the $\vec{a}_1$ direction

N_2 cells in the $\vec{a}_2$ direction, and

N_3 cells in the $\vec{a}_3$ direction.

Call

$$\vec{R}_N = N_1 \vec{a}_1 + N_2 \vec{a}_2 + N_3 \vec{a}_3 \tag{354}$$

If we impose periodic boundary conditions (PBC), we obtain

$$\psi_{\vec{k}}(\vec{r} + \vec{R}_N) = \psi_{\vec{k}}(\vec{r}) \tag{355}$$

namely,

$$e^{i\vec{k}\cdot\vec{R}_N} = 1 \tag{356}$$

This equation determines the allowed values of $\vec{k}$; it implies that

$$\vec{k} \cdot \vec{R}_N = 2\pi s \qquad (s = \text{integer}) \tag{357}$$

The wave vector can be expressed as follows

$$\vec{k} = k_1 \hat{b}_1 + k_2 \hat{b}_2 + k_3 \hat{b}_3 \tag{358}$$

where $\hat{b}_i$ = unit vector in the $\vec{b}_i$ direction.

Therefore

$$\vec{k} \cdot \vec{R}_N = (k_1 \hat{b}_1 + k_2 \hat{b}_2 + k_3 \hat{b}_3) \cdot (N_1 \vec{a}_1 + N_2 \vec{a}_2 + N_3 \vec{a}_3)$$

$$= N_1 k_1 \vec{a}_1 \cdot \hat{b}_1 + N_2 k_2 \vec{a}_2 \cdot \hat{b}_2 + N_3 k_3 \vec{a}_3 \cdot \hat{b}_3$$

$$= 2\pi s \tag{359}$$

This relation is satisfied if

$$\begin{cases} k_1 = \dfrac{2\pi s_1}{N_1 \vec{a}_1 \cdot \hat{b}_1} \\ k_2 = \dfrac{2\pi s_2}{N_2 \vec{a}_2 \cdot \hat{b}_2} \\ k_3 = \dfrac{2\pi s_3}{N_3 \vec{a}_3 \cdot \hat{b}_3} \end{cases} \tag{360}$$

with s_1, s_2 and s_3 positive or negative integers or zero.

Claim

The values of s_1 greater than N_1,

the values of s_2 greater than N_2, and

the values of s_3 greater than N_3

are redundant.

Proof

Let us assume that $s_i < N_i$ and let us replace s_i by $s_i + N_i$. Then

$$k_1 = \frac{2\pi\, s_1}{N_1\, \vec{a}_1 \cdot \hat{b}_1} + \frac{2\pi\, N_1}{N_1\, \vec{a}_1 \cdot \hat{b}_1} = \frac{2\pi\, s_1}{N_1\, \vec{a}_1 \cdot \hat{b}_1} + \frac{2\pi|\vec{b}_1|}{\vec{a}_1 \cdot \vec{b}_1}$$

$$= \frac{2\pi\, s_1}{N_1\, \vec{a}_1 \cdot \hat{b}_1} + |\vec{b}_1|$$

$$k_2 = \frac{2\pi\, s_2}{N_2\, \vec{a}_2 \cdot \hat{b}_2} + |\vec{b}_2|$$

$$k_3 = \frac{2\pi\, s_3}{N_3\, \vec{a}_3 \cdot \hat{b}_3} + |\vec{b}_3|$$

and

$$\vec{k} = \frac{2\pi\, s_1}{N_1\, \vec{a}_1 \cdot \hat{b}_1}\,\hat{b}_1 + \frac{2\pi\, s_2}{N_2\, \vec{a}_2 \cdot \hat{b}_2}\,\hat{b}_2 + \frac{2\pi\, s_3}{N_3\, \vec{a}_3 \cdot \hat{b}_3}\,\hat{b}_3$$

$$+ \vec{b}_1 + \vec{b}_2 + \vec{b}_3$$

But

$$e^{i(\vec{b}_1+\vec{b}_2+\vec{b}_3)\cdot\vec{R}_N} = 1$$

Therefore the values of $s_1 > N_1$, $s_2 > N_2$ and $s_3 > N_3$ are redundant. Q.E.D.

We can then limit k_1, k_2 and k_3 to the following values:

$$k_1 = \frac{2\pi\, s_1}{N_1(\vec{a}_1 \cdot \hat{b}_1)} \qquad s_1 = 1, 2, \ldots, N_1$$

$$\text{or } \pm 1, \pm 2, \ldots, \pm \frac{N_1}{2}$$

$$k_2 = \frac{2\pi s_2}{N_2(\vec{a}_2 \cdot \hat{b}_2)} \qquad s_2 = 1, 2, \ldots, N_2 \quad \text{or } \pm 1, \pm 2, \ldots, \pm \frac{N_2}{2}$$

$$k_3 = \frac{2\pi s_3}{N_3(\vec{a}_3 \cdot \hat{b}_3)} \qquad s_3 = 1, 2, \ldots, N_3 \quad \text{or } \pm 1, \pm 2, \ldots, \pm \frac{N_3}{2}$$

The number of states with k_1 in $(k_1 + dk_1)$, k_2 in $(k_2 + dk_2)$ and k_3 in $(k_3 + dk_3)$ is given by

$$ds_1\, ds_2\, ds_3 = \frac{N_1\, N_2\, N_3}{(2\pi)^3} [(\vec{a}_1 \cdot \hat{b}_1)(\vec{a}_2 \cdot \hat{b}_2)(\vec{a}_3 \cdot \hat{b}_3)] \times dk_1\, dk_2\, dk_3 \tag{361}$$

The infinitessimal volume element in $\vec{k}$ space is given by

$$d^3\vec{k} = dk_1\, dk_2\, dk_3(\hat{b}_1 \cdot \hat{b}_2 \times \hat{b}_3) \tag{362}$$

Therefore

$$ds_1\, ds_2\, ds_3 = \frac{N_1\, N_2\, N_3}{8\pi^3} d^3\vec{k} \frac{(\vec{a}_1 \cdot \hat{b}_1)(\vec{a}_2 \cdot \hat{b}_2)(\vec{a}_3 \cdot \hat{b}_3)}{\hat{b}_1 \cdot \hat{b}_2 \times \hat{b}_3} \tag{363}$$

But

$$\frac{(\vec{a}_1 \cdot \hat{b}_1)(\vec{a}_2 \cdot \hat{b}_2)(\vec{a}_3 \cdot \hat{b}_3)}{\hat{b}_1 \cdot \hat{b}_2 \times \hat{b}_3} = \frac{(\vec{a}_1 \cdot \vec{b}_1)(\vec{a}_2 \cdot \vec{b}_2)(\vec{a}_3 \cdot \vec{b}_3)}{\vec{b}_1 \cdot \vec{b}_2 \times \vec{b}_3} \tag{364}$$

We know that

$$\vec{a}_i \cdot \vec{b}_j = 2\pi\, \delta_{ij}$$

and

$$\vec{b}_1 \cdot \vec{b}_2 \times \vec{b}_3 = \text{volume of unit cell of reciprocal lattice} = 8\pi^3/\Omega_a$$

where

Ω_a = unit cell of "direct" lattice

Therefore

$$\frac{(\vec{a}_1 \cdot \hat{b}_1)(\vec{a}_2 \cdot \hat{b}_2)(\vec{a}_3 \cdot \hat{b}_3)}{\hat{b}_1 \cdot \hat{b}_2 \cdot \hat{b}_3} = \frac{8\pi^3}{\frac{8\pi^3}{\Omega_a}} = \Omega_a \tag{365}$$

and

$$ds_1\ ds_2\ ds_3 = \frac{N_1\ N_2\ N_3}{8\pi^3}\ d^3\vec{k}\ \Omega_a = \frac{V}{8\pi^3}\ d^3\vec{k} \tag{366}$$

because

$N_1\ N_2\ N_3\ \Omega_a = V$ = volume of the crystal

The volume of the (first) Brillouin Zone is also $8\pi^3/\Omega_a$. The number of allowed $\vec{k}$ values in this zone is

$$\frac{8\pi^3}{\Omega_\alpha}\frac{V}{8\pi^3} = \frac{N\Omega_a}{\Omega_a} = N \tag{367}$$

where $N = N_1\ N_2\ N_3$.

VI.G. Interaction of Photons with Collective Excitations

The relevant quantity in the creation or annihilation of a quantum of collective excitation via the absorption or emission of a photon is [29]

$$\langle\psi_u\Big|\sum_i \frac{q_i}{m_i}\ e^{i\vec{k}_\alpha \cdot \vec{r}_i}(\vec{\pi}_\alpha^\sigma \cdot \vec{p}_i)\Big|\psi_\ell\rangle \tag{368}$$

where $\vec{k}_\alpha$ = wave vector of the photon, $\vec{\pi}_\alpha^\sigma$ = (unit) polarization vector, and the sum extends to all the electrons in the optically active atom.

In our case

$$\psi_\ell = |u_1\ u_2\ \dots\ u_N\rangle \tag{369}$$

$$\psi_u = \frac{1}{\sqrt{N}} \sum_s e^{i\vec{k}\cdot\vec{R}_s} \left| u_1 u_2 \ldots v_s \ldots u_N \right> \tag{370}$$

We can now write

$$\left<\psi_u\right| \sum_i \frac{q_i}{m_i} e^{i\vec{k}_\alpha\cdot\vec{r}_i} (\vec{\pi}^\sigma_\alpha \cdot \vec{p}_i) \left|\psi_\ell\right> = \left<\psi_u\right| \sum_i C_i \, e^{i\vec{k}_\alpha\cdot\vec{r}_i} \left|\psi_\ell\right>$$

$$= \frac{1}{\sqrt{N}} \sum_s e^{-i\vec{k}\cdot\vec{R}_s} \left<u_1 u_2 \ldots v_s \ldots u_N\right| \sum_i C_i \, e^{i\vec{k}_\alpha\cdot\vec{r}_i} \left|u_1 u_2 \ldots u_s \ldots u_N\right>$$

$$= \frac{1}{\sqrt{N}} \sum_i \sum_s e^{-i\vec{k}\cdot\vec{R}_s} \left<v_s\right| C_i \, e^{i\vec{k}_\alpha\cdot\vec{R}_s} \, e^{i\vec{k}_\alpha\cdot\vec{r}\,'_i} \left|u_s\right> \delta_{is}$$

$$= \frac{1}{\sqrt{N}} \sum_s e^{i(\vec{k}_\alpha-\vec{k})\cdot\vec{R}_s} \left<v\right| \sum_i C_i \, e^{i\vec{k}_\alpha\cdot\vec{r}\,'_i} \left|u\right>$$

$$= \sqrt{N} \left<v\right| \sum_i C_i \, e^{i\vec{k}\cdot\vec{r}\,'_i} \left|u\right> \delta_{\vec{k}_\alpha, \vec{k}+\vec{K}_s} \tag{371}$$

where $C_i = \frac{q_i}{m_i} \vec{\pi}^\sigma_\alpha \cdot \vec{p}_i$ and $\vec{r}_i = \vec{R}_i + \vec{r}\,'_i$. Setting $\vec{K}_s = 0$ (no umklapp processes)

$$\left<\psi_u\right| \sum_i C_i \, e^{i\vec{k}_\alpha\cdot\vec{r}_i} \left|\psi_\ell\right> = \sqrt{N} \left<v\right| \sum_i C_i \, e^{i\vec{k}_\alpha\cdot\vec{r}\,'_i} \left|u\right> \delta_{\vec{k},\vec{k}_\alpha} \tag{372}$$

This selection rule is illustrated in Figure 24 which indicates that only excitations with $\vec{k} = \vec{k}_\alpha \simeq 0$ can be created in absorption and only excitations with $\vec{k} = \vec{k}_\alpha \simeq 0$ can produce the emission of a photon. We observe also that no dispersion effect can be seen because the dispersion curve of the collective excitations and that of the photons cross at *one* point.

The $\vec{k} = \vec{k}_\alpha$ rule is relaxed if more than one collective excitation is involved in the radiative process. For example, in absorption

$$\vec{k}_\alpha = \vec{k}_1 + \vec{k}_2 \tag{373}$$

would correspond to the creation of a collective excitation of wave vector $\vec{k}_1$ and of a collective excitation of wave vector $\vec{k}_2$;

$$\vec{k}_\alpha = \vec{k}_1 - \vec{k}_2 \tag{374}$$

would correspond to the creation of a collective excitation of wave vector $\vec{k}_1$ and the annihilation of a collective excitation of wave vector $\vec{k}_2$.

For a more extensive treatment of collective excitations in solids the reader is referred to reference [29] which is an introductory treatment of this subject and to the book in reference [30].

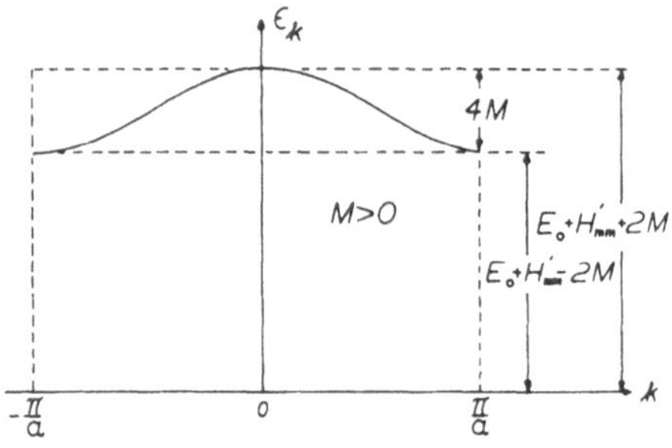

Fig. 24. Radiative processes and collective excitations.

ACKNOWLEDGEMENTS

The author wishes to thank Dr. R. K. Watts for the benefit of discussions on energy transfer processes during the 1974 NATO

Advanced Study Institute in Erice and the Plenum Publishing Corporation for the permission to reproduce some figures that appear in R.K. Watt's article in the proceedings of the above mentioned meeting.

REFERENCES

1. H. Eyring, J. Walter and G. F. Kimball, Quantum Chemistry, Wiley, New York (1944), p. 351.
2. H. Margenau, Phys. Rev. 38, 347 (1931).
3. H. Hargenau, Rev. Mod. Phys. 11, 1 (1939).
4. B. C. Carlson and G. S. Rushbrooke, Proc. Camb. Phil. Soc. 46, 626 (1950).
5. P. M. Levy, Phys. Rev. 177, 509 (1969).
6. A. J. Freeman and R. E. Watson, Phys. Rev. 127, 2058 (1962).
7. E. Clementi, IBM J. of Res. and Dev. 9, 2 (1965).
8. B. DiBartolo, Optical Interactions in Solids, Wiley, New York (1968), p. 456.
9. R.C. Powell, B. Di Bartolo, B. Birang and C.S. Naiman, in Optical Properties of Ions in Crystals, H. M. Crosswhite and H. W. Moos, eds., Interscience, New York (1967), p. 207.
10. T. Miyakawa and D. L. Dexter, Phys. Rev. B1, 2961 (1970).
11. F. Auzel, in Radiationless Processes, B. Di Bartolo, ed., Plenum Press, New York (1980), p. 213.
12. R. K. Watts, in Optical Properties of Ions in Solids, B. Di Bartolo, ed., Plenum Press, New York and London (1975), p. 307.
13. F. Perrin, Compt. Rend. 178, 1978 (1928).
14. O. Stern and M. Volmer, Physik Z. 20, 183 (1919).
15. Th. Förster, Z. Naturforsch. 4a, 321 (1949).
16. Th. Förster, Discussions Faraday Soc. 27, 7 (1959).
17. M. D. Galanin, Zh. Eksperim. i Teor. Fiz. 28, 485 (1955) [English translation: Soviet Phys. JETP 1, 317 (1955)].
18. M. Inokuti and F. Hirayama J. Chem. Phys. 43, 1978 (1965).
19. F. Reif, Fundamentals of Statistical and Thermal Physics, McGraw Hill, New York (196[illegible]) [illegible]. 483.
20. M. Yokota and O. Tanimoto, J [illegible]ys. Soc. of Japan 22, 779 (1967).
21. R. K. Watts, notes from 197[illegible] NATO ASI lectures.
22. A. I. Burshtein, Soviet Phys[illegible]cs JETP 35, 882 (1972).
23. M. V. Artamova, Ch. M. Brisk[illegible]a, A. I. Burshtein, L. D. Zusman and A. G. Skleznev, S[illegible]iet Physics JETP 35, 457 (1972).
24. N. Krasutsky and H. W. Moos, Phys. Rev. B8, 1010 (1973).
25. W. B. Gandrud and H. W. Moos, I. Chem. Phys. 49, 2170 (1968).
26. R. K. Watts and H. J. Richter Phys. Rev. 6, 1584 (1972).
27. J. T. Karpick and B. Di Bartol[illegible], J. of Luminescence 4, 309 (1971).
28. L. A. Riseberg, Phys. Rev. A7,1971 (1973).

29. B. Di Bartolo, in Collective Excitations in Solids, B. Di Bartolo, ed., Plenum Press, New York and London (1983), p. 19.

30. B. Di Bartolo, ed., Collective Excitations in Solids, Plenum Press, New York and London (1983).

MATHEMATICAL METHODS FOR THE DESCRIPTION OF ENERGY TRANSFER

V. M. Kenkre

Department of Physics and Astronomy
University of Rochester
Rochester, New York 14627

ABSTRACT

Some modern mathematical methods developed for the investigation of energy transfer are described. They are based primarily on master equations and are particularly useful for the description of coherent motion, capture, annihilation, and related phenomena involving quasiparticles such as Frenkel excitons.

I. INTRODUCTION

I.A. Preliminary Remarks

This article describes a unified framework of mathematical methods developed in recent years for the description of energy transfer in solids occuring via the motion of excitons. It is hoped that the article will fulfill two functions: the description of some modern theoretical approaches of transport theory of interest not only to exciton dynamics and energy transfer but to the broader area of quasiparticle transport, and the presentation of an overview,from the theoretical viewpoint, of Frenkel exciton motion in molecular crystals.

Although the applicability of these mathematical methods extends over a wide area, the systems of direct interest to these developments are molecular crystals. Examples are crystals of aromatic hydrocarbons such as anthracene, napththalene, and tetrachlorobenzene. The special characteristics of these systems are that the entities occupying the lattice sites in the crystal, the molecules, have complex internal structure and motion (whence intramolecular motions arise); that intermolecular interactions are weak relative to most inorganic solids; that anisotropy can prevail as a result of the non-spherical shape and orientation of the molecules; and that dynamic disorder is of paramount importance in transport phenomena. These characteristics force the transport theorist to abandon traditional methods of analysis that have been used with success for many years in fields such as that of electron transport in metals, and to look for fundamentally new formalisms. The traditional methods employ kinetic treatments in k-space. They are based on the theory of bands which are slightly perturbed by interactions with phonons or other sources of scattering which can therefore be treated as small corrections. However, in molecular crystals, the bandwidth of the moving quasiparticle, the thermal energy k_BT, phonon energies, and other interaction energies can all acquire magnitudes comparable to one another. The new methods that are described below are based on master equations, usually in real space. These master equations are of the so-called "generalized" kind as well as of the simple kind. The existence of disorder which is not a small perturbation on crystalline properties, and the fact that molecular crystals often retain the properties of the individual constituent molecules, lead to the use of real space transport equations. On the other hand, the fact that the disorder is dynamic rather than static (which would be the case for amorphous sytems), the system being still perfectly crystalline at zero temperature, leads to translationally invariant master equations being used for the analysis.

The quasiparticle whose motion brings about the process of energy transfer in molecular crystals is the Frenkel exciton. Differences of opinion exist about the convenience of the terminology used around the phrase "Frenkel exciton." Some authors (see elsewhere in this book) prefer to mean by that phrase a Bloch state of the electronic excitation of the molecules in the crystal, following early usage [1]. Other authors [2-6] look upon the Frenkel exciton as a quasiparticle (in analogy with the electron) which may occupy a delocalized Bloch state, a localized Wannier state, or any other allowable state. We find the latter usage conceptually more natural and practically more convenient and therefore employ it in this article. Thus, excitation transfer is identical to Frenkel exciton transport in this article and, if one were to consider systems with sufficient static disorder to make quasimomentum a very poor quantum number, we would still describe excitation transfer as the motion of a Frenkel exciton albeit among the sites of a disordered array.

As is well known, the subject of energy transfer is of special importance because of its obvious connections to other disciplines such as biology [7]. Energy transfer in molecular crystals derives its particular importance both from the fact that it raises basic issues about transport as mentioned above, and from the well-known fact that a molecular crystal is a solid state physicist's experimentally realizable first approximation to a complex biological system.

The present article stresses mathematical methods for the transport description, although special effort has been made to relate the contents to experimental observations.

I.B. Processes and Questions of Interest

Optical absorption can produce electronic excitations in the crystal, i.e., it can create Frenkel excitons. These excitons lead a rather eventful life before they die their radiative or radiationless death. They may undergo vibrational relaxation which may be of a simple kind as when the excited molecule relaxes among its intramolecular modes, or of a relatively dramatic kind as when an excimer, involving a drastic interaction of two (or more) molecules, is formed. They may decay through luminescence, which may be fluorescence as in the case of singlets, or phosphorescence as in the case of triplets. They may undergo internal conversion, i.e., a transition from one singlet manifold to another, or intersystem crossing as when a singlet changes into a triplet. The excitons may move from molecular site to molecular site, i.e., bring about energy transfer from one spatial location to another. If during this motion they come under the influence of traps in the crystal, which may be there either inadvertently or precisely because they were put there to detect motion, the excitons may be captured. They may also come under the influence of one another during motion and undergo mutual annihilation. The latter process is particularly striking when the moving excitons are triplets because the product of the mutual annihilation is often the formation of singlets which luminesce differently - much faster and at higher frequencies producing blue rather than red light which is typical of triplets.

A variety of questions are of interest in this field. The extent and speed of energy transfer depends on the magnitude of the diffusion constant of the excitons. Measurements of this central quantity have been made by many experimentalists over the last three decades but serious problems of interpretation remain. The value of the diffusion constant of singlet excitons in a prototype crystal such as anthracene is therefore still unknown even at room temperature although opinions abound. There is thus a disparity of several orders of magnitude in the reported values of this quantity. The temperature dependence of the diffusion constant and the nature of the underlying processes are also under question. In particular, the coherence issue, concerning whether excitons move in a wave-like coherent manner or a diffusive incoherent manner, continues to be widely debated. The validity of the simple picture of an exciton thermalizing before each transfer event, i.e., the question of whether energy transfer occurs after, before, or during relaxation, also continues to be under study. So does the connection between exciton motion and optical spectra. The validity of simple kinetic schemes for understanding exciton trapping and mutual annihilation, the role of the capture process (as differentiated from the motion process) in the former, and the existence of time-dependent versus time-independent rates of energy transfer constitute other important questions in this field.

I.C. Some Experiments

Of the large variety of experiments that have been carried out in the area of energy transfer in molecular crystals we depict schematically in Fig. 1, four kinds which use direct probes into

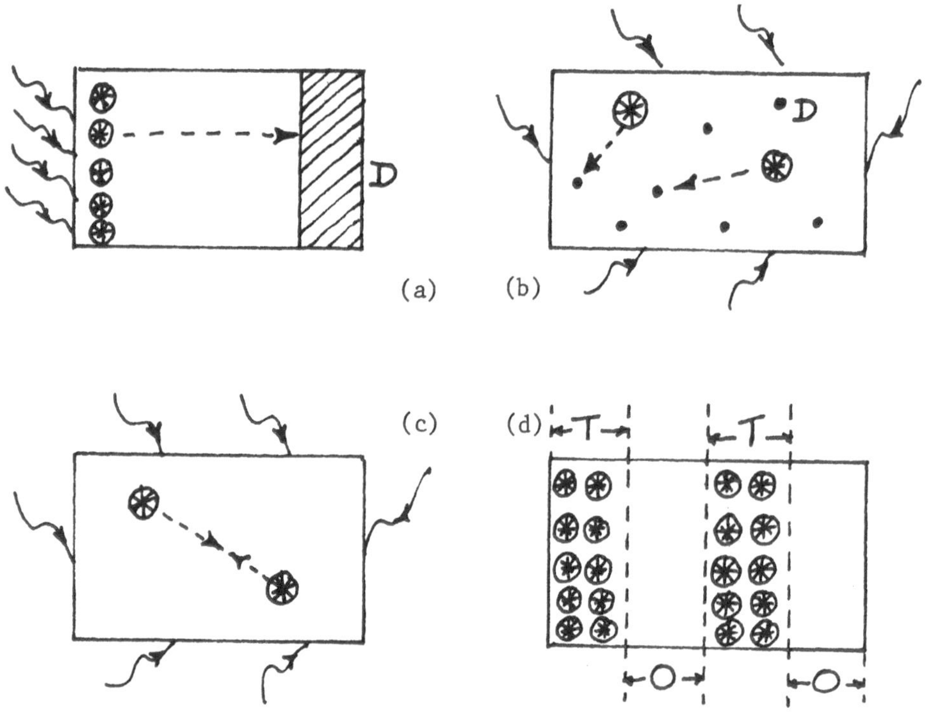

Fig. 1. Schematic depiction of four kinds of experiment for the measurement of energy transfer via exciton motion. Circled asterisks ⊛ represent excitons, wavy lines with arrows show illumination and dotted lines with arrows represent the process which allows the measurement. In (a) and (b), D represents the detector material whose luminescence is monitored, and in (d), T and O respectively depict transparent and opaque regions of the gratings.

energy transfer. Conceptually, the simplest experiment to measure energy transfer consists of starting excitons at one location in the crystal and detecting them at another. The two ends of a crystal serve as these well-defined locations in one of the oldest experiments [8] in this field. Shown in (a) in Fig. 1, the experiment uses a detector coating of a material which luminesces at a different frequency as the host crystal, illumination being at the opposite end of the crystal. Another, and more popular, kind of capture experiment is represented in Fig. 1(b) and involves homogeneous illumination of the entire crystal which is doped with other molecules serving as traps. Fig. 1(c) shows experiments which use no external agencies to probe into the motion of the excitons but employ the excitons themselves as the detectors. Their mutual annihilation is the probe process. In these capture and annihilation experiments, observables may be time-independent as in the case of steady-state quantum yields, or time-dependent as in the case of luminescence intensities. In capture observations they may refer either to the host or the guest, i.e., the traps. Fig. 1(d) represents the Ronchi grating experiments in which triplet excitons are created in the bulk of the crystal in spatially alternating regions by covering the crystal by an array of alternating opaque and transparent strips during illumination, and detecting their motion from the delayed fluorescence signal arising from their mutual annihilation. In a modern modification of this experiment which uses picosecond observations and studies singlet motion, laser beams are crossed to create a sinusoidal population of excitons, and the diffraction of a third laser beam off this population used to detect the decay of the amplitude of the inhomogeneity in a time-dependent manner.

We have not described many other important experiments used in the investigation of energy transfer simply because they are not as closely related to the mathematical methods to be developed in this article. Many excellent reviews exist on various aspects of the experiments [9-13] and should be consulted for further information.

I.D. Outline Of This Article

The basic transport equation to be used in most of the analysis, viz, the generalized master equation, is presented in section II. It contains the motivation for non-Markoffian, i.e, memory-possessing, transport equations and an explanation of their particular suitability for the systems and questions under study. Central to these transport equations are their "memory functions." Procedures for the calculation of these memory functions and explicit results for various models and interactions are exhibited in section III. Calculations of experimentally observable quantities are presented in section IV. They are directed specifically at capture and grating observations. Miscellaneous mathematical methods appropriate to energy transfer and concluding remarks form section V.

II. THE BASIC TRANSPORT INSTRUMENT: THE EVOLUTION EQUATION

II.A. Introduction and the Coherence-Incoherence Problem

Insight into the physics of the basic evolution equation to be used in the sequel can be gained by studying briefly some historical aspects of the subject. In 1932 Perrin attempted to use the Schroedinger equation among sharp molecular site states to describe excitation transfer in the context of experiments on fluorescence depolarization [14] and found clear disagreement with observations. That the problem lay in the evolution equation itself, and not in the specific transport mechanism assumed by Perrin - diplole-dipole interactions - was shown a number of years later by Foerster [15]. He recognized that the levels among which the motion transitions were occuring were not sharp but rather 'broadened' into groups of states as a result of bath (i.e., reservoir) interactions. By using a Master equation with transition rates given by the Fermi Golden Rule, with the same dipole-dipole interactions assumed by Perrin, Foerster was able to obtain excellent agreement with experiment. However, as further experiments were carried out at various temperatures with various environments and on various systems, departures from the Foerster theory were observed. The Schroedinger equation and the Master equation were clearly understood to be valid in the two extreme limits, called coherent and incoherent respectively. But one was faced with two non-trivial tasks: how to give a unified description which would reduce to the two limits and would furthermore be capable of treating the intermediate range, and how to ascertain practically which limit is applicable to a given experimental system.

The simplest way of appreciating the coherence-incoherence issue is to consider motion of the exciton in a system of just 2 sites, 0 and 1, which would have equal energies in the absence of the intersite interaction. If the latter is V, one solves a simple Schroedinger equation and shows that the probability $P_0(t)$ that the initially occupied site is occupied by the exciton at time t, is

$$P_0(t) = \cos^2(Vt) \tag{1}$$

Here and henceforth we put $\hbar=1$. Equation (1) shows oscillations, and a reversible or ringing character. However, if the 2 sites provide smeared-out (rather than sharp) levels, i.e., if each site represents a group of an extremely large number of states as a result of bath interactions, the familiar procedure is to take for the evolution equation the Master equation, the rates of transfer between the sites F being given by the Fermi Golden Rule

$$F = 2V^2/\alpha \tag{2}$$

where $1/\alpha$ contains an appropriate density of states factor in addition to other proportionality constants. The result for the probability of the initially occupied site is then

$$P_0(t) = (\tfrac{1}{2})[1 + \exp(-2Ft)] \tag{3}$$

and shows a non-oscillatory decay and an irreversible approach to equilibrium in contrast to (1). Equation (1) depicts coherent motion, while equation (3) describes incoherent motion. It is clear that these motions have entirely different character relative to each other.

A more realistic system is an infinite linear chain wherein the exciton moves via nearest-neighbour matrix elements V in the coherent case and nearest-neighbour transport rates F in the incoherent case. The evolution equation for coherent motion is the Schroedinger equation for the amplitude $c_m(t)$:

$$i\frac{dc_m}{dt} = V(c_{m+1} + c_{m-1}) \tag{4}$$

To solve (4) one multiplies it by exp(ikm) and sums over all sites m, i.e., performs a discrete Fourier transform. This is a standard mathematical procedure for the solution of translationally invariant equations such as (4). Denoting the discrete Fourier transforms by superscripts k as in

$$c^k = \sum_m c_m e^{ikm} \tag{5}$$

the interconnected equations (4) are transformed into N unconnected equations for the individual c^k's, where N is the number of sites in the crystal (infinite in the present case). Thus,

$$i\frac{dc^k}{dt} = (2V\cos k)c^k \tag{6}$$

with the immediate solution $c^k(t) = c^k(0)\exp(-i2Vt\cos k)$. If, initially, the exciton occupies a single site which, without loss of generality we shall call 0, the $c^k(0)$'s are all equal to 1 from Eq. (5). The inversion of (5) through

$$c_m(t)=(1/N)\sum_k \exp(-i2Vt\cos k)e^{-ikm} \tag{7}$$

and multiplication by the complex conjugate of c_m then give $P_m(t)$, the probability of occupation of site m. In the limit $N\to\infty$ the right hand side of (7) equals $(1/2\pi)$ times an integral over a continuous k-variable from $-\pi$ to π. One immediate obtains for this infinite chain,

$$P_m(t) = J_m^2(2Vt) \tag{8}$$

where J is the ordinary Bessel function. The probabilities exhibit oscillations as in (1), although the infinite size of the system considered destroys Poincaré recurrences evident in the 2-site result (1).

The evolution equation in the incoherent case is the Master equation

$$\frac{dP_m}{dt} = F(P_{m+1} + P_{m-1} - 2P_m) \tag{9}$$

rather than (4). This probability equation can also be solved with the use of discrete Fourier transforms. Proceeding as above,

$$P_m = (1/N) \sum_k \exp[-4Ft\sin^2(k/2)]e^{-ikm} \tag{10}$$

which is analogous to (7) in the coherent case and results in

$$P_m(t) = [\exp(-2Ft)]I_m(2Ft) \tag{11}$$

where I_m is the modified Bessel function. Unlike (8), this incoherent result shows a non-oscillatory decay.

The profound difference in the nature of the motion depicted respectively by the coherent probability propagators (8) and their incoherent counterparts (11) is also reflected clearly in the mean-square-displacement $\langle x^2 \rangle$. With a as the lattice constant, i.e., the distance between nearest neighbour sites on the linear chain, one has the general result

$$\langle x^2 \rangle = a^2 \langle m^2 \rangle = a^2 \sum_m m^2 P_m = -a^2 \left[\frac{d^2 P^k}{dk^2}\right]_{k=0} \tag{12}$$

The discrete Fourier transform of (8) is

$$P^k = J_0[4Vt\sin(k/2)] \tag{13}$$

whereas that of (11), which occurs in the process of the derivation of (11), is

$$P^k = \exp[-4Ft\sin^2(k/2)] \tag{14}$$

On combining (13), (14) with (12), one sees that the mean-square-displacement is bilinear in t for the coherent case,

$$\langle x^2 \rangle = (\sqrt{2}Va)^2 t^2 \tag{15}$$

but linear in t for the incoherent case,

$$\langle x^2 \rangle = 2(Fa^2)t \tag{16}$$

The quantities in parentheses in (15) and (16) should be familiar from standard treatments. In the incoherent case, it is the diffusion constant of the exciton, $D = Fa^2$. In the coherent case it is the square of the average over the band of the group velocity of the exciton. To recover the latter result, observe that the $c^k(t)$'s in (6) are nothing other than amplitudes in the Bloch representation, and the factor 2Vcosk in the exponent in the solution of (6) is but the band energy E_k in the tight-binding scheme. The group velocity, given by a times the k-derivative of E_k, therefore has the band average $\sqrt{2}Va$.

Fig. 2 shows the self-propagator, i.e., P_0, the probability of occupation of the initially occupied site for the purely coherent case and the completely incoherent case obtained respectively from (8) and (11). These plots as well as the expression (15) and (16) for the mean-square-displacement $\langle x^2 \rangle$ given above make clear the strong differences between coherent and incoherent motion. While we have seen that it is trivial to describe these two extreme limits, the construction of a unified framework to treat them both as well as the intermediate range presents a challenging problem. A result of the solution of this problem is seen in Fig. 2, where the intermediate self-propagator is also plotted for two given arbitrary degrees of coherence. The corresponding expression and its derivation will be found in IV. For now the form of the problem and the relevant questions should be amply clear. What is the general expression for the probability propagator which has the general unifying behaviour shown in Fig. 2 and which reduces to (8) for a system wherein the exciton does not suffer any scattering but to (11) when the scattering is so strong that the exciton motion has the aspect of a random walker? What is the general evolution equation whose respective limits are (4) and (9)? What is a practical prescription to extract the degree of coherence, i.e., the degree of the departure from the two extreme limits, for a given realistic crystal? What are the observable effects of this departure in practical experiments?

It is worth commenting in passing that the discrete Fourier transform technique explained above can be used to obtain explicit solutions in the case of long-range interactions V_{mn} or F_{mn} as well as for higher-dimensional systems. To treat the former one merely forms the Fourier transforms of V_{mn} or F_{mn} since translational invariance demands that the V's or F's are functions of the differences m-n, and proceeds exactly as shown in the case of nearest-neighbour transfer. The generalization to higher-dimensional crystals is also straightforward. The indices m, k, etc. then represent vectors of appropriate dimensionality and expressions such as km in

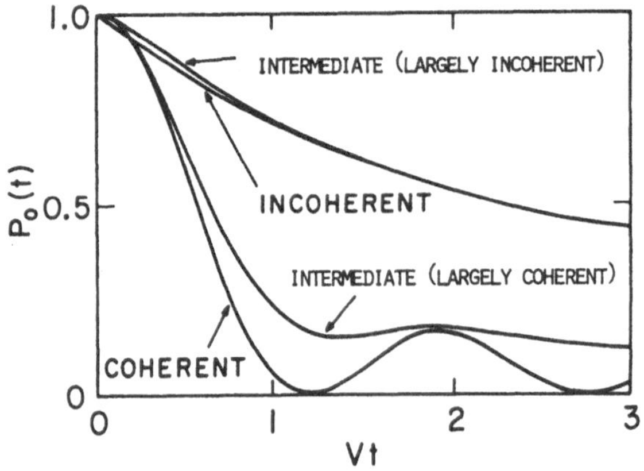

Fig. 2. The profound difference between coherent and incoherent motion shown through plots of the self-propagator, i.e., the probability of occupation of the initially occupied site, displayed as a function of (a dimensionless) time. Shown are the purely coherent case, the perfectly incoherent case and two intermediate cases, one being almost coherent and the other almost incoherent.

(5), (7), (10) represent dot products. Furthermore, for nearest-neighbour interactions in simple cubic lattices, allowing for arbitrary anisotropy, i.e., different V's or F's in different directions, the probability propagators P_m are simply products of the one-dimensional ones given in (8) and (11).

II. B. Motivation for the GME

The most natural and convenient solution of the unification problem posed in section II.A. is found in the method of the generalized master equation (GME). The essential characteristic of the GME is that it is non-Markoffian, i.e., an integro-differential equation with kernels which are non-local in time. These kernels, which are known as memory functions, give the GME its particular suitability for the analysis of the coherence-incoherence issue. In order to understand this suitability consider the two functions $y_c=\cos(wt)$ and $y_i=\exp(-Kt)$. The former oscillates and the latter decays. The single equation

$$\frac{dy(t)}{dt} + \int_0^t dt'z(t-t')y(t') = 0 \tag{17}$$

gives both y_c and y_i as solutions in the respective extreme limits $z(t)=w^2$ and $z(t)=K\delta(t)$ of the "memory kernel" $z(t)$. Furthermore, for an extremely large class of z's the solution of (17) behaves like y_c and y_i at long times. To appreciate this quantitatively, take $z(t)=w^2\exp(-\alpha t)$. One then has

$$\frac{d^2y}{dt^2} + \alpha\frac{dy}{dt} + w^2y = 0 \tag{18}$$

which shows that one recovers from (17) y_c and y_i as extreme solutions for $\alpha \to 0$ and $\alpha \to \infty$, $w \to \infty$, $w^2/\alpha = K$ respectively. These correspond to the above two choices of the memory kernel z: a constant and a δ-function respectively. Also the solution of (17), i.e. of (18) behaves like y_c and y_i respectively for times which are much smaller than, and much larger than, $1/\alpha$. Equation (18) is only a special case of (17). Generally the time for comparison ($1/\alpha$ in the simple case of the exponential z(t) corresponding to (17)) will be the characteristic time over which the kernel z decays. The initial value of z equals the square of the frequency of the oscillation of y_c while its time integral from t=0 to $t=\infty$ equals the decay constant of y_i. The unification of the oscillatory behaviour of the cosine and the decay behaviour of the exponential can thus be done with the help of (17).

We have seen the simplest possible example of the unification of oscillatory and decay behaviour via an evolution equation which is non-Markoffian. It should be immediately clear that such equations would be of value in the coherence-incoherence issue since oscillations and decay are indeed characteristic of coherent and incoherent motion respectively.

The above example corresponds to an actual physical system: the quantity y could be the amplitude of a damped harmonic oscillator with frequency w and damping constant α. Although this example does not contain site-to-site motion, one can easily include it by replacing w^2y in (18) by $-c^2(\partial^2y/\partial x^2)$. One now has the well-known wave equation and diffusion equation as the extreme (coherent and incoherent) limits which are unified by the single telegrapher's equation

$$\frac{\partial^2y}{\partial t^2} + \alpha\frac{\partial y}{\partial t} = c^2\frac{\partial^2y}{\partial x^2} \tag{19}$$

or what is the same, a memory-possessing diffusion equation with exponential memory. A perfect memory corresponds to wave-like behaviour with its oscillations and speed c, while a perfectly absent memory (a δ-function) to diffusive behaviour. The general solutions of (19) are well known to combine wave-like and diffusive behaviour. They exhibit transport coherence for times short with respect to the decay time of the memory and incoherence at large times.

All the power of evolution equations possessing memory functions would of course be useless to the issue under analysis if such equations were not natural to the exciton transport problem. In fact the telegrapher's equation (19) is of little direct use to excitons because the wave equation does not describe exciton motion in the coherent limit. The dispersion relation is quite different. However, memory-possessing evolution equations do turn out to be completely natural to exciton transport. Indeed they are actually unavoidable in the process of the derivation of the Master equation, the basis of Foerster's analysis of incoherent motion. We will therefore be able to harness their unification properties for the description of transport with arbitrary degree of coherence.

II.C. Derivation and Validity of the GME

Extensive details of the derivation and validity of the generalized master equation have been given in other reviews of the author [5,16] and will not be repeated here. Only a brief description follows.

The starting point for the evolution is the Von Neumann equation for ρ, the density matrix of the exciton along with whatever bath (phonons, imperfections, etc.) it is in interaction with. One defines projection operators $\mathcal{P}$ which diagonalize and coarsegrain the density matrix. The coarsegraining eliminates the bath coordinates and the diagonalization is in the representation of site-local states. Acting on the full density matrix ρ, the operator $\mathcal{P}$ thus yields a reduced probability vector which describes the probabilities of site occupation by the exciton. The procedure is exact, although it involves the elimination of the coordinates of a part of the total system - the bath - and although it involves only the diagonal part of the reduced density marix. The Von Neumann equation is

$$i\frac{\partial\rho}{\partial t} = [H,\rho] = L\rho \tag{20}$$

where L is the Liouville operator which, acting on any operator, produces the commutator of the full Hamiltonian H with that operator. The respective application of the projection operator $\mathcal{P}$ and of its complement $(1-\mathcal{P})$ to (20), followed by the elimination of $(1-\mathcal{P})$ from the equation involving $\mathcal{P}$ through the simple substitution of the formal solution of $(1-\mathcal{P})$, gives

$$\frac{dP\rho(t)}{dt} = -\int_0^t dt' P L e^{-i(t-t')(1-P)L}(1-P)LP\rho(t')$$

$$- iPLe^{-it(1-P)L}(1-P)\rho(0) \qquad (21)$$

This equation is always exact. Furthermore, if the full density matrix at t=0 equals the projected density matrix, i.e., $\rho(0) = P\rho(0)$, the last term in (21) is zero at all times. One then has an evolution equation for $P\rho$, i.e., for the site-occupation probabilities of the exciton, which is closed in $P\rho$ (i.e., in the probabilities), and is non-Markoffian in nature. Since P has no off-diagonal elements in the site representation, the tetradics in (21) reduce to square matrices and one obtains the probability equation

$$\frac{dP_m(t)}{dt} = \int_0^t dt' \sum_n [W_{mn}(t-t')P_n(t') - W_{nm}(t-t')P_m(t')] \qquad (22)$$

This is the generalized master equation.

The W's are the memory functions. Their functional form depends on the extent of coarsegraining as well as on the interactions in the Hamiltonian. A detailed examination of how the memory functions vary on varying the level of coarsegraining may be found elsewhere [17]. What is important to realize is that, if the initial state is such that the last term in (21) vanishes, the GME is an exact consequence of microscopic dynamics. Even if the last term in (21) does not vanish at all times but does in some practical time range, the GME becomes exact in that time range. The GME is more general than the Master equation, the basis of traditional theories such as Foerster's, in that its W's are not δ-functions. In fact the Master equation can be derived only when the approximation of replacing $W(t)$ by

$$\delta(t)[\int_0^\infty ds W(s)]$$

is made in the GME. To address the coherence issue, and to describe the exciton transport problem in a general way, one need merely refrain from making this approximation. The powers of non-Markoffian equations pointed out in section IIB are then at our disposal in an entirely natural manner.

The range of validity of the GME is decided by the validity of the passage from (21) to (22), i.e., the validity of neglecting the last term in the former. The condition $\rho(0) = P\rho(0)$ means, first, that the initial state is an outer product of an exciton state and a bath state and, second, that the exciton is initially site diagonal. The latter condition is extremely restrictive and would seldom apply in exciton physics since the creation of excitons usually involves optical absorption and, therefore, an initial state that is by no means localized. An initial Bloch state would be much more appropriate than an initial site-local state. Fortunately it can be shown [5,18] that a sufficient condition for the vanishing of the first term in (21) is that $L(1-P)\rho(0)=0$ rather than $(1-P)\rho(0)=0$. Initial Bloch state occupation or generally the initial occupation of a delocalized state does indeed result in $L(1-P)\rho(0)=0$ for crystals. The GME is thus valid for the practically occuring case of initially delocalized excitons as well as for initial localization.

To obtain the memory functions explicitly, it is necessary to evaluate the first term on the right hand side of (21). Some exact evaluations will be described in section III. Here we give the exact definition of the projection operator and a general approximate expression for the memory functions when the site-to-site interaction is small enough to be treated perturbatively.

The general definition of the projection operator P is

$$\langle\xi|PO|\mu\rangle = [\sum_{\xi\epsilon m}\langle\xi|O|\xi\rangle][\sum_{\xi\epsilon m}1]^{-1}Q_\xi\delta_{\xi,\mu} \quad (23)$$

where the ξ,μ are eigenstates of the full exciton-bath system, O is any operator, Q is arbitrary except for being subject to the condition

$$\sum_{\xi\epsilon m}Q_\xi = \sum_{\xi\epsilon m}1$$

to ensure that P is idempotent. The summation of ξ in the "grain" m is the coarsegraining operation and involves the elimination of the bath coordinates. The Q's do not affect the expresssions for the memory functions in an exact calculation but do affect them in calculations involving approximations. If the Hamiltonian is $H_0 + V$, the part H_0 being site-diagonal, the memory functions are given perturbatively by

$$\mathcal{W}_{mn}(t) = 2\sum_{\xi\epsilon m}\sum_{\mu\epsilon n}[Q_\mu/g_n]|\langle\xi|V|\mu\rangle|^2\cos[(E_\xi-E_\mu)t] \quad (24)$$

$$\mathcal{W}_{nm}(t) = 2\sum_{\xi\epsilon m}\sum_{\mu\epsilon n}[Q_\xi/g_m]|\langle\xi|V|\mu\rangle|^2\cos[(E_\xi-E_\mu)t] \quad (25)$$

where $g_m = \sum_{\xi\epsilon m}1$ and $g_n = \sum_{\mu\epsilon n}1$.

Equations (24) and (25) are coarsegrained generalizations [16,17] of memory expressions given originally by Zwanzig [19].

II. D. Solution of Foerster's Problem

If the interaction V responsible for exciton motion is dipole-dipole in nature, its dependence on the intersite distance R is given by $1/R^3$. The rate of transfer calculated through the Golden Rule is proportional to the square of V and has therefore the distance dependence $1/R^6$. However, if the motion is coherent, the oscillations of probability would be characterized by a frequency which is proportional to V at least for a two-site system. It was often thought, therefore, that the distance dependence of transfer rates in the coherent limit should be $1/R^3$. Foerster sharpened this form of the coherence question by presenting a plot of the transfer rate versus the intersite distance R. He stated that the dependence was $1/R^6$ for large R, i.e., for weak enough V for perturbation theory in terms of the Golden Rule to be valid, and that one might argue that the dependence was $1/R^3$ for small enough R which would make the V overwhelm the bath broadening. He further hoped that a theory of transfer rates which could bridge the two limits would be available [20].

The unification of transfer rates which Foerster had hoped for is possible in a natural way through the GME. No details of the memory functions are necessary. To stress the extreme simplicity of the argument, assume that the memory functions are of the separable form $\mathcal{W}_{mn}(t) = F_{mn}\phi(t)$, the F's being the transition rates in the corresponding Master equation, which one would arrive at on replacing $\phi(t)$ by a δ-function. One first realizes that it is necessary to define a transfer rate unambiguously and not merely take it to be a transition rate in one case and a frequency in the other. To that end one calculates the mean-square-displacement $\langle x^2\rangle$ of the exciton for an initial localized condition, finds the time for which it becomes equal to the square of the lattice constant, and defines the reciprocal of that time as the rate of transfer w. Obviously this definition is sensitive to the time taken by the exciton to move one site irrespective of the degree of coherence. It is straightforward to see that this definition leads to

$$\int_0^{1/w} dt \int_0^t dt'\phi(t') = [\sum_m m^2 F_m]^{-1} \equiv [\overline{m^2}]^{-1} \tag{26}$$

The unification is apparent from (26). For incoherent transfer is a δ-function, the first integration gives a constant and the rate w is proportional to $\overline{m^2}$ and for nearest neighbour F's to R^{-6}. For coherent transfer ϕ is a constant, the first integration gives t, and w^2 is proportional to $\overline{m^2}$ and thus the rate w is proportional to R^{-3}. To calculate an explicit expression for the R-dependence of the rate, assume a simple expression for $\phi(t)$ such as an exponential

$$\phi(t) = \alpha e^{-\alpha t} \tag{27}$$

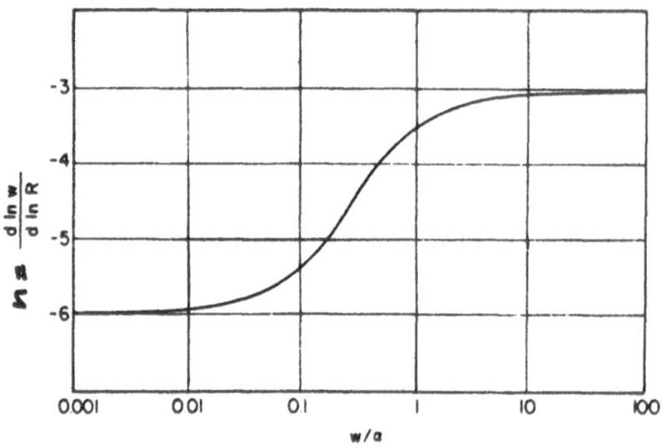

Fig. 3. The solution of Foerster's problem demonstated through a plot of the exponent of Rin the transfer rate for the case of dipolar interaction, as a function of the transfer rate in units of α. The inflection point is at $\omega/\alpha \approx$ 0.22.

along with a simple expression such as $(2V^2/\alpha)(\delta_{m,n+1}+\delta_{m,n-1})$ for F_{mn}, representative of nearest-neighbour interactions on a linear chain. Substitution in (26) gives the implicit expression

$$(\alpha/w) + \exp[-(\alpha/w)] - 1 = (\alpha^2/2V^2) \tag{28}$$

With V = constant $\cdot$ R^{-3}, a plot of the rate w versus the intersite distance R can be given from (28) and Foerster's problem solved explicity [21]. In Fig. 3 is shown a plot of the exponent n in w = constant $\cdot$ R^n, defined [21] as

$$n = \frac{d\ell n(w)}{d\ell n(R)} = 6\left(\frac{w}{\alpha} - \frac{1}{1-e^{-\alpha/w}}\right) \tag{29}$$

This plot has found use in analyzing excitation transfer in some biological systems [21,22].

II.E. General Remarks about the GME

The characteristic features of the GME are that it is an exact consequence of the microscopic dynamics for some initial conditions, that its memory functions may be obtained from knowledge of the microscopic interactions at least in principle, and that its non-Markoffian nature makes it especially adapted to the analysis of the coherence-incoherence issue. The structure of the GME approach is as follows. One calculates the memory functions of the GME from the microscopic interactions whenever possible. These calculations may be exact as is the case in a small number of model systems, or they may involve standard perturbation techniques. The memory functions may also be obtained in some cases directly from experimental observations which do not involve transport, an example being optical spectra. With the memory functions as an input, one calculates probability propagators. These appear directly in various experimental observables. A connection is thus established between memory functions and observables. Memory functions which are long-lived generally correspond to coherent motion and little or no bath interactions, although factors other than the latter also affect the decay. Rapidly decaying memory functions signal incoherent transport. The decay characteristics of the memories are reflected in the propagators and therefore in the observables. There are experiments such as the grating ones shown in Fig. 1d which probe directly the Fourier transforms of the propagators and are therefore highly sensitive to the degree of coherence in the system. While any realistic non-pathological system behaves coherently at short enough times and incoherently at long enough times, it is thus possible to measure the degree of coherence quantitatively by analyzing observations with the help of the GME framework outlined above.

III. MEMORY FUNCTIONS: EXPLICIT CALCULATIONS

III.A. Outline

Exact evaluation of the memory functions for a realistic system is obviously out of the question since such an evaluation would be tantamount to an exact solution of the dynamics of the full complex system. Model calculations are therefore undertaken as elsewhere in physics with the hope that in simplifying the mathematical problem the model does not sacrifice the essential features of the system under consideration. Such exact model calculations are to be found in III.B. and III.C. below. It is also necessary to perform approximate calculations of realistic systems which defy exact solution. Here the hope is that the approximation procedures employed do not destroy the essential features of the system. Section III.D. provides an example. An attractive result in the GME theory is that it is sometimes possible to obtain the memory functions for a real system directly from observations in a different realm. This is described in section III.E.

III.B. Exact Results for Pure Crystals

The general expression for the memory function $\mathcal{W}_{nm}(t)$ for motion in a crystal of arbitrary dimensionality and size (but obeying periodic boundary conditions) is [5,23]:

$$\mathcal{W}_{mn}(t)=-\int d\varepsilon e^{\varepsilon t} \sum_k \{ e^{-ik(m-n)} / \sum_q [\varepsilon+i(V^{k+q}-V^q)]^{-1} \} \qquad (30)$$

where the ε-integration is on the Bromwich contour, where m,n are direct lattice vectors, k, q are vectors in the first Brillouin zone of the reciprocal lattice of the crystal, km represents a dot product, the k and q summations are within the first Brillouin zone, and V is the discrete Fourier transform of the interaction matrix elements $V_{mn} = V_{m-n}$, the only peculiarity of the expressions relative to standard usage being that k,q,m,n are dimensionless in (30) as also elsewhere in this review.

Translational invariance, i.e., a true crystalline environment, is the only requirement to obtain (30). The proof is as follows. The Schroedinger equation for transport in the crystal is

$$\frac{dc_m(t)}{dt} = -i \sum_n V_{mn}\, c_n(t) \tag{31}$$

the site energies being taken to be zero without loss of generality. Equation (4) representing motion on a linear chain is a particular case of (31) and so is the 2-site equation whose solution leads to (1). Discrete Fourier transforms in the manner of section II.A. lead to the solution of $c^k(t)$ and thence through a Fourier inversion to $c_m(t)$. For the initial condition that the exciton occupies a single site, which we label zero, the solution gives, when multiplied by its complex conjugate,

$$P_m(t) = (1/N) \sum_{k,q} e^{-it(V^k - V^q)}\, e^{-im(k-q)} \tag{32}$$

If we define the quantities $\mathcal{A}_{mn}$ as equal to $-\mathcal{W}_{mn}$ for $m \neq n$, with $\mathcal{A}_{mm} = \sum_n \mathcal{W}_{nm}$, the GME (22) takes on the form

$$\frac{dP_m(t)}{dt} + \int_0^t dt' \sum_n \mathcal{A}_{mn}(t-t')\, P_n(t') = 0 \tag{33}$$

A discrete Fourier transform, a Laplace transform, and the initial condition stated above which leads to $P^k(0)=1$, yield from (33)

$$\tilde{\mathcal{A}}^k(\varepsilon) = [1/\tilde{P}^k(\varepsilon)] - \varepsilon \tag{34}$$

The calculation of the memory functions is now immediate on substituting the transform of (32) in (34) and using the relation

$$\tilde{\mathcal{W}}^k(\varepsilon) = -[\varepsilon + \tilde{\mathcal{A}}^k(\varepsilon)] \tag{35}$$

which follows from the above definition of the $\mathcal{A}$'s. The derivation of (32) is thus complete without the need to disentangle the projection operator expression (21).

It is possible for one to have two misconceptions about this derivation: that it is useless because it assumes knowledge of the probability solutions which it is the function of the GME and of its memory functions to arrive at, and that it cannot be correct since it claims to obtain the N(N-1) quantities $\mathcal{W}_{mn}$ from the N quantities P_m. The latter misconception is easily removed by observing that in a translationally invariant system (which alone is under consideration) the quantities $\mathcal{W}_{mn}$ are functions of m-n, there being thus only N independent $\mathcal{W}$'s or A's. The other misconception will disappear in section III.C. when this calculation will be put to use to obtain results which are extremely hard to get without its help and for which the probability solutions are certainly not known beforehand.

Particular cases of the exact general result (30) are presented in the table below. The interaction is characterized by the single matrix element V in all cases and is of nearest-neighbour range in all cases but the last one. In that last case the interaction is V between <u>any</u> two sites. A system which can be said both to have such a universal range in its interaction and to be of the nearest-neighbour kind, as in the case of the others presented in the table, is the trimer (3 sites). All the systems shown above obey periodic boundary conditions, i.e., have no ends or surfaces. The crystals being all pure (no bath interactions), the motion is perfectly coherent in all cases.

The result for the dimer (2 sites) shows the constant memory familiar from the pedagogical examples given in section II.A. Introduction of bath interactions can indeed be shown to cause the decay of this memory. In section III.C. we shall see that the

Table I

NO. OF SITES IN THE CRYSTAL	RANGE OF INTERACTION	MEMORY FUNCTIONS
2	-	$\mathcal{W}_{12} = \mathcal{W}_{21} = 2V^2$
3	-	$\mathcal{W}_{12} = \mathcal{W}_{23} = \mathcal{W}_{31} = 2V^2\cos(tV\sqrt{3})$
4	nearest-neighbour between sides 1 and 2, 2 and 3, 3 and 4, and 4 and 1.	$\mathcal{W}_{12} = \mathcal{W}_{23} = \mathcal{W}_{34} = \mathcal{W}_{41} = 2V^2\cos(tV2\sqrt{2})$ $\mathcal{W}_{13} = \mathcal{W}_{24} = 4V^2\sin^2(tV\sqrt{2})$
linear chain)	nearest-neighbour	$\mathcal{W}_{mn}(t) = \frac{1}{t}\frac{d}{dt}\left[J^2_{m-n}(2Vt)\right]$
N	equal among all sites	$\mathcal{W}_{mn}(t) = 2V^2\cos[tV\sqrt{N(N-2)}]$

exponential decay referred to in II.A. is quite physical in origin. However, it must not be concluded that coherent motion is always accompanied by constant memory functions. The result for the trimer already shows that the memory generally oscillates in the case of coherent motion. The frequency of this oscillation happens to be zero for a dimer. For crystals of finite size, true decay of memory functions does not occur unless some degree of incoherence is introduced. But the infinite chain result shows that the memories can decay even for purely coherent motion as a consequence of the destruction of Poincaré cycles brought about by the infinite size of the system. An alternative form of $\mathcal{W}_{mn}(t)$ for the infinite linear chain is given below

$$\mathcal{W}_{mn}(t)=2V^2[J^2_{m-n+1}+J^2_{m-n-1}+2J_{m-n-1}J_{m-n+1} -2J^2_{m-n}-J_{m-n}(J_{m-n+2}+J_{m-n-2})] \tag{36}$$

The J's are all Bessel functions of argument 2Vt.

III.C. Exact Results for an SLE

A transport equation that has often appeared [2,3,6,24] in the analysis of exciton motion as well as in other transport contexts is the stochastic Liouville equation (SLE). A form of the SLE is

$$\frac{\partial\rho_{mn}}{\partial t} = -i\sum_r(\tilde{V}_{mr}\rho_{rn} - \tilde{V}_{rn}\rho_{mr}) - (1-\delta_{m,n})\,\alpha\rho_{mn} + \delta_{m,n}\sum_r(\gamma_{mr}\rho_{rr}-\gamma_{rm}\rho_{mm}) \tag{37}$$

where ρ is the exciton density matrix; m,n,etc., represent site-localized states as always in this review; $\tilde{V}$'s are the intersite interaction matrix elements; α represents scattering and is the rate at which the off-diagonal elements of ρ (in the m,n representation) decay; and the γ's are additional rates of incoherent transfer. In microscopic derivations such as Silbey's [3], one naturally attaches the following meaning to these various quantities: $\tilde{V}$'s are proportional to the bandwidth of the exciton dressed with phonons, in other words to the bandwidth of the excitonic polaron; α arises from scattering of phonons and other sources; and the γ's are phonon-assisted rates. An exact calculation of the memory functions appearing in the GME corresponding to (37) is possible [5,25] and leads to (22) with the $\mathcal{W}$'s given by

$$\mathcal{W}_{mn}(t) = \mathcal{W}^c_{mn}(t)e^{-\alpha t} + \gamma_{mn}\delta(t) \tag{38}$$

In (38) the quantities $\mathcal{W}^c$ are the purely coherent memory functions corresponding to (37) in the absence of α and of the γ's. The expression for them is given in (30) above.

The proof of (38) is facilitated by the introduction of some operator manipulations involving the projection operators $\mathcal{P}$. One rewrites the SLE as

$$i\frac{\partial\rho}{\partial t} = L_c\rho + L_i\rho + L_a\rho \tag{39}$$

where L_c represents the first term in the right side of (37) which describes coherent motion, L_i represents the second term in (37) which describes the primary source of incoherence, and L_a represents the third term in (37) which describes the "assisted" transport. The application of the diagonalizing projection operator $\mathcal{P}$ to (39) in the manner described in section II.C. leads to a slightly modified form of (21):

$$\frac{d\mathcal{P}\rho(t)}{dt} = \mathcal{P}L_a\mathcal{P}\rho(t) - \int_0^t dt'\mathcal{P}(L_c+L_i)e^{-i(t-t')(1-\mathcal{P})(L_c+L_i)}$$

$$\times\ (1-\mathcal{P})(L_c+L_i)\mathcal{P}\rho(t') \tag{40}$$

where we have dropped the initial term involving $(1-\mathcal{P})\rho(0)$ as in section II.C. With the definition

$$\mathcal{O}'' = (1-\mathcal{P})\mathcal{O} \tag{41}$$

for any operator $\mathcal{O}$, the identity

$$\{\exp[-it(1-\mathcal{P})(L_c+L_i)]\}\mathcal{O}''$$

$$= [1+(-it)(L_c''-i\alpha) + \ldots]\mathcal{O}'' = e^{-\alpha t}\exp[-it(1-\mathcal{P})L_c] \tag{42}$$

follows for any off-diagonal operator $\mathcal{O}''$. This remarkably simple result is a consequence of the fact that $L_i'' = (1-\mathcal{P})L_i$, acting on any off-diagonal operator, merely multiplies it by $-i\alpha$. Equation (42) when substituted in (40), immediately produces the first term in the memory function result (38). The other part of the memory function in (38) follows directly from the term L_a in (40) and is the complete contribution of L_a.

This calculation of the memory functions for the SLE (37) is exact. It illustrates the method of direct computation with projection operators and clarifies the questions raised in section III.B. about the usefulness of the calculation of the coherent memories $\mathcal{W}^c$. Although the latter are obtained from knowledge of the probability solutions in the coherent case, the result (38) for the full memory functions in the presence of the scattering α and of the "phonon-assisted" transport signified by the γ's has been obtained without such knowledge. The probability solutions in the presence of the α and the γ's do not bear a very simple relation to those in their absence and must be computed by solving the GME after the result (38) is obtained. On the other hand the memory functions in the two cases are simply related. One merely multiplies the coherent memory functions by the exponential $e^{-\alpha t}$ and adds the δ-function terms $\gamma_{mn}\delta(t)$ to obtain the expresssion valid in the SLE case.

The physics underlying the SLE is that of the coexistence of two channels of transport: band or coherent transport represented by the V's which is interrupted by scattering events controlled by α, and diffusion-type or incoherent transport represented by the γ's. It is extremely satisfying that this coexistence is reflected so clearly in the expresssion for the memory function as a sum of two terms. The effect of scattering appears as a decay with time constant $1/\alpha$ superimposed on whatever time dependence the coherent memory function $\mathcal{W}^c$ has. Such a clean separation of the contributions of the two scattering mechanisms also appears in transfer rates or diffusion constants obtained from the SLE [5,6] since they merely involve the integration of the memories from $t = 0$ to $t = \infty$.

III.D. Perturbative Evaluation for Linear Exciton-Phonon Coupling

The Hamiltonian H given by

$$H = E_o \sum_m a_m^\dagger a_m + \sum_{m\neq n} V_{mn} a_m^\dagger a_n + \sum_q \omega_q b_q^\dagger b_q$$

$$+ \sum_{m,q} g_q \omega_q (b_q + b_{-q}^\dagger) a_m^\dagger a_m e^{iqm} \tag{43}$$

where a and b destroy respectively an exciton and a phonon, and where ω_q is the phonon frequency, is an important and useful model for the description of exciton transport in realistic systems. While the evaluation of memory functions is trivial when the exciton-phonon coupling term - the last term in (43) - is negligible or relatively small, the case when it dominates requires a transformation to be carried out prior to the application of a perturbative formula such as (24). The transformation [3,26] is designed to eliminate the coupling and it is said to dress the excitons with phonons, giving rise to excitonic polarons. The perturbative formula (24) is then applied to the residual interaction which is treated as a small quantity.

The transformation is given by the relation

$$Z = \{\exp[\sum_{m,q} g_q (b_q - b_{-q}^\dagger) a_m^\dagger a_m e^{iqm}]\} \; z \; \{\exp[-\sum_{m,q} g_q (b_q - b_{-q}^\dagger) a_m^\dagger a_m e^{iqm}]\} \tag{44}$$

where Z is the transformed operator corresponding to any operator z. It gives rise to the excitonic polaron operators A_m given by

$$A_m = a_m \exp[- \sum_q g_q (b_q - b_{-q}^\dagger) e^{iqm}] \tag{45}$$

Their form obviously justifies the statement that the polaron consists of the bare particle surrounded by a cloud of phonons. The new (displaced) phonon operators B_q are given by

$$B_q = b_q + \sum_m g_q a_m^\dagger a_m e^{-iqm} \tag{46}$$

The Hamiltonian H is now expressed as

$$H = \sum_m [E_0 - \sum_q g_q^2 \omega_q] A_m^\dagger A_m + \sum_q \omega_q B_q^\dagger B_q + \sum_{m \neq n} V_{mn} A_m^\dagger A_n e^{\alpha_m^\dagger} e^{\alpha_n} \tag{47}$$

where

$$\alpha_m = \sum_q g_q (b_q - b_{-q}^\dagger) e^{iqm}.$$

Application of the perturbative formula (24) with the last term of (47) as the perturbation leads [27] in the case of a dimer to the memory function expression

$$\mathcal{W}_{mn}(t) = 2|V_{mn}|^2 \exp\{-\sum_{r,s} [h_{rs}(t) - h_{rs}(0)]\} \tag{48}$$

where r and s each take the values m and n, and h_{rs} is given by

$$h_{rs}(t) = -\sum_q 4g_q^2 \sin^2 q(r-s)[N_q e^{i\omega_q t} + (N_q+1)e^{-i\omega_q t}] \tag{49}$$

N_q being the average number of phonons given by the Bose distribution $[\exp(\omega_q/k_B T)-1]^{-1}$.

Equation (48) is the generalization of the pure dimer results displayed in the table of section III.B. to the case of exciton-phonon interactions as described in the Hamiltonian of (43) or (47).

III.E. Evaluation from Spectra

The rate of transfer F_{mn} for singlet exciton transport is given by the theories of Foerster [15] and Dexter [28] in a form which is extremely convenient from a practical point of view. In these theories the F's are proportional to the spectral overlap of the emission of the donor and the absorption of the acceptor. The great advantage of such a prescription is that, when valid, it allows one to bypass model calculations and assumptions and to connect exciton transport directly to another experimental realm, viz, optical spectra. In situations wherein the Foerster-Dexter mechanism of transport is valid but the Master equation formalism underlying the Foerster-Dexter theory is not, the memory functions can be obtained through a simple generalization [5,29] of their prescription. The generalization is based on the fact that an expression such as (24) for the memory function is a straightforward generalization of the corresponding Golden Rule expression for the transfer rates F:

$$F_{mn} = 2\pi \sum_{\xi\epsilon m} \sum_{\mu\epsilon n} [Q_{\mu}/g_n] \, | \langle\xi|V|\mu\rangle|^2 \; \delta(E_{\xi} - E_{\mu}) \tag{50}$$

The only difference between (24) and (50) is that $\cos(E_{\xi}-E_{\mu})t$ appears in the former where $\pi\delta(E_{\xi}-E_{\mu})$ appears in the latter.[5] Indeed the latter can be obtained from (24) by replacing each cosine by a δ-function in t, times the integral from t=0 to t =∞ of the cosine. This is the Markoffian approximation necessary to convert the GME into the ordinary Master equation and is responsible for the fact that, while the GME is able to describe transport at short times, the Master equation is not. To gain the capability of providing a short-time (i.e. coherence) description while retaining the basic mechanism of transfer, one need therefore make only the necessary modifications in the Foerster-Dexter formula and obtain

$$\mathcal{W}_{mn}(t) = \text{constant} \cdot \frac{1}{R^6_{mn}} \int_{z=-\infty}^{+\infty} dz \, \cos(zt) \int_{\omega-0}^{\infty} d\omega \, \frac{A(\omega-z)E(\omega+z)}{(\omega-z)^3(\omega+z)} \tag{51}$$

The constant factor in (51) is unimportant for the present discussion. The quantities A and E are the absorption and emission spectra respectively and R is the intersite distance, the sixth power being characteristic of the diple-dipole interaction. The prescription implied by (51) is as follows. One obtains F_{mn} as given by the Foerster-Dexter prescription, renames it $f_{mn}(0)$, recalculates it after displacing the two spectra on the frequency axis by z/2 and

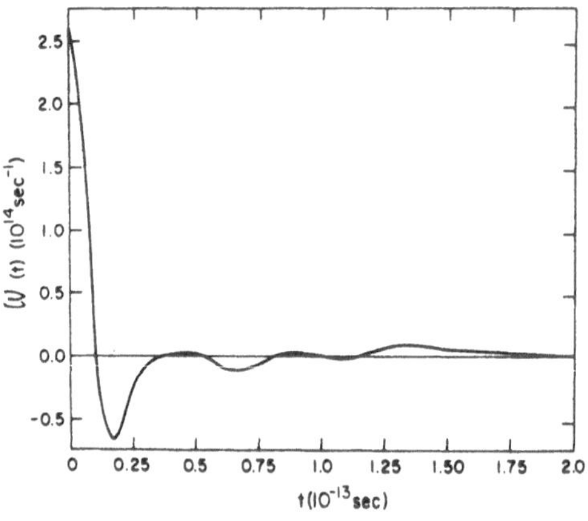

Fig. 4. The time dependence of the memory function for exciton transport among anthracene molecules in cyclohexane solution at room temperature obtained from the spectral prescription of (51). Ordinates are chosen in a way to normalize the memory function.

-z/2 respectively, renames the results $f_{mn}(z)$ and repeats for all values of z. The Fourier cosine transform of the function $f_{mn}(z)$ thus obtained is the memory function $\mathcal{W}_{mn}(t)$.

An intimate relation thus exists between spectra and memory functions. Narrow spectra generally correspond to long-lived memories and therefore to coherent behaviour in the transport whereas bath interactions cause incoherence as well as spectral broadening. Well-known restrictions exist on the applicability of the Foerster-Dexter theory, particularly when inhomogeneous broadening is important. These restrictions have been discussed elsewhere in this book and must certainly not be ignored in the use of the above procedure. However, when the restrictions do not apply, (51) provides a direct method of extracting memory functions from experiment. As an example of the application of (51), Fig. 4 shows the time-dependence of the memory function for exciton motion from one anthracene molecule to another in cyclohexane solution at room temperature.

IV. CALCULATION OF OBSERVABLES

IV.A. Prelude: Calculation of Propagators

The previous sections of this article have set up the basic evolution equation, the GME, and shown how its memory functions are obtained. Now we shall use that equation to address experiment. Of central relevance to the experimental quantities, particularly in the context of the experiments described schematically in Fig. 1, are the probability propagators ψ, which are nothing other than special solutions of the GME for initially localized conditions or, what is the same, the Green functions of the GME. Thus, by $\psi_m(t)$ is meant the probability that the exciton occupies the site m at time t, given that it occupied site 0 and time 0. Although for a disordered system with no translational invariance, the propagator would depend explicitly both on the site of initial occupation and that of later interest, the crystalline nature of our system makes the propagator a function of the single index m, the difference between the indices representing the two locations in question. Some of the observations require knowledge of $\psi_m(t)$, while others are related to its discrete Fourier transform $\psi^k(t)$ and yet others probe their respective Laplace transforms $\tilde{\psi}_m(\varepsilon)$ and $\tilde{\psi}^k(\varepsilon)$. Thus, the grating experiments depicted in Fig. 1d are directly sensitive to the time dependence of the propagator in the Fourier domain, i.e., to $\psi^k(t)$ and the capture experiments shown in Fig. 1b probe the self-propagator in the Laplace domain, i.e. $\tilde{\psi}_0(\varepsilon)$. The GME itself is an integro-differential difference equation with a variable upper limit and a difference t-kernel in the time integration. The t-structure of the equation suggests the use of the Laplace transform for its solution while the crystalline nature, i.e., the properties in m,n-space, suggest the discrete Fourier transform.

Thus, purely calculational considerations focus one's attention on the propagator in the Laplace and Fourier domain. It is a delightful accident that some of the experiments probe the transformed propagators directly and thereby save the theorist the often troublesome - and always tedious - task of inverting the transforms.

One begins then with the GME (22) and uses the procedure already outlined in section III.A.: One calculates the $A_{mn}(t)$'s from the memory functions $W_{mn}(t)$'s, obtains the Fourier transform $A^k(t)$, and uses its Laplace transform in the following general relation between propagators and memory functions which is a trivial consequence of the GME itself:

$$\tilde{\psi}_m(\varepsilon) = (1/2\pi)^d \int dk e^{-ikm}[\varepsilon + \tilde{A}^k(\varepsilon)]^{-1} \tag{52}$$

Here d is the number of dimensions of the crystal, taken to be infinite in extent, and the integration is in d-dimensional k-space. In arriving at (52) one also encounters the propagator in the Fourier domain:

$$\tilde{\psi}^k(\varepsilon) = [\varepsilon + \tilde{A}^k(\varepsilon)]^{-1} \tag{53}$$

Equations (52) and (53) show how the various characteristics that memory functions possess enter into the behaviour of the propagators, i.e., the solutions of the GME, and therefore into that of observable quantities. Thus, if the memory functions are short-lived, the $\tilde{A}$'s are largely constant in ε-space, the ψ^k's are exponential, and the behaviour of the ψ_m's is the same as from a Master equation. It exhibits no coherence. For highly coherent systems on the other hand, $\tilde{A}$'s are far from constant in ε-space, the time dependence of the propagators is profoundly different from that of solutions of a simple Master equation and can indeed exhibit oscillations characteristic of coherence.

In order to study the effects of coherence on observables we shall choose the simplest possible evolution capable of describing exciton transport of arbitrary degree of coherence. It has the form

$$\frac{\partial \rho_{m\,n}}{\partial t} = -iV(\rho_{m+1\,n} + \rho_{m-1\,n} - \rho_{m\,n+1} - \rho_{m\,n-1}) - (1-\delta_{m,n})\alpha\rho_{m\,n} \tag{54}$$

It was first used for exciton transport by Avakian et al. [30] and can be considered to be a particular case of the SLE (37). In the absence of the scattering α, the off-diagonal elements of the density matrix do not decay and the motion of the exciton is purely coherent. The evolution is exactly the same as that of (4) if the system is taken to be a linear chain, and the propagators are given by (8). One can show that, if the scattering is very large and justifies the limit $\alpha \to \infty$, $V \to \infty$, $2V^2/\alpha = F$, (54) reduces to the Master equation (9) and the motion is perfectly incoherent. The underlying picture in (54) is "band motion" with a bandwidth proportional to V and scattering at the rate α. This can be made particularly clear by transforming (54) to k-space. The diagonal part of ρ in the k-representation follows the "Boltzmann" equation

$$\frac{\partial \rho^{kk}}{\partial t} = (\alpha/N) \sum_q (\rho^{qq} - \rho^{kk}) \tag{55}$$

which shows scattering with equal scattering rates (α/N) among all k-states, N being the number of sites in the crystal.

It is possible, in principle, to analyze the observables with any GME with any degree of complexity by following the procedure outlined earlier. However, to understand the coherence issue, it is important not to be distracted by other elements of the evolution which are not essential to the issue. All the calculations in this section will be made, therefore, from (54). The GME corresponding to (54) has the memory functions [5]

$$\mathcal{W}_{mn}(t) = [\frac{1}{t}\frac{d}{dt} J^2_{m-n}(2Vt)]e^{-\alpha t} \tag{56}$$

as is clear from (36). The corresponding $\tilde{A}^k(\varepsilon)$ is given by

$$\tilde{A}^k(\varepsilon) = [(\varepsilon + \alpha)^2 + 16V^2\sin^2(k/2)]^{\frac{1}{2}} - \varepsilon \tag{57}$$

The discrete Fourier transform of the square of the ordinary Bessel function with respect to the space index is given quite simply. Thus

$$\sum_{m=-\infty}^{+\infty} J^2_m(x)e^{ikm} = J_0(2x\sin|k/2|) \tag{58}$$

It is this simple result that allows the effortless passage from (56) to (57). Equation (53) yields, on Laplace inversion,

$$\psi^k(t) = e^{-\alpha t} J_0(bt) + \int_0^t du\alpha e^{-\alpha(t-u)} J_0(b\sqrt{t^2-u^2}) \tag{59}$$

where $b = 4V\sin|k/2|$. The derivation of (59) uses a well-known theorem in Laplace-transform theory which allows inversion of transforms of the form $\tilde{f}[(\varepsilon^2 + c^2)^{\frac{1}{2}}]$ in terms of known transforms of $\tilde{f}(\varepsilon)$. Here f is any function and c is some ε-independent quantity. To obtain the real-space propagators one must evaluate the Fourier-inverse of (59). For this purpose one uses (58) in reverse. Then

$$\psi_m(t) = J^2_m(2Vt)e^{-\alpha t} + \int_0^t du e^{-\alpha(t-u)} J^2_m(2V\sqrt{t^2-u^2}) \tag{60}$$

Equations (59) and (60) constitute the solutions of the GME

corresponding to (54) and, with their counterparts in the Laplace domain, enter directly into the description of experimental observables. They exhibit oscillations when the degree of coherence is high, i.e., when α is small, and show incoherent behaviour when α is large. The former case is obtained trivially by taking the limit of small α in (59), (60); the latter is obtained by expressing the square root in (57) in the form of a Binomial expansion, and retaining the lowest power of V/α. The cases (13) and (14) are thus recovered as extreme limits of (60). The solutions for intermediate degree of coherence displayed along with those extreme limits in Fig. 2 are given by (60).

IV.B. Application to Grating Experiments

The grating experiment of Fig. 1d consists of creating a periodic inhomogeneity in the spatial distribution of the excitons and measuring the time evolution of that inhomogeneity [31]. The spatial inhomogeneity is produced simply by covering the crystal with an array of alternating opaque and transparent strips during illumination. The excitons under study are triplets. The measurement of the time evolution can therefore be made by monitoring the delayed fluorescence which arises as a result of the formation of singlets through the mutual annihilation of the triplets. It is possible to make the illumination strength small enough to make the annihilation a negligible perturbation on the evolution of the exciton distribution and yet large enough to make the signal clearly discernible. The characteristics of the motion of the excitons is reflected in the time dependence of the delayed fluorescence. Exponential decay after illumination would be typical of incoherent motion. Oscillatory features would characterize a high degree of coherence.

The evolution equation is the GME (22) corresponding to (54) but with its right hand side augmented by two terms: a radiative term $-P_m(t)/\tau$ where τ is the exciton lifetime and a term $S_m(t)$ which describes the spatial and temporal dependence of the illumination. The former term merely multiplies the GME solutions by $\exp(-t/\tau)$, or replaces ε in the Laplace domain by $\varepsilon' = \varepsilon + 1/\tau$. The source term $S_m(t)$ introduces an additional driven contribution in the solution. Thus the solutions in the Laplace domain are now

$$\tilde{P}_m(\varepsilon) = \sum_n \tilde{\psi}_{m-n}(\varepsilon')P_n(0) + \sum_n \tilde{\psi}_{m-n}(\varepsilon')\tilde{S}_n(\varepsilon) \tag{61}$$

The Ronchi grating experiment consists of three parts. In the build-up part, in which the delayed fluorescence signal builds up to its saturation value, $P_n(0)$ is identically zero as there are no excitons initially, and (61) gives, through the Fourier transform,

$$\tilde{P}^k(\varepsilon) = (1/\varepsilon)i_0 g^k \tilde{\psi}^k(\varepsilon') \tag{62}$$

Here the illumination is switched on to a constant value which, when multiplied by the appropriate absortion coefficient, equals i_o, and has a spatial dependence g_m characteristic of the ruling geometry. The second part of the experiment is concerned with the steady-state-value of the signal. The limit of $P^k(t)$ as $t\to\infty$ is given from (62) as

$$\lim_{t\to\infty} P^k(t) = i_o g^k \tilde{\psi}^k(1/\tau) \tag{63}$$

The third part of the experiment consists of observing the time dependent decay of the signal from the steady state value. To calculate this decay, one returns to (61). There is no driving term S now, but the $P_n(0)$ are given by (63). Thus, for the decay

$$\tilde{P}^k(\varepsilon) = [i_o g^k \tilde{\psi}^k(1/\tau)]\tilde{\psi}^k(\varepsilon') \tag{64}$$

It suffices for illustrative purposes to study only this decay stage (64). We see that, except for the time-independent quantity in the square brackets in (64), the probability solution is given in the time domain by $\exp(-t/\tau)$ times that given in (59) above.

The measured quantity is the delayed fluorescence signal which is proportional to

$$\sum_m [P_m(t)]^2$$

and therefore to

$$\sum_k P^k(t)P^{-k}(t)$$

following a standard Fourier result. It is therefore clear that, if the illumination were to excite a single Fourier component k, the delayed fluorescence signal would be given essentially by squaring the result (59) for $\psi^k(t)$. It is thus that the grating experiment contains a direct probe of the time dependence of the Fourier transform of the propagator.

The actual Ronchi grating experiment does not populate a single Fourier component but several ones with amplitudes given by the Fourier transform of the square wave, since a mask is employed. The expression for g_m and, therefore, for g^k in (62)-(64) is written down trivially. Straightforward calculations [32] lead then to explicit expressions for the normalized decay signal of delayed fluorescence. Careful experimentation requires that this signal be subtracted from one in the absence of the mask in order to eliminate spurious contributions. The difference signal thus obtained is given by $\Delta\phi(t)$ with

$$\Delta\phi(t) = e^{-2t/\tau} E(t) \tag{65}$$

$$E(t) = [1 + \sum_{\ell=1}^{\infty} A_\ell]^{-1} [\sum_{\ell=1} A_\ell \, [\psi^{k_\ell}(t)]^2 - 1\}] \tag{66}$$

$$A_\ell = 8[\pi^2(2\ell-1)^2]^{-1}[(1/\tau)\tilde{\psi}^{k\ell}(1/\tau)]^2 \tag{67}$$

The quantity k_ℓ appearing in (61)-(63) is the dimensionless wave-vector given by

$$k_\ell = 2\pi(a/x_0)(2\ell-1) \tag{68}$$

with a as the lattice constant of the crystal and x_0 as the period of the ruling, i.e., of the mask. The dynamics of the exciton is reflected in the ψ's appearing in (66) and (67). The grating period decides which Fourier components appear in the expression.

The time dependence of E(t), equivalently that of the delayed fluorescence difference signal which differs from E(t) only by the factor $\exp(-2t/\tau)$, is completely controlled by the time dependence of the Fourier transform of the propagator. For incoherent motion the latter is an exponential as given by (14) and -E(t) rises from 0 to the value

$$[1+\sum_\ell A_\ell]^{-1}[\sum_\ell A_\ell]$$

monotonically. For purely coherent motion, i.e., when the exciton suffers no scattering, -E(t) rises in an oscillatory fashion. The

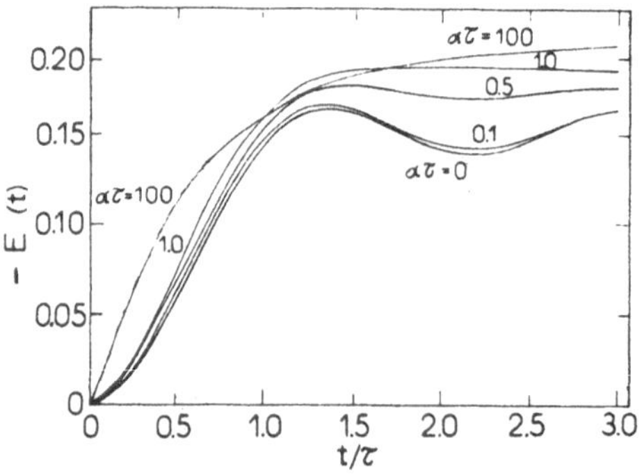

Fig. 5. The delayed fluorescence signal for the decay stage of the Ronchi ruling experiment times the factor $\exp(2t/\tau)$ plotted as a function of time for several degrees of exciton coherence. The effect of coherence is seen in the characteristic shape near the origin and in oscillations. See equation (66).

spatial inhomogeneity disappears monotonically in the former case but overshooting with resultant oscillations occurs in the latter. Fig. 5 shows the evolution of -E(t) for various degrees of coherence typified by several values of $\alpha\tau$. The latter parameter is the ratio of the exciton lifetime to the time between scattering events.

The measured quantity is $\Delta\phi(t)$ rather than E(t). The clear oscillations seen in E(t) are not always seen in $\Delta\phi(t)$ because the exponential factor in (65) generally suppresses them. However, oscillations are not the only characteristic of coherence. It is manifested also in the shape near the origin - concave versus convex - of the time dependence. Quantitative analysis for representative crystals has shown [32] that this latter effect would be quite discernible in realistic systems. Fig. 6 shows the actual measurable quantity $-\Delta\phi$, rather than the quantity E(t), plotted for the extreme limits of pure coherence and complete incoherence for several values of the quantity, ℓ_T/x_o, which is the ratio of the exciton transport length ℓ_T to the ruling period x_o. The transport length ℓ_T is a generalization, to arbitrary degree of coherence, of the well-known diffusion length. For the present system it is given by [32]

$$\ell_T = 2a\ (V/\alpha)(\alpha\tau - 1 + e^{-\alpha\tau})^{\frac{1}{2}} \tag{69}$$

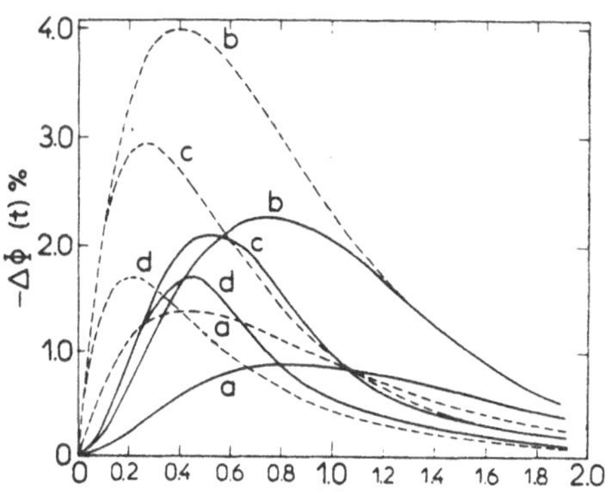

Fig. 6. Delayed fluorescence decay signal $-\Delta\phi(t)$ plotted as a function of the dimensionless time t/τ for the extreme cases of pure coherence and complete incoherence. Curves a, b, c, d, refer, respectively, to the values 0.05,0.15,0.35,0.45 of ℓ_T/x_o, the ratio of the transport length to the ruling period. Solid lines represent the purely coherent case, and the dashed lines the completely incoherent case. Curves of the latter kind have been already observed experimentally [31].

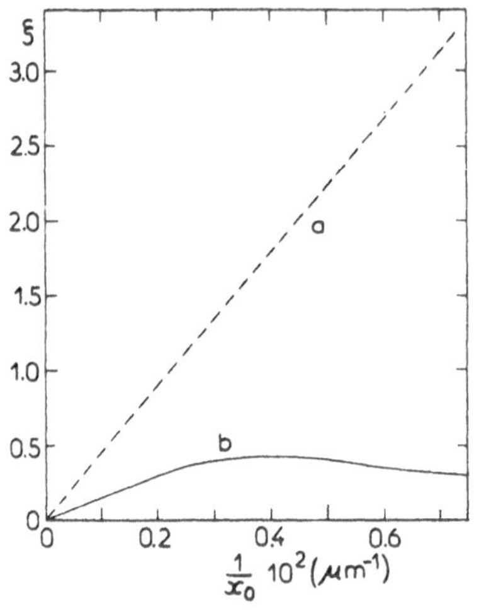

Fig. 7. Plot vs $(1/x_0)$, the spatial frequency of the rulings, of the dimensionless quantities $\xi_{coh} = 2\pi\sqrt{2}(\ell_T)_{coh}/x_0$ for the purely coherent case (curve a) and $\xi_{inc} = \pi\sqrt{2}(\ell_T)_{inc}/x_0$ for the completely incoherent case (curve b), as obtained by graphically inverting the relevant expressions for the same "observed" signal $-\Delta\phi(t)$. The latter is obtained by fitting curves such as those in Fig. 6. The straight-line behaviour a indicates that the correct theory has been used to interpret the measurements. The clear departure of curve b from straight-line behaviour shows that the theory used for b is incorrect. The straight line a corresponds to a transport length ℓ_T=50 μm.

The measurable shown in Fig. 6 is quite different in the coherent and incoherent cases. Attempts to fit one with any of a family of curves of the other kind show [32] such poor results that experimental differentiation would be quite unambiguous. We exhibit this consequence graphically in Fig. 7.

The question of coherence has been debated for a long time in the literature. However, clear experimental methods for its measurement have not been developed. A careful study along the lines outlined in the present section has recently shown [32-34] that the Ronchi grating experiment is an excellent candidate for this task. This had not been realized earlier, although grating experiments have had a long history [31]. It is expected that such experiments will be carried out in the near future.

IV.C. Capture Experiments

The capture experiment represented by Fig. 1b employs homogeneous bulk illumination of the host crystal under study which is doped with guest molecules. These capture the excitons moving in the host and luminesce at a frequency different from that of the host. Monitoring the luminescence from the host and/or guest allows one to measure exciton motion in the host. An enormous amount of information has been gathered over the years through the use of this technique [9,13,35,36].

The evolution equation is the GME (22) once again with two terms appended to its righthand side: $-P_m(t)/\tau$ to describe decay as in IV.B. and a term which represents capture by the guests or traps. The simplest form of the latter term is

$$-c \sum_r{}' P_r \delta_{m,r}$$

and means that whenever the exciton is at one of the trap-influenced host sites r (over which the primed summation runs), its probability decays into the trap from the host at rate c. To solve the GME with these terms, a new mathematical technique needs to be introduced. This is the defect technique of Montroll [37]. The Laplace transform of the augmented GME gives an equation similar to but different from (61)

$$\tilde{P}_m(\varepsilon) = \tilde{\eta}_m(\varepsilon') - c\sum_r{}' \tilde{\psi}_{m-r}(\varepsilon')\tilde{P}_r(\varepsilon) \tag{70}$$

Here, as in (61), $\varepsilon' = \varepsilon + 1/\tau$ and η is the homogeneous solution, i.e., the first term on the righthand side of (61).

The experimental observable is the total illumination intensity which is proportional to

$$n_H(t) = \sum_m P_m(t),$$

the probability that the host is excited. Summation of (70) over all host sites m gives

$$\tilde{n}_H(\varepsilon) = \frac{1}{\varepsilon'} [1 - c\sum_r{}' \tilde{P}_r(\varepsilon)] \tag{71}$$

and shows that n_H is simply related to the probability that the trap-influenced host region is excited. However, the latter cannot be generally evaluated since, from (70),

$$\sum_r \tilde{P}_r(\varepsilon) = \frac{\rho}{\varepsilon'} - c\sum_s{}' \tilde{\nu}_s(\varepsilon')\tilde{P}_s(\varepsilon) \tag{72}$$

where the initial illumination has been assumed explicitly homogeneous as in usual experiments, where ρ is the trap concentration, i.e., the ratio of the trap-influenced host sites to the total number of host sites, and ν is defined as

$$\nu_s(t) = \sum_r{}' \psi_{r-s}(t) \tag{73}$$

An exact evaluation is possible for small ρ since then one can use the single-trap solution of (71)-(73). The ν-function is then equal to the self-propagator $\psi_0(t)$ and (71) is easily evaluated. An important observable is the steady-state yield ϕ_H defined as the ratio of the number of photons emerging radiatively from the host to that put initially through illumination and is given by

$$\phi_H = (1/\tau) \int_0^\infty dt\; n_H(t) = [\tilde{n}_H(\varepsilon)]_{\varepsilon=0} \tag{74}$$

for the simple case when the radiative lifetime equals the total lifetime. For the single-trap case one gets

$$\phi_H = 1 - \frac{\rho\tau}{(1/c\tau)+(1/\tau)\tilde{\psi}_0(1/\tau)} \tag{75}$$

Thus, in the capture experiment, characteristics of exciton motion influence the observable, in this case ϕ_H, through the self-propagator $\tilde{\psi}_0(1/\tau)$. To study the effect of coherence, (60) may be used as in IV.B. The propagator to be calculated is the m=0 case of (60) evaluated in the Laplace domain. The result is [38]

$$\tilde{\psi}_0(\varepsilon) = \frac{\alpha}{[(\varepsilon^2+2\varepsilon\alpha)(\varepsilon^2+2\varepsilon\alpha+16V^2)]^{1/2}} + \frac{(2/\pi)}{[(\varepsilon+\alpha)^2+16V^2]^{1/2}}\mathbb{K}(k)$$

$$+ \frac{1}{[(\varepsilon+\alpha)^2+16V^2]^{1/2}} \cdot \frac{(2/\pi)}{(\varepsilon^2+2\varepsilon\alpha+16V^2)^{1/2}}\,\Pi(a_1^2,k) \tag{76}$$

where

$$a_1^2 = 16V^2(\varepsilon^2+2\varepsilon\alpha+16V^2)^{-1}, \tag{77}$$

$$k = 4V[(\varepsilon+\alpha)^2+16V^2]^{-1/2} \tag{78}$$

and $\mathbb{K}$ and Π are elliptic integrals of the first and third kinds respectively defined through

$$\mathbb{K}(b) = \int_0^1 dx(1-x^2)^{-1/2}(1-b^2x^2)^{-1/2} \tag{79}$$

$$\Pi(a_1^2,b) = \int_0^1 dx(1-x^2)^{-\frac{1}{2}}(1-b^2x^2)^{-\frac{1}{2}}(1-a_1^2x^2)^{-1} \tag{80}$$

In the coherent limit $\alpha \to 0$, (76) gives

$$(1/\tau)\tilde{\psi}_0(1/\tau) = (2/\pi)\ (1+16V^2\tau^2)^{-\frac{1}{2}}\mathbb{K}[4V\tau(1+16V^2\tau^2)^{-\frac{1}{2}}] \tag{81}$$

whereas in the incoherent limit $\alpha\to\infty$, $V\to\infty$, $2V^2/\alpha = F$, one has

$$(1/\tau)\tilde{\psi}_0(1/\tau) = (1+4F\tau)^{-\frac{1}{2}} \tag{82}$$

For exciton transport $V\tau >> 1$ and $F\tau >> 1$ usually applies: the excitons cover a distance of many lattice constants before they die. It is then possible to express (81) and (82) more simply. The elliptic integral reduces to a logarithmic expression and (81) gives

$$(1/\tau)\tilde{\psi}_0(1/\tau) \approx (1/4\pi V\tau)\ \ell n\ (256V\tau) \tag{83}$$

as the key quantity in this experiment for coherent transport. On the other hand, for incoherence, (82) gives

$$(1/\tau)\ \tilde{\psi}_0(1/\tau) \approx 1/(2\sqrt{F\tau}) \tag{84}$$

These results show that the quantity examined by these capture experiments is ℓ_T/a, the ratio of the transport length (see (69)) to the lattice constant. For higher concentration of traps the probed quantity can be shown to be ℓ_T/ℓ_θ where ℓ_θ is the distance between traps.

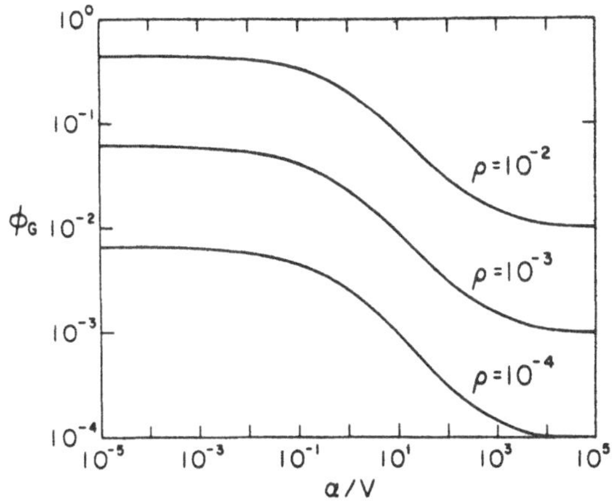

Fig. 8. Guest yield plotted as a function of the (in)coherence parameter α/V, i.e., the ratio of the lattice constant to the mean free path, for several values of the trap concentration.

To treat the multi-trap case one returns to (72) and by assuming s-independence of ν_s writes an average form of (73):

$$\nu_s(t) = \nu(t) = \sum_m \psi_m(t) p_m \qquad (85)$$

Here p_m is the probability that the mth host-site is trap-influenced given that the 0th is; it will be called the trap pair correlation function. The problem is now immediately solved, the generalization of (75) to the multi-trap case being

$$\phi_H = 1 - \frac{\rho\tau}{(1/c\,\tau)+(1/\tau)\tilde{\nu}(1/\tau)} \qquad (86)$$

For random placement of traps, it follows that [39]

$$\nu(t) = \rho + (1-\rho)\psi_0(t) \qquad (87)$$

Fig. 8 is the guest yield, i.e., $1-\phi_H$ obtained from (86) and (87), the exciton dynamics being given by (54), plotted to show the effect of coherence. The coherence parameter is V/α, which is proportional to the mean free path of the exciton in units of the lattice

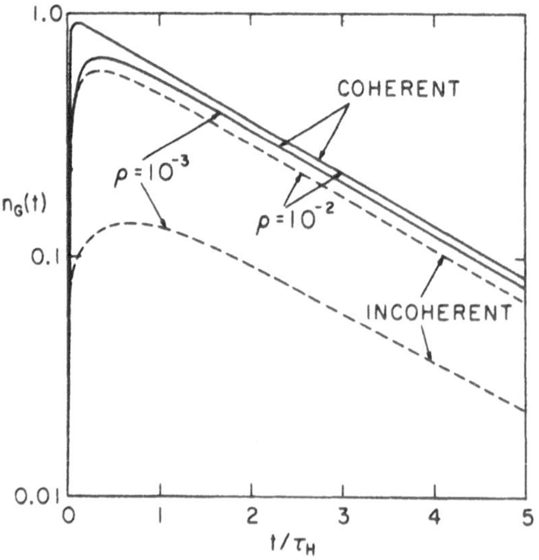

Fig. 9. Guest luminescence intensity plotted as a function of time for two values of the trap concentration to show the explicit effect of transport coherence. The extreme limits of completely coherent motin and completely incoherent motion are shown. The parameter values are $V\tau=1.8\times10^5$, $F\tau=1.8\times10^4$, and $c\tau=10^5$.

constant. Higher degree of coherence is seen to result in more efficient trapping but no dramatic differences comparable to those in Figs. 5-7 in the grating context occur here. Fig. 9 shows the time dependence of guest luminescenc obtained from the solution of (71) by using numerical inversion of the Laplace transform but again no striking difference occurs [39].

The fact that such a pronounced difference exists between the qualitative effects of coherence for capture experiments on the one hand and grating observations on the other suggests that one compare the two analyses. In both kinds of experiment one begins with

$$\frac{dP_m(t)}{dt} + \frac{P_m(t)}{\tau} + \int_0^t dt' \sum_n A_{mn}(t-t')P_n(t') = 0 \tag{88}$$

In the grating analysis one adds an annihilation term which gives rise to the delayed fluorescence signal but neglects it in finding solutions of (88). In the capture analysis one adds a capture term which, however, cannot be neglected. This difference is in keeping with the experimental situation. No information can be gathered about exciton motion if c in (70) is put equal to zero.

The initial condition on the exciton population is different for the two kinds of experiment. In the grating case it is inhomogeneous but periodic. In the capture case it is homogeneous. It is certainly possible to make it inhomogeneous in the latter case but the experiments carried out so far use homogeneous illumination or populations near one end. The latter case [5] is not treated in the present article.

The observed signal measures

$$\sum_k P^k P^{-k} \text{ , i.e., } \sum_k \psi^k \psi^{-k} g^k g^{-k}$$

in the grating experiment but is sensitive to $\sum_m \psi_m P_m$ in the capture observations. Here g_k is the Fourier transform of the spatial dependence - square wave - of the initial illumination and is controlled by the ruling period. The corresponding control in the capture case is exercised by the trap concentration ρ on the trap pair correlation function p_m. Thus, for random placement of traps,

$$p_m = \rho + (1-\rho)\delta_{m,0} \tag{89}$$

The control in the capture case is, however, much weaker than in the grating case because it is not as systematic. The rulings have a periodic arrangement. The doping by guest or trap molecules is random. This is the reason that dramatic manifestations of coherence are possible in the grating case but not in capture experiments. The quantity to measure in these experiments is the diffusion length of excitons if the extent of transport is under study while it is the mean free path if the quality of transport (i.e., degree of coherence) is under study. The measuring unit employed in capture observations of the kind discussed is the inter-trap distance and is a random quantity. The measuring unit in grating observations is the ruling period which is a fixed quantity. In modern singlet grating experiments it appears possible to make the grating measurement unit even more systematic as it involves the selection of a single Fourier component of the P's by illuminating the crystal with crossed laser beams [5,40] instead of through a ruling.

V. MISCELLANEOUS METHODS AND CONCLUSIONS

V.A. Methods for Cooperative Trap Interactions

The mathematical technique of the GME and the issue of exciton transport coherence are well matched to each other, play a central role in the subject of energy transfer, and have been explained in sections II-IV. However, the subject poses other important issues and requires other useful mathematical methods. Although space considerations do not allow us to treat them all, an attempt is being made to describe one of them and briefly mention some of the others. The present section contains techniques used in capture situations with a high concentration of traps with particular reference to the difficult question, seldom discussed in the literature, of the effect of cooperative interactions among the traps, i.e., among the guest molecules in sensitized luminescence observations.

The interesting feature of this treatment is that it combines the methods of non-equilibrium statistical mechanics, i.e. transport theory, with techniques of equilibrium statistical mechanics, particularly Ising model arguments. Although time dependent observables can be analyzed as well as steady state quantities [39,41,42], attention will be restricted to the latter for simplicity. Central to this analysis are (86) which relate the observable, the yield, to the ν-function, and (85) which expresses the ν-function in terms of the p_m and the ψ_m. The powerful feature of (85) is that the observables can be calculated by combining two distinct parts of the system under study: (i) the dynamic part which involves the motion of the exciton in the pure host as described by the propagators ψ_m, and (ii) the static part which involves the placement of the traps in the doped crystal as described by the pair correlation function p_m. Calculation of the propagators has been illustrated in earlier sections. A procedure for obtaining the pair correlation function p_m follows.

If one represents the crystal under investigation by a lattice gas, one can exploit the well-known analogy between the latter and the Ising model, and use known results for spin correlation functions in the Ising model. The host sites are treated as the lattice gas sites, and are considered "occupied" if they are trap-influenced and "unoccupied" otherwise. The interaction between any two trap-influenced host sites is characterized by an energy which equals infinity if the two sites coincide, $-\Delta$ if they are nearest neighbours, and zero in all other cases. It follows that the quantities Δ and $2\rho-1$, where ρ is the trap concentration, respectively correspond to the magnetic interaction $4J$ and magnetization M of the Ising model. The spin-spin correlation function $\langle\sigma_o\sigma_m\rangle$ is related to the p_m we seek, through

$$p_m = 1 - (1/4\rho)(1-\langle\sigma_o\sigma_m\rangle) \tag{90}$$

We shall restrict ourselves to a one-dimensional system. Ising model arguments and the above correspondence then give

$$p_m = \delta_{m,0} + (1-\delta_{m,0})[\rho+(1-\rho)x^{|m|}] \tag{91}$$

The details of the computation of $\langle\sigma_o\sigma_m\rangle$ may be obtained from extensions of standard textbook calculations and therefore have not been

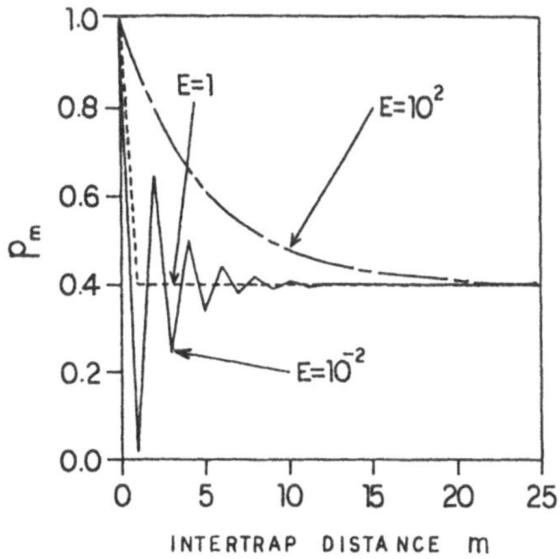

Fig. 10 The trap pair correlation function p_m plotted as a function of the dimensionless intertrap distance m for attractive, repulsive, and no interaction among the traps ($E=10^2$, 10^{-2}, 1 respectively). The value of the concentration has been arbitrarily made 0.4.

shown here. The quantity x in (91) is defined by

$$x = (y - 1)/(y + 1) \tag{92}$$

and can take values between -1 and 1 depending on the nature of the trap-trap interaction and the value of the concentration. Here y is given by

$$y = [1-4\rho(1-\rho)(1-E)]^{\frac{1}{2}} \tag{93}$$

and E, the trap-trap interaction parameter, equals $\exp(\Delta/k_BT)$ where T is a characteristic temperature. For no interactions Δ vanishes, E equals 1, x vanishes since y equals 1, and (91) reduces to the random placement result (87). Attractive interactions lead to the formation of clusters while repulsive ones generally cause desert regions to form. Fig. 10 shows the pair correlation function for three values of E showing typical spatial oscillations for repulsive interactions and slow decay for attractive ones.

The ν-function is obtained by combining (91) with (85):

$$\nu(t) = \rho + (1-\rho)\sum_{m=-\infty}^{\infty} x^{|m|}\psi_m(t) \tag{94}$$

and reduces to (87) for random trap placement. The sum in (94) can be evaluated in simple cases. Thus, for incoherent motion with nearest-neighbour transfer rates F as in (9) or (11), the ν-function gives, for the relevant quantity appearing in (86)

$$(1/\tau)\tilde{\nu}(1/\tau) = \rho + (1-\rho)[\tanh(\xi/2)][\tanh\{(\xi+\mu)/2\}]^{-\mathrm{sgn}(x)} \tag{95}$$

The factors $\tanh(\mu/2)$ and $\tanh(\xi/2)$ respectively equal y and $(1/\tau)\tilde{\psi}_0(1/\tau)$ where ψ_0 is, as before, the self-propagator. The behaviour of ν, and therefore that of the observables, is controlled by the interplay of the two quantities ξ and μ. They characterize the motion of the exciton and the interaction among the traps respectively and are given by

$$\xi = \cosh^{-1}[1 + (1/2F\tau)] \tag{96}$$

$$\mu = \ell n\ (1/|x|) \tag{97}$$

Simple physical meaning can be ascribed to these two quantities in certain limits. Thus, for excitons in molecular crystals usually $F\tau >> 1$ with the consequence that ξ equals $(4F\tau)^{-\frac{1}{2}}$ to an excellent approximation. Except for a factor of $\sqrt{2}$, ξ, therefore, equals the ratio of the lattice constant a to the diffusion length $\sqrt{2Fa^2\tau} = \sqrt{D\tau}$. It can be similarly shown [41] that, except for unimportant proportionality constants, μ equals the ratio of the lattice constant to the effective distance over which the trap-trap interaction extends. The effective distance is defined as one over which the pair correlation function p_m falls off by some characteristic amount. The reciprocal of μ therefore measures how far the effective trap-trap interaction extends, whereas the reciprocal of ξ measures how far the exciton can travel in its lifetime.

Fig. 11 shows the result of the application of this theory to compute the observable, the guest quantum yield ϕ_G which is obtained from (86). Specifically,

$$\phi_G = \rho\tau[(1/c\tau) + (1/\tau)\tilde{\nu}(1/\tau)]^{-1} \tag{98}$$

with (95)-(97). One draws the conclusion that trapping is more efficient for repulsive interactions.

The present technique can be used for arbitrary degree of transport coherence (whose effect will be felt through the propagators appearing in (87)) and also for arbitrary strength of the capture rate. However, it involves the averaging approximation inherent in the use of the ν-function theory [5,39]. An exact

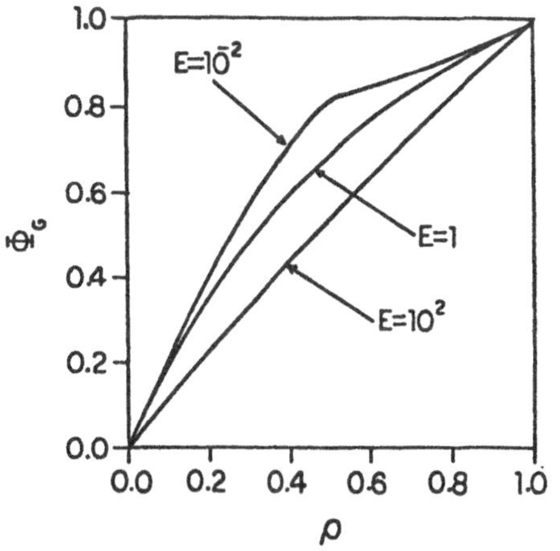

Fig. 11. The guest (trap) yield ϕ_G plotted as a function of the guest concentration ρ for the three values of trap-trap interactions corresponding to Fig. 10. Repulsive interactions are seen to lead to enhanced efficiency of capture, whereas attractive interactions inhibit capture.

solution which avoids all approximations is possible [42] if one is willing to sacrifice the ability to cover arbitrary degree of coherence and capture. For incoherent one-dimensional exciton motion with nearest-neighbour rates F and nearest-neighbour trap-trap interactions characterized by as above, the exact expression for the guest yield for the case of infinite capture rates is

$$\phi_G = \rho_e \sum_{N=1}^{\infty} \rho_e (1-\rho_e)^{N-1}$$

$$[\tanh(\xi/2)]^{-1}\{\tanh[(N+1)\xi/2]-\tanh(\xi/2)\} \tag{98}$$

where ρ_e is an "effective" trap concentration which equals $\rho(1-x)$, with x as in (92).

V.B. Conclusion

There is a large number of important questions and techniques which it has not been possible to describe in this review. One of those techniques addresses mutual annihilation of excitons. The procedure [5] is to consider the evolution of the system point representing two interacting (annihilating) excitons in a space of twice the number of dimensions as the real crystal. The annihilation problem reduces then to the capture problem explained in section IV. Another question concerns the applicability of traditional expressions for the so-called energy transfer rate and the time dependence of the latter. For many years sensitized luminescence observations had been interpreted in terms of a simple kinetic scheme involving an energy transfer rate which described the transfer of energy from the host to the guest [9]. Later experimentss reported a time dependence of this rate [13] and various theoretical and experimental investigations of this time dependence began. The most recent developments on this question are as follows. The systems that were reported earlier [13] to exhibit time dependence in the energy transfer rate have been found [36] to have a time-independent rate. The earlier observatinns appear to have been an experimental artifact. However, time dependence in the rate is seen [43] for one-dimensional systems. The unified theoretical framework described in this article is able to reconcile quantitatively both of these observations (time-dependent rates in the experiments of ref. [43] and time-independent rates for the system of ref. [36]), as is clear from detailed fits carried out recently [44]. A related new development [45] is the concern that usual exciton capture experiments probe capture rather than motion parameters and a feeling is developing that a large amount of information gathered over the years may be of little relevance to exciton motion.

Another issue in this field is that of the interplay of energy transfer with vibrational relaxation. The natural mathematical technique to study it is to employ two interlocked master equations

[5], one for the relaxation and the other for transfer. It is possible to investigate in this way the dependence of energy transfer on the wavelength of initial excitation which could be observable if relaxation occurs on a scale comparable to or slower than transfer.

A number of other techniques exist including those involving a return to k-space equations such as the Boltzmann equation. The interested reader is referred to several reviews by the author for detailed [5] as well as perspective [46-48] descriptions and also to other reviews cited in this article.

ACKNOWLEDGEMENTS

I thank Profesor B. DiBartolo for inviting me to Erice and for conducting a truly delightful summer school. I thank Professor V. Ern and Dr. A. Fort for the invitation to, and hospitality at, Strasbourg where a part of this review was completed. It is a pleasure to acknowledge my debt to my students and other collaborators, most of all to Paul E. Parris. This work was supported in part by the National Science Foundation under grants DMR-8111434 and INT-8210098.

As this article went to press, news arrived of the passing away of an esteemed friend and greatly renowned scientist who, in his active life, had made numerous pioneering contributions to the mathematics of energy transfer. Indeed, many of the techniques described in this article owe their origin to him. It is difficult to imagine the community of statistical mechanics without Elliott Montroll. This article is dedicated to his memory.

REFERENCES

1. D. L. Dexter and R. S. Knox, Excitons (Interscience, New York, 1965).
2. H. Haken and G. Strobl in The Triplet State, ed. by A. B. Zahlan (Cambridge University, Cambridge, 1967).
3. R. Silbey, Ann. Rev. Phys. Chem. 27, 203 (1976).
4. R. S. Knox in Collective Excitations in Solids, ed. by B. DiBartolo (Plenum, New York, 1981).
5. V. M. Kenkre in Exciton Dynamics in Molecular Crystals and Aggregates, ed. G. Hoehler (Springer-Verlag, Berlin, 1982).
6. P. Reineker in Exciton Dynamics in Molecular Crystals and Aggregates, ed. G. Hoehler (Springer-Verlag, Berlin, 1982).
7. R. S. Knox in Primary Processes of Photosynthesis, ed. by J. Barber (North-Holland, Amsterdam, 1977), p. 55.
8. O. Simpson, Proc. Royal Soc. London A, 238, 402 (1956).

9. H. C. Wolf in Adv. in Atomic and Molecular Physics, vol. 3, ed. by D. R. Bates, I. Esterman (Academic Press, New York, 1967).
10. D. Burland and A. Zewail, Adv. Chem. Phys., 50, 385 (1980).
11. C. B. Harris and D. A. Zwemer, Ann. Rev. Phys. Chem., 29, 473 (1978).
12. A. H. Francis and R. Kopelman in Excitation Dynamics in Molecular Solids, Topics in Applied Physics, ed. by W. M. Yen and P. M. Selzer (Springer, Berlin, Heidelberg, New York, (1981).
13. R. Powell and Z. Soos, J. Lumin., 11, 1 (1975).
14. F. Perrin, Ann. Physique, 17, 283 (1932).
15. Th. Foerster, Ann. Phys. (Leipzig) (b), 2, 55 (1948).
16. V. M. Kenkre in Statistical Mechanics and Statistical Methods in Theory and Application, ed. by U. Landman (Plenum, New York, (1977).
17. V. M. Kenkre, Phys. Rev. B. 11. 3406 (1975).
18. V. M. Kenkre, J. Stat. Phys. 19, 333 (1978).
19. R. W. Zwanzig in Lectures in Theor. Phys., ed. by W. Downs and J. Downs (Gordon and Breach, Boulder, Colorado, 1961).
20. Th. Foerster in Modern Quantum Chemistry, Part III, ed. by O. Sinanoglu (Academic Press, New York, 1965), p. 93.
21. V. M. Kenkre and R. S. Knox, Phys. Rev. Lett. 33, 803, 1974.
22. K. D. Philipson and K. Sauer, Biochem., 11, 1180 (1972).
23. V. M. Kenkre, Phys. Rev. B, 18, 4064 (1978).
24. R. P. Hemenger, K. Lakatos-Lindenberg, and R. M. Pearlstein, J. Chem. Phys., 60, 3271 (1974).
25. V. M. Kenkre, Phys. Lett. 65A, 391 (1978).
26. C. B. Duke and T. Soules, Phys. Lett. A, 29, 117 (1969).
27. V. M. Kenkre and T. S. Rahman, Phys. Lett., 50A, 170 (1974).
28. D. L. Dexter, J. Chem. Phys. 21, 836 (1953).
29. V. M. Kenkre and R. S. Knox, Phys. Rev. B, 9, 5279 (1974).
30. P. Avakian, B. Ern, R. E. Merrifield, and A. Suna, Phys. Rev. 165, 974 (1968).
31. V. Ern and M. Schott in Localization and Delocalization in Quantum Chemistry, ed. by O. Chalvet (D. Reidel Publishers, Dordrecht-Holland, 1976), vol. II, p. 249.
32. V. M. Kenkre, V. Ern, and A. Fort, Phys. Rev. B, 28, 598 (1983).
33. V. M. Kenkre, A. Fort, and V. Ern, Chem. Phys. Lett. 96, 658, (1983).
34. A. Fort, V. Ern, and V. M. Kenkre, Chem Phys.,80, 205 (1983).
35. H. Auweter, A. Braun, U. Mayer and D. Schmid, Z. Naturforsch. 34A, 761 (1979).
36. A. Braun, U. Mayer, H. Auweter, H. C. Wolf, and D. Schmid, Z. Naturforsch. 37a, 1013 (1982).
37. See e.g., E. W. Montroll and B. West, J. Stat. Phys. 13, 17 (1975).
38. V. M. Kenkre and Y. M. Wong, Phys. Rev. B 23, 3748 (1981).
39. V. M. Kenkre and P. E. Parris, Phys. Rev. B 27, 3221 (1983).

40. J. Salcedo, A. E. Siegman, D. D. Dlott, and M. D. Fayer, Phys. Rev. Lett. 41, 131 (1978).
41. V. M. Kenkre, P. E. Parris, and S. M. Phatak, Physica A, to be published (1984).
42. P. E. Parris, S. M. Phatak, and V. M. Kenkre, J. Stat. Phys., to be published (1984).
43. D. D. Dlott, M. D. Fayer, and R. D. Wieting, J. Chem. Phys. 69, 2753 (1978); 67, 3808 (1977).
44. P. E. Parris, Ph.D. Thesis, University of Rochester (1984).
45. V. M. Kenkre and D. Schmid, Chem. Phys. Lett. 94, 603 (1983).
46. V. M. Kenkre in Proceedings of the Fourth International Seminar on Energy Transfer, ed. J. Pantoflicek (Prague, 1981)p. 54.
47. V. M. Kenkre, J. Stat. Phys. 30, 293 (1983).
48. V. M. Kenkre in Electronic Excitations and Interaction Processes in Organic Molecular Aggregates, ed. P. Reineker, H. C. Wolf, and H. Haken (Spring-Verlag, Berlin, 1983).

ENERGY TRANSFER IN INSULATING MATERIALS

G. Blasse

Physical Laboratory, University of Utrecht
3508 TA Utrecht, The Netherlands

ABSTRACT

In many important luminescent materials energy transfer plays an important role. First we review shortly energy transfer between two ions with an extension to macroscopic energy transfer. The results are applied on some special cases, especially cross relaxation. Multistep energy migration including donor-donor and donor-acceptor transfer is summarized in a next paragraph. A number of important classes of materials in which energy migration plays a role is discussed. They are divided into two groups, viz. those in which the energy migration has a strong temperature dependence and those in which it has a weak temperature dependence. Examples in the former group are YVO_4, $CaWO_4$ and certain Ce^{3+} compounds. Examples in the latter group are transition metal ions (MnF_2), uranium compounds and rare earth compounds. It is shown that the migration processes depend strongly on the nature of the rare earth ions and the type of impurities. The reason for this is discussed and some recent experiments are summarized. Finally fluorescence line narrowing and energy migration in glasses is dealt with shortly.

I. INTRODUCTION

Energy transfer is not only an interesting and intriguing physical phenomenon, but has also found application in many materials of importance. In 1942 the well-known lamp phosphor $Ca_5(PO_4)_3(F,Cl)$: Sb^{3+},Mn^{2+} was discovered. Excitation is into the Sb^{3+} ion, which transfers the excitation energy partly to the Mn^{2+} ion. In this way a white emission results. In 1941 the U.S. production of fluorescent lamps was 21 million lamps, in 1970 260 million. Other codoped materials have also been applied for some time, e.g. $CaSiO_3$:Pb^{2+},Mn^{2+} [1].

In 1964 the phosphor $YVO_4:Eu^{3+}$ was introduced as the red component of a color television picture tube. In this material energy transfer occurs from the broad-band emitting vanadate groups (VO_4^{3-}) to the Eu^{3+} ions. However, also transfer among the VO_4^{3-} groups appeared to play an important role.

In the new three-color fluorescent lamps the green phosphor is $(Ce,Tb)MgAl_{11}O_{19}$. Here energy transfer from Ce^{3+} to Tb^{3+} occurs. Energy transfer among the Ce^{3+} ions appeared to be of no importance [2]. Further energy transfer finds application in sensitizing solid state and glass lasers and in luminescent solar concentrators.

It is therefore not surprising that through the years the phenomenon of energy transfer has been studied extensively.

Following the early work of Botden [3], Dexter presented in 1953 his theory of energy transfer in solids [4]. This paper had an enormous influence up till today. The introduction of pulsed lasers (time-resolved spectroscopy) made a more detailed study possible. A recent book gives a good review of the state of the art [5].

It is the purpose of this chapter to review the characteristics of materials in which energy transfer plays a role. We restrict ourselves to insulating materials, i.e. we exclude energy transfer by electric charge carriers. This is treated in the chapter by Klingshirn. Also we will not dwell into theory, but restrict ourselves to a short summary in a form suitable for further treatment. A similar approach has been presented elsewhere [6].

II. SINGLE-STEP ENERGY TRANSFER

Dexter has considered energy transfer from a donor or a sensitizer ion (S) to an acceptor or activator ion (A). This may occur if the energy differences between the ground and excited states of S and A are equal (resonance condition) and if there exists a suitable interaction between both systems [4]. The interaction may be either an exchange interaction (if the wave functions overlap) or an electric or magnetic multipolar interaction. In practice the resonance condition can be tested by considering the spectral overlap of the S emission and the A absorption spectra.

The Dexter result looks as follows

$$P_{SA} = \frac{2\pi}{\hbar}|\langle\Psi_S\Psi_A^*|H_{SA}|\Psi_S^*\Psi_A\rangle|^2 \int g_S(E)g_A(E)dE.$$

Here the integral presents the spectral overlap, H_{SA} the interaction Hamiltonian and Ψ_S^* and Ψ_S, Ψ_A^* and Ψ_A the excited and ground state wave functions of S and A, respectively. The distance dependence depends on the interaction mechanism. For exchange interaction it is exponential, for the other interactions of the type r^{-n}.

The final results of Dexter's calculations can be applied rather generally, which makes them useful up till today. It has been shown by Orbach, however, that certain transfer processes are omitted in spectral overlap treatments [7]. The reason for this lies with the character of the ion-phonon coupling. If the ion-phonon interaction takes place at only a single site, the Dexter approach is equivalent to that in [7]. If the interaction takes place at both sites in the transfer process, there is no obvious connection with the optical linewidth, so that the transfer rate cannot be obtained from spectral overlap considerations. The reader is referred to [7] for detailed treatment. Since in our case the precise character of the ion-phonon coupling is usually unknown, we will mainly use Dexter's theory.

In general, in a real crystal there is a random distribution of sensitizers and activators. An excited sensitizer can interact with all unexcited activators and it is necessary to account for the distribution in SA separations. This problem was treated by Förster and later by Inokuti and Hirayama [8]. They obtained the following expression for the decay of S in the presence of A:

$$I(t) = I(o)\exp\left[\left(-\frac{t}{\tau_o}\right)-\Gamma\left(1-\frac{3}{S}\right)\frac{c_A}{c_o}\left(\frac{t}{\tau}\right)^{3/S}\right],$$

where τ_o is the decay constant of S in the absence of A, c_A is the concentration of A, c_o is the critical activator concentration and S = 6, 8 or 10 depending on the electric multipole interaction. For exchange interaction their result reads

$$I(t) = I(o)\exp\left[\left(-\frac{t}{\tau_o}\right)-\gamma^{-3}\frac{c_A}{c_o}\,g\left(\frac{e^{\gamma}t}{\tau_o}\right)\right],$$

where $\gamma = 2r_0/L$ with r_0 the critical distance and L an effective Bohr radius. Note that I(t) is not an exponential in the presence of A. In this treatment only SA transfer is considered and SS transfer is assumed not to occur. These equations have been encountered in many experimental situations.

Let us now consider some examples. For $VO_4^{3-} \rightarrow Eu^{3+}$ transfer the resonance condition is satisfied (the blue vanadate emission overlaps several Eu^{3+} absorption lines). For backtransfer this is not the case (the red Eu^{3+} emission does not overlap the uv vanadate absorption), so that it does not take place. The nature of the interaction in the case of the $VO_4^{3-} \rightarrow Eu^{3+}$ transfer has not yet been clarified convincingly.

In $(Ce,Tb)MgAl_{11}O_{19}$ the absence of $Ce^{3+} \rightarrow Ce^{3+}$ transfer is due to the vanishing spectral overlap of the Ce^{3+} emission and absorption spectra, so that the resonance condition is not fulfilled. The Ce^{3+} emission overlaps several Tb^{3+} absorption lines, so that $Ce^{3+} \rightarrow Tb^{3+}$ transfer is possible.

A case where the interaction is vanishing, is transfer between Eu^{3+} ions in $YAl_3B_4O_{12}$ [9]. At low temperatures only the $^5D_0$ and $^7F_0$ levels are occupied. These are connected by an optical transition which is strictly forbidden in this host lattice. Due to the vanishing interaction the Eu^{3+} - Eu^{3+} transfer does not occur, although the resonance condition is fulfilled (see also below).

A special case is cross relaxation in rare earth ions of which we will consider a few examples. In this case the excited donor transfers only part of its excitation energy to an acceptor (which may be identical with the donor).

An illustrative example is the Sm^{3+} ion, where cross relaxation between identical ions quenches the Sm^{3+} luminescence above a certain critical concentration. The Sm^{3+} ion shows emission from the $^4G_{5/2}$ level (see Fig. 1). For high enough Sm^{3+} concentrations the following transfer may occur:

$$Sm(^4G_{5/2}) + Sm(^6H_{5/2}) \rightarrow 2\ Sm(^6F_{9/2}),$$

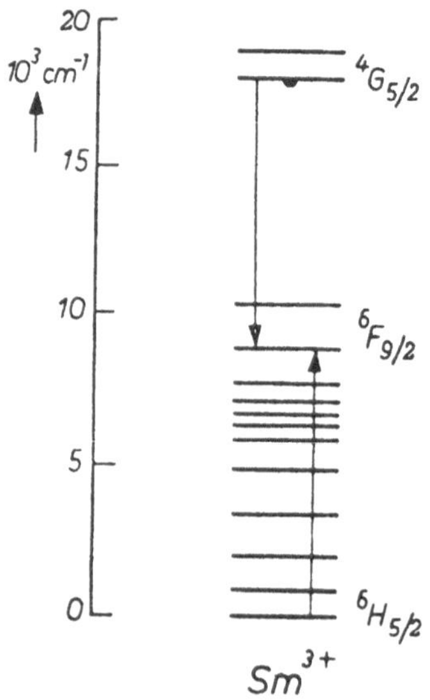

Fig. 1. Energy level scheme of the Sm^{3+} ion.

so that the orange $^4G_{5/2}$ emission is quenched. As a matter of fact the transitions $^4G_{5/2} \rightarrow {}^6F_{9/2}$ and $^6H_{5/2} \rightarrow {}^6F_{9/2}$ should match each other. The critical distance for this transfer is estimated to be some 20 Å. As a consequence Sm^{3+} compounds usually do not luminesce. An exception is $(bu_4N)_9SmW_{10}O_{36}$, but here the shortest Sm^{3+} - Sm^{3+} separation is > 18 Å [10]. Consequently the quantum efficiency is high.

An application of this cross relaxation phenomenon in the case of Sm^{3+} is found in the direct observation of a structural detail

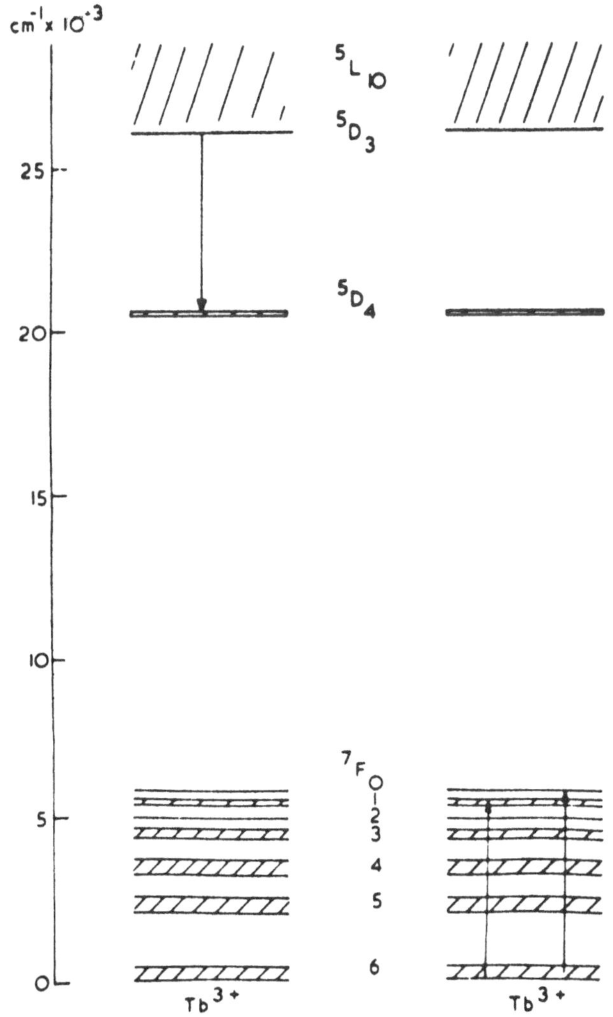

Fig. 2. Energy level scheme of the Tb^{3+} ion.

of solid electrolytes based on the fluorite structure [11]. These have composition $M^{2+}_{1-x}M^{3+}_{x}F_{2+x}$ (M^{2+} = Ca, Sr, Ba and M^{3+} = lanthanide). The M^{3+} ions occur in the lattice as clusters. This fact determines the ionic conductivity. Samples $Ca_{1-x}Sm_xF_{2+x}$ do not show any luminescence at all in contradiction with samples $NaY_{1-x}Sm_xF_4$ where quenching occurs for $x > 0.02$. The absence of emission shows directly that the Sm^{3+} ions occur in pairs, so that the excitation energy is lost by cross relaxation even if the Sm^{3+} concentration is extremely low.

A cross relaxation of the type described for Sm^{3+} has also been observed for Dy^{3+}. A comparable process quenches the higher level emission of the Eu^{3+} and the Tb^{3+} ions and is of great importance for practical application. The energy level diagram of the Tb^{3+} ion has been given in Fig. 2. Emission occurs from the $^5D_3$ and the $^5D_4$ level (blue and green, respectively). At low Tb^{3+} concentrations both emissions are usually observed. For increasing Tb^{3+} concentration the $^5D_3$ emission is quenched by

$$Tb^{3+}(^5D_3) + Tb^{3+}(^7F_6) \rightarrow Tb^{3+}(^5D_4) + Tb^{3+}(^7F_1).$$

This transfer occurs by electric-dipole interaction [12]. Its critical distance is about 13 A. Note that this cross relaxation results in green $^5D_4$ emission. If one is interested in the preparation of green Tb^{3+}-activated phosphors, the Tb^{3+} concentration cannot be too low. Due to the high price of terbium, this is rather unfavourable. In certain blue-emitting X-ray phosphors (e.g. LaOBr-Tb) the Tb^{3+} concentration should be low to avoid the green Tb^{3+} emission.

A way to keep the Tb^{3+} concentration low and to obtain still green emission is to quench the $^5D_3$ emission with the Dy^{3+} ion. Finally $^5D_4$ emission from Tb^{3+} results. The exact transfer mechanism is not yet known [13].

A similar situation exists for the Eu^{3+} ion which has emission from $^5D_2$ (blue), $^5D_1$ (green) and $^5D_0$ (red). The Eu^{3+} ion is used as an activator in redemitting phosphors, where only $^5D_0$ emission is required. This is obtained by increasing the Eu^{3+} concentration, so that the other emissions are quenched by cross relaxation. Due to the high europium price this is economically not very favourable.

In all these examples only pairs of ions play a role in the total transfer process. In many cases, however, this is more complicated. In concentrated materials the energy may migrate through the lattice looking for a place where it can end its life. Such situations will be the subject of the remainder of the next section.

III. MULTISTEP ENERGY TRANSFER

Up till now we have considered only the microscopic, single step donor-acceptor energy transfer. It is possible, however, that donor-donor transfer plays also a role. Excitation energy may migrate among the donor species before being transferred to an acceptor. A good survey of earlier and recent attacks of this problem has been given by Huber [14]. From this survey we derive the following for our purpose.

Consider the time evolution of $P_n(t)$, the probability that species n is excited and all other atoms are in their ground state:

$$dP_n(t)/dt = -\left(\gamma_R + X_n + \sum_{n'\neq n} W_{nn'}\right)P_n(t) + \sum_{n'\neq n} W_{n'n}P_{n'}(t).$$

The first term on the right-hand side corresponds to processes which bring the species n back to the ground state: γ_R is the radiative probability, X_n is the transfer rate to acceptors and $\sum_{n'} W_{nn'}$ gives the transfer rate from species n to other donor species n'. The second term describes the reverse process. For simplicity back transfer from the acceptors is neglected.

The energy difference between ground and excited state, $E_{n'}$, will vary from donor to donor due to perturbations from impurities, strains, etc. This yields the inhomogeneous line broadening observed under broad-band excitation. There are two techniques to follow the excitation energy migration in the donor system, viz. fluorescence line narrowing (FLN) and the time evolution of the donor luminescence in the presence of acceptors.

In FLN a pulsed, narrow band light source (a laser) excites those donors whose resonance frequencies span a small part of the inhomogeneous line. After the pulse, the luminescence evolves as shown schematically in Fig. 3. Broad-band luminescence arises to donors which were not excited directly. The decay of the narrow component yields information on the microscopic transfer process.

A well-known example is the case of $La_{0.8}Pr_{0.2}F_3$ [15]. In Fig. 4 we give the time evolution of the emission of the $^3P_0 \rightarrow (^3H_6)_1$ transition on the Pr^{3+} ion. Excitation is at 12 cm^{-1} higher energy than the line centre. Note that the line decreases in time, whereas the background luminescence increases. This shows that energy transfer occurs within the Pr^{3+} subsystem and that the temperature is high enough (14 K) to make the transfer process independent of the energy mismatch. From these experiments we can find the ratio R(t):

$$R(t) = \frac{\text{narrow band intensity at time t}}{\text{total intensity at time t}}.$$

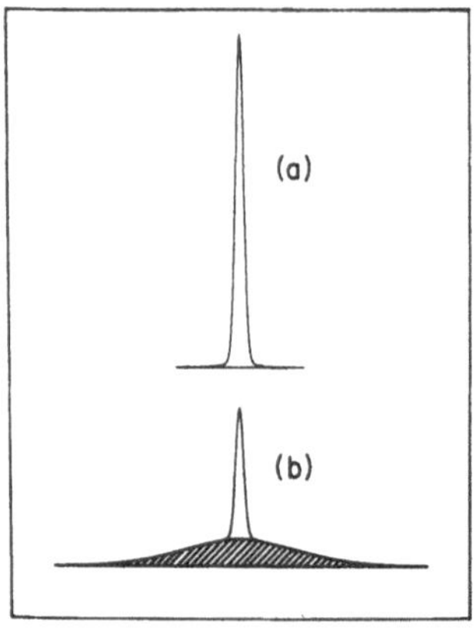

Fig. 3. Schematic diagram of the time development of the luminescence in a FLN experiment. (a) t = 0, (b) t > 0. The shaded area corresponds to luminescence from ions which were not initially excited [14].

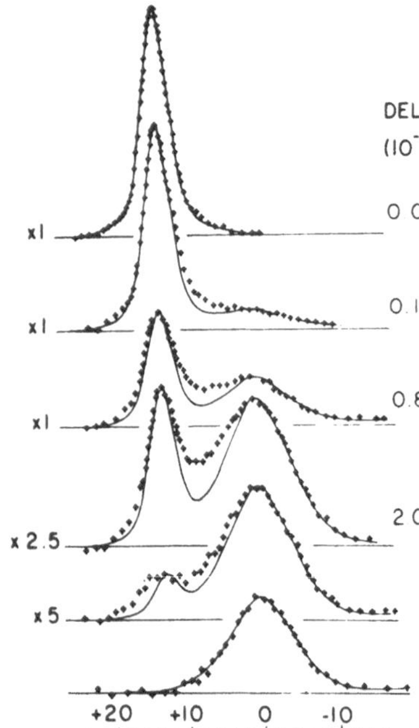

Fig. 4. Time resolved emission spectra for the $^3P_0 \rightarrow (^3H_6)_1$ luminescence in $La_{0.8}Pr_{0.2}F_3$. Excitation is 12 cm^{-1} above the line centre. T = 14 K [15].

Theoretical expressions for R(t) have been derived in the literature. In this way it becomes possible to derive transfer characteristics from a comparison between experiment and theory. For the case of Fig. 4, for example, it has been found that electric dipole-dipole transfer is dominant in the system $La_{0.8}Pr_{0.2}F_3$ and that the nearest-neighbour transfer rate is $0.4 \times 10^6 s^{-1}$ (14 K).

The time evolution of the donor fluorescence upon broad-band excitation is an old problem in luminescence. By measuring the time dependence of the donor fluorescence it is possible to obtain information about the donor-donor and donor-acceptor transfers by analysing the decay curve.

Since the integrated intensity at a time t is proportional to the number of excited donors at that time, $N_D(t)$, the decay can be described by

$$N_D(t) = N_D(0) \exp(-\gamma_R t) f(t).$$

Here $N_D(0)$ is the number of excited donors at the time the pulse is turned off and f(t) is the fraction of excited donors if the radiative lifetime (γ_R^{-1}) would be infinite. The function f(t) depends on time as described above for $P_n(t)$, if $\gamma_R = 0$.

Exact solution is possible for two extreme cases, viz. no donor-donor transfer at all (the case considered in chapter II) and very rapid donor-donor transfer. The behaviour of f(t) in between these two cases is extremely complicated.

In the limit of no donor-donor transfer at all we obtain

$$f(t) = \Pi_\ell [1-C_A+C_A \exp(-X_{o\ell}t)]. \tag{1}$$

This is a generalization of the results obtained by Inokuti and Hirayama [8]. C_A gives the acceptor concentration and $X_{o\ell}$ the transfer rate from a donor at site 0 to an acceptor at site ℓ. The value of $1-C_A$ gives the probability to find no acceptor on site ℓ. If site ℓ is occupied by an acceptor, it contributes a factor $\exp(-X_{o\ell}t)$ to $\exp(-X_n t)$. Here X_n is the total donor-acceptor transfer rate for the nth donor. Our equation represents therefore an average of $\exp(-X_n t)$ over all configurations of acceptors.

In the case of rapid transfer the donor-donor transfer takes place so quickly that for t > 0 all donors have equal probability to be excited. f(t) has now a very simple form, viz.

$$f(t) = \exp(-C_A \sum_\ell X_{o\ell}t). \tag{2}$$

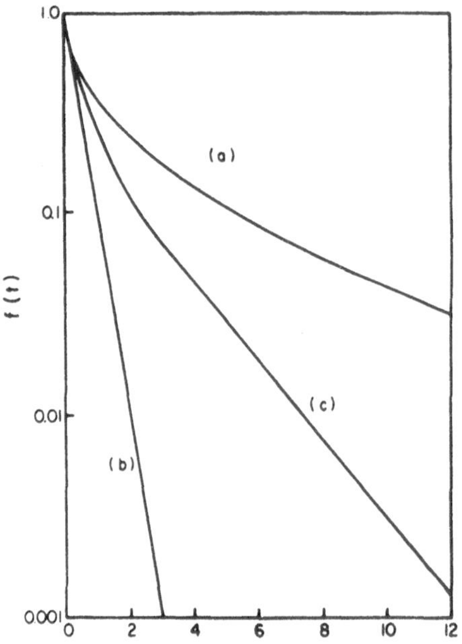

Fig. 5. Schematic plot of f(t) vs t. (a) donor-donor transfer absent, (b) rapid transfer, (c) intermediate between (a) and (b).

In Fig. 5 we present some schematic plots of f(t) vs t. In general f(t) is initially nonexponential, but becomes exponential after a certain time. In the rapid transfer case [(b) in Fig. 5] $N_D(t)$ is exponential in the whole time region. In the absence of donor-donor transfer $N_D(t)$ becomes exponential after long times only, with a slope equal to the radiative decay time (compare chapter II). Also in the intermediate case $N_D(t)$ is initially nonexponential, but becomes exponential in the limit $t \to \infty$. The slope, however, is steeper than in the absence of donor-donor transfer [(c) in Fig. 5].

Several theories have been presented in the literature, e.g. a hopping model [16] and a diffusion model [17]. Especially the latter solution has become popular if the diffusion is not fast enough to maintain the initial distribution of excitation (diffusion-limited transfer). The following expression was found

$$N_D(t) = N_D(0)\exp(-\gamma_R t)\exp[-\frac{4}{3}\pi^{\frac{3}{2}}C_A(Ct)^{\frac{1}{2}}(\frac{1+10.87x + 15.50x^2}{1+8.743x})^{\frac{3}{4}}]$$

Here C is the interaction parameter for donor acceptor transfer, and $x = DC^{-1/3}t^{2/3}$ where D is the diffusion constant. For $t \to \infty$ an exponential time dependence is predicted with decay rate τ_D^{-1} = 11.404 $C_A C^{1/4} D^{3/4}$. Here the diffusion is assumed to be isotropic. For one- and two-dimensional diffusion, however, f(t) has the asymptotic limits $[4\pi(C_A/a)^2 Dt]^{-1/2}$ and $(4\pi C_A a^{-2} Dt)^{-1}$, respectively, where a is the lattice constant [14],[15]. A more fundamental theory of energy transfer has been given by Huber [14],[15].

Note finally that back transfer from acceptor to donor was neglected up till now. It is possible, however, to incorporate back transfer in the existing theories [14]. For our purpose the formulations given above are sufficient. We will distinguish three cases, viz. (a) no donor-donor transfer; (b) diffusion-limited energy migration, i.e. donor-donor transfer is less probable than donor-acceptor transfer; (c) fast diffusion or trapping-limited energy migration, i.e. donor-donor transfer is more probable than donor-acceptor transfer. We will apply these models especially to compounds with high concentrations of luminescent ions, i.e. high donor concentrations.

IV. CHARACTERISTICS OF MATERIALS

In the following sections we review experimental results which contribute to our knowledge of energy migration in solid insulators. A large part of these results is of a qualitative nature only. Nevertheless this is instructive and often of large value in predicting transfer phenomena in new materials. The number of quantitative results is increasing, especially after the introduction of laser spectroscopy.

This section divides the materials to be discussed into two categories depending on whether the energy migration in the lattice exhibits a strong or a weak temperature dependence. A special section is devoted to glasses.

A strong temperature dependence is found if the excitation energy is self trapped after each step due to the relaxation of the surrounding lattice. In this category we find tungstates and compounds of Ce^{3+} and Bi^{3+}. A weak temperature dependence is encountered if the excited state shows only a very small amount of relaxation. Energy migration persists down to low temperatures. Examples are compounds of Mn^{2+}, U^{6+} and the rare earth ions.

The spectra in both cases are different. In the former they consist of broad bands, in the latter they show vibrational structure and zero-phonon lines. In the case of the rare earth ions the latter are dominating.

It is an easy task to make this more quantitative [6]. At low temperatures the transfer probability between identical ions is proportional with the spectral overlap of the zero-phonon lines in emission and in absorption. Therefore the transfer probability vanishes if the spectra do not show vibrational structure with zero-phonon lines.

V. STRONG TEMPERATURE DEPENDENCE

Let us first consider the classic phosphor $CaWO_4$. The luminescent centre is a WO_4^{2-} group. The optical transitions are broad band and of the charge-transfer type. They involve considerable lattice relaxation. In $CaWO_4$ the concentration of luminescent groups is high and the question whether excitation energy migrates among the lattice or not, has been the subject of many investigations. In view of the absence of a zero-phonon line energy transfer between tungstate groups is expected to be negligible at low temperatures. At higher temperatures, however, it is conceivable that energy migration occurs. Since thermal quenching of the luminescence is another possibility at high temperatures, it is necessary to distinguish between the two processes. This can be done as follows without using any complicated instrumentation. In $CaWO_4$-Sm^{3+} the Sm^{3+} emission upon host lattice excitation can be studied as a function of temperature. At low temperatures the Sm^{3+} emission intensity is very low, but it increases above a certain temperature ($\sim$250 K). This indicates that above this temperature the tungstate excitation energy starts to migrate. The thermal quenching can be studied in the very similar, but diluted system $CaSO_4$-W where no energy transfer is possible due to the low tungsten concentration. The quenching occurs also above 250 K, which makes the situation rather complicated.

In $CaMoO_4$ the situation is a little different, because thermal quenching occurs above 150 K. In the composition $CaMoO_4$-Sm^{3+} the molybdate emission intensity decreases above 150 K, but no increase of the Sm^{3+} emission intensity is observed. This is only observed at still higher energies.

A strikingly different situation occurs in $PbWO_4$ and YVO_4 where energy migration among the tungstate and vanadate sublattice starts at much lower temperatures than the thermal quenching temperature does. Fig. 6 gives a survey of these results. The situation encountered in the case of $PbWO_4$ and YVO_4 makes host lattices of this type promising for efficient phosphors. This has been realized, for example, in the case of YVO_4-Eu^{3+}. Excitation at room temperature into the vanadate group results in efficient Eu^{3+} emission, because the vanadate excitation energy migrates through the lattice until it is trapped by the activator. Other host lattices of this type are Ba_2CaWO_6 and La_2MgTiO_6 [6].

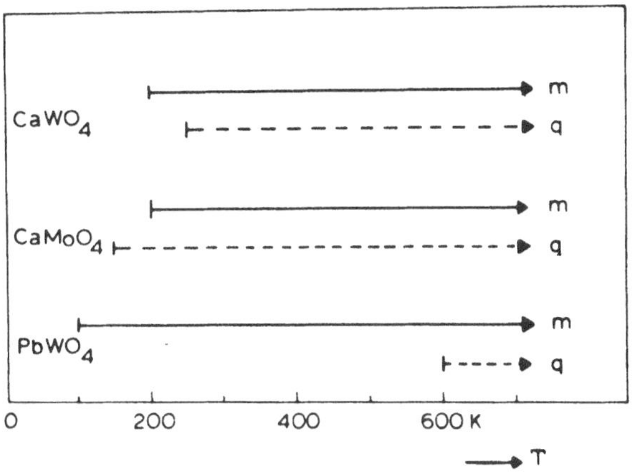

Fig. 6. Onset of energy migration (m) and thermal quenching (q) for several materials.

Nevertheless La_2MgTiO_6-Eu^{3+} is not an efficient luminescent material. This is due to the fact that the lattice provides its own traps due to a certain amount of crystallographic disorder between the Mg^{2+} and Ti^{4+} ions. This results in pairs of titanate octahedra which emit at much longer wavelength than the intrinsic titanate octahedra. The real activator, Eu^{3+}, cannot compete successfully with these defect centres for the migrating energy.

Surprisingly enough; not much quantitative data are available on these, often important, materials. Some work has been performed on YVO_4-Eu^{3+} (for a review, see [6]). The critical distance for energy transfer between the vanadate groups is about 8 Å, to be compared with the shortest V-V distance of 3.6 Å yielding a hopping time of about 10^{-7} s at 300 K. The interaction mechanism is not yet known. Values for the vanadate → Eu^{3+} transfer probability in the literature vary from 10^4-10^8 s^{-1}. This illustrates the unsatisfactory situation for materials of this type.

Many compounds of Ce^{3+} and Bi^{3+} belong also to the present class of materials [6]. A good example is CeF_3 which shows at 300 K and lower temperatures high Ce^{3+} efficiency. This is also the case in the system $La_{1-x}Ce_xF_3$. The relevant spectra are broad band, so that no energy migration is possible. Consequently CeF_3-Tb does not yield much Tb^{3+} emission upon Ce^{3+} excitation. This is not the case with $CeBO_3$-Tb, but here the Ce^{3+} emission and excitation overlap considerably.

It is interesting to note the similarity between $CeBO_3$-Tb and

YVO_4-Eu on one hand and between CeF_3-Tb and (for example) $YNbO_4$-Eu on the other hand. In the former case energy migration in the host lattice occurs at room temperature, resulting in efficient phosphors. In the latter case, however, energy migration is negligible.

Two well-known Ce^{3+} compounds activated with Tb^{3+} have been investigated more in detail, viz. CeP_5O_{14}-Tb [18] and $CeMgAl_{11}O_{19}$-Tb [2]. However, Ce^{3+}-Ce^{3+} energy transfer was not observed.

It is here the place to mention an unresolved problem in energy transfer between two species, viz. $Ce^{3+} \rightarrow Eu^{3+}$ transfer does not occur, nor does vanadate $\rightarrow Tb^{3+}$ occur. Not only is the transfer absent, also the luminescent centres seem to quench each other. It has been proposed that quenching occurs via a charge-transfer state which could be at low energy in the cases mentioned (viz. Ce^{4+}-Eu^{2+} and V^{4+}-Tb^{4+}, respectively) [19].

Some Bi^{3+} and Pb^{2+} compounds belong also to the present class of materials. The situation with compounds of that type is complicated, however. The reason for this is that some of them behave as semiconductors, for example Bi_2O_3, $CsPbCl_3$ and $Cs_3Bi_2Br_9$ [6]. Others belong to the materials with energy migration with weak temperature dependence, for example $Cs_2NaBiCl_6$. Large Stokes shifts without any indication for energy migration at low temperatures have been encountered in others, for example $Bi_4Ge_3O_{12}$, $Bi_2Al_4O_9$, $PbSO_4$ and PbF_2. The observation of several types of energy transfer processes in Bi^{3+} and Pb^{2+} compounds has the same origin as the observation of several types of emission spectra, i.e. exciton recombination, zero-phonon line with vibrational structure and broad band spectra.

VI. WEAK TEMPERATURE DEPENDENCE

VI.A. Transition Metal Compounds

Manganese fluoride, MnF_2, has become an outspoken and well-studied example of an inorganic material in which exciton diffusion persists down to very low temperatures [20]. Rare earth ions (Eu^{3+} and Er^{3+} have been used as activators. The emission of the Mn^{2+} (3 d^5) ion corresponds to the $^4T_1$-$^6A_1$ transition which is strongly forbidden. The radiative lifetime of the excited $^4T_1$ level in MnF_2 is about 30 ms. The relaxation around this excited state is not very large and a zero-phonon line has been observed (Fig. 7). In MnF_2 the excitation does not stay on the same ion, but can travel readily through the sublattice of in-resonance Mn^{2+} ions. Even in the purest MnF_2, the emission originates, therefore, from Mn^{2+} ions associated with impurities and from defects. Common impurities are

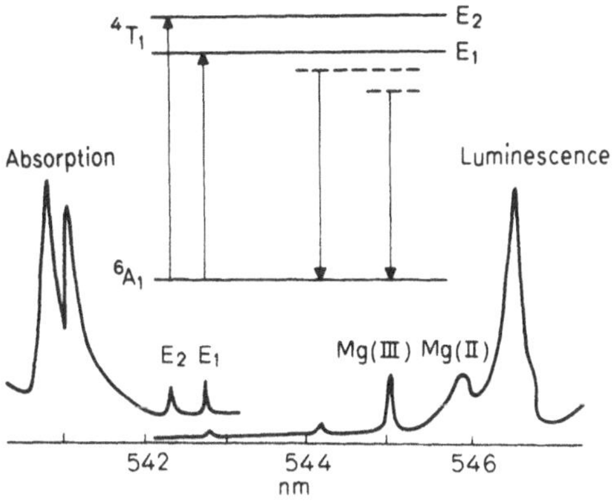

Fig. 7. Part of the absorption and emission spectrum of a MnF_2 crystal at LHeT. Mg(II) or Mg(III) refer to Mn^{2+} ions with a second or third nearest neighbour Mg^{2+} ion.

Mg^{2+}, Zn^{2+} and Ca^{2+} ions which are always present at concentrations of a few parts per million. They occupy regular cation sites in the lattice and perturb the surrounding Mn^{2+} ions, lowering their energy levels relative to those of the unperburbed (intrinsic) Mn^{2+} ions. The diffusing excitons can now be trapped by the perturbed Mn^{2+} ions. At low temperatures this excitation cannot return to the exciton state; the excited, perturbed Mn^{2+} ions decay radiatively with a spectrum characteristic of the particular trap. Examples of such traps are Mg(3 nn, 48 cm^{-1}), Mg(2 nn, 77 cm^{-1}), Zn(3 nn, 36 cm^{-1}), Zn(2 nn, 66 cm^{-1}) and Ca(1 nn, 300 cm^{-1}). Here we have indicated a Mn^{2+} ion in the first, second or third nearest-neighbour site around Mg, Zn or Ca and its trap depth. Deeper traps are also present and are effective as killer sites, i.e. traps from which no emission occurs, but where the excitation is lost nonradiatively.

Below 4 K the traps are effective. Around 4 K, however, the shallower traps [like Zn(3 nn), Mg(3 nn)] begin to lose their trapped excitation energy by thermally activated back-transfer to the exciton level. From here the energy may be trapped by deeper traps. Finally all the emitting traps are emptied and only the deep, non-emitting traps are operative. As a consequence the luminescence has been quenched. These quenching traps may be Ni and/or Fe ions (Fig. 8).

It has been found that the net low-temperature transfer rate from the exciton level to all traps amounts to $8 \times 10^3 s^{-1}$ [20]. Since the radiative decay rate of the exciton is 30 s^{-1}, transfer to traps is about 270 times more probable than intrinsic exciton emission. In fact the ratio of trap to intrinsic emission has been

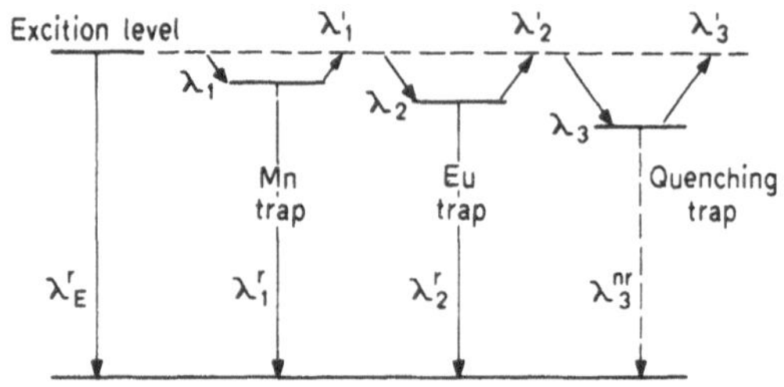

Fig. 8. Model used to describe the populating of Mn traps, Eu traps and quenching traps in MnF_2-Eu^{3+}.

measured to be about 250. The transfer rate increases rapidly with temperature. This has been explained in the following way. The $^4T_1$ state of the Mn^{2+} ion is split by the combined effect of the orthorhombic crystal field, spin-orbit coupling and exchange. The two lowest levels, E1 and E2 are 17 cm^{-1} apart, E1 being the lower one. At higher temperatures the E2 state becomes thermally populated. The E2 exciton state transfers more readily to the traps (transfer rate 6 x 10^6 s^{-1}). This higher transfer rate from the E2 level has been ascribed to a more favourable spectral overlap resulting from lifetime broadening due to fast relaxation down to E1 [20].

The luminescences of MnF_2-Eu^{3+} and MnF_2-Er^{3+} have also been studied. The Eu^{3+} emission usually originates from the $^5D_0$, $^5D_1$ and $^5D_2$ level of this ion. The E1 level is above the $^5D_0$ level, but below the $^5D_1$ and $^5D_2$ level. Excitation into the $^5D_1$ and $^5D_2$ level does not yield emission from these levels. It is assumed that the $^5D_1$ and $^5D_2$ level relax by energy transfer to the E1 level of a neighbouring Mn^{2+} ion, and that this ion in turn transfers its energy to the $^5D_0$ level of the Eu^{3+} ion. Excitation into the MnF_2 host lattice results in a considerable amount of Eu^{3+} emission indicating that the Eu^{3+} ion acts indeed as a trap. Whereas the luminescence of pure MnF_2 is quenched above 50 K, the Eu^{3+} emission is quenched above 100 K. This occurs by back transfer from the $^5D_0$ level of Eu^{3+} to the exciton levels which transport the energy to killer sites. The trap depth of the Eu^{3+} ion is some 1300 cm^{-1}. Since the emitting levels of the Er^{3+} ions are more than 3000 cm^{-1} below E1, Er^{3+} traps remain effective at much higher temperatures and their emission is observed at room temperatures.

Finally we note that transfer between nearest-neighbour Mn^{2+} ions in MnF_2 does not occur within the exciton lifetime.

All Mn^{2+}-Mn^{2+} transfer mentioned above are between next-nearest Mn^{2+} neighbours. The reason for this is that MnF_2 is antiferromagnetic in the temperature region involved. The Mn^{2+} ions are on two sublattices with antiparallel magnetic moment. Energy transfer between Mn^{2+} ions with antiparallel spin moments is a relatively improbable process.

Some properties of the exciton dynamics of MnF_2 crystals have been elucidated through time-resolved studies of the emission at different parts of the exciton band after selective laser excitation [21]. In some cases phase memory was found to be retained up to ~ 1 µsec but the details of the exciton dynamics require further knowledge of the exciton scattering mechanisms.

Similar effects have been observed for other manganese(II) fluorides e.g. $KMnF_3$, $RbMnF_3$, and $BaMnF_4$ doped with rare earth activators. An interesting result for $KMnF_3$:Eu^{3+} crystals is that the activator ions perturb the neighbouring Mn^{2+} ions which act as traps for the host excitons which then transfer the energy to the Eu^{3+} ions [22].

For Mn^{2+} compounds with linear manganese chains a different behavior has been found. In tetramethyl ammonium manganese cloride the manganese spins are correlated antiferromagnetically. This prevents energy transfer between nearest chain-neighbours. Transfer between next-nearest neighbours has to occur over a large distance. As a consequence the emission of this compound exists of intrinsic Mn^{2+} emission at low temperatures. The energy becomes mobile above about 50 K [23].

A considerable amount of work has been performed on the luminescence of the ions $Co(CN)_6^{3-}$ and $Cr(CN)_6^{3-}$ [24]. The interesting point is that quantitative measures for the relaxation around the excited ion have been derived from the vibrational structure of the spectra. In $K_3Co(CN)_6$ the $Co(CN)_6^{3-}$ ion emits a broad band peaking at 14000 cm^{-1}. The corresponding absorption band peaks at 26000 cm^{-1}. This is a very large Stokes shift. There is a weak vibrational structure. The maximum intensity lies at about the 10th member of the progression in the totally symmetric Co-C vibration. The zero-phonon line is not observable. The Co-C distance in the excited state is 0.11 A longer than in the ground state. This points to a strong relaxation in the excited state. The luminescence quenches below room temperature. It is generally agreed that there is no energy migration in $K_3Co(CN)_6$ in view of the negligible spectral overlap between the emission and absorption bands. The thermal quenching is ascribed to internal nonradiative decay in the $Co(CN)_6^{3-}$ ion.

In the system $K_3Co_{1-x}Cr_x(CN)_6$ the $Cr(CN)_6^{3-}$ ions can be excited selectively. Concentration quenching of the $Cr(CN)_6^{3-}$ luminescence occurs between x = 0.05 and x = 0.1. This has been ascribed to energy migration among the $Cr(CN)_6^{3-}$ ions. At lower Cr concentrations the $Cr(CN)_6^{3-}$ luminescence has a high quantum efficiency up to room temperature. This behaviour, completely different from that of $Co(CN)_6^{3-}$ is due to small relaxation around the excited state. The change in the Cr-C distance upon excitation is only some 0.02 A. The zero-phonon line in the emission spectrum dominates the spectrum. Only in $[(C_6H_5)_4P]_3Cr(CN)_6$ the distance between the Cr^{3+} ions has become so large that energy migration is no longer possible.

The large difference in the relaxation of the two ions (with all its consequences) is due to the difference in the nature of the crystalfield transitions involved in the emission: in the $Co(CN)_6^{3-}$ ion ${}^3T_{1g} \rightarrow {}^1A_{1g}$ and in $Cr(CN)_6^{3-}$ ion ${}^2E_g \rightarrow {}^4A_{2g}$. Note, finally, that in $K_3Co_{1-x}Cr_x(CN)_6$ no energy transfer from the cobalt to the chromium ion has been observed. Due to the large relaxation of the cobalt system it is out of resonance with the chromium ion levels.

VI.B. Hexavalent Uranium Compounds

The study of hexavalent uranium compounds illustrates what types of centres are involved and which factors determine their concentration. The U^{6+} ion in solids occurs as the linear uranyl ion (UO_2^{2+}) or as the octahedral uranate group (UO_6^{6-}).

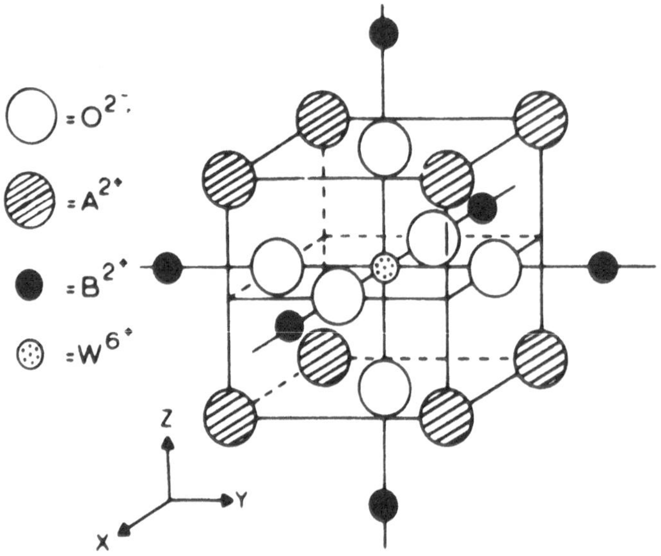

Fig. 9. The crystal structure of an ordered perovskite A_2BWO_6.

Let us first consider the ordered perovskite Ba_2CaUO_6 where UO_6^{6-} groups are present. The crystal structure is shown in Fig. 9. The shortest U-U distance is about 6 Å. At low temperatures this compound luminesces efficiently, but above 20 K the luminescence is quenched [25]. The emission spectrum shows several zero-phonon lines. These do not coincide with the zero-phonon line in the excitation spectrum, which indicates that the emission, even at the lowest temperatures, does not originate from the intrinsic uranate group. The emission lines depend on sample history and quench subsequently with increasing temperature.

The explanation is as follows. After excitation the energy migrates through the lattice among the intrinsic uranate groups. This energy is trapped by uranate groups near defects which have energy levels at somewhat lower values than the intrinsic groups. From the "defect" uranate groups the energy is emitted as luminescence. The trap depth in Ba_2CaUO_6 is some 100 cm^{-1}. If the temperature is increased, the traps are emptied (see Fig. 10) and the energy reaches the killer centres. In Ba_2CaUO_6 the trap concentration is very high, viz. about 1 at. %. A slight disorder in the ordered perovskite structure (probably between Ba and Ca) is responsible for this disorder. Due to the disorder the migrating energy is trapped and cannot reach the killer sites. Note that the deviation from the perfect crystal structure is responsible for the fact that Ba_2CaUO_6 luminesces at all.

This is different in $MgUO_4$ [26]. Even at the lowest temperatures (1.4 K) the quantum efficiency is not higher than 10%. The number of traps is large as follows from the number of zero-phonon lines in emission. More important, however, the ratio of killers and traps must be larger than in Ba_2CaUO_6, so that the greater part of the migrating energy is lost at killer sites. It can

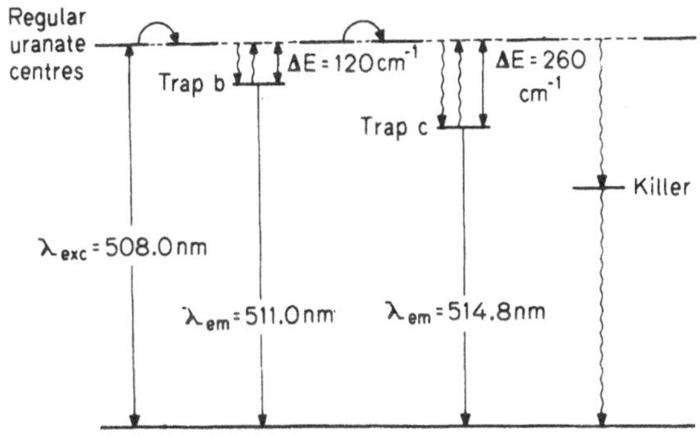

Fig. 10. Schematic energy level diagram of Ba_2CaUO_6

also be shown what the nature of these killer centres is, viz. U^{5+} ions. These give an intervalence charge-transfer absorption, which picks up the excitation energy. In fact $MgUO_4$ is slightly oxygen deficient. The strongly oxygen-deficient Y_6UO_{12} does not luminesce at all.

Uranyl compounds are usually more stoichiometric. In fact $Cs_2UO_2Cl_4$ shows efficient luminescence at 300 K. The killer concentration is low and the traps are no longer effective. At LHeT, however, also trap emission is observed. The trap concentration is so low that at low temperatures the migrating excitons meet each other whereby they annihilate themselves (biexciton decay) [27].

In the corresponding $Cs_2UO_2Br_4$ the killer concentration is higher and quenching occurs below room temperature. The trap concentration is some 100 ppm and the trap depth 40 cm^{-1} [28]. Fig. 11 shows the time dependence of the intrinsic and the trap emission at low temperatures, showing energy migration among the intrinsic groups to the "defect" groups (the traps). Excitation is by selective excitation into the intrinsic uranyl groups.

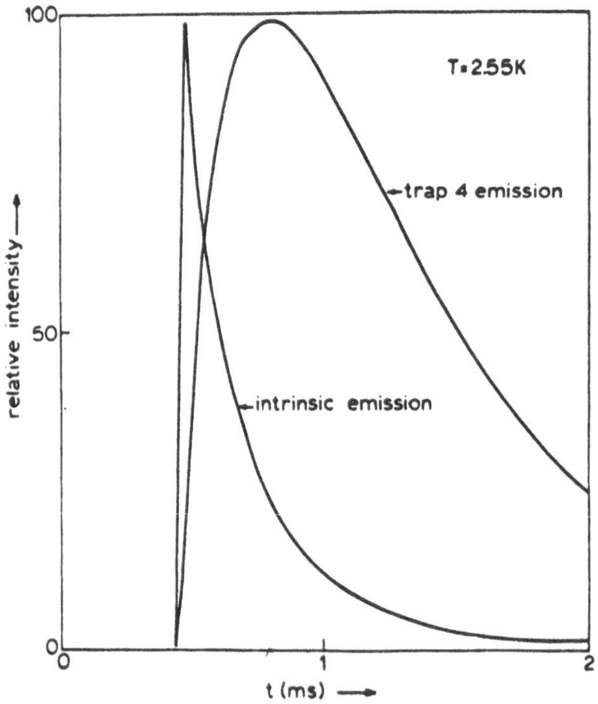

Fig. 11. Time dependence of intrinsic and trap emission of $Cs_2UO_2Br_4$ after excitation into the intrinsic centres.

The diffusion constant for the hexavalent uranium compounds is about $10^{-8} cm^2 s^{-1}$ with a resulting hopping time of about 10^{-8} s (10 K). Note that the spectra of the U^{6+} ion in oxides show always a zero-phonon line, so that energy migration down to very low temperatures is possible.

VI.C. Trivalent Rare Earth Compounds

Compounds of the rare earth ions have been investigated in detail. Their sharp line spectra (mainly zero-phonon lines) are very suitable for this purpose. High-concentration rare earth compounds may be used as minilasers. The surprise has been that in these compounds concentration quenching of the luminescence may be absent. For phosphors it is sometimes important to have a low critical concentration of the activator in view of its expensive price.

It has become use to refer to undiluted rare earth compounds with efficient luminescence as stoichiometric (laser) materials. We want to stress that the term "stoichiometric" used in this sense is incorrect. High concentration materials may well be non-stoichiometric, whereas diluted compositions can be stoichiometric. This will be illustrated below.

The different possibilities for energy migration in high-concentration materials will first be illustrated on a study of rare

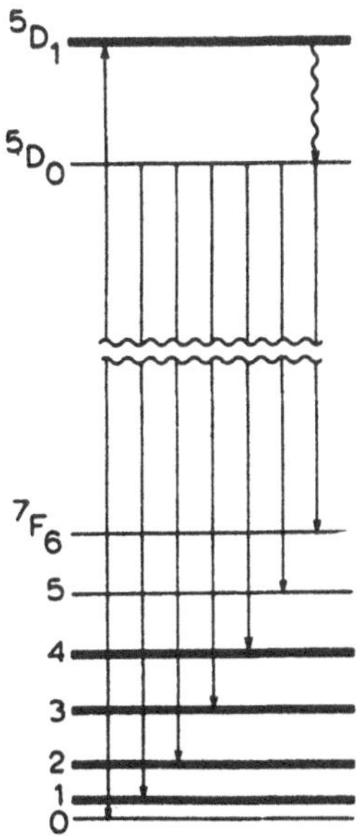

Fig. 12. Energy level scheme of the Eu^{3+} ion.

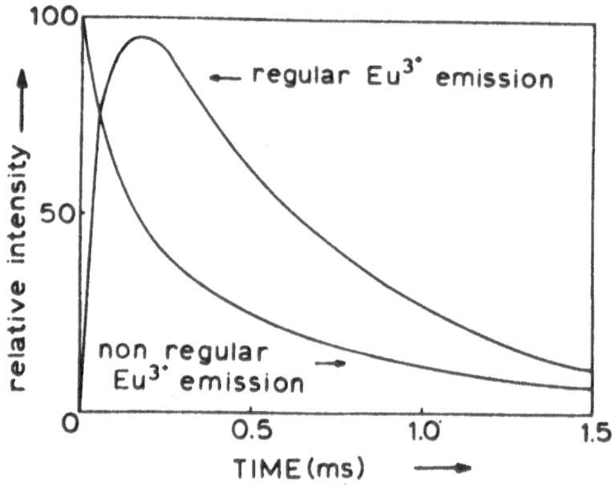

Fig. 13. Time dependence of non-regular and regular Eu^{3+} luminescence of $EuAl_3B_4O_{12}$ after excitation into the non-regular ions.

earth luminescence in the huntite structure ($YAl_3B_4O_{12}$). We choose this example, because diluted as well as concentrated materials, crystals as well as powders were investigated and because the results depend on the choice of the rare earth ion. Secondly we will deal with some other examples studied in more detail.

Our first example is $EuAl_3B_4O_{12}$ [9]. In the huntite structure the rare earth ions occupy trigonal prisms, so that no inversion symmetry is present. The shortest distance between the Eu^{3+} ions is about 6 Å, which is a relatively long distance. At room temperature the single crystals show only weak luminescence, whereas the powders emit efficiently. The energy level scheme of Eu^{3+} is given in Fig. 12. The emission spectrum of $EuAl_3B_4O_{12}$ in the $^5D_0$-$^7F_1$ region shows not only the two lines expected for the trigonal prismatic coordination. There are also a couple of other weak lines. These lines have their own excitation lines. They can be excited selectively and their time dependence can be followed (Fig. 13). The extra lines have been ascribed to Eu^{3+} ions which occupy sites different from the regular Eu^{3+} sites. These non-regular sites transfer excitation energy to the regular sites, so that the non-regulars do not act as optical traps. Consequently, $EuAl_3B_4O_{12}$ is not stoichiometric. At least one of the non-regular Eu^{3+} ions is a Eu^{3+} ion on an Al^{3+} site.

Fig. 14 shows the decay curves of the regular Eu^{3+} ions of $EuAl_3B_4O_{12}$ crystals at different temperatures. At LHeT the curve is identical to that for diluted samples $Y_{1-x}Eu_xAl_3B_4O_{12}$. This means that the decay occurs only by radiative transitions in the excited centres. No energy transfer occurs. At higher temperatures

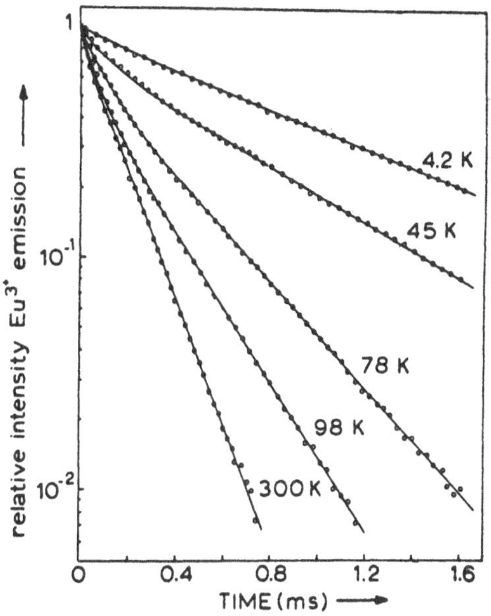

Fig. 14. Decay curves of regular Eu^{3+} luminescence of $EuAl_3B_4O_{12}$.

the decay is much faster than expected for radiative decay. Energy migration among the Eu^{3+} sublattice occurs and brings the excitation energy to killer sites. An analysis of these curves shows that the excitation diffuses slowly through the lattice. We are dealing with diffusion-limited energy migration. During its life time the excitation energy performs at 300 K some 1400 transfers from Eu^{3+} to Eu^{3+}, reaching a diffusion length of 230 Å. In crystals this is long enough to have a reasonable probability to reach a killer site, but in powders it is too short. This makes the powders efficient phosphors. The killer sites appear to be Mo^{3+} on Al^{3+} sites. The molybdenum is incorporated from the flux (K_2SO_4-MoO_3).

Why is there no migration at low temperatures? At low temperatures the Eu^{3+} ion in the ground state occupies the $^7F_0$ level (see Fig. 12), and in the excited state the $^5D_0$ level. The transition between these levels (0-0) is strongly forbidden and is not observed in the spectra. Consequently the interaction strength vanishes. If the temperature increases, the $^7F_1$ level becomes populated. The interaction strength increases and migration starts. The diffusion coefficient has an activation energy of about 200 cm^{-1}, equal to the energy difference between the $^7F_0$ and $^7F_1$ levels as observed from the spectra.

A linear crystal field at the site of the Eu^{3+} ion induces a certain transition probability in the $^5D_0$-$^7F_0$ transition. This is, for example, the case for $NaEuTiO_4$. It has been observed recently that in this compound energy migration among the Eu^{3+} sublattice persists down to the lowest temperatures [30].

In the $EuAl_3B_4O_{12}$ crystals the transfer probability for Eu^{3+}-Eu^{3+} is much smaller than for Eu^{3+}-Mo^{3+} (killer) as required for diffusion-limited energy migration. This is different in the systems (Y,Ce,Tb)$Al_3B_4O$and (Y,Bi,Tb)$Al_3B_4O_{12}$ [31]. Excitation is into the Ce^{3+} or Bi^{3+} ions. The energy migrates rapidly over the Ce^{3+} or Bi^{3+} ions which have allowed transitions. The probability for transfer from Ce^{3+}(Bi^{3+}) to Tb^{3+} is smaller than that for transfer among the excited ions. This is the case of trap-limited migration. A similar situation was encountered in (La,Bi,Tb)MgB_5O_{10} [32]. In designing efficient phosphors such effects are of large importance.

The situation in $TbAl_3B_4O_{12}$ is different from $EuAl_3B_4O_{12}$ at low temperatures [33]. Let us have a look at the emission spectra (Fig. 15). Immediately after the excitation pulse into the regular Tb^{3+} ions (Fig. 15a), the emission is equal to that of a diluted sample (Y,Tb)$Al_3B_4O_{12}$ (Fig. 15c). But after longer times it changes dramatically (Fig. 15b). This is due to energy migration among the Tb^{3+} sublattice which is here possible because the transition involved ($^5D_4$-$^7F_6$) is not so strongly forbidden as in the case of the Eu^{3+} ion. The other difference with $EuAl_3B_4O_{12}$ is that the non-regular Tb^{3+} ions, also present here, act as optical traps. It is their emission which we observe some time after the pulse. An analysis shows that we are dealing with trap-limited diffusion. At higher temperatures the Tb^{3+} traps are too shallow (trap depth 25 cm^{-1}) to be effective. Migration ends on the Mo^{3+} ions. This migration appears to be diffusion limited. However, the diffusion constant decreases with increasing temperature. This has been ascribed to a decreasing spectral overlap which is not compensated by thermal population of higher crystal-field levels. In $Tb_3Al_5O_{12}$ the reverse behaviour has been observed [34]. These studies present a reliable overall picture of the transport processes in a wide temperature region.

The material which induced the extensive research of concentrated rare earth systems is undoubtedly NdP_5O_{14} in view of its unusually weak concentration quenching. This quenching is linearly dependent on the Nd^{3+} concentration and the luminescence decays exponentially for all temperatures and Nd^{3+} concentrations. Several mechanisms have been put forward to explain these effects, for example, a crystal-field overlap model, a surface quenching model and cross relaxation [6], [39]. However, recently it has been shown that it is possible to use existing energy-transfer theories to explain the behaviour of the system $La_{1-x}Nd_xP_5O_{14}$ [35]. In this study

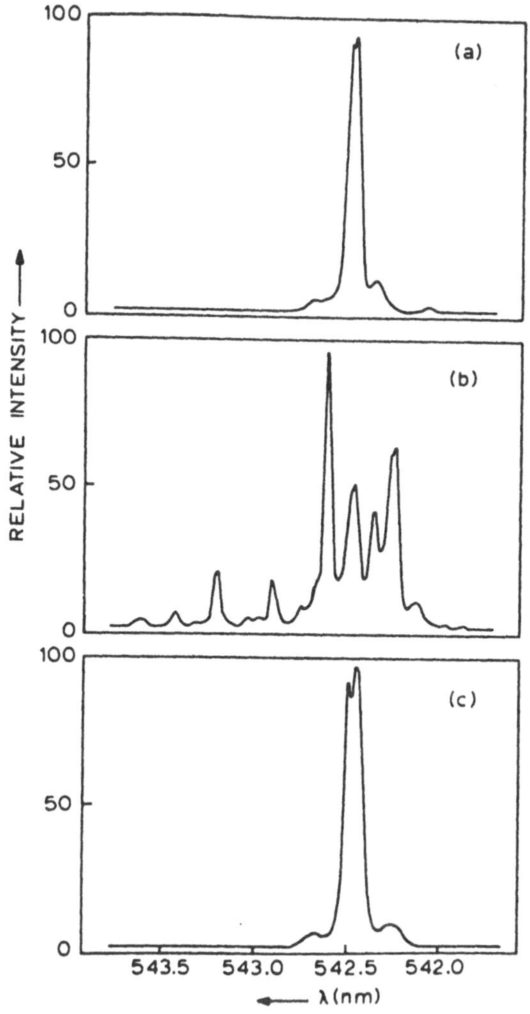

Fig. 15. Emission spectra of $TbAl_3B_4O_{12}$.

the author applied resonant and high-resolution excitation and measured the time-resolved fluorescence line narrowing for several Nd^{3+} concentrations. Donor-donor transfer appeared to be absent at low temperatures (up to 20 K for x = 210 and x = 0.75 and up to 6 K for x = 1). At higher temperatures donor-donor transfer starts.

Eq.(1) can be approximated by

$$f(t) \approx \exp(- tx \sum_{\ell} X_{0\ell}) \quad (3)$$

for small trapping rates. This results in an exponential decay rate given by

$$\tau^{-1} = \tau_o^{-1} + x \sum_{\ell} X_{0\ell}$$

which shows a linear dependence on the concentration [35].

At room temperature rapid donor-donor transfer is observed. This shows that the donor-donor transfer rate increases rapidly with temperature which means that this transfer is phonon-assisted. In this case a T^3 or higher-power dependence is expected [7].

The exponential decay can also be explained. For room temperature it follows from eq. 2. For low temperatures it follows from eq. 3 by realizing that we are involved with the early time regime, i.e.

$$\tau_o < X_{0\ell}^{-1}.$$

For X_{01} (the nearest-neighbour trapping rate) a value of $1.1 \times 10^3 s^{-1}$ was found, which is comparable with the radiative decay rate. Note that trapping corresponds to cross relaxation whereby the donors act as their own acceptors. For the nearest neighbour donor-donor transfer rate at 5 K a value of $7 \times 10^3 s^{-1}$ was found in two independent ways, viz. from the fluorescence line narrowing and by extrapolation from the 300 K value of the diffusion constant. This value is comparable with the decay rate too. Note that the temperature dependence of the donor-donor transfer is responsible for complete different behaviour at low and high temperatures.

Another model system in this type of studies is $La_{1-x}Pr_xF_3$ (see also section III). Also here trapping is due to cross relaxation, so that all donors can act as trapping centres for excitation [36],[37]. In fact the quantum efficiency as well as the luminescence lifetime decrease with increasing Pr^{3+} concentration for the levels emitting in the visible.

The cross-relaxation behaviour can be measured by observing the luminescence decay after pulsed excitation. Hegarty et al. [36] were able to interpret the results for $La_{0.8}Pr_{0.2}F_3$ by assuming that 12 % of the Pr^{3+} ions are in non-quenching sites and contribute an exponential to the decay. The remaining Pr^{3+} ions are in quenching sites. Cross relaxation occurs directly in steps after transfer within the inhomogeneous line. However, the assumption of two types of Pr^{3+} ions seems to be a little artificial.

Another solution of this problem has been presented by Vial and Buisson [37]. They criticise the use of the Inokuti-Hirayama formula, because of the use of a single trapping mechanism. Earlier comments were made by Heber et al. who proposed to use a discrete instead of a continuous model [38]. Also Hegarty et al. use a single mechanism.

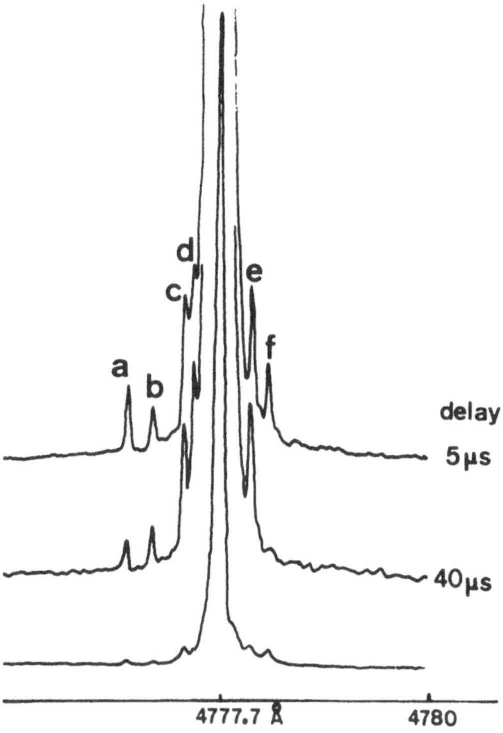

Fig. 16. The lower curve is the absorption spectrum of LaF_3-Pr^{3+} (0.1 %) around the $^3H_4 \rightarrow {}^3P_0$ transition. The two other curves are time-resolved excitation spectra of the Pr^{3+} luminescence for two different delays [37].

Vial and Buisson have measured directly pair relaxation in diluted crystals and use the obtained results to explain the data for concentrated compositions in an elegant way.

Their measurements are reproduced in Fig. 16. The main line is the $^3H_4$-$^3P_0$ transition on Pr^{3+} in a 0.1 % Pr^{3+} sample. The satellites are due to pairs of ions. The pairs have a shorter decay (see Table 1). The cross relaxation rate W_c follows from $W_c = \tau^{-1} - \tau_0{}^{-1}$, where τ_0 is the radiative lifetime. From these data the authors deduce that the interaction between the Pr^{3+} ions is of the superexchange type.

What is of interest here is that they can explain in this way the luminescence decay in heavily doped crystals. In eq.(1) $X_{0\ell}$ gives the cross relaxation rate for ions of a certain class of pairs with the central ion. The contribution to f(t) of the class ℓ is $(1-x)^{N_\ell}$ for $t >> X_{0\ell}^{-1}$ and $(1)^{N_\ell}$ for $t << X_{0\ell}^{-1}$. Here N_ℓ is the number of equivalent ions which make a certain class of pairs with the central ion.

Table 1. Decay time τ and cross relaxation rate W_c for various pairs in LaF_3-Pr^{3+} (0.1 %) [37].

Line	τ (μs)	W_c (μs^{-1})
a	22	2.5×10^{-2}
b	45.5	1×10^{-3}
c	>47	$<5.4 \times 10^{-4}$
d	>47	$<5.4 \times 10^{-4}$
e	>46.5	$\sim 5.4 \times 10^{-4}$
f	8.5	9.7×10^{-2}

In the experimental time interval (100 - 200 μs) only the pairs a and f contribute, and well by $(1-x)^{N_f+N_a}$. Therefore, the luminescence decay can be described by

$$\Phi(t) \simeq (1-x)^{N_f+N_a} e^{-t/\tau_0}.$$

In the experimental results [36] the exponential variation with τ_0 is indeed observed (and ascribed to Pr^{3+} in non-quenching sites). By comparing this equation with the experimental results it follows that $N_f+N_a \simeq 9$, so that only cross relaxation between one ion and its possible 9 first neighbours has to be considered. Crystallographically $N_f=6$ and $N_a=4$ seems to be an obvious choice. In this way the decay can be described assuming only two types of pairs where cross relaxation occurs (see Fig. 17). These results show convincingly the importance of studying the diluted and the concentrated systems simultaneously.

Auzel [39] has developed a criterion to predict self quenching of Nd^{3+} luminescence in concentrated materials. This author stresses that the energy mismatch in the cross-relaxation process is of large importance. As a matter of fact this mismatch is determined by the crystal field splitting of the Nd^{3+} energy levels involved, viz. $^4F_{3/2}$, $^4I_{15/2}$, $^4I_{9/2}$. The crystal field parameters B^k_q determine a quantity

$$N_v = [\sum_{k'\neq 0,q} (B^k_q)^2]^{1/2}.$$

If $N_v < 1800$ cm^{-1}, we have materials with low quenching. We illustrate this with some data: (La,Nd)Cl_3 $N_v = 928$ cm^{-1}, NdP_5O_{14} $N_v = 1406$ cm^{-1}, (La,Nd)F_3 $N_v = 1841$ cm^{-1}, Li(Y,Nd)F_4 $N_v = 2194$ cm^{-1} and $(Y,Nd)_3Al_5O_{12}$ $N_v = 3575$ cm^{-1} (ref. [39]). The criterion $N_v < 1800$ cm^{-1} corresponds to $\Delta E_1 < 470$ cm^{-1}, where ΔE_1 is the total crystal field splitting of the $^4I_{9/2}$ level. Auzel tried to correlate these criteria to simple physical properties and found

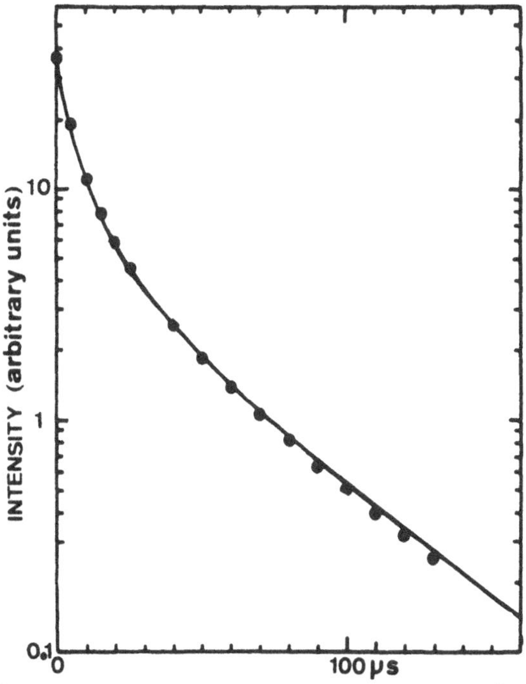

Fig. 17. The $^3P_0$ luminescence decay of a LaF_3-Pr^{3+} (20 %) sample. Points are experimental points from Hegarty et al. [36]. The curve is the theoretical expression given in the text.

also $T_{melting}$ < 1200 °C. From the material presented, his criterion seems to be reliable, although it is based on the Dexter's spectral overlap condition (compare the comment in chapter II).

We will close this paragraph with an application to luminescent materials. If energy migrates through a host lattice, we have available a suitable host lattice for an efficient phosphor. A condition is that the excitation energy can be introduced into this lattice by an allowed transition and that it can be trapped by an activator which takes care of the emission. An example has been given by De Hair [40], viz. $GdAl_3B_4O_{12}$-Bi,Dy. Excitation is into the Bi^{3+} ion via the allowed $^1S_0 \rightarrow {}^3P_1$ transition. Transfer occurs into the Gd^{3+} sublattice in which energy migrates rapidly [41], until it is trapped by the Dy^{3+} ion from which emission occurs.

Recently we have studied the system $(Y,Ce,Gd,Tb)F_3$ [42]. Samples GdF_3-Ce,Tb are efficient green phosphors. The luminescence and energy transfer processes can be given schematically as follows

$$\overset{exc.}{\rightarrow} Ce^{3+} \rightarrow (Gd^{3+})_n \rightarrow Tb^{3+} \overset{em.}{\rightarrow}.$$

In YF_3-Ce,Tb energy transfer from Ce^{3+} to Tb^{3+} is not complete, showing the influence of the Gd^{3+} intermedium. It turns out that about 20 % Gd^{3+} is enough to "connect" the Ce^{3+} and Tb^{3+} ions. In GdF_3-Ce^{3+} excitation into the intrinsic Ce^{3+} ion results in Ce^{3+} defect emission:

$$\stackrel{\text{exc.}}{\rightarrow} Ce^{3+} \rightarrow (Gd^{3+})_n \rightarrow Ce^{3+}(O^{2-}) \stackrel{\text{em.}}{\rightarrow}.$$

These phenomena are probably rather general and comparable with those described above. In $GdNbO_4$, for example, luminescence is quenched at room temperature as follows:

$$\stackrel{\text{exc.}}{\rightarrow} NbO_4^{3-} \rightarrow (Gd^{3+})_n \rightarrow NbO_3\square^-.$$

In the trifluoride system these phenomena are not strikingly temperature dependent in view of the pronounced spectral overlaps involved. It cannot be exluded that attractive new phosphors will be developed on this principle.

VII. FLUORESCENCE LINE NARROWING IN GLASSES

Laser-excited luminescence spectroscopy has contributed considerably to our understanding of glass structure. For an extensive review the reader is referred to Ref. [43]. Since glass is a strongly disordered medium, the environment of each ion in a glass is different and the local fields at individual ion sites vary strongly. The spectra exhibit, therefore, considerably inhomogeneous broadening unless the technique of fluorescence line narrowing (FLN) is used. In crystals this broadening may be about 1 cm^{-1} in the case of the narrow line transitions, but in glasses the same transition may be broadened by some 100 cm^{-1}. Consequently, FLN may be of the greatest importance as a microscopic probe of the local environment in a glass. A typical example of such a study is the work by Brecher and Riseberg [44] on the Eu^{3+} ion in sodium-barium-zinc silicate glass. With the FLN measurements of site-to-site variations in energy levels and transition probabilities, the authors succeeded in finding a geometric model of the glass structure. In this study energy transfer was excluded by studying diluted samples short after the excitation pulse.

A good example of energy migration in a glass studied by FLN is the case of Yb^{3+} in a silicate glass [45]. The time-resolved spectra are given in Fig. 18. The emission corresponds to the transition between the lowest levels of the $^2F_{5/2}$ and $^2F_{7/2}$ manifolds. Following the laser pulse, the initial line-narrowed luminescence decays due to radiative decay and nonradiative transfer to other Yb^{3+} ions without pronounced broadening. However, luminescence from the acceptor ions results in a rise of the inhomogeneously broadened emission profile, demonstrating that transfer occurs to nearby, but spectrally different sites. The transfer was attributed to multipolar processes with an effective range of about 1 nm.

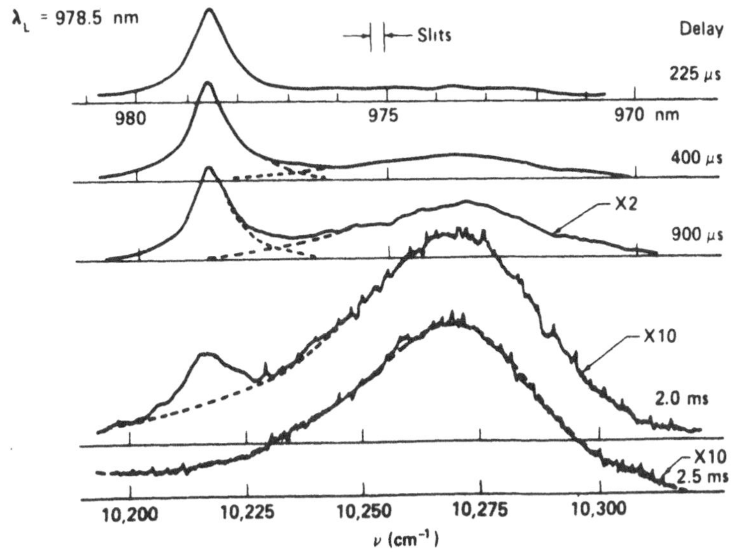

Fig. 18. Time evolution of the resonant line-narrowed $^2F_{5/2} \rightarrow {}^2F_{7/2}$ luminescence of 0.5 % Yb^{3+} in silicate glass at 100 K [45].

Quantitative comparison between transfer rates and the time development of the luminescence with theoretical expressions is very complicated for glasses, because the inhomogeneities are large. As a consequence there is a site-to-site variation of energy levels, line strengths and linewidths. This makes a quantitative description of energy transfer in a glass very difficult.

ACKNOWLEDGMENT

The author is indebted to Everdina M. Bos, who typed the manuscript.

REFERENCES

1. K. H. Butler, Fluorescent lamp phosphors, The Pennsylvania State University (1980).
2. J. L. Sommerdijk, J. A. W. van der Does de Bye, P. H. J. M. Verberne, J. Luminescence 14, 91 (1976).
3. Th. P. J. Botden, Philips Research Reports 7, 197 (1952).
4. D. L. Dexter, J. Chem. Phys. 21, 836 (1953).

5. Laser Spectroscopy of Solids (Eds. W. M. Yen and P. M. Selzer), Springer Verlag, Berlin, 1981.
6. R. C. Powell and G. Blasse, Structure and Bonding 42, 43 (1980).
7. See e.g. T. Holstein, S. K. Lyo and R. Orbach, chapter 2 in [5].
8. M. Inokuti and F. Hirayama, J. Chem. Phys. 43, 1978 (1965).
9. F. Kellendonk and G. Blasse, J. Chem. Phys. 75, 561 (1981).
10. G. Blasse and F. Zonnevijlle, Recueil Trav. Chem. Pays Bas 101, 434 (1982).
11. G. Blasse and G. J. Dirksen, J. Electrochem. Soc. 127, 978 (1980).
12. D. J. Robbins, B. Cockayne, B. Lent and J. L. Glasper, Solid State Comm. 20, 673 (1976).
13. Liu Xing-ren, Changchun Institute of Physics, private communication.
14. D. L. Huber, chapter 3 in [5].
15. D. L. Huber, D. S. Hamilton and B. Barnett, Phys. Rev. B16, 4642 (1977).
16. A. I. Burshtein, Sov. Physics-JETP 35, 882 (1972).
17. M. Yokota and I. Tanimoto, J. Phys. Soc. Japan 22, 779 (1967).
18. B. Blanzat, J. P. Denis and R. Reisfeld, Chem. Phys. Letters 51, 403 (1977).
19. G. Blasse, Chem. Phys. Letters 56, 409 (1978).
20. B. A. Wilson et al., Phys. Rev. B19, 4238 (1979); Phys. Rev. Letters 41, 268 (1978).
21. R. M. Macfarlane and A. C. Luntz, Phys. Rev. Letters 31, 832 (1973).
22. M. Hirano and S. Shionoya, J. Phys. Soc. Japan 28, 926 (1970).
23. H. Yamamoto and D. S. McClure, Chem. Phys. 22, 79 (1977).
24. See e.g. D. Oelkrug, M. Radjaipour and E. Eitel, Spectrochim. Acta 35A, 167 (1979).
25. D. M. Krol and G. Blasse, J. Chem. Phys. 69, 3124 (1978).
26. K. P. de Jong, D. M. Krol and G. Blasse, J. Luminescence 20, 241 (1979).
27. D. M. Krol, Chem. Phys. Letters 74, 515 (1980).
28. D. M. Krol and A. Roos, Phys. Rev. B23, 2135 (1981).
29. F. Kellendonk and G. Blasse, J. Chem. Phys. 75, 561 (1981).
30. P. A. M. Berdowski and G. Blasse, to be published.
31. F. Kellendonk, T. van den Belt and G. Blasse, J. Chem. Phys. 76, 1194 (1982).
32. M. Saakes and G. Blasse, to be published.
33. F. Kellendonk and G. Blasse, J. Phys. Chem. Solids 43, 481 (1982).
34. J. P. van der Ziel, L. Kopf and L. G. Van Uitert, Phys. Rev. B6, 615 (1972).
35. M. M. Broer, D. L. Huber, W. M. Yen and W. K. Zwicker, Phys. Rev. Letters 49, 394 (1982).
36. J. Hegarty, D. L. Huber and W. M. Yen, Phys. Rev. B25, 5638 (1982).
37. J. C. Vial and R. Buisson, J. Physique 43, L745 (1982).

38. H. Dornauf and J. Heber, J. Luminescence 22, 1 (1980); H. Siebold and J. Heber, J. Luminescence 22, 297 (1981).
39. F. Auzel, Mater. Res. Bull. 14, 223 (1979), in Radiationless Processes (Ed. B. Di Bartolo), Plenum Press, New York, 1980, p. 213.
40. J. Th. W. de Hair, J. Luminescence 18/19, 797 (1979).
41. F. Kellendonk and G. Blasse, Phys. Stat. Sol. (b) 108, 541 (1981).
42. G. Blasse, Phys. Stat. Sol. (a) 73, 205 (1982).
43. M. J. Weber, chapter 6 in [5].
44. C. Brecher and L. A. Riseberg, Phys. Rev. B13, 81 (1976).
45. M. J. Weber, J. A. Paisner, S. S. Sussman, W. M. Yen, L. A. Riseberg and C. Brecher, J. Luminescence 12/13, 729 (1976).

Wahrheit und Klarheit sind komplementär
E. Mollwo

ENERGY TRANSFER IN SEMICONDUCTORS

C. Klingshirn

Physikalisches Institut der Universität
Robert-Mayer-Str. 2-4
D 6ooo Frankfurt am Main

ABSTRACT

From the rather large and complex field of energy transfer processes in semiconductors we selected for this article the following topics: The transfer of energy from a photon-field in vacuum through the surface into a semiconductor, the propagating modes of light in a semiconductor (the socalled polaritons), the transfer of energy from exciton-polaritons to the phonon field and finally the transfer of energy from one exciton-polariton mode to others by nonlinear interaction.

I. INTRODUCTION, OR THE PHYSICAL PROBLEM OF LOOKING THROUGH A WINDOW

Interaction processes of "real" particles in vacuum or of quasiparticles in condensed matter are connected with energy and momentum transfer. Since there exist many different quasiparticles in semiconductors, it is obvious, that the field of energy transfer processes in semiconductors will be a rather wide and complex one. Therefore only selected aspects may be presented in a contribution of finite length. Here, we restrict ourselves to the following topics: the transfer of energy from a photon field through the surface into a semiconductor, the propagating modes of light in a semiconductor (the socalled polaritons), the transfer of energy from exciton-polaritons to the phonon field and from one exciton-polariton mode to another by nonlinear interaction. Other aspects of energy-transfer in semiconductors e.g. the energy conversion in solar cells [1] are treated in this volume by other authors. Some important ideas concerning the efficiency of these devices are reviewed in [2]. A comprehensive presentation of photoconductivity is found e.g. in [3], aspects of photoluminescence and exciton migragtion are reviewed e.g. in [4,5].

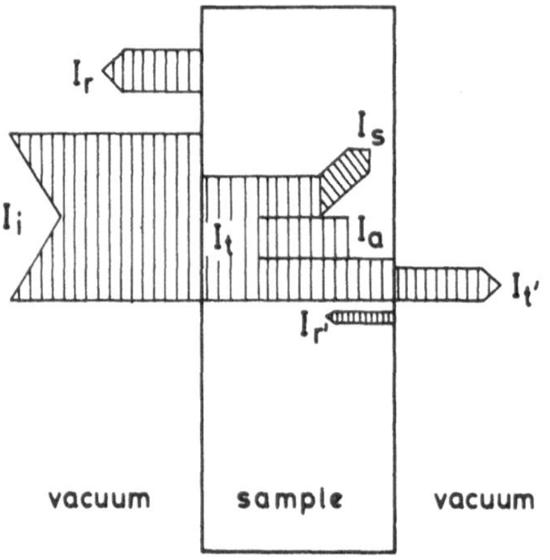

Fig. 1. Schematic presentation of various processes which may occur, when a light beam falls on a slab of matter. I_i incident intensity, I_r and $I_{r'}$ reflected intensities, I_t and $I_{t'}$ transmitted intensities, I_a absorbed intensity and I_s scattered intensity. Interference effects e.g. between the amplitudes of waves I_t and $I_{r'}$ are neglected.

The simple and common phenomenon of looking through a window involves already many aspects of the topics selected for this contribution. To illustrate this and to bring the problem in a physically conceivable form, the process is shown schematically in Fig. 1:
An electromagnetic wave or in other words a photon field comes from the left-hand side. The incident intensity is called I_i (for a more precise definition of "intensity" see II.1). At the surface of the medium a part I_r is reflected and the rest I_t transmitted. The propagating modes of light in the medium are no longer pure electromagnetic waves or simple photons but polaritons as will be discussed in some detail in section II.4. A certain part I_a of the transmitted light is absorbed in the medium on the way from one side of the sample to the other. This means, energy is transformed into other degrees of freedom of the system, often into heat. Another part I_s is diffused in some other directions, either with or without frequency shift. Raman and Brillouin scattering are examples of scattering with frequency shift (see III.4,5), the spectrally unshifted component in the scattered of diffused light is called the Rayleigh scattered part. A third fraction finally reaches the other side of the sample and here again some part is reflected, $I_{r'}$, and some part transmitted, $I_{t'}$. In this example of looking through a window the part $I_{t'}$ finally may reach the eye and produce a picture on the retina of what is on the other side of the window.

In the following we want to discuss a physically well defined situation. So we assume to have vacuum around our medium instead of air. The glass of the window, which is a rather complex, amorphous, supercooled melt is replaced by a perfectly crystallized semiconductor without any impurities, vacancies, dislocations, etc. Since energy transfer processes between dopants are extensively treated in other articles of this volume, we restrict ourselves in the following to the ideal semiconductor, and give only as an appendix some comments about impurities (Appendix 3).

In Chapter II we deduce the dielectric function of a semiconductor from a simple, mechanical model. From the dielectric function, we deduce the polariton concept and apply it to phonons and excitons. The transfer of energy through the surface is investigated in some detail.

In Chapter III we discuss the energy transfer between various systems of quasiparticles with emphasis on the interaction between excitons and phonons.

In Chapter IV finally we present some aspects of energy transfer between various exciton-polariton modes by nonlinear interaction.

In the Appendix we give additional aspects of subjects which are relevant to the field but which would interrupt the logic development of the subject if incorporated in the main text.

To keep the list of references limited, we generally restrict ourselves to the citation of review articles which then contain exhaustive lists of the original literature.

II. ENERGY TRANSFER FROM AN EXTERNAL PHOTON-FIELD INTO A SEMICONDUCTOR

II.A. Photons in Vacuum

To start with, we review in the following the properties of an electromagnetic field in vacuum. In the classical description, the electromagnetic phenomena are governed by Maxwell's equations. They read in their general form

$$\vec{\nabla} \times \vec{E} = -\dot{\vec{B}} \qquad \vec{\nabla} \times \vec{H} = \vec{j} + \dot{\vec{D}} \qquad (1a,b)$$

$$\vec{\nabla} \cdot \vec{D} = \rho \qquad \vec{\nabla} \cdot \vec{B} = 0 \qquad (2a,b)$$

$$\vec{D} = \varepsilon_o \vec{E} + \vec{P} \qquad \vec{B} = \mu_o \vec{H} + \vec{M} \qquad (3a,b)$$

where

$\vec{E}$ = electric field; $\vec{H}$ = magnetic field

$\vec{D}$ = electric displacement; $\vec{B}$ = magnetic induction

ρ = charge density; $\vec{j}$ = current density

$\vec{P}$ = polarization density of the medium

$\vec{M}$ = magnetization density of the medium

ε_o = permittivity of vacuum

μ_o = permeability of vacuum

The Nablaoperator $\vec{\nabla}$ is given by $\vec{\nabla} = [(\partial/\partial x), \quad (\partial/\partial y), \quad (\partial/\partial z)]$ and the operator "dot" means differentiation with respect to the time (e.g., $\dot{A} = \partial/\partial t\ A$).

Equations (1a,b) tell us how temporally varying electric and magnetic fields generate each other. Equations (2a,b) show that the charge density ρ is the source of the electric displacement and that the magnetic induction is source-free. Equations (3a,b) are the material equations. In many cases $\vec{P}$ and $\vec{M}$ are proportional to $\vec{E}$ and $\vec{H}$, respectively. In these linear cases (3a,b) simplify to

$$\vec{D} = \varepsilon\, \varepsilon_o\, \vec{E} \qquad \text{and} \qquad \vec{B} = \mu\, \mu_o\, \vec{H} \tag{4a,b}$$

with ε known as dielectric "constant" or relative permittivity and μ as relative permeability. In the following we discuss only non-magnetic materials, i.e. $\mu = 1$. The quantity ε is indeed a function of ω for periodic fields, as will be shown in Sections II.2 and II.3.

In vacuum we have

$$\vec{P} = 0 \Rightarrow \varepsilon = 1 \tag{5}$$

$$\vec{M} = 0 \Rightarrow \mu = 1 \tag{6}$$

$$\vec{j} = 0 \qquad \rho = 0 \tag{7a,b}$$

This simplifies Eqs. (1a,b) to

$$\vec{\nabla} \times \vec{E} = -\mu_o\, \dot{\vec{H}} \qquad \vec{\nabla} \times \vec{H} = \varepsilon_o\, \dot{\vec{E}} \tag{8a,b}$$

Applying $\vec{\nabla}\times$ to Eq. (8a) and $\partial/\partial t$ to Eq. (8b) gives

$$\vec{\nabla} \times (\vec{\nabla} \times \vec{E}) = -\mu_o \vec{\nabla} \times \dot{\vec{H}} \qquad \vec{\nabla} \times \dot{\vec{H}} = \varepsilon_o \vec{E} \tag{9a,b}$$

From (9a,b) we find

$$-\mu_o \varepsilon_o \ddot{\vec{E}} = \vec{\nabla} \times (\vec{\nabla} \times \vec{E}) = \vec{\nabla}(\vec{\nabla} \cdot \vec{E}) - \vec{\nabla}^2 \vec{E} \tag{10}$$

The first term on the righthand side of (10) is zero because of (2a) and (7b) and we are left with the well-known wave equation

$$\vec{\nabla}^2 \vec{E} - \mu_o \varepsilon_o \ddot{\vec{E}} = 0 \tag{11}$$

The solutions are transverse, electromagnetic waves. For the case of plane waves we can write

$$\vec{E} = \vec{E}_o \, e^{i(\vec{k}\cdot\vec{r}-\omega t)} \tag{12}$$

We find from (1a)

$$\vec{B} = \omega^{-1} \vec{k} \times \vec{E} = \vec{B}_o \, e^{i(\vec{k}\cdot\vec{r}-\omega t)} \tag{13}$$

where

$\vec{k}$ = wave-vector $|\vec{k}| = k = 2\pi/\lambda$

λ = wavelength

ω = angular frequency

For a non-absorbing medium $\vec{k}$ is real. For an absorbing one, $\vec{k}$ is a complex quantity, where Re{k} describes the wave propagation and Im{k} the absorption. (See e.g. Section II.5). If not stated otherwise we denote in the following by $\vec{k}$ always the real part.

The wavevector $\vec{k}$ is always (also in medium) perpendicular to the wavefront defined by $\vec{D}$ and $\vec{B}$

$$\vec{D} \perp \vec{k} \perp \vec{B} \perp \vec{D} \tag{14}$$

In vacuum (and isotropic materials) we have

$\vec{D} \parallel \vec{E}$ and $\vec{B} \parallel \vec{H}$

The phase velocity is defined by

$$v_p(\omega) = \frac{\omega}{k} \tag{15}$$

and the group velocity by

$$v_g(\omega) = \frac{d\omega}{dk} \tag{16}$$

In vacuum we find

$$v_p = v_g = c = (\varepsilon_o \mu_o)^{-1/2} \tag{17}$$

c being the vacuum speed of light.

The momentum of the electromagnetic field per volume $\vec{\pi}$ is given by

$$\vec{\pi} = \vec{D} \times \vec{B} \tag{18}$$

The Poynting vector $\vec{S}$ gives the energy flux density

$$\vec{S} = \vec{E} \times \vec{H} \tag{19}$$

Since $\vec{S}$ is a rapidly varying function of time (see (12) and (13)) one generally considers its time average

$$\langle S \rangle = \frac{1}{2} |\vec{E}_o \times \vec{H}_o| \tag{20}$$

The quantity $\langle S \rangle$ is often referred to as light intensity I or (average) energy flux density.

In the picture of quantum electrodynamics, the electromagnetic field is quantized in a way that the energy of this field may be increased or decreased only by integer multiples of $\hbar\omega$, resolving thus the dualism of wave and particle. The quanta of energy $\hbar\omega$ are called photons. The Hamiltonian for the photons reads

$$H = \sum_{\vec{k}} \hbar\omega_{\vec{k}} \, c^+_{\vec{k}} \, c_{\vec{k}} \tag{21}$$

where the creation and annihilation operators $c_{\vec{k}}$ and $c^+_{\vec{k}}$, respectively, obey the usual commutation relations for Bosons. The sum runs over all possible values of $\vec{k}$ and two polarizations for every $\vec{k}$. The wavevector $\vec{k}$ is used as a label for the various photon modes. $c^+_{\vec{k}} \cdot c_{\vec{k}}$ is the number operator. The energy in a single photon-mode $\vec{k}$ is then given by

$$c_{\vec{k}}^{+} c_{\vec{k}} | n_{\vec{k}} \rangle = (n_{\vec{k}} + \frac{1}{2})\, \hbar\omega_{\vec{k}} \quad ; \quad n_{\vec{k}} = 0,\ 1,\ 2,\ \ldots \tag{22}$$

The zero point energy $(1/2)\hbar\omega_{\vec{k}}$, which is responsible for the stimulated emission, is generally neglected in energy considerations. The momentum of a photon is given by

$$\vec{P} = \hbar\, \vec{k} \tag{23}$$

The energy flux density of the photon-field is given by

$$\langle S \rangle = \sum_{\vec{k}} N_{\vec{k}}\, \hbar\omega_{\vec{k}}\, v_{\vec{k}} \tag{24}$$

where $N_{\vec{k}}$ is the density of photons $\vec{k}$, and $v_{\vec{k}}$ their energy velocity with $v_{\vec{k}} = c$ in vacuum. The momentum per volume reads

$$\langle \vec{\pi} \rangle = \sum_{\vec{k}} N_{\vec{k}}\, \hbar\vec{k} \tag{25}$$

The relation between eigenenergy E and wavevector $\vec{k}$ is called in the following "dispersion relation." By this relation we shall characterize the various (quasi-) particles. The reason for this choice is that there are conservation laws for both the (quasi-) momentum $\hbar\vec{k}$ and the energy $E = \hbar\omega$.

For photons the dispersion relation is given by $E = cp = c\hbar k$ (Figure 2). As in the classical limit we find

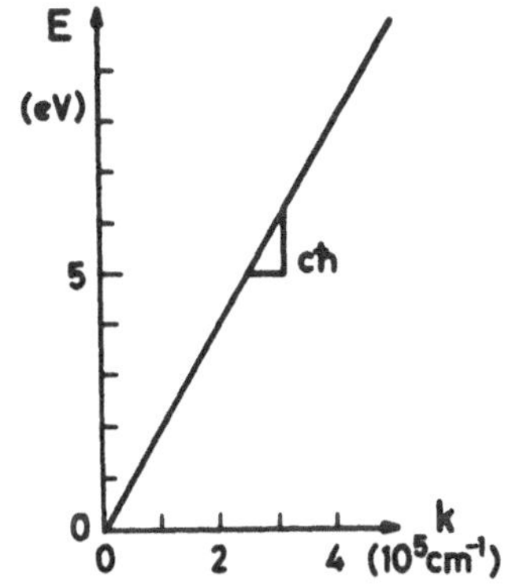

Fig. 2. The dispersion relation E(k) for photons in vacuum.

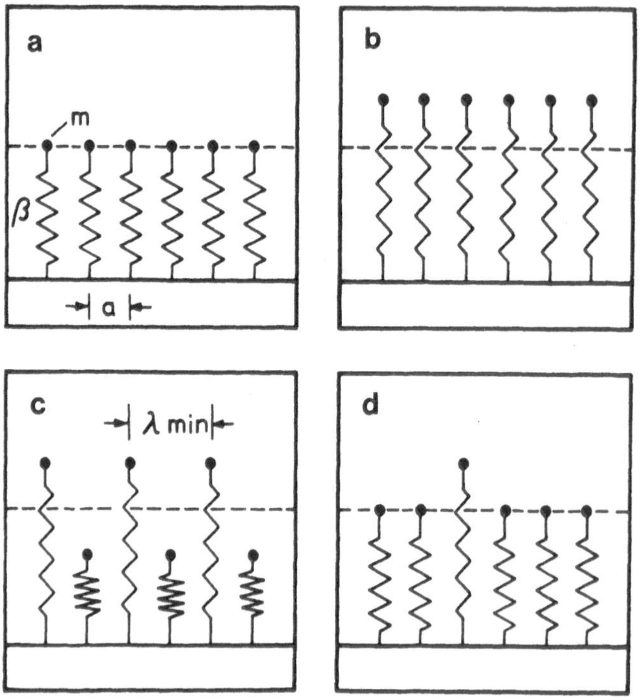

Fig. 3. An ensemble of uncoupled harmonic oscillators, representing excitations in a semiconductor. According to [6]
(a) all oscillators in their equilibrium positions
(b) all oscillators elongated in phase ($\lambda = \infty$)
(c) all oscillators elongated in antiphase ($\lambda = 2a$)
(d) a "wave-packet"

$$v_p = \frac{1}{\hbar} E/k = c \quad ; \quad v_g = \frac{1}{\hbar}\frac{dE}{dk} = c \tag{26}$$

II.B. A Mechanical Model for a Medium

The aim of this subjection is to present a simple mechanical model from which we may deduce in the next subsection the dielectric function $\varepsilon(\omega)$ of semiconductors. In a semiconductor (insulator) the forces between the ions are in first approximation harmonic. The electrons, which are responsible for the chemical binding are situated energetically in the valenceband, which is completely filled for T = OK and which is separated from the conduction band by a finite energy gap E_g. The transitions between the bands may be considered as "effective oscillators." It may thus be a reasonable approximation to represent a semiconductor by an ensemble of various harmonic oscillators. For simplicity, we assume in the moment that these oscillators are not coupled to each other, that they have all the same eigenfrequency ω_o', and that we have one of these oscillators per unit cell.

The mechanical model which represents the oscillators by masses m bound to the rigid lattice-site by a spring with force constant β looks like Figure 3a in the one-dimensional case, with $\omega_o' = (\beta/m)^{1/2}$.

If we elongate the oscillators in phase, Figure 3b, corresponding to the limit $\lambda \Rightarrow \infty$ or $\vec{k} \Rightarrow 0$, the whole ensemble will oscillate with the frequency ω_o'. If we excite the oscillators in antiphase (Figure 3c), the same will be true. The latter case corresponds to the shortest physically reasonable wavelength $\lambda_{min} = 2a$ or to the maximum k-vector $k_{max} = \pi/a$ ($\hat{=}$ boundary of first Brillouin-zone). For all intermediate cases $0 \leq k \leq \pi/a$ the eigenfrequency is also ω_o'. Thus a dispersion-relation results which is a horizontal line (Figure 4). A wavepacket produced e.g. by exciting only on of the oscillators (Figure 3d) will not propagate since the group velocity $v_g = d\omega/dk$ vanishes in this system, or, in other words, since there is no coupling between the model-oscillators. This shows one of the physical shortcomings of our model. Some consequences of the case

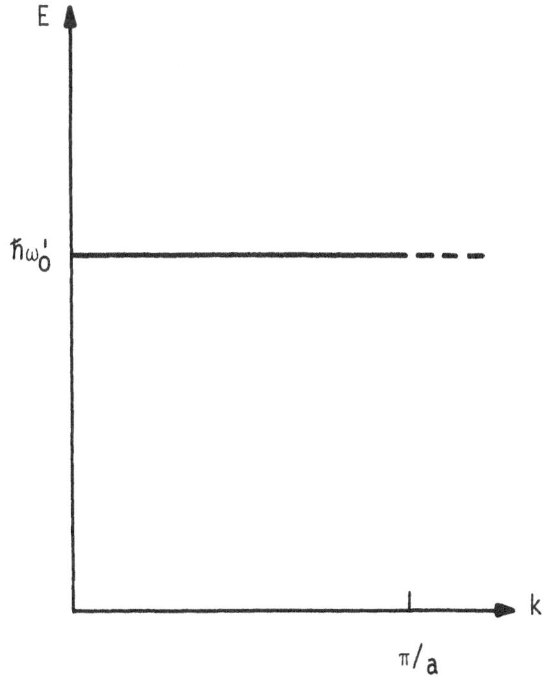

Fig. 4. The dispersion relation of waves in an ensemble of uncoupled oscillators which have all the same eigenfrequency ω_o'.

$d\omega/dk \neq 0$ will be discussed in Sections II.G and II.H.

Now we want to couple our oscillators to the electromagnetic field of the light. This can be done by assuming that every mass is connected with a charge e. For neutrality-reasons we have to assume that at every equilibrium position there is a fixed charge - e. An electric field will exert a force on the mass and elongations of the oscillators by the amount x are now connected with a dipole-moment e x. Still we assume that the oscillators interact only with an external electromagnetic field but not with each other.

If now a plane electromagnetic wave polarized in x-direction travels through the medium in z-direction given by

$$\vec{E} = (E_o, 0, 0)\, e^{i(k_z z - \omega t)} \tag{27}$$

the differential equation for the motion of the oscillators at every lattice site is given by

$$m\ddot{x} + \gamma\, m\dot{x} + \beta\, x = e\, E_o\, e^{-i\omega t} \tag{28}$$

To be physically realistic, we introduced a small damping term γ in Eq. (28). Equation (28) is the differential equation of an externally driven, damped oscillator. After switching on the external force $e\, E_o\, e^{-i\omega t}$ at a time $t_o = 0$, the reaction of the system consists of two parts, shown by Eq. (29)

$$x(t) = x_o\, e^{-i(\omega_o'^2 - \gamma^2/2)^{1/2} t}\, e^{-t\gamma/2} + x_p\, e^{-i\omega t} \tag{29}$$

The first term of the righthand side of (29) is a damped oscillation with the eigenfrequency of the system. It disappears after $t \gg \gamma^{-1}$. The second part describes the forced oscillation with the frequency of the perturbation.

The amplitude x_p is given by the resonance formula

$$x_p = \frac{e/m}{\omega_o'^2 - \omega^2 - i\omega\gamma}\, E_o \tag{30}$$

This oscillation is connected with a dipole moment ex_p and we can define a polarizability $\hat{\alpha}(\omega)$.

$$\hat{\alpha}(\omega) = \frac{ex_p}{E_o} = \frac{e^2/m}{\omega_o'^2 - \omega^2 - i\omega\gamma} \tag{31}$$

II.C. The Dielectric Function

The polarization density of the model substance is given by

$$P_x = N\,\hat{\alpha}\,E_o = \frac{Ne^2}{m}\,\frac{1}{\omega_o'^2 - \omega^2 - i\omega\gamma}\,E_o \tag{32}$$

From this we conclude with (3a) for the electric displacement

$$\vec{D} = \varepsilon_o\,\vec{E} + \vec{P} = \varepsilon_o\left(1 + \frac{Ne^2/m\,\varepsilon_o}{\omega_o'^2 - \omega^2 - i\omega\gamma}\right)\vec{E} \tag{33}$$

Comparing (33) with Eq. (4a) we find

$$\varepsilon(\omega) = 1 + \frac{Ne^2/m\,\varepsilon_o}{\omega_o'^2 - \omega^2 - i\omega\gamma} \tag{34}$$

The quantity $\chi(\omega) = \varepsilon(\omega) - 1$ is introduced in the literature as susceptibility.

A quantum mechanical treatment will yield the same result, except that the term $Ne^2/m\,\varepsilon_o$ is replaced by a term f' proportional to the square of the dipol-matrix element between ground- and excited states of the oscillators

$$f' \sim |\langle f|H_{Dipol}|i\rangle|^2 \tag{35}$$

yielding thus

$$\varepsilon(\omega) = 1 + \frac{f'}{\omega_o'^2 - \omega^2 - i\omega\gamma} \tag{36}$$

Until now, we silently assumed that the macroscopic field $\vec{E}$ of the electromagnetic wave travelling through the medium is equal to the local electric field $\vec{E}^{loc}$ acting on a single oscillator. This is true only in dilute systems like gases. In condensed matter the electric fields produced by the other dipols act on the dipol under consideration, too. Taking into account this effect leads to Eq.

(37) instead of Eq. (36). The equation (37) is known as Clausius-Mosotti formula. Its derivation is given in every textbook on optics, e.g. [7],[8].

$$\frac{\varepsilon(\omega) - 1}{\varepsilon(\omega) + 2} = \hat{\alpha}(\omega) = \frac{Ne^2/3m\,\varepsilon_o}{\omega_o'^2 - \omega^2 - i\omega\gamma} \tag{37}$$

It can be shown that (37) can be rewritten in the form (36), however, with a shifted eigenfrequency ω_o given by

$$\omega_o^2 = \omega_o'^2 - \frac{Ne^2}{3m\,\varepsilon_o} = \omega_o'^2 - f'/3 \tag{38}$$

In a semiconductor we can only measure the eigenfrequency ω_o. It is not possible to produce a "dilute" semiconductor. So we just continue to use Eq. (36), however, with ω_o instead of ω_o'.

To make our model somewhat more realistic, we assume that we have in every unit cell oscillators with various eigenfrequencies ω_{oj}. Equation (36) then becomes

$$\varepsilon(\omega) = 1 + \sum_j \frac{f_j'}{\omega_{oj}^2 - \omega^2 - i\omega\gamma_j} \tag{39}$$

For frequencies $\omega << \omega_o$ the resonance term in (36) and (39) gives a constant contribution

$$\varepsilon_b = \frac{f'}{\omega_o^2} \tag{40}$$

For $\omega >> \omega_o$ the contribution of the resonance term tends to zero. If we are spectrally in the vicinity of a single resonance j' and the other resonances $j \neq j'$ are reasonably far away, Eq. (39) can be approximated by

$$\varepsilon(\omega) = \varepsilon_b\left(1 + \frac{f_{j'}'/\ _b}{\omega_{oj'}^2 - \omega^2 - i\omega\gamma_{j'}}\right) \tag{41}$$

where ε_b is the background dielectric constant. It describes the contribution of all oscillators with frequencies higher than $\omega_{oj'}$. The quantity

$$f_{j'} = f_{j'}'/\varepsilon_b \tag{42}$$

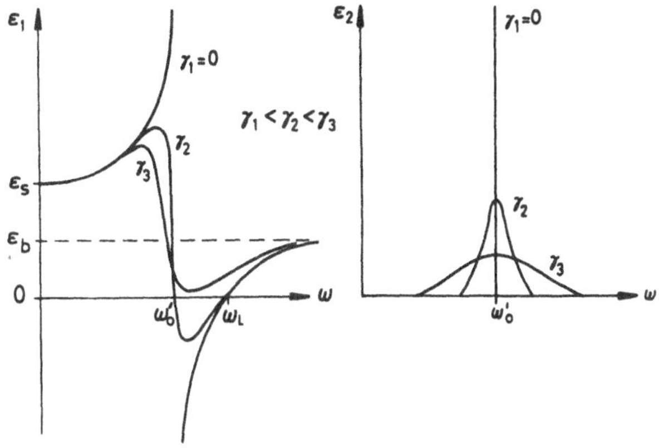

Fig. 5. Schematic representation of the frequency dependence of real and imaginary parts $\varepsilon_1(\omega)$ and $\varepsilon_2(\omega)$ of the dielectric function in the vicinity of a resonance for various values of the damping γ.

is called oscillator-strength $f_{j'}$ of the resonance j'.

In the following we regard for simplicity a single resonance j' well-separated from the other resonances, and we drop the index j'. The dielectric function $\varepsilon(\omega)$ can be separated into a real and imaginary part

$$\varepsilon(\omega) = \varepsilon_b\left(1 + \frac{f}{\omega_o^2 - \omega^2 - i\omega\gamma}\right) = \varepsilon_1(\omega) + i\varepsilon_2(\omega)$$

$$= \varepsilon_b\left(1 + \frac{(\omega_o^2 - \omega^2)\, f}{(\omega_o^2 - \omega^2)^2 + \omega^2\gamma^2}\right) + i\varepsilon_b \frac{\omega\gamma f}{(\omega_o^2 - \omega^2)^2 + \omega^2\gamma^2} \tag{43}$$

In Figure 5 we show qualitatively the frequency dependence of $\varepsilon_1(\omega)$ and $\varepsilon_2(\omega)$ for various values of the damping parameter γ. For $\gamma \Rightarrow 0$ ε_1 exhibits a singularity at ω_o and ε_2 becomes a δ-like function centered at ω_0. With increasing γ $\varepsilon_1(\omega)$ has always finite values and $\varepsilon_2(\omega)$ broadens to a Lorentzian-like curve.

Two comments about $\varepsilon(\omega)$ shall be added: We treated $\varepsilon(\omega)$ as a scalar-function until now. This is true for isotropic materials only. Semiconductors with cubic symmetry are, e.g., isotropic within a good approximation. For anisotropic materials $\varepsilon(\omega)$ is a tensor. As a consequence $\vec{E}$ and $\vec{D}$ are for certain orientations no longer parallel. For these cases the vectors of energy propagation $\vec{S}$ (Poynting vector) $\vec{S} = \vec{E} \times \vec{H}$ and for the field-momentum $\vec{\pi} = \vec{D} \times \vec{B} \,||\, k$ are no longer parallel and we encounter the wealth of phenomena connected with optical birefringence (see, e.g. [7],[8]). These aspects are beyond this review on energy transfer, however, and we shall use in the following the isotropic approximation throughout.

Without real charges Eq. (2a) reads

$$\vec{\nabla} \cdot \vec{D} = 0 \qquad \text{or} \qquad \varepsilon_o \, \vec{\nabla} \, \varepsilon(\omega) \cdot \vec{E} = 0 \tag{44}$$

In vacuum $\varepsilon(\omega) \equiv 1$ and the only solution is $\vec{\nabla} \cdot \vec{E} = 0$ yielding transverse waves.

In matter, we have this solution as the main one, too. As can be seen from Figure 5, Eq. (44) can be fulfilled for small damping γ also by $\varepsilon(\omega) = 0$. For $\varepsilon = 0$ a longitudinal eigenmode exists with frequency ω_L which is a pure polarization mode with $\vec{E} \,||\, \vec{k}$; $\vec{D} = 0$, $\vec{B} = \vec{H} = 0$; $\vec{P} = -\varepsilon_o \vec{E}$.

We deduce from (41)

$$\varepsilon(\omega_L) = 0 \Rightarrow \omega_L^2 = \omega_o^2 + f \tag{45}$$

the eigenfrequency ω_o in turn corresponds to the resonance for transverse oscillations ω_T. So we can rewrite Eq. (41)

$$\varepsilon(\omega) = \varepsilon_b \left(1 + \frac{\omega_L^2 - \omega_T^2}{\omega_T^2 - \omega^2 - i\omega\gamma}\right) \tag{46}$$

and we find immediately the Lyddane-Sachs-Teller relation

$$\frac{\varepsilon_s}{\varepsilon_b} = \frac{\omega_L^2}{\omega_T^2} \tag{47}$$

where ε_s and ε_b are the static and background dielectric constants, i.e. the values of $\varepsilon(\omega)$ according to Eq. (41) for frequencies much smaller or much larger than ω_T, respectively. One finds

$$\varepsilon_s = \varepsilon_b(1 + f/\omega_T^2) \qquad (48)$$

If an oscillator couples to the electromagnetic light field (i.e. $f \neq 0$) a finite longitudinal-transverse splitting and a finite difference between ε_s and ε_b are necessary consequences as seen from Eqs. (47) and (48) and vice versa.

The term "small damping" used already several times can be defined more precisely as

$$\gamma < h^{-1} \Delta_{LT} = \omega_L - \omega_T \qquad (49)$$

Δ_{LT} being the longitudinal-transverse splitting.

The complex dielectric function $\varepsilon(\omega)$ is connected with the complex index of refraction $\tilde{n}(\omega)$

$$\tilde{n}(\omega) = n(\omega) + ik(\omega)$$

$$\varepsilon(\omega) = \tilde{n}^2(\omega) \qquad (50)$$

and

$$\varepsilon_1(\omega) = n^2(\omega) - k^2(\omega)$$

$$\varepsilon_2(\omega) = 2n(\omega)\ k(\omega) \qquad (51)$$

The set of Eq. (51) is an implicit representation of $n(\omega)$ and $k(\omega)$ since $\varepsilon_1(\omega)$ and $\varepsilon_2(\omega)$ are known. In Figure 6 we represent $n(\omega)$ and $k(\omega)$ schematically for various damping values γ.

The comment about the tensor-character of $\varepsilon(\omega)$ given above results in an orientation dependence of $n(\omega)$. With the known complex index of refraction we are now able to describe the propagation of light in matter.

The wavevector in vacuum and the complex wavevector in medium $\vec{k}_V$ and $\vec{k}_M$, respectively, are connected by

$$\vec{k}_M = \vec{k}_V\ \tilde{n}(\omega) \qquad (52)$$

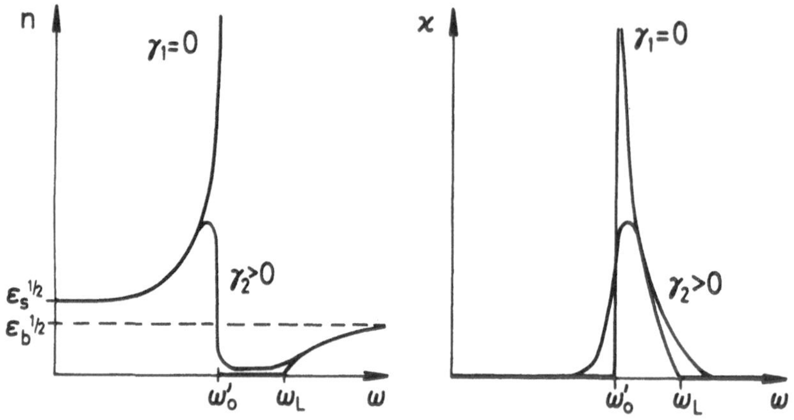

Fig. 6. Schematic representation of the real and imaginary parts $n(\omega)$ and $k(\omega)$ of the complex refractive index in the vicinity of a resonance for two different values of γ.

If we have in vacuum a plane electromagnetic wave (again $\vec{E} \parallel$ x-axis and $\vec{k} \parallel$ z-axis)

$$\vec{E}(z,t) = (E_{vx}, 0, 0)\, e^{i(k_v z - \omega t)} \tag{53}$$

the corresponding wave in the medium reads as:

$$\vec{E}_M = (E_{Mx}, 0, 0)\, e^{i(k_v \tilde{n}(\omega) z - \omega t)}$$

$$= (E_{Mx}, 0, 0)\, e^{i(k_v n(\omega) z - \omega t)}\, e^{-k_v k(\omega) z} \tag{54}$$

$k(\omega)$ obviously describes the absorption of the light. Since the intensity is proportional to the amplitude squared one gets

$$I \sim E^* E \sim e^{-2k_v k(\omega) z} \tag{55}$$

On the other hand we have

$$I(z) = I_o\, e^{-\alpha(\omega) z} \tag{56}$$

$\alpha(\omega)$ being the absorption coefficient. From the comparison of Eqs. (55) and (56) we find

$$\alpha(\omega) = 2k_v \; k(\omega) = \frac{1}{c} \, 2\omega \; k(\omega) = \frac{4\pi}{\lambda_v} \; k(\omega) \tag{57}$$

In principle, we are able to describe with Eqs. (52) to (57) and (10) or (13) for the magnetic field the propagation of light in any medium if $\varepsilon(\omega)$ or $\tilde{n}(\omega)$ are known. However, it is worthwhile to think a little bit more what we mean by the term "light propagating through a medium."

II.D. Polaritons

In Section II.1 we have seen that the light field in vacuum is a pure electromagnetic wave. The quantization yields the photons. If light propagates through a medium, we have always a superposition of an electromagnetic wave and the driven vibrations of the oscillators in the medium. In other words, we have a mixture of an electromagnetic wave and a polarization wave according to

$$\vec{D} = \varepsilon_o \, \vec{E} + \vec{P} = \varepsilon(\omega) \; \varepsilon_o \; \vec{E} \tag{58}$$

The Hamiltonian of this mixed state of electromagnetic field and polarization can be diagonalized (see, e.g. [6],[9]) and the quanta are called polaritons.

The energy of light travelling through matter propagates as polaritons. This is true for insulators, semiconductors and metals, and it is true from the far IR over the visible to the far UV. Only for frequencies ω that are well above all eigenfrequencies of ω_{oj} of the medium (Eq. (39)) we find $\varepsilon(\omega) \equiv 1$. This occurs in the spectral region of γ-quanta and means that there is virtually no more coupling between electromagnetic field and matter. It is only in this region that pure photons can propagate through matter. For all lower frequencies we have in matter only polaritons and no photons.

Now, we want to characterize the polaritons by their dispersion relation (E(k). In vacuum the total energy and the total momentum are conserved. In matter the total energy is conserved (e.g. in an interaction process) the total wavevector $\sum_i \vec{k}_i$ of the involved particles is conserved in a crystalline solid only modulo integer multiples of vectors of the reciprocal lattice. For this reason, the quantity is referred to as "quasi-momentum" in crystals.

Until now, we have $\varepsilon(\omega)$. To find $\omega(k)$ or $E(k) = \hbar\omega(k)$ we need a relation between $\varepsilon(\omega)$ and k. This relation is given by the so-called polariton equation introduced by Huang [10]

$$\varepsilon(\omega) = \frac{c^2 k^2}{\omega^2} \tag{59}$$

This equation is valid for complex ε and $\vec{k}$. If we consider for the moment only the real parts of ε, $\tilde{n}$ and $\vec{k}$, Eq. (59) is easily intelligible by

$$n(\omega) = \frac{c}{c_M(\omega)} = \frac{ck}{\omega} \quad ; \qquad n^2(\omega) = \varepsilon_1(\omega) = \frac{c^2 k^2}{\omega^2} \tag{60}$$

where c_M and k are the phase velocity of light and the wavevector in the medium.

From Eqs. (59) and (41) we find

$$\frac{c^2 k^2}{\omega^2} = \varepsilon(\omega) = \varepsilon_b\left(1 + \frac{f}{\omega_o^2 - \omega^2 - i\omega\gamma}\right) \tag{61}$$

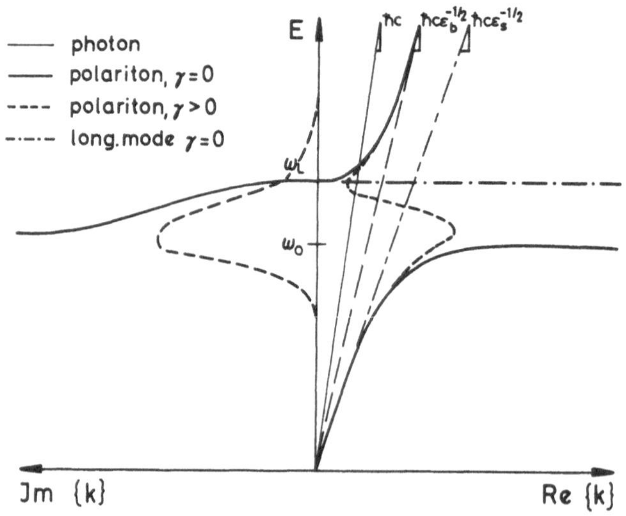

Fig. 7. Schematic representation of the polariton dispersion E(k) for a single resonance, without and with damping.

The right- and lefthand sides of (61) are an implicit representation of E(k). In Figure 7 we show the dispersion for the real and imaginary parts of k for $\gamma = 0$ and $\gamma > 0$. First we discuss the case $\gamma = 0$. The dispersion starts for small energy as a straight line with a slope $dE/dk = \hbar c \varepsilon_s^{-1/2}$. Below the transverse eigenfrequency ω_0 the dispersion bends over to a horizontal line. This part of the dispersion curve is called lower polariton branch (LPB). Between the transverse and longitudinal eigenfrequencies, there exists no propagating mode in the material, k being purely imaginary. At ω_L exists the longitudinal eigenmode as discussed in Section II.3. For $\omega \geq \omega_L$ starts the upper polariton branch (UPB) which bends over to a straight line with slope $\hbar c \varepsilon_s^{-1/2}$. The parts of the polariton-dispersion which coincide with lines with slope $\hbar c \varepsilon_s^{-1/2}$ and $\hbar c \varepsilon_b^{-1/2}$ respectively are called "photon-like" since the dispersion of photons is also a straight line. However, this name is somehow misleading, because we are dealing with polaritons in the "photon-like" part of the dispersion, too. This is immediately obvious by comparison with the dispersion of pure photons which is also shown in Figure 7 and which has a slope of $\hbar c$, differing thus from the "photon-like" parts of LPB and UPB by a factor $\varepsilon_{s,b}^{-1/2}$. For $\gamma > 0$ the dispersion is modified as shown by the dashed lines. Between ω_0 and ω_L occurs a propagating mode with anomalous dispersion, which is strongly damped, however.

The k-vectors of upper and lower polariton branches get some imaginary part (i.e. some sbsorption; see Eq. (57)) which decreases with increasing distance from the resonance.

To summarize this subsection, we have seen that the energy flux of a light field is carried by photons in vacuum and by polaritons in the medium. In the next subsection, we shall discuss what happens at the interface between vacuum and medium.

II.E. What Happens at the Surface

In Fig. 8 are schematically shown the wavefronts of a plane electromagnetic wave, travelling in vacuum and impinging on the surface of a semiconductor. A part of the incident intensity is reflected, the other part is transmitted through the surface and propagates as a polariton field.

The data of the incident wave (electric and magnetic field amplitudes $\vec{E}_i$, $\vec{B}_i$ and wavevector $\vec{k}_i$) are given and the values for the reflected and the transmitted wave have to be calculated. Since we have two unknown waves, we need two boundary conditions. They can be deduced from Maxwell's equations and state that the components of the electric field $\vec{E}_p$ parallel to the surface and the normal components of the magnetic induction $\vec{B}_n$ must be steady through the surface, i.e.

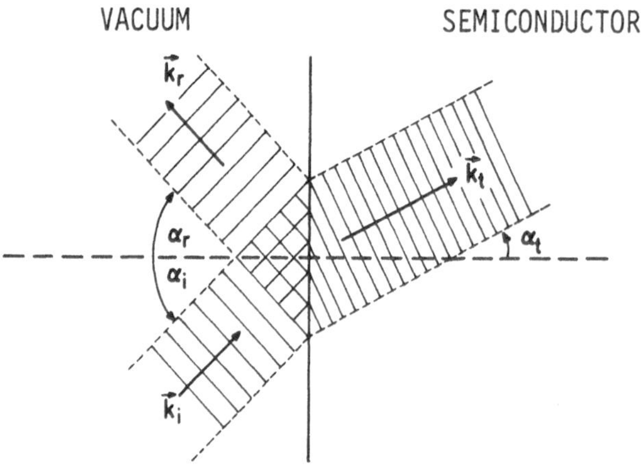

Fig. 8. Schematic representation of the reflection and refraction of a plane wave at the interface vacuum-semiconductor. The wave-vectors and a "snapshot" of the wavefronts are shown. Diffraction which occurs at the boundaries of the beams is neglected.

$$\vec{E}_{pi} + \vec{E}_{pr} = \vec{E}_{pt} \quad ; \qquad \vec{B}_{ni} + \vec{B}_{nr} = \vec{B}_{nt} \tag{62}$$

where the indices i, r and t refer to the incident, reflected and transmitted wave, respectively. The quantitative treatment yields the following results (see, e.g., [7],[8]). The laws of reflection and refraction can be expressed as:

$$\frac{\sin \alpha_i}{\sin \alpha_t} = n(\omega) \qquad \alpha_i = \alpha_\tau \tag{63}$$

$n(\omega)$ being the real part of the refractive index of the medium. They can also be written in terms of the real parts of the wave-vectors. They read:

$$k_i = k_r$$

$$k_t = n(\omega)\ k_i \tag{64}$$

Since the problem has translational invariance parallel to the surface (modulo integer multiples of lattice vectors parallel to the surface), the parallel components $\vec{k}_p$ of the wavevector are conserved (modulo integer multiples of reciprocal lattice vectors parallel to the surface), thus

$$\vec{k}_{p_i} = \vec{k}_{p_r} = \vec{k}_{p_t} \tag{65}$$

For the normal components we find then from (64) and (65)

$$\vec{k}_{ni} = -\vec{k}_{nr}$$

$$k_{nt}^2 = k_i^2 \, n^2(\omega) - k_{p_t}^2 \tag{66}$$

Furthermore, the reflectivity and the transmittivity for the field amplitudes

$$r = \frac{E_r}{E_i} = r(\alpha_i, \tilde{n}(\omega), \theta)$$

$$t = \frac{E_t}{E_o} = t(\alpha_i, \tilde{n}(\omega), \theta) \tag{67}$$

may be deduced by applying (62).

The quantities r and t depend on the angle of incidence α_i on the complex refractive index $\tilde{n}(\omega)$ and on the angle θ defined as the angle between the direction of $\vec{E}_i$ and the plane of the main section. The main section contains $\vec{k}_i$ and $\vec{k}_t$. With $\theta = 0°$ we describe linearly polarized light, with $\vec{E}$ in the main section, and with $\theta = 90°$ light with $\vec{E}$ perpendicular to the main section. The explicit representation of Eq. (67), known as Fresnel's formulas, is somehow lengthy and is found, e.g., in [7],[8].

For reflectivity under normal incidence $r(\alpha_i = 0, \tilde{n}, \theta)$ simplifies to

$$r(\alpha_i = 0) = \frac{\tilde{n}(\omega) - 1}{\tilde{n}(\omega) + 1} \tag{68}$$

The reflectivity R and transmittivity T of the surface for the light intensities are given by

$$R = r^* \cdot r \quad , \quad T = t^* \cdot t \quad , \quad R + T = 1 \tag{69}$$

For $\alpha_i = 0$ we find from (68) and (50)

$$R = \frac{[n(\omega) - 1]^2 + k^2(\omega)}{[n(\omega) + 1]^2 + k^2(\omega)} \tag{70}$$

Figure 9 gives a summary of some of the main aspects concerning Eq. (70). First we discuss the case for $\gamma = 0$. Figure 9a shows the refractive index of the LPB and the UPB; Figure 9b shows on the same energy scale the reflectivity $R(\omega)$.

Coming from lower energies, R starts with a value given by

$$R = [(\varepsilon_s^{1/2} - 1)(\varepsilon_s^{1/2} + 1)]^2$$

Approaching the resonance frequency, R increases and reaches the value $R = 1$ for $\omega = \omega_0$. Since we have no propagating modes in the medium for $\omega_0 < \omega < \omega_L$ we obtain total reflection in this region ($R = 1$). For $E > \hbar\omega_L$ R drops very rapidly and is zero for the value where $n(\omega) = 1$. Above this value R increases gradually and reaches finally the high frequency limit

$$R = [(\varepsilon_b^{1/2} - 1)/(\varepsilon_b + 1)]^2$$

The part of the light intensity which is not reflected is transmitted

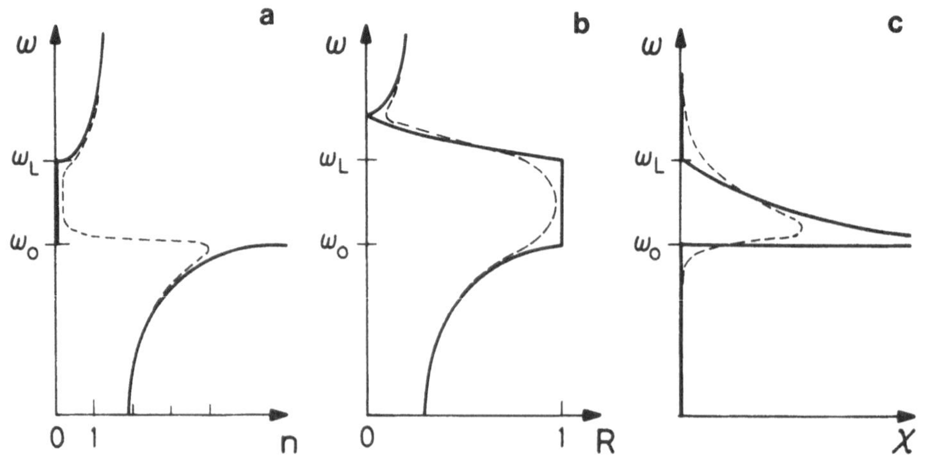

Fig. 9. Real part of refractive index $n(\omega)$, reflectivity $R(\omega)$ and imaginary part $k(\omega)$ in the vicinity of a resonance. Full line: without damping; dashed line: weak damping.

through the surface, penetrates into the sample and propagates according to Figures 9a and c.

For $\gamma > 0$ we obtain the dashed lines in Figure 9. This means that the reflection structure is somewhat smeared out and reaches only maximum values slightly below 1. On the other hand, the polariton-modes are damped outside the resonance region.

Now we know in principle everything concerning the energy transfer of a light field through the interface between vacuum and a medium. However, we must bear in mind that all the above results have been obtained using the simple mechanical model of chapter II.2.

In the subsection II.6. and 7. we shall inspect the main constituents of a semiconductor, namely the electronic and the ionic systems and shall see to which extent they can be described by the above developed model and/or which corrections have to be added.

II.F. Phonon Polaritons

In this subsection we discuss the phonons as an example for an oscillator as used in our model. Phonons are the quanta of the normal modes of the lattice vibrations. These normal modes are collective excitations of the ions or atoms forming a semiconductor.
The concept of collective excitations and of quasi-particles as the quanta of these collective excitations has been presented in some detail in the preceeding summerschool of this series [11] or e.g. in [12].

To recall the properties of phonons, Figure 10 shows some typical normal modes of a diatomic chain, i.e. a chain with two atoms per unit cell. If the atoms of different type are elongated in phase between two nodes of the wave, we call these modes acoustic phonons. There are two transverse and one longitudinal mode for every $\vec{k}$. If the different kinds of atoms are elongated in opposite directions, we have optical phonon modes. The number N of optical normal modes for a given $\vec{k}$ is

$$N = 3m - 3 \tag{71}$$

where m is the number of atoms per unit cell.

Figure 11a shows the dispersion of the phonons of the diatomic chain. The two transverse modes, polarized either in the plane of drawing or perpendicular to it are degenerate in this model. An extension to a three dimensional lattice gives no qualitative differences.

If the different atoms in the unit cell carry a charge, i.e. if

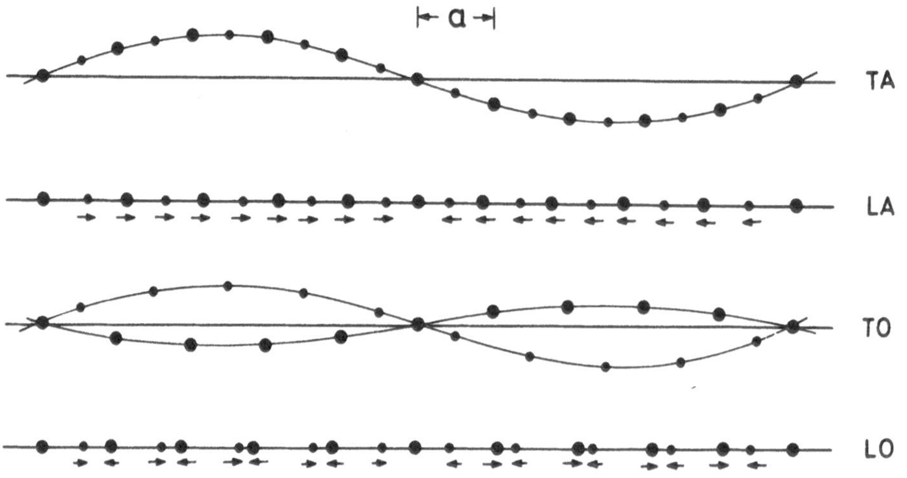

Fig. 10. Schematic representation of acoustic (A), optic (O), transversal (T) and longitudinal (L) normal modes of a diatomic chain, i.e. a chain with two different atoms in the unit cell with masses m and M, respectively. The masses are connected by identical springs with force constant β. The dispersion relation resulting from this model is given by

$$E(k) = \{\beta(\frac{1}{M} + \frac{1}{m})\}^{1/2} \{1 \pm (1 - \frac{4Mm \sin^2 ka}{(M+m)^2})\}^{1/2}$$

The - sign refers to the lower (acoustic), the + sign to the upper (optic) branch (see Figure 11a). This model differs considerably from the one of Figure 3. In Figure 3 every mass was attached by a spring to its lattice site, without coupling to the neighboring oscillators, resulting in the dispersion of Figure 4. Here we have only coupling between neighboring atoms resulting in the completely different dispersion of Figure 11a.

the chemical binding is at least a partially ionic one, some of the optical normal modes may be connected with a polarization wave

$$\vec{P} = \vec{P}_o \, e^{i(\vec{k}\cdot\vec{r}-\omega t)}$$

They then couple to the electromagnetic field, ($f > 0$), resulting in a finite longitudinal transverse splitting Δ_{LT} at $k = 0$. (See Eq. (47)). In Figure 11a we assume a case with finite Δ_{LT}.

In contrast to the model of Chapter II.2, the ions forming the la-tice are coupled to each other. A necessary consequence of coupling between the oscillators is a finite group-velocity $v_g \neq 0$ or a finite width 2B of the band. This width 2B is shown in Figure 11a for the TO-branch. The dependence of the eigenfrequency ω_0 of the oscillators on $\vec{k}$

$$\omega_o = \omega_o(\vec{k}) \tag{72}$$

is called spatial dispersion. The model of Chapter II.2 was thus one without spatial dispersion, the phonons exhibit spatial dispersion.

If we look a little bit closer to the phonons, we shall see, however, that neglecting the spatial dispersion of the optical phonons is a reasonable approximation: The eigenenergies of optical phonon modes are situated in the range from 10 meV to 100 meV i.e. in the IR-part of the spectrum. The corresponding photon-wavelength in vacuum is between 100 μm and 10 μm. If we use as an average $\lambda_v = 30$ μm we find $k_v \simeq 2 \cdot 10^3$ cm^{-1}. The first Brillouin zone shown in Figure 11a extends to $\pi/a \simeq 10^8$ cm^{-1}. If we want to plot the dispersion of the photons in Figure 11a we would get a line which practically coincides with the ordinate.

If we change the scale of the abscissa in a way to be able to show the photon dispersion, the k-dependence of the eigenfrequency is negligible, and we may apply the model without dispersion.

Figure 11b thus gives the E(k) relation of the phonon-polariton. The phonon-polariton is said to be photon-like for $\hbar\omega < \hbar\omega_0$ (LPB) and for $\hbar\omega > \hbar\omega_L$ (UPB). For $\hbar\omega \simeq \hbar\omega_0$ the dispersion curve bends over and the polariton is said to become more and more phonon-like. For $\hbar\omega \gtrsim \hbar\omega_L$ the UPB starts phonon-like and gradually becomes photon-like again.

In order to show that this phonon-polariton concept is valid for the reflection and transmission of the light at the interface vacuum-semiconductor, we show in Fig. 12 reflection spectra for various

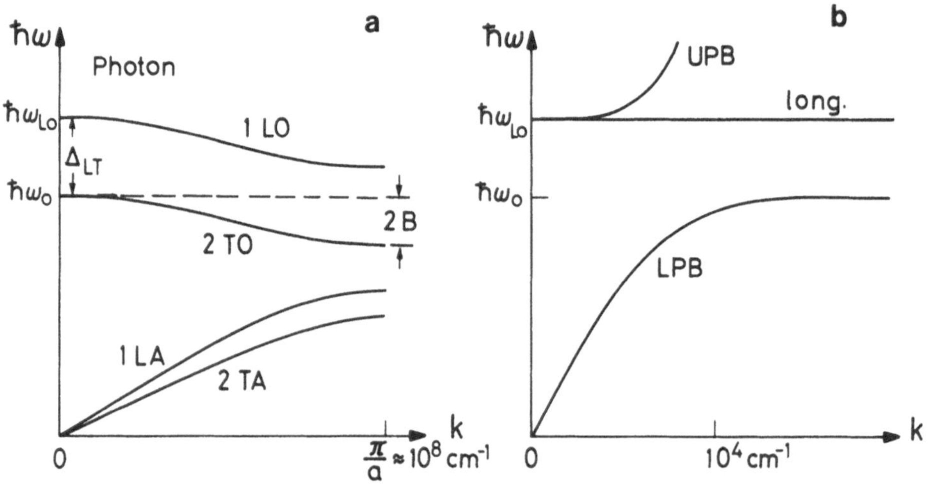

Fig. 11. Dispersion of acoustic and optic phonon modes of a diatomic chain in the first Brillouin-zone (a); dispersion of the lower and upper polariton branches and of the longitudinal phonon mode (b). Note the different scales for (a) and (b) on the abscissa.

III - V semiconductors from [13]. The reflection structures coincide well with the prediction of the polariton concept (see Fig. 9). The fine structures seen in some of the reflection spectra are due to the more numerous and complex branches of optical phonons in these materials as compared to Fig. 11. The spectral regions with high reflectivity $R \gtrsim 0.8$ are called Reststrahlen-banden for the following reason. If a "white" IR-spectrum is reflected several times at a material a considerable intensity is reflected only in the region $\hbar\omega_T \leq \hbar\omega \leq \hbar\omega_L$. To give some numbers: If we have $R \simeq 0.2$ outside the Reststrahlen-bande and $R = 0.9$ inside we have after four reflection processes $R_o = (0.2)^4 = 1.6 \cdot 10^{-3}$ outside and $R_i = (0.9)^4 = 0.65$ inside, corresponding to $R_i/R_o \simeq 0.4 \cdot 10^{+3}$.

The subsection about phonon polaritons is concluded by two comments:

If a material has perfect homeopolar binding like the elementary semiconductors Ge or Si, there is only weak coupling between the optical phonons and the electromagnetic field. The L - T splitting is zero for $k = o$. Still it must be remembered that the photon-like branch describes a polariton-state. In this case, however, only the driven oscillations of the electronic system are contributing.

The slope of the dispersion of the acoustic phonons is given for small k by the velocity of sound v_s. Because of $v_s/c \simeq 10^{-5}$ the dispersion of the acoustic phonons almost coincides in Fig. 11b with the

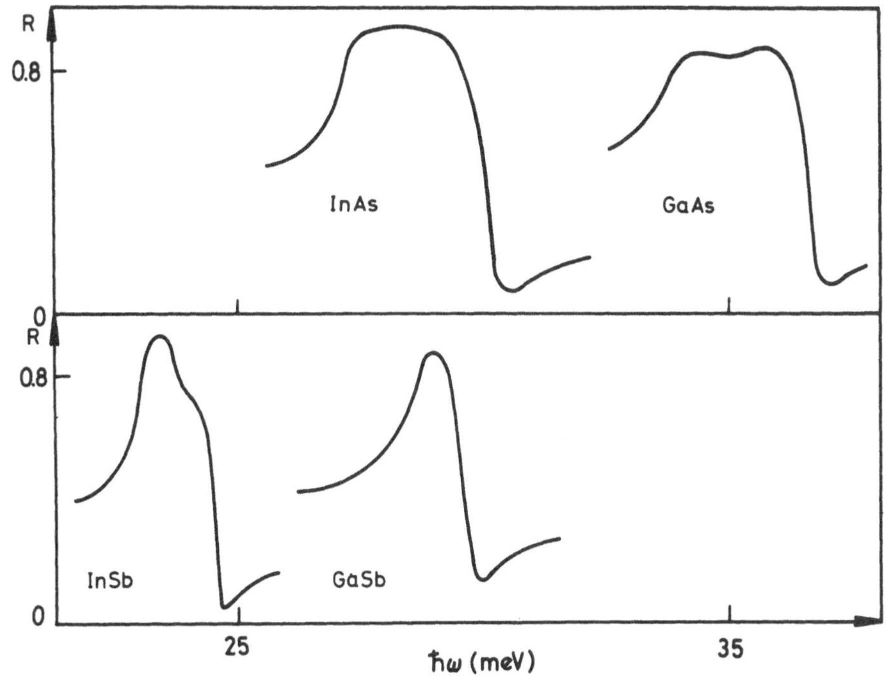

Fig. 12. Reflection spectra of phonon polaritons in various III-V semiconductors. From [13].

abscissa. Because of the large discrepancy of k-vectors for phonon-polaritons and acoustic phonons of the same energy it is obvious that there is no direct interaction between these two modes. A weak interaction is possible in scattering-processes involving several particles (III.D.).

More literature on phonon polaritons and on Raman-like scattering experiments showing rather directly the dispersion of phonon polaritons are found e.g. in [14].

II.G. Excitons: Oscillators with Spatial Dispersion

In this section, we investigate the properties of light in the vicinity of electronic resonances. We use here throughout an idealized electronic bandstructure as shown by Fig. 13. We assume that conduction and valence bands are isotropic, parabolic and nondegenerate, and that the maximum of the valence band and the minimum of the conduction band occur at k = 0, i.e. at the Γ-point. With this assumption, we may use the effective mass approximation, m_e and m_h the effective masses of electrons and holes, respectively.
The modification brought about by a realistic band structure and the case of indirect gap materials is outlined in appendix 1.

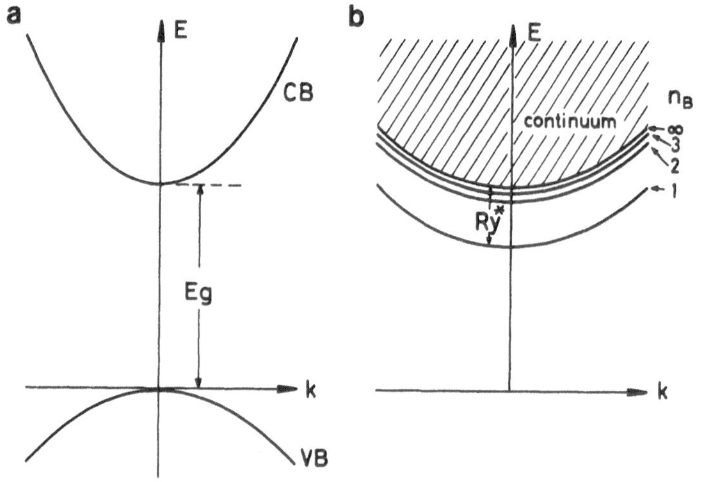

Fig. 13. Band structure (a) and exciton dispersion (b) for an idealized, direct gap semiconductor.

If we excite an electron in a semiconductor across the forbidden energy gap e.g. by absorbing a photon we create simultaneously two quasi particles: the electron in the conduction band and the unoccupied place in the valence band. Instead of regarding all the electrons (about 10^{22} cm^{-3}) in the va' nce band with one missing, it is more convenient to introduce the quasi-particle concept of the holes (see e.g. [12],[15]). The hole is a quasi-particle with positive charge. Wavevector and spin are opposite to those of the electron which has been removed from the valence band.
The negatively charged crystal electron and the positive hole (or defect electron) interact with each other due to their charges. They may form bound states in analogy to the hydrogen- or the positronium atom characterized by the main quantum number n_B (see Fig. 13b). These bound states are called exciton states. The excitons are the energetically lowest excitations of the electronic system of an ideal semiconductor at T = o K. They correspond to the electronic oscillators which we have to introduce in the polariton concept for the spectral region around the band gap. The properties of the excitons in the limit of our idealized band structure are given by:

$$E_x(n_B, \vec{k}) = E_g - Ry^* \frac{1}{n_B^2} + \frac{\hbar^2 k^2}{2M} \quad ; \quad n_B = 1, 2, 3, \ldots \tag{73}$$

$$M = m_e + m_h$$

$$\mu = m_e m_h/(m_e + m_h)$$

with the modified Rydberg energy Ry* given by

$$Ry^* = 13.6 \text{ eV } \mu/\varepsilon_1^2 \tag{74}$$

The symbols in Eqs. (73) and (74) have the following meanings: E_g = Bandgap at k = 0; n_B = main atomic quantum number; M = translational mass of the exciton; μ = reduced mass of the exciton.

Typical values for Ry* in semiconductors are

$$1.0 \text{ meV} \lesssim Ry^* \lesssim 200 \text{ meV} \tag{75}$$

The Bohr-radius a_B of the $n_B = 1$ groundstate is ranging from

$$200 \text{ Å} \gtrsim a_B \gtrsim 5 \text{ Å}$$

This means that a_B is generally considerably larger than the lattice constant a. In this limit the excitons are called Wannier-excitons. Their wavefunction is given by

$$\phi(\vec{k}, n_B, \ell, m) = \frac{1}{\sqrt{\Omega}} e^{i\vec{k}\vec{R}} \rho_e(\vec{\tau}_e) \rho_h(\vec{\tau}_h) \phi_{n_B,\ell,m}(\vec{\tau}_e - \vec{\tau}_h) \tag{76}$$

$1/\sqrt{\Omega}$ is the normalization factor, $\vec{R}$ the vector of the center of mass $\vec{R} = (m_e \vec{r}_e + m_h \vec{r}_h)/(m_e + m_h)$.

$\rho_e/h(\vec{\tau}_{e,h})$ represents Wannier functions of electrons or holes centered at $\vec{r}_e$ and $\vec{r}_h$, respectively. For the relation between Bloch- and Wannier functions, see e.g. [12],[15]. The envelope function $\phi_{n_B,\ell,m}$ is the modified Hydrogen-function with the usual quantum numbers n_B, l, m. Modified means that everywhere the free electron mass m_0 has to be replaced by μ and the permittivity of vacuum ε_0 by $\varepsilon_0 \cdot \varepsilon_1$. In Figure 13b the E(k) relations of the excitons are summarized. Above the ionization-limit of the discrete exciton states ($n_B \Rightarrow \infty$) the continuum states start. They considerably modify the density of states for $E > E_g$ as compared to a simple square root function. The influence of the Coulomb interaction extends roughly over a region of 10 Ry* into the bands. Many aspects of excitons are summarized e.g. in the following review articles and the literature cited therein [11],[16]-[21]. For phonons, the bandwidth 2B (see Figure 11a) is about 10 meV and we were able to neglect the spatial dispersion in the discussion of the optical properties of phonon polaritons. For the excitons, the bandwidth 2B is given by the sum of the widths of conduction and valence band, 2B being thus of the order of 10 eV. In this case, the spatial dispersion is so large that it must not be neglected even for wavevectors comparable to those of the photon. Photon wavevectors in the spectral region around 2 eV are roughly 10^5 cm^{-1}. In our mechanical model of Figure 3, a spatial dispersion can be introduced by a coupling between the oscillators, e.g. in the way shown in Figure 14 (see also [6]).

Before we investigate in the next section the optical and energy transport properties of excitonic polaritons in detail, it should be stated that the exciton concept is a very important one for the optical properties of semiconductors and insulators. Almost all absorption and emission processes of light in the electronic system are via exciton states, i.e. they always involve an electron and an empty state (hole) and the Coulomb-interaction between these two quasiparticles. To illustrate this, we enumerate some other types of excitons:

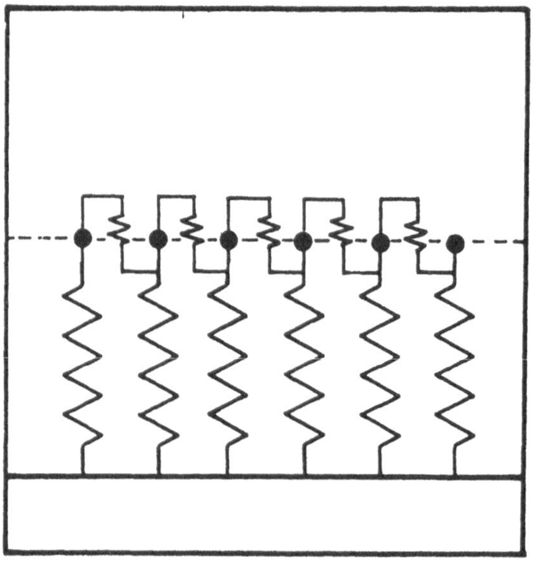

Fig. 14. Mechanical model for an ensemble of oscillators with spatial dispersion. Compare Fig. 3a

In semiconductors one has apart from the free (Wannier) excitons presented above also bound excitons, i.e. excitons bound and localized at some defect or impurity. There are monocentric bound excitons like donor or acceptor bound excitons or polycentric bound excitons including donor-acceptor pairs (see, e.g. Appendix 3). In insulators, rare solid gases (see [20],[22]) and organic molecular crystals (see, e.g. [20]), the electrons and holes involved in an optical transition are often located at the same lattice site forming so-called Frenkel-excitons. The self-trapping of a carrier in insulators with strong electron-phonon interaction gives rise to the formation of self-trapped excitons which may coexist with free excitons (see, e.g. [20]). The discussion will be limited in the following to Wannier-excitons.

II.H. The Exciton Polariton and the Problem of Additional Boundary Conditions

Excitons may couple to the electromagnetic field, yielding polaritons. In our simplified model the selection rules are roughly the following. If the band to band transition is dipole-allowed, excitons states with opposite electron and hole spin (singletts) and s-envelope function (l = 0) will couple strongly to the electromagnetic field thus forming polaritons. The longitudinal transverse splitting depends on n_B [16],[17]:

$$\Delta_{LT}(n_B) \sim |M_{cv}|^2 \frac{1}{n_B^3} \quad ; \quad M_{cv} = \langle C|H_{dipole}|v\rangle \tag{77}$$

M_{cv} is the band to band dipole matrix element. Triplet states (electron and hole spin parallel) and states with p,d ... envelope function do not couple to the electromagnetic field. They remain pure excitons and do not form polaritons. The longitudinal modes are also generally pure excitons.

In the following we discuss only the $n_B = 1$ exciton state, bearing in mind that the same results are true for the higher states, however with reduced oscillator-strength (Eq. (77)). In order to take into account the spatial dispersion, we have to introduce the $\vec{k}$-dependence of the eigenfrequency. For the 1s excitons we find with Eq. (41):

$$\varepsilon(\omega_1 \vec{k}) = \varepsilon_b\left(1 + \frac{f}{\left(\omega_o + \frac{\hbar k^2}{2M}\right)^2 - \omega^2 - i\omega\gamma}\right) \tag{78}$$

with $\omega_o = (E_g - Ry^*)/\hbar$.

In principle f and γ may also depend on $\vec{k}$ (and, in fact, do so in real semiconductors) but we ignore here this additional complication.

By Eq. (78) ε now becomes a function of the two independent variables ω and $\vec{k}$. The term

$$(\omega_o + \frac{\hbar k^2}{2M})^2$$

is generally approximated by

$$(\omega_o + \frac{\hbar k^2}{2M})^2 \simeq \omega_o^2 + \frac{\hbar k^2 \omega_o}{M} \qquad (79)$$

neglecting the terms proportional to k^4.

The transverse and longitudinal eigenenergies may be deduced for $\gamma = 0$ from (78)

$$\varepsilon(\omega_T, \vec{k}) = \infty \Rightarrow \omega_T^2(\vec{k}) = (\omega_o + \frac{\hbar \vec{k}^2}{2M})^2$$

and

$$\varepsilon(\omega_L, \vec{k}) = 0 \Rightarrow \omega_L^2(\vec{k}) = \omega_o^2 + f + \frac{\omega_o \hbar k^2}{M} = \omega_{oL}^2 + \frac{\hbar k^2}{M} \omega_o \qquad (80)$$

The dispersion of the mixed state of electronic excitation and electromagnetic field is found again with the help of the polariton equation (59) yielding

$$\frac{c^2 k^2}{\omega^2} = \varepsilon_b (1 + \frac{f}{\omega_o^2 + \frac{\hbar k^2 \omega_o}{M} - \omega^2 - i\omega\gamma}) \qquad (81)$$

The quanta of this mixed state are called exciton-polaritons.

Equation (81) is quadratic in k^2. The solutions for $E = \hbar\omega(k)$ are shown for $\gamma = 0$ and real and imaginary k in Figure 15a. The dispersion of the exciton-polariton starts as in the case of the phonon-polariton with a photon-like lower polariton branch. For $E \lesssim \hbar\omega_o$, the LPB bends over to an exciton-like dispersion, the transition region being generally called bottleneck.

The longitudinal exciton branch starts at $\hbar\omega_{oL}$. The upper

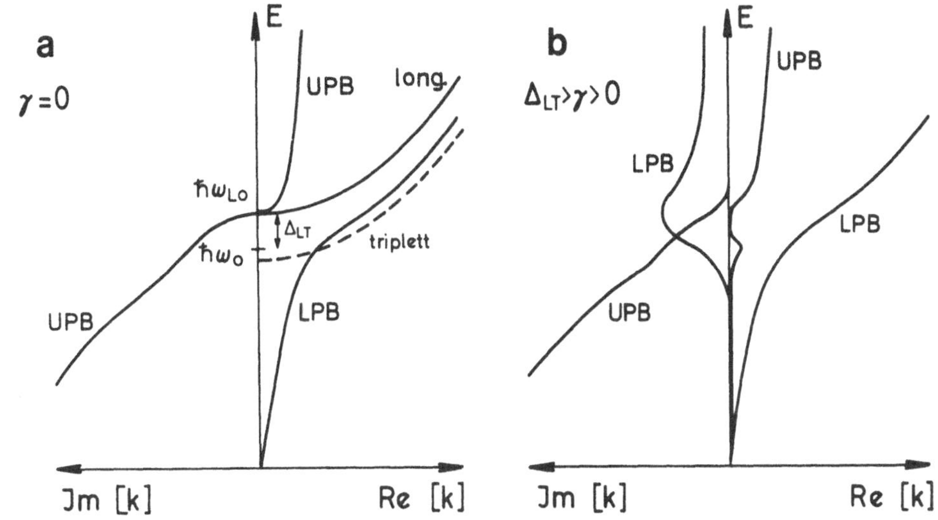

Fig. 15. Dispersion of the lower and upper polariton branches, of the longitudinal and of the triplett exciton without damping (a) and for small damping (b). In (b) the pure exciton states have been omitted for simplicity.

polariton branch has a purely real wavevector for $E \geq \hbar\omega_{oL}$ and a purely imaginary one for $E < \hbar\omega_{oL}$. The latter case describes not a propagating mode but an exponentially decaying one. The dispersions of the triplet excitons and of the longitudinal exciton mode which do not interact with light are shown, too.

For small damping the picture is modified according to Figure 15b. Approaching the resonance region, the LPB is damped ($Im\{k\} > 0$). The UPB gets a small real k for $\omega < \omega_{oL}$, connected, however, with very large damping, which exhibits a tail for $\omega > \omega_{oL}$. For increasing damping, the small peak of the UPB around $\hbar\omega_o$ increases and may reach the LPB. This is the beginning of strong damping $\hbar\gamma > \Delta_{LT}$ (see e.g. [23]-[25]. For strong damping, we still have an oscillator with spatial dispersion and the exciton-like branch for large k, but the optical properties like reflection and absorption can then be described to a good approximation by the model of II 5 without spatial dispersion. Though the dispersion of the exciton-polariton is exciton-like only in the vicinity of the resonance, a theoretical analysis shows, that the wave function of the polariton is predominantly exciton-like over a rather large region of E given by the inequality [25]:

$$|E - \hbar\omega_o| \lesssim (\Delta_{LT} \cdot \hbar\omega_o)^{1/2} \tag{82}$$

This allows in the evaluation of transition-matrix elements the use of the exciton-wavefunction for the same wave vector as the polariton which is involved.

To discuss the reflection and transmission of a light beam at the interface vacuum-semiconductor, we come back to Fig. 15. There are two main differences as compared to the case without spatial dispersion: There is no forbidden region between ω_o and ω_{oL}, i.e. we have for every frequency at least one propagating mode in the semiconductor. This reduces even for $\gamma = 0$ the maximum values of R from $R_{max} \simeq 1$ (Figure 12) to about $0.2 \lesssim R_{max} \lesssim 0.6$.

For every frequency, there are in fact at least two possible modes in the medium. For $\omega \geq \omega_{oL}$ there are two propagating modes (LPB, UPB); for $\omega < \omega_{oL}$ there is one propagating mode (LPB) and one exponentially decaying mode which, however, has a finite amplitude at the semiconductor-side of the surface. In the case of oblique incidence there may even be some coupling to the longitudinal mode (Fig. 16) for some crystal symmetries. For a given incident field amplitude, Maxwell's equations give only two boundary conditions (see eq. (62)) for the reflected and the transmitted amplitudes. In order to know how the amplitude in the sample is distributed over the various modes, one needs an additional boundary condition (abc). Many abc's have been discussed in the literature. The most widely used ones are that either the excitonic contribution to the polarization should be zero at the surface

$$P_{ex}(x = 0) = 0 \tag{83}$$

or the derivative normal to the surface. Linear combinations of both abc's are discussed, too. Often the abc is connected with the idea that there is an exciton-free layer on top of the semiconductor, described by ε_b. This "dead layer" has a thickness d of at least the exciton Bohr-radius and takes into account that the center of mass of the exciton cannot approach infinitely close to the surface because of the finite diameter of the exciton.

It is a common feature of all abc that the energy of light travels almost exclusively on the LPB for $E < \hbar\omega_o$. In the region $E \gtrsim \hbar\omega_{oL}$ both modes share comparable parts of the intensity and for $E >> \hbar\omega_{oL}$ the light intensity goes over on the UPB.

Recent reviews of the abc-problem are found in, e.g., [20] and [26]. In [27] it is assumed that there is a detuning of the exciton eigenfrequency close to the surface because of electric surface potentials.

The different abc give different calculated reflection spectra, if the other parameters are kept constant. However, it was until now not possible to deduce the "correct" abc from a detailed analysis

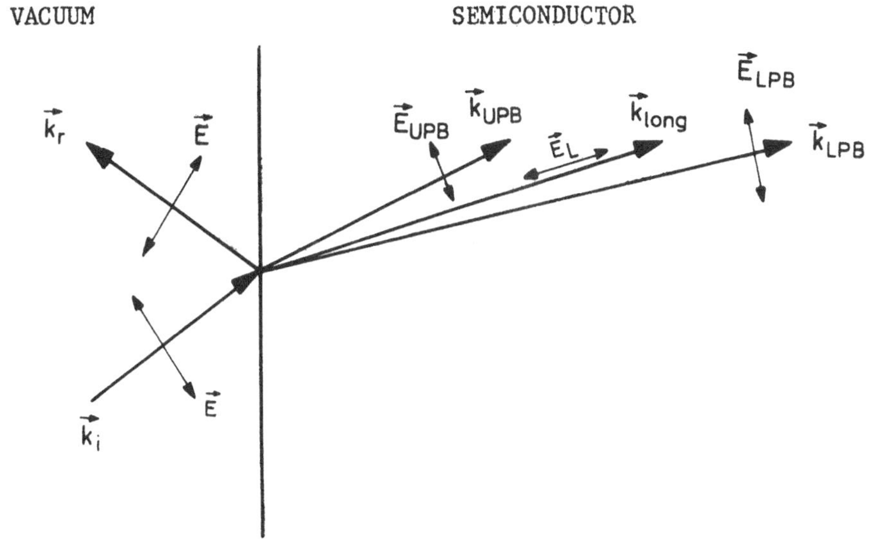

Fig. 16. Wave vectors and electric fields for reflection and refraction of an incident electromagnetic wave at the boundary between vacuum and a medium with spatial dispersion.

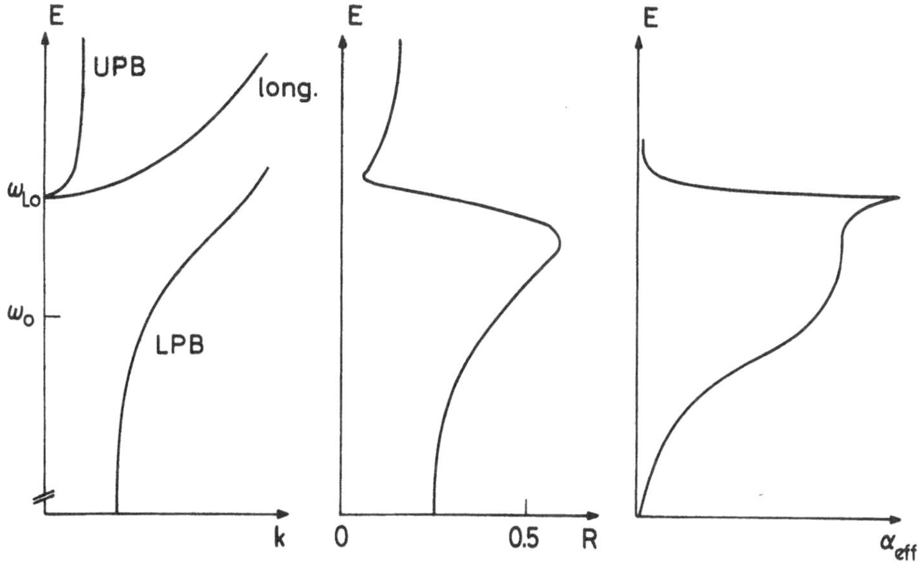

Fig. 17. Polariton dispersion, reflection spectrum and effective absorption coefficient for an excitonic resonance.

of the reflection spectra. With almost every abc, it is possible to get a reasonably good fit to the experiment by a suitable choice of the other parameters like ε_b, ω_o, Δ_{LT}, M, γ and the thickness of the dead layer. If these parameters are determined by an independent method, e.g. resonant Brillouin scattering, it has been found that none of the abc was able to describe consistently all experimental findings [24].

This is a highly unphysical situation. The reflection spectrum of a given sample of a semiconductor may depend on a large number of parameters but definitely not on the abc, the experimentator is bearing just in mind.

In fact, the problem of the abc is an artifical one. If we calculate the dielectric function and the polariton dispersion, we do this for the bulk of the crystal, i.e. we are investigating the volume modes. Then we want to use this model to describe the properties of the interface between vacuum and a semi-infinite sample. The price we have to pay for this not appropriate use of the bulk dielectric function is the problem of abc. If the optical properties would be calculated from the very beginning for a semi-infinite material including the surface, the problem of more or less arbitrary abc would not occur. However, calculation of the bandstructure, the Coulomb interaction etc. for the semi-infinite space is a very hard problem, too. Until now, only approximative but promising attempts have been made to tackle this problem e.g. [28]. The "state of the art" is to use one of the abc's, generally the Pekar-Hopfield one, which uses eq. (83) together with an exciton-free layer on top of the sample. On this background, it is astonishing that a great deal of the knowledge about exciton polaritons come from the investigation of reflection spectra.

In Fig. 17 we summarize the results of energy transfer through the surface, reflection, and the damping of the transmitted modes. Fig. 17a gives the polariton dispersion for $\gamma = 0$, Fig. 17b the resulting calculated reflection spectrum and Fig. 17c finally the effective absorption coefficient α. It should be noted that the damping of the various modes is different for the same energy. This means that the intensity travelling on the different modes decays spatially into the sample with different absorption coefficients resulting thus in a nonexponential decay on the total intensity.

II.I. Experimental Proofs for the Concept of Excitonic Polaritons

In recent years, several linear and nonlinear spectroscopic methods have been found which allow to really investigate the volume modes of light in semiconductors including their dispersion $E(\vec{k})$, i.e. these methods offer the possibility of direct spectroscopy in k-space. The most prominent of these techniques are two-photon Raman scattering (see IV.B.) resonant Brillouin scattering (see III.D.) the time of flight measurements (see below), the transmission through a Fabry-

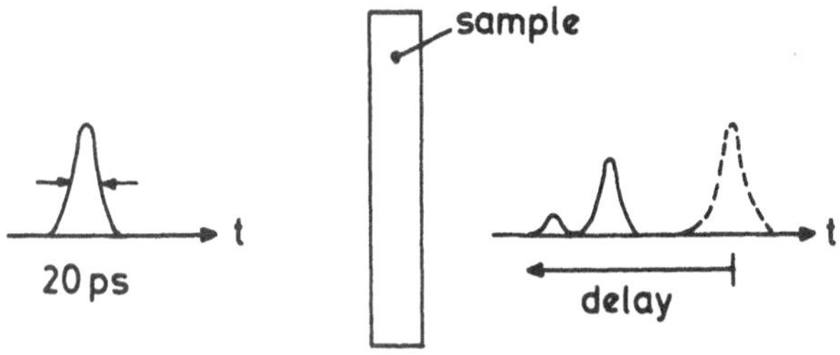

Fig. 18. Schematic drawing of the setup for time of flight measurements.

Perot resonator (see below) or the refraction of light by a thin prism [29]. A recent review of these methods will be published in [30]. The principle of the time of flight measurements is the following:
A spectrally narrow, tunable, short pulse is created by a picosecond laser. Typical values of the temporal FWHM are about τ_L = 20 psec and the spectral halfwidth E is limited by the uncertainty relation

$$\tau_L \cdot \Delta E \simeq h \tag{84}$$

This pulse travels through vacuum and is transmitted through a thin slab of matter (Fig. 18). When it appears on the other side of the sample, it has a certain time delay as compared to a pulse which travelled only through vacuum. From this delay, it is easily possible to deduce the velocity of energy propagation v_e in the medium, if the geometrical thickness of the sample is known. According to the theoretical investigations of Birman [31] one finds

$$v_e \simeq v_g = \frac{1}{h}\frac{dE}{dk} \tag{85}$$

except with some minor deviations in the regions of very high absorption. If E(k) is known for one energy e.g. by a precise measurement of the refractive index, the dispersion-curve can be constructed from the measured energy dependence of v_e with the help of Eq. (85). For $E > h\omega_{0L}$ two propagating modes exist. If for a given damping the sample thickness is chosen suitable, both modes will contribute to the energy transport to the opposite surface. Since the group-velocity is different for the LPB and the UPB, two transmitted pulses may appear. This situation is depicted in Fig. 18. The time of flight method has been applied successfully to various semiconductors like CuCl [32],[33], GaAs [34] or CdSe [35],[36]. Fig. 19 shows results for CuCl from [33].

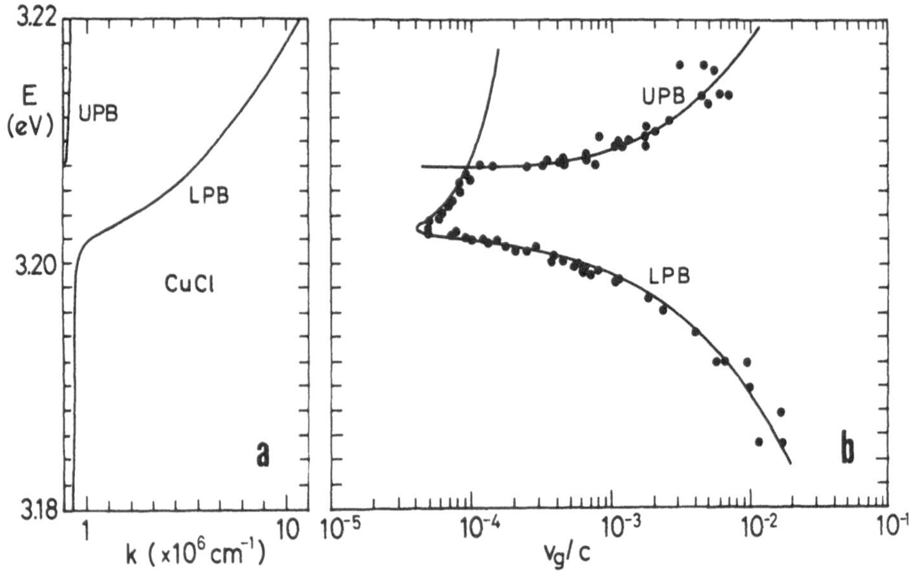

Fig. 19. Calculated dispersion of lower and upper polariton branches of the lowest free exciton resonance in CuCl (a). Measured values (•) for the energy velocity and calculated curves (———) for the group velocity (b). From [33].

On the lefthand side are the dispersion curves of LPB and UPB calculated with the parameters $M = 2.5\ m_0$, $\varepsilon_b = 5.0$, $\hbar\omega_0 = 3.2025$ eV and $\Delta_{LT} = 5.5$ meV, $\gamma = 0$. The solid lines on the righthand side of Figure 19 are calculated curves $v_g(E)$ deduced from the dispersion relation. They coincide very well with the measured data of $v_e(E)$, giving thus direct evidence for exciton polaritons as the modes which carry the energy of light through the semiconductor. It should be noted that the energy transport velocity may become smaller than 10^{-4} c in the vicinity of the resonance.

A problem is to prepare samples of appropriate thickness. In the vicinity of the resonance, the absorption coefficient may be as high as $10^6\ cm^{-1}$. To get a measurable transmitted intensity, samples with a thickness

$$10^{-1}\ \mu m \lesssim d \lesssim 1\ \mu m$$

are required. High quality samples of this thickness are difficult to prepare and to handle.

Before considering the specific implications of a Fabry-Perot etalon made from a medium with spatial dispersion i.e. with two or more propagating modes for one energy, we regard some general features of such an optical resonator [7],[8]. A Fabry-Perot (FP) consists of two plan-parallel, partly reflecting surfaces with a geometrical distance d. We assume that the medium inside is described by a refractive index $n(\omega)$. Outside is vacuum. If a light beam falls on the FP, we get at every surface partial reflection and partial transmission. The partial waves in the FP may interfere, if the coherence length is sufficiently long. For normal incidence, we get constructive superposition for waves with

$$2 \cdot d \cdot n(\omega) = m \cdot \lambda_m \quad , \qquad m = 1, 2, 3, \ldots \tag{86}$$

λ_m is the wavelength in vacuum connected with the transition maximum m situated energetically at $\hbar\omega_m$. Constructive interference in the FP means a high field amplitude in the resonator and correspondingly a transmission maximum. In the case of destructive interference, a transmission minimum results. Transmission maxima of a FP coincide with reflection minima and vice versa. It is possible to translate the condition of eq. (86) for transmission maxima in the language of dispersion curves. Maxima occur for polariton energies $\hbar\omega_m$ with wavevectors k_m in the medium given by

$$k_m = m\frac{\pi}{d} \quad ; \qquad m = 1, 2, 3, \ldots \tag{87}$$

It is thus obvious that the transmission maxima are equidistant in k-space, the separation k between adjacent maxima being

$$\Delta k = k_{m+1} - k_m = \frac{\pi}{d} \tag{88}$$

Quantitatively, one gets from a summation over all amplitudes for the transmittivity of the energy flux T_{PF} and the reflectivity R_{PF} the following relations for normal incidence and a nonabsorping medium:

$$T_{PF} = I_T/I_i = \frac{1}{1 + F\sin^2(\delta/2)} \tag{89}$$

$$R_{PF} = I_R/I_i = \frac{F\sin^2(\delta/2)}{1 + F\sin^2(\delta/2)} \tag{90}$$

$$T_{PF} + R_{PF} = 1 \tag{91}$$

The phase argument δ is given by

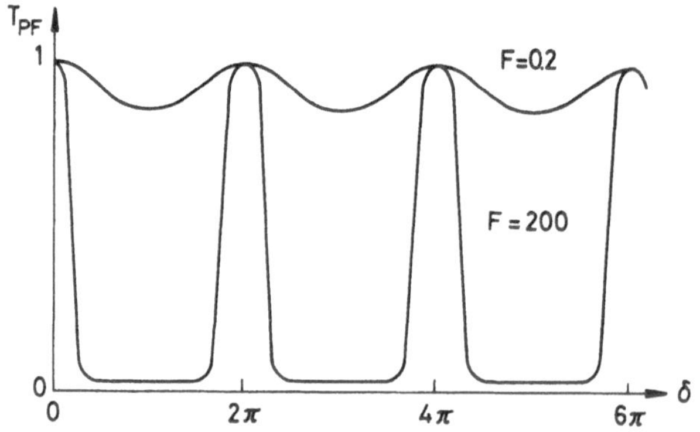

Fig. 20. Transmission through a Fabry-Perot as a function of the phase shift δ for two different values of the finesse F. From [8].

$$\delta = 2d\, k_{vac}\, n(\omega) = 2d\, k \tag{92}$$

It is the phase shift connected with one round-trip in the resonator.

The finesse F is defined with the help of the reflectivity of a single surface rr* as

$$F = \frac{4rr^*}{(1 - rr^*)} \tag{93}$$

Figure 20 gives examples to the transfer of energy through a FP for two different values of F. T_{PF} is plotted as a function of δ.

The transfer of energy through a FP made from a semiconductor has been investigated by several authors for CdSe and CdS (see, e.g. [36],[37]) and for CuCl [38]. Here we review the results for CuCl which can directly be compared with those found from the time of flight method.

The measured reflectivity R_{PF} and transmittivity T_{PF}, and the calculated polariton dispersion are plotted in Fig. 21 on a common energy scale for a 0.15 m thick CuCl platelet. The sample has high quality, i.e. small damping γ. For $E \leq \hbar\omega_{oL}$ there is only one propagating mode. The transmission maxima calculated from Eq. (87) are indicated on the dispersion curve by crosses. Their positions coincide perfectly with the observed transmission maxima.

For $E > \hbar\omega_{oL}$ an LPB-wave and an UPB-wave are created from the incident wave. Both travel through the medium to the opposite side. The phase shift δ_{UPB} of the UPB-wave is much smaller than δ_{LPB} because of the smaller k. Both waves are partly transmitted and partly reflected at the opposite side. But now, the UPB-wave produces both a reflected LPB and UPB wave, and so does the LPB wave. As a result, there are already after half a round-trip two LPB-waves which may interfere, one has a phaseshift δ_{LPB} and the other δ_{UPB}, as mentioned with $\delta_{LPB} >> \delta_{UPB}$. As a consequence, the transmission maxima are still equally spaced for $E > \hbar\omega_{oL}$ but with a Δk which is roughly given by

$$\Delta k_{LPB} = \frac{2\pi}{d} \tag{94}$$

as shown by the crosses on the righthand side of Figure 21. The doubling of Δk when going from $E \leq \hbar\omega_{oL}$ to $E > \hbar\omega_{oL}$ is a consequence of the spatial dispersion only. A quantitative analysis of the spectral positions of the transmission maxima allows us to reconstruct the polariton dispersion E(k).

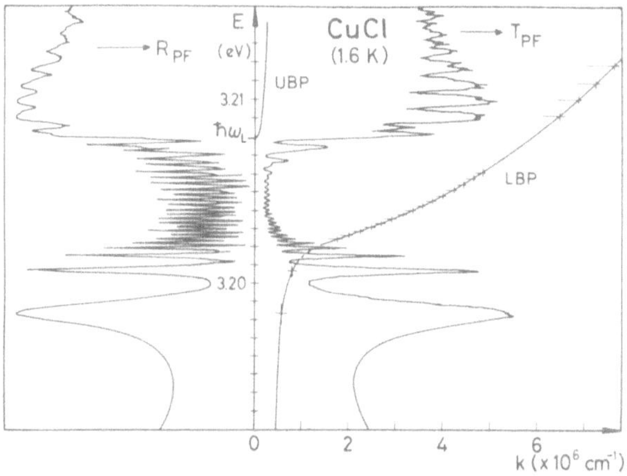

Fig. 21. Reflection and transmission of a 0.15 μm thick Fabry-Perot etalon made from CuCl. From [38].

References on review articles which give more information on the dispersion and the optical properties of exciton polaritons [11], [16]-[38]. In Appendix 1, where we shortly discuss modifications of the polariton properties brought about by the bandstructure of real semiconductors, additional references may be found.

III. ENERGY TRANSFER FROM THE EXCITON-POLARITON TO THE PHONON-FIELD

III.A. Review of Energy Transfer Processes in Semiconductors

In section II we have treated various types of quasiparticles. There are various possible definitions for the terminus "quasiparticle". We regard them here as the quanta of the collective excitations in a solid. There is a rich variety of such quasiparticles. Most of them are to a good approximation Bosons like phonons, plasmons, polaritons, magnons, etc. Others obey Fermi-statistics, like crystal-electrons, polarons, etc.

The energy transfer processes can now be described as interaction between different or equal quasiparticles either among themselves or with lattice imperfections or localized excitations. To elucidate this concept further, we present in the following part of this subsection some examples from various fields of semiconductor-physics.

In a first example, we regard a simple resistor R connected with an electrical power source (Fig. 22). This resistor consumes an electrical power $P = UI = U^2R^{-1}$ and transfers it into heat. This energy transfer process can be attributed on a microscopic scale to the interaction between (crystal) electrons and/or holes with the phonons.

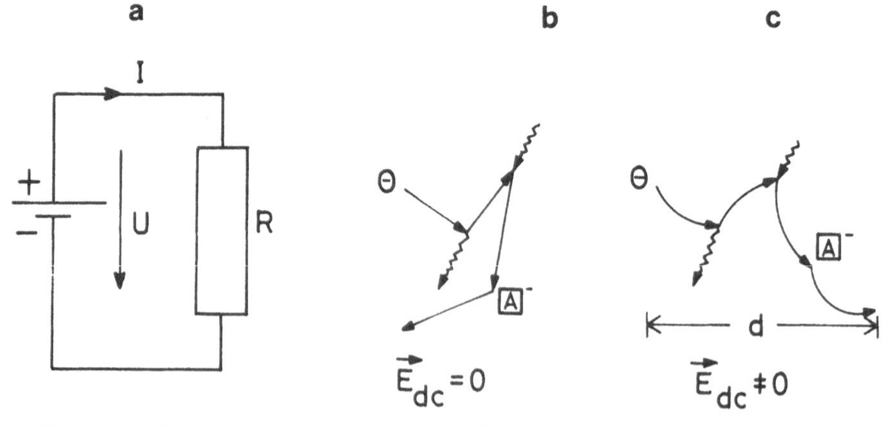

Fig. 22. An ohmic resistor R, transferring electric energy into heat (a). Schematic representation of the "path" of a crystal electron without (b) and with applied dc electric field (c).

In Fig. 22b the "path" of a crystal-electron in the sample is schematically depicted without applied voltage. After a free path of a certain length the electron is scattered e.g. by absorbing or emitting a phonon or by collision with an impurity (here a charge acceptor A^-). In the latter case only the direction of k, but not the amount of k, is changed because of $m_{A^-} >> m_{el}$. The mean kinetic energy of the electron is constant and given by $\langle E \rangle = 3/2\ k_BT$ for a nondegenerate electron system. This means that the amount of energy the electron gains and looses in the scattering processes, cancel each other on the average. If now an electric dc field E is applied to the resistor, the electron is accelerated during the free path according to $\hbar\dot{\vec{k}} = e\,\vec{E} = m_{el}\,\dot{\vec{r}}$. The path of the electron looks now like that shown in Figure 22c. As a result the electron has superimposed on its thermal motion a drift velocity v_D given by

$$\vec{v}_D = \mu_D\,\vec{E} \quad ; \qquad \mu_D = e\tau/m_{el} \tag{95}$$

μ_D is the drift mobility, τ the mean time of flight between two collisions. For the applicability of this relaxation time model it is necessary that the drift velocity v_D is much smaller than the thermal

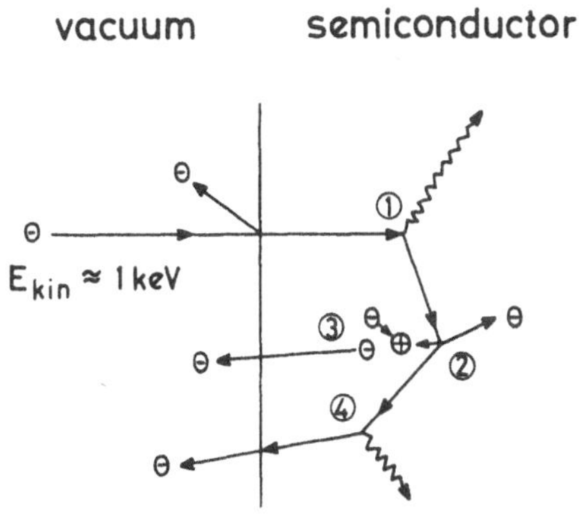

Fig. 23. Schematic representation of some energy loss and scattering mechanisms of electrons in a semiconductor. 1: creation of a valence-band plasmon. 2: creation of an electron-hole pair. If the hole is in a deeper valence-band subsequently an Auger process (3) or an X-ray emission may occur. 4: creation of a phonon.

velocity v_{th}. If the electron has travelled along a distance d it has got an amount of electrical energy e $\vec{E} \cdot \vec{d}$ and has transferred it to the phonon-system, i.e. into heat since the single creation and absorption processes of phonons are not correlated but random.

In the next example, we regard an electron which is accelerated in vacuum by a voltage U of about 10^3 V and then hits the surface of a semiconductor. The electron is transmitted through the surface with a certain probability, and then travels through the sample as a crystal electron in a very high conduction band. It will relax down to lower conduction bands by interaction with other quasiparticles. Some possible processes are shown schematically in Fig. 23. The first process shown is the creation of a valenceband plasmon (a short discussion of plasmons will be given in appendix 2) or it may create an electron-hole pair. If the hole is in a deep valence-band, it may recombine with an electron of a higher valence-band by either emitting an X-ray quantum or an Auger-electron. Furthermore, the electron may be scattered by absorbing or emitting a phonon. Eventually, the electron may come back to the surface, is remitted and may be detected by an energy sensitive detector. Fig. 24 shows schematically the energy-spectrum of these secondary electrons. There is a peak of electrons elastically scattered from the surface (for a review of these see e.g. [39] and the literature cited therein). Towards lower energies there are some structures in the N(E) spectrum. They are due either to electrons which have suffered a single energy-loss (e.g. by emitting a plasmon) or to Auger electrons [40]. The former have a fixed energetic distance ΔE from E_0, the latter ones occur at a fixed energy independent of E_0. Finally, there is a broad unstructured peak of so-called "true secondaries". These are electrons which have undergone many scattering processes and thus contain no information about an individual process.
Electrons interact strongly with many quasiparticles in a semiconductor because of their electric charge. Neutrons can interact - apart from strong interaction - only via their magnetic moment with the spins of electrons and nuclei. This interaction is much weaker. If monoenergetic neutrons fall on a semiconductor of about 1 cm^3 size, most of them will interact not at all or only once with a quasiparticle creating e.g. a phonon or, in a magnetic material, a magnon. By measuring energy and wavevector of the incident and scattered neutrons, it is possible to reconstruct the dispersion curve of the quasiparticle under investigation.

As the last example of this subsection, we consider a monochromatic light beam falling on the surfaces of a semiconductor. A part is reflected, and one (or two) polariton-modes are excited in the sample as discussed in some detail in chapter II. A polariton may now be scattered by creating e.g. a phonon. Depending on the scattering angle, it may return to the surface and here again one (or two) reflected polariton-modes are created and a part is transmitted through the surface. These "scattered" photons now contain the information about

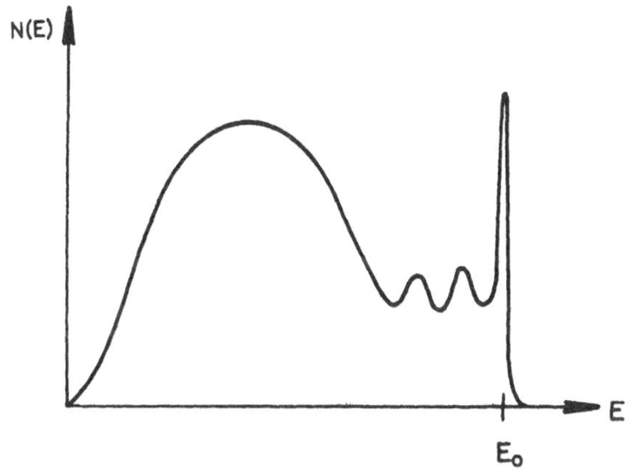

Fig. 24. Schematic drawing of the energy distribution N(E) of secondary electrons, emitted from the surface of a semiconductor or metal. One observes (quasi-) elastically scattered electrons at E_0. E_0 is assumed to be around 10^3 eV. Then there are some structures which are due either to scattered electrons, which have suffered only one or two discrete energy losses or to Auger electrons. Since the former ones occur at a fixed energetic distance from E_0 and the later ones at a fixed energy, they can easily be separated by modulation techniques. The broad unstructured peak is due to "true" secondaries.

the interaction process of the polariton with the phonons in the sample. The following part of this chapter will be devoted to a detailed study of this example.

III.B. Interaction Mechanisms Between Excitons and Phonons

In this subsection the most important interaction processes between exciton-polaritons and phonons are discussed. For polar ones we should more precisely use the term phonon-polaritons. In many interaction processes of exciton-polaritons ($k \geq 10^5$ cm^{-1}) the wavevectors of the involved phonon-polaritons are of the same order of magnitude. The phonon-polaritons then are to a high degree phononlike. Cases where the phonon-polariton dispersion shows up explicitely are discussed e.g. in [14].

An interaction mechanism which is possible both for polar and non-polar optical phonons is the deformation-potential scattering. The basic idea is the following: the band parameters, especially the energetic positions of the bands are functions of the lattice parameters. The dependence of these energies on the lattice deformation is described by the deformation-potential. A phonon can now be considered as a periodic distortion of the lattice. Via the deformation-potential this is connected with a periodic modulation of the band gap and this is felt by the electron and hole forming the exciton. With the argument that the exciton polariton wavefunction is exciton-like over a rather large spectral region eq. (82), we can use the exciton wavefunction for the same k as the exciton-polariton involved. The interaction between one polariton (wavevector $\vec{k}$, main quantum number n_B) with a field of phonons of wavevector $\vec{q}$ and occupation number $N_{\vec{q}}$ is given by [25],[41] for $qa_B \ll 1$:

$$\langle n_B, \vec{k}, N_{\vec{q}}\, H_{\mathrm{Def.\ Pot.}} | n_B, \vec{k}', N_{\vec{q}} \pm 1 \rangle$$

$$= \left(\frac{h}{2\mu_i N\omega_o} \right)^{1/2} \frac{E_c - E_v}{a} \left(N_{\vec{q}} + \frac{1}{2} \pm \frac{1}{2} \right)^{1/2}; \quad \vec{k}' = \vec{k} \pm \vec{q} \tag{96}$$

The minus sign in $\pm$ describes absorption of a phonon, the plus sign the (stimulated) emission. E_c and E_v are the deformation-potentials of conduction and valence bands, respectively. The case $E_c - E_v = 0$ means that both bands are shifted by the same amount. The relative gap would be constant and the interaction zero. N gives the density of unit cells, ω_o the energy of the phonons. μ_i is here the reduced mass of positively and negatively charged ions and a the lattice constant.

Polar optical phonons are connected with a macroscopic electric field. The electron and the hole in the exciton may be influenced by this field. The interaction with the field of LO-phonons is predominant (Fröhlich coupling).

The transition matrix element between two exciton states with equal main and angular quantum numbers is given by (see, e.g. [25])

$$<n_B, \ell, \vec{k}, N_{\vec{q}} | H_{\text{Fröhlich}} | n_B, \ell, \vec{k}', N_{\vec{q}} \pm 1$$

$$= \frac{e}{q} \left[\frac{2\pi \hbar\omega_{Lo}}{V} \left(\frac{1}{\varepsilon_b} - \frac{1}{\varepsilon_s} \right) \right]^{1/2} (q_e - q_h)(N_{\vec{q}} + \frac{1}{2} \pm \frac{1}{2})^{1/2} ;$$

$$k = \vec{k} \pm \vec{q} \qquad (97)$$

V is the volume of the unit cell, q_e and q_h a measure for the charge-distribution of electrons and holes in the exciton. For 1s excitons $q_{e/h}$ reads [25]:

$$q_{e/h} = [(1 + (\frac{m_{e/h}}{m_e + m_h}) \frac{q\, a_B}{2})^2]^{-2} \qquad (98)$$

ε_s and ε_b are the low- and high-frequency dielectric constants of the phonon-resonance, respectively. The other quantities have their usual meaning. The matrix element is zero for $\varepsilon_b = \varepsilon_s$. In this case the oscillator strength of the phonon is zero (see Eqs. (47) and (48)) and it is not connected with an electric field.

The matrix element vanishes for $q_e = q_h$. In this case, the charge-distributions of electrons and holes coincide, so every fraction of the exciton is neutral and has in first order no interaction with an electric field.

The Fröhlich-interaction is proportional to q for $q \ll a_B^{-1}$. Therefore this interaction is sometimes called a forbidden one, though it may be the dominant one in polar materials. For q = 0 the electric field of the phonon degenerates to a homogeneous electrostatic field and the Stark effect of the 1s-exciton is zero in linear, first-order approximation.

The latter statement is true, however, only for scattering processes between exciton states with the same parity (see Figure 25). Transitions between exciton states with different parity are allowed for q = 0. For large q ($q \gg a_B^{-1}$) the electric field of the LO-phonon oscillates many times over the diameter of the excitons and the opposite contributions tend to cancel each other. Therefore $|M|$

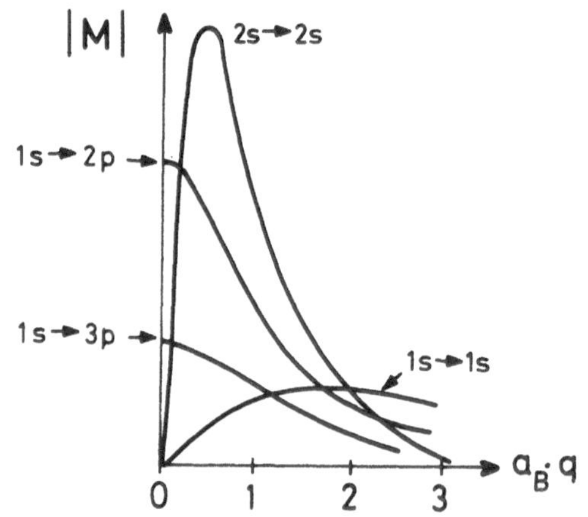

Fig. 25. Dependence of the matrix element M of the Frohlich interaction on the wavevector of the LO-phonon q, for transitions between exciton states of equal parity (1s → 1s or 2s → 2s) or different parity (1s → 2p or 1s → 3p). From [42].

decreases for large q. The decrease is faster for states with higher n_B due to their larger radius. Similar arguments hold for $qa_B >> 1$ in the case of deformation potential scattering, too.

The interaction between exciton-polaritons and acoustic phonons is generally some orders of magnitude smaller than with optic ones. Again there exists the possibility of deformation-potential interaction. Since all atoms in a unit cell are elongated in phase for acoustic phonons, the important quantity is here the variation of the amplitude from one unit cell to the next, not the amplitude itself. The deformation-potential interaction is generally stronger for LA than for TA phonons. For our idealized semiconductor with isotropic, nondegenerate bands, the deformation-potential interaction would be zero for TA phonons [41]. In piezoelectric materials an acoustic phonon may produce an electric field which in turn interacts with the electron-hole pair in the exciton. As to the piezoeffect itself, the piezocoupling between acoustic phonons and exciton-polaritons is strongly direction-dependent. Fig. 26 gives an example. The piezocoupling is generally stronger for TA than for LA phonons.

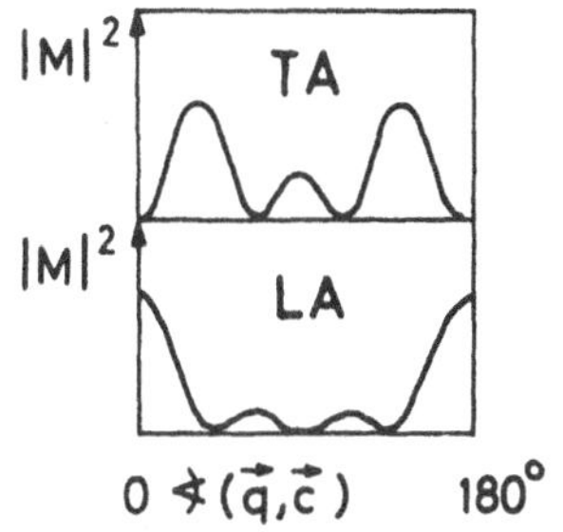

Fig. 26. Dependence of the matrix element squared of the piezo-coupling on the angle between phonon-wavevector $\vec{q}$ and crystallographic axis $\vec{c}$ for transverse and longitudinal, acoustic phonons. From [43].

More information about exciton-phonon coupling may be found, e.g. in [20],[25],[41] and the references cited therein.

III.C. LO-Phonon Assisted Luminescence

In this and the following subsections, various aspects of exciton-phonon interaction are discussed. Here we assume that we have a population of exciton-like polaritons which is in thermal quasi-equilibrium. It can thus be described by a temperature T_x using Boltzmann-statistics for not too high densities and too low temperatures. T_x may be different from the lattice temperature. Deviations of the exciton distribution towards Bose-statistics can be observed only under extreme conditions [86].

The distribution of exciton-like polaritons is then described by their kinetic energy E_k

$$N(E_k) \sim \begin{cases} E_k^{1/2}\, e^{-E_k/k_B T_x} & \text{for } E_k > E_o \\ 0 \quad \text{otherwise} & E_k = \dfrac{\hbar^2 k^2}{2M} \end{cases} \qquad (99)$$

Eq. (99) gives a good approximation for the exciton-like part of the polariton dispersion. In the transition-region from exciton-like to photon-like behaviour (bottleneck) there occurs a tail in the population to lower energies which is not described by Eq. (99). If the exciton-like polaritons reach the surface, they will be partly transmitted and appear as a luminescence photon in vacuum. This direct luminescence of the free excitons is generally rather weak for two reasons: first, the damping (i.e. the $\mathrm{Im}\{\varepsilon(\omega)\}$) is rather large around the resonance (see Fig. 17) and thus only exciton-like polaritons from a thin layer may reach the surface. Second, the reflectivity is rather high in this spectral region and so only a small fraction of the polaritons is transmitted through the surface as photons. More information about the diffusion of excitons and their luminescence is found e.g. in [4],[11],[44] and references given therein. Another luminescence process of the exciton-like polariton involves some energy transfer to the phonon field: an exciton-like polariton may be scattered onto the photon-like part of the dispersion curve by emitting one or more LO-phonons (Fig. 27). The transition rate is somehow restricted by the small density of final states on the photon-like part of the dispersion (see also III.2.). This is compensated, however, by the fact that all exciton-like polaritons in the volume of the sample can contribute, since the photon-like polaritons are almost undamped and may reach the surface. Due to their samller index of refraction, they are also easily transmitted through the surface. As a consequence, the LO-phonon satellites of the free excitons are generally more intense than the zero-phonon emission.

The lineshape of the band arising from the emission of m LO-phonons is given by [45].

$$I_m(\hbar\omega) = \begin{cases} E_k^{1/2}\, e^{-E_k/k_B T_x}\, W_m(E_k) & \text{for } E_k > 0 \\ 0 \quad \text{otherwise} & \end{cases} \qquad (100)$$

with

$$\hbar\omega = E_o - m\hbar\omega_{LO} + E_k$$

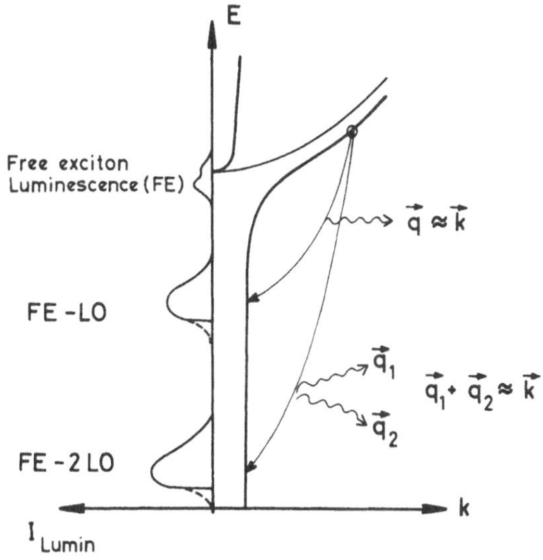

Fig. 27. Luminescence from exciton-polaritons under emission of one or two LO-phonons. Right hand side: schematic drawing of the scattering process, left hand side: resulting luminescence bands.

W_m is the square for the transition matrix element for the emission of m LO-phonons. Often W_m can be approximated by a power law

$$W_m(E_k) = E_k^{l_m} \tag{101}$$

Then one finds for the position of the emission-maximum of band m:

$$\hbar\omega^m_{max} = E_o - m\hbar\omega_{LO} + (l_m + \frac{1}{2})\, k_B\, T_x \tag{102}$$

The ratio $Q_{m1,m2}$ of the integrated emission bands m_1 and m_2 is given by

$$Q_{m1\ m2} = \int I_{m1} \cdot d\omega / \int I_{m2}\, d\omega = T(\ell_{m1} - \ell_{m2}) \tag{103}$$

As explained in III.2, the Fröhlich interaction between excitons and phonons is proportional to the wavevector q of the LO-phonon. Already for temperatures $T_x > 3K$, the average k-vector of the exciton-like

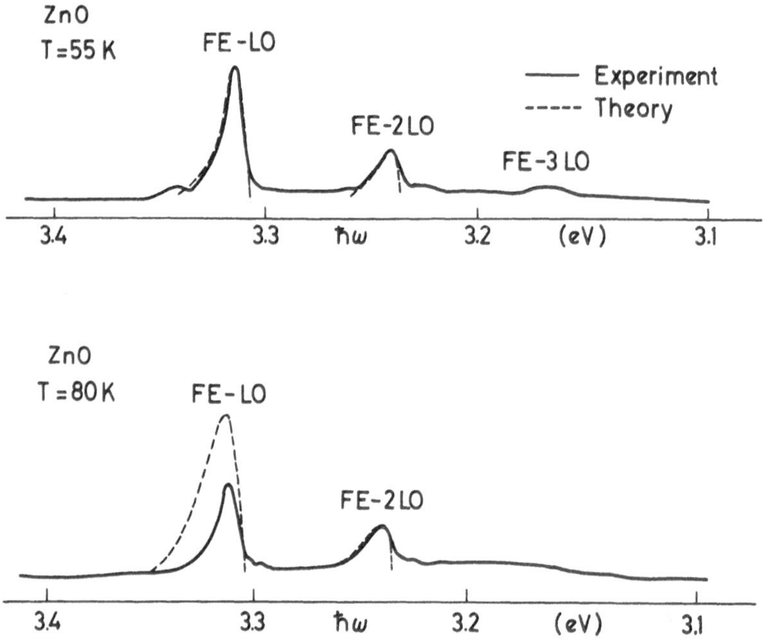

Fig. 28. Emission spectra of ZnO for two different temperatures showing LO-phonon replica of the free exciton luminescence. From [46].

polaritons in the initial state is considerably larger than in the photon-like final state. Thus we may write for m = 1

$$\vec{k} \simeq \vec{q} \Rightarrow W_1 \sim k^2 \sim E_k \Rightarrow \ell_1 = 1$$

$$I_1(\hbar\omega) \sim (\hbar\omega + \hbar\omega_{LO} - E_o)^{3/2}\, e^{-(\hbar\omega + \hbar\omega_{LO} - E_o)/k_B T_x} \tag{104}$$

$$\hbar\omega^1_{max} = E_o - \hbar\omega_{LO} + \frac{3}{2} k_B T_x$$

For m = 2 there is only a rule for the sum of the wavevectors of the two LO-phonons

$$\vec{k} = \vec{q}_1 + \vec{q}_2 \tag{105}$$

$\vec{q}_1$ and $\vec{q}_2$ may spread over the whole Brillouin zone. As a consequence W_2 does not depend on E_k. Thus

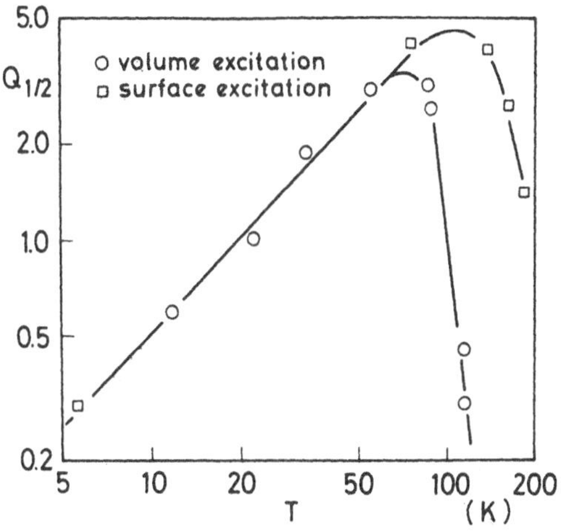

Fig. 29. The ratio of the integrated intensities of the first and the second LO-phonon sideband $Q_{1/2}$ as a function of temperature for volume and surface excitation. From [46].

$$I_2(\hbar\omega) = E_k^{1/2}\ e^{-E_k/k_BT_x} = (\hbar\omega + 2\hbar\omega_{LO} - E_o)\ e^{-(\hbar\omega+2\hbar\omega_{LO}-E_o)/k_BT_x} \tag{106}$$

Comparing Eq. (106) with Eq. (99), one sees that $I_2(\hbar\omega)$ directly reflects the distribution of the exciton-like polaritons and the integrated intensity $\int I_2(\hbar\omega)\, d\omega$ is directly proportional to the density of excitons.

Finally one gets from (103)

$$Q_{1/2} \sim T \tag{106}$$

The II-VI semiconductor ZnO is used to compare the above results to experiment:

Fig. 28 shows two emission-spectra for 55K and 80K, respectively. The first three LO-phonon replica are seen. For 55K the shape of the spectra of the first two satellites is well described by Eqs. (104) and (105). The small tail to lower energies is due to the population of the dispersion in the bottleneck region which is not considered by Eq. (99). For T = 80K the first LO-satellite is already influenced by the damping of the photon-like polariton, which extends to lower energies with increasing temperature (Urbach-Martienssen rule, see, e.g. [87]).

Fig. 29 shows the temperature dependence of $Q_{1/2}$. For temperatures up to about 70K, $Q_{1/2}$ follows the linear temperature dependence predicted by Eq. (107). For higher T, $Q_{1/2}$ decreases because of the increasing damping in the spectral region of I_1, which is not considered in the model-calculation above. This deviation is obviously more pronounced for two-photon excitation, where exciton-like polaritons are created homogeneously in a sample of about 2 mm thickness, as compared to one-photon band-to-band excitation, where a layer of a few μm depth is excited only.

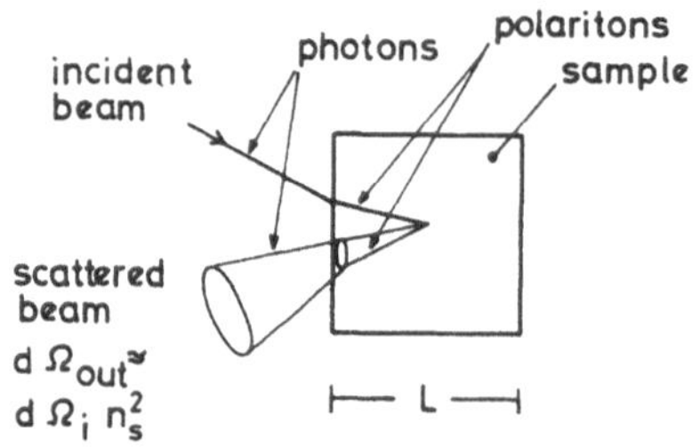

Fig. 30. Geometric configuration for the calculation of the scattering efficiency according to Eq. (108).

III.D. Resonant Brillouin Scattering

Before we discuss in this subsection an aspect of the energy transfer from exciton-polaritons to acoustic phonons, namely the Brillouin-scattering, a formula is presented for the scattering efficiency which is applicable both to Brillouin- and Raman-scattering.

The geometry is shown in Figure 30. A monochromatic beam of photons with intensity I_i and frequency ω_i falls on the surface of the semiconductor. The scattered intensity is collected from a solid angle $d\Omega_{out}$ which for $d\Omega_{out} << 4\pi$ is connected with the inner solid angle $d\Omega_{in}$ by

$$d\Omega_{out} = d\Omega_{in}\ n^2(\omega_s) \tag{108}$$

(ω_s) being the frequency of the scattered light. One finds [41], [49],[50]:

$$\eta = \frac{1}{I_i}\frac{dI_s}{d\Omega_{out}} = T(\omega_i)\ T(\omega_s)\ \frac{k_s^2\ L_{eff}|<k_i|H_{sc}|k_s>|^2}{4\pi^2\hbar^2\ n^2(\omega_s)\ v_g(\omega_i)\ v_g(\omega_s)} \tag{109}$$

where $T(\omega_i)$ and $T(\omega_s)$ are the probabilities that the incident beam is transmitted through the surface into the sample and the scattered light out of the sample. L_{eff} is an effective scattering length given by

$$L_{eff} = \begin{cases} \text{Length of the sample} \\ [\alpha(\omega_i) + \alpha(\omega_s)]^{-1} \end{cases} \quad \text{(whatever is shorter)} \tag{110}$$

The other symbols have their usual meaning.

In Fig. 31 the Brillouin scattering process is presented for various spectral regions. In agreement with most experiments, a backscattering-geometry is assumed. The slopes of the transition arrows are given by $\hbar v_a$, v_a being the velocity of the acoustic phonon branch involved in the scattering (see Fig. 10).

In the region 1 we are below resonance. The dispersion is almost a vertical line and Stokes and anti-Stokes lines are situated symmetrically to the incident laser. The intensity ratio depends on the sample temperature. For $E_o \lesssim \hbar\omega_i \lesssim E_{lo}$ (region 2) the anti-Stokes shift is larger than the Stokes shift due to the transition from a photon-like to an exciton-like dispersion. For $\hbar\omega_i > E_{lo}$ (region 3), there are two polariton branches for the ingoing and two for the scattered polaritons, resulting in a total of four possible Stokes and four anti-Stokes lines. Since the frequency shift ΔE is directly proportional to Δk according to

$$\Delta E = \hbar v_a\ \Delta k = \hbar v_a|\vec{k}_i - \vec{k}_s| \tag{111}$$

resonant Brillouin scattering allows for direct spectroscopy in momentum space. This scattering has been observed in several direct

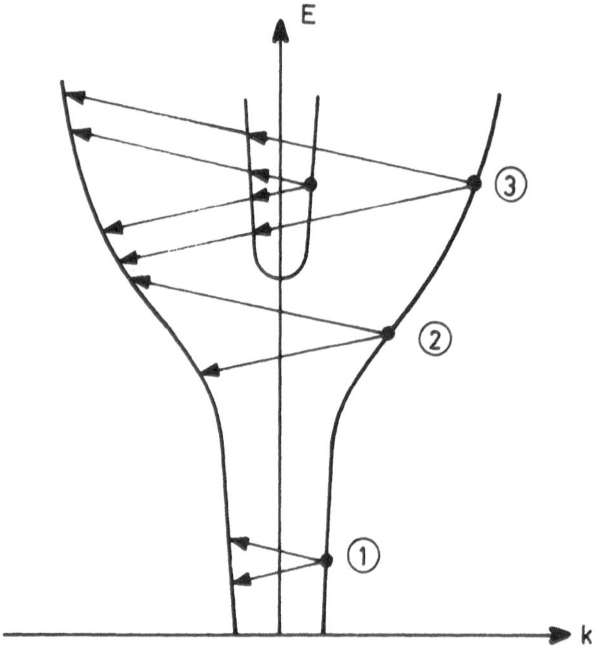

Fig. 31. Brillouin scattering in backward configuration for different parts of the dispersion of exciton-polaritons.

and indirect gap semiconductors (see, e.g. [18],[25],[41],[47]). If scattering with different acoustic phonon modes is possible, the different sound velocities have to be taken into account in Figure 31.

In Figure 32 results are shown for GaAs. The Stokes and anti-Stokes shifts are plotted along an energy scale for $\hbar\omega_i$. Dots are experimental points, solid lines are calculated. The data agree with the predictions given above and close coincidence between experiment and theory shows that the properties of the exciton-polariton in GaAs are well understood. The discussion of RBS will be completed by two comments: RBS allows in a direct way to demonstrate the uncertainty relation $\Delta P_x \cdot \Delta x \simeq \hbar$.

If the material is strongly absorbing for ω_i and/or ω_s the effective scattering length L_{eff} is small (see Eq. (110)). The place where the acoustic phonon is created is known for the direction normal to the surface with a precision $\Delta x \simeq L_{eff}$. A momentum broadening $\hbar\Delta q_x$ results in

$$\Delta q_x \simeq L_{eff}^{-1} \simeq \mathrm{Im}\{k\} \qquad (112)$$

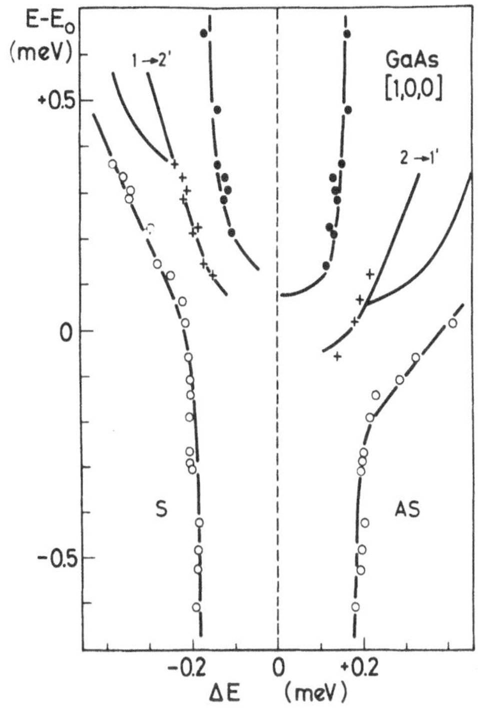

Fig. 32. Stokes (S) and Anti-Stokes (AS) shift in resonant Brillouin scattering of GaAs involving LA phonons. Solid line: theory; crosses, points and circles: experiment from [49].

Equation (112) is valid for the incident and the scattered light beam resulting in [49]

$$\frac{\Delta h\omega_q}{h\Delta_q} = \frac{\mathrm{Im}\{k_i + k_s\}}{\mathrm{Re}\{k_i + k_s\}} \tag{113}$$

Figure 33 gives an experimental example of this so-called opacity broadening.

As is clear from Figure 31, in region 3 different polariton branches yield different (anti-) Stokes shifts. From an analysis of the intensities, information about the abc should follow [31]. Unfortunately, the efficiency is determined by many, partly unknown quantities (see Eq. (109)).

It has been tried to find out which is the correct abc by simultaneously measuring the reflection, transmission and RBS-spectra. This has been done to our knowledge only once in Ref. [24] for CdS.

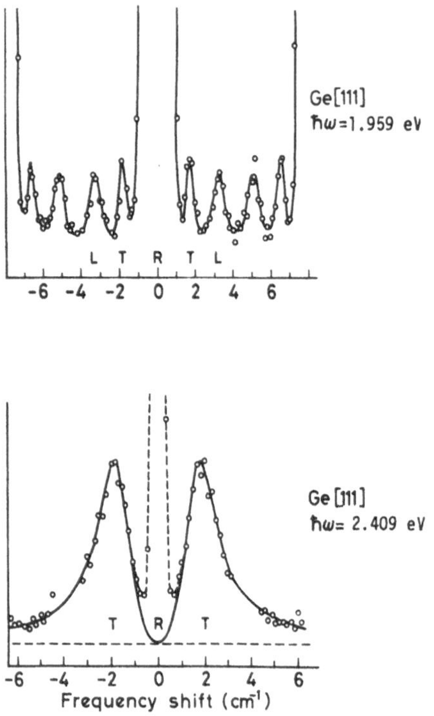

Fig. 33. Brillouin scattering from Ge in backward configuration for two different photon energies. The increase of the absorption coefficient with photon energy results in an increasing opacity broadening seen for the scattering with TA-phonons. From [50].

The result was that none of the abc was able to explain all data. The reason for this result has been already discussed in Section II.G.

III.E. Raman Scattering

Raman scattering means scattering of an exciton-polariton under emission (or absorption) of one or more optical phonon-polaritons. The process is shown schematically in Fig. 34 for backward and forward scattering geometry. Since the eigenenergies of optical phonons are of the order of 10 meV to 100 meV, at low temperatures the Stokes emission is observed only. Since the dispersion of phonon-polaritons is rather flat except for the photon-like region, the energy-shift for Raman scattering

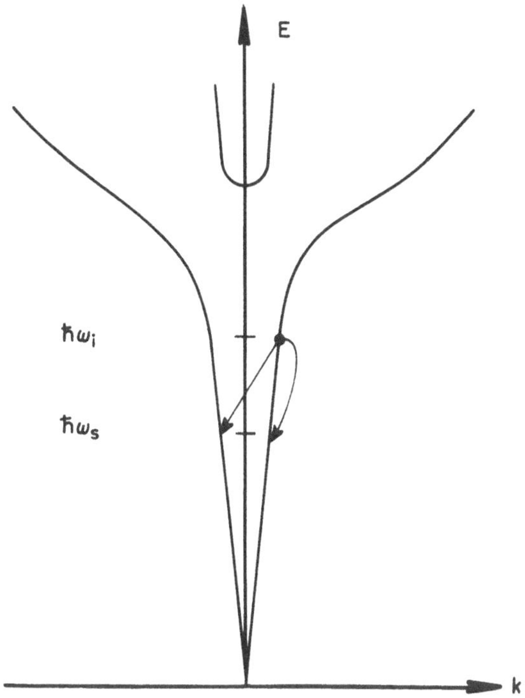

Fig. 34. Raman-scattering in forward and backward configuration.

$$\Delta E_R = \hbar\omega_i - \hbar\omega_s = \hbar\omega_{LO/TO}$$

is independent of the scattering geometry for changes in wavevector $|\Delta\vec{k}| = |q|$ larger than 10^4 cm^{-1}. In a nonresonant Raman scattering experiment, i.e. if $\hbar\omega_i$ and $\hbar\omega_s$ are several Δ_{LT} below resonance, the values of q are about 8.10^5 cm^{-1} and 2.10^4 cm^{-1} for backward and forward scattering, respectively. In the case of the Fröhlich interaction, the transition matrix element is proportional to q (see III.B.). So there should be an enormous increase in the scattering efficiency when going from forward to backward scattering. In an experiment on CdS only a factor of two has been found [51]. It has been argued that localized bound exciton states are involved as virtually excited intermediate states. Due to the localization, the k-conservation is relaxed, resulting in a rather weak dependence of the scattering efficiency on the scattering geometry.

Raman scattering below resonance gives mainly information on Raman-active optical phonon modes. For one phonon scattering, information is obtained about the eigenenergies at the Γ-point ($q \simeq 0$). If more LO phonons are involved in a single scattering process, only the sum of their wavevectors must be small and points with a high

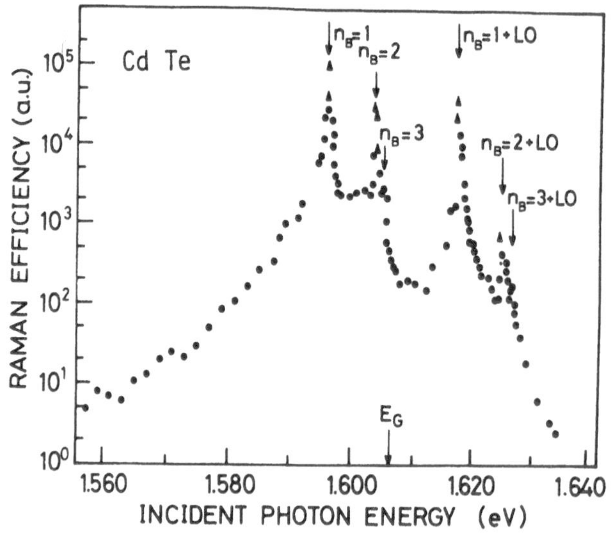

Fig. 35. Excitation spectra of resonant Raman-scattering in CdTe showing the resonances when either the incoming or the scattered light falls energetically on the exciton resonances. From [49],[53].

density of states far in the Brillouin zone show up (see, e.g. [41], [52] and references therein).

In contrast, Raman-scattering experiments at resonance give information about the exciton-polariton and the mechanism of energy transfer, too. In the following two examples of this type are presented.

As seen from Eq. (109) the scattering efficiency is inversely proportional to the group velocities of ingoing and scattered light in the medium

$$\eta \sim v_g(\omega_i)^{-1} \cdot v_g(\omega_s)^{-1} \qquad (114)$$

This dependence has nicely been shown e.g. in CdTe [49],[53]. As seen in Figure 19 (n_B = 1 state for CuCl), the group-velocity decreases considerably at the resonance. As explained in II.7, such resonances occur also for the higher exciton states ($n_B = 2, 3, \ldots$), however, with decreasing oscillator strength.

Figure 35 gives the scattering efficiency of the 1 LO Raman-scattering as a function of $\hbar\omega_i$ in CdTe. Distinct resonances occur when $\hbar\omega_i$ coincides with the n_B = 1, 2 and 3 states of the exciton series. These peaks are due to the minima of $v_g(\omega_i)$. Shifted by the energy of an LO-phonon to the blue, three similar maxima occur. In this case, the energy of the scattered light coincides with the excitonic resonances and the structure is caused by the minima of $v_g(\omega_s)$.

As already mentioned above, the phonon wavevectors in Raman-processes involving more than one phonon may be rather large, enhancing thus the transition probability in the case of Fröhlich interaction. In Fig. 36 a multiresonant, multiphonon Raman process is shown schematically. The incident photon energy is situated above the band gap and the polariton states are imbedded in the excitonic continuum. The 1 LO Raman process has a rather small probability due to

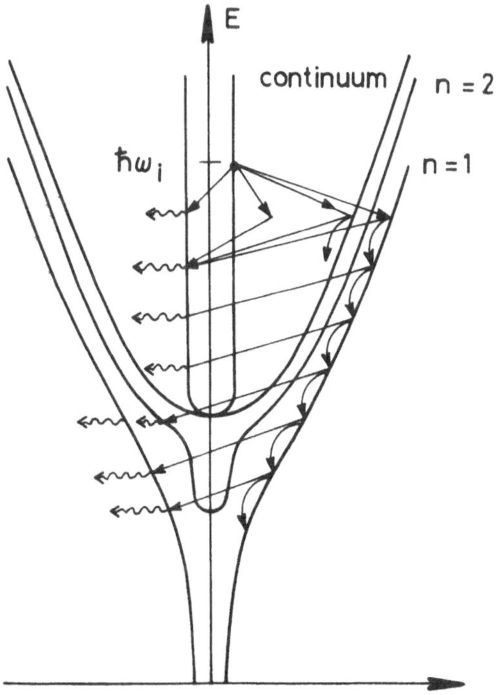

Fig. 36. Schematic drawing for a multiresonant, multiphonon Raman-scattering process.

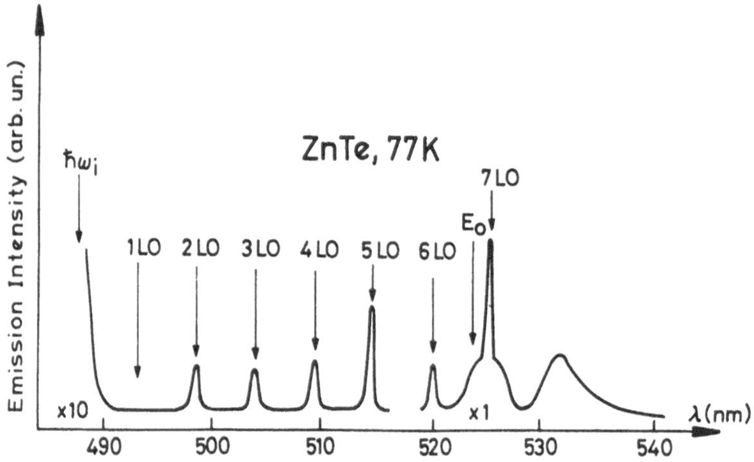

Fig. 37. Multiphonon-Ramanemission spectrum of ZnTe. From [45].

the small q of the phonon. For higher Raman processes (in Fig. 36 depicted up to m = 7) always many scattering processes are possible involving large q phonons. As sketched for the 2 LO Raman emission continuum states or discrete branches (represented here by $n_B = 1$ $n_B = 2$) may be reached. Fig. 37 gives an experimentally measured spectrum from [45], corresponding to the schematic drawing in Fig. 36. The one LO Raman line is below detection limit. The other Raman lines are enhanced both by the group velocity terms Eq. (114) and by the q dependence of the Fröhlich interaction. Below the $n_B = 1$ resonance, the Raman lines are too weak to be detected. The reasons for this decrease are the high group velocity and the small q-vector possible in the phonon-emission processes. The m = 7 Raman line is superimposed on a broader, luminescence-like emission band. A LO-phonon replica of this band is seen around λ = 533 um. This transition from Raman scattering to hot luminescence and to "normal" photoluminescence will be the subject of the last section of this chapter. An exhaustive treatment of light-scattering phenomena in solids and the energy transfer processes connected with them, is found e.g. in [48]. Some aspects of scattering of polaritons under simultaneous emission of optical and acoustic phonons are found e.g. in [41],[48].

III.F. Resonant Raman Scattering, Hot Luminescence, Thermalisation and Photoluminescence

In this subsection, we discuss comprehensively resonant Raman scattering, hot luminescence, thermalisation and photoluminescence.

In all of these processes, energy is transferred from exciton polaritons to the phonon field. In a resonant Raman scattering process with emission of only one phonon (e.g. Fig. 34) the properties of the scattered polariton are known, if the incident polariton and the properties of the phonon are known. In other words, there is a close correlation or even some phase coherence between incident and scattered polariton.

In Raman processes where several phonons are emitted (see Fig.36) the phase coherence between incident and scattered polaritons is lost, because there are many possible scattering-paths which are superimposed. Due to the weak dispersion of optical phonons, there is still an energy-correlation between incident and scattered polariton given by

$$\hbar\omega_s = \hbar\omega_i - m\,\hbar\omega_{LO} \tag{115}$$

and the spectral width of the Raman emission is not very much broader than that of the incident laser (see e.g. Fig. 37).

Now it may happen that there occurs - apart from emission of LO-phonons - emission of acoustic phonons or scattering between exciton polaritons. These processes lead to an energetically broad distribution of the polaritons and consequently to a broad emissionband. The shape of these emissionbands depends on the energy of excitation, and a lineshape analysis shows that the distribution is either nonthermal or corresponds to an effective temperature, which is considerably higher than the lattice-temperature and which depends on $\hbar\omega_{exc}$. This type of emission is called "hot luminescence". The two broad emissionbands on the low energy side of the spectrum in Fig. 37 could be due to hot luminescence. If the lifetime of the exciton-polaritons is sufficiently long, they may come into thermal equilibrium with the lattice. They relax down first mainly by emission of LO phonons. This process becomes ineffective, if the final state of the scattering is below the resonance (see III.2.). Then further relaxation and randomization of the distribution take place by emission (and absorption) of acoustic phonons. A complete thermalization is reached if the distribution of the polaritons on the exciton-like part of the dispersion can be described by the lattice temperature. The distribution of the polaritons is then completely independent of the excitation energy. The emission from such a state has been treated in III.3. It is simply called (photo-) luminescence. As a conclusion of this discussion, we may state that resonant one-phonon and multi-phonon Raman scattering, hot luminescence and photoluminescence from a thermalized distribution are related phenomena. The correlation to the excitation conditions is decreasing from Raman scattering to photoluminescence by an increasing energy transfer to the lattice or to other degrees of freedom. More information about this subject may be found e.g. in [49].

IV. ENERGY TRANSFER BETWEEN VARIOUS EXCITON-POLARITON MODES BY NONLINEAR INTERACTION

In the preceding chapters II and III, predominantly the energy transfer from one system of quasi particles to another one has been discussed. It is also possible to get an energy transfer from one eigenstate to another one of the same type of quasiparticles. This process is generally due to nonlinearities. To illustrate this concept, we recall shortly the case of phonons which is treated in every textbook on solid state physics (see, e.g. [11],[12],[15]) and then we shall discuss in more detail various aspects of the nonlinear interaction between exciton polaritons.

IV.A. Nonlinear Interaction Between Phonons

In an idealized model for the lattice vibrations one assumes that all forces are harmonic, i.e. the forces acting between the atoms (or ions) of the lattice are assumed to be strictly proportional to the elongation. This assumption leads to a good overall description of the properties of phonons, e.g. concerning the specific heat, but also to some conclusions which do not agree with the experience, like:

- there is no thermal expansion of matter
- phonons do not influence each other
- the heat conductivity is proportional to the geometric dimension of the sample for all temperatures.

It is well known that the harmonic approximation of the vibration potential is strictly valid only for very small amplitudes. For larger amplitudes deviations occur.

As a result, phonons start to interact with each other and the average "lattice constant" changes with the amplitude, i.e. with temperature. Due to the nonlinear interaction, one high energy phonon may decay into two phonons of lower energy or two incident phonons with energies $\hbar\omega_{1/2}$ and wavevectors $\vec{q}_{1/2}$ may form a third phonon ($\hbar\omega_3$, $\vec{q}_3$). In both processes energy and momentum are conserved, e.g.

$$\vec{q}_1 + \vec{q}_2 = \vec{q}_3 \quad ; \quad \hbar\omega_1 + \hbar\omega_2 = \hbar\omega_3$$

If $\vec{q}_3$ is outside the first Brillouin-zone, the transfer of a momentum $\hbar\vec{g}$ to the lattice as a whole (where $\vec{g}$ is a vector of the reciprocal lattice) brings phonon "3" back into the first Brillouin-zone. This type of Umklapp-process is, e.g., responsible for the finite heat-conductivity of reasonably pure materials at elevated temperatures $T \gtrsim \theta_{Debey}$.

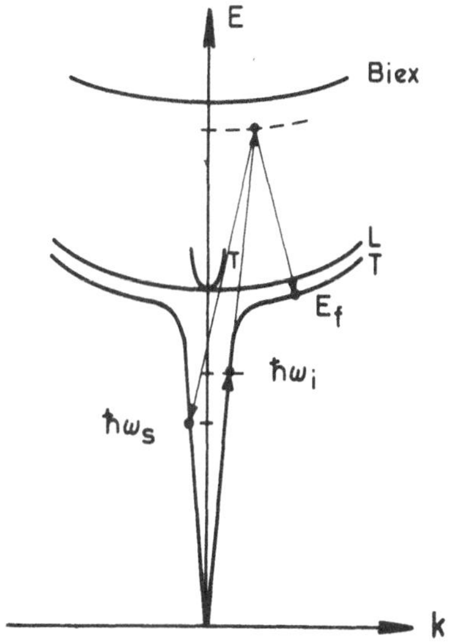

Fig. 38. Two-"photon" Raman-scattering via virtually excited biexcitons.

IV.B. Two-Photon Raman Scattering

In this subsection, we describe a process which is known in the literature as two-photon Raman scattering (TPRS) or hyper-Raman scattering (HRS). Before giving details of the process, a further quasiparticle is introduced, namely the biexciton (or excitonic molecule). It is a bound state of two electrons and two holes. If the exciton is an analogon to the hydrogen atom, then the biexciton corresponds to the H_2 molecule. The existence of this biexciton has been proved in various semiconductors. Recent reviews of this subject are found e.g. in [9],[30],[54]. The dispersion of the biexciton is given by

$$E_{biex}(\vec{k}) = 2E_x(k = 0) - E^b_{biex} + \frac{\hbar^2 k^2}{2 \cdot 2M} \qquad (116)$$

where $E_x(k = 0)$ is the eigenenergy of the lowest free exciton state, which is generally a triplet state (see Fig. 15a). E^b_{biex} is the binding energy of the excitonic molecule and the third term on the right hand side of Eq. (116) gives the kinetic energy, under the assumption that the translational mass of the biexciton is twice that of the exciton. This assumption is roughly verified by experiment [55]-[57]. In the TPRS-process, two polaritons from a spectrally narrow, incident laser beam virtually excite a biexciton. Virtual excitation means that the biexciton is created at an energy which deviates frwm the eigenenergy by an amount ΔE. This is possible due to the uncertainty relation for a time τ given by

$$\Delta E \cdot \tau \simeq \hbar \qquad (117)$$

After this time, the virtually excited state has to disappear. As shown schematically in Fig. 38 for a backward scattering geometry, a possible decay process is into a photonlike polariton and an exciton-like polariton (or a longitudinal exciton). Energy and momentum are conserved during this process:

$$\left.\begin{aligned} 2\hbar\omega_i &= \hbar\omega_s + E_f(\vec{k}_f) \\ 2\vec{k}_i &= \vec{k}_s + \vec{k}_f \end{aligned}\right\} \qquad (118)$$

The index i refers to the incident polaritons, s means the photonlike

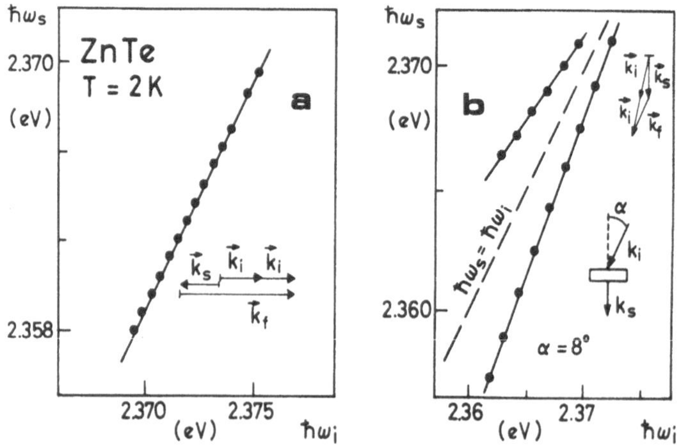

Fig. 39. The relation between the photon energy of the incident laser light ($\hbar\omega_i$) and the two-photon Raman-emission $\hbar\omega_s$, for backward (a) and near forward (b) scattering geometry in ZnTe. Points: experimental data; solid lines: theory. From [58].

outgoing polariton and f the other particle in the final state. For backward scattering k_f is rather large, and E_f corresponds to a state on the exciton-like part of the dispersion. The energy variation with k_f is weak in this part, $\hbar\omega_s$ shifts with a slope of two compared to $\hbar\omega_i$. These slope two curves (Fig. 39a) are the experimental proof for the two-polariton character of TPRS. In near-forward scattering experiments the wavevectors of both particles are on the photon-like part of the dispersion. Both particles in the final state may be detected outside the sample, and a typical result are two TPRS-lines situated energetically above and below $\hbar\omega_i$ as shown in Fig. 39b. TPRS has been observed, e.g., in CuCl, CuBr, CdS, ZnSe or ZnTe. See, e.g. [9],[30],[54],[58] and the literature cited therein.

It should be noted that a strong population at the polariton-state $\hbar\omega_i$, $\vec{k}_i$ will result in an additional polariton-branch with a resonance-like dispersion an an energy $\hbar\omega_{an}$.

$$\hbar\omega_{an} = E_{biex}(k) - \hbar\omega_i \qquad (119)$$

due to two-polariton transitions to the real biexciton state. In other words, shining an intense laser-beam on a sample may result in variations of the refractive index and the appearance of additional absorption bands by nonlinear interaction. The transition to the biexciton is only one example of numerous processes of this kind [9],[30],[54],[57],[59]. We shall use this knowledge in Sections IV.C and IV.D.

IV.C. Degenerate Four Wave Mixing

In IV.2. we considered the case that the two polaritons creating the biexciton virtually have parallel k-vectors and equal energy. This is not a necessary condition. They may have different energies or may propagate into different directions. Another peculiar situation is to excite the virtual, intermediate state by two counter-propagating polaritons of equal energy. The wavevector of this state is then zero. An infinite number of decay processes becomes possible which must fulfill the conditions:

$$\vec{k}_s + \vec{k}_f = 0 \quad ; \quad \hbar\omega_s + \hbar\omega_f = 2\hbar\omega_i \qquad (120)$$

If the two polaritons $\vec{k}_s$ and $\vec{k}_f$ are on the same polariton branch, one gets from Eq. (120):

$$\hbar\omega_s = \hbar\omega_f = \hbar\omega_i \qquad (121)$$

If they are on different branches $\hbar\omega_f \neq \hbar\omega_s$, but Eq. (120) still has

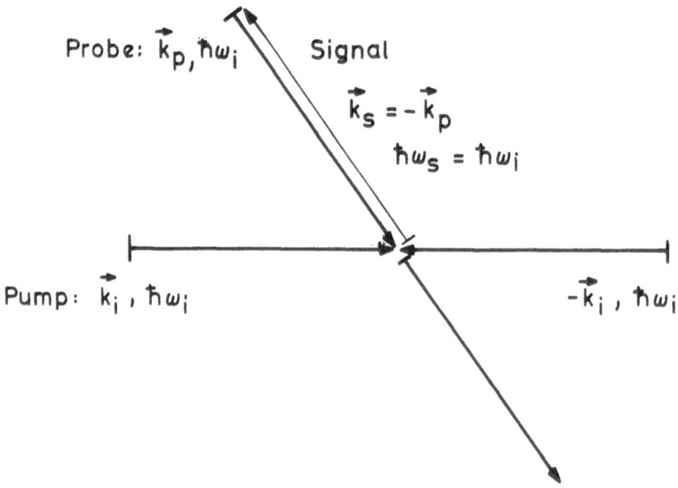

Fig. 40. Geometrical arrangement for degenerate four wave mixing.

to be fulfilled. We consider in the following only the case that ingoing and outgoing polaritons have the same energy. One of the decay mechanisms possible in this case may be selected by stimulation: Additional to the two counter-propagating "pump" beams, a probe beam with the same photon energy is sent on the sample at some arbitrary angle. This probe beam stimulates the decay of the virtually excited biexciton. As a consequence, a second polariton is created which is identical to those of the probe beam and another one propagating in the opposite direction. (See Fig. 40). Though at least five polaritons are involved in this nonlinear interaction process, it is referred to in the literature as "degenerate four wave mixing" DFWM [30],[61]-[63] or sometimes as "three wave mixing".

The process shown in Fig. 40 has some similarity with the reflection from a mirror. But it is a special type of mirror. One generally speaks of a phase-conjugate mirror or phase-conjugate reflection [62]. The difference to a usual mirror is worked out in Fig. 41. As already discussed in II.E., the law of reflection reads:

$$\vec{k}_{pi} = \vec{k}_{pr} \quad ; \quad \vec{k}_{ni} = -\vec{k}_{nr} \tag{122}$$

and if there is some phase-delay in the incident plane wave, the reflected wave will have the same phase-structure. For a PC-mirror, the law of reflection reads

$$\vec{k}_i = -\vec{k}_r \tag{123}$$

and if there is a phase delay in some part of the incident wave, there will be just the opposite phase-structure in the "reflected" beam. This phenomenon which gave the name, is important for the reconstruction of distorted wavefronts. Various other aspects of DFWM are discussed, e.g., in [61]-[63].

IV.D. Laser Induced Gratings

As we have seen in Section IV.B, an intense light beam propagating through a medium may produce a variation $\Delta\tilde{n}$ of the real and/or imaginary part of the refractive index. In Figure 42 we assume that two plane polariton waves intersect in the medium under an angle α. Both waves have the same energy $\hbar\omega_i$ and amount of the wavevector k. As is known from simple wave-theory, the constructive interference occurs on planes indicated by dotted lines in Figure 42 which form an angle of $\alpha/2$ with the wavevectors. Along these planes $\Delta\tilde{n}$ reaches its maximum value. In the region of destructive interference $\Delta\tilde{n}$ is small. As a result, we get a plane amplitude- or phase-grating, depending whether $\mathrm{Im}\{\Delta\tilde{n}\}$ or $\mathrm{Re}\{\Delta\tilde{n}\}$ is predominant. (LIG = laser induced grating). If we shine now a third beam on the sample, diffraction will occur. If this beam 3 has the same energy and opposite wavevector to beam 1 it is easy to show that one of the diffracted beams is counter-propagating to beam 2. This means we have the same situation as in DFWM discussed in Section IV.C.

It is obvious that we discussed in these three subsections almost identical processes: TPRS is the spontaneous process of nonlinear frequency mixing via a virtually excited intermediate biexciton state, DFWM is the same process stimulated by a third external beam. However, while in TPRS phasematching (i.e., $\vec{k}$-conservation) depends on the angular and energetic configurations, it is automatically achieved in DFWM and LIG. Whereas TPRS and DFWM stress the quasiparticle concept, the laser-induced gratings are the adequate description of the same phenomenon in the wave-picture. Higher diffraction orders in LIG correspond to higher nonlinearities of DFWM, i.e. processes involving more than five polaritons. In CuCl this higher-order emission is easily observed in the vicinity of the biexciton resonance [64].

Though the three effects TPRS, DFWM and LIG are very closely related, they have been investigated by various research groups which did not take too much notice of each other. It was only

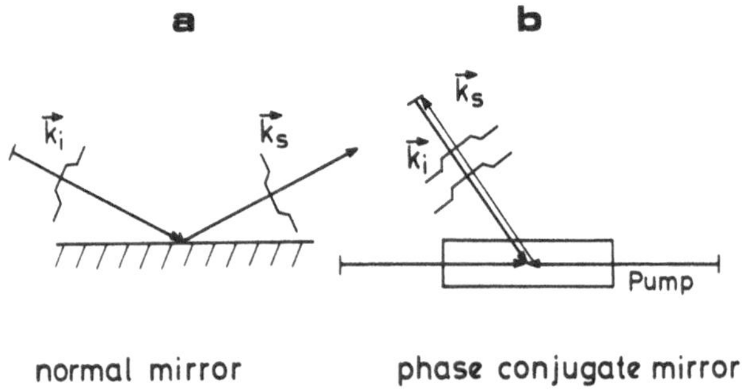

Fig. 41. Wavevectors and a distorted wave-front reflected from a usual mirror (a) and a phase-conjugate mirror (b).

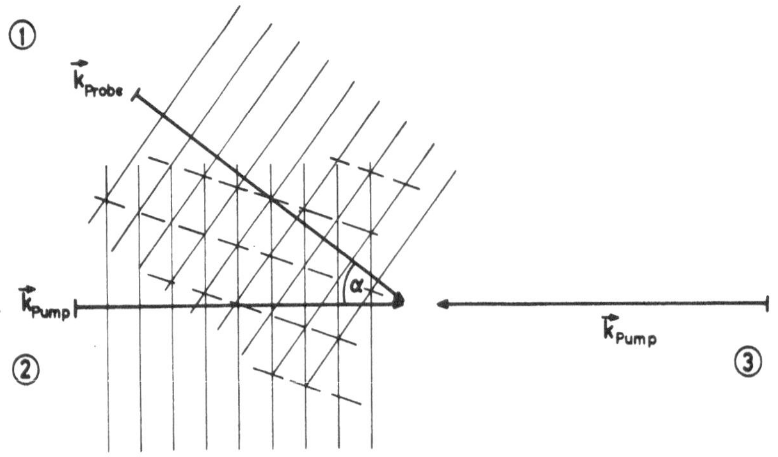

Fig. 42. Creation of and diffraction from a laser-induced grating.

rather recently that the close connection has been realized and now one tries to describe these phenomena from a unified point of view.

As a concluding remark of this chapter, it has to be stressed that the excitation of biexcitons is by far not the only possible process to produce nonlinearities which result in DFWM or LIG. The formation of an electron-hole plasma is, e.g., connected with a considerable renormalization of the bandgap and consequently with large $\Delta\tilde{n}$ [65]. The bleaching of an optical transition also gives rise to phenomena of the type discussed above. For some recent reviews of this field, see e.g. [30], [57]-[65].

V. CONCLUSION

As already stated in the introduction, the field of energy transfer processes in semiconductors is a very extensive and complex one. Some examples have been mentioned in the introduction. In this contribution, some aspects of the propagation of light in semiconductors have been stressed. The most important message of this article is that polaritons are the propagating modes of light in semiconductors. To summarize this, we have plotted the dispersion of polaritons and of some other quasiparticles from the IR to the X-ray region in a double-logarithmic scale in Figure 42. We have used a sum-formula for the dielectric function similar to Eq. (39) but properly taking into account the k-dependence of the eigenenergies, i.e.,

$$\frac{c^2 \vec{k}^2}{\omega^2} = \varepsilon(\omega, \vec{k}) = 1 + \sum_{j=1}^{3} \frac{f_j}{\omega_{oj}^2(\vec{k}) - \omega^2} \qquad (124)$$

Damping is neglected for simplicity. One optically active phonon resonance is included in Eq. (124) and two electronic resonances. The energetically lower one represents an exciton resonance and the higher one all other band-to-band excitations. Note that the upper polariton branch of the phonon polariton is simultaneously the lower polariton branch of the exciton polariton.

APPENDICES

Appendix A. Exciton-Polaritons in Real Semiconductors

Until now, we considered semiconductors with an idealized electronic bandstructure only (see Section III.A). In this appendix we shall summarize the main properties of the bandstructures of the

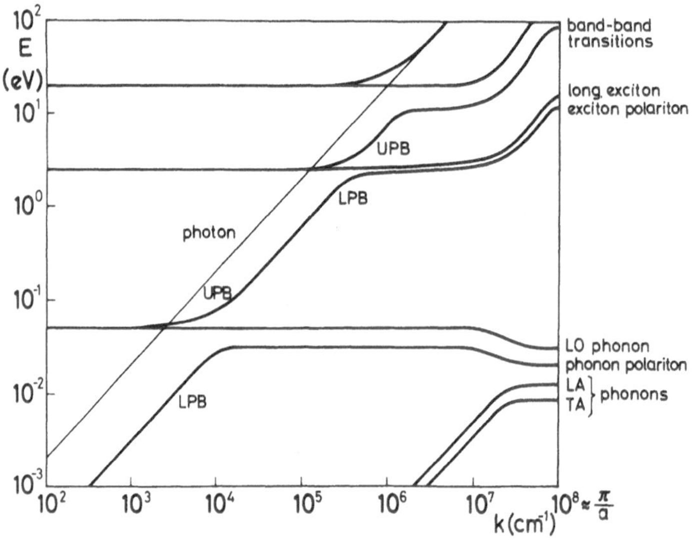

Fig. 43. Schematic dispersion of photons, acoustic phonons and of polaritons in a semiconductor. Three resonances are included, representing the phonon polariton, the exciton polariton and the band-to-band transitions.

most common semiconductors. An extensive review of the corresponding (and other) data of almost all semiconductors is found in [66].

In Table 1 we summarize the more widely known semiconductors of the families of I-VII, II-VI, III-V and the elementary semiconductors of group IV. All of these materials are tetrahedrally coordinated. The crystal-structures are either the diamond structure, the zinc-blende structure or the wurtzite-structure.

For ionic binding the lowest conduction band is formed from the lowest unoccupied s-levels of the kations and the valence-band is formed from the highest occupied p-levels of the anions with a more or less pronounced admixture of kation d-levels. The covalent binding is due to sp^3 hybridization. The bonding and antibonding states form the balence and conduction-bands, respectively (see, e.g., [9]).

Figure 39 in Volume I of Ref. [12] gives a nice schematic

Table 1. Synopsis of bandstructure, lattice- and binding typed of the most widely known semiconductors.

B. Blende-type lattice
W. Wurtzite-type lattice
D. Diamond lattice.

Familiy of Semiconductor	I - VII	II - VI	III - V	IV (-IV)
Direct gap	CuCl, CuBr, CuJ (B)	ZnS, ZnSe, ZnTe, CdTe (B) ZnS, ZnO, CdS, CdSe (W)	GaAs, InSb (B) GaN (W)	
Indirect gap	AgBr (B)		GaP (B)	C, Si, Ge (D) SiC (B)
Type of chemical Binding	Ionic	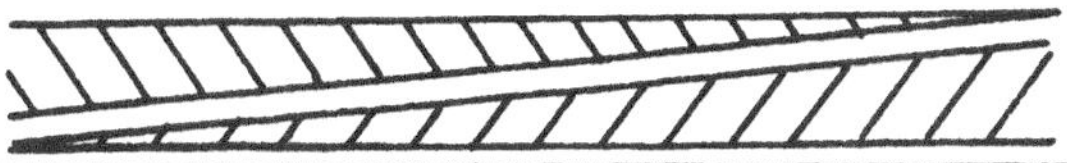		Covalent

drawing of the bandstructure of tetrahedrally coordinated semiconductors for diamond and blende type lattice. The details of the bandstructure show up in the dispersion of excitons and excitonic polaritons. In the following, some typical cases are shortly reviewed, starting with direct gap materials. The conduction band is isotropic to a good approximation and often parabolic in the vicinity of the Γ-point. The valence-band is split by spin-orbit interaction at the Γ-point into a fourfold and a twofold degenerate band. The fourfold band is generally the upper one and splits for $k = 0$ into a light and a heavy hold band. This results in heavy and light exciton- and polariton-branches, as shown e.g. in Figure 8 of Ref. [58]. Warping and k-linear terms bring about some additional complications. For CuCl, the twofold valence band is the upper one due to a negative spin-orbit splitting [67]. Thus CuCl corresponds closely to our idealized model, as can be seen by comparing, e.g., Figure 15a and Figure 19.

In a wurtzite-type lattice, the crystal field and spin-orbit interaction result in a splitting of the p-levels into three twofold degenerate bands, which are often strongly anisotropic with respect to the crystallographic c-axis [9]. The exciton- and polariton-dispersion for the typical wurtzite-type material CdS are shown, e.g., in [9],[19],[20],[29],[30],[57] and [68]. The second valence band has a k-linear term, resulting in an additional polariton branch. In ZnO, the ordering of the two upper valence bands is interchanged because of a negative spin-orbit coupling [44],[69]. Some of the

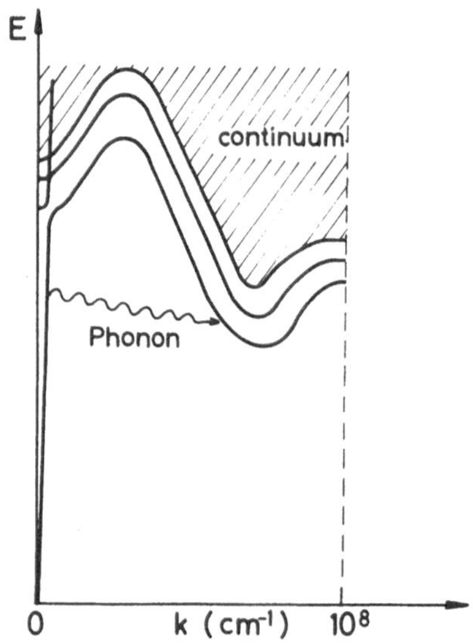

Fig. 44. Schematic drawing of exciton and polariton states for a semiconductor with an indirect gap. The whole first Brillouin zone is shown. If the direct band-to-band transition is dipole-allowed at k = 0, a polariton is formed in the way discussed in Chapter II. Excitons in the absolute minimum can be created only under momentum transfer to another quasiparticle, e.g. by creation or annihilation of a phonon.

resonance are optically allowed only for one orientation of the electric field relative to the crystallographic c-axis, resulting in strong birefringence and dichroism in the resonance regions. The frequency dependences of $\varepsilon(\omega)$ and of the polaron corrections, the exchange interaction and other phenomena often lead to a considerable deviation of the exciton-states from a simple Rydberg series equation [73]. More detailed information on the dispersion of excitons and polaritons in the direct gap materials is found, e.g., in [9],[20] and [30]. In indirect gap materials, the valence band is qualitatively the same as in blende-type direct gap materials. In contrast, the conduction band minimum is situated not at the Γ-point but somewhere else in the Brillouin zone. In Ge it is situated, e.g., at the L-point and in Si in the Δ-direction rather close to the X-point resulting in several equivalent minima. The resulting exciton and polariton dispersions are shown schematically in Figure 44. The abscissa covers here the whole first Brillouin zone. Absorption sets in, when phonon-assisted transitions to the indirect minimum become energetically allowed. At the direct gap, a polariton is formed in the usual way. The damping of this resonance is larger than in direct gap materials, however, since polaritons in the vicinity of the Γ-point may easily relax in the indirect minimum by phonon emission. More details about excitons and polaritons in indirect gap material are found, e.g., in [12], [18] and in the literature cited therein.

Appendix B. Surface Polaritons

In Chapter II, we treated only the volume polariton modes. From Maxwell's equations it can be deduced that there exists a surface polariton mode to every volume polariton. A surface polariton mode can propagate only along the interface between different materials. Here we consider the interface between a semiconductor and vacuum only. The field amplitudes decay exponentially on both sides of the interface. If we choose the coordinates so that the x-y plane represents the surface and $\vec{k}$ is parallel to the x-axis, then the electric field $\vec{E}$ of the surface polariton is in the x-z plane and the magnetic induction is parallel to the y-axis. The surface polariton mode exists if the following conditions are fulfilled:

$$\mathrm{Re}\{\varepsilon_M(\omega)\} < 0$$

and

$$|\mathrm{Re}\{\varepsilon_M(\omega)\}| > \varepsilon_{vac} = 1 \qquad (125)$$

where $\varepsilon_M(\omega)$ is the dielectric function of the semiconductor. We do not go through the Maxwell theory to prove all the above statements

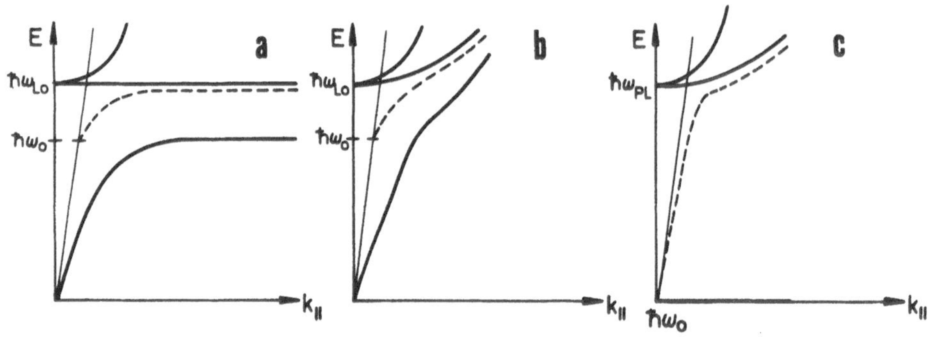

Fig. 45. The dispersion of bulk and surface polariton modes and of photons for various quasiparticles in a semiconductor. Heavy solid line: bulk polaritons; dashed line: surface polaritons; light solid line: photons.

(see e.g. [69]). Looking, however, at the law of refraction (Eqs. (64-66))

$$k_{nt}^2 = \varepsilon_M(\omega)\, k_i^2 - k_p^2 \tag{126}$$

One immediately finds for $\varepsilon_M(\omega) < 0$ the result $k_{nt}^2 < 0$, i.e. a polariton mode decaying exponentially into the medium.

In Figure 45 we give the dispersion of volume and surface polaritons for phonons, excitons and plasmons, neglecting damping for simplicity. The dispersion of the surface polaritons for the first two examples is intelligible from Figures 6, 7 and 15a, and Eq. (125). Plasmons have not yet been treated and we shall give some comments. Plasmons are the quanta of longitudinal collective excitations of the electron-system. There are two types of plasmons. The more common ones which are treated in every textbook [11],[12], [15] are connected with the electrons of a partially occupied conduction band. Their optical properties can be described by an oscillator having a transverse eigenfrequency which is identical to zero for all k and a longitudinal eigenfrequency given for k = 0 by the plasma-frequency

$$\omega_{PL_o} = \left(\frac{N_{el}\, e^2}{m_{eff}\, \varepsilon\, \varepsilon_o} \right)^{1/2} \tag{127}$$

and showing a quadratic dispersion for k > 0. The lower polariton branch coincides then with the abscissa (Figure 45c) and the upper polariton branch starts at ω_{PL_o} and then turns over to a photon-like

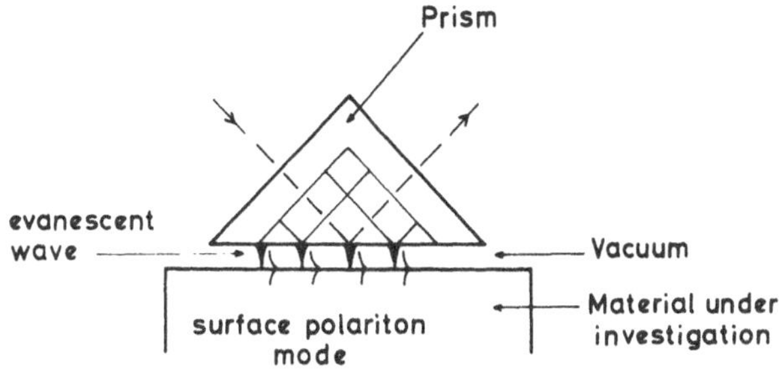

Fig. 46. Arrangement for the excitation of surface polariton modes by attenuated total reflection.

dispersion. With Figure 9 and Figure 45c, it is obvious that a material with a partially occupied band will show a reflectivity close to 1 for $\hbar\omega < \hbar\omega_{PL_0}$ and a reflection minimum around $\hbar\omega_{PL_0}$. For metals, the plasma energy turns out to be in the UV. For semiconductors, $\hbar\omega_{PL}$ can come to the IR-spectral region if a high density of free carriers is produced, e.g., by high doping. (See, e.g., the standard example in [70].)

In a more classical approach, the high reflectivity of metals can also be explained by the shortcircuiting of the electric field of an incident electromagnetic wave by the high conductivity connected with a partially occupied band. The reflected wave is produced by the surface currents which are necessary to shortcircuit the field. From this point of view it is immediately obvious that collective excitations of a completely filled band will not result in such spectacular optical properties as mentioned above. The quanta of the collective excitation of a completely filled band are called valence band plasmons. They can be treated in a certain analogy to excitons, but they are situated energetically not below the band gap but above an energy given by the energetic distance from the bottom of the valence band to the top of the conduction band. For more information of valence band plasmons see e.g. [71].

Returning to Fig. 45 it is obvious that the dispersion of the surface plasmons intersects neither the upper or lower polariton-branch nor the photon-dispersion. This means, there is no possibility of direct optical coupling to the surface polariton modes. Surface plasmons can be excited by electrons and are observed in electron energy-loss spectroscopy (see III.1. or [72]). There exist, however, some tricks, to transfer energy by optical means to surface polaritons. The most widely used is the attenuated total reflection (ATR).

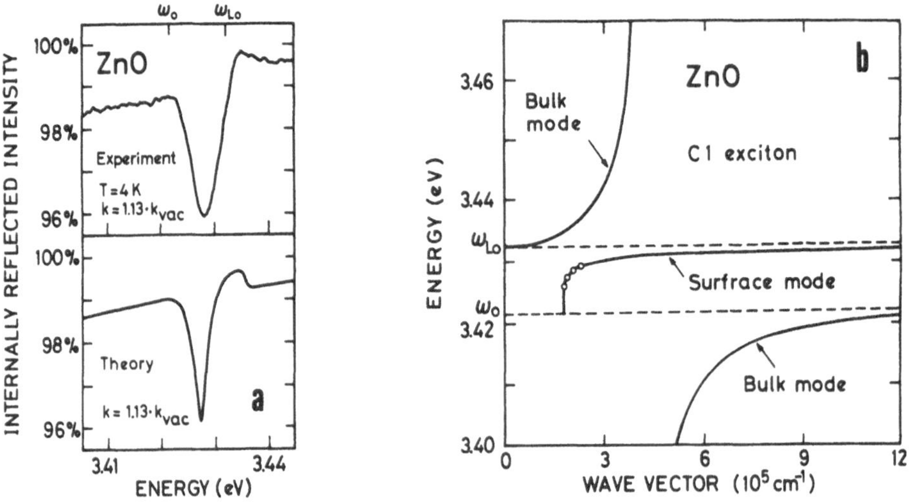

Fig. 47. Attenuated total reflection signal of an exciton surface polariton in ZnO. Experiment and theory (a), dispersion of bulk and surface modes (b). From [73].

In ATR a parallel, monochromatic light-beam with frequency ω_i is sent into a prism with refractive index $n_p(\omega)$ in the way shown in Figure 46. If the internal angle of incidence θ is above a critical value θ_T, total reflection occurs. θ_T is given by

$$\sin\theta_T = \frac{1}{n_p(\omega_i)} \tag{128}$$

In the case of total reflection, an evanescent wave is "leaking" into the vacuum. The frequency of this wave is ω_i and its wave-vector k_p can be varied by varying θ from 90° to θ_T between

$$\frac{\omega_i}{c} n_p(\omega_i) \geq k_p \geq \frac{\omega_i}{c} \tag{129}$$

The distance over which the evanescent wave decays in the vacuum is comparable to the wavelength. If the medium in which we want to investigate surface polaritons is brought sufficiently close to the base of the prism, the evanescent wave couples to the surface polariton modes, if energy and wavevector $\vec{k}_p$ coincide. In this case, energy is transferred to the surface polariton modes and the reflectivity falls below 100%. This method of attenuated total reflection has been applied successfully to all surface modes shown in Figure 45. Figure 47 gives an example for exciton surface polariton modes in the semiconductor ZnO. For a recent review and for other possibilities to detect surface polariton modes like diffraction from surface gratings or nonlinear mixing, see e.g. [69].

Appendix C. The Role of Impurities

The discussion of energy transfer in semiconductors has so far been limited to intrinsic aspects. But even the purest semiconductors available still contain about 10^{15} - 10^{16} imperfections per cm^3. By intentional doping, this number can be increased by several orders of magnitude. These imperfections include various types of point defects, like vacancies or interstitial atoms or dopants, both interstitial and substitutional ones. Furthermore, there are line defects (dislocations) or two-dimensional defects, like small angle grain-boundaries. In the following, we shall shortly mention some aspects of point defects. For a recent review of this field see e.g. [74]. The transfer of energy from donor to acceptor-states is discussed in great detail for many materials in other contributions of this volume, So this point will only shortly be touched in this chapter (see below).

Point defects also play an important role in electrical conductivity of semiconductors both by influencing the concentration and mobility of free carriers.

Besides the Umklapp-processes, the scattering at impurities limits the heat conductivity, especially at low temperatures. Here also isotopes act as impurities because of their different mass. The main aspect investigated in the following is the influence of point defects on the optical properties of semiconductors, i.e. on the propagation and absorption of light.

Impurities like isoelectronic traps (I), neutral acceptors (A^0), neutral donors (D^0), charged donors (D^+) and charged acceptors (A^-) may bind an electron-hole pair forming a bound exciton complex (BEC). The binding energy of the exciton to the complex is decreasing from IX complexes over A^0X and D^0X to D^+X and A^-X. The A^-X complex is generally unbound, i.e. it is for this BEC energetically more favourable to form a neutral acceptor and a free electron. The wavefunction of a bound exciton can be described in simplest approximation by a superposition of free exciton wavefunctions $\phi_x(n_B, \vec{k})$

$$\phi_{BEC} = \sum_{\vec{k}} a_{\vec{k}} \, \phi_x(n_B, \vec{k}) \tag{130}$$

Equation (130) shows that the k-selection is relaxed for a BEC, because it is a localized state, without translational invariance.

If we want to represent a bound exciton state in a E(k) diagram, we can do this by a horizontal line, the thickness of which represents $|a_{\vec{k}}|$ (Figure 48). Obviously, it is a type of oscillator without spatial dispersion. The singlet states of the BEC are generally coupling to the electromagnetic field. The oscillator strength per impurity atoms of a BEC is considerably larger as compared to the oscillator-strength per unit cell of free excitons [79]. On the other hand, the density of the impurity centers is in "pure" materials several orders of magnitude smaller than that of unit cells, resulting in moderate values of absorption

$$10^{1}\ \mathrm{cm}^{-1} \leq \alpha_{\mathrm{BEC}} \leq 10^{3}\ \mathrm{cm}^{-1}$$

$$10^{4}\ \mathrm{cm}^{-1} \leq \alpha_{\mathrm{free\ exciton}} \leq 10^{6}\ \mathrm{cm}^{-1} \qquad (131)$$

As a consequence, BEC gives rise to narrow absorption-bands. The resonance-structure in the polariton dispersion is generally too weak to

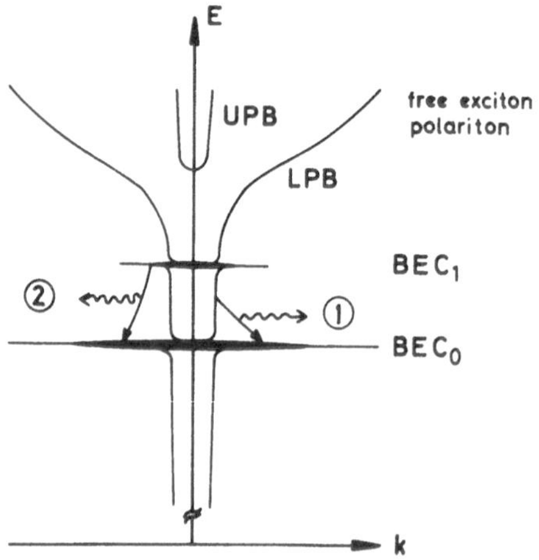

Fig. 48. Dispersion for a free exciton polariton and of bound exciton complex (BEC). The upper BEC is regarded as an excited state of the lower one. The transitions are described in the text.

show up, e.g. in reflection experiments (see, e.g., the small reflection structure around 2.545 eV in the 5K spectra of Figure 2 of Ref. [75] as an exception).

There are many optical phenomena connected with BEC, which have an analogon in the optical properties of free exciton-polaritons. As an example, they can act as virtually excited intermediate states in Raman scattering, or they can give luminescence without emission of phonons or under emission of one or several optical phonons. In contrast to the case discussed in III.3., the lineshape is not determined by the thermal distribution of BEC, because they populate only an energetically narrow state. The width of the zero-phonon lines is therefore small at low temperatures and increases roughly proportional to T^2 due to a detuning of the eigenenergy by thermal lattice vibrations e.g. [76]. The shape of the phonon replica is influenced by the dispersion of the optical phonons and the spread of the BEC-wavefunction in k-space.

The BEC can also decay under emission of acoustic phonons in a photonlike polariton which is observed as luminescence light. In contrast to free exciton-polaritons, acoustic phonons with various wavevectors may participate, resulting in a rather broad "acoustic wing" on the long wavelength side of the emission, the spectral shape of which may be used as a temperature probe for the population of acoustic phonons [9],[77]. In the excitation spectra of BEC, a similar acoustic sideband appears on the high energy side, i.e. a photonlike polariton decays in the sample into a BEC and an acoustic phonon (see, e.g., [78],[79]).

A lot of information about excited states of BEC has been collected rather recently by photoluminescence excitation spectroscopy (see, e.g., for ZnO [21],[79],[80], for CdS [78],[79],[81],[82]. This phenomenon can be nicely described in terms of energy transfer. In Figure 48 we plot the dispersion of the LPB together with two energy levels of a BEC, the lower one being the groundstate, the upper one the excited state. If a spectrally narrow, tunable (laser-) beam ($\hbar\omega_i$) is sent on the sample, it will be absorbed if $\hbar\omega_i = E_{BEC_0}$. In addition, strong light scattering (or resonance-fluorescence) occurs. For $\hbar\omega_i > E_{BEC_0}$, polaritons on the LPB are created. They may relax to a small extent by acoustic phonon emission to the BEC_0 producing some luminescence in this spectral region ①. If $\hbar\omega_i = E_{BEC_1}$ a large number of BEC_1 will be created. They relax to BEC_0 by transferring energy, e.g. to the acoustic phonon system (process ②). As a consequence, a peak appears in a plot of the BEC_0-luminescence intensity as a function of $\hbar\omega_i$ for constant excitation intensity whenever $\hbar\omega_i$ falls on an optically allowed excited state of the BEC. Such excitation spectra are found, e.g., in [71],[79]-[82].

If a semiconductor contains donors and acceptors, D° and A°

complexes may be formed by illuminating the sample with photons of an energy larger than the bandgap. If the localized wavefunctions of electrons and holes of the D° and A° complexes overlap at least a little bit, the electron-hole pair may recombine under emission of a photon-like polariton of energy $\hbar\omega$

$$\hbar\omega = E_g - E_D^b - E_A^b + \frac{e^2}{4\pi\,\varepsilon\,\varepsilon_o}\,\frac{1}{R_{DA}} - m\hbar\omega_{LO} \tag{132}$$

where E_g is the energy of the gap, E_D^b and E_A^b are the binding energies of electrons and holes measured relative to the bottom of the conduction band and the top of the valence band, respectively. R_{DA} is the geometric distance between the donor and acceptor. The term containing R_{DA} describes the Coulomb-energy between D^+ and A^- centers after the recombination. The last term on the righthand side gives the participation of optical phonons. The zero-phonon line corresponds to m = 0. The D° A°-system can be regarded as a polycentric bound exciton.

The matrix element for the optical transition depends strongly on the overlap of electron- and hole-wavefunctions and thus on R_{DA}. With increasing doping and excitation level, the average distance R_{DA} decreases, resulting in an increase of the Coulomb term in Eq. (132) and a blue-shift of the emission maximum. This blue-shift with increasing excitation is characteristic for the donor-acceptor pair recombination.

If the concentration of lattice defects is increased further, the overlap of the wavefunctions will increase, too. As a consequence, the BEC will no longer form energetically narrow states but will broaden to impurity-bands. The transition from BEC to impurity-bands has been investigated extensively e.g. in the system $CdS_{1-x}Se_x$ in [83],[84] by increasing the content of Se. One observes localized and free states. The mobility edge is found to coincide with the high energy edge of the emission band. Energy transfer between localized states and coupling to acoustic phonons are investigated by excitation spectroscopy, resulting at least in a good qualitative knowledge of energy transfer processes in this system.

ACKNOWLEDGEMENTS

The author would like to thank many colleagues at the Universities of Frankfurt, Erlangen and Strasbourg for stimulating discussions about the subject treated in this article. He is especially grateful to Prof. Mohler, Dr. Bohnert, Dr. Kempf and Prof. Haug (Frankfurt) and to Dr. Hönerlage (Strasbourg) for a critical reading of the manuscript. Part of the work reported here has been supported by the Deutsche Forschungsgemeinschaft.

REFERENCES

1. P. Landsberg, this volume.
2. P. Würfel and W. Ruppel, IEEE Trans. Electron Devices ED27, 745 and 877 (1980).
3. H. J. Hoffmann and F. Stöckmann, Advances in Solid State Physics, Vol. XIX, 271, ed.: J. Treusch, Vieweg, Braunschweig (1979).
4. I. Broser, in Physics and Chemistry of II-VI Compounds, p. 487, eds.: M. Aven and J. S. Prener, North Holland Publishing Co., Amsterdam (1967).
5. Modern Problems in Condensed Matter Sciences, Series eds.: V. M. Agranovich and A. A. Maradudin; Vol. 3, Electronic Excitation Energy Transfer in Condensed Matter, eds.: V. M. Agranovich and M. D. Galanin, North Holland Publishing Co., Amsterdam (1982).
6. J. J. Hopfield and D. G. Thomas, Phys. Rev. 132, 563 (1963).
7. R. W. Pohl, Optik und Atomphysik, 11th edition, Springer (1963).
8. E. H. Hecht and A. Zajac, Optics, Addison Wesley Publishing Co., Reading, MA (1974).
9. C. Klingshirn and H. Haug, Physics Reports 70, 315 (1981).
10. M. Born, K. Huang, Dynamical Theory of Crystal Lattices, University Press, Oxford (1954).
11. Collective Excitations in Solids, ed.: B. DiBartolo, Proc. of the NATO Advanced Study Institute on Collective Excitations in Solids, Erice (1981), NATO ASI Series B, Vol. 88, Plenum Press (1983).
12. O. Madelung, Festkörpertheorie, Vols. I, II and III, Heidelberger Taschenbücher, Vols. 104, 109 and 126, Springer (1972).
13. M. Hass and B. W. Henvis, J. Phys. Chem. Solids 23, 1099 (1962).
14. C. H. Henry and J. J. Hopfield, Phys. Rev. Lett. 15, 964 (1965); R. Claus, L. Merten and J. Brandmüller, Springer Tracts in Mod. Physics 75 (1975); R. Claus, Phys. Stat. Sol. B100, 9 (1980).
15. See, e.g., Ch. Kittel, Introduction to Solid State Physics, John Wiley & Sons, Inc., New York (1966); J. M. Ziman, Principles of the Theory of Solids, Cambridge University Press, London (1972); N. W. Ashcroft and N. D. Mermin, Solid State Physics, Holt Saunders International Editions (1976).
16. S. Nikitine, Progr. in Semiconductors 6, 235 (1962).
17. R. S. Knox, Theory of Excitons, Solid State Physics Supplement 5, eds.: F. Seitz and D. Turnbull, Academic Press (1963).
18. Advances in Solid State Physics: D. Bimberg, Vol. XVII, 195 (1977); R. G. Ulbrich and C. Weisbuch, Vol. XVIII (1978); U. Rössler, Vol. XIX, 77 (1979); D. Fröhlich, Vol. XXI, 363 (1981).

19. Topics in Current Physics, Vol. 14, Excitons, ed.: K. Cho, Springer (1979).
20. Modern Problems in Condensed Matter Sciences, Series eds.: V. M. Agranovich and A. A. Maradudin; Vol. 2, Excitons, eds.: E. I. Rashba and M. D. Sturge, North Holland Publishing Co., Amsterdam (1982).
21. C. Klingshirn, Lecture Notes in Physics 177, 214 (1983), ed.: G. Landwehr, Springer (1983).
22. R. Grasser and A. Scharmann, in Ref. [11], p. 317.
23. M. I. Strashnikova, Ukrainian Physical Journal 24, 440 (1979) (in Russian).
24. M. Rosenzweig, Ph.D. Thesis, Berlin (1982).
25. C. Weisbuch and R. Ulbrich, in Topics in Applied Physics 51, Light Scattering in Solids III, p. 221, eds.: M. Cardona and G. Güntherodt, Springer (1982).
26. A. Stahl and Ch. Uihlein, Advances in Solid State Physics XIX, 159 (1979).
27. J. Lagois, Phys. Rev. B23, 5511 (1981).
28. I. Balslev and A. Stahl, Phys. Stat. Sol. B111, 531 (1982).
29. I. Broser, R. Broser, E. Beckmann and E. Birkicht, Sol. State Commun. 39, 1209 (1981).
30. B. Hönerlage, R. Levy, J. B. Grun, C. Klingshirn and K. Bohnert, to be published in Physics Reports.
31. J. L. Birman, in Ref. [20], p. 83.
32. Y. Segawa, Y. Aoyagi and S. Namba, Sol. State Commun. 32, 229 (1979).
33. Y. Masumoto, Y. Unuma, Y. Tanaka and S. Shionoya, J. Phys. Soc. Japan 47, 1844 (1979).
34. R. G. Ulbrich and G. W. Fehrenbach, Phys. Rev. Lett. 43, 963 (1979).
35. T. Itoh, P. Lavallard, J. Reydellet and C. Benoit à la Guillaume, Sol. State Commun. 37, 925 (1981); P. H. Duong, T. Itoh and P. Lavallard, Sol. State Commun. 43, 879 (1982).
36. V. A. Kiselev, B. S. Razbirin and I. N. Uraltsev, Phys. Stat. Sol. B72, 161 (1975), and I. V. Makarenko, I. N. Uraltsev and V. A. Kiselev, ibid. 98, 773 (1980).
37. J. Voigt, M. Senoner and I. Rückmann, Phys. Stat. Sol. B75, 213 (1976).
38. T. Mita and N. Nagasawa, Sol. State Commun. 44, 1003 (1982).
39. K. Heinz, K. Müller, in Springer Tracts in Modern Physics, Vol. 91, Springer (1982).
40. R. Weibmann, K. Müller, Surf. Sci. Rep. 1 (1981).
41. P. Y. Yu, in Ref. [19], p. 211.
42. D. S. Bulyanista, Sov. Phys. Semicond. 4, 1081 (1970).
43. D. Berlincourt, J. Jaffe, L. R. Shiozawa, Phys. Rev. 129, 1009 (1963).
44. R. Kuhnert, R. Helbig and K. Hümmer, Phys. Stat. Sol. B107, 83 (1981).
45. S. Permogorov, in Rev. [20], p. 177, and the literature cited therein.

46. C. Klingshirn, Ph.D. Thesis, Erlangen (1975), and C. Klingshirn, Phys. Stat. Sol. B71, 547 (1975).
47. E. S. Koteles, in Ref. [20], p. 83.
48. Light Scattering in Solids, Vol. I, ed.: M. Cardona (1975); Vols. II to IV, eds.: M. Cardona and G. Güntherodt (1982), Topics in Applied Physics, Springer.
49. C. Weisbuch and R. Ulbrich, in Ref. [48], Vol. III, p. 207.
50. J. R. Sandercock, in Ref. [48], Vol. III, p. 173.
51. A. A. Klochikhin, S. A. Permogorov and A. N. Reznitsky, Sov. Phys. Sol. State 18, 1304 (1976).
52. M. Cardona, in Ref. [48], Vol. I.
53. A. Nakamura and C. Weisbuch, Sol. State Commun. 32, 301 (1979).
54. J. B. Grun, B. Hönerlage and R. Lévy, in Ref. [20], p. 459.
55. R. Lévy, B. Hönerlage and J. B. Grun, Phys. Rev. Lett. 44, 1355 (1980).
56. T. Mita, K. Sâtome and M. Ueta, J. Phys. Soc. Japan 48, 496 (1980).
57. V. G. Lyssenko, K. Kempf, K. Bohnert, G. Schmieder, C. Klingshirn and S. Schmitt-Rink, Sol. State Commun. 42, 401 (1982).
58. W. Maier, G. Schmieder and C. Klingshirn, Z. Physik B50, 193 (1983).
59. K. Kempf, G. Schmieder, G. Kurtze and C. Klingshirn, Phys. Stat. Sol. B107, 105 (1981).
60. H. J. Eichler, Advances in Solid State Physics, Vol. XVIII, 241 (1978).
61. A. Miller, D. A. B. Miller and S. D. Smith, Advances in Physics 30, 697 (1981).
62. M. Ducloy, Advances in Solid State Physics, Vol. XXII, 35 (1982).
63. H. Haug, Advances in Solid State Physics, Vol. XXII, 149 (1982).
64. A. Maruani and D. S. Chemla, Phys. Rev. B23, 841 (1981).
65. A. L. Smirl, T. F. Bogess, B. S. Wherrett, G. P. Perryman and A. Miller, Phys. Rev. Lett. 49, 933 (1982), and A. L. Smirl, to be published in Semiconductor Processes Probed by Ultrafast Laser Spectroscopy, ed.: R. Alfano, Academic Press, New York (1983).
66. Landolt-Börnstein, New Series Group III, Volumes 17a-17e, eds.: O. Madelung, M. Schulz and H. Weiss, Springer (1982).
67. M. Cardona, Phys. Rev. B129, 580 (1963); A. Goldmann, Phys. Stat. Sol. B81, 9 (1977).
68. G. Blattner, G. Kurtze, G. Schmieder and C. Klingshirn, Phys. Rev. B25, 7413 (1982).
69. Modern Problems in Condensed Matter Sciences, Series eds.: V. M. Agranovich and A. A. Maradudin; Vol. 4, Surface Polaritons, eds.: D. L. Mills and V. M. Agranovich, North Holland Publishing Co., Amsterdam (1981).
70. W. G. Spitzer and H. Y. Fan, Phys. Rev. 106, 882 (1957).
71. I. Egri, in Ref. [11], p. 643.
72. H. Raether, Surf. Sci. 8, 233 (1967).

73. J. Lagois and B. Ficher, in Ref. [19], p. 183.
74. P. J. Dean and D. C. Herbert, in Ref. [19], p. 55; P. J. Dean, in Ref. [11], p. 247.
75. K. Bohnert, G. Schmieder and C. Klingshirn, Phys. Stat. Sol. B98, 175 (1980).
76. Ch. Solbrig and E. Mollwo, Sol. State Commun. 5, 625 (1967).
77. J. Shah, R. F. Leheny and W. F. Brinkman, Phys. Rev. B10, 659 (1974).
78. R. Baumert and J. Gutowski, Phys. Stat. Sol. B107, 707 (1981).
79. C. Klingshirn, W. Maier, G. Blattner, P. J. Dean and G. Kobbe, J. Crystal Growth 59, 352 (1982).
80. G. Blattner, C. Klingshirn, R. Helbig and R. Meinl, Phys. Stat. Sol. B107, 105 (1981).
81. J. Puls and J. Voigt, Phys. Stat. Sol. B94, 199 (1979), and J. Puls, H. Redlin and J. Voigt, Phys. Stat. Sol. B107, K71 (1981).
82. R. Baumert, I. Broser, J. Gutowski and A. Hoffmann, Phys. Stat. Sol. B116, 261 (1983), and Phys. Rev. B27, 6283 (1983).
83. E. Cohen and M. D. Sturge, Phys. Rev. B25, 3828 (1982).
84. S. Permogorov, A. Reznitsky, S. Verbin, G. O. Müller, P. Flögel and M. Nikiforova, Phys. Stat. Sol. B113, 589 (1982), and S. Permogorov, A. Reznitsky, S. Verbin and V. Lysenko, Sol. State Commun. 47, 5 (1983).
85. I. V. Kukushkin, V. D. Kulakovskii and V. B. Timofeev, J. Lumin. 24/25, 393 (1981); A. Mysyrowicz, D. Hulin and C. Benoit à la Guillaume, J. Lumin. 24/25, 629 (1981); N. Peyghambarian, L. L. Chase and A. Mysyrowicz, Phys. Rev. B27, 2325 (1983).
86. S. Schmitt-Rink, H. Haug and E. Mohler, Phys. Rev. B24, 6043 (1981).

TRIPLET EXCITATION TRANSFER STUDIES IN ORGANIC CONDENSED MATTER VIA COOPERATIVE EFFECTS

V. Ern

Laboratoire de Spectoscopie, Associé au CNRS,
Université Louis Pasteur, 5 rue de l'Université
67000 Strasbourg, France

ABSTRACT

The fundamental elementary excitations of most molecular organic condensed matter are the Frenkel excitons, neutral excited states which transport electronic excitation energy without transport of charge. Of special interest are the metastable paramagnetic triplet excitations which are known to travel during their lifetime over macroscopic distances. These excitations exhibit cooperative effects, notably triplet-triplet pair annihilation which leads to the creation of a higher energy singlet and to the emission of detectable delayed fluorescence. The scope of this presentation is to show how this mutual annihilation process can be put to use to extract information on the transport of the excitation energy. The advantage of such approach is that the migrating exciton itself is thus used as a probe to detect the excitation motion in the crystal. A direct technique using a spatially inhomogeneous, periodic, excitation of the crystal will be described and it will be shown how diffusion constants and coherence effects can be extracted from the delayed fluorescence data.

I. INTRODUCTION

Transfer of excitation energy in organic condensed matter has been subject of an intensive study in the last decade. Different theoretical approaches and experimental techniques have been used to tackle the problem. An overall, concise review of the subject can be found in the recently published text of Pope and Swenberg [1]. More details on specific theoretical models of motion like the use of Generalized master equations (GME), Stochastic Louiville equations (SLE), percolation, spin-relaxation, as well as of techniques for measuring exciton transfer, for instance via optical lineshapes, magnetic resonance or quantum yields of sensitized luminescence in purposely doped crystals, can be found in several review articles [2]-[11] and in the references given therein.

Because of the naturally limited extension of this presentation we shall restrict ourselves to present here in some detail one specific, direct approach to study exciton motion in neat organic solids. The approach relies on the detection of the signal due to the mutual annihilation of two excitons which takes place when the two excitations, due to their motion, come sufficiently close to each other so that a fusion in a new, higher energy excitation, can take place. The advantage of this technique lies in the fact that no impurities must be added to the lattice to act as detectors of motion, and, as will be seen below, at sufficiently low intensities of excitation, that is, at low free exciton densities, the annihilation process itself can be neglected in the kinetics, and one can view this approach as one in which the migrating exciton itself is used as a probe to detect the motion of the excitation in the crystal. The technique requires only the initial exciton density to have a known spatial distribution, the spatial inhomogeneities having characteristic distances comparable to the distance travelled by the exciton during its lifetime in the crystal.

The approach using the biexcitonic process is best put to use to study the motion of triplet excitons in organic solids, since in many materials the outcome of triplet-triplet annihilation can lead to an excited singlet state, and hence to an easily detectable delayed fluorescence signal [1]. As will be seen below, the time-dependence of this signal contains direct information on the excitation transport and no additional theory (which already would require some model for the motion) for the triplet-triplet annihilation rate constant γ_T is needed. Singlet-singlet annihilation [1] on the other hand doesn't typically lead to a new emitting species and motion parameters, e.g. the diffusion constants, can be deduced only indirectly from the measurement of the singlet annihilation rate constant γ_S which, for instance, can be obtained from studies of fluorescence quantum yield efficiencies as a function of the exciting intensity [1].

Whenever it is possible to detect delayed fluorescence there are several advantages to use the direct experimental approach described here to get triplet motion parameters, and test any general theory of exciton transport in a neat organic crystal. First, triplet excitons are long-lived excitations so that during their lifetime they can scan macroscopic distances in the crystal, even in the completely incoherent, diffusive, case. Moreover, due to the low $S_o \rightarrow T_1$ absorption coefficient the triplet exciton population can be excited throughout the bulk of the sample so that ill-defined surface quenching effects, which usually play an important role for singlet excitons, become negligible. Also, the paramagnetic nature of the triplet excitation allows ESR experiments [12],[5],[11] on the excited states as well as the possibility to modulate the triplet-pair annihilation rate with an external magnetic field [13],[14],[1]. The theoretical understanding of the observed ESR linewidths [5] and of the magnetic field dependence of $\gamma_T(\vec{H})$ [14] will typically require some knowledge of the magnitude and of the anisotropy of triplet exciton transport. This is especially the case in the $\gamma_T(\vec{H})$ experiments in which an external magnetic field is used as a handle to probe the pair-annibhilation process itself. A priori direct, independent, measurements, as will be described here, of triplet's transport parameters, for instance, hopping rates in a given crystallographic plane, are essential to reduce the number of parameters needed for an unambiguous fit of the experiments by the theory [14],[15].

This presentation is laid out as follows. In section II a brief introduction to crystal band states and to the delayed flourescence mechanism will be given. Section II will present an analysis of the direct experimental approach, using a spatially periodic excitation of the crystal, to detect triplet motion effects via the delayed fluorescence signal. In section IV, the approach will be applied to the study of the macroscopic exciton diffusion equation and in section V, practical experimental techniques as well as examples of extraction of triplet-exciton-diffusion tensor components will be given. Finally, section VI will present briefly recent work using generalized master equations (GME) which shows that, in principle, coherence effects in the transport can be detected using exactly the same experimental techniques as up to now applied to study the completely incoherent diffusive triplet motion.

II. INTRODUCTION TO MOLECULAR CRYSTAL BAND STATES

As compared to inorganic solids a molecular crystal can be considered as an ordered ensemble of weakly interacting asymmetric units consisting either of groups or of a single organic molecule. As an example, let's consider the case of many organic solids, notably those of the best studied aromatic hydrocarbons molecules, which crystallize in the monoclinic space group $C_{2h}^5(P2_1/a)$ containing two, four or eight translationally enequivalent molecules in the unit cell [16]. Figure 1 schematically depicts the molecular packing in the ab plane of such a crystal for the case of two translationally inequivalent molecules in the unit cell. The two differently oriented molecules are related by the two-fold symmetry b axis, as required by the point group C_{2h} of the unit cell. As shown, the structure can also be regarded as consisting of two equal interpenetrating lattices of differently oriented molecules. For the general case of N molecules packed in N/h unit cells each containing q=1,2,....h translationally inequivalent molecules, the crystal hamiltonian can be written as [17],[19]

$$H = \sum_{nq} h_{nq} + \frac{1}{2} \sum_{nq} \sum_{n'q'} (1 - \delta_{nn'}\delta_{qq'}) V(nq,n'q') \tag{1}$$

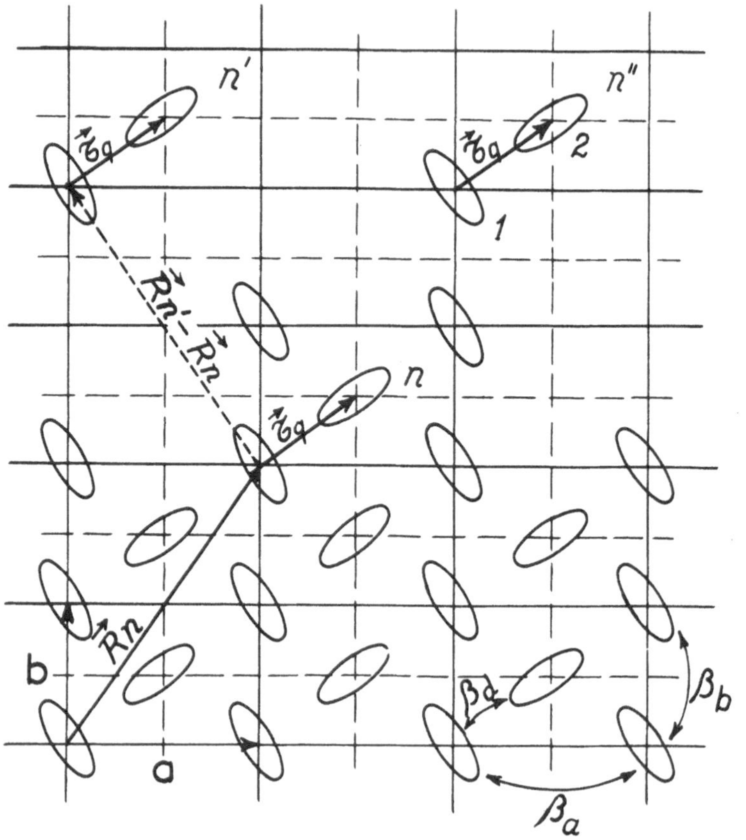

Fig. 1. Schematic representation of a typical molecular packing in an organic crystal. The particular case shown corresponds to the projection down c_5^{+}, that is, the ab plane of the monoclinic space group C_{2h} with h = 2 molecules per unit cell. The heavy line lattice is the q = 1 sublattice, the dashed line one, the q = 2 sublattice. All unit cell positions $\vec{R}_n$ are referred to the site (q = 1) on the extreme lower left. For such a choice of the origin, the molecules on the q = 1 sublattice are specified by the vector $\vec{\tau}_q = 0$, while the ones on the q = 2 sublattice by the vector $\vec{\tau}_q = (\vec{a} + \vec{b})/2$. The transfer matrix elements (see text below) between the nearest-neighbor molecules are indicated as β_a, β_b, and β_d.

where h_{nq} is the hamiltonian for the molecule situated at the qth site of the nth unit cell, and V(nq,n'q') represents the interaction energy between molecules on sites (nq) and (n'q') due to nuclear-nuclear, electron-nuclear, and electron-electron interactions. The Kronecker deltas ensure that in the double sum the interaction of a molecule with itself is excluded. In the Frenkel exciton limit [17]-[19] one assumes that the second intermolecular interaction sum in (1) constitutes a small perturbation to the first sum in which h_{nq} is taken to be the hamiltonian of the free isolated molecule. This is a reasonable assumption for most molecular crystals, especially for the relevant low-energy lying states, since one expects that the electronic wavefunctions of a molecule, e.g. the π-orbitals do not appreciably extend beyond few Å's. The intermolecular interaction energies are thus usually much smaller (say less than few percent) than a typical electronic energy ε^f of the fth excited state of the isolated molecule. In this tight-binding approximation the first sum in (1) is the hamiltonian of an aggregate of N noninteracting molecules placed at the positions, and in orientations, as prescribed by the crystal structure, that is, of an "oriented gas", and the intersite interaction term is treated by well-known methods of perturbation theory. The eigenfunctions of the oriented gas hamiltonian are thus the basic functions for the perturbation treatment of (1).

The wavefunctions for the ground state of the oriented gas are simply given by

$$\phi^\circ = \underline{\underline{A}} \prod_{nq} \chi^\circ_{nq} \tag{2}$$

where χ°_{nq} is the ground state wavefunction of the free molecule, that is $h_{nq}\chi^\circ_{nq} = \varepsilon^\circ\chi^\circ_{nq}$, ε° being the molecular ground state energy, and where the operator $\underline{A}$ is the one which antisymmetrizes the product with respect to intermolecular exchange of a pair of electrons. The excited states for the oriented gas hamiltonian are now obtained when one of the molecules is taken to be in one of its excited states. The zero-order wavefunction for hamiltonian (1) corresponding to the excitation in its fth state of the molecule situated on the qth site of nth unit cell is given by

$$\phi^f_{nq} = \underline{\underline{A}}\chi^f_{nq}\prod'_{n'q'}\chi^\circ_{n'q'} \tag{3}$$

where $h_{nq}\chi^f_{nq} = \varepsilon^f\chi^f_{nq}$, ε^f being the molecular energy of the fth excited state, and $\prod'$ means that in the product the molecular ground state function for the site (nq) has now been excluded. While there is only one ground state function (2), there are N possible localized, usually called single site, excitation functions (3). They form a N-fold degenerate set since any one of the molecules of the crystal can be in the fth excited state giving the same zero-order (or oriented gas) crystal energy. Turning on the intermolecular interaction as a perturbation in (1) will lift this degeneracy and mix the localized functions (3) giving rise to N delocalized crystal, or exciton, states, whose energies are spread over a band and which will belong now to the entire crystal.

The perturbation problem can now be solved by first constructing the so-called one-site exciton functions for each sublattice which, due to the translational symmetry, can be written as the normalized Bloch sums

$$\psi_q^f(\vec{k}) = (h/N)^{\frac{1}{2}} \sum_n \phi_{nq}^f e^{i\vec{k}\cdot(\vec{R}_n + \vec{\tau}_q)} \quad , \tag{4}$$

q = 1,2....h, where $\vec{R}_n$ is the lattice vector specifying a given unit cell (see Fig.1) and $\vec{k}$ is the exciton wavevector spanning the first Brillouin zone. The entire crystal problem is now solved by switching on the interaction between the molecules on different sublattices. The exciton bands deriving from a fth molecular state are now obtained by solving the h×h secular determinant

$$|M_{qq'}^f - E(\vec{k})\delta_{qq'}| = 0 \tag{5}$$

where the matrix elements

$$M_{qq'}^f = \langle\psi_q^f|H|\psi_{q'}^f\rangle \tag{6}$$

can be reduced to the sums

$$M_{qq'}^f = e^{i\vec{k}\cdot(\vec{\tau}_{q'} - \vec{\tau}_q)} \sum_{n'} \beta_{qq'}^{n'} e^{i\vec{k}\cdot\vec{R}_{n'}} \tag{7}$$

in which the nth reference unit cell has been taken as the common origin ($\vec{R}_n = 0$) for all distances. The key quantities in (7) are the values of

$$\beta_{qq'}^{n'} = \langle\phi_{0q}^f|H|\phi_{n'q'}^f\rangle \quad , \tag{8}$$

the so-called exciton transfer matrix elements, which determine the spread of the exciton energy band dispersion. For q' = q one has the transfer matrix elements between the translationally equivalent molecules in a given sublattice, while for q' ≠ q between all the translationally inequivalent ones which are in the same (n' = 0) or different unit cells (n' ≠ 0). When the transfer matrix elements are known the band structure calculation is simply obtained by constructing matrix elements $M_{qq'}^f$ via sums (7) and then solving the h×h secular problem (5) which will yield the $E_i^f(\vec{k})$ i = 1,2....h exciton band branches derived from a given molecular excited state f. In practice, a considerable simplification of the secular problem is obtained through symmetry considerations when $\vec{k}$ has special values or when $\vec{k}$ lies along a special direction or in a special symmetry plane of the Brillouin zone [19],[20][21]. Once the exciton bands $E(\vec{k})$ are known, a dey exciton transport parameter, anmely its mean square velocity, can be obtained by calculating the termally weighted average of the square of the exciton wave-packet group velocity

$$\vec{v}(\vec{k}) = \hbar^{-1}\vec{\nabla}_{\vec{k}} E(\vec{k})$$

over the entire Brillouin zone.

The transfer matrix elements (8) can be calculated explicitly [18],[22] via (2) and (3) if the one-electron (molecular orbital) approximation is used for the molecular wavefunctions χ^f. In this approximation one has [22] for an excited singlet state

$$ {}^1\beta^{n'}_{qq'} = J^{n'}_{qq'} + K^{n'}_{qq'} \tag{9}$$

and

$$ {}^3\beta^{n'}_{qq'} = K^{n'}_{qq'} \tag{10}$$

for the triplet state if the molecular spin orbit coupling is neglected [22],[23]. In the usual notation the $J^{n'}_{qq'}$ are the <u>intermolecular</u> Coulomb integrals and $K^{n'}_{qq'}$ the <u>intermolecular</u> exchange integrals, the integrations being carried over the spatial coordinates of two electrons. The exchange integrals appear in the problem because of the antisymmetrization operator $\underline{\underline{A}}$ in (3), that is, of Pauli's principle applied to pair-electron exchange between different sites. The Coulomb integrals do not contribute to triplet states because of the orthogonality of the spin functions which can be integrated separately from the spatial part if spin-orbit coupling in the molecule is neglected. In practice, this coupling is indeed negligible [18] for transfer matrix elements calculations, notably for aromatic hydrocarbons [24], however, its inclusion becomes essential for understanding optical singlet-triplet transition intensities and polarisations [25].

It will be instructive at this point to make some qualitative remarks on the effects that the difference between Eqs (9) and (10) makes on the band structure calculations and the behavior of singlet and triplet excitons. Unlike the exchange integrals $K^{n'}_{qq'}$, the intermolecular Coulomb integrals $J^{n'}_{qq'}$ are strongly dependent on the electronic structure of the isolated molecule and can be reduced using a multiple expansion to products of integrals over electrons on the <u>same</u> molecule. The lattice sums (7) with transfer matrix elements given by (9) are slowly convergent and contributions from distant neighbors must be taken into account, especially for $\vec{k} \sim 0$ [17]-[19]. Long-range interactions must thus be included in singlet exciton band structure calculations. On the other hand, the intermolecular exchange integrals (10) decrease rapidly (nearly exponentially [22],[18] with distance and the lattice sums (7) converge rapidly. Hence, consideration of just few nearest-neighbor interactions is usually sufficient to obtain a good approximation to triplet exciton band structures. Clearly, the price of this simplification is the fact that the exchange integrals are very sensitive to the behavior of the molecular wavefunctions in the region between the molecules and also the relative intermolecular orientation which determines the degree of overlap. Hence, a good knowledge of the crystal structure and of the wave functions at great distances from the molecules are needed for a first-principles band structure calculation. The smallness of the exchange integrals makes the result more sinsitive to the assumptions involved in the numerical computation, and to effects like non-orthogonality and admixture of charge transfer states [23]. In the usual case of aromatic crystals the electrons involved in the lower triplet transitions are π electrons whose orbitals are perpendicular to the plane of the molecules. One expects the biggest overlap, and therefore, the biggest transfer matrix elements between neighboring molecules whose planes cover the most each other in a nearly parallel way. So a simple inspection of the crystal structure can allow sometimes a qualitative prediction of the anisotropy of the triplet exciton bandwidths and hance of triplet migration.

The fact that transfer matrix elements for singlet excitions are typically two to three orders of magnitude bigger than those for triplet excitions, should not mislead the reader into thinking that the transport length for singlets will be correspondingly larger than for triplet excitions. Indeed, the excitation transfer time between two sites is 10^2 - 10^3 times faster for singlet excitions, but triplet exciton lifetimes ($\sim 10^{-2}$ sec) are typically 6 orders of magnidude longer than those for singlet excitions ($\approx 10^{-8}$ sec) and, as pointed out first by Sternlicht et al. [26], the degree of transfer during the lifetime for triplet excitions is usually considerabley greater than that for singlet excitions. As has been first suggested by Avakian et al. [27], triplets do indeed travel macroscopic distances, several thousands of lattice spacings, and clearly represent a buld property of the organic solid state.

It will be instructive for what will follow in this presentation to close this section with a simple example of triplet exciton band structure for the packing shown in Fig. 1. As mentioned before, it corresponds to the ab plane of the C_{2h}^5 structure with two molecules per unit cell, like anthracene, naphthalene and others. The triplet transfer matrix elements between molecules lying in adjacent ab (ab is usually the cleavage plane) planes can be considered as negligible. The relevant nearest-neighbor matrix elements will be donoted by β_a, β_b, and β_d, as shown in Fig. 1. This last matrix element corresponds to the interaction between the two translationally inequivalent lattices. The $\vec{k}$-dependent part of lattice sums (7) is simply given by

$$\begin{aligned} M_{11}(\vec{k}) &= M_{22}(\vec{k}) = 2\beta_b\cos(\vec{k}\cdot\vec{b}) + 2\beta_a\cos(\vec{k}\cdot\vec{a}) \\ M_{12} &= M_{21} = 4\beta_d\cos(\tfrac{1}{2}\vec{k}\cdot\vec{a})\cos(\tfrac{1}{2}\vec{k}\cdot\vec{b}) \quad , \end{aligned} \tag{11}$$

since each molecule is surrounded by 8 nearest-neighbors, namely, two at $\pm\vec{b}$, two at $\pm\vec{a}$ and four at $\frac{1}{2}(\pm\vec{a}\pm\vec{b})$. With matrix elements (11) the eigenvalue problem (5) is trivial and the dispersion relations for the two exciton band branches are given by

$$E_{\pm}(\vec{k}) = 2\beta_a\cos(\vec{k}\cdot\vec{a}) + 2\beta_b\cos(\vec{k}\cdot\vec{b}) \pm 4\beta_d\cos(\tfrac{1}{2}\vec{k}\cdot\vec{a})\cos(\tfrac{1}{2}\vec{k}\cdot\vec{b}). \quad (12)$$

Figure 2 illustrates a typical behavior of the two branches given

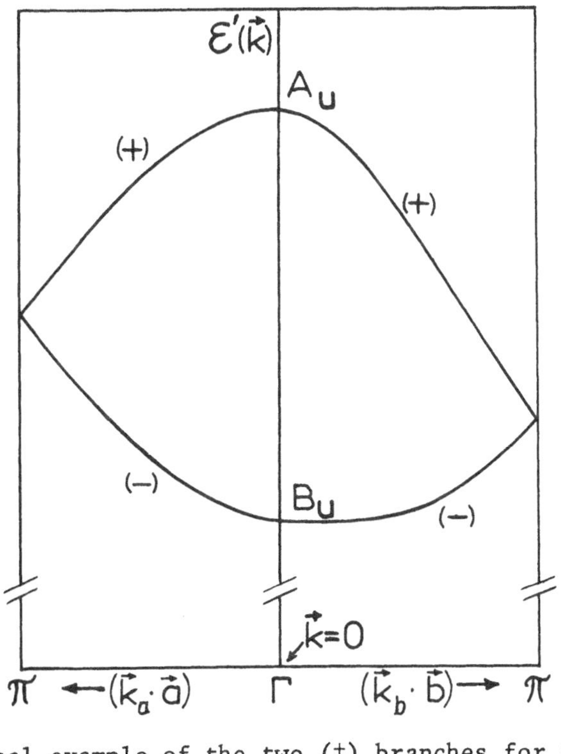

Fig. 2. Typical example of the two (±) branches for the lowest triplet exciton band state of a crystal of space group C_{2h}^5 with two molecules per unit cell. The bands are drawn for $\beta_d \approx 2\beta_b > 0$, and $\beta_a << \beta_b$. Note the brake in the energy scale. Typical bandwidths are of the order of 10-20cm^{-1}, while the center of gravity of the bands lies above the ground state ($\vec{k} = 0$) point at an energy which roughly corresponds to the molecular first triplet excited state, e.g., typically, 2×10^4cm^{-1}.

by (12) for $\vec{k}$ along the a and b directions for the first BZ in the reciprocal lattice. A general treatment of the space group C_{2h}^5 can be found in Ref. (21). Note that, contrary to inorganic solids, the crystal ground state is not a band but a single state at $\vec{k} = 0$ (Γ point of the BZ). In Fig. 2 the two $\vec{k} = 0$ triplet states have A_u and B_u symmetries [20],[23], so that they are optically accessible from the A_g ground state. The separation of the A_u, B_u $\vec{k} = 0$ band states is usually called the factor group or Davydov splitting and can be observed spectroscopically [25],[28],[29]. For the case shown, this splitting Δ is the total bandwidth along the $\vec{k}//\vec{a}^*$ direction, and according to (12) is $\Delta \simeq 8\beta_d$, so that β_d can be directly extracted from spectroscopic date even at room temperature [25],[28],[29].

As stressed before, the quantity of interest for transport studies is the exciton's mean-square velocity. As bandwidths for triplets are typically much smaller than k_BT the Boltzmann weighting factor can be neglected. For instance, from (12) one immediately gets

$$\begin{aligned} \langle v_a^2 \rangle &= \hbar^{-2}(\beta_d^2 + 2\beta_a^2)a^2 \\ \langle v_b^2 \rangle &= \hbar^{-2}(\beta_d^2 + 2\beta_b^2)b^2 \end{aligned} \tag{13}$$

as the mean-square exciton velocities along the $\vec{a}$ and $\vec{b}$ crystallographic axes, a and b being the lattice constants.

For a one-dimensional crystal having just one strong nearest-neighbor matrix element β along one direction of lattice constant a, one simply has

$$E(k) = 2\beta\cos(ka) \quad , \tag{14}$$

the bandwidth being 4β, and the exciton's mean-square velocity

$$\langle v^2 \rangle = 2\beta^2 a^2 \quad , \tag{15}$$

so that in the purely coherent limit of motion the exciton's transport length L_T, that is, the distance travelled by the exciton along the chain during its lifetime τ will be simply given by

$$L_T = \sqrt{2}\beta a\tau \quad . \tag{16}$$

III. DIRECT APPROACH FOR STUDY OF TRIPLET TRANSPORT VIA DELAYED FLUORESCENCE

From the previous section it is apparent that triplet excitons can be viewed as mobile pseudoparticles which during their lifetime τ can travel in the coherent (absence of scattering) limit a distance $L_T = \langle v^2 \rangle^{\frac{1}{2}} \tau$. As temperature is raised, scattering by phonons diminishes this transport length , the motion tending to the diffusive (completely incoherent) limit. Even in this limit the transport length $L_T = (2D\tau)^{\frac{1}{2}}$, D being the diffusion constant, can be considered as macroscopic (e.g. several microns). Thus, even at a low exciton concentration excited in a crystal ,there will be a non-negligible probability that two excitons will meet, e.g. reside on two neighboring sites. This leads to the formation of virtual pair-states (T,T) which can simply dissociate or undergo a fusion, that is, a pair-annihilation process which in turn can lead to creation in the crystal of a higher energy singlet or triplet state [k],[14]. The path leading to emission of delayed fluorescence is depicted schematically in Fig. 3. For more details about fusion and dission of excitons the reader is referred to Ref. (1) and more specifically to the works of Merrifield [13] and Suna [14].

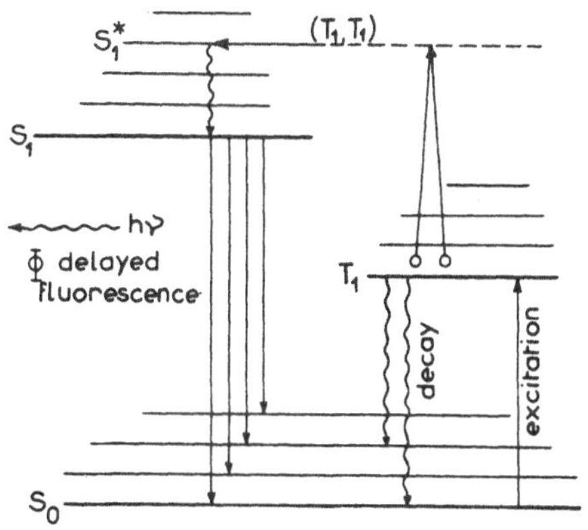

Fig. 3. Schematic representation of triplet-triplet annihilation leading to emission of delayed fluorescence.

In Fig.3 the first narrow excited triplet and singlet bands are depicted as single energy levels. The heavy line levels are the purely electronic states, while the lighter lines the vilbronic bands associated with these states. Upon $S_0 \rightarrow T_1$ excitation and creation of a population in the T_1 band, the triplets can disappear by direct radiative (phosphorescent) or radiationless decay with an overall lifetime τ (typically 10^{-2}sec) and by formation of pair states (T_1, T_1) a fraction of which undergoes a spin-allowed fusion process into a near-resonant vibrationally excited first singlet state S_1^*, which in turn, very rapidly (10^{-12}sec) relaxes in the lowest S_1 band. An excited first singlet population (typical lifetime $\sim 10^{-8}$sec) is thus built in the crystal leading to an emission of fluorescence. This fluorescence differs from the usual prompt fluorescence by the fact that its buildup and decay are governed by the slowest rate in the overall process, that is, by the triplet exciton lifetime. Hence the name of delayed fluorescence. In many aromatic molecular crystals the relative energies of the T_1 and S_1 states are such that the double of the energy of T_1 lies well above the S_1 energy so that singlet creation via triplet-triplet fusion can be considered as a relatively common process and delayed fluorescence emission can be observed in many materials upon excitation of the lowest triplet state.

We shall discuss now in which way observations of the delayed fluorescence emission can be put to use to extract information on the transport properties of the triplet exciton itself [30]-[33]. Let's suppose that a spatially inhomogeneous, time-dependent triplet exciton population has been created in the crystal by a direct $S_0 \rightarrow T_1$ excitation. The spatial and temporal evolution of the concentration $n(\vec{r},t)$ [cm^{-3}] of this population will be in general governed by the kinetic equation

$$\frac{\partial n}{\partial t} = G(\vec{r},t) - \frac{n}{\tau} + \left.\frac{\partial n}{\partial t}\right|_{ann} + \left.\frac{\partial n}{\partial t}\right|_{mot} \quad . \tag{17}$$

On the right-hand side, $G(\vec{r},t)$ represents the spatially inhomogeneous time-dependent source term, the second term the exciton's decay with lifetime τ. The third term is the loss of excitons due to mutual annihilation and the fourth term gives the change in time of the exciton concentration at a point $\vec{r}$ in space due to exciton motion. The mutual annihilation term is assumed, for not too high concentrations [14], to have the form

$$\left.\frac{\partial n}{\partial t}\right|_{ann} = -\gamma_{tot} n^2(\vec{r},t) \quad , \tag{18}$$

that is, proportional to the square of the concentration via a total annihilation rate constant γ_{tot}.

Solution of (17) for a given source term $G(\vec{r},t)$ and different initial conditions $n(\vec{r},0)$ will yield $n(\vec{r},t)$ from which the observable delayed fluorescence signal is obtained as

$$\Phi(t) = \frac{1}{2} f\gamma_{tot}\int_V n^2(\vec{r},t)d^3\vec{r} \tag{19}$$

where the integration is performed over the excited volume V of the crystal. Factor 1/2 in (19) takes care of the fact that two triplets produce only one fluorescing singlet, while factor $f<1$ accounts for the fact that only a fraction of pair-annihilations leads to singlet states which in general will emit fluorescence with some quantum efficiency $\eta \leqslant 1$. The signal $\Phi(t)$ contains information on the motion of excitons due to the fact that the square of the concentration must be integrated in (19) which forces the integral to depend on the actual spatial triplet exciton distribution at any instant t. Note that this would not be the case for the phosphorescent signal due to the direct decay [30],[33].

For several experimental and practical reasons detailed below, it is convenient to perform transport experiments based on (17) and (19) under the following conditions:

a) To avoid complications in obtaining the solution, which would be typically numerical, of the non-linear equation (17) it is convenient to perform the experiments using sufficiently low exciting intensities so that the annihilation term (18) that is, $\gamma_{tot}n^2$, becomes negligible as compared to the direct decay and motion terms in (17). With such experimental precaution analytical solutions of (17) can be used in practice, which, not only eases the computation of (19) but also avoids the necessity of fitting the observed function $\Phi(t)$ with additional parameters like the exciting light intensity, the mutual annihilation rate γ_{tot}, and the ground state singlet-first excited triplet absorption coefficient, quantities which are known to be difficult to obtain with great accuracy[1]. As will be seen later, in the low-exciton concentration limit the observable time evolution of $\Phi(t)$ will only depend on the easily measurable triplet lifetime τ and on the exciton transport parameters of interest which are contained in the last term of (17).
b) The mutual annihilation process leads typically (e.g. $\gamma_{tot} \simeq 10^{-11}cm^3sec^{-1}$) to weak delayed fluorescence signals, and restriction a) will require that the excited crystal volu-

me V in (19) is made as big as possible so that adequate signal-to-noise ratio for the signal $\Phi(t)$ is achieved. Thus, in practice, sufficiently big samples of the material, typically ∿1cm x 1cm x 0.2cm platelets, and excitation with an expanded laser beam are required.

c) As seen in section 3, triplet exciton band structures can be highly anisotropic so that, typically, one should expect a highly anisotropic exciton motion. Thus, in order to explore this anisotropy and to extract, for instance, tensorial motion parameters, it is necessary to perform the experiments using in (17) a source function $G(\vec{r},t)$ which creates a strong spatial exciton-concentration gradient only along one, easily controlable, direction, say $\hat{x}_0$, with respect to the crystallographic axes.

d) For the purpose stated in c) the best spatially inhomogeneous excitation of the crystal would be that through a narrow slit of some width x_0 so that a sharp excitation concentration gradient is created along the $\hat{x}_0$direction. This direction can be varied with respect to the crystal axes by cutting or cleaving samples in the form of platelets according to chosen crystallographic planes, and then rotating the slit around the exciting beam axis so as to explore the motion anisotropy in each plane. However, there is a strong drawback to such single slit-excitation. It is clear that in order to obtain noticeable effects due to motion from the last term in (17) the slit width x_0 must be comparable to exciton's transport length L_T during its lifetime. In practice this will mean that the excited volume of the crystal will be small, in contradiction with the requirement stated in b).

e) A good compromise between conditions a) through d) above can be met by exciting the crystal not through just one slit but through an array of periodically alternating open and opaque strips with spatial period x_0, that is, through what is usually known as a Ronchi ruling or grating. Such rulings are available with spatial periods x_0 ranging from fractions of a cm to fractions of a micron. In this way, a sufficiently big volume of the crystal can be excited, through the ruling, for instance, with an expanded to ca. 1cm diameter laser beam. At the same time one has a spatially inhomogeneous distribution whose basic length x_0 can be chosen to be comparable to exciton's transport length so that motion effects can be detected in (19) where the signal $\Phi(t)$ is collected by a photomultiplier from the whole excited volume.

f) Finally, to ease the solution of (17) it is convenient to have a source term of the form

$$G(x,t) = \alpha_0 i_0 g(x) f(t), \tag{20}$$

that is, to separate the spatial and temporal dependencies. In (20) α_0 is the $S_0 \rightarrow T_1$ absorption coefficient [1], i_0 the exciting beam intensity, and $g(x)$ and $f(t)$ are normalized to unity functions determining the space-time-dependence, respectively, of the source term per unit volume and unit time. As will be seen below, in the experimental setup such separation of variables is very simply achieved, for instance by using a chopped or pulsed laser beam which excites the crystal through a stationary Ronchi ruling of known period x_0. An example of the case of a moving ruling can be found in Ref. [32].

By meeting experimental conditions a) through f) above, equation (17) can be written with $\mu \equiv 1/\tau$ as

$$\frac{\partial n}{\partial t} = \alpha_0 i_0 g(x) f(t) - \mu n + \left.\frac{\partial n}{\partial t}\right|_{mot} \tag{21}$$

in which $g(x)$ can be any spatially periodic function. For a Ronchi ruling whose open strips can be, in general, a fraction $r \leqslant 0.5$ of the period x_0 as shown in Fig.4, one can write $g(x)$ in terms of its Fourier components as

$$g(x) = \sum_{\ell=-\infty}^{\infty} S_\ell e^{i\ell\eta x} , \tag{22}$$

with coefficients

$$S_0 = r, \quad \text{for } \ell = 0$$
$$S_\ell = S_{-\ell} = \sin(\ell\pi r)/\ell\pi, \qquad \ell=1,2,3... \tag{23}$$

the parameter

$$\eta = 2\pi / x_0 \tag{24}$$

being just the spatial angular frequency. For the typically used Ronchi gratings with equal open and opaque strips one gets only contributions from the odd spatial harmonics, that is

$$S_0 = \frac{1}{2} , \qquad S_\ell = S_{-\ell} = \sin(\ell\pi/2)/\ell\pi \qquad \underset{(\text{odd})}{\ell=1,3,5...} \tag{25}$$

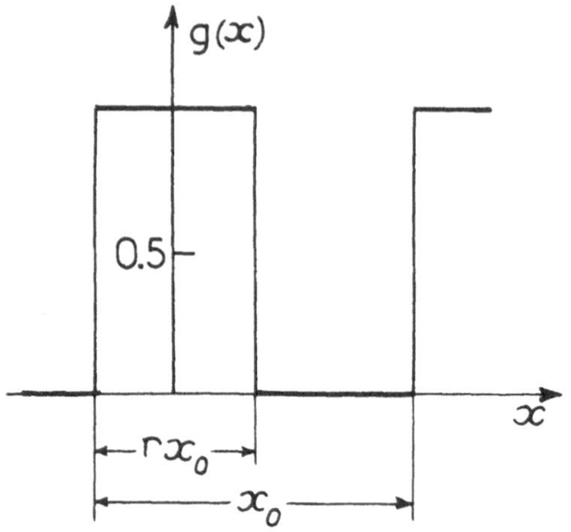

Fig. 4. The spatial modulation function g(x) introduced by a Ronchi ruling.

If some amount of uniform stray light Δi is assumed to be present in the shadow ($g(x) = 0$) regions due to diffraction and scattering from the illuminated strips the coefficients S_ℓ, $\ell \neq 0$ will be diminished to [30] .

$$S'_\ell = \sigma S_\ell \qquad \ell = \text{odd} \tag{26}$$

with $\sigma \equiv 1 - 2\Delta i/i_0 \leqslant 1$, reflecting the loss of contrast in the perfect shadow pattern of the ruling. The experiments require good optical quality crystals and surfaces, typical values for $\Delta i/i_0$ achieved in practice being of the order or less than 5% [30]-[33].

A spatially periodic excitation of the crystal can also be obtained by using two crossed laser beams whose interference will give the pattern

$$g(x) = \frac{1}{2}(1 + \cos\eta x) \quad , \tag{27}$$

that is, $S_0 = S_1 = S_{-1} = \frac{1}{2}$, $S_\ell = 0$ for $\ell>1$ in (22) the η given by (24), the spatial period x_0 being now determined by the beam-crossing angle θ and the wavelength λ of the excitation by Bragg's formula $x_0 = \lambda/2\sin(\frac{1}{2}\theta)$. Such spatially periodic excitation has been successfully applied to study, via delayed fluorescence, the diffusion of excited molecules in solution [34]. An advantage over Ronchi rulings of distribution (27) is that smaller periods x_0 can be obtained, a disadvantage being the fact that the observable effects due to transport become considerably smaller as only one spatial harmonic, and hence a less sharp exciton concentration gradient is used to probe exciton's motion. The spatially periodic excitation (27) has been recently applied using two crossed pulsed laser beams to study singlet exciton diffusion [35],[36], a third laser beam being used to probe the scattering by the cecaying pattern. Ref. [36] describes in detail the precautions that should be taken to interpret such experiments in terms of singlet transport.

IV. THE TRIPLET EXCITON MACROSCOPIC DIFFUSION EQUATION

At not too low temperatures, due to the strong exciton phonon interaction in organic solids [1], exciton motion is expected to be completely incoherent so that a macroscopic diffusion equation is expected to be valid to describe exciton transport in a neat organic crystal. The motion term in (21) witll be then given by Fick's law

$$\left.\frac{\partial n}{\partial t}\right|_{mot} = D_x \frac{\partial^2 n}{\partial x^2} , \tag{28}$$

where D_x is the component of the triplet exciton diffusion tensor along the $\hat{x}_0$ direction of the rulings periodicity, which is perpendicular to the ruling lines. If the chosen direction $\hat{x}_0$ for the ruling periodicity makes the angles θ_j $j = 1,2,3$ with the tensor's principal axes, one has

$$D_x = \sum_{j=1}^{3} D_{jj} \cos^2\theta_j , \tag{29}$$

where D_{jj} $j = 1,2,3$ are the principal values of the triplet diffusion tensor whose knowledge allows to map out the diffusion anisotropy in any crystallographic plane of the crystal.

With (28) and (22) equation (21) can be written as

$$\frac{\partial n}{\partial t} = \alpha_0 i_0 \sum_{\ell=-\infty}^{\infty} S_\ell e^{i\ell\eta x} f(t) - \mu n + D_x \frac{\partial^2 n}{\partial x^2} \tag{30}$$

which represents the macroscopic triplet exciton diffusion equation under a spatially periodic excitation. For vanishing or spatially periodic initial conditions as used in the experiments to be described later,one can write the solution of (30) in the form

$$n(x,t) = \sum_{\ell=-\infty}^{\infty} n_\ell(t) e^{i\ell\eta x} \tag{31}$$

in which the time-dependence of the spatial amplitudes of exciton-concentration harmonics is determined by

$$\frac{dn_\ell}{dt} = \alpha_0 i_0 S_\ell f(t) - a_\ell \mu n_\ell \quad , \tag{32}$$

in which the quantities

$$a_\ell = 1 + \ell^2 \xi^2 \tag{33}$$

are factors which, for each spatial harmonic, increase due to diffusion, the exciton's decay rate $\mu = \tau^{-1}$. In (33) the adimensional motion parameter ξ is defined as

$$\xi \equiv \sqrt{2}\pi L_T / x_0 \quad , \tag{34}$$

that is, it scales the exciton transport, here diffusion, length during its lifetime

$$L_T = (2D_x \tau)^{\frac{1}{2}} \tag{35}$$

to the "yardstick" used to probe the transport, which here is the ruling periodicity x_0 . For any time-dependence $f(t)$ of the exciting beam, solutions of equations (32) can be in general written as

$$n_\ell(t) = \alpha_0 i_0 S_\ell \int_0^t e^{-a_\ell \mu t'} f(t - t')dt' + n_\ell(0)e^{-a_\ell \mu t} \quad , \tag{36}$$

or more compactly in the Laplace transform form

$$\tilde{n}_\ell(\varepsilon) = (\varepsilon + a_\ell \mu)^{-1}\{\alpha_0 i_0 S_\ell \tilde{f}(\varepsilon) + n_\ell(0)\} \tag{37}$$

where the tildes denote the Laplace transforms in the time domain, rate ε being the Laplace variable.

The delayed fluorescence signal (19) can now be obtained by squaring (31) and integrating over one ruling period. One gets

$$\Phi(t) = K \sum_{\ell=0}^{\infty} (2 - \delta_{\ell 0}) n_\ell^2(t) \quad , \tag{38}$$

the signal being thus determined by the sum of the squares of the spatial amplitudes of the exciton concentration harmonics given by (36). The time-dependence of $\Phi(t)$ is sensitive to exciton motion through the a_ℓ's which, in turn, depend on the transport parameter ξ. For a given x_0 in an experiment, the best fit of the observed time-dependence of delayed fluorescence will yield a value of ξ, from which the triplet-exciton-diffusion tensor component D_γ is extracted via (34) and (35). As will be seen in the next section such extraction can be carried out without a priori knowledge of the proportionality factor K in (38) which in general contains unknown crystal and experimental parameters.

Before going into details of the work needed in actual experiment, it will be instructive to point here the basic interest in obtaining this tensor for an organic material. First of all, a direct measurement of the principal values D_{jj} $j = 1,2,3$ of the triplet exciton diffusion tensor will be of interest to verify calculated exciton band structure anisotropies [23],[28] and for obtaining the intermolecular exciton hopping rates. In the hopping model one has

$$D_{ij} = \frac{1}{2} \sum_{n'q'} F_{nn'}^{qq'} (\vec{R}_{n'}^{q'} - \vec{R}_n^q)_i (\vec{R}_{n'}^{q'} - \vec{R}_n^q)_j \tag{39}$$

where $F_{nn'}^{qq'}$ is the hopping rate between a molecule in the nth unit cell for sublattice q and some other molecule in unit cell n' and sublattice q', and where $(\vec{R}_{n'}^{q'} - \vec{R}_n^q)_{i,j}$ is the distance between the pair projected, respectively on the i and jth axes of the tensor component. As stressed before, the nearest-neighbor approximation is typically adequate for triplet excitons so that (39) is in practice simplified by considering hopping rates in just one or a few unit cells. For instance, for the lattice shown in Fig.1 having a triplet band structure given by (39) one has in the ab plane

$$\begin{aligned} D_{aa} &= (F_a + \tfrac{1}{2}F_d)a^2 \\ D_{bb} &= (F_b + \tfrac{1}{2}F_d)b^2 \end{aligned} \tag{40}$$

and $D_{c'c'} \ll D_{aa}, D_{bb}$. In (40) the hopping rates are labelled with the same subscripts as those used for the nearest-neighbors triplet transfer matrix elements shown in Fig.1. In a crystal like anthracene one expects $F_a \ll F_d$ so that a measurement of D_{aa} , the triplet diffusion tensor component along the crystal a axis, can yield immediately F_d, the hopping rate between the inequivalent molecules in the unit cell [30],[37],[38]. If D_{bb} is also measured one can get F_b from $F_b = (D_{bb}/b^2) - (D_{bb}/a^2)$.

An independent determination of the principal intermolecular hopping rates is of primary interest to test basic theories like thos of the magnetic field effects on the triplet mutual annihilation rate constant [14], of the ESR linewidths [5], and of the exciton dynamics itself. For instance, in the SLE approach to exciton motion of Haken and Stroble [2],[5], the hopping rate F_d for the lattice considered above will be determined by two terms $F_d = \hbar^{-1}(\beta_d^2\Gamma^{-1} + 2\gamma d)$, where the first term β_d^2/Γ, sometimes called the local [37] or phonon hindered term, can be determined from spectroscopic data on the $S_o \rightarrow T_1$ absorption band: β_d from the Davydov splitting, and Γ from the (0,0) absorption linewidth [5],[28[,[37]-[41], so that γd, the strength of the correlation in the fluctuations of $\beta_d m$ which determines the nonlocal or "phonon assisted" contribution to diffusion in this theory can be estimated from the diffusion measurements. For more details on all thes theoretical aspects the reader is referred to the references cited above.

V. EXPERIMENTAL DETERMINATION OF THE TRIPLET EXCITON DIFFUSION TENSOR

We shall focus now our attention on basic experiments which allow determination of the triplet diffusion tensor using the formalism of the previous section, that is, via detection of delayed fluorescence from a crystal excited with a spatially periodic pattern of light. The starting point is equation (36) in which f(t), the time modulation of excitation, can be any function of time. This function will be now chosen so that the experiments can be performed in the easiest and most reliable way. In practice, three approaches have been developed for this purpose: A) Time-dependent buildup and decay transient experiments [30]. B) Phase-lag steady-state experiments using a periodic in time excitation [31]-[33], and C) Steady-state experiments in which the crystal is excited through a ruling using a constant intensity excitation [27], As has been remarked early [30] steady-state experiments C) can lead to gross overestimation of the diffusion constant since such experiment cannot simply distinguish if the presence of the excitons in the non-illuminated regions of the ruling is due to their motion or to spurious effects like diffraction or scattering of the exciting light into the dark regions. In the time-dependent experiments of type A) and B), the relatively slow spreading of the excitons from the illuminated into the dark regions can be distinguished from the instantaneous effects due to presenceof stray light in these regions. Moreover, the experiments are performed in such a way that spurious stray-light effects can only lead to a decrease of the effect attribuable exclusively to exciton motion and the interpretation of the experiment can never lead to overestimations of the extent of exciton's travel. In what follows we shall thus only describe the transient and phase-lag experiments A) and B). The experimental setup for both experiments is shown schematically in Fig. 5. The experiments will be illustrated in what follows using rulings of equal open and opaque strips, that is, with coefficients S_ℓ given by (25) and taking a stray-light factor of $\sigma = 1$. Additional experimental details can be found elsewhere [30]-[32].

V.A. Time-Dependent Buildup and Decay Transient Experiments

In Fig.5 the switches S are set on positions 1. The time-modulation consists of either a sudden, step-function illumination of the sample for the buildup study, or of a sudden cut-off of a constant excitation so that the decay of $\Phi(t)$ from its steady-state value can now be followed. The signal-to-noise ratio of the delayed fluorescence signal is improved with a transients-averaging computer triggered by the exciting light signal. The time-dependent signals are normalized by dividing by the saturation value for buildup and by the initial value for decay experiments, respectively. From each normalized to unity time-dependence $\Phi_N(t)$ obtained in this way for a given ruling, of known period x_0, the normalized time-dependence $\Phi_0(t)$ in the absence of motion effects is substracted. These curves are obtained in the same way but just with a blank glass substituting the ruling in the beam path so that $L_T << x_0 (\xi \simeq 0)$, the diffusion effect becoming negligible. The difference curves $\Delta\Phi(t) \equiv \Phi_N(t) - \Phi_0(t)$ are plotted or printed for analysis and extraction of the motion parameter ξ.

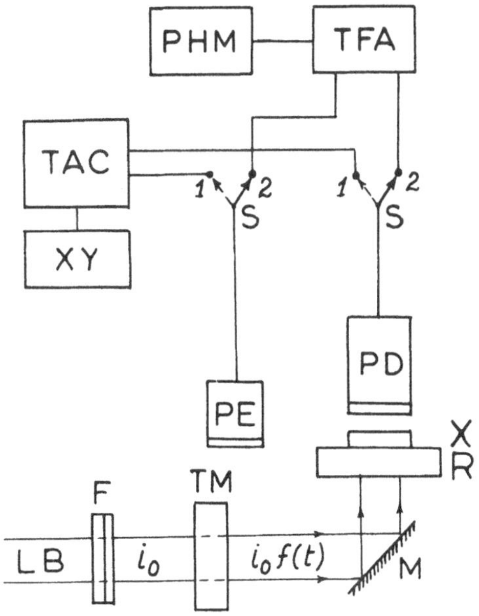

Fig. 5. Experimental setup to study triplet transport via time-dependent delayed fluorescence. LB expanded laser beam, F filters, TM time-molulator, R Ronchi ruling, X crystal platelet, PD and PE photo-multipliers detecting through appropriate filters the time-dependent delayed fluorescence and exciting light signals, respectively; S switches, TFA two-channel Fourier analyser, PHM digital phase meter, TAC signal averaging computer, XY recorder or printer. Rülings can be replaced without disturbing crystal position in the beam.

1. Buildup of Delayed Fluorecence. For this case one has $f(t) = \theta(t)$, the Heaviside function, the initial conditions being $n_\ell(0) = 0$. From (36) one gets

$$n_\ell(t) = \alpha_0 i_0 \tau S_\ell a_\ell^{-1}(1 - e^{-a_\ell \mu t}) \quad . \tag{41}$$

The delayed fluorescence buildup time-dependence is then by (38) of the form

$$\Phi^b(t) \propto \sum_{\ell=0}^{\infty} A_\ell (1 - e^{-a_\ell \mu t})^2 \tag{42}$$

where

$$A_\ell \equiv S_\ell^2 (2 - \delta_{\ell 0}) \frac{1}{S_0^2 (1 + \ell^2 \xi^2)^2} \quad . \tag{43}$$

The steady-state saturation value $S \equiv S(\xi)$ for the buildup is given by

$$S \equiv \Phi_N^b(\infty) \propto \sum_{\ell=0}^{\infty} A_\ell = 1 + \sum_{\ell=1}^{\infty} A_\ell \tag{44}$$

and depends also on the motion parameter ξ defined by (34). Note that for the $\ell = 0$ term, $A_0 = 1$, $a_0 = 1$, so that, as expected, the spatially uniform component of the ruling's periodicity doesn't contribute to the diffusion effect. The normalized buildup of delayed fluorescence is customarily written in the form

$$\Phi_N^b(t) = S^{-1}\{(1 - e^{-\mu t})^2 + \sum_{\ell=1}^{\infty} A_\ell (1 - e^{-a_\ell \mu t})^2\} \quad . \tag{45}$$

As explained before the function of interest to the experiment is the time-dependence of the difference $\Delta\Phi^b(t)$ between (45) and $(1 - e^{-\mu t})^2$ the normalized time-dependence obtained in the absence of rulings ($\xi \simeq 0$). Specifically, for the typical experiments with rulings with equal open and opaque strips (25), that is, $r = 1/2$, one has $A_\ell \neq 0$ for $\ell \neq 0$ odd only, so that $A_0 = 1$,

$$A_\ell = 8\pi^{-2} \frac{1}{\ell^2 (1 + \ell^2 \xi^2)^2} , \qquad \ell = 1,3,5\ldots \text{ (odd)} \tag{46}$$

and the difference buildup function can be written as

$$\Delta\Phi^b(t) = (1 - e^{-\mu t})^2 E^b(t) \tag{47}$$

with

$$E^b(t) = S^{-1} \sum_{\ell = \text{odd}} A_\ell (B_\ell^2(t) - 1), \tag{48}$$

where

$$B_\ell(t) \equiv (1 - e^{-(1 + \ell^2\xi^2)\mu t})/(1 - e^{-\mu t}) \tag{49}$$

Eqs (47) - (49) have been written in the form must suitable for computer calculations. The calculated functions are compared to the experimental $\Delta\Phi^b(t)$'s obtained for rulings of known periods x_0, the best fits yielding ξ and hence L_T from which D_x is extracted.

2. Decay of Delayed Fluorescence. For the decay experiments one has as initial conditions the value $n_\ell(\infty) = \alpha_0 i_0 \tau S_\ell a_\ell^{-1}$ as given by (41) and f(t) = 0. Eq. (36) yields immediately

$$n_\ell(t) = \alpha_0 i_0 \tau S_\ell a_\ell^{-1} e^{-a_\ell \mu t} . \tag{50}$$

Proceeding now in the same way as above one gets for difference function $\Delta\Phi^d(t)$ for the decay

$$\Delta\Phi^d(t) = e^{-2\mu t} E^d(t) \tag{51}$$

with

$$E^d(t) = S^{-1} \sum_{\ell = \text{odd}} A_\ell (D^2(t) - 1), \tag{52}$$

where

$$D(t) \equiv e^{-\ell^2\xi^2\mu t}. \tag{53}$$

Expressions (47) and (51) for coefficients A_ℓ as given by (46) are plotted in Fig.6 for several values of the motion parameter ξ as a function of time expressed in units of lifetime. An experimental example of measured differences of normalized delayed fluorescence is shown in Fig. 7.

In order to perform the experiments unde the most favorabel conditions some remarks on the behavior of the $\Delta\Phi(t)$ functions will be useful. It is important to note that the behavior of the $\Delta\Phi(t)$'s being determined by the product of two time-dependent fuctions in (47) and (51), passes for each value of the motion parameter ξ through a maximum. The diffusion effects increase as the parameter ξ or, equivalently the ratio L_T/x_0 increases, and attain their biggest values for $\xi \simeq 0.9$ when the diffusion length is about 1/5 or the ruling period x_0 used to probe the motion, that is, about one

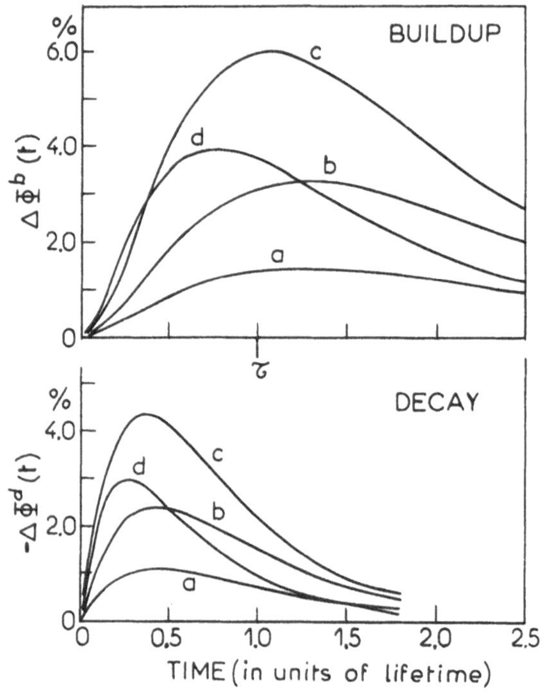

Fig. 6. Plot as a function of time in units of lifetime of the buildup and decay difference functions $\Delta\Phi(t)$ given by (47) and (51). Curves a,b,c,d, correspond to values of motion parameter ξ = 0.2, 0.4, 0.9, and 1.5, respectively, or equivalently, for diffusion length to ruling period ratios L_T/x_0 of 0.045, 0.09, 0.20, and 0.35, respectively.

half of the width of a dark region. The maximum effects occur at times in the vicinity of the triplet exciton lifetime τ(at t $\simeq$ 1.06 τ for the buildup and at t $\simeq$ 0.36 τ for the decay, respectively). As physically expected, when the parameter ξ increases further, that is, as the exciton diffusion length L_T becomes comparable or larger than the ruling period, the observable effects begin to decrease, with the maximum of the effect occuring at increasingly shorter times, since now the excitons diffusing out of a given illuminated strip can reach during their lifetime the neighboring illuminated regions, and the triplet exciton concentration becomes more rapidly uniformized over the whole crystal. In practice it is convenient to use ruling periods x_0 which, typically, lie in the range 2 to 20 times the diffusion length L_T to be extracted from the data. For a given ruling the effect in the buildup is always bigger and appears at longer times from

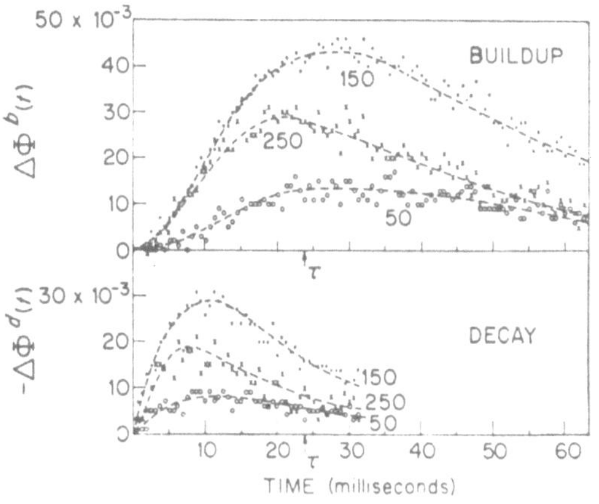

Fig. 7. Experimental behavior (points) of the normalized delayed fluorescence differences observed for an anthracene crystal having a triplet exciton lifetime of τ = 24.0 msec. as indicated by the arrow. The numbers labelling each curve correspond to the spatial frequencies x_0^{-1} of the rulings used to probe the motion, expressed in number of lines/inch. (From Ref. [30]).

the start of the process than that for the decay. It is thus advantages, whenever possible, to perform the experiments detecting changes in the buildup of the delayed fluorescence. Spurious effects dur to presence of stray light in the dark region when taken into account by using coefficients (26) instead of (25) in (43) lead to a nearly uniform decrease of the observable $\Delta\Phi(t)$ effects. Effects as those shown in Fig. 6 and 7 can be only interpreted as due to triplet diffusion. an overestimation of the transport length being not possible [30],[33]. The experimental behavior shown in Fig. 7 is for an ab plane anthracene platelet, the ruling lines being set parallel to the crystal a axis so that motion along the crystal b axis is being studied. Analysis of the data [30] yield a diffusion length of 30 $\pm$ 3μm and by (35) the value D_{bb} = 1.9 $\pm$ 0.3 x 10^{-4} cm^2sec^{-1} for the triplet diffusion tensor component along b in anthracene.

V.B. Phase-Lag Steady-State Experiments

Transient experiments described above provide a direct physical way to extract exciton diffusion parameters. However, the experiments become cumbersome and time-consuming if one wishes to map out the complete anisotropy of the triplet diffusion tensor of a crystal at different temperatures. It was soon realized [31] that some information can be more quickly extracted from measurements of the changes of phase-angles of the delayed fluorescence signal harmonics with respect to those of a periodic in time modulation of the exciting light. Because of the intrinsic triplet lifetime the delayed fluorescence waveform under periodic illumination lags that of the exciting light. Since motion effects, appearing when a ruling is introduced in the beam, speed-up both buildup and decay of delayed fluorescence, this phase-lag is reduced, the changes depending thus, for each harmonic, on the diffusion parameter ξ.

In the experimental setup shown in Fig.5 the switches S are set on position 2. The time-modulation of the laser beam is made to be periodic in time, that is, in general, proportional to

$$f(t) = \sum_{n=-\infty}^{\infty} f_n e^{in\omega t} , \qquad (5.4)$$

where ω is the angular frequency. This excitation signal, detected with photomultiplier PE, and the resulting delayed fluorescence signal $\Phi(t)$ which will be also a periodic function of time when the steady-state regime is reached, that is, of the form [31].

$$\Phi(t) \propto \sum_{p=-\infty}^{\infty} H_p e^{ip\omega t} , \qquad (55)$$

are fed to a two-channel Fourier analyzer TFA, The phase-lag of a given harmonic of $\Phi(t)$ with respect to that of the excitation (54) is measured with the digital phase meter PHM with a high degree of accuracy, Typically 0.2^0. As written in (55) the amplitudes H_p are in general complex quantities, their arguments being the phase-lags of interest to the experiment.

The complex amplitudes H_p are obtained by specializing (36) for the periodic in time excitation f(t) given by (54) and one has then

$$n_\ell(t) = \alpha_0 i_0 S_\ell \sum_{n=-\infty}^{\infty} f_n e^{in\omega t} \int_0^t e^{-(a_\ell \mu + in\omega)t'} dt' + n_\ell(0) e^{-a_\ell \mu t} . \qquad (56)$$

In the periodic in time steady-state regime, reached in the limit $t \to \infty$, that is, for $t >> \tau$ so that contributions from the initial conditions vanish in (56), the integral is then the Laplace transform of unity with variable $a_\ell \mu + i n\omega$, one gets

$$n_\ell(t) = \alpha_0 i_0 S_\ell \sum_{n=-\infty}^{\infty} f_n (a_\ell \mu + i n\omega)^{-1} e^{i n\omega t} \quad , \tag{57}$$

each spatial amplitude being thus a periodic function of time. By squaring (57) and using (38) one obtains the delayed fluorescence signal in the disired periodic in time form (55) in which the complex amplitude H_p for the pth harmonic can be written as

$$H_p \propto \sum_n \sum_{n'} f_n f_{n'} \sum_{\ell=0}^{\infty} A_\ell^{nn'} \quad , \tag{58}$$

with the condition $n + n' = p$, and in which the quantities $A_\ell^{nn'}$ are given by

$$A_\ell^{nn'} \equiv S_\ell^2 (2 - \delta_{\ell 0}) \frac{1}{S_0^2 (1 + \ell^2\xi^2 + i n\omega\tau)(1 + \ell^2\xi^2 + i n'\omega\tau)} \quad . \tag{59}$$

Expressions (59) can be viewed as a generalization of the quantities A_ℓ defined by (43). Because of the restriction $n + n' = p$ in the double sum (58) its numerical calculation, and extraction of its phase ϕ_p , as a function of the motion parameter ξ or of $\omega\tau$ does not offer in practice particular difficulties.

The experiments are usually performed using the, strongest, first harmonic (p - 1) in (55). For a square-wave time-modulation [33], [38] of the exciting beam one has $f_0 = 1/2$, $f_{\pm n} = 1/n\pi$, n odd only, and by (58) the first harmonic amplitude of delayed fluorescence is then given by

$$H_1 \propto \sum_{\ell=0}^{\infty} A_\ell^{10} \quad . \tag{60}$$

If, as done above for the transient experiments, one considers the case of rulings of equal open and opaque strips, one has explicitly using (59)

$$H_1 \propto \frac{1}{1 + i\omega\tau} + \frac{8}{\pi^2} \sum_{\ell = \text{odd}} \frac{1}{\ell^2 (1 + \ell^2\xi^2)(1 + \ell^2\xi^2 + i\omega\tau)} \tag{61}$$

whose argument $\phi_1 \equiv \arg(H_1)$,

$$\phi_1 = \tan^{-1}(\mathrm{Im}H_1/\mathrm{Re}H_1) \tag{62}$$

$\mathrm{Im}H_1$, $\mathrm{Re}H_1$ being the imaginary and real parts of (61), can be readily computed. The first term in (61) determines for this harmonic the phase-lag in the absence of the motion effect

$$\phi_0 \equiv -\tan^{-1}(\omega\tau) \quad , \tag{63}$$

and which is measured with no ruling ($\xi \to 0$) in the beam. This angle can be used to obtain the triplet lifetime τ. As in the transient experiments motion effects stem exclusively from the second term in (61) to which contribute the $\ell \neq 0$ harmonics of the spatial periodicity. Hence, the observed decreases in the phase-lag

$$\Delta\phi \equiv \phi_1 - \phi_0 = \arg(H_1/(1 + i\omega\tau)) \tag{64}$$

when a ruling of period x_0 is inserted in the beam, are determined

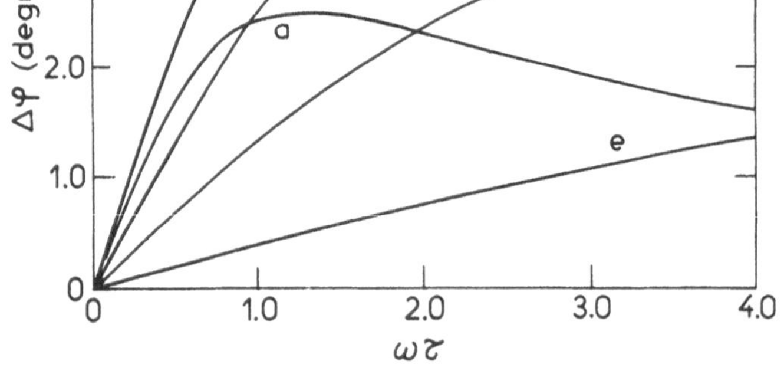

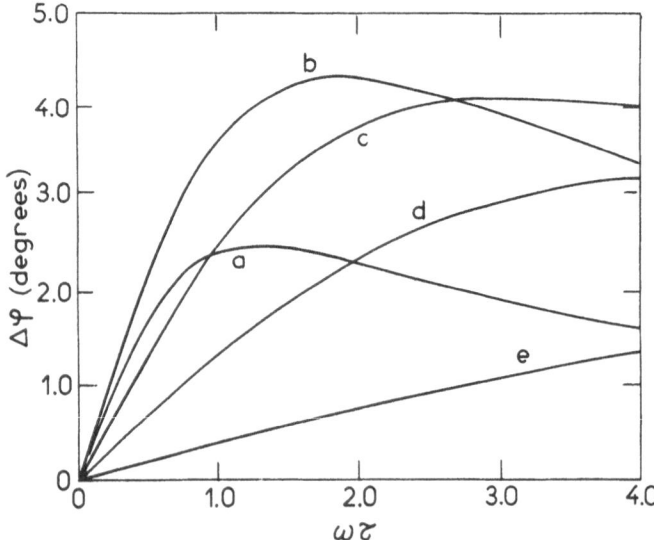

Fig. 8. Phase-lag differences due to exciton motion given by Eq. (64) as a function of $\omega\tau$ for different values of the diffusion parameter ξ = 0.5, 1.0, 1.5, 2.0, 3.0 for curves labelled (a) to (e), respectively.

by the motion parameter ξ from which the triplet diffusion tensor component is extracted via (34) and (35). Fig. 8 and 9 show the expected observable phase difference due to diffusive motion as given by (64) with (61). Fig. 8 shows the $\Delta\phi$'s as a function of $\omega\tau$ for several values of the motion parameter ξ. It is seen that to obtain the biggest motion effects (e.g. $\sim 4.5^o$) it is convenient to use chopping angular frequencies such that $\omega\tau$ lies between 1 and 2, and using rulings of periods x_0 so that the diffusion parameter $\xi = \sqrt{2}\pi L_T/x_0$ for the particular srystal studied lies in the neighborhood of $\xi \approx 1$. In Fig. 9 function (65) is plotted as usually done to fit experimental data, that is, for a given $\omega\tau$ as function of the ruling spatial frequency x_0^{-1} and using the diffusion length L_T as the fitting parameter. For the figure shown $\omega\tau$ = 2.0 with diffusion lengths of 15, 20, 30, 40 μm for the curves (a) through (d), respectively.

The precision typically achieved in the stationary phase-lag experiments is illustrated in Fig. 10. The figure shows measurements of phase-lag differences (points) observed for an anthracene

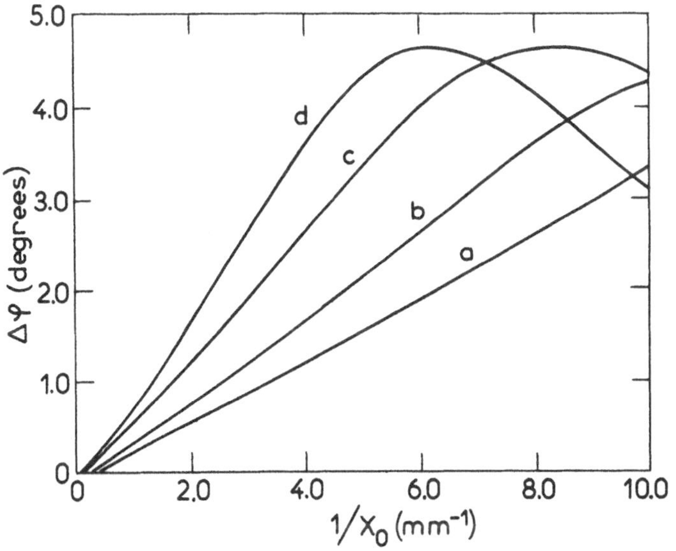

Fig. 9. Phase-lag differences given by (64) for $\omega\tau$ = 2 and plotted as a function of the ruling spatial frequencies for different values of the diffusion length L_T = 15, 20, 30, 40 μm for curves labelled (a) to (d), respectively.

crystal at room temperature having a triplet lifetime $\tau = 16.9 \times 10^{-3}$ sec [31]. The ruling lines are set perpendicular to the crystal a axis along which the diffusive transport will be measured. The solid curves are the expected behaviors given by (64) plotted as in Fig. 9 for the indicated values of the diffusion length. From the best fit one deduces a diffusion length along the a axis of this crystal of L_T^a - 22.0 ± 1.5 μm so that $D_{aa} = 1.5 \pm 0.2 \times 10^{-4}$ cm^{-2} sec^{-1}. Fig. 11 illustrates the results of a complete measurement of the anisotropy of the triplet diffusion length and of the diffusion tensor as described by (29) in the ab plane of a *p*-terphenyl crystal [38]. The direct technique via delayed fluorescence described in this section to measure components of the triplet exciton diffusion tensor has been applied to several organic solids, namely anthracene [30],[31],[28],[37],[12],[14],[5], naphthalene [40],[41],[5],[15],[42],1-4 dibromonophthalene [40],[43],[45], tetracene [46]-[48], trans-stilbene [40], *p*- terphenyl [38],[49], and pyrene [50]-[53], The by-no-means exhaustive list of references given for each material include also those for related studies in which triplet diffusion tensor values have

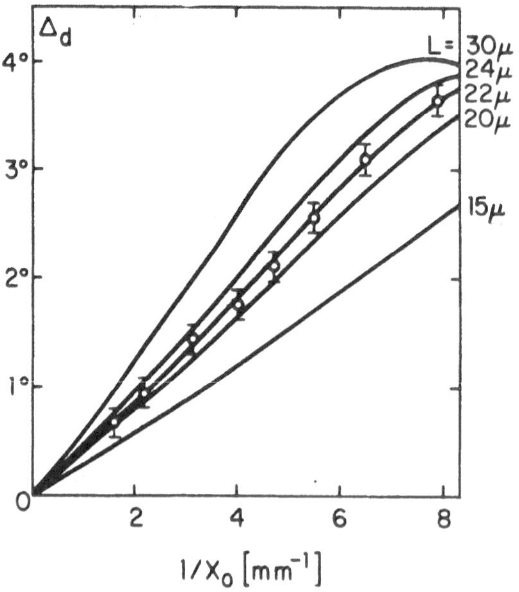

Fig. 10. Measured phase-lag differences due to diffusion in an anthracene crystal (from Ref. [31]).

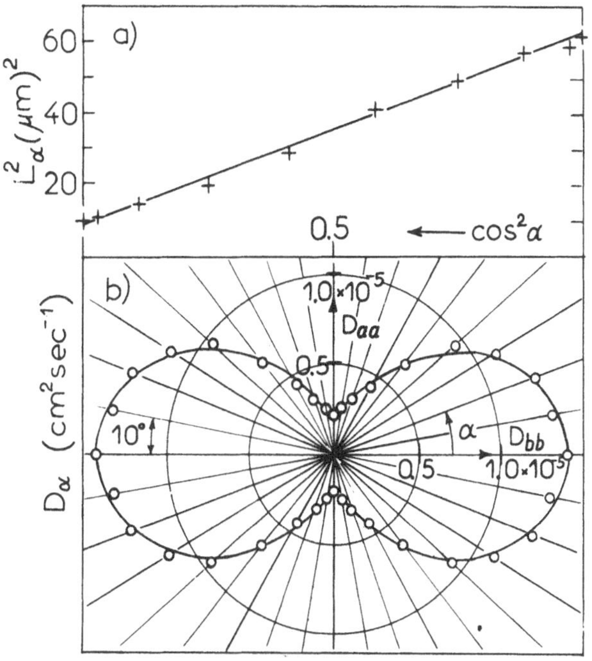

Fig. 11. Anisotropy of diffusion length and diffusion tensor in the ab plane of *p*-terphenyl (from Ref. [38].

been used to test theories of exciton dynamics in organic matter, to fit experimental behaviors of other triplet properties, as well as those in which triplet hopping rates have been inferred indirectly.

VI. POSSIBILITY OF DETECTING COHERENCE EFFECTS IN TRIPLET TRANSPORT

The direct approach to study triplet-exciton motion via delayed fluorescence described in Section 4 has been developed in Section 5 and 6 specifically for the case when a macroscopic diffusion equation is applicable to describe triplet motion in a near organic crystal. Recently, the theory of this experimental approach has been extended [54] to the situation in which the motion would not be completely incoherent as in the diffusive case, that is, when coherence effects are present in the transport. The microscopic theory of exciton transport used in Ref. [54] is that of generalized master equation (GME) [4],[55], which is capable of providing a unified analysis of coherent and incoherent motion [4],[56],[57]. The theory allows, in principle, to gradually move as the conditions, for instance, the interaction with the phonon bath, change from the completely coherent case when the exciton transport length is given by (16) to the completely incoherent (diffusive) case in which the exciton transport length is given by (35). The theoretical basis of the powerful GME approach and an analysis of its applicability to a variety of situations including singlet motion can be found in the presentation of V.M. Kenkre [58].

For the experimental situation described in detail in Section 4 the starting point of the theory is, instead of the macroscopic equation (21), the motion equation

$$\frac{\partial P_m}{\partial t} = \alpha_0 i_0 g_m f(t) - \mu P_m + \frac{\partial P_m}{\partial t}\Big|_{mot} \tag{65}$$

for the probability $P_m(t)$ of finding an exciton at the mth site of the crystal at time $\underline{t}$. The motion contribution in (65) is given by

$$\frac{\partial P_m}{\partial t}\Big|_{mot} = \sum_{m'} \int_0^t \{W_{mm'}(t - t')P_{m'}(t') - W_{m'm}(t - t')P_m(t')\}dt' \quad , \tag{66}$$

where the $W_{mm'}(t)$'s are the standard GME memory functions [4], the solutions of (66) being usually written in the form [4],[56],[57]

$$P_m(t) = \sum_{m'} \Psi_{mm'} P_m(0) \tag{67}$$

where the $\Psi_{mm'}(t)$'s are the propagator-functions between the sites $\underline{m}$ and the site $\underline{m'}$, the $P_{m'}(0)$'s being the initial (t = 0) conditions for the probabilities of occupation [4],[54]. In (65) g_m is the discrete site-function determined by the spatially periodic excitation, and f(t), the normalized time-dependence of the exciting beam, the parameters α_0, i_0, τ having similar meanings as in Section 4. To illustrate the appearance of coherence effects, the delayed fluorescence signals of interest to the time-dependent experiments described in Section 6 will be obtained explicitly for one dimensional crystals in the nearest-neighbor approximation, that is crystals whose triplet exciton's band structure and mean-square velocity are given by expressions (14) and (15). For motion studied along the crystalline chain of lattice spacing $\underline{a}$ the site function characteristic of the ruling geometry will be given by

$$g_m = \sum_{\ell=-\infty}^{\infty} S_\ell e^{i\ell\eta m} \quad , \tag{68}$$

where now η stands for the adimensional wavevector

$$\eta \equiv 2\pi a/x_0 \tag{69}$$

the coefficients S_ℓ being given, as before, by expressions (23).

(25), or (26) depending on the periodic spatial geometry used.

With (67) and (68) the solution of the augmented GME equation (65) for any time-dependence f(t) for the exciting beam can be written as

$$P_m(t) = \alpha_0 i_0 \sum_{\ell=-\infty}^{\infty} S_\ell e^{i\ell\eta m} \int_0^t \Psi^{\ell\eta}(t')e^{-\mu t'} f(t-t')\,dt' + e^{-\mu t} \sum_{m'} \Psi_{m-m'}(t)P_{m'}(0) \qquad (70)$$

where rate $\mu \equiv \tau^{-1}$, as before, and $\Psi\ell\eta(t)$ stand for the discrete Fourier transforms of the propagator for the wavevectors $\ell\eta$. It should be noted that because of the translational symmetry of the crystal the GME memories and propagators $\Psi_{mm'}$ can be considered as single-indexed quantities $\Psi_{m-m'}(t) = \Psi_{m'-m}(t)$, depending only on the distance (m - m') separating sites m and m'. The time-dependence of the delayed fluorescence signal of interest to the experiments of Section V is obtained as

$$\Phi(t) \propto \sum_m P_m^2(t) \quad , \qquad (71)$$

that is, proportional to the sum of the squares of the probabilities $P_m(t)$'s as given by (70) for the specific time-dependencies f (t) and initial conditions used in Section V. To make the results immediately comparable to the graphs presented in Section V for the diffusive case, all expressions will be directly written for the case of rulings of equal open and opaque strips (r = 1/2), that is $S_0 = 1/2$, $S_\ell = S_{-\ell} = \sin(\ell\pi/2)/\ell\pi$, ℓ = odd (Eq (25)) and separating in the sums over the spatial harmonics the uniform, $\ell = 0$, term which as before doesn't contribute to the effects on the time-dependence of $\Phi(t)$ due to the exciton transport. Stray light effects can be accounted for by using coefficients (26) instead of (25).

VI.A. Buildup and Decay Transient Experiments

The experiments are performed in the same way as described in Section V. The experimental signals sought are, as before, the differences $\Delta\Phi(t) = \Phi_N(t) - \Phi_0(t)$ between the normalized delayed fluorescence with and without the ruling in the exciting beam.

For the buildup of delayed fluorescence one has $f(t) = \theta(t)$ as before and $P_m(0) = 0$ for all m'. For a ruling with equal open and opaque strips one gets from (70)

$$P_m^b(t) \propto (1 - e^{-\mu t}) + 2 \sum_{\ell=\text{odd}} \frac{\mu}{\ell\pi} \cos(\ell\eta m) \int_0^t \Psi^{\ell\eta}(t')e^{-\mu t'}dt' \qquad (72)$$

whose steady-state saturation value is given by

$$P_m(\infty) \propto 1 + 2 \sum_{\ell = \text{odd}} \frac{\mu}{\ell\pi} \cos(\ell\eta m)\tilde{\Psi}^{\ell\eta}(\mu) \quad . \tag{73}$$

In (73) $\tilde{\Psi}^{\ell\eta}(\mu)$ stands for the Laplace transform of variable $\mu=\tau^{-1}$ of the Fourier transform of the propagator function appropriate to the motion under study. The expressions for $\Delta\Phi^b(t)$ of interest to the experiment can be now written in a manner which parallels Eqs (46) to (47). Squaring (72) and (73) and performing the sum (7) one gets

$$\Delta\Phi^b(t) = (1 - e^{-\mu t})^2 E^b(t) \tag{74}$$

with

$$E^b(t) = S^{-1} \sum_{\ell = \text{odd}} A_\ell\{B_\ell^2(t) - 1\} \ , \tag{75}$$

as before, but where the functions

$$B_\ell(t) = \int_0^t \Psi^{\ell\eta}(t')e^{-\mu t'}dt'/(1 - e^{-\mu t})\ \tilde{\Psi}^{\ell\eta}(\mu) \quad , \tag{76}$$

and coefficients

$$A_\ell \equiv \frac{8}{\ell^2\pi^2}\{\mu\tilde{\Psi}^{\ell\eta}(\mu)\} \tag{77}$$

depend specifically on the propagators, or equivalently on the GME memory functions, which describe the motion under study. The quantity S in (75) stands, as before, for the steady-state saturation value

$$S \equiv 1 + \sum_{\ell = \text{odd}} A_\ell \tag{78}$$

in which the A_ℓ's are given now by their generalized form (77)

For the decay of delayed fluorescence one has f(t) = 0, the initial conditions $P_m'(0)$ being given by (73). From (70) one gets

$$P_m^d(t) \propto e^{-\mu t} + 2 \sum_{\ell = \text{odd}} \frac{\mu}{\ell\pi} \cos(\ell\eta m)\tilde{\Psi}^{\ell\eta}(\mu)\Psi^{\ell\eta}(t)e^{-\mu t} \quad , \tag{79}$$

and the difference delayed fluorescence decay signals are given by

$$\Delta\Phi^d(t) = e^{-2\mu t}E^d(t) \tag{80}$$

with

$$E^{d}(t) = S^{-1} \sum_{\ell=\text{odd}} A_{\ell}\{D_{\ell}^{2}(t) - 1\} \quad , \tag{81}$$

where one has now

$$D_{\ell}(t) \equiv \Psi^{\ell\eta}(t) \; , \tag{82}$$

with A_{ℓ} and S given by (77) and (78)

Exact analytic expressions for the propagators have been given [4],[54]-[58] for the case of one-dimensional crystals in the nearest neighbors approximation for three types of motion: completely coherent, completely incoherent, and an intermediate case for which the exciton-phonon interaction is taken into account via a scattering rate α and whose memory function is assumed to decay in time as $W^{\alpha}_{mm'}(t) - W^{c}_{mm'}(t)e^{-\alpha t}$, $W^{c}_{mm'}(t)$ being the GME memory of the completely coherent motion [4].

For the case of completely coherent motion the propagator for a one-dimensional crystal with triplet exciton band structure (14) and nearest-neighbor transfer matrix element β is $\Psi_m(t) = J_m^2(2\beta t)$ where J_m is the mth ordinary Bessel function, its Fourier transform for wavevector $\ell\eta$ being

$$\Psi_c^{\ell\eta}(t) = J_0(4\beta\sin(\tfrac{1}{2}\ell\eta)t) \; . \tag{83}$$

By the difinition (69) the $\eta\ell$'s can be considered as very small quantities since the lattice constant a typically ranges from 5 to 10 Å while the ruling periods x_0 as used in the experiments are at least of the order of several microns. The continuum limit approximation can be considered here as applicable so that

$$\sin(\tfrac{1}{2}\ell\eta) \simeq \tfrac{1}{2}\ell\eta = \ell\pi a/x_0 \quad . \tag{84}$$

As discussed in Section IV, it is physically logical to present the results by defining an adimensional motion parameter, to be extracted from the experiment, which scales the exciton transport length (given here by (16) to the ruling period x_0 used to probe the effects of transport. If, instead of (34), one defines for the coherent case

$$\xi_c = 2\sqrt{2}\pi L_T^c/x_0 \quad , \tag{85}$$

where $L_T^c = \sqrt{2}\beta a\tau$ (Eq. 16), the Fourier transforms (83) needed for (82) and (81) can be simply written in the continuum limit (77) as

$$\Psi_c^{\ell\eta}(t) = J_0(\ell\xi_c\mu t) \quad , \tag{86}$$

and its Laplace transform needed to obtain the coefficients (77) as

$$\tilde{\Psi}_c^{\ell\eta}(\mu) = \tau/(1 + \ell^2\xi_c^2)^{1/2} \quad . \tag{87}$$

For the completely incoherent (hopping or diffusive) case the propagator is $\Psi_m(t) = \exp(-2Ft)I_m(2Ft)$ where I_m is the mth modified Bessel function and F the nearest-neighbor hopping rate. For the one-dimensional crystal under study this motion can be also characterized from (39) by a diffusion constant along the chain D = Fa and the transport(diffusion)length given by (35) can also be written as

$$L_T = (2D\tau)^{1/2} = (2F\tau)^{1/2}\, a. \tag{88}$$

The Fourier transform of the propagator is here $\Psi_{in}^{\ell\eta} = \exp(-4F\sin^2(\frac{1}{2}\ell\eta)t)$ which in the continuum limit can be written as

$$\Psi_{in}^{\ell\eta}(t) = e^{-\ell^2\eta^2 Ft} = e^{-\ell^2\xi^2\mu t} \quad , \tag{89}$$

and its Laplace transform, to obtain coefficients (77)

$$\tilde{\Psi}_{in}^{\ell\eta}(\mu) = \tau/(1 + \ell^2\xi^2) \quad . \tag{90}$$

In (89) and (90) the adimensional motion parameter for the completely incoherent case

$$\xi \equiv \sqrt{2}\pi L_T/x_0 \tag{91}$$

with L_T given by (88), has exactly the same meaning as the parameter (34) defined for the macroscopic diffusion equation study in Section IV and V. It is a simple matter for the reader to verify that (89) and (90) for the GME completely incoherent case lead via (74) and (80) to experimental signals $\Delta\Phi^b$ (t) and $\Delta\Phi d$ (t) having exactly the same time-dependences as those obtained in Section V for the macroscopic diffusion equation and plotted in Fig. 6.

Finally, for the intermediate coherence case characterized by an exponentially decaying memory with scattering rate α the Fourier transform of the propagator in the continuum limit approximation (84) is given by [54]

$$\Psi_\alpha^{\ell\eta}(t) = J_0(\ell\xi_c\mu t)e^{-\alpha t} + \alpha\int_0^t e^{-\alpha(t-t')}J_0(\ell\xi_c\mu(t^2-t'^2)^{\frac{1}{2}})dt' \quad , \tag{92}$$

whose Laplace transform is

$$\tilde{\Psi}_\alpha^{\ell\eta}(\mu) = \tau/\{((1 + \alpha\tau)^2 + \ell^2\xi_c^2)^{\frac{1}{2}} - \alpha\tau\} \quad , \tag{95}$$

and the transport length, calculated from the mean-square displacement[4] (t = τ) , is given by $L_T^\alpha = \sqrt{2}L_T^c(\alpha\tau + e^{-\alpha\tau} - 1)^{\frac{1}{2}}/\alpha\tau$. Eq. (93) and L_T^α tend to the coherent values (87) and (16) for $\alpha\tau \to 0$ (no scattering), and to the completely incoherent ones (90) and (88) for the hopping limit $F \to 2\beta^2/\alpha$ as β/μ and α/μ tend to infinity [4], [54], [28].

The experimental signals $\Delta\Phi^b(t)$ and $\Delta\Phi^d(t)$ given by (74) and (80) can now be written in analytical form for the three kinds of motion described above using (86), (89), or (92) in (76) and (82) for the buildup and decay, respectively, and by calculating coefficients (77) with expressions (87), (90), or (93). In practice, sums (75), (81) and (78) converge rapidly so that in a numerical computation inclusion of just the first ten spatial harmonics is more than amply sufficient to obtain the time-dependence of $\Delta\Phi(t)$ with accuracy. Also, for the buildup case calculation of (76) by a numerical integration doesn't pose special difficulties.

The fact that the delayed fluorescence experiments described above will indeed have a sensitivity to the type of triplet motion, well above the typical signal-to-noise ratio found in practice, is illustrated in Figs.12 and 13 for the case of the buildup $\Delta\Phi^b(t)$ differences given by Eq. (76). Figure 12 compares a family of expected signals $\Delta\Phi^b(t)$ for the purely coherent motion (solid line curves) with those for the completely incoherent case (dashed line curves), these last ones having the same behavior as already displayed in Fig.6. To facilitate the comparison of the differences in the temporal behavior for the two types of motion, in the two families, curves obtained with the same L_T/x_0 ratio have been labelled with the same lower case letters. In practice, this means that either with a given ruling of period x_0 the $\Delta\Phi^b(t)$ behavior is compared for two crystals having same transport lengths but in which motion is of a different type or, alternatively, if the experiments are performed with the same crystal, the ruling periods x_0 being adapted so that L_T/x_0 remains the same as motion changes from one type to the other. For instance, if one considers the pair of curves labelled (b), that is, for a L_T/x_0 value which lies in the range for which maximum motion effects due to the presence of the ruling occur in the $t \simeq \tau$ region, a coherent motion leads to $\Delta\Phi^b(t)$ values in the vicinity of 12%, a value which under no circumstances can be observed in the completely incoherent case (see discussion in Section V). Similar considerations apply to the other pairs so that coherence is unmistakably distinguishable in the data plotted in this way. Fig.13 illustrates the evolution of the coherent behavior into the completely incoherent one by showing a family of curves computed using the intermediate case propagator formulae (92) and (93) with L_T/x_0 = 012 taken to be the same for different values of $\alpha\tau$ as shown. The $\alpha\tau = 0$ case corresponds to the purely coherent motion while for $\alpha\tau \geqslant 10^3$ the behavior becomes already experimentally undistinguishable from the completely incoherent diffusive case.

The behavior of $\Delta\Phi^d(t)$ for the decay of delayed fluorescence

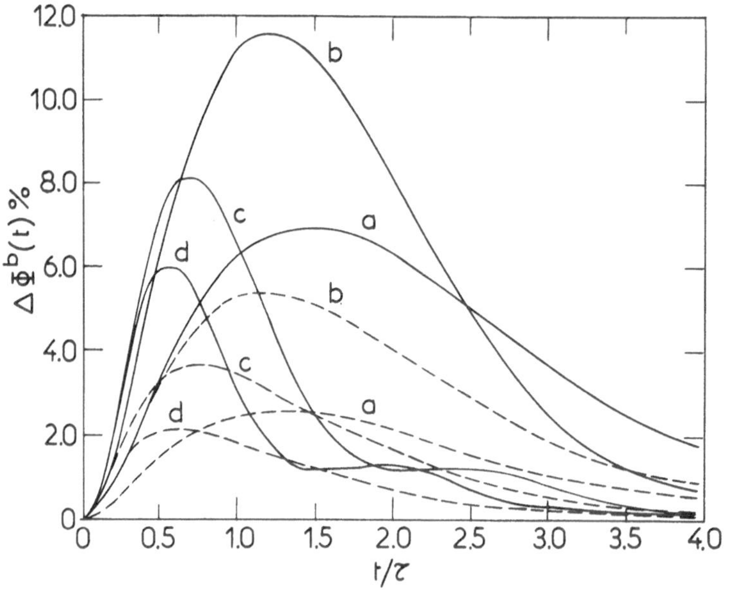

Fig. 12. Delayed fluorescence buildup changes $\Delta\Phi^{b}(t)$ for coherent (solid line curves) and diffusive (dashed curves) triplet motions. The pairs labelled a,b,c,d are for same ratio L_T/x_0 = 0.075, 0.15, 0.30, and 0.45, respectively.

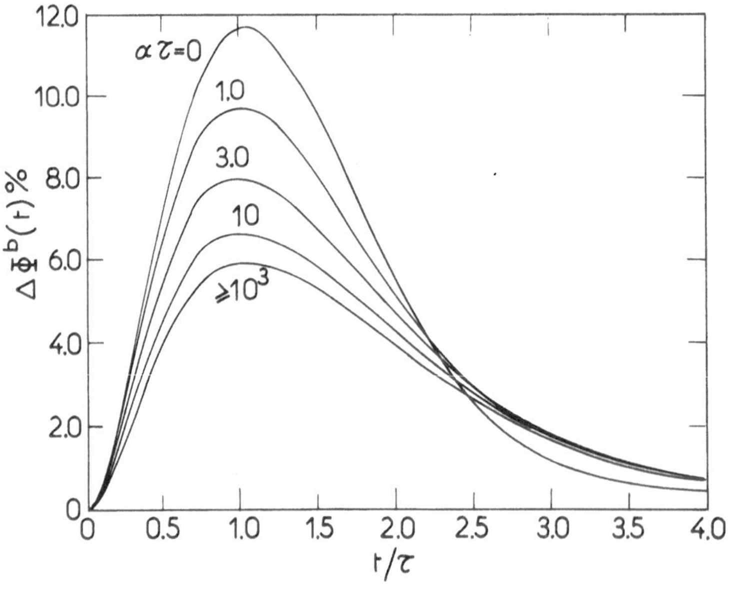

Fig. 13. Buildup $\Delta\Phi^{d}(t)$ changes using intermediate coherence propagator (92). Ratio L_T/x_0 = 0.2 is the same for all $\alpha\tau$.

corresponding to the cases shown for buildup in Figs. 12 and 13 has been given in Ref. [54] leading, as physically expected, to the same conclusions as above. In Ref. [54] the reader can also find an extended discussion showing that, as usually done in practice, the extraction by best fit of motion parameter ξ from measurements with a series of rulings of periods x_0 for a given sample will yield, as required by definition of ξ, a straight line for a $\xi = f(1/x_0)$ plot, only if the correct theory has been applied to fit the experimental $\Delta\Phi(t)$ data. It should be noted that the mathematical structure of Eqs. (74) and (80) is such that buildup effects due to motion, notably for the coherent case, are bigger than those obtained for the decay. This makes buildup experiments, whenever possible, preferable as better signal-to-noise ratios are obtained for the data. However, as manifestations of coherence produce opposite effects in buildup and decay, it is advisable to perform at the same time both types of observations to double-check the self-consistency of the fit.

VI.B. Phase-Lag Steady-State Experiments

As discussed in Section V these time-dependent experiments constitute the preferable experimental tool to obtain information on triplet motion, notably when an extensive study of the motion anisotropy is desired. The extension to this experiment of the GME approach given above is immediate. Following the reasoning developped in Section V and taking as starting point, instead of (36), the GME solution (70) one gets with the periodic in time excitation (54) the occupation probabilities given by

$$P_m(t) = \alpha_0 i_0 \sum_{n=-\infty}^{\infty} f_n e^{in\omega t} \sum_{\ell} S_\ell e^{i\ell\eta m} \int_0^t e^{-(\mu + in\omega)t'} \Psi^{\ell\eta}(t')\, dt' + \sum_{m'} \Psi_{m-m'}(t) P_{m'}(0) e^{-\mu t} . \qquad (94)$$

When, as discussed in Section V, the periodic in time steady-state regime is reached one has

$$P_m(t) = \alpha_0 i_0 \sum_{n=-\infty}^{\infty} f_n e^{in\omega t} \sum_{\ell} S_\ell e^{i\ell\eta m} \tilde{\Psi}^{\ell\eta}(\mu + in\omega) \qquad (95)$$

in which $\tilde{\Psi}^{\ell\eta}(\mu + in\omega)$ is the Laplace transform with variable $\mu + in\omega$ of Fourier transforms of the propagator appropriate for the motion under study.

The periodic in time delayed fluorescence signal given by (71) with (95) can now be written in form (55) in which the amplitudes of the temporal harmonics can be written as (58)

$$H_p \propto \sum_n \sum_{n'} f_n f_{n'} \sum_{\ell=0}^{\infty} A_\ell^{nn'} , \qquad (96)$$

with condition $n + n' = p$, but in which now the quantities $A_\ell^{nn'}$ have the generalized form[59]

$$A_\ell^{nn'} \propto \frac{S_\ell^2(2 - \delta_{\ell 0})}{S_0^2} \tilde{\Psi}^{\ell\eta}(\mu + in\omega)\ \tilde{\Psi}^{\ell\eta}(\mu + in'\omega) \qquad (97)$$

which, for each spatial harmonic of the ruling, depend on the particular type of transport through the Laplace transform of variable $\mu + in\omega$ of Fourier transform for wavevector $\ell\eta$ of the propagator appropriate to the motion.

As discussed in Section V, the experimental quantity of interest is the phase-lag, that is, the argument of (96). Considering again the typically used experimental case of the first harmonic of delayed fluorescence, and specializing the expressions for rulings of equal open and opaque strips one has now, instead of (61), the expression

$$H_1 \propto \frac{1}{1 + i\omega\tau} + \frac{8}{\pi^2}\mu^2 \sum_{\ell = \text{odd}} \frac{1}{\ell^2} \tilde{\Psi}^{\ell\eta}(\mu)\ \tilde{\Psi}^{\ell\eta}(\mu + i\omega) \quad , \qquad (98)$$

which using (87), (90), and (93) can be written, respectively,

$$H_1^c \propto \frac{1}{1 + i\omega\tau} + \frac{8}{\pi^2} \sum_{\ell = \text{odd}} \frac{1}{\ell^2 K_\ell(0) K_\ell(\omega\tau)} \qquad (99)$$

with

$$K_\ell(0) \equiv (1 + \ell^2\xi_c)^{\frac{1}{2}} \qquad (100)$$

$$K_\ell(\omega\tau) \equiv \{(1 + i\omega\tau)^2 + \ell^2\xi_c^2\}^{\frac{1}{2}}$$

for the completely coherent motion with the adimensional motion parameter ξ_c defined by (85), as

$$H_1^{in} \propto \frac{1}{1 + i\omega\tau} + \frac{8}{\pi^2} \sum_{\ell = \text{odd}} \frac{1}{\ell^2 C_\ell(0) C_\ell(\omega\tau)} \qquad (101)$$

with

$$C_\ell(0) \equiv 1 + \ell^2\xi^2 \qquad (102)$$

$$C_\ell(\omega\tau) \equiv 1 + i\omega\tau + \ell^2\xi^2$$

for the completely incoherent, diffusive, transport, ξ being the adimensional incoherent motion parameter defined by (91), and as

$$H_1^\alpha \propto \frac{1}{1 + i\omega\tau} + \frac{8}{\pi^2} \sum_{\ell = \text{odd}} \frac{1}{\ell^2 I_\ell(0) I_\ell(\omega\tau)} \qquad (103)$$

with

$$I_\ell(0) \equiv \{(1 + \alpha\tau)^2 + \ell^2\xi_c^2\}^{\frac{1}{2}} - \alpha\tau$$
$$I_\ell(\omega\tau) \equiv \{(1 + i\omega\tau + \alpha\tau)^2 + \ell^2\xi_c^2\}^{\frac{1}{2}} - \alpha\tau \qquad (104)$$

for the intermediate coherence case with scattering rate α. In the limits $\alpha\tau \to 0$ and $\alpha\tau \to \infty$ one retrieves the completely coherent and diffusive behaviors (99) and (101), respectivley.

The changes (64) in the phase-lag due to exciton motion have been obtained [59] for the first harmonic of the delayed fluorescence signal with the harmonic amplitudes (99), (101) and (103) for the three types of motion. Fig. 14 illustrates the big difference, in amplitude and frequency domain behavior, to be expected in the $\Delta\phi$ data obtained in the completely coherent and diffusive cases. The solid line curves correspond to the coherent motion, the dashed ones to the completely incoherent case. These last ones exhibit, of course, the same behavior as that already shown in Fig.8. Pairs labelled in Fig.14 with same letter correspond to same L_T/x_0 ratio for the two types of motion. The effects of coherence become especially noticeable, well above the typical signal-to-noise ratios, if in the experiment one uses modulation frequencies and ruling periods such that $\omega\tau$ lies in the 2.0 to 5.0 and L_T/x_0 in the 0.1 to 0.6 range for the crystal under study. Fig.15 illustrates the intermediate coherence behavior as predicted by (103). The $\Delta\Phi$ behavior becomes experimentally undistinguishable from the diffusive case when the scattering rate is such that $\alpha\tau \gg 10^3$. As already stressed in Section V, phase-lag experiments constitute the preferable tool to probe triplet exciton motion via delayed fluorescence. As seen in Fig.14 coherence makes the observable effects bigger with a maximum of $\sim 8°$, a value which never can be observed in this geometry with diffusive motion. A typical, 5 to 10%, spurious stray light effect, if taken into account by using coefficients (26) in (77), will reduce nearly uniformly the size of the $\Delta\Phi$'s plotted as functions of $\omega\tau$, but the different positions of the maxima would still allow to discriminate between coherent and diffusive effects [59].

The model discussed in this section must be now generalized to three dimensional crystals containing several translationally inequivalent molecules in the unit cell, at least for the simple case of materials with two molecules per unit cell in the C_{2h}^5 symmetry as anthracene and naphthalene. A first step in this direction has already been taken by an analysis of the case of a simple orthorombic crystal [60]. The analysis given above for the one-dimensional case could be experimentally tested in materials in which one of the triplet exciton transfer elements is known to be much bigger than the other ones, so that, motion can be considered as nearly one-dimensional [9]-[11], [40], [43], [44]. For these materials

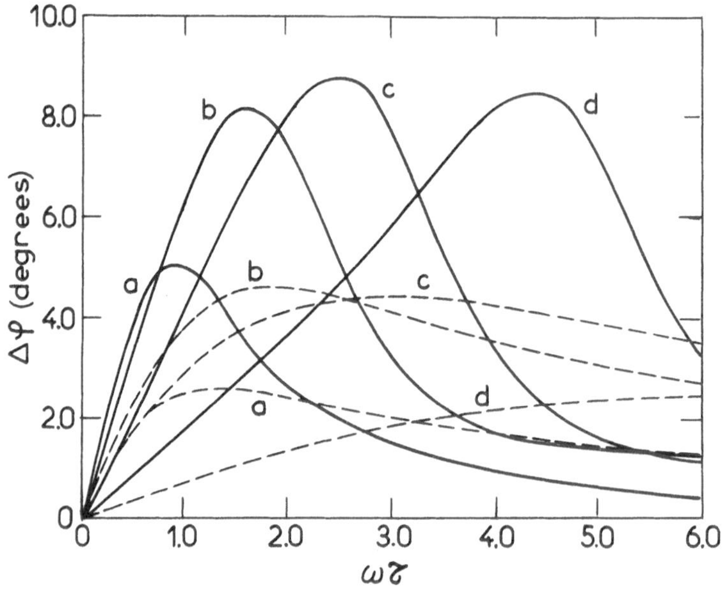

Fig. 14. Phase-lag differences as predicted for the coherent (solid line curves) and diffusive (dashed curves) motions. Same letter labels curves with same L_T/x_0 ratio. Pairs a,b,c,d are for ratios L_T/x_0 = 0.11, 0.22, 0.33, and 0.55, respectively.

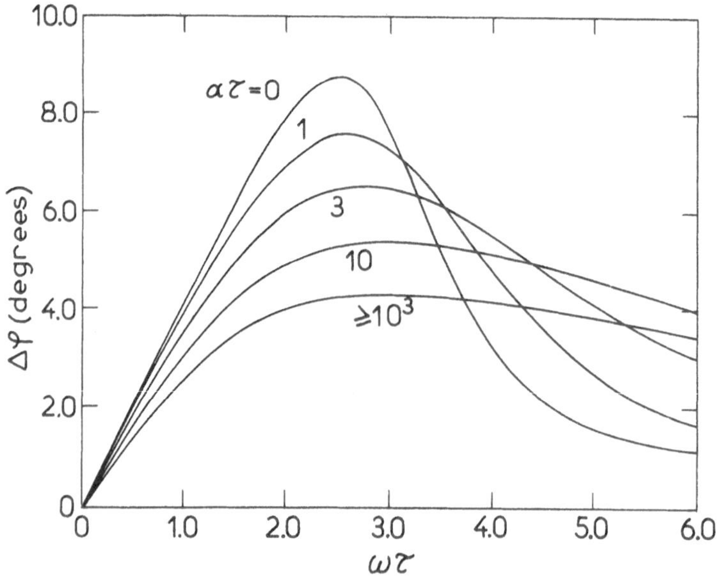

Fig. 15. Phase-lag differences for the intermediate coherence motion case plotted with same L_T/x_0 = 0.35 ratio for all curves with the values of $\alpha\tau$ as shown.

the challenge is to be able to grow highly purified and crystallographically perfect crystals of appropriate size so that secondary factors, for instance, trapping at low temperature, would not mask coherence effects in the time-dependence of delayed fluorescence.

REFERENCES

1. M. Pope and C.E. Swenberg, "Electronic Processes in Organic Crystals", Oxford University Press, New York (1982), and references therein.
2. H. Haken and G. Strobl, in "The Triplet State", A.B. Zahlan ed., Cambridge University Press, Cambridge (1967).
3. R. Silbey, Ann. Rev. Phys. Chem. 27, 203 (1976).
4. V.M. Kenkre, in "Exciton Dynamics in Molecular Crystals and Aggregates", G. Höhler ed., Springer Verlag, Berlin (1982).
5. P.Reineker, in "Exciton Dynamics in Molecular Crystals and Aggregates", G. Höhler ed., Springer Verlag, Berlin (1982).
6. H.C. Wolf, in "Advances in Atomic and Molecular Physics", D.R. Bates and I. Estermann eds., Vol.3, Academic Press Inc. New York (1967).
7. A.H. Francis and R. Kopelman, in "Excitation Dynamics in Molecular Solids", Topics in Applied Physics, W.M. Yen and P.M. Selzer eds., Springer Verlag, Berlin (1981).
8. H.C.Brenner, in "Triplet State ODMR Spectroscopy", R.H. Clarke ed., John Wiley, Interscience, New York (1982).
9. C.B. Harris and D.A. Zwemer, Ann. Rev. Phys. Chem. 29, 473 (1978)
10. D.M. Burland and A.H. Zewail, Adv. Phys. Chem. 50, 385 (1980); D.D. Smith and A.H. Zewail, J. Chem. Phys. 71, 3533 (1979).
11. A.J. van Strien and J. Schmidt, Chem. Phys. Letters 86, 203 (1982); A.J. van Strien and J. Schmidt, in "Physics and Chemistry of Molecular Structures", I. Zschokke-Gränacher ed., D. Reidel Publishing Co., Dordrecht, Holland (in press).
12. D. Haarer and H.C. Wolf, Mol. Cryst. Liquid Cryst. 10, 359 (1970)
13. R.C. Johnson and R.E. Merrifield, Phys. Rev. B1, 896 (1970).
14. A. Suna, Phys. Rev. B1, 1716 (1970).
15. See, for example, L. Altwegg and I. Zschokke-Gränacher, Phys. Rev. B20, 4326 (1979); J. Rosenthal, L. Yarmus and N.F. Berk, Phys. Rev. B24:1103 (1981), Phys. Rev. B23, 5673 (1981); L. Altwegg and I. Zschokke-Gränacher, Phys. Rev. B24, 1106 (1981) and references therein.
16. A.I. Kitaigorodskii, "Organic Chemical Crystallography", Consultants Bureau, New York (1961).
17. R.S. Knox, "Theory of Excitons", in "Solid State Physics", F. Seitz and D. Turnbull eds., Supplement 5, Academic Press, New York (1963); D.P. Craig and S.H. Walmsley, "Excitons in Molecular Crystals", W.A. Benjamin, Inc., New York (1968); A.S. Davydov, "Theory of Molecular Excitons", Plenum Publishing Co., New York (1971).
18. S.A. Rice and J. Jortner, in "Physics and Chemistry of the Organic Solid State", D. Fox, M.M. Labes and A. Weissberger eds.,

Vol.3, Interscience Publishers-John Wiley, New York (1967); M. Philpott, Adv. Chem. Phys. 23,227 (1973).
19. G.W. Robinson, Ann. Rev. Phys. Chem. 21,429 (1970).
20. R. Kopelman, J. Chem. Phys. 47,2631 (1967); S.D. Colson, R. Kopelman and G.W. Robinson, J. Chem. Phys. 47,27 (1967); J. Hoshen, R. Kopelman and J. Jortner, Chem. Phys. 10.185 (1975).
21. V.Ern, Chem. Phys. 25,307 (1977).
22. R.E. Merrifield, J. Chem. Phys. 23,402 (1955).
23. A. Tiberghien and G. Delacôte, J. Phys. 31,637 (1970); A. Tiberghien, G. Delacôte and M. Schott, J. Chem. Phys. 59,3762 (1972); D.P. Craig and B.S. Sommer, Chem. Phys. Letters 22, 239 (1973).
24. S.P. McGlynn, T. Azumi and M. Kinoshita, "Molecular Spectroscopy of the Triplet State", Prentice-Hall, Inc., Englewood Cliffs, N.J. (1969).
25. R.H. Clarke and R.M. Hochstrasser, J. Chem. Phys. 46,4532 (1967) R.M. Hochstrasser, J. Chem. Phys. 47,1015 (1967); R.M. Hochstrasser and T. Lin, J. Chem. Phys. 49,4929 (1968).
26. H. Sternlicht and H.M. McConnell, J. Chem. Phys. 35,1793 (1961) H. Sternlicht, G.C. Nieman and G.W. Robinson, J. Chem. Phys. 38,1326 (1963).
27. P. Avakian and R.E. Merrifield, Phys. Rev. Lett. 13,541 (1964).
28. P.Avakian, V. Ern, R.E. Merrifield and A. Suna, Phys. Rev. 165,974 (1968).
29. D.M. Hanson and G.W. Robinson, J. Chem. Phys. 43,4174 (1965).
30. V. Ern, P. Avakian and R.E. Merrifield, Phys. Rev. 148,862 (1966).
31. V. Ern, Phys. Rev. Lett. 22,343 (1969).
32. V. Ern, A. Suna and R.E. Merrifield, J. Appl. Phys. 42,2770 (1971).
33. V. Ern and M. Schott, in "Localization and Delocalization in Quantum Chemistry", O. Chalvet, D. Reidel Publishing Co., Dordrecht, Holland (1976).
34. B.Nickel, Berichte der Bunsen. Gesell. 76,582 (1972); ibid. 76,584 (1972).
35. J.R. Salcedo, A.E. Siegman, D.D. Dlott and M.D. Fayer, Phys. Rev. Lett. 41,131 (1978).
36. K.A. Nelson, R. Casalegno, R.J. Dwayne Miller and M.D. Fayer, J. Chem. Phys. 77,1144 (1982).
37. V. Ern, A. Suna, Y. Tomkievicz, P. Avakian and R.P. Groff, Phys. Rev. B5,3222 (1972).
38. A. Fort and V. Ern, Chem. Phys. Lett. 74,519 (1980).
39. H. Port and D. Rund, Chem. Phys. Lett. 69,406 (1980).
40. V. Ern, J. Chem. Phys. 56, 6259 (1972).
41. E. Schwarzer, Thesis Dissertation, University of Stuttgart, Germany (1974); H. Port, Thesis Dissertation, University of Stuttgart, Germany (1974).
42. M. Chabr and I. Zschokke-Gränacher, J. Chem. Phys. 64, 3093 (1976).
43. R.M. Hochstrasser and C.D. Whiteman, J. Chem. Phys. 56, 5945

(1972).
44. R. Schmidberger and H.C. Wolf, Chem. Phys. Lett. 16,40 (1972); ibid. 25,185 (1974).
45. V. Ern, H. Bouchriha, M. Schott and G. Castro, Chem. Phys. Lett. 29,453 (1974).
46. J.B. Aladekomo, S. Arnold and M. Pope, Phys. Stat. Sol. B80, 333 (1977).
47. N.F. Berk, J. Rosenthal, L. Yarmus and C.E. Swenberg, Phys. Stat. Sol. B83,K1 (1977).
48. H.Bouchriha, V. Ern, J.L. Fave, C. Guthmann and M. Schott, Phys. Rev. B18,525 (1978); J. Physique 39,257 (1978).
49. V. Ern and A. Fort, Mol. Cryst. Liq. Cryst., 10C,1 (1983).
50. S. Arnold, J.L. Fave and M. Schott, Chem. Phys. Lett. 28,412 (1974).
51. S. Arnold, W.B. Whitten and A.C. Damask, Phys. Rev. B3,3452 (1971); W.B. Whitten, S. Arnold and C.E. Swenberg, J. Chem. Phys. 60,4219 (1974).
52. H. Port and K. Mistelberger, J. Lum. 12/13,351 (1976).
53. N.F. Berk, W. Bizzaro, J. Rosenthal and L. Yarmus, Phys. Rev. B23,5661 (1981); N.F. Berk, J. Rosenthal and L. Yarmus, Phys. Rev. B28,4963 (1983).
54. V.M. Kenkre, V. Ern and A. Fort, Phys. Rev. B28,598 (1983).
55. V.M. Kenkre and R.S. Knox, Phys. Rev. B9,5279 (1974).
56. V.M. Kenkre and Y.M. Wong, Phys. Rev. B23,3748 (1981); V.M. Kenkre and P.E. Parris, ibid. B27,3221 (1983).
57. V.M. Kenkre, Phys. Rev. B22,2089 (1980); Z. Phys. B43,221 (1981).
58. V.M. Kenkre, article in present volume.
59. A. Fort, this Laboratory, in preparation for publication.
60. A. Fort, V. Ern and V.M. Kenkre, Chem. Phys. 80,205 (1983).

ENERGY TRANSFER IN SOLID RARE GASES

N. Schwentner

Institut für Atom- und Festkörperphysik
Freie Universität Berlin
Arnimallee 14, 1000 Berlin 33, Germany West

E. E. Koch

Deutsches Elektronen Synchrotron
Notkestr. 85, 2000 Hamburg 52, Germany West

J. Jortner

Department of Chemistry
Tel Aviv University
6997 Tel Aviv, Israel

ABSTRACT

Three types of energy transport are essential in condensed rare gases: migration of free excitons, transfer by localized centers and mass diffusion of electronically excited centers. The first two processes are observed in solid rare gases, a combination of the last two processes in liquid rare gases. Free and localized exciton states are described and the balance between localization and migration of free excitons is analyzed because it determines the transfer range of free excitons. Photoelectron and luminescence experiments are discussed which monitor the migration of free excitons. Förster-Dexter type of energy transfer for localized centers is illustrated by host to guest and guest to guest electronic and vibrational transfer. The competition between transfer and nonradiative relaxation is crucial and examples for transfer prior to electronic or vibrational relaxation are presented. The long lifetime of triplet states especially in liquid helium causes transfer ranges of the order of centimeters. Finally the application of emission from selftrapped excitons in rare gas crystals for vacuum ultraviolet solid state lasers is shown. The efficiency is determined by losses due to energy transport. At these high excitation densities an extremely hot electron plasma is created in the crystal which induces additional transport and scattering processes.

I. INTRODUCTION

The atoms in rare gas crystals are densely packed by Van der Waals forces in a fcc structure. The structural properties follow from the gas phase pair potentials. The crystals are insulators and are transparent far into the vacuum ultraviolet spectral region. The onset of absorption is dominated by strong exciton bands
The group velocities of free excitons in solid rare gases are large compared to those of other typical Van der Waals crystals like for example organic molecular crystals. Therefore a fast wave like or diffusive transport of energy by free excitons is feasible as has been pointed out quite early by Gould and Knox [1]. An excited rare gas atom is highly reactive in contrast to an inert ground state atom because the excited electron resembles the lone electron of an alkali atom and the remaining hole in the valence shell corresponds to the partly filled shell of an halogen atom. In the gas phase strongly bound excimers between the excited atom and a ground state atom are formed. The reactive excited atoms in the rare gas crystal induce local structure changes which are similar to excimers or to bubbles. These structural changes destroy the free exciton states and localized excitons are formed. The various types of excitons provide a case study for the competition of wave like and diffusive energy transport by free excitons, localization of excitons and transport by localized excitons. Furthermore the structural and electronic properties of rare gases change rather smoothly in going from the gas phase through the liquid phase to the solid phase. Therefore the influence of mass diffusion, density and structural order or disorder on energy transport can be studied for the whole range of densities.

Rare gas crystals are easily doped with atoms or molecules by codeposition on a cold substrate. The electronic and vibrational states are in general preserved in the rare gas matrix due to the inertness and softness of the matrix. The coupling to the matrix is weak in many cases and radiationless depopulation of excited electronic and vibrational guest states is comparatively slow. The resulting long lifetimes of the excited states present ideal conditions for energy transfer from even high excited electronic and vibrational host or guest states to guests. By an appropriate choice of guest and matrix the dependence of energy transfer on the initially prepared state, the coupling to the surrounding and on the phonon spectrum of the matrix can be studied in a systematic way.

The emission bands of pure solid rare gases are nearly identical with the gas phase excimer bands which are used in commerical excimer laser for the VUV. Rare gas crystals are the only materials which could be used for solid state VUV lasers. Doped rare gas crystals could provide in principle a way to cover the range from $\hbar\omega$ = 17 eV to the visible with solid state lasers. The efficiency depends on energy transport and energy dissipation in the excited electronic states of the crystal and on the transfer to the dopands.

II. ELEMENTARY EXCITATIONS OF RARE GAS CRYSTALS

II.A. Lattice Vibrations

The excess energy in radiationless relaxation processes is dissipated into phonons. The fcc lattice with a basis of one atom allows only acoustic phonons with one longitudinal and two transversal branches. The weak Van der Waals bonding results in soft crystals with low phonon energies of the order 5-10 meV (Fig. 1). The structural changes in excited electronic states of rare gas crystals will be responsible for local modes. The compression of the surrounding at an excited bubble state causes a breathing mode. The stretching modes in an excimer center will have energies which are similar to the free excimer vibrational energies. Atomic dopands introduce in general even in the ground state local modes due to the different masses, sizes and Van der Waals binding energies. For molecular guests the internal vibrational modes can be considered as local modes. Rotation of guest molecules is often hindered and librational modes will appear in addition. The spectra of local modes cannot be presented in a general way like in Fig. 1 since they depend strongly on the special combination of host, guest and electronic state.

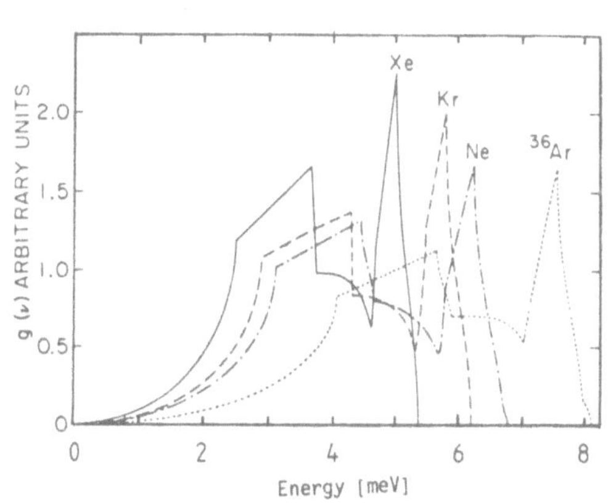

Fig. 1. Density of phonon states of rare gas crystals versus phonon energy [2], [3].

II.B. Resonant Electronic States

A description of the physical properties of a solid starts with the total Hamiltonian for all ions, electrons and their mutual interactions. The resonant electronic states follow from band structure calculations in which a periodic potential is assumed given by the ions at their equilibrium positions and in addition the one electron approximation is used. The band structure calculations yield the electron energy E(k) versus the wave vector k of the electron. An example for Xe (Fig. 2) shows the valence bands around -10 eV, originating from the outer atomic p levels. The upper two levels with a total angular momentum $j = 3/2$ of the hole are degenerated at the center of the Brillouin zone Γ. At Γ the $j = 3/2$ and the lower lying $j = 1/2$ bands are separated by a spin orbit splitting of $\approx$ 1.3 eV. The considerable dispersion i.e. width of the individual valence bands of the order of 1 eV has been confirmed experimentally by photoelectron spectroscopy [7]. These large widths (Table 1) indicate a significant overlap of the p wave

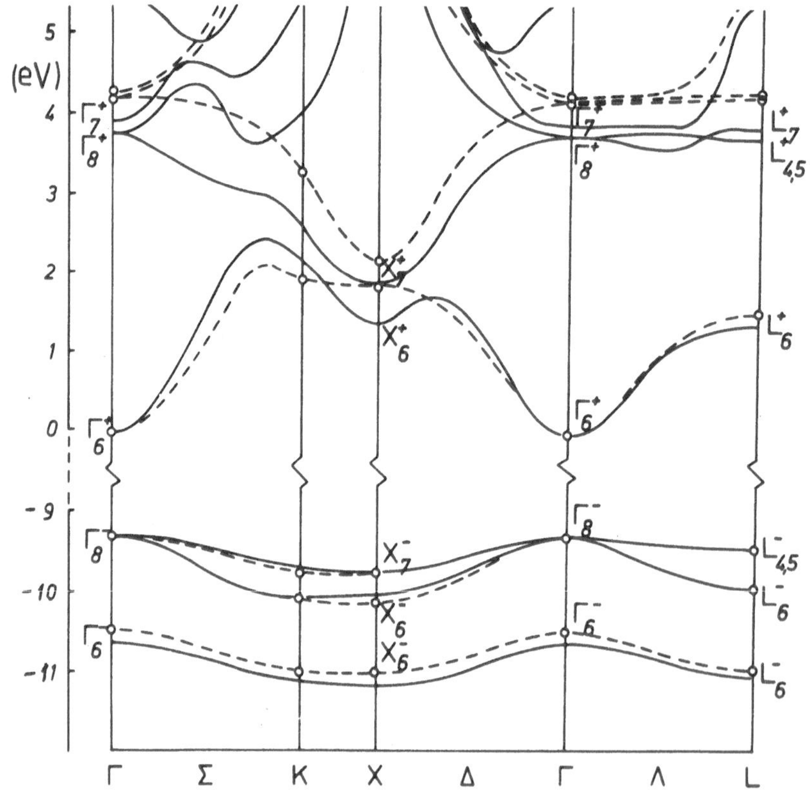

Fig. 2. Two band structure calculations for Xe |4,5| compiled by Rössler [6].

Table 1. Band structure parameters like band gap energy E_G, valence band width VBW, light and heavy hole effective masses m_h^l, m_h^h and electron effective mass m_e. Also exciton binding energy B and the lattice relaxation energy for eximer centers E_{LR} are given.

		Ne	Ar	Kr	Xe
E_G (eV)[a]		21.58	14.16	11.61	9.33
VBW (eV)		0.4	0.7	0.9	0.9
m^l	in free [c]	6.16	1.93	1.21	
m_h^h	electron [c]	24.8	11.6	7.66	
m_e	mass units	0.802[c]	0.488[c]	0.418[c]	0.35[d]
B	(eV)	5-6.93[a]	2.36[e]	1.53[e]	1.02[e]
E_{LR}	(eV)[b]	2.0	1.86	1.38	0.85

[a]Ref. [11]

[b]Ref. [20]

[c]A.B. Kunz and D.J. Miklisch, Phys. Rev. 8, 779 (1973)

[d]Ref. [4]

[e]V. Saile, W. Steinmann and E.E. Koch in Extended Abstracts, V. International Conference on VUV Radiation Physics, Montpellier Vol. 1, p. 199 (1977)

functions of neighbouring atoms in the crystal and establish a rather high group velocity of these resonant hole states. The lowest empty conduction band states are separated from these occupied valence bands by a large band gap E_G of the order of 10 eV (Table 1). Near Γ the lowest conduction band is free electron like (Fig. 2) and has s-symmetry. The contributions of additional bands complicate the picture at higher energies. The calculations and experiments differ in details (Fig. 2) but they all give this general scheme and the trends in the band gap energies, valence band widths and effective masses of the light m_h^l and heavy m_h^h holes and of the electrons m_e at Γ (Table 1).

The partly screened Coulomb interaction between an electron in the conduction band and a hole in the valence band leads to a hydrogen like series of bound states within the band gap. Two such

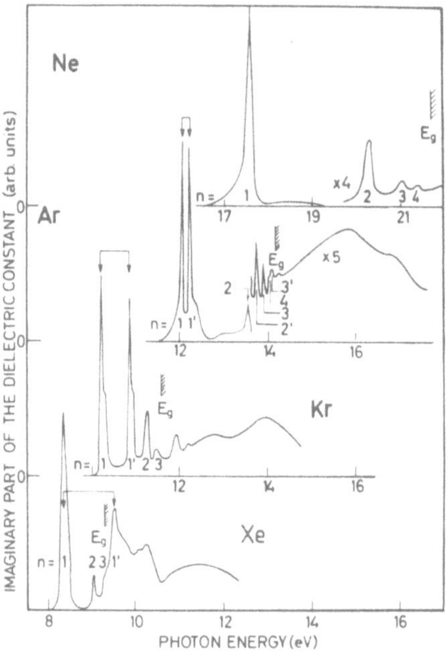

Fig. 3. Absorption bands of solid Ne, Ar, Kr, Xe. Eg: band gap; n and n' are j = 3/2 and j = 1/2 exciton bands [8].

series of excitations converging to the bottom of the conduction band are observed in absorption and reflection spectra (Fig. 3). These two series are separated by the spin orbit splitting of the j = 3/2 and j = 1/2 hole states. The wave function of an exciton can be constructed from valence band and conduction band wave functions [6],[9].The continuum approximation of the Wannier model is appropriate for excitons with a separation of the hole and electron which is large compared to the nearest neighbour separation of the atoms. Within the effective mass approximation it predicts for the excitation energies E_n:

$$E_n = E_G - \frac{B}{n^2} + \frac{\hbar^2 k^2}{2M} \qquad n = 1,2,3 \ldots \tag{1}$$

with an exciton "Rydberg constant"

$$B = \frac{\mu e^4}{2\hbar^2 \varepsilon^2} \tag{2}$$

ε is the dielectric constant, μ is the reduced effective mass defined by

Table 2. Energies of bulk and surface excitons and of the quantum defect δ (eV).

		Xe 5p [a]		Kr 4p [a]		Ar 3p [a]		Ne 2p [b]	
	j =	3/2	1/2	3/2	1/2	3/2	1/2	3/2	1/2
	n = 1	8.37	9.51	10.17	10.86	12.06	12.24	17.36	17.50
	2	9.07		11.23	11.92	13.57	13.75	20.25	20.36
bulk	3	9.21		11.44	(12.21)	13.87	14.07	20.94	21.02
	4			11.52		13.97		21.19	21.29
	5							21.32	
	n = 1	8.21		9.95	10.68	11.71	11.93		17.15
				10.02		11.81			
surface	2			11.03			12.99		
							13.07		
	δ	-0.03-0.06 [c]		0.08-0.17 [c]		0.21-0.28 [c]		0.28 [b]	

[a] see [e] Table 1

[b] Ref. [11]

[c] L. Resca, R. Resta and S. Rodriguez, Sol. State Comm. 26, 849 (1978)

$$\frac{1}{\mu} = \frac{1}{m_e} + \frac{1}{m_h} \tag{3}$$

and M is the total exciton mass

$$M = m_e + m_h \tag{4}$$

The last term in (1) describes the center of mass motion of the exciton as a plane wave with wave vector k and group velocity $\hbar k/M$. In optical spectra only $k \underset{\sim}{\sim} 0$ excitons are excited because of the small wave vector of the light compared to the extension of the Brillouin zone. The hydrogen like term B/n^2 fits the exciton series for $n \geq 2$ (Fig. 3) nicely with the binding energies B given in Table 1. The radius of the electron orbit for the $n = 1$ excitons is smaller than the nearest neighbour separation and the Frenkel model should be used in this case. This model, which is more complicated, starts with an excited atom at a given lattice site. A calculation for the dispersion of the $n = 1$ and $n' = 1$ excitons in Ar has been carried out by Knox [9],[10] within the framework of the Frenkel model (Fig. 4). The $n = 1$ exciton state in optical spectra corresponds to the lower Δ_5 transverse branch, the $n' = 1$ exciton to the upper Δ_5 transverse branch near $k = 0$. The absolute energies are about 3 eV to low due to convergence problems in the calculation. But this picture shows that in optical spectra only some points of the three-dimensional exciton dispersion curves are accessible and several branches are completely missing in optical spectra. Many attempts have been made to derive a consistent description of all exciton bands (Table 2) for this situation which is intermediate between the Wannier and Frenkel case [11]. One approximation uses a central cell correction in (1) for the $n = 1$ excitons, another introduces a quantum defect δ in (1) of the form (Table 2)

$$E_n = E_G - B/(n-\delta)^2 \tag{5}$$

Besides these bulk exciton states also surface exciton states are observed with energies about 0.2 eV smaller than the bulk states [11] (Table 2). These exciton states are restricted to the surface and can act as sinks in energy transport processes to the surface.

II.C. Localized Electronic States

Up to now a periodic potential for the electrons due to a periodic arrangement of the ions has been assumed. An electronic excitation in an insulator causes local distortions of the electron distribution which induces also structural changes. To get a feeling for the new equilibrium configurations which are favoured by electronically excited solid rare gases we make use of the smooth change of the electronic properties with density. We start with the interaction of a free electron in the conduction band with the surrounding neutral atoms. The energy $V_0 = T+U$ of an excess electron at the bottom of the conduction band relative to the vacuum level consists of an attractive polarisation energy U and a repulsive kinetic energy term T arising from multiple scattering at nearest neighbours [12]. This energy V_0 is required to inject an electron from the vacuum into the condensed rare gas. A positive value of V_0 indicates a dominating repulsive interaction and the electron can lower its energy by pushing away neighbouring atoms thus creating an empty sphere, i.e. a bubble. In this way the electron becomes localized. The trend to form bubbles decreases in going from the light rare gases to the heavy rare gases and for Kr and Xe the

Table 3. Electron polarisation energy V_o of solid rare gases together with energetic and structural parameters of rare gas molecules. D_+ molecular ion dissociation energy, r_k: nearest neighbor separation in a crystal; r_o, r_*, r_+: internuclear distance of rare gas molecules in the ground state, excited state and of molecular ions respectively (compiled in Ref. [15]).

	He	Ne	Ar	Kr	Xe
V_o (eV)	1.05 (liquid)	1.3	0.4	-0.3	-0.4
D_+ (eV)	2.67	1.2	1.25	1.15	0.99
r_K (Å)		3.156	3.755	3.992	4.335
r_o (Å)	2.96	3.102	3.761	4.006	4.361
r_* (Å)	1.04	1.79	2.42		3.04
r_+ (Å)	1.06	1.75	2.43	2.79	3.04

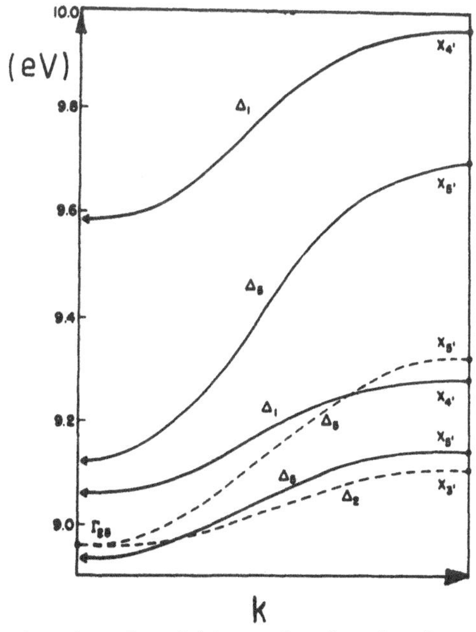

Fig. 4. Exciton bands of solid Ar in (100) direction. Δ_5: transverse exciton; Δ_1: longitudinal excitons; dashed lines: forbidden branches due to $^3P_2$ atomic states [9[,[10].

attractive polarisation term dominates (Table 3). The bubbles around excess electrons in liquid He are of considerable size (∿ 30 Å) and have been studied extensively [13].

A rare gas ion Rg^+ is chemically aggressive and tends to form strongly bound excimer ions Rg_2^+ with dissociation energies D_+ of the order of 1-2 eV (Table 3). The bond length r_+ of the excimer ion is about 30 % smaller than the position of the ground state Van der Waals minimum r_o and the nearest neighbour separation r_k of the crystal in the ground state (Table 3). An ion in the crystal lattice i.e. a hole tries also to form an excimer ion by attracting one of the nearest neighbours and thus becomes localized as an R_2^+ center. An R_2^+ hole tries to capture an electron resulting in an R_2^* excimer center in the crystal. The local structure at this center depends on the mainly repulsive interaction of the orbiting electron with surrounding atoms and on the attractive $R-R^+$ interaction. The orbiting electron is bound in Rydberg states R_2^*. The electron on its large orbit has little influence on the structure of the R_2^* center. The dissociation energies and the equilibrium separations r_* remain essentially that of the excimer ion R_2^+ (Table 3). The Rydberg states of the electron form a series converging to the ionisation energy similar but more complicated like the free exciton series. A repulsive interaction of the electron with the surrounding will cause bubbles around the R_2^* centers similar to the excess electron bubbles. In addition also atomic excited states Rg^* within bubbles are observed.

Liquid He is a good example for illustrating Rg^* and Rg_2^* centers in bubbles. The arrows in Fig. 5 indicate experimentally investigated transitions in atomic He^* and molecular He_2^* bubble centers. The sizes and shapes of the bubbles are derived from the shift of transition energies relative to the gas phase values and from the width of the bands together with theoretical predictions. Bubble diameters of 10-20 Å are found compared to an average interatomic spacing in liquid He of 3 Å. The bubble diameters for the atomic 2s ^3S, 2s ^3P, 3s ^3S and 3s ^1S states are 12 Å, 12 Å, 22 Å, and 26 Å, respectively. Furthermore the bubble diameter can be reduced by external pressure [14], [15].

In solid rare gases we have to deal with free electrons and localized electrons in bubbles, free holes and localized holes in Rg^+ and Rg_2^+ centers, free excitons and localized excitons in Rg^* and Rg_2^* configurations. Typical recombination times are of the order of 10^{-9} s or shorter. Thus free excitons and localized excitons in one center Rg^* and two center Rg_2^* configurations are the most important species. The contribution of free and localized centers to energy transport is rather different. Therefore the time scales and the efficiencies for localization of free excitons in the two types of centers have to be discussed.

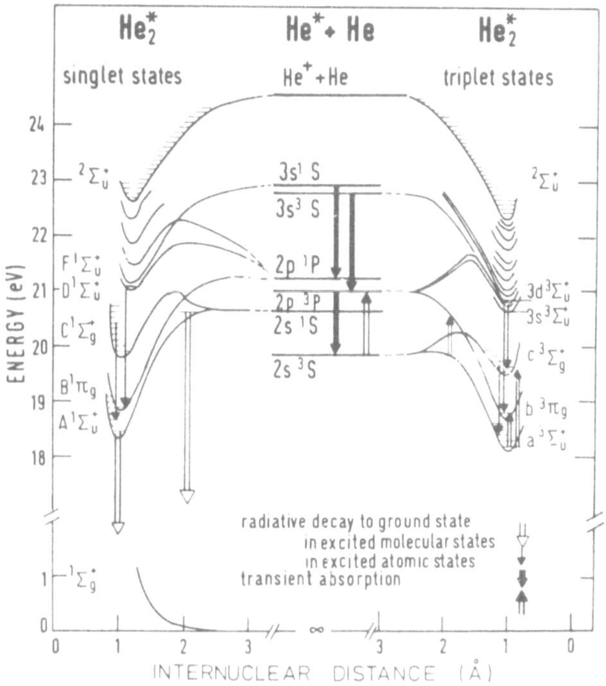

Fig. 5. Potential curves and observed transitions of liquid He compiled in [15].

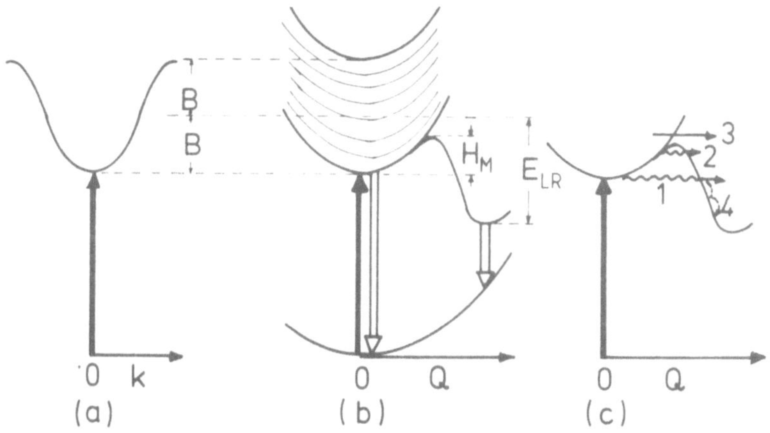

Fig. 6. Scheme for a) dispersion of free excitons versus wavevector k, b) selftrapping versus configuration coordinate Q and c) tunneling through selftrapping barrier [17], [18].

II.D. Localization (Self-Trapping) of Excitons

The fate of an exciton will be analyzed in subsequent steps. Is the initially created exciton free or localized? Does a free exciton remain free or does it become localized before it decays radiatively or by an energy transfer process? Finally we ask for a microscopic description of the localized states. This field has been treated theoretically by several groups [9], [15], [20].

1. Exciton-Phonon Scattering. The exciton-phonon scattering time τ and the line width Γ of an exciton are determined by the balance between the local nature characterized by D and the wave like nature characterized by B' [18].

$$\frac{2\Gamma}{\hbar} = \frac{1}{\tau} = \frac{\pi}{\hbar}\,\frac{D^2}{B'} \tag{6}$$

The width of the exciton dispersion curve 2B'(Fig. 4) corresponds essentially to the valence band width (Table 1). D^2 represents the statistical fluctuations of the exciton energy due to nuclear motion. D^2 follows from the energy gain E_{LR} by a lattice relaxation into a new local equilibrium structure.

$$D^2 = \begin{cases} E_{LR} \cdot \hbar\omega_p & \text{for } kT \ll \hbar\omega_p \\ 2E_{LR} \cdot kT & \text{for } kT \gg \hbar\omega_p \end{cases} \tag{7}$$

$\hbar\omega_p$ is a mean phonon energy (Fig. 1). The motion of the exciton is slow compared to the thermal fluctuations in the case of strong scattering $D \gg B'$ and the exciton will be localized. In the weak scattering case $D \ll B'$ the exciton is fast and stays only a small fraction of time on each lattice site on the time scale of nuclear motion. Thus the effect of nuclear motion is smeared out and the exciton remains free. E_{LR} values have been estimated semiempirical from the Stokes shift of luminescence bands for Rg_2^* centers (Table 1). A comparison of the D values with 2B' shows that the heavier rare gase like Xe correspond to the weak scattering limit ($D \approx 0.07 \ll 2B' \approx 0.9$) whereas Ne with $D \approx B'$ is still on the verge of weak scattering.

The local or wave like character follows also from the line shape of exciton absorption bands. The line shape is Lorentzian and the half width Γ is given by (6) for the case of weak scattering. Strong scattering results in a Gaussian line shape with the half width fwhm given by 2.355D. Unfortunately the line shapes are partly obscured by longitudinal excitons, overlap of spin orbit partners and surface excitons. The line shape seems to be Lorentzian with a fwhm of $\sim$ 80 meV for the n = 1 excitons of Xe, Kr and Ar [11]. Thus the heavier rare gases belong also according to this criterion to the weak scattering case with phonon scattering times of the order of $\tau \sim 10^{-15}$ s. The larger line width of $\sim$ 200 meV for Ne [11] perhaps indicates a transition towards the strong scattering case.

2. Self-Trapping. Does an initially free exciton become self-trapped during its lifetime? A free exciton remains stable from the point of view of energy minimalization if the energy gain due to the free wave like propagation exceeds the energy gain given by the lattice relaxation $B' > E_{LR}$ or vice versa (Fig. 6). This criterion allows localization even if the scattering criterion predicts the excitation of initially free excitons. But besides $E_{LR} > B'$ it is necessary for localization that the localization time is shorter than the radiative lifetime of the exciton (10^{-6} to 10^{-9} s). The energy balance $E(\alpha)$ is given by [8]:

$$E(\alpha) = B'\alpha^2 - E_{LR}\,\alpha^3 \tag{8}$$

for an exciton whose extension $r \sim \frac{1}{\alpha}$ is described by a Gaussian wave function with orbital exponent α. $\alpha = 0$ ($r \to \infty$) represents a free exciton state and $\alpha = 1$ a localized exciton. $E(\alpha)$ increases with α near $\alpha = 0$ because more energy is required to squeeze the free exciton compared to the energy gain delivered by the lattice deformation. But after crossing a maximum of height H_M the energy is reduced below $E(\alpha = 0)$ provided $E_{LR} > B'$.

$$H_M = 4B'^3/27\,E_{LR}^2 \tag{9}$$

The barrier H_M prolongates the localization time of excitons and a coexistence of free and localized excitons becomes possible. The efficiency ratio of for example energy transfer by free or localized electrons follows from the rate constant for crossing the barrier scaled by the transfer rate. The calculation of quantitative rates for localization by tunneling 1, thermoactivated tunneling 2, over barrier transitions 3 and reverse tunneling in competition to relaxation of hot localized excitons 4 is a difficult task (Fig. 6). Approximate expressions [17],[21[contain a prefactor representing a mean phonon frequency and the transparency Tr. $Tr = \exp(-2\,S)$ results for process 1 in Fig. 6 with a quasiclassical action S and $Tr \sim (1-T/\Theta)^{-\alpha}$ for process 2 with a density α of vibrational modes and the Debye temperature Θ. This treatment and also the elaborate calculations of Nasu and Toyozawa [22] yield a high tunneling rate for the solid rare gases compared to the radiative lifetimes. The tunneling rate should decrease with the exciton band width, i.e. in going from Ne to Xe. Thus luminescence originates mainly from localized excitons and the rather spurious contributions from free excitons should be strongest for Xe. But for fast energy transfer processes the contribution of free excitons can be quite important.

3. Microscopic Picture. A quantitative description requires a microscopic picture for the localized states in addition to the general criteria delivered by the contiuum models discussed above. A microscopic model [23],[24], for alkalihalide and rare gas crystals predicts one center and two center self-trapped excitons. The two center state is oriented along the (110) axis. The essential vibrational modes for coupling these centers to the lattice are the breathing mode Q_1 for both centers and the stretching mode in addition for the two center state. The similarity of these centers with the atomic and the excimer bubble centers is obvious.

Luminescence experiments yield broad emission bands (Fig. 7) for solid Ar, Kr and Xe width quantum efficiencies near 1. These bands are nearly identical concerning the transition energies and shapes with the well known gas phase excimer bands [25]. Time resolved luminescence experiments [26], [27] allow even to separate the contributions of the dipole allowed transition $^1\Sigma_u^+ \rightarrow {}^1\Sigma_g^+$ and of the dipole forbidden transition $^3\Sigma_u^+ \rightarrow {}^1\Sigma_g^+$ due to the two lowest excimer states (Fig. 7). The $^3\Sigma_u^+$ state is responsible for the temperature dependent long lifetimes and the $^1\Sigma_u^+$ state for the short lifetimes (Fig. 7). Transient absorption experiments [27], [28] starting from the lowest excited state $^3\Sigma_u$ after the lattice relaxation to excimer centers reveal that also the next higher states in these centers are similar to molecular Rg_2^* states in agreement with calculations [19]. The exciton Rydberg (2) and the continuum states are strongly influenced by the surrounding crystal atoms [27].

In solid Ar, Kr and Xe also spurious signals (Fig. 8) from free excitons (a), atomic bubble centers (b), hot molecular centers (c) with quantum efficiencies of 10^{-2} - 10^{-3} have been reported. The free exciton emission of Xe is quite well established and has been analyzed also in terms of a polariton model [31]. Recently also the temperature dependence of its lifetime has been studied [32]. The assignments of the (a), (b) and (c) emissions for Kr and Ar have to be considered as preliminary because for example also contributions due to surface excitons have been identified [32]. The emission bands of solid Ne (Fig. 8) seem to depend in a complicated way on the excitation conditions. There is no doubt that the main emission band (b) around 16.8 eV is due to an excited atomic bubble state. The nearest neighbour separation in the bubble increases by about 1 Å i.e. 30% to 50% compared to the configuration of the ground state [33], [34], [35]. Transient absorption experiments [36], [37] show that in addition to this fast bubble formation also 3 to 4 vacancies are accumulated at these atoms on a

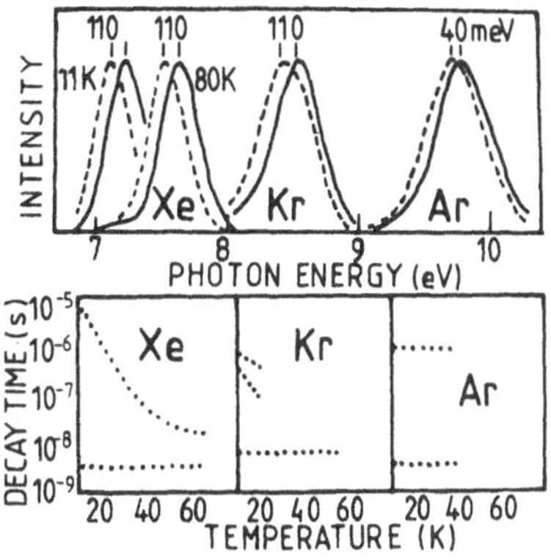

Fig. 7. Results of time resolved luminescence spectroscopy for the vibrationally relaxed selftrapped exciton emission bands [26],[27].

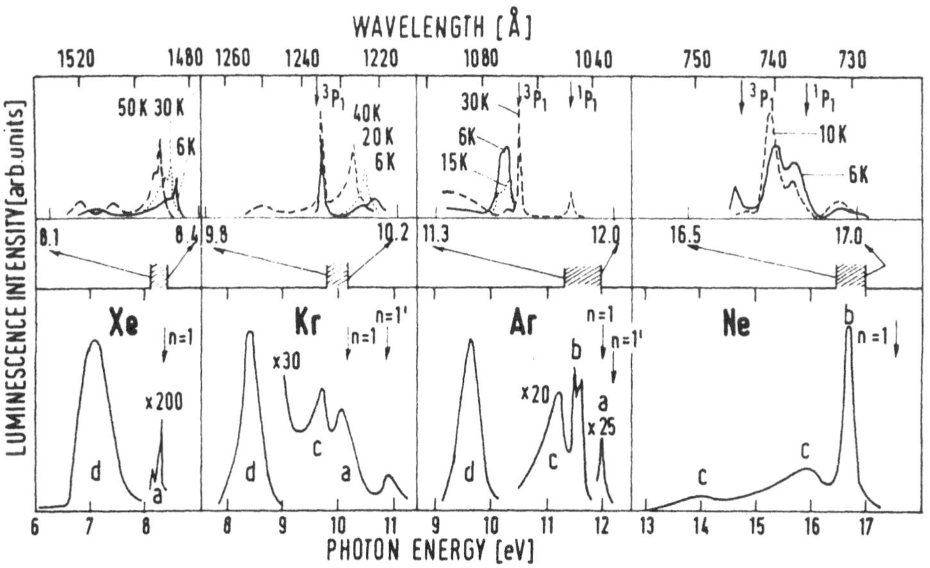

Fig. 8. Lower part: complete emission spectra of Xe [20], Kr, Ar [20]and Ne [29] with a: free excitons, b: atomic centers, c: vibrationally excited excimer centers and d: vibrationally relaxed excimer centers. Upper part: temperature dependence of a and b [10] compiled in [15].

time scale of 10^{-4} s - 10^{-6} s [35], [36. The temperature dependence of a growth rate proves that the vacancies are thermally created and these migrating vacancies are captured by the center. The lower lying emission bands (c in Fig. 8) originate from hot excimer centers. Vibrational relaxation in the excimer center is retarded around the 3 to 5 vibrational levels because the vibrational spacing increases due to the anharmonicity and becomes large compared to the phonon frequencies of the lattice [15],[16],[37].

III. ELECTRONIC STATES OF GUEST ATOMS AND MOLECULES IN RARE GAS MATRICES

III.A. Transition Energies

In this very brief discussion it will be distinguished between Rydberg transtions and valence transitions. In a Rydberg transition of a free atom or molecule an electron is excited to a state with higher main quantum number and its orbit will increase accordingly. The higher Rydberg states converge to the ionisation limit I_g. In the matrix these ionisation energies will be reduced to E_G^i due to the polarisation energies of the hole P^+ and the electron V_o [38]

$$E_G^i = I_g + P^+ + V_o \tag{10}$$

The Rydberg of the free states will be reduced to B^i by the shielding of the Coulomb interaction $(1/\varepsilon^2)$ and by introducing the reduced effective mass instead of the free electron mass like in (7). The reduced effective masses of the guest centers and the host excitons are very similar because the dominant contribution comes in both cases from the dispersion of the host conduction band [39]. Therefore a relation similar to (1)and(5) holds also for the excitation energies E_n^i of guest Rydberg states (Table 4)

$$E_n = E_G^i - B^i \,/\, (n-\delta_i)^2 \tag{11}$$

with $B^i \sim B$ (see eq. 2). The quantum defect δ_i accounts for the deviations from the Coulomb potential within the core of the guest ion and for the small orbit of the first Rydberg states.

The valence transition represent a change of the electronic configuration within the molecule. Therefore the valence transition energies are only weakly changed by the surrounding matrix. The shifts are of the order of 0.01 - 0.1 eV. In general an increasing red shift in going from a Ne matrix to Xe matrix is observed mainly because of the increasing polarisability of the matrix with atomic weight. The relative intensities of different valence transitions are often strongly modified due to a reduced symmetry and additional spin orbit mixing by the matrix.

Table 4. Exciton energies E_n^i of Xe, Kr and Ar guest atoms in Ne, Ar and Kr matrices together with the impurity binding energies B^i and quantum defects δ^i (in eV).

Guest		Matrix Ne[a]	Matrix Ar[b]	Matrix Kr[b]
Xe	n = 1	9.10	9.22	9.01
	n'= 1	10.06	10.53	
	n = 2	11.31	9.97	9.76
	n = 3	12.00	10.25	
	n = 4	12.19	10.36	
	n = 5	12.43		
	n'= 2	12.60	10.80	
	n'= 3	13.34		
	B^i	5.00-6.93	2.36	1.53
	δ^i	0.43[d]		
Kr	n = 1	10.60	10.79	
	n'= 1	11.22	11.36	
	n = 2	13.32		
	n'= 2	13.94		
	n = 3	14.05		
	n = 4	14.48		
	B^i	5.00-6.93	2.36	
	δ^i	0.31[d]		
Ar	n = 1	12.48		
	n'= 1	12.74		
	n = 2	14.84		
	n'= 2	15.25		
	n = 3	15.51		
	n = 4	15.75		
	n'= 3	15.90		
	B^i	5.00-6.93		
	δ^i	0.40[d]		

[a] Ref. [41]
[b] G. Baldini, Phys. Rev. 137A, 508 (1965)
[c] see Tab. 1
[d] L. Resca and R. Resta, Phys. Rev. B19, 1683 (1979)
W. Böhmer, R. Haensel, N. Schwentner and E. Boursey Chem. Phys. 49, 225 (1980)

III.B. Lattice Relaxation and Line Shapes

The optical line shape of transitions localized at guest centers follows from the lattice rearrangement of the matrix around the excited centers [40]. The situation can be illustrated in a configuration coordinate diagram. The separation between the guest and the first shell of matrix atoms might be an appropriate configuration coordinate. The potential surfaces are represented by parabola which are displaced for different electronic states. The size of the displacement determines the strength of the interaction of a specific guest electronic transition with the host lattice vibrations. This strength can be defined by the Huang-Rhys coupling constant

$$S = E_{LR}/\hbar\omega_p \tag{12}$$

which is the number of phonons $\hbar\omega_p$ dissipated during the lattice relaxation around the excited state. Simple analytic expressions are obtained in the linear approximation and for one phonon energy $\hbar\omega_p$. The absorption and emission spectra in the weak coupling limit $S \lessapprox 1$ consist of a zero phonon line followed by one and multiphonon contributions. The intensity I of the zero phonon line relative to the rest of the bands is given by

$$I(T) = \exp(-S \coth(\hbar\omega_p/2\ kT)) \tag{13}$$

The strong coupling limit $S > 1$ yields a Gaussian line shape in absorption and emission with a Stokes shift E_S and widths H(T).

$$E_S = 2\ S\ \hbar\omega_p \tag{14}$$

$$H(T) = \sqrt{8 \ln 2\ S \coth(\hbar\omega/2\ kT)}\ \hbar\,\omega_p \tag{15}$$

Valence transitions usually fall into the class of weak coupling. For Rydberg transitions the interaction of the excited electron with the matrix is strong due to the large orbit. Broad bands with widths of the order of .1 eV and Stokes shifts of the order of 0.5 eV are observed. The interaction of the electron is often repulsive as is indicated by V_o and bubbles around excited Rydberg states with considerable displacements of the neighbouring atoms are formed [41].

IV. ELECTRONIC AND VIBRATIONAL RELAXATION

To pin down the initial state for energy transfer it is necessary to know the fate of the excited states. It is well known that high lying electronic states in solids decay in most cases very fast to the lowest excited electronic states or even to the ground state by converting the excess energy nonradiatively into local phonons and delocalized lattice phonons. Since we are mainly interested in examples where energy transfer can compete with relaxation, the time scale of relaxation has to be examined. Relaxation to a new equilibrium configuration by a continuous change of the configuration coordinate within the same electronic state (Fig. 9) corresponds to a successive emission of the excess energy in single phonon steps. This process is expected to occur on a time scale given by the vibrational frequency and is therefore very fast (10^{-12} - 10^{-13} s). The decay to the next lower electronic state separated by an energy gap ΔE requires the emission of $N = \Delta E/\hbar\omega_p$ phonons in one step. The rate constant W_{nr} for this multiphonon process follows from [15],[16],[43].

$$W_{nr} = A \exp\{-S(2v+1)\} \{(v+1)/v\}^{N/2} \quad I_N\{2S(v(v+1))\}^{1/2} \tag{16}$$

$$A = 2\pi |V_{ab}|^2 / (\hbar^2 \omega_p)$$
$$v = \{\exp(\hbar\omega_p/kT)-1\}^{-1} \tag{17}$$

with the modified Bessel functions $I_N(Z)$ and the electronic matrix element V_{ab}. For large coupling strength S (see eq. 12) it can be shown that the rate decreases exponentially with N and increases with S [43].

$$W_{nr}(T \to 0) = \frac{A}{\sqrt{2\hbar S}} \exp(-S) \exp(-\gamma N) \tag{18}$$

$$\gamma = \ln (N/S) - 1 \tag{19}$$

Conversion of vibrational energy of guest molecules into matrix phonons follows the same scheme [16].

First we discuss the relaxation processes following electron-hole recombination in a pure crystal. In the gas phase recombination of an R_2^+ center with an electron leads to dissociation to an excited atom R^* and several successive recombination-dissociation-recombination processes due to crossings of bonding and repulsive potential curves [44]. A similar series of processes is expected in the R_2^* states of a localized exciton in a crystal because the excimer centers in the crystal show up similar crossings of bonding and repulsive potential curves (Fig. 9). It is interesting to note that dissociation of an excimer center can lead either to an atomic bubble state or even a free exciton again. Since these relaxation processes involve continuous changes of configuration coordinates it is expected that the relaxation down to the lowest excited electronic state of an excimer center is fast. Only vibrational relaxation within this excimer centers can be retarded when the vibrational spacing exceeds the phonon energies as in the case for

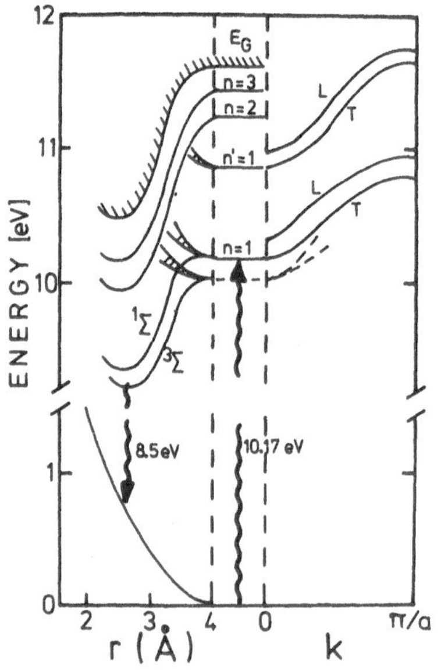

Fig. 9. Scheme for exciton states in solid Kr [42].
Right: estimated dispersion of the n=1 and n' =1 excitons versus wave vector **k**; T and L indicate transversal and longitudinal excitons; dashed are optically forbidden branches.
Center: exciton series at k=0 as observed in absorption
Left: estimated energy of excimer-type localized excitons versus internuclear separation of an excited Kr atom and a neighbouring Kr atom.

Ne. Little experimental information is available for the intermediate steps in the relaxation cascase. Decay time measurements show that the whole cascade is completed within the radiative lifetime of about 10^{-9} s. Therefore these intermediate steps are not accessible to experiments using radiative decay of excitons. Fast energy transfer experiments give additional information about these early stages.

Next we ask for electronic relaxation in guest centers. Energy gaps of the order of 1 eV appear between the Rydberg levels of excited guest atoms. A detailed investigation [41] of the relaxation processes for Xe, Ar and Kr guest atoms in Ne matrix and for Xe and Kr guest atoms in Ar matrix showed that for sufficiently large gaps the relaxation times exceed the radiative lifetimes considerably. Thus energy transfer from high excited species becomes feasible. Furthermore the nonradiative relaxation rate can be controlled via the temperature of the sample [41]. In molecules the electronic relaxation cascade is complicated by an overlap of vibrational progressions belonging to different electronic states. An interesting oscillation between two electronic states via close lying vibrational states has been reported for example for CN in Ar matrix [45]. The gaps are still large enough to hinder relaxation and to allow for radiative decay. Thus in molecules vibrational and electronic relaxation are often mixed. A clear cut case that vibrational relaxation in an excited electronic state can be very slow provides N_2 in rare gas matrices. Selective excitation of vibrational levels of the lowest excited state A $^3\Sigma_u^+$ of N_2 in the matrices shows that more than 50 % of the intensity remains in the initially populated vibrational level on the time scale of radiative decay [46]. This demonstrates that vibrational relaxation in this case is much slower than 10^{-3} s. More details will be discussed later.

Several examples of slow vibrational relaxation in the electronic ground state are known [45]. The most spectacular case of CO in rare gase matrices will be also discussed later on [47]. For vibrational relaxation it has been shown that especially in the case of hydride molecules conversion into rotational levels plays an important role [45], [48].

V. ENERGY TRANSFER

V.A. Concepts

The investigation of exciton states has been stimulated by the idea of transport in solids not connected to transport of free charges [9]. A wealth of information has been collected for proto type materials like organic crystals and for application purposes in solid state laser systems. This volume represents a profound review on the current activities. Several contributions have been devoted to the difficult task of a reliable description of energy transfer processes. In this part only some basic equations are collected which have been used in the description of experiments. The examples of energy transfer in solid rare gases have been selected mainly to illustrate the competition between localization and energy transfer of free excitons and to demonstrate that electronic states and vibrational states which lie above the lowest excited states also contribute to energy transfer.

Three time scales the scattering time τ_{sc}, the thermalization time τ_{th} and the localization time τ_{loc} govern the migration of free excitons. τ_{sc} determines the mean free path Λ of a coherent wave like free exciton before its momentum and energy are changed by scattering with phonons or imperfections. The group velocity $v_g(k)$ of an exciton with wave vector k follows from $v_g = \frac{1}{\hbar}\frac{dE}{dk}$ and corresponds in the effective mass approximation near Γ to $\hbar k/M$ [see (1)] or to $v_g = \sqrt{2E/M}$. According to the scattering times $\tau_{sc} \sim 10^{-13}$ s derived from the line shape of exciton absorption and for $v_g \sim 10^7$ cm/s, which is typical for an exciton with a kinetic energy E of some tenths of an eV, a mean free path $\Lambda \sim 10^{-8}$ cm is calculated from

$$\Lambda(k) = v_g(k) \cdot \tau_{sc}(k) \ . \tag{20}$$

This mean free path is of the order of the lattice spacings. The excitons are scattered at each lattice site and the migration of free excitions will not be wave like but diffusive.

The localization times τ_{loc} are not really known. The rise times in time resolved luminescence experiments indicate that $\tau_{loc} \leq 10^{-9}$ s. After localization the energy of the exciton lies about 1 eV (E_{LR} see Table 1) below the energy of the free, i.e. resonant exciton states (Fig. 9). The localized states are off-resonant by this energy compared to any excited state with the ground state equilibrium configuration. Due to this off-resonance and the lattice distortions involved in the localization process it is clear that these localized excitons are really spatially fixed. Therefore only free excitons are able to migrate and energy transfer by migration is restricted to the very short time before localization.

Each main quantum number n of a free exciton state is connected with several exciton bands (transversal, longitudinal, allowed and forbidden substates, spin orbit splitting, see Figs. 4 and 9). Phonon scattering lowers the exciton energy by a small amount but the momentum transfer can have any value of the Brillouin zone, because of the flat phonon dispersion curves (Fig. 1). An exciton with n = 3 for example can be scattered within the same branch from higher to lower kinetic energy (interstate scattering). Alternatively it can be scattered from a small k value in n = 3 to an approximately energetically resonant state of n = 2 or n' = 1 with high k value as is suggested by the level scheme in Fig. 9 (intrastate scattering). In this case a n = 3 exciton with low kinetic energy is converted to an exciton with lower n but high kinetic energy. The sum of all scattering times involved in scattering an exciton state with a given main quantum number and kinetic energy down to the bottom of the lowest n = 1 branch can be defined as the thermalization time τ_{th} of this exciton state. After reaching this bottom the kinetic energy will be stabilized to the lattice temperature by successive phonon absorption and emission events.

For $\tau_{th} \ll \tau_{loc}$ we can expect a rather simple description for the energy transfer processes. In this case we deal essentially only with one type of excitons. They belong to the lowest branches and have only small kinetic energies of almost 1 meV given by the lattice temperature. The spatial range of these free excitons is limited by the localization time and can be described by a diffusion length l or a diffusion constant D by

$$l = \sqrt{D\,\tau_{loc}} \tag{21}$$

The diffusion constant can be expressed by a hopping time τ_H and a hopping distance a(lattice spacing) or by a mean group velocity $\langle v_g^2 \rangle$ and Λ.

$$D = \frac{a^2}{6\tau_H} \tag{22}$$

$$D = \sqrt{\langle v_g^2 \rangle}\ \Lambda \tag{23}$$

Experiments indicate a dependence of the transfer range on the initially prepared exciton state in solid rare gases. This case is more complicated but also more interesting because it opens the possibility for a systematic investigation of migration properties of higher excited excitons. General equations have been derived [9]. But an application to the available experimental results including the different scattering pathways is still missing. The experiments suggest an even more complicated case. Also the higher exciton states can become localized (Fig. 9). Therefore an initially free exciton with n > 1 may be localized in a state with n > 1 which is energetically still resonant with a free state of lower n. Perhaps these localized excitons are detrapped by following repulsive excimer potential curves (Fig. 9) and converted again to free excitons. With this speculation we leave the free excitons and come to energy transfer by localized states.

The spectrum of donors for energy transfer by localized states consists of localized matrix excitons and of excited dopands. In the simplest case of excited singlet states energy transfer originates from the dipol-dipol interaction between donor and acceptor states [49], [50]. The transitions probability W(r) from the donor (D) to the acceptor (A) with separation r is given by this Förster-Dexter mechanism.

$$W(r) = \frac{1}{\tau} \left(\frac{R_q}{r}\right)^6 \tag{24}$$

The transfer radius R_q represents the D-A separation where the probability $1/\tau$ for decay of the donor state (in the absence of the acceptor) equals the probability for energy transfer. R_q can be expressed by the electronic transition moments $\mu(A)$ for acceptor absorption and $\mu(D)$ for donor emission, a spectral overlap function F and a numerical constant α.

$$R_q = (\alpha\ \mu\ (A)^2\ \mu\ (D)^2\ F \cdot \tau)^{1/6} \tag{25}$$

$$F = \int f_{Aa}\ (E)\ f_{De}\ (E)\ dE \tag{26}$$

$f_{Aa}(E)$ and $f_{De}(E)$ are normalized acceptor absorbtion and donor emission line shapes. These equations describe the interaction of one D-A pair. It is necessary to average for the different D-A separations. Försters results [49] are a good approximation for a statistical distribution and for concentrations low enough to inhibit migration within the donors or migration within acceptors and backtransfer. The time evolution of the donor emission intensity $n_o(t)$ represents the donor quenching after pulsed excitation

$$n_D(t) = n_D(t{=}o)\ \exp\ (-t/\tau - 2b\ \sqrt{t}) \tag{27}$$

Transfer to the acceptor is contained in the $2b\ \sqrt{t}$ term which includes the concentration N_A of acceptors and R_q.

$$2\ b = \frac{4}{3}\ \sqrt{\pi^3/\tau}\ R_q^{\ 3}\ N_A \tag{28}$$

Time integration yields the total quenching given by the ratio of the doped donor totalemission yield $\bar{n}$ to the pure donor yield $\bar{n}_o$.

$$\bar{n}/\bar{n}_o = 1 - \sqrt{\pi}\ q \cdot \exp(q^2)\ \left[1 - \mathrm{erf}\ (q)\right] \tag{29}$$

$$q = \frac{2}{3}\ \sqrt{\pi^3}\ R_q^{\ 3}\ N_A \tag{30}$$

The many interesting aspects involved in the spatial averaging are discussed elsewhere in this book. We want to mention an additional transfer process for liquids. Diffusion of an excited center through the liquid to an acceptor has to be taken into account because of the high mobility of the constituents of the liquid. This mass diffusion can be described by a bilinear reaction rate. In the case of a combined process including mass diffusion and dipol-dipol energy transfer an approximate treatment [51] yields a rate constant k_{ET} for energy transfer which contains R_q and the relative mass diffusion coefficient D *.

$$K_{ET} = 0.51\ 4\pi\ (R_q^6/\tau)^{1/4}\ D^{*\ 3/4} \tag{31}$$

The physical consequences of energy transfer by free excitons or by localized states can be classified in two types of transitions:

(a) Bound-bound transitions where the energy acceptor is produced in a bound excited state which is located below the ionization energy of the acceptor. The excited state of the acceptor can subsequently decay either radiatively or nonradiatively.

(b) Bound continuum transitions. When the energy of the donor exceeds the solid state ionization potential of the energy acceptor, the energy transfer process will result in the ionization of the acceptor. Such ionization processes bear a close analogy to Penning ionization in the gas phase. From the foregoing classification it is apparent that two general techniques can be utilized to interrogate the dynamics of energy transfer. The consequences of transfer resulting in bound-bound transitions can be explored by the techniques of luminescence spectroscopy. On the other hand energy transfer resulting in ionization can be investigated by photoelectron emission studies monitoring the photoelectron yield and the photoelectron energy distribution.

V.B. Migration of Free Excitons

1. Transfer to Guests. To explore electronic energy transfer by "free" excitons one has to utilize sensitive interrogation methods in order to probe processes occuring on the 10^{-12} s time scale. Photoelectron emission yield and energy distribution measurements provide an adequate tool. The electron emission process is prompt ($\lesssim 10^{-16}$s) on the time scale of nuclear motion. The kinetic energy of an electron which has been released in an ionizing energy transfer process contains therefore the information on the initially transferred energy. Light emission from an acceptor in contrasts needs 10^{-9} s or longer. The memory on the initially transferred energy can be destroyed within this time by electronic and lattice relaxation at the acceptor. Fig. 10a shows an energy level scheme for Xe atoms doped into an Ar matrix. The energy even of the lowest free matrix exciton is sufficient to ionize a Xe guest atom. Fig. 10b illustrates the expected differences in the kinetic energy distributions of electrons emitted from the Xe guests I) when the energy of free n=2 matrix excitons is transferred, II) when n=2 excitons relaxe to n=1 excitons before transfer, III) when the free n=2 exciton becomes localized before transfer and IV) when an n'=1 exciton (j=1/2 hole state) relaxes to an n=1 exciton (j=3/2 hole state) before transfer [52]. Evidently the kinetic energies of the emitted electrons immediately reflect the status of the matrix exciton at the

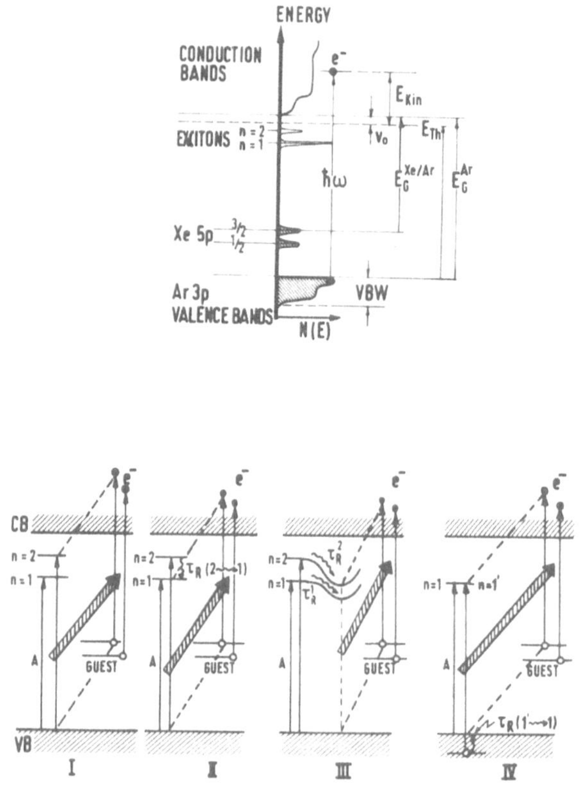

Fig. 10. (a) Energy level scheme for Xe in Ar
(b) Scheme for energy transfer and relaxation
I: Transfer by free excitons to guest;
II: relaxation of free n=2 to free n=1 exciton states and subsequent transfer,
III: localization of n=2 and n=1 and subsequent transfer;
IV: relaxation of j=1/2 hole to j=3/2 hole (n'=1 → n=1) and subsequent transfer [52].

moment of transfer. Only electrons located above the vacuum level which is marked by V_o are able to leave the sample. The signal is not masked by electrons from the matrix itself because the matrix excitons lie below the vacuum level (Fig. 10). Electron energy distribution curves (EDC's) from thin films of solid Ar doped with 1 % Xe are shown in Fig. 11. The samples have been excited with photon energies covering the range of n=1 up to n'=2 matrix excitons. The counting rate is plotted versus kinetic energy. The spectra are shifted upwards proportional to the increase in the exciting photon energy. The maxima B and C are due to the two spin orbit splitted Xe 5p states in the band gap of Ar (Fig. 10). Inspection of Fig. 11 shows that the energies of the maxima B and C are the same as if the Xe atoms would have been excited directly by photons. This is

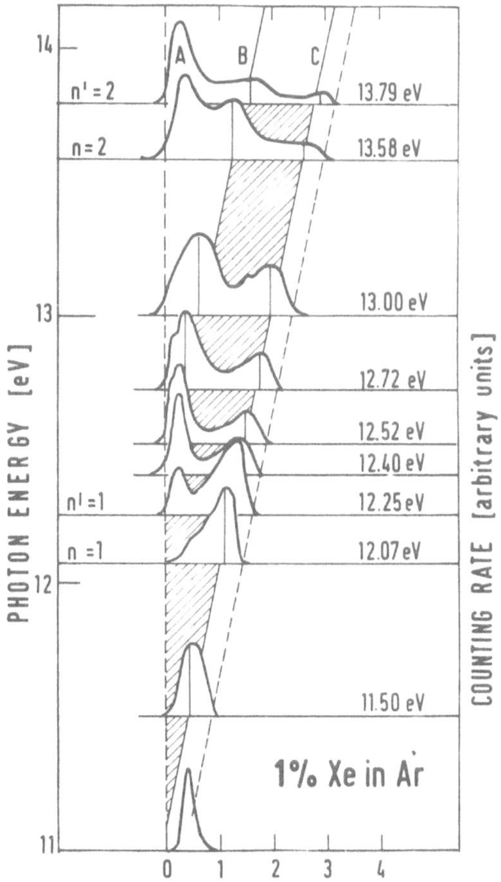

Fig. 11. Electron energy distribution curves for 1 % Xe in solid Ar. The base lines are shifted proportional to the ex-exciting photon energies |from 52|.

easily seen by comparison with the diagonal lines which correspond to the predicted positions of maxima B and C for transfer of the total energy of each matrix exciton, i.e. before any relaxation process. A contribution due to direct absorption of light by Xe atoms is negligible (< 0.1 %) for photon energies in the matrix exciton region because of the low Xe concentration and the high matrix absorption coefficient. The measured kinetic energies for the population of matrix n=1 excitons prove that the total energy of free n=1 excitons is transferred. For localized n=1 excitons the transferred energy would be 2 eV lower and the final state would even be located below the vacuum level. The difference in kinetic energy for the spectra for n=1' and n=1 excitons also excludes relaxation from n=1' to n=1 before energy transfer. For the n=2 and n=2' excitons of the Ar matrix again the maxima of the Xe 5p states appear in the EDC's at kinetic energies indicating energy transfer from unrelaxed n=2 and n=2' excitons. There is an additional maximum near the vacuum level which is attributed to energy transfer from selftrapped n=2 excitons. This maximum cannot be explained by transfer after relaxation to n=1 and n'=1 excitons.

The situation is quite different for Xe in Ne. The EDC's in Fig. 12 are taken for photon energies corresponding to population of Ne n=2, n=1 excitons and at a somewhat lower energy for direct excitation of the Xe guest atoms ($h\nu$ = 16 eV). In the latter spectrum emission from the two spin orbit split Xe 5p states situated

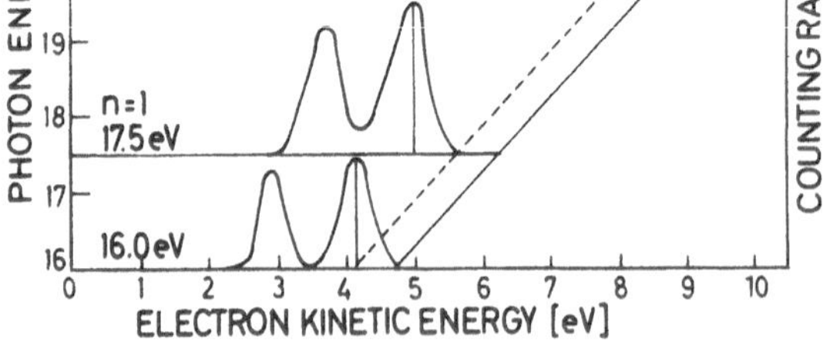

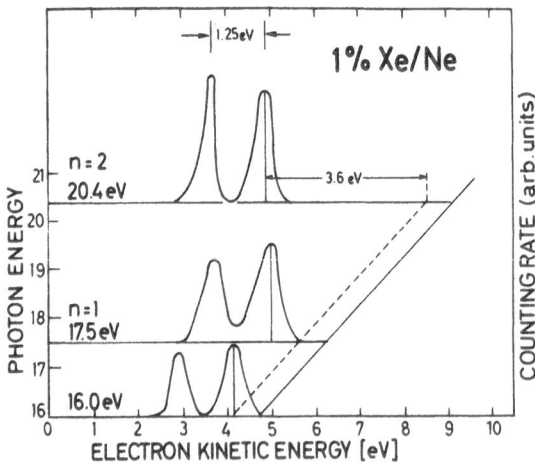

Fig. 12. Electron energy distribution curves for 1 % Xe in solid Ne.

in the band gap of Ne is observed. The Xe 5p states are well reproduced when the energy is transferred from the Ne n=1 exciton ($h\nu$ = 17.5 eV) to the Xe atoms. However, there is a shift of $\simeq$ 0.7 eV of the total spectrum to lower kinetic energies indicating some relaxation either to the one center selftrapped exciton or to the high vibrational levels of the two center selftrapped exciton before energy transfer. The striking observation is that in the EDC obtained by exciting the Ne n=2 excitons more than 3 eV are missing. The electrons have the same energy distribution as for excitation of the n=1 excitions. This observation indicates a complete relaxation of the n=2 excitons to n=1 excitons and dissipation of an energy of more than 3 eV before transfer.

These data illustrate the power of EDC measurements yielding direct information about the state of relaxation before transfer. A time scale derived from these experiments for the competition of relaxation and transfer processes is given in Table 5. Similar experiments have been reported for C_6H_6 guest molecules in solid rare gas matrices [53].

These experiments demonstrate that energy transfer can be fast enough to compete with electronic relaxation and with exciton localization. To derive the range of energy transfer it is necessary to vary the separation between exciton and acceptor. The concentration dependence of exciton induced impurity ionization is one possibility. Photoelectron yield spectra for Xe atoms in Ar matrix are shown in Fig. 13 for Xe concentrations decreasing from 1 % down to 0.005 % [38]. The yield corresponds to the total amount of emitted electrons and represents for each photon energy the area of an energy distribution curve (for example Fig. 11). The immediate information on the transferred energies is lost in the yield spectra but there is still a threshold energy E^i_{Th}. The transferred energy has to be sufficient for ionization of the guest atoms and for exciting the electrons also above the vacuum level.

$$E^i_{Th} = \begin{cases} E^i_G & \text{for } V_o > 0 \\ E^i_G - V_o & \text{for } V_o < 0 \end{cases} \tag{32}$$

For Xe in Ar the free excitons but not the localized excitons fulfill this condition and only energy transfer events by free excitons contribute to the yield. At photon energies below 12 eV the matrix is transparent and electrons in Fig. 13 originate from the Au substrate and from the Xe guests. Both contributions are small at all Xe concentrations up to 1 %. At photon energies between 12 eV and 13.9 eV light is absorbed by excitons of the Ar matrix. Hence the yield from the Au substrate and from direct absorption by Xe guest atoms is reduced. Nevertheless the yield increases. The observed large yield in this range originates from the creation of host

Table 5. Time hierarchy for the competition of energy transfer and relaxation. The time constants τ (in s) have the following meaning: τ_1: radiative decay of R_2^* or R^* centers of the matrix: τ_2; vibrational relaxation of R_2^* centers; τ_3 localization of excitons; τ_R $(i \rightarrow j)$; electronic relaxation of exciton i to exciton j; τ_T: energy transfer. (After Ref. [52]).

	Time Hierachy from Electron Emission Experiments	Time Hierarchy Including Lifetime Measurements
1% Xe in Ar		
n = 1	$\tau_1 > \tau_2 > \tau_T$	$\tau_1 \sim 10^{-5} - 10^{-9} > \tau_2 \sim 10^{-9} > \tau_T$
n = 1'	$\tau_R\ (1' \rightarrow 1) > \tau_T$	
n = 2	$\tau_1 > \tau_R(2 \rightarrow 1) > \tau_T \sim \tau_3$	$\tau_1 \sim 10^{-5} - 10^{-9} > \tau_R\ (2 \rightarrow 1) > \tau_T \sim \tau_3 \sim 10^{-12}$
1% Xe in Ne		
n = 1	$\tau_2 > \tau_R > \tau_T > \tau_3$	$\tau_2 \sim 10^{-5} > \tau_1 \sim 10^{-6} - 10^{-8} > \tau_T > \tau_3 \sim 10^{-12}$
n = 2	$\tau_2 > \tau_1 > \begin{cases} \tau_R(2 \rightarrow 1) \\ \tau_3 \end{cases}$	$\tau_2 \sim 10^{-5} > \tau_1 \sim 10^{-6} - 10^{-8} > \tau_T > \begin{cases} \tau_R(2 \rightarrow 1) \\ \tau_3 \sim 10^{-12} \end{cases}$

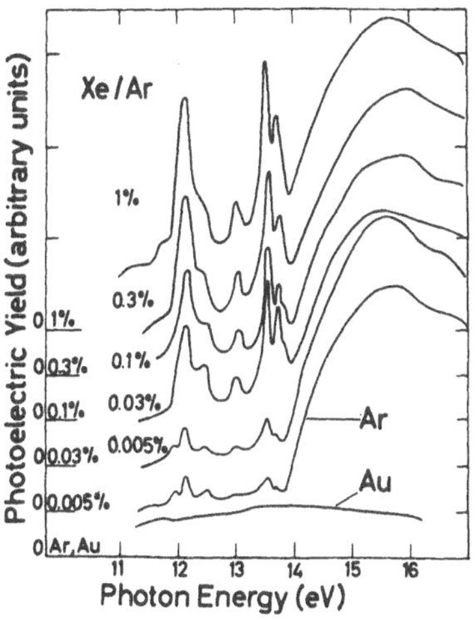

Fig. 13. Electron yield of Xe doped Ar films versus the exciting photon energy. Parameter is the Xe concentration. Ar and Au indicate pure Ar and Au substrate yields from [38].

excitons (n=1, 1', 2 ...) and energy transfer of free Ar excitons to the guest atoms. The small maxima in pure Ar are due to energy transfer of Ar excitons to the Au substrate. Above $h\nu$ = 13.9 eV direct photoemission of the Ar matrix causes a further steep increase of the yield. In Table 6 the matrix exciton energies and the energy of the emission bands are compared for various systems together with the threshold energy E^i_{Th} necessary for electron emission from the guest levels. The occurrence of energy transfer is marked. The essential result from these experiments is that in all solid rare gases matrices electron emission starts when the energy of primarily excited excitons just exceeds the threshold for photoemission of the guest atom. For Ar and Kr matrices the luminescence emission bands lie below the threshold and only transfer by free excitons leads to electron emission. The high yield for Xe in Ar (Fig. 13) proves that at high concentrations this is the dominant transfer process.

The range of energy transfer has been determined by modelling the dependence of the yield on concentration, photon energy and

Table 6. Comparison of energies which are transferable by the matrix either in the selftrapped exciton states ($E_{R_2}{}^*$) or in the free exciton states n=1 ($E_{n=1}$) and n=1' ($E_{n=1'}$) with the lowest excitation energies ($E^i_{n=1}$) and ionisation energies E^i_{Th} of guest atom. E_{Th}: ionisation energy of the matrix. + and - means observation or absence of energy transfer in electron emission spectra (energies in eV).

Matrix	Guest	E_{Th} [e]	E^i_{Th}	$E^i_{n=1}$ [f]	$E_{R_2}^*$	$E_{n=1}$	$E_{n=1'}$	
Ne		20.3			16.80	17.36	17.50	
	Xe		11.60	9.10	+	+	+	c,d
	Kr		13.48	10.60		+	+	c
	Ar		15.05	12.48		+	+	c
Ar		13.8			9.80±0.44	12.06	12.24	
	Xe		10.2	9.22	-	+	+	b,d
	Kr		12.2	10.79	-	-	+	b
Kr		11.9			8.45±0.32	10.17	10.86	
	Xe		10.3	9.01	-	-	+	a,b

[a] Ref. [64]
[b] Ref. [38]
[c] Ref. [39]
[d] Ref. [52]
[e] Ref. [7]
[f] Tab. 4

sample thickness by a diffusion equation for the density of free excitons [38]. The diffusion equation has been solved with appropriate boundary conditions which lead to a dead layer, i.e. an exciton free region at the surface. Thus the transfer has to compete with localization and also with quenching near the surface. The influence of surface quenching is very sensitive to the film thickness and to the spectral dependence of the matrix absorption coefficient. A consistent fit of all spectra allows a determination of the diffusion length and provides a test for the boundary conditions which have been used. These boundary conditions are an interesting problem in itself. The diffusion length for Ar of l = 120 Å exceeds the lattice constant considerably. Also for the other rare gases diffusion length of the order of 100 Å up to several thousand Å (Ne) have been found (Table 7).

Table 7. Diffusion lengths for energy transfer to guest atoms and to boundaries (in Å).

	Photoelectron Emission		Luninescence	
Excitons	Guest Atom	Boundary	Guest Atom	Boundary
Ne n = 1,1',2	2500±500 a	observed a		
Ar n = 1,1' n = 2,2',3	120 b	observed b		50 g
Kr n = 1 n = 1' n = 2	observed b	10-150 150-250 d 250-350	300 f	200-250f 200 g,h
Xe n = 1,2	170 c	300 e	25-260 f	500 f

[a]Ref.[39]
[b]Ref.[38]
[c]Ref.[64]
[d]Ref.[42]
[e]Ref.[57]
[f]Ref.[65]
[g]Ch. Ackermann, R. Brodmann, G. Zimmerer and U. Hahn, J. Luminesc. 12/13, 315 (1976)
[h]Ref.[54]

2. Transfer to Boundaries. The guest atoms can be replaced by boundaries as acceptors for energy transfer. Even in the doped samples the surface played a role, but a passive role in acting as a sink for excitons. Several experiments have been reported where luminescence of rare gas solids has been quenched by energy transfer to surfaces which have been contaminated by residual gas [54],[55],[56]. The quenching efficiency is large for a small penetration depth of the exciting photons. Therefore minima are observed in the luminescence yield for regions of high absorption coefficients, i.e. in the centers of exciton absorption bands. The interpretation is complicated by the fact that free and localized excitons can contribute to the transfer. The experimental results can be explained in terms of diffusion of free excitons as well as by Förster Dexter type transfer (Tables 7,8).

Table 8. Energy transfer radii Rq (in Å) for electronic energy transfer to guest atoms and boundaries in solid and liquid (1) rare gas matrices.

Host	Temperature	Guest			Boundary
	K	Species	Rq(exp.)	Rq(cal.)	Rq(exp.)
Ar	6-20	Xe	18[a)]		
		Kr	6[b)]		
Kr	60	Xe	17[a)]	10[a)]	
	110	Xe	25[a)]	15[a)]	
(1)	120	Xe	24[a)]	21[a)]	
	5	C_6H_6	21-22[c)]		
					25-29[c)]
					22[d)]
Xe	5-15	C_6H_6	24-29[c)]		40[c)]

[a)] Ref. [62]
[b)] Ref. [63]
[c)] Ref. [65]
[d)] Ref. [54]

Detailed information can be obtained when the boundary plays an active role and delivers a signal proportional to the transfer efficiency. Excitons in rare gas films on a metal substrate can migrate to the substrate and ionize the substrate. The ejected electrons can be detected [42],[57],[58]. The incident light in the scheme of Fig. 14 is represented by a wavy beam I_0. Part of it is reflected (I_0 R), part of it creates excitons in the film, which move to the substrate (hatched beam), eject electrons (straight beam) some of which penetrate the film and are detected as Y_{ET} [42]. The transmitted light causes a background of photoelectrons from the substrate (Y_{Au}). The thickness dependence of this yield has been used to derive diffusion lengths for Xe [57], Kr [42] and Ne [58]. Kr delivers a very pronounced signal and is therefore treated as an example.

Yield spectra for the films on an Au substrate are shown in Fig. 15 for a number of film thicknesses d. The Kr films are transparent up to photon energies of 9.8 eV. The small photoelectron yield below 9.8 eV represents the escape probability for electrons

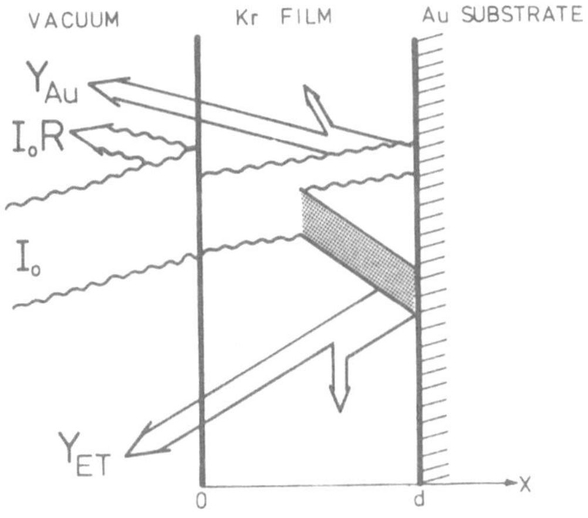

Fig. 14. Contribution of Y_{Au} and Y_{ET} to the electron yield of a Kr film on an Au substrate. The crosshatched beam indicates energy transfer by excitons, the straight beams escape of electrons and the wavy beams the light pass [42].

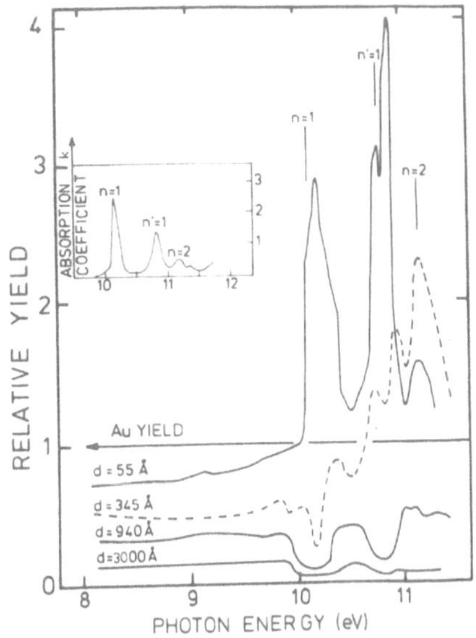

Fig. 15. Electron yield versus photon energy for several Kr film thicknesses d normalized to the Au substrate yield. The absorption coefficient k of Kr is shown in the inset [42].

from the Au substrate. The strong increase of Y for photon energies in the Kr exciton region above 10 eV is due to the energy transfer to the substrate. The yield Y_{ET} in this region depends on the diffusion length l, the thickness d, the absorption coefficient k and the escape length L of the electrons. L and k have been determined independently. The absolute value of the yield indicates an efficiency for electron emission due to energy transfer of 0.3 - 0.5 electrons per exciton compared with the Au direct photo yield efficiency of 0.03 - 0.05 electrons per photon. A similar high efficiency is observed in the Penning ionization of metal surfaces by metastable rare gas atoms. The interaction of the exciton with the metal substrate is a surface effect yielding the electron at the metal surface. This explains the high efficiency compared to the ordinary photoeffect where the penetration dept of the light into the metal reduces the efficiency. The most important result concerning exciton migration is a pronounced dependence of the diffusion length on the energy of the initally prepared exciton (Fig. 16). The diffusion length increases from about 30 Å in the

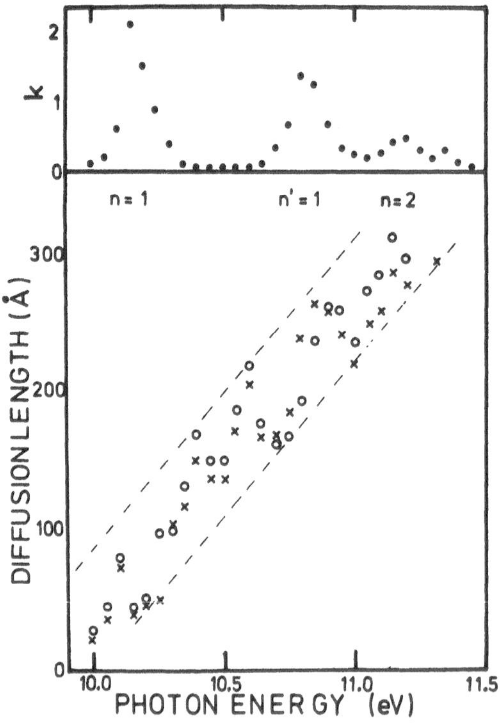

Fig. 16. Bottom: diffusion length versus the excitation energy of the excitons for solid Kr. Dashed lines indicate the trend.
Top: absorption coefficient showning n=1, n'=1 and n=2 exciton peaks.

low energy tail of the n=1 exciton to l = 300 Å for the n=2 exciton. The value l $\sim$ 30 Å can be explained by the interaction of a metal surface with a dipole, i.e. a localized exciton in front of it [42]. The monotonic increase of l with the exciton energy is due to a memory of the transfer process on the primarily excited exciton state. This increase reflects the diffusion of free excitons while they are scattered down within the original exciton band, scattered to lower exciton bands and become thermalized and finally localized (Fig. 9). The larger diffusion length of higher exciton states can be ascribed to a longer lifetime as a free exciton because more cumulative scattering events for thermalizations are necessary and to an increase in the diffusion coefficient. A theoretical interpretation of the monotonic increase of l within each exciton band and also for switching to higher exciton states (Fig. 16) would be very interesting.

Finally for extreme thin films with thicknesses d of only 15 Å energy transfer of surface excitons to the substrate has been observed [42]. This transfer died out already at d = 55 Å. The surface excitons are restricted to the surface plane due to the lower energy compared to bulk excitons (Table 2). Therefore they cannot penetrate into the bulk and diffusion is impossible for these states. The energy transfer is mediated by a dipole interaction of the exciton with the metal which yields a transfer range of the order of 30 Å.

Another type of experiments is closely related to energy transfer to surfaces. For solid rare gases extremely high erosion yields Y_{er} of the order of $Y_{er} \sim 10$ ejected atoms per electron for keV electrons [59] and up to $Y_{er} \sim 10^3$ for MeV charged particles [60] have been reported. One model explains the high erosion yield by a thermal spike in the vicinity of the track of the penetrating high energy particle because about 2 eV of the kinetic energy of electrons produced by the stopping of the particle are converted to nuclear motion in the dissociative recombination of localized holes with these electrons (Fig. 9). Many atoms are evaporated at the crossing point of the particle track with the surface due to this thermal spike [60]. An alternative model [59] explains the yield by an efficient transport of energy, which has been deposited in form of excitons along the track of the particle, to the sample surface. The transport is attributed to the diffusion of free exciton thus involving the diffusion length. The excitons are trapped as excimer centers at the surface. They decay radiatively to the repulsive ground state (Fig. 9) and one atom for each excimer is ejected with a large energy. The kinetic energy of the ejected atom is given by the repulsive energy of the ground state of about 1 eV.

V.C. Energy Transfer Between Localized Centers

1. Electronic Energy Transfer of Self-Tapped Excitons to Guest Centers. In luminescence experiments the threshold criteria which have been discussed for photoelectron spectra are not relevant and energy transfer by free or localized excitons will lead to light emission from the acceptor. In the past energy transfer observed in luminescence experiments has been attributed mainly to transfer from localized excimer centers. This assumption will be more valid for low acceptor concentrations because the time scale for energy transfer will exceed the lifetime of free excitons due to the larger mean donor-acceptor separation. The Förster-Dexter mechanism is responsible for energy transfer between matrix excimer centers R_2^* and guest centers. The efficiency depends mainly on the spectral overlap of the excimer emission band and the guest absorption spectrum (24-26). The spectral overlap is large for Xe guest atoms in solid Kr and Ar matrices [61].

The transfer radius [61],[62],[63] is derived from the concentration dependence of the donor yield η (29). A typical plot of η for the Xe in Kr system is portrayed in Fig. 17. The position and width of the Rydberg type impurity bands of Xe atoms in the Kr matrix change with temperature (15). This variation is reflected in the experimental and calculated Rq values (Table 8) because of its

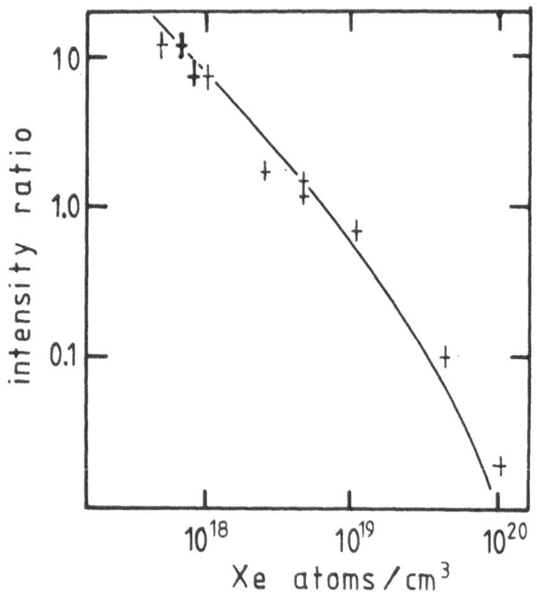

Fig. 17. Ratio of Kr donor emission yield to Xe acceptor emission yield versus acceptor concentration. Crosses: experiment; solid line: fit with Förster theory [62].

influcence on the spectral overlap (26).

Electronic energy transfer to molecular impurities in RGS was studied utilizing the benzene molecule as a prototype [64],[53],[65],[66], The luminescence intensity of the benzene acceptor [66] is much higher for excitation of host excitons than for direct excitation of benzene molecules in the transparent region of the host. Evidently, energy transfer from the host to guest molecules takes place, an observation, made for all rare gas matrices. The surface competes with the acceptor centers in quenching the donor centers especially for excitation processes with a small penetration depth. Ackermann [65] established that the ratio of host emission intensity to guest emission intensity (Fig. 18) is quite insensitive to changes

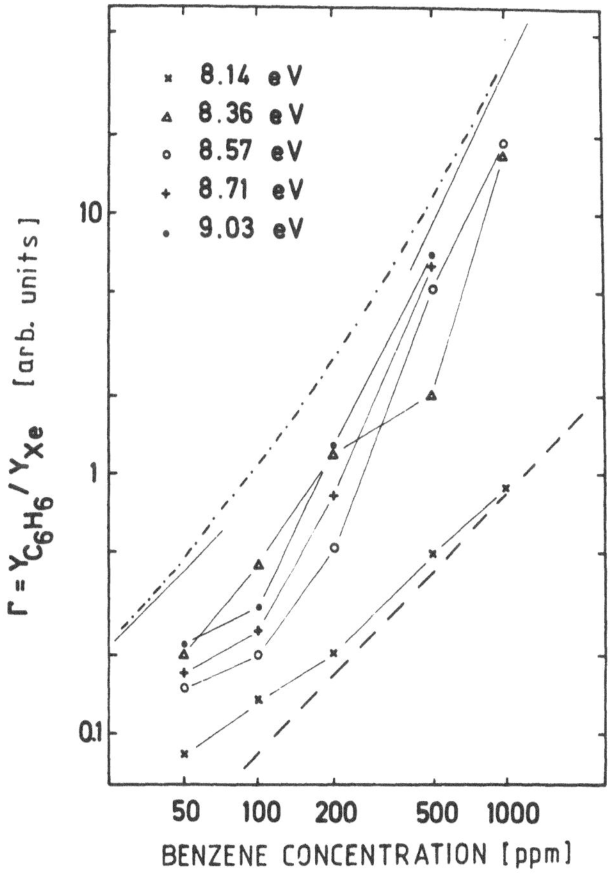

Fig. 18. Ratio of C_6H_6 acceptor emission yield to Xe donor emission yield versus acceptor concentration. Solid lines: experiment; dashed line: prediction for diffusion model; dashed-dotted curve: prediction for Förster model [65].

in the surface quenching rate. A careful study of the concentration and thickness dependence of host and guest emission efficiencies in the excitonic region of the rare gas host indicated that for benzene in Xe and Kr matrices there is an important contribution of self-trapped excitons to the energy transfer for all host exciton states yielding a critical Förster Dexter radius of 28-29 Å in a Xe matrix and 21-22 Å in a Kr matrix |Table 8|.

2. Elecctronic Energy Transfer Between Guest Centers . For matrix to guest energy transfer it is difficult to sort out the contribution of electronic energy transfer from free exciton states of the host and that from trapped excitons. Energy transfer from a localized donor to a molecular acceptor in solid rare gases was explored in the study of energy transfer from Xe atoms to benzene molecules when both partners had been embedded in an Ar matrix [65]. Both centers are fixed in the Ar lattice and exciton diffusion is excluded. The transfer mechanism is exclusively of the Förster Dexter type. Again the intensity in the emission bands of both Xe and C_6H_6 has been measured after the excitation of Xe exciton states. More details of the line shape show up in the excitation spectra of the three emission bands at 1640 Å, 1460 Å excitation spectra of the three emission bands at 1640 Å, 1460 Å and 1250 Å of Xe atoms in Ar matrix. The analysis results in a Förster Dexter radius of 24 Å.

By exciting selectively different electronic states of a donor guest atom it is possible to compare the relaxation rate of higher excited donor states with the transfer efficiency to acceptors. The Rydberg character of the n=1 and n'=1 excited states of Kr guest atoms in a Ne matrix causes a large line width and a Stokes shift of about 0.5 eV of the transitions to the ground state. The n'=1 state in emission is accidently resonant with the n=1 state in absorption due to this Stokes shift. Therefore a Förster Dexter type energy transfer between a n'=1 Kr excited state in a relaxed Ne configuration and another Kr atom in the ground state will be very efficient. The time evolution of the n'=1 Kr emission and the n=1 Kr emission after population of n'=1 states can be nicely fitted with the Förster Dexter equations (27,28) yielding a value of R_q = 21 Å [41].

3. Vibrational Energy Transfer Between Guest Molecules. Migration of vibrational energy in doped rare gas solids has been discussed in recent review [45]. A spectacular example is the upconversion of v=1 vibrational levels of CO in Ar up to v=7 which has been reported by Dubost and Charneau [47] and which has been analyzed in many theoretical papers [45].

CO molecules have been trapped in solid Ne and Ar matrices at ∿ 8 K with concentrations of typically 1:1000. The first vibrational state in the electronic ground state of $^{12}C^{16}O$ has been populated from the vibrational ground state by excitation with a Q-switched frequency doubled CO_2 laser pulse. The emission spectra consist of several lines which can be attributed to transitions between vibrational levels v = 0 up to v = 7 with $\Delta v = 1$. The transitions are separated due to unharmonicity in the CO potential curve. The surprising observation is this energy upconversion from v =1 up to v = 7. The complication of additional fine structure due to energy transfer to $^{13}C^{16}O$ and $^{12}C^{18}O$ isotopes will not be discussed here [47]. First for an interpretation of this energy upconversion the lifetime of the v = 1 state is essential. Because of the small matrix phonon energies and the large vibrational energies of the CO molecule radiationless transitions are unimportant (18). The lifetime is determined by the radiative lifetime which is long and lies in the order of msec. During this lifetime the vibrational energy migrates with a high rate constant due to a Förster Dexter-like energy transfer through the crystal from one CO molecule to another. In this way several vibrational v = 1 energy packets can reach one CO molecule simultaneously and form a higher excited vibrational state v' > 1. The very small energy mismatch for example of 2 (E (v=1) - E (v=0)) > E (v=2) - E (v=0) due to the small anharmonicity of the potential can be overcome by emission of a lattice phonon. At the low temperatures of the experiment (4 K) phonon emission is some orders of magntidude more probable than phonon absorption. As a consequence the opposite process of decay of a higher excited vibrational level into two lower vibrational states is much less probable. Therefore, a transient population of higher vibrational levels occurs and even a population inversion between population of v = 2 and v =1 is observed. From the measured rise times and decay times of all the emission bands any population of higher vibrational states by two photon absorption or radiative energy transfer has been excluded. Based on the available potential curves a quantitative description of the population rates of different vibrational states and also of the time dependence of the population has been given [45],[47].

Vibrational energy transfer plays also an important role in electronically excited states of guest molecules. Electronic and vibrational relaxation in excited states of a guest molecule can be accelerated by a distribution of the excess energy into ground state vibrational quanta of neighbouring guest molecules as has been observed for N_2 molecules in solid rare gas matrices [46]. N_2 in rare gas matrices is an exceptional example because vibrational relaxation in its first excited state $A^3\Sigma_u^+$ is extreme slow. Therefore, despite the long radiative lifetime of its dipole forbidden transition to the ground state luminescence also from vibrationally excited $A^3\Sigma_u^+$ states is observed [67]. One of the reasons is the weak coupling of the excited singlet [68] and triplet [46] states to the matrix phonons which has been derived from the line shape of absorption and emission spectra (13-15). The second reason is the large vibrational spacing of 165 meV in the $A^3\ \Sigma_u^+$ state which requires for nonradiative relaxation a dissipation of about 33 matrix phonons. This multiphonon process is inefficient according to the energy gap law (18). Each histogram in Fig. 19 represents the population distribution in the vibrational levels v' = 0,...,7 of the $A^3\ \Sigma_u^+$ state of N_2 in Xe matrix for excitation of that vibrational level which has been marked by a horizontal arrow. The left hand distribution for 0.5 % N_2/Xe (Fig. 19) corresponds to the case of essentially isolated N_2 molecules. For v' = 0,...6 most (> 50 %) of the intensity is emitted from the initially populated level and only a small fraction relaxes down to the next lower levels, illustrating the slow multiphonon relaxation rate. The distribution for excitation energies exceeding v' = 6 is given by the radiative transition $B^3\ \Pi_g$ (v' = 0) → $A^3\ \Sigma_u^+$ (v'') with its Franck Conden distribution. At 2 % N_2 in Xe the statistical probability that two N_2 molecules occupy neighbouring lattices sites reaches 30 %. Evidently the initially excited vibrational levels are depopulated much more effectively. Furthermore relaxation steps corresponding to a dissipation of two vibrational quanta dominate the distribution. The ground state vibrational energy of N_2 is 287 meV. Thus two quanta of the $A^3\ \Sigma_u^+$ state (2 x 165 meV) can be converted into a ground state vibration and the remaining excess energy of 43 meV has to be dissipated in 8 matrix phonons. This energy transfer to a neighbouring N_2 molecule reduces the number of emitted matrix

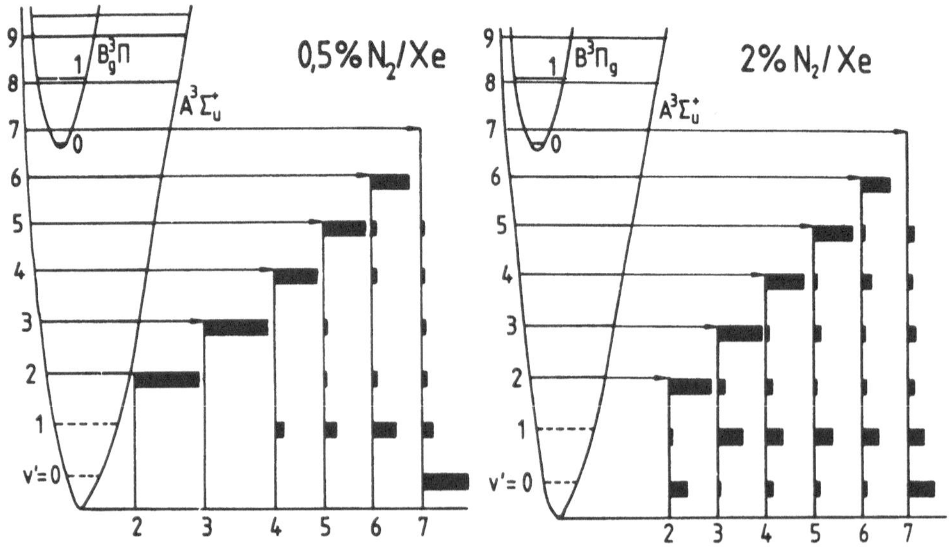

Fig. 19. Population distributions for 0.5 % N_2 and 2 % N_2 in solid Xe matrices at 5 K. Each vertical arranged histogram represents one population distribution. The arrow indicates the initially excited vibrational level.

phonons from 33 to 8. The exponential terms in the energy gap law (18) favour a low number of phonons even for a very weak coupling between the two molecules with $S \lesssim 10^{-2}$.

V.D. Energy and Mass Transport in Liquid Rare Gases

Electronic excitation of liquid rare gases leads to the formation of atomic and excimer bubble centers. These centers can migrate in the liquid by mass diffusion due to the high mobility of the atoms in the liquid. The two lowest long living excited states in liuqid He are the atomic 2s ^3S and the excimer $a^3\Sigma_u^+$ state (Fig. 5). These two states are most important for energy transport by mass diffusion. Transport of energy over distances of 1 cm without appreciable attenuation has been observed in superfluid He between 0.3 and 0.6 K [69]. At higher temperatures the range is reduced. The transport range by these centers increases with the lifetime of these triplet states. The lifetimes have been probed by delayed absorption spectra of 2s $^3S \rightarrow 2p\ ^3p$ and a $^3\Sigma_u^+ \rightarrow b\ ^3\Pi_g$ transitions after pulsed excitation [70]. The lifetime of a $^3\Sigma_u^+$ is mainly limited by the collision between two a $^3\Sigma_u^+$ excimers and decomposition into two ground state He atoms and one higher excited excimer.

The time dependent concentration M of metastable states is described by second order kinetics

$$1/M = 1/M_o + \alpha(T) \cdot t \qquad (33)$$

with the steady state concentration M_o and the bilinear reaction rate $\alpha(T)$. M_o is given by the number I_o of inicially produced metastable molecules $M_o = (I_o/\alpha(T))^{1/2}$. To provide a quantitative estimate we note that the lifetime of the $a^3\Sigma_u^+$ state due to the bilinear quenching is about 1 nsec at a beam current of 1 μA and the lower limit of the radiative lifetime is at least as long as 0.1 sec. A most important experimental observation is that the bilinear reaction rate $\alpha(T)$ increases with decreasing temperature. $\alpha(T)$ is inversely proportional over the higher part of the temperature range to the number density of rotons given by $e^{-\Delta/T}$ with $\Delta = 8.6$ K [70].This interesting temperature dependence reveals that the reaction rate is diffusion limited and determined by scattering with rotons. The energy gap of rotons is pressure dependent. The change of $\alpha(T, P)$ with external perssure P has been measured [70]. The observed decrease of $\alpha(T, P)$ with pressure agrees with the predictions based on the pressure dependence of Δ.

Some direct implications of the annihilation process were recorded. The afterglow observed in the VUV emission of the $A^1\Sigma_u^+$ (Fig. 5) is due to repopulation of this state in the annihition process of two metastable $a^3\Sigma_u^+$ states as follows from the similarity in the rate constants for annihilation $\alpha(T, P)$ and for repopulation. The destruction of metastable molecules feeds the $A^1\Sigma_u^+$ channel as well as higher lying states.

Similar results for triplet-triplet annihilation have been obtained for solid He_3 [71]. In solid He the molecular absorption band for the $a^3\Sigma_u^+ \rightarrow b^3\Pi_g$ transition is similar to that in the liquid phase. Also the time dependence of the $A^1\Sigma_u^+$ afterglow, i.e. the bilinear reaction rate $\alpha(T)$ is the same. The common reaction rate is surprising, because α is determined by diffusion of metastable molecules and the transport properties for other species like positive and negative ions are about four orders of magnitude smaller in the solid than in the liquid phase. Thus, mass transport in the solid is excluded. Short range electron exchange interaction provides an attractive mechanism for triplet migration in solid He. If this is indeed the excitation transport mechanism, the triplet-triplet annihilation in solid He bears a close analogy to triplet-triplet annihilation in aromatic crystals.

Finally, we mention briefly electronic energy transfer in doped liquid rare gases. Elelctronic energy transfer between Ar_2^* and Kr_2^* donors and Xe acceptor in liquid rare gases was studied by Chesnowsky et al. [72]. The donor emission yield η of Kr_2^* for the liquid Xe/Kr system exhibits a linear dependence on the Xe impurity concentration. This behaviour is different from that of the Xe/Kr solid alloys and can be accounted for in terms of a simple kinetic scheme with a rate constant $k_{ET}\tau_0 = 6 \times 10^{-19}$ cm^3 which, with the value $\tau_0 \approx 10^{-6}$ s^{-1} for the radiative decay of the Kr_2^* state, results in $k_{ET} = 6 \times 10^{-13}$ cm^3 s^{-1} (see equ. (31)). This rate includes mass diffusion and Förster Dexter type of transfer. With the date $\tau_0 = 10^{-6}$ s^{-1} and the relative diffusion coefficient D^* of the donor acceptor pair $D^* = 10^{-6}$ cm^2 s^{-1} and the experimental value for k_{ET} a transfer radius $R_q = 24$ Å in the liquid was calculated with (31). This value is in agreement with the R_q values in Table 8. Thus the energy transfer process between localized states in the liquid involves a long range electronic energy transfer process coupled with diffusive motion.

VI. HIGH EXCITATION DENSITIES

VI.A Laser Applications

The broad bandwidth emission continua from noble gas excimers, extending from 60 nm in the case of He to 172 nm for Xe favour these gases for short wavelength lasers. The bound excimer states decay radiatively to the strongly repulsive part of the ground state potential curve. The repulsive ground state ensures an extremely fast depopulation by dissociation (10^{-13}s) thus facilitating inversion. It is responsible for the large width of the emission bands which should allow a broad tunability region. The particular problems encountered in the construction of short wave length laser follow from an λ^{-5} scaling law for the pumping power per unit volume which is required to produce a gain coefficient of unity per unit length. In 1965 it ahs been recognized that noble gases emit essentially the same excimer bands in the solid, liquid and gaseous phase [25]. The idea to exploite the higher density in the condensed phases for obtaining a sufficient high density of excimers lead to the first successful operation of an excimer laser by the Basov group [73] in 1970. Amplified stimulated emission from electron beam excited liquid Xe at 178 nm has been observed. This project has not been continued because of the low efficiency which has been attributed mainly to reabsorption of the light. The gas phase excimer lasers made their way [44]. Despite of the progress in gas phase excimer lasers and in frequency multiplication it is still difficult to cover the region of wavelengths shorter than about 180 nm by lasers. Any additional way should be pursued. Since two years we try to obtain stimulated emission in the region below 180 nm from Ar, Kr and Xe crystals. Most of our emphasis has been concentrated on Ar crystals which radiate around 126 nm. Several experimental observations indicate that we succeeded in obtaining stimulated emission from Ar crystals. The expenses in pumping power and in equipment for achieving a comparable radiation power seem to be lower than for the Ar gas phase laser. Considering this success and the relatively little efforts devoted up to now to solid phase lasers it seems to be promising to continue also this way [74].

The emission bands (Fig. 7) are very similar in the gaseous, liuqid and solid phase. The short living component (Fig. 7) is due to an dipole allowed transition of a $^1\Sigma_u^+$ excimer state to the ground state. This allowed transition is the only laser active band because the cross section for stimulated emission of the dipole forbidden radiative decay of the close lying $^3\Sigma_u^+$ state is two orders of magnitude smaller. The cross section σ_s for stimulated emission follows from

$$\sigma_s = \frac{1}{8\Pi\, n_r^2} \quad \frac{\lambda^2}{\tau_r \Delta\nu} \tag{34}$$

with the index of refraction n_r, the lifetime τ_r and the linewidth $\Delta\nu$. The optical data yield a value for solid Ar of $\sigma_s = 0.7 \times 10^{-17}$ cm^2 which is close to the gasephase value of 1×10^{-17} cm^2.

The light flux from spontaneous emission in an strongly excited crystal will induce also stimulated emission. The intensity

I (t) due to spontaneous emission and due to amplified spontaneous emission within the cone dΩ into the amplifying direction is given by

$$I(t) = \frac{d\ h}{\sigma_s \tau_r} (\exp(n(t)\ \sigma_s\ L)-1)\ d\Omega \tag{35}$$

with the excimer density n(t), the penetration depth d, the length L and the height h of the crystal. The gain α due to amplification is given by

$$\alpha = n(t)\ \sigma_s\ L \tag{36}$$

This gain is obtained for a single pass through the excited crystal. If mirrors are used than this gain increases by the number of passes. This number is limited by the mirror reflectinity and the decay time of the centers compared to the transit time of the light in the resonator. In order to obtain a gain of ≈ 1 in single pass for a crystal with a length of 1 cm it is necessary to create a density of about $n=10^{17}$ excimers/cm^3.

The growth even of single crystals from noble gases is well known. The demands for laser crystals are less severe since only optically clear but free standing crystals with a volume of about 1 cm^3 and a front surface of 2 x 1 cm^2 are necessary. A pyrex box is mounted below a copper block which can be cooled down to 10 K by a closed cycle refrigerator.

With the copperblock pressed versus the box and gas flowing through the pipe at the bottom we grow a crystal within about 30 minutes. By shifting the refrigerator and the crystal upwards with a bellow sealed translation stage we bring the free standing crystal into the optical path of the ultrahigh vacuum sample chamber.

The crystals are excited by an electron gun delivering electron pulses of up to 10 J (600 keV, 5 kA, 3 ns) through a 0.025 mm thick Ti foil into the sample chamber. The penetration depths of the electrons are about 0.24, 0.12 and 0.08 mm in solid Ar, Kr and Xe respectively.

The electron beam cross section is larger than the crystal surface, therefore only about 0.22 J are really deposited in the crystal. Nevertheless this energy is enough to create the necessary density for amplification of the order of 10^{17} excimers/cm^3. Amplification has been proved by the following observations.
a) A first indication for amplification is a decrease by 10% of the fwhm of the $^1\Sigma$ band for an increase of the deposited energy from 0.09 J to 0.22 J yielding α = 0.7 for the highest excitation

power. This interpretation is confirmed by the fact that the $^3\Sigma$ line shape remains unchanged.
b) Most of the $^1\Sigma$ contribution is emitted synchronously with the exciting Febetron pulse and much faster than a convolution of the exciting pulse with τ_r would predict. This part contains an essential part of amplified emission on a background of spontaneous emission. From the enhancement of the intensity a gain coefficient $\alpha = 1$ is estimated.
c) The intensity in the $^1\Sigma$ contribution increases progressively in contrast to the $^3\Sigma$ contribution which increases linear or even slightly sublinear with deposited energy. The enhancement of the $^1\Sigma$ component is caused by an increasing collimation of the $^1\Sigma$ radiation into the amplifying direction. The $^3\Sigma$ radiation is emitted isotropically because of the much smaller cross section for stimulated emission. This experiment yields a gain which increases from $\alpha = 0.4$ to $\alpha = 1.3$ for enlarging the deposited energy from 0.09 J to 0.22 J.
d) The measured absolute photon fluxes for a deposited power of 0.22 J correspond to maximal intentsities of 8×10^{20} photons/s into $d\Omega = 10^{-2}$ sterad. From equation (36) with the known value for σ_s we can derive at least a density $n = 5 \times 10^{16}$ cm^{-3} of $^1\Sigma$ excimers and a gain $\alpha = 0.3$. The total number of $^1\Sigma$ excimers $N_{tot} = 4 \times 10^{15}$ in the crystal follows from the time sourse and the excited volume. A peak power of 1300 watts due to amplified spontaneous emission has been detected in 10^{-2} sterad. Furthermore a total of 4×10^{15} photons due to $^1\Sigma$ excimers is emitted yielding a peak power into all directions of 1×10^6 watts. Finally the long component yields a density of $8 \times 10^{16}/cm^3$ $^3\Sigma$ excimers.
e) With a laser resonator consisting of a focusing Al mirror coated with MgF_2 and a MgF_2 plate a divergence of the amplified radiation of ~ 1 mrad and an additional amplification by a factor of 3 has been obtained. Fig. 20 shows a comparison of the efficiencies for an Ar gas laser [75] and a solid state laser starting with the electron gun. Evidently the crystal can be more efficient concerning the conversion of deposited energy into photons. The measured efficiency in solid Ar is about 12% corresponding to 6% $^1\Sigma$ photons and 6% $^3\Sigma$ photons. In this context losses and energy transfer come into play.

VI.B. Loss Processes and Electron Plasma

A typical Ar gas excimer laser [44],[75] requires gas pressures between 10 - 100 bars in a stainless steel tube and excitation by a Pulserad electron source (2 MeV, 15 KA, 20 ns, 500 J). A delicate compromise is necessary between the gas pressure, the thickness of the tube wall and the penetration depth given by the energy of the electrons. The conversion of the energy deposited in the gas into light is limited mainly due to reabsorption of the VUV light by excimers and by collisions between two excimers. The collison of two excimers leads to one ionised excimer and two ground state

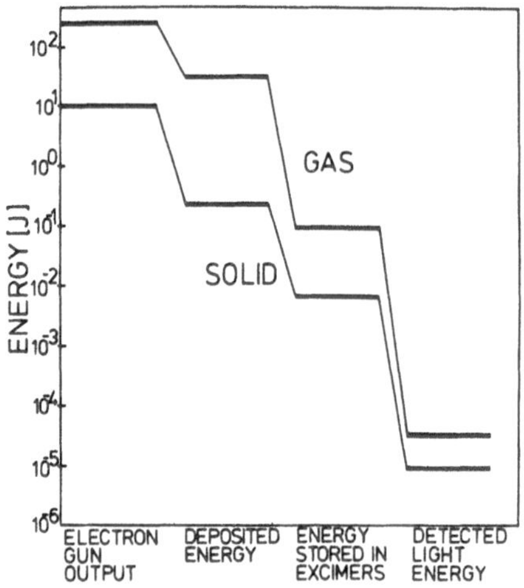

Fig. 20. Approximate conversion efficiencies for Ar gas [75] and Ar crystals [74].

atoms i.e. annihilation of one excimer. The higher efficiency of the Ar crystals at excitation densities of about 10^{17} excimers/cm^3 has to be attributed to different conditions concerning excimer-excimer annihilation.

In part V it has been explained that the excimers in the crystal are spatially localized due to the lattice distortions and the energy relaxation involved in the selftrapping process. Therefore collisions between the excimers in the crystal do not take place and the only process leading to excimer-excimer annihilation would be Förster Dexter type of energy transfer. The mean separation between two excimers at a density of 10^{17} excimers/cm^3 is about 100 Å which is larger than the expected Förster Dexter radii (Table 8) and quenching due to this process is not important on the time scale of the pulsed stimulated emission of 10^{-9}s.

But severe annihilation processes can be expected from the precursor states of the excimers. The high energetic primary electrons are converted to many pairs of free electrons and holes according to a mean energy for electron hole pair creation of 19.3 eV. for Ar. Recombination of slow electrons with free and localized holes will populate the whole manifold of free and localized excitons. The free exciton can travel distances of the order 100 Å and more before localisation (Table 7). Thus collison processes between free excitons and between free excitons and excimer centers will reduce the efficiency. Therefore fast localisation which suppresses the migration of excitonswill enhance the light output of rare gas crystals at high excitation densities.

In addition the dense cloud of free electrons can cause succesive ionisation processes of excitons and excimer centers and electron-electron collisions. In Kr and Xe crystals a flat continuum emission is observed extending from the infrared, through the visible and down to the ultraviolet spectral region [74]. This emission is only present during the excitation time of 3×10^{-9} s. In Ar crystals it is absent or at least one order of magnitude smaller than in Kr. This emission is due to free-free Bremsstrahlung and free-bound recombination radiation. The emission coefficient ε_ν for this radiation does not depend on the photon energy and is given by [76].

$$\varepsilon_\nu \ (\mathrm{W\ s\ sr^{-1}\ m^{-3}}) = 5.44 \times 10^{-52} \ \frac{N_e \cdot N_i}{\sqrt{T}} \qquad (37)$$

with the electron and ion densities N_e, N_i in (m^{-3}) and the temperature T in (K).

The measured absolute fluxes prove that in Xe and Kr crystals during the excitation pulse of 3×10^{-9} s a dense plasma of electrons with a density of 10^{18} electron/cm^3 and a distribution of kinetic energies corresponding to a temperature of 7000-8000 K is maintained. The crystal lattice remains at its low temperature of 10-60 K. This radiation is similar to the plasma radiation observed in high pressure arcs. The black spot and the the arrows Fig. 21) illustrate the characteristics of this plasma in rare gas crystalsin comparison with other plasma sources [77]. This plasma in rare gas crystalspresents an extraordinary case of an electron temperature of several thousand K and an electron density comparable to general plasma sources but with the ions fixed a lattice sites at very low temperatures. The scattering properties of these hot electrons and the interaction with the free excitons and excimers which are also present at high densities will be an interesting field in the future.

This work has been partly supported by the Bundesministerium für Forschung und Technologie.

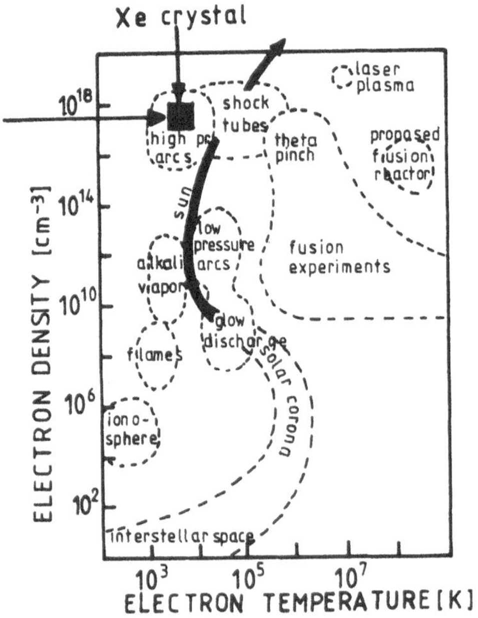

Fig. 21 Electron densities and electron temperatures for typical plasmas [77]. The arrows and the black dot indicate the plasma parameter for high excited Xe crystals.

REFERENCES

1. A. Gold and R. Knox, J. Chem Phys. 36, 2805 (1962)
2. H. Bilz and W. Kress in Phonon Dispersion Relations in Insulators, Springer, Berlin (1979)
3. B.M. Powell and G. Dolling in Rare Gas Solids Vol. II, eds. M.L. Klein and J.A. Venables, Academic Press, London (1977)
4. U. Rössler, phys. stat. sol. 42, 345 (1970)
5. M.H. Reilly, J. Phys. Chem. Solids 28, 2067 (1967)
6. U. Rössler in Rare Gas Solids Vol. I. ed. M.L. Klein and J.A. Venables, Academic Press London (1976)
7. N. Schwentner, F.-J. Himpsel, V. Saile, M. Skibowski, W. Steinmann and E.E. Koch, Phys. Rev. Lett, 34, 528 (1975)
8. B. Sonntag in Rare Gas Solids Vol. II eds. M.L. Klein and J.A. Venables, Academic Press, London (1977)
9. R.S. Knox in Theory of Excitons Solid State Physics suppl. 5, Academic Press, New York (1963)
10. R.S. Knox, Phys. Chem. Solids 9, 265 (1959)
11. V. Saile and E.E. Koch, Phys. Rev. B20, 784 (1979)
    V. Saile, Appl. Optics 19, 4115 (1980)
12. B.E. Springett, M.H. Cohen and J. Jortner, J. Chem. Phys. 48, 2720 (1968)
13. M.W. Cole, Rev. Modern Phys. 46, 451 (1974)
14. W.A. Fitzsimmons in Atomic Physics 3 ed. J. Smith and G.K. Walters, Plenum, New York (1973)
    J.C. Hill, O. Heyby and G.K. Walters, Phys. Rev. Lett 26, 1213 (1971)
    F.J. Soley and W.A. Fitzsismmons, Phys. Rev. Lett 32, 988 (1974)
    A.P. Hickman, W. Steets and N.F. Lane, Phys. Rev. B12, 3705 (1975)
15. N. Schwentner, Appl. Optics 23, 4104 (1980)
    N. Schwentner, E.E. Koch and J. Jortner, Internal Report DESY F41, HASYLAB 80/81, Deutsches Elektronen Synchrotron Hamburg (1980)
16. J. Jortner in Vacuum Ultraviolet Radiation Physics eds. E.E. Koch, R. Haensel and C. Kunz, Vieweg Pergamon, Braunschweig (1974)
17. E.I. Rashba in Excitons eds. E.I. Rashba and M.P. Sturge, North Holland (1982)
18. Y. Toyozawa, Prog. Theor. Phys. 26, 29 (1961)
    Y. Toyozawa in Vacuum Ultraviolet Radiation Physics eds. E.E. Koch, R. Haensel and C. Kunz, Vieweg Pergamon, Braunschweig (1974)
19. K.S. Song and L.J. Lewis, Phys. Rev. B19, 5349 (1979)
20. I. Ya Fugol, Adv. in Phys. 27, 1 (1978)
21. E.I. Rashba in Defects in Insulating Crystals ed. V.M. Tuchkevich and K.K. Shvarts, Springer, Berlin (1981)
22. K. Nasu and Y. Toyozawa, J. Phys. Soc. Jap. 50, 235 (1981)
23. K.S. Song, J. Phys. Soc. Jap. 26, 1131 (1969)
24. K.S. Song and C.H. Leung, J. Phys. C. Solid State Phys. 14,

L 359 (1981)
25. J. Jortner, L. Meyer, S.A. Rice and E.G. Wilson, J. Chem. Phys. 42, 4250 (1965)
26. G. Zimmerer, J. Luminescence 18/19, 875 (1975)
27. O. Dössel, H. Nahme, R. Haensel and N. Schwentner, J. Chem. Phys. in print (1983)
28. T. Suemoto and H. Kanzaki, J. Phys. Soc. Jap. 46, 1554 (1979)
29. F. Schuberth and M. Creuzburg, phys. stat. sol (b) 71,797 (1975)
30. F. Coletti and A.M. Bonnot, Chem. Phys. Lett.55, 92 (1978)
31. R. Kink, A. Lohmus and M. Selg, phys. stat. sol (b) 107, 479 (1981) and 96, 101 (1979)
32. G. Zimmerer, P. Gürtler and E. Roick to be published
33. P.L. Kunsch and F. Coletti, J. Chem. Phys. 70 726 (1979)
34. F.V. Kusmartsev and E.I. Rashba, Czech. J. Phys. B32, 54 (1982)
35. K.S. Song, Emery, and C.H. Leung to be published
36. T. Suemoto and H. Kanzaki, J. Phys. Soc. Jap. 49, 1039 (1980) and 50, 3664 (1981)
37. Y. Yakhot, Chem. Phys. 14, 441 (1976)
38. Z. Ophir, B. Raz, J. Jortner, V. Saile, N. Schwentner, E. E. Koch, M. Skibowski and W. Steinmann, J. Chem Phys. 62, 650 (1975)
39. D. Pudewill, F.-J. Himpsel, V. Saile, N. Schwentner, M. Skibowski and E.E. Koch, phys. stat. sol (b) 74, 485 (1976)
40. A.A. Maradudin, Solid State Physics 18, 273 (1966)
41. U. Hahn and N. Schwentner, Chem. Phys. 48, 53 (1980)
U. Hahn. R. Haensel and N. Schwentner, phys. stat. sol. (b) 109, 233 (1982)
42. N. Schwentner, G. Martens and H.W. Rudolf, phys. stat. sol. (b) 106, 183 (1981)
43. J. Jortner, J. Chem Phys. 64, 4860 (1976)
44. Ch. K. Rhodes in "Topics in Applied Physics" Vol. 30, Springer, New York (1979)
45. V.E. Bondybey in Adv. Chem. Phys. 41, 269 (1982) and XLVII, Part 2, 521 (1981)
J. Manz, Chem. Phys. 24, 51 (1977)
46. R. Haensel, H. Kühle, N. Schwentner and H. Wilcke Proceedings of the VII Vacuum Ultraviolet Radiation Conference, Jerusalem (1983)
47. H. Dubost and R. Charneau, Chem. Phys. 12, 407 (1976)
48. F. Legay in Chemical and Biological Applications of Lasers Vol II, Academic, New York (1977)
49. Th. Förster, Z. für Naturforschung 4a, 321 (1949)
50. D.L. Dexter, J. Chem. Phys. 21, 836 (1953)
51. M. Yokota and O. Tanimoto, J. Phys. Soc. Jap. 22, 779 (1967)
52. N. Schwentner and E.E. Koch, Phys. Rev. 14, 4687 (1976)
53. N. Schwentner, E.E. Koch, Z. Ophir and J. Jortner, Chem. Phys. 34, 281 (1978)
54. Ch. Ackermann, R. Brodmann, U. Hahn, A. Suzuki and G. Zimmerer phys. stat. sol. (b) 74, 579 (1976)

55. T. Nanba and N. Nagasawa, J. Phys. Soc. Jap. 36, 1216 (1974)
56. G. Zimmerer in Luminescence of Inorganic Solids ed. B. DiBartolo, Plenum, New York (1978)
57. Z. Ophir, N. Schwentner, B. Raz , M. Skibowski and J. Jortner, J. Chem. Phys. 63, 1072 (1975)
58. D. Pudewill, F.-J. Himpsel, V. Saile, N. Schwentner M. Skibowski, E.E. Koch and J. Jortner, J. Chem. Phys. 65, 5226 (1976)
59. P. Børgesen, J. Schou, H. Sørensen and C. Classen, Appl. Phys. A29, 57 (1982)
60. R.E. Johnson and M. Inokuti, Nucl. Instr. and Meth. to be published
61. O. Chesnovsky, B. Raz and J. Jortner, Chem.Phys. Lett 15, 475 (1972)
62. O. Chesnovsky, A. Gedanken, B. Raz and J. Jortner, Chem. Phys. Lett 22, 23 (1973)
63. A. Gedanken, B. Raz and J. Jortner, J. Chem. Phys. 59, 1630 (1973)
64. Z. Ophir, B. Raz and J. Jortner, Phys. Rev. Lett. 33, 415 (1974)
65. Ch. Ackermann, Thesis, Universität Hamburg (1976) and DESY, Internal Report F41-76/04 (1976)
66. S.S. Hasnain, T.D.S. Hamilton, I.H. Munro, E. Pantos and I.T. Steinberger, Phil. Mag. 35, 1299 (1977)
67. D.S. Tinti and G.W. Robinson, J. Chem. Phys. 49, 3229 (1968)
68. P. Gürtler and E.E. Koch, J. Mol. Structure 60, 287 (1980) and Chem. Phys. 49, 305 (1980)
69. C.M. Surko and F. Reif, Phys. Rev. 175, 229 (1968)
70. J.W. Keto, M. Stockton and W.A. Fitzsimmons, Phys. Rev. Lett. 28, 792 (1972)
J.W. Keto, F.J. Soley, M. Stockton and W.A. Fitzsimmons Phys. Rev. 10, 887 (1974)
71. F.J. Soley, R.K. Leach and W.A. Fitzsimmons, Phys. Lett. 55, 49 (1975)
72. O. Chesnovsky, B. Raz and J. Jortner, J. Chem.Phys. 59, 5554 (1973)
73. N.G. Basov, V.A. Danilychev, Yu.M. Propov and D.D. Khodkevich, JETP Lett. 12, 329 (1970)
74. N. Schwentner, O. Dössel and H. Nahme in Laser Techniques for Extreme Ultraviolet Spectroscopy AIP Conference Proceedings 90 ed. Mc Ilrath and Freeman, American Institute of Physics, New York (1982)
75. W.G. Wrobel, H. Röhr, K.-H. Steuer, Appl. Phys. Lett. 36, 113 (1980)
76. H. Maecker and T. Peters, Z. Physik 139, 448 (1954)
77. W.B. Kunkel and M.N. Rosenbluth in Plasmaphysics in Theory and Application ed. W.B. Kunkel, McGraw-Hill Book Company New York (1966)

ENERGY TRANSFER AND LOCALIZATION IN RUBY

G.F. Imbusch

Department of Physics
University College
Galway, Ireland

ABSTRACT

The efficiency of energy transfer between donor ions and acceptor ions in a solid may be strongly affected by whether or not the excitation on the donor ions can transfer rapidly among the donors, that is, whether the excited donor state is a delocalized or a localized state. According to a theorem of Anderson if the interaction between donors is sufficiently short range the donor excitation will be localized below a critical concentration of donors and will be delocalized above this concentration. Whether or not this Anderson transition occurs in ruby has been hotly debated. In this article we review the experiments which have attempted to elucidate the nature of the energy transfer in ruby and to decide whether an Anderson transition occurs there.

I. THE LOCALIZATION OF OPTICAL EXCITATION IN A SOLID

When an ion is excited to an energy state of optical excitation it may decay radiatively by emission of a photon, or it may decay nonradiatively either by multi-phonon emission or by transferring all or part of its excitation to a nearby ion. The decay may occur by a combination of some or all of these processes. We can visualize a situation where a solid contains two types of dopant ion (A and B) each with distinct absorption and fluorescent transitions. In a typical experiment we would selectively excite a fraction of the A ions. If these were the only dopant ions they would decay radiatively. Now because of the presence of the B ions the excitation on the A ions may also decay by nonradiative energy transfer to B ions. We call the A ions the "donors" and the B ions the "acceptors". The occurrence of this energy transfer would be noticed as a reduction in the donor fluorescence and a concomitant appearance of acceptor fluorescence. The decay rate of the excited A ions may also be affected by the transfer.

The transfer mechanisms which can cause transfer between two adjacent ions have been described by many authors [1], [2], [3]. To relate these microscopic transfer mechanisms to experimentally observed effects one must apply sophisticated statistical analyses. An additional complication is that donor → donor transfer may be occurring at the same time as donor → acceptor transfer and this may greatly affect the efficiency of the observed donor → acceptor transfer. Indeed, it may be felt that since donor → donor transfer is a "resonant" process it would tend to occur more efficiently than donor → acceptor transfer where there is invariably an energy mismatch. One must invoke a phonon-assisted transfer mechanism for donor → acceptor transfer and the energy difference between the donor and acceptor states is released in the form of a phonon or phonons. Hence to analyze a particular experiment in which an energy transfer occurs between two dissimilar types of ions, one must consider whether a resonant transfer is occurring also among similar ions.

On further analysis, however, the conditions for a true resonant process may be difficult to achieve. Strains in the material can shift the positions of the energy levels of ions, and random microscopic strains may mean that the energy levels of two adjacent donor ions may be shifted in energy relative to each other. Whether or not this strain-induced energy shift is greater than the homogeneous linewidth of the levels then becomes of paramount importance; if there is no overlap within the homogeneous linewidth there is no resonant transfer.

We see that the nature of the strains in the material is an important consideration. Random microscopic strains are much more disruptive than macroscopic strains where the strain regions are large and the difference in strain between adjacent sites is small. In our problem we are concerned with the strains in the crystal in the vicinity of the dopant ion. If another dopant ion is nearby, close enough for transfer to occur, it is likely that the presence of this nearby dopant ion causes a local strain in the host crystal. Thus the introduction of random microscopic strains may be an inherent part of the doping process. Whereas true resonant donor → donor transfer could be orders of magnitude faster than the phonon-assisted donor → acceptor transfer which might involve the emission of a 100 cm^{-1} phonon, an energy mismatch among adjacent donor ions greater than the homogeneous linewidth would mean that the donor → donor transfer, too, would be a phonon-assisted process. And since in this case we are dealing with phonons of energy 1 cm^{-1} or smaller, the donor → donor transfer rate could be orders of magnitude smaller than the donor → acceptor transfer rate.

So far we have looked at the situation from the single ion point of view. But if a similar dopant ion is nearby, close enough for a strong near-resonant interaction to occur between the ions, then it is more correct to look at the coupled ions as constituting a single quantum system. We can include other nearby similar ions in this analysis, eventually ending up with a single extended quantum system. The excited state of this extended quantum system is an "extended state", and donor → acceptor transfer would be visualized as a transfer of the excitation of this extended state to an acceptor ion within the region of the extended state.

The question of whether the donor excitation stays localized on single donor ions or becomes delocalized and exists in the form of an extended state is extremely interesting and should greatly affect the process of donor → acceptor transfer. Conversely, the experimental study of donor → acceptor transfer may shed light on the question of whether the excitation on the donor ion system is localized or delocalized.

To return to the donor → donor case, a theoretical problem involving the transport of excitation in a lattice of random centres was considered by Anderson in an important paper [4]. He considered a three-dimensional lattice in which the active centres could be randomly distributed and the energy of a centre varies randomly from site to site. This randomness in energy gives an inhomogeneous width to the observed energy level, and this inhomogeneous width, ΔE, is one of the parameters of the model. The interaction between centres j and k which can transfer excitation between the centres is written $V_{jk}(r_{jk})$. The probability that excitation resides on centre j at time t is written $a_j(t)$. It is assumed that $a_j(0) = 1$ for a particular centre j, that is, that this centre is excited at t = 0, and the value of $a_j(t)$ at a later time is calculated. If $\langle|aj(\infty)|^2\rangle_{AV} \neq 0$ the excitation is said to be localized. If $\langle|aj(\infty)|^2\rangle_{AV} = 0$ the excitation is said to be delocalized. Anderson's conclusion is that if $V_{jk}(r)$ falls off faster than 1/r3 then there is a critical concentration of centres below which transport cannot take place and the excitation stays localized. The criterion for localization is

$$\frac{\Delta E}{V} \geqslant \sim 2 \qquad (1)$$

where V is the average interaction between adjacent centres. V will increase as the average separation between adjacent centres decreases, that is, as the concentration of centres increases until it reaches the critical concentration above which the criterion no longer holds and the excitation is delocalized. As the concentration increases through the critical concentration there should be an abrupt transition from a localized to a delocalized state. An unambiguous demonstration of this Anderson transition has not been easy to find, as it is difficult to find a system which satisfies the basic assumptions of Anderson's theory.

Anderson's ideas were extended by Mott [5] and others [6],[7] in the case where the concentration is above the critical value. Consider the distribution of centres which makes up the inhomogeneous line. Above the critical concentration fast transfer of excitation should occur among centres whose energies are at or near the centre of the inhomogeneous line but excitation may stay localized on the centres whose energies occur far out in the wings of the line. The states which make up the inhomogeneous line will then consist of delocalized states in the centre and localized states in the wings, and the boundaries between these states are called "mobility edges". This situation is sketched schematically in Fig. 5 and will be discussed in greater detail later. The demonstration of mobility edges would be indicative of an Anderson transition.

In order for Anderson localization to occur the interaction between ions must be of shorter range than dipole-dipole (which varies as $1/r^3$) and must be much less than the inhomogeneous width. Hsu and Powell [8] claim that such is the case for $CaWO_4 : Sm^{3+}$ where the interaction between Sm^{3+} ions is electric quadrupole-quadrupole, and their data suggest an absence of transfer at low temperatures where the interaction strength is much less than the inhomogeneous linewidth. Likewise it has been claimed [9] that the excitation on Cr^{3+} ions in ruby undergoes a transition from a localized state to a delocalized state at a concentration around 0.3 at.%. The possibility that an Anderson transition might be occurring in ruby has led to an intense reexamination of the ruby system by a variety of experimental and theoretical techniques. These studies have been fruitful in deepening our understanding of the excitation transfer process in inorganic insulating materials and in clarifying theoretical aspects of the Anderson localization problem.

II. ENERGY TRANSFER IN RUBY - EARLY EXPERIMENTS

Ruby consists of Al_2O_3 in which some Cr^{3+} ions are substituted for Al^{3+} ions. These Cr^{3+} ions have broad and strong absorption bands in the visible, and all Cr^{3+} ions excited by optical pumping end up on the metastable ^2E state from which the ions decay radiatively; the quantum efficiency is close to unity (Fig.1). The lifetime of this metastable state is about 4 ms and the transition between ^2E and the ground state consists mainly of two sharp lines, R_1 and R_2, of which only the R_1 line is observed in emission at liquid helium temperatures. The R_1 and R_2 lines also appear in absorption As a result R_1 photons emitted from some excited Cr^{3+} ions can be reabsorbed by other unexcited Cr^{3+} ions ("trapping"), leading to a

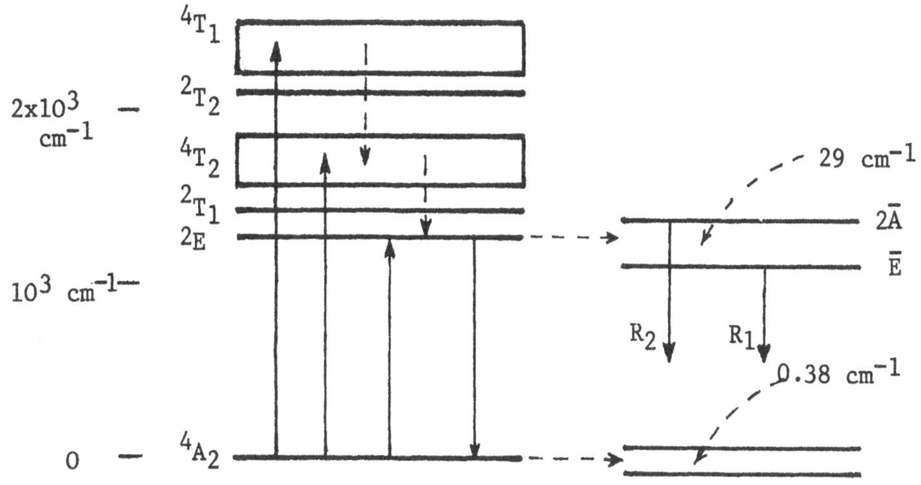

Fig. 1. Low lying energy levels of Cr^{3+} in ruby. The splitting of the ^2E and $^4A_2$ states are shown out of scale. The solid vertical arrows indicate radiative transitions, the broken vertical arrows indicate nonradiative transitions.

resonant radiative transfer of optical excitation which is pronounced at low temperatures in heavily doped single crystals of mm or larger dimensions [10]. This reabsorption of luminescence causes a lengthening of the observed luminescence decay time of ruby crystals.

As the concentration of Cr^{3+} ions increases the likelihood increases that two Cr^{3+} ions will enter sites near enough to each other to interact strongly and form a distinct luminescence centre - an exchange coupled pair. A rich spectrum of such pairs has been found in ruby [11]-[15]. Much attention has focussed on third nearest neighbour pairs and fourth nearest neighbour pairs as these have particularly strong luminescence features. The N_1 line at 7041 A comes from the third nearest neighbour pairs, and the N_2 line at 7009 A comes from the fourth nearest neighbour pairs.

Because the intensities of the N_1 and N_2 lines seemed more intense than was expected on the basis of statistical probabilities of the occurrence of exchange coupled pairs, Schawlow et al. [11] were led to suggest that these pairs could be excited by energy transfer from excited single ions as well as by optical excitation by broadband light sources. This energy transfer was studied, mainly at 77K, by Imbusch [16]. Two possibly energy transfer mechanisms are (i) radiative transfer, in which the pairs would absorb the single ion luminescence before it leaves the crystal, and (ii) nonradiative transfer from excited single ions to pairs. The nonradiative transfer process, which acts as an additional decay mechanism for the single ions, should result in a decrease in the observed lifetime of the single ion emission. Radiative transfer to pairs, on the other hand, should not affect the decay time of the single ions. The observed R_1 decay time as a function of concentration is seen in Fig.2. As the chromium concentration is increased the density of exchange-coupled pairs increases and transfer from single ions to pairs should become more probable. In the mm size single crystal samples trapping by unexcited single ions causes an initial increase in the measured lifetime, but at concentrations above 0.3 at.% there is a rapid drop in lifetime. By using tiny amounts of microcrystalline ruby (powdered samples) the trapping is eliminated and the lifetime stays constant until around 0.3% where it begins to show a sharp decrease, confirming the nonradiative energy transfer from single ions to pairs.

Additional evidence of transfer is seen by studying the decay

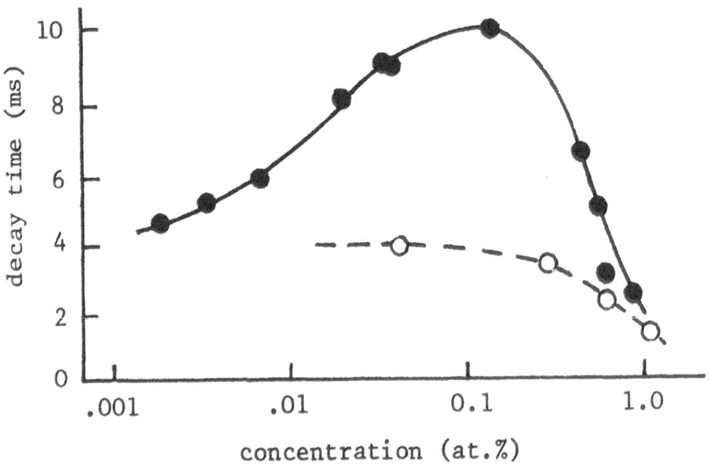

Fig. 2. Experimental values of R_1 decay time at 77 K as a function of concentration. The solid circles are for mm-size samples in which trapping occurs. The open circles are for powdered samples free from trapping.

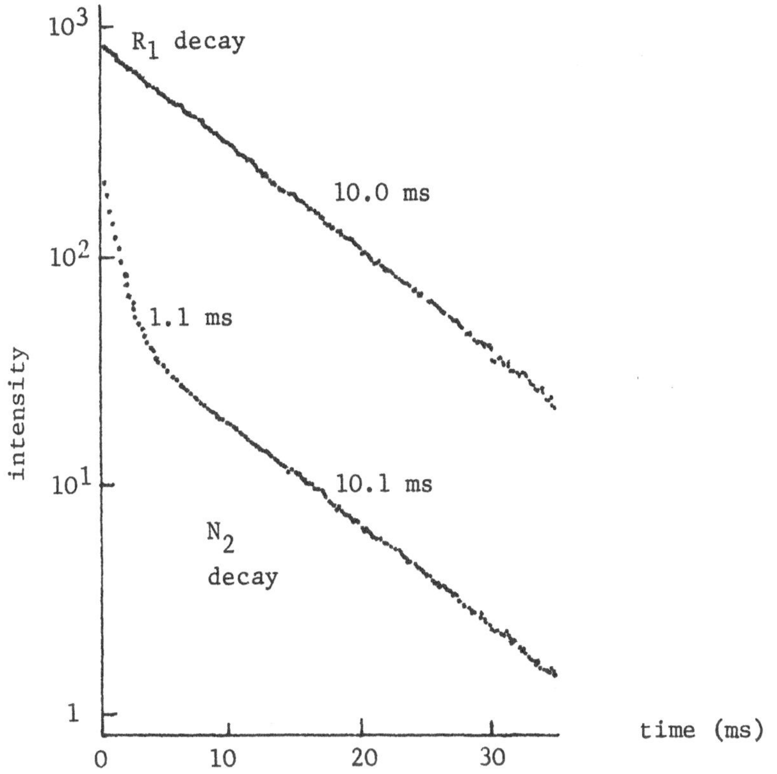

Fig. 3. Fluorescence decay patterns of the R_1 and N_2 lines of ruby (0.15 at .%) at 77K.

patterns of the single ion and pair luminescence. Fig.3 shows the decay patterns after pulsed excitation of the R_1 and N_2 lines at 77K in a mm-size ruby crystal of ≃ 0.2% concentration of chromium [16]. The R_1 decay is single exponential with a decay time of 10.0 ms - lengthened because of trapping. The N_2 decay is a double exponential with an initial fast decay time of 1.1 ms and a later component with the slower time of 10.1 ms, and this later component has, within experimental error, exactly the same decay time as the single ions. The intrinsic decay time of the excited fourth nearest neighbour pair centre is 1.1 ms, hence the 1.1 ms component of the decay is emitted by pairs pumped directly by the broadband flash lamp. The 10.1 ms decay is emitted by pairs excited by energy transfer from single ions. Now the fact that the later component of the N_2 decay has exactly the same decay rate as the R_1 line suggested that at liquid nitrogen temperatures where these measurements were made the pairs were being excited by energy transfer not just by nearly single ions but by the main body of single ions. This could occur if there were a fast resonant transfer among the single ions so that they acted collectively. That becomes clear if the rate equations are examined. The situation is sketched in Fig.4. The single ions act collectively

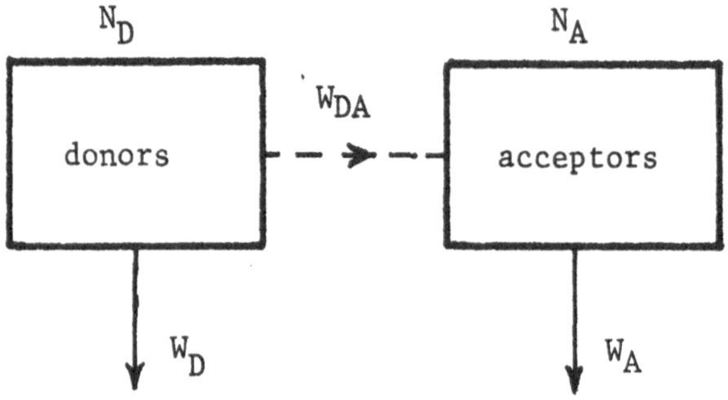

Fig. 4. Representation of the donor (single ion) to acceptor (ion pair) nonradiative energy transfer where the single ions act collectively. There is no back transfer from pairs to single ions at low temperatures.

as a donor system, the ion pairs are the acceptors. The donors have a collective radiative decay rate, W_D, and a collective nonradiative transfer rate to the acceptors, W_{DA}. The intrinsic decay rate of the acceptors is W_A. At liquid nitrogen temperatures and below there is no back transfer. Let N_D and N_A be the numbers of excited donors and excited acceptors, respectively. Then

$$\frac{dN_D}{dt} = -W_D N_D - W_{DA} N_D \tag{2}$$

$$\frac{dN_A}{dt} = W_{DA} N_D - W_A N_A \tag{3}$$

The solution of Eq.2 is

$$N_D(t) = N_D(0) \exp\left[-(W_D + W_{DA})t\right] \tag{4}$$

where $N_D(0)$ is the number of excited donors at $t = 0$, immediately after pulsed excitation. Inserting expression (4) into expression (3) and solving for N_A we obtain

$$N_A(t) = N_A(0)\exp\left[-W_A t\right] + N_D(0)\,\frac{W_{DA}}{W_A-(W_D+W_{DA})}\,\{\exp\left[-(W_D+W_{DA})t\right]-\exp\left[-W_A t\right]\} \tag{5}$$

where $N_A(0)$ is the number of excited acceptors at t = 0 immediately after pulsed excitation. We assume that $W_A > W_D + W_{DA}$, as is the case for the ruby system where the intrinsic pair decay time is shorter than the observed single ion decay time. The first term of Eq. 5 gives the decay, by radiation, of these initially excited acceptors. The second term gives the number of acceptors excited by transfer from the donors. This number rises from zero at t = 0, peaks at some later time, and then decays at exactly the same decay rate as the donor system.

Since the intensity of luminescence (I) from a centre varies as the number of such excited centres, the donor intensity decays as a single exponential rate $(W_D + W_{DA})$ and the acceptor intensity should decay with a double exponential character, an initial fast decay at W_A and a slower decay at $(W_D + W_{DA})$. This is exactly the behaviour seen in Fig.3.

The occurrence of trapping is a complicating factor here since trapping could act as a resonant transfer mechanism and cause the single ions to act collectively. However, measurements on powdered samples (which were essentially free from trapping) by Imbusch showed that in all cases at liquid nitrogen temperatures the slow component of the N_2 decay has the same decay time as the main body of single Cr^{3+} ions. Further, this same behaviour was later reported in powdered samples at liquid helium temperatures [17],[18].

With increasing concentration one expects W_{DA} to increase and one expects to find a decrease in the lifetime of the single ions, $\tau = (W_D + W_{DA})^{-1}$. This is found in the ruby lifetime data above ≃0.3 at.% as Fig.2 shows. The decrease in observed single ion lifetime confirms that the single → pair transfer is a nonradiative process, since radiative transfer would not affect the observed lifetime of the single ions.

Assuming that the single ion system acts collectively Imbusch [16] concluded that a fast (∿ microsecond transfer time) resonant nonradiative transfer occurred among the single ions. The mechanism for this transfer had to be faster than dipole-dipole and was possibly a quadrupole-quadrupole mechanism. Birgeneau [19] made accurate calculations of the quadrupole-quadrupole matrix elements and showed that they were much too small to cause a microsecond transfer. He suggested that an exchange interaction could provide the mechanism for such fast resonant transfer. Now the exchange interaction is of very short range and this is one of the requirements for Anderson localization to occur below some critical concentration. Lyo [9], using reasonable values for the exchange interaction and using Anderson's analysis, estimated that the critical concentration below which localization would occur should be 0.3 - 0.4 at.%

Not all workers were willing to accept this picture of the energy transfer in ruby. Gerlovin [20] reported that his decay measurements made on thin polycrystalline ruby samples showed departures from a single exponential R_1 line and he concluded that fast resonant transfer was not occurring among the single ions. Heber and Murmann [21] performed heat pulse measurements on ruby samples at low temperatures and also concluded that single ion → single ion transfer is much slower than single ion → pair transfer. Siebold and Heber [22] also failed to find evidence of fast resonance transfer among the single Cr^{3+} ions in $LaA\ell O_3 : Cr^{3+}$.

The early experimental work on energy transfer in ruby was carried out using broadband optical pumping sources and was mainly done at liquid nitrogen temperatures. From 1970 onwards more sophisticated experimental techniques using narrowband laser pump sources were used, and the theoretical analysis focussed on the question of the nature of the single ion → single ion transfer at liquid helium temperatures where the complication due to thermal effects are minimized and the homogeneous broadening is much smaller than the inhomogeneous broadening.

III. THE SEARCH FOR MOBILITY EDGES IN THE RUBY R_1 LINE

An ingeneous experiment to look for the o-currence of Anderson transition in ruby at liquid helium temperatures was carried out by Koo et al. [23]. Their idea was to use a tunable narrow band laser to excite ions in distinct regions of the inhomogeneous R_1 line so as to discriminate between localized and delocalized states and detect a mobility edge (Fig.5(a)). A cw ruby laser whose output bandwidth was much narrower than the R_1 linewidth was tuned by a combination of magnetic field and temperature and pumped by an argon ion laser. This was used to excite ions in different regions of the R_1 linewidth. The ruby laser output could be chopped by means of a mechanical shutter.

In the centre of the line where the excitation is expected to be delocalized fast resonant transfer among the single ions should lead to an efficient pumping of the exchange coupled pairs and consequently to an enhanced fluorescence from the pairs. Exciting the localized states, on the other hand, should not enhance transfer to pairs. In their experiment Koo et al. tuned the narrow band laser from the centre of the R_1 line out to the wings on the low energy side of the R_1 line and they monitored the ratio of N_2 emission to R_1 emission (N_2/R_1). The N_2/R_1 value should stay constant over the region of delocalized states but should drop abruptly at the mobility edge, as is shown schematically in Fig.5(b). Their data, taken with great care and using many precautions showed evidence of an abrupt break which, they felt, indicated the existence

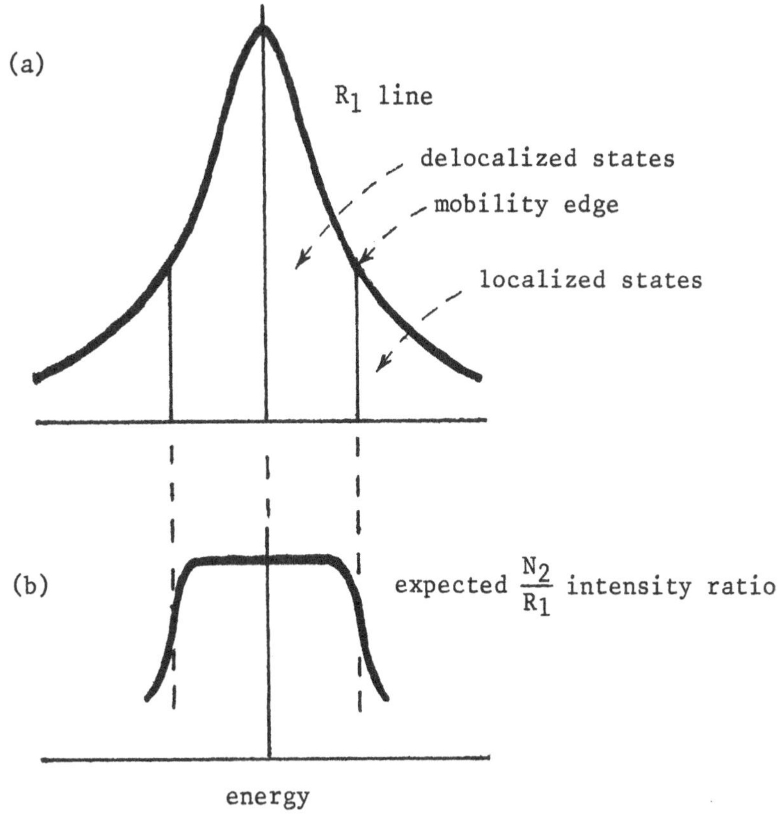

Fig. 5. (a) Schematic representation of the division of the R_1 states into localized and delocalized states separately a mobility edge. (b) The expected N_2/R_1 ratio as the exciting laser beam is tuned through the R_1 line.

of a mobility edge and was consistent with an onset of localization at a critical concentration of around 0.8%.

Doubts were expressed [17] at whether the results of the Koo et al. experiment could only be interpreted as evidence for a mobility edge between localized and delocalized states; other explanations were considered possible. It was felt that a search for the appearance of a symmetrical N_2/R_1 pattern on either side of the R_1 line should be made, Koo et al. had only looked at the low energy side. Because of the significance of this experiment it was repeated again by Chu et al. [24] using a cw dye laser which could be tuned across the entire R_1 line. Their N_2/R_1 patterns were not identical on the high and low energy sides of line centre and did not exhibit clearly the abrupt break found by Koo et al. So the question as to whether a transition between localized and delocalized states occurs in ruby at low temperatures remained in doubt.

IV. FLUORESCENCE LINE NARROWING AND HOLE BURNING EXPERIMENTS IN RUBY

The R_1 line has an observed width of around 1 cm^{-1} at helium temperatures. This is inhomogeneous broadening due to random strains in the crystal. The homogeneous width (ΔE_{homo}) of the line is much narrower than this, being around 0.003 cm^{-1} ($\simeq$ 100 MHz). The observed lineshape is a profile made up of very many narrow emissions (or absorptions) from ions in different strain environments. This is shown schematically in Fig. 6. If a very narrow band laser with bandwidth ΔE_{laser} is used as a pumping source it can excite the ions in a particular strain environment. When the pumping source

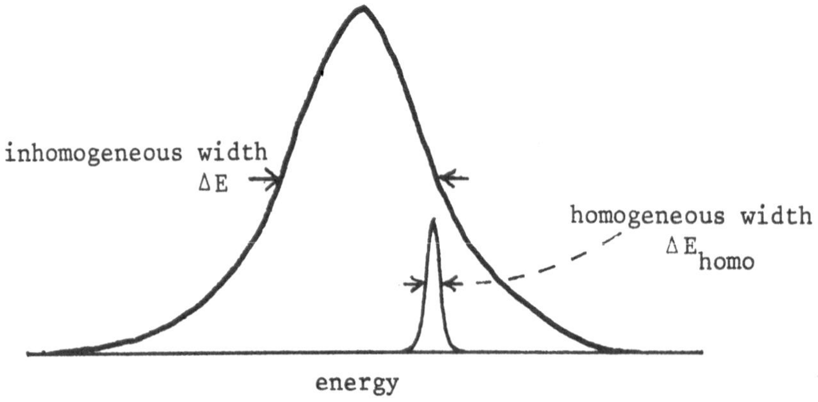

Fig. 6. The homogeneously broadened line is a profile of very narrow emissions (or absorptions) each from ions in sites of identical strain.

is turned off one expects to see fluorescence only from those ions excited by the laser, and the observed linewidth is

$$\Delta E = \Delta E_{laser} + 2\,\Delta E_{homo} \quad (6)$$

Such a fluorescence signal can be much narrower than the R_1 bandwidth. This is the phenomenon of fluorescence line narrowing (FLN).

Fluorescence line narrowing in ruby was first demonstrated by Szabo [25] who demonstrated directly the homogeneous bandwidth of the R_1 line [26]. Szabo also demonstrated optical hole burning in the ruby R_1 line [27]. He used these techniques to gain information on broadening mechanisms affecting the ruby line. Szabo and Kroll also demonstrated [28] that information about homogeneous broadening could be obtained from measurements on the ruby R_1 line in the time domain such as photon echoes and optical free induction decay.

Because of the 0.38 cm^{-1} ground state splitting in the ruby R_1 line the FLN signal appears as a double line of 0.38 cm^{-1} separation as Fig. 7. shows. Sometimes additional lines may appear.

FLN in the ruby R_1 line was used by Selzer et al. to study energy transfer. The narrow band laser excites ions in a particular strain field. If a mechanism exists for transferring excitation from this excited set of ions to ions in different strain regions (and hence with different excited state energy) and if the transfer can occur within the radiative decay time of the excited ions one should see a gradual broadening of the FLN signal. Such spectral diffusion was observed by Selzer et al. [29], [17] in the FLN signal in ruby. Some of their experimental results are shown in Fig. 8. What is interesting is that the spectral diffusion does not appear as a gradual broadening of the narrow FLN lines but appears as a gradual rise of the entire inhomogeneous emission line.

Spectral diffusion is a phonon-assisted transfer mechanism and the diffusion rate is seen to be independent of the energy mismatch - since the entire inhomogeneous line grows. Further, the observed spectral transfer rate shows a linear dependence on temperature [29] up to around 50 K. This linear dependence on temperature of the transfer rate and its independence of energy mismatch are consistent with a direct one-phonon assisted process, as the calculations of Holstein et al. have shown [30]. The appearance of the full inhomogeneous emission line can also be considered as evidence for the occurrence of microscopic strains in ruby.

The slow non-resonant single ion → single ion transfer demonstrated in these experiments is quite different from the resonant ion → ion transfer discussed previously.

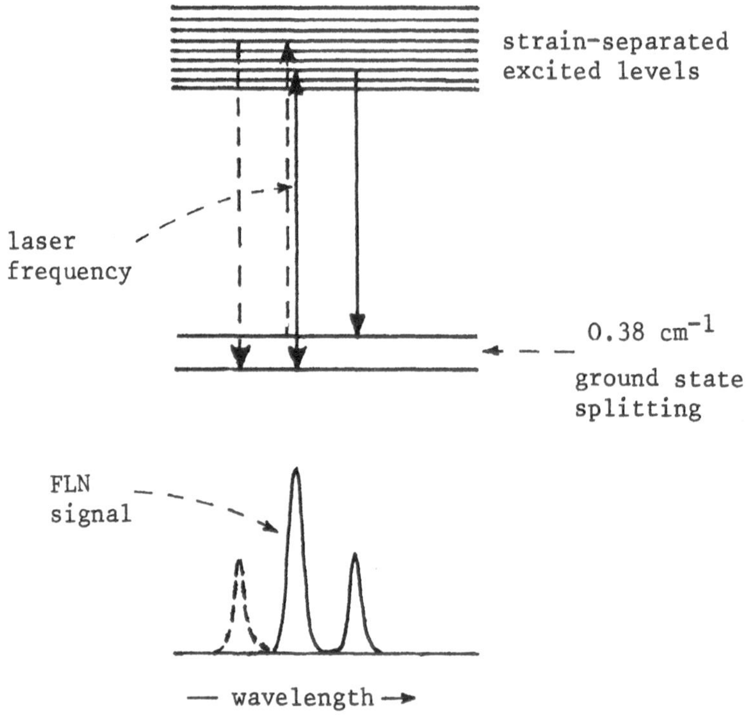

Fig. 7. The excited states of ions in different strain fields have different energies. The narrow band laser pulse excites one set of ions in identical strain fields (upward solid arrow) which emit two sharp fluorescence lines as shown (FLN signal). The laser beam may also excite a separate set of ions - as shown by the broken upward arrow. These ions will contribute a third narrow fluorescence line at lower wavelength.

In this same series of experiments Selzer et al. demonstrated the existence of single ion → pair energy transfer in a very clear way. When the narrowband laser pulse (≃10 ns duration) was tuned to the centre of the R_1 line and the R_1 fluorescence monitored a single exponential decay was always obtained. When the laser was kept tuned to the R_1 line centre but the N_2 line was monitored the fluorescence consisted of a rise followed by an exponential decay,

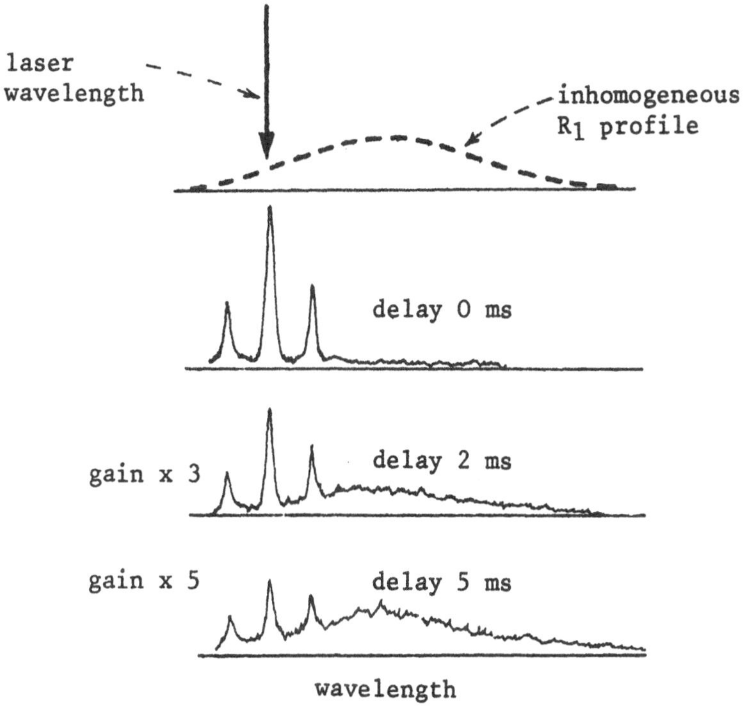

Fig. 8. FLN signals from a 0.9 at.% ruby at 10 K. The laser excitation is on the high energy side of line centre. The delay indicates the time after the pulsed laser excitation.

and this decay had the same decay time as the R_1 line. This is consistent with the transfer of excitation from the main body of Cr^{3+} ions to the pairs, and was taken by Selzer et al. as supporting the idea of a fast resonant transfer among the single ions.

By shifting the laser energy by about 2 cm^{-1} so as to coincide with an excited pair level in near coincidence with the R_1 level Selzer et al. were able to pump the pairs directly and by monitoring the N_2 fluorescence they obtained the expected fast intrinsic decay pattern of the pairs. These N_2 decay patterns are shown in Fig. 9.

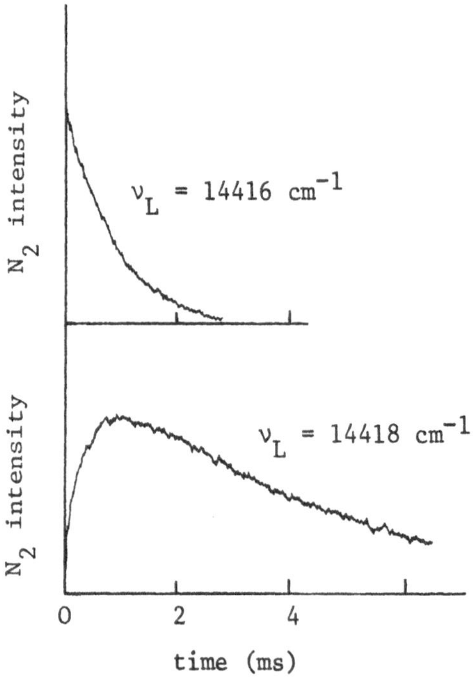

Fig. 9. Decay patterns of the N_2 line in a 0.51 at.% sample at 5 K. In the upper trace the laser was tuned to a pair obsorption. In the lower trace the laser was tuned to the centre of the R_1 line.

Whereas Selzer et al. interpreted their FLN experiments as fovouring the occurrence of fast resonant transfer among single Cr ions, Szabo, on the other hand, interpreted his experimental results as militating against the notion of such fast resonance transfer. Some more direct methods of examining transfer were needed to resolve the question of the existence of an efficient resonant ion → ion transfer in ruby.

V. DEGENERATE FOUR WAVE MIXING EXPERIMENTS IN RUBY: AN ATTEMPT TO DIRECTLY MEASURE THE ENERGY MIGRATION DISTANCE

Attempts wer made by Eichler et al. [31], Liao et al. [32], and Hamilton et al. [33] to obtain a direct experimental measurement of the single ion → single ion energy migration distance in ruby using the technique of degenerate four wave mixing. In these experiments two laser beams (derived from the same laser) are focussed onto a thin ruby sample where they produce a spatial periodic variation in the density of excited Cr^{3+} ions. This produces a spatial periodic modulation of the refractive index of the material which acts as a diffraction grating and can cause diffraction of a third laser beam. When the two laser beams which create the grating are switched off the grating decays with a time constant which depends on the radiative decay time of the excited Cr^{3+} ions and on the diffusion of excitation among the Cr^{3+} ions. In principle this affords a direct method of measuring the diffusion of excitation among the ions.

In the experiments of Liao et al. two laser beams derived from the same laser of wavevectors $\vec{k}_1$ and $\vec{k}_2$ are exactly counter-propagating ($\vec{k}_2 = -\vec{k}_1$) while the third laser beam, also derived from the same laser, of wavevector $\vec{k}_3$ makes a small angle ($\theta \simeq 3^o$) with laser beam $\vec{k}_1$. (Fig. 10). If $\vec{k}_2$ is switched off and $\vec{k}_1$ and $\vec{k}_3$ focussed on the sample then a grating is formed, as shown schematically in Fig. 10. The spatial periodicity of the grating is

$$d = \frac{\lambda}{2 \text{ Sin } \theta/2} \qquad (7)$$

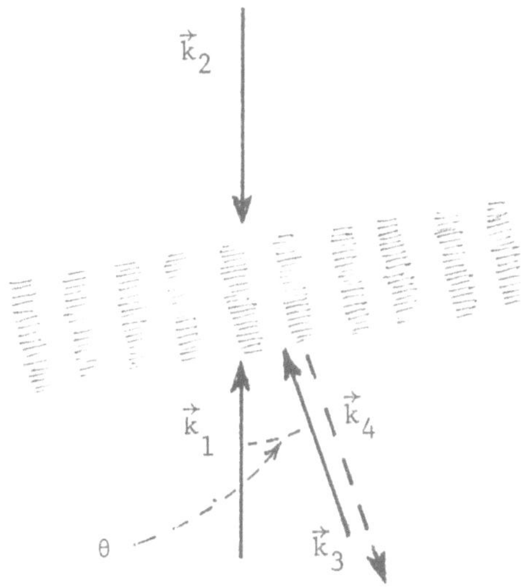

Fig. 10. Representation of the grating formed by laser beams $\vec{k}_1$ and $\vec{k}_3$. Beam $\vec{k}_2$ is diffracted by the grating resulting in a beam coherently scattered in the direction $\vec{k}_4$

where λ is the laser wavelength and θ is the angle between the two laser beams forming the grating ($\simeq 3^{\circ}$ in this case).

If $\vec{k}_1$ and $\vec{k}_3$ are now switched off and $\vec{k}_2$ focussed on the crystal it will be diffracted by the grating and according to the Bragg law we expect the first diffracted beam to be along the $\vec{k}_4$ direction, i.e. exactly backwards along the $\vec{k}_3$ direction, as shown in the figure. The gradual fading of the grating means that the backward diffracted beam decreases in intensity as

$$I_4(t) = I_4(0) \exp\left[-t/\tau\right] \tag{8}$$

where the decay time τ is given by

$$1/\tau = \frac{2}{\tau_R} + 2D\left(\frac{4\pi}{\lambda}\,\mathrm{Sin}\,\frac{\theta}{2}\right)^2 \tag{9}$$

where τ_R is the fluorescence lifetime of the Cr^{3+} ions and D is the diffusion constant describing the transfer of excitation among the Cr^{3+} ions.

In a second experiment $\vec{k}_2$ and $\vec{k}_3$ are used to create a grating whose spatial periodicity is given by expression (7) with $\theta \simeq 177^{\circ}$. These beams are then switched off and beam $\vec{k}_1$ is focussed on the crystal. The diffracted beam is once again along $\vec{k}_4$. The decrease in intensity of beam $\vec{k}_4$ is again monitored, and this is expected to decay as described by expressions (8) and (9) with $\theta = 177^{\circ}$. By comparing the backward diffracted lifetimes obtained with the two different gratings ($\theta = 3^{\circ}$ and $\theta = 177^{\circ}$) a value of D should be obtained. Very little difference was found between the two lifetimes and only an upper limit for D could be obtained. For a chromium concentration of 1.55 at.% the upper limit for D is 1.7×10^{-9} cm^2 s^{-1}.

One can attempt to relate this diffusion constant to an average transfer rate by the following very simple approach. We relate the diffusion constant to the average separation between adjacent <u>resonant</u> ions, ℓ, and the average transfer time between adjacent resonant ions, τ_0 by the equation

$$\ell^2 = 6\,D\,\tau_0 \tag{10}$$

The average separation between adjacent ions in the 1.55 at.% sample is about 10A but the transfer takes place between resonant ions (i.e. those whose energies are within the same homogeneous width) and two such adjacent resonant ions should be considerably farther apart than 10A. If the homogeneous width is taken as 100 MHz while the inhomogeneous width is taken as 1 cm^{-1} one estimates an average separation between adjacent resonant ions at around 60A for the 1.55 at.% sample. Taking this as ℓ we calculate the transfer time as $1/\tau_0 \approx 3 \times 10^4 \ s^{-1}$. No more than about 100 transfers would occur in this material within the radiative lifetime, indicating a diffusion distance of less than 600 A. From their experiments on many samples Liao et al. feel that the upper limit for the diffusion distance is 300 A. Hence there is no detectable smearing out of the grating because of energy transfer within the fluorescence decay time of the ions. This is also the conclusion of Hamilton et al. It should be noted, however, that the upper limit for the diffusion distance is greater than the average separation between pair centres in the 1.55 at.% sample, which is around 60 A.

It is clear that a finer probe than that supplied by these experiments is necessary to elucidate the nature of the energy transfer in ruby at the atomic level.

VI. ELECTRIC FIELD EXPERIMENTS IN RUBY

A very ingeneous experiment was devised and carried out by Chu et al. [34] to study $Cr^{3+} \rightarrow Cr^{3+}$ transfer at the level of individual Cr ions. This experiment takes advantage of the fact that individual Cr^{3+} ions in the sapphire lattice occupy one of two inequivalent sites, A and B sites, as shown in Fig.11. At each site there is an odd component of electric field due to the surrounding ions, this internal field acts along the crystallographic axis, but acts in opposite directions at the A and B sites. If an external field, E_0 is applied along the axis it adds to the internal field at one site but subtracts from the internal field at the other site. Because of the different resultant fields acting at the two sites Cr^{3+} ions in the A and B sites have their energies shifted relative to each other when an external electric field is applied. The resultant "pseudo Stark splitting" [35] of the R_1 line is quite large; the R_1 lines from the A and B site ions can be separated by around 1 cm^{-1}, enough to move ions out of resonance with each other.

The experiment of Chu et al. can be visualized with the aid of Fig. 12. (a): Without the external field the R_1 lineshapes of the A and B ions are superimposed upon each other. (b): When the external field is applied the A and B transitions are separated. A subset of the A ions is resonantly excited by a fast pulse from a narrow bandwidth dye laser. A FLN signal is seen immediately after the excitation. This is shown in the figure as a single shaded narrow line, whereas in practice it consists of three or four

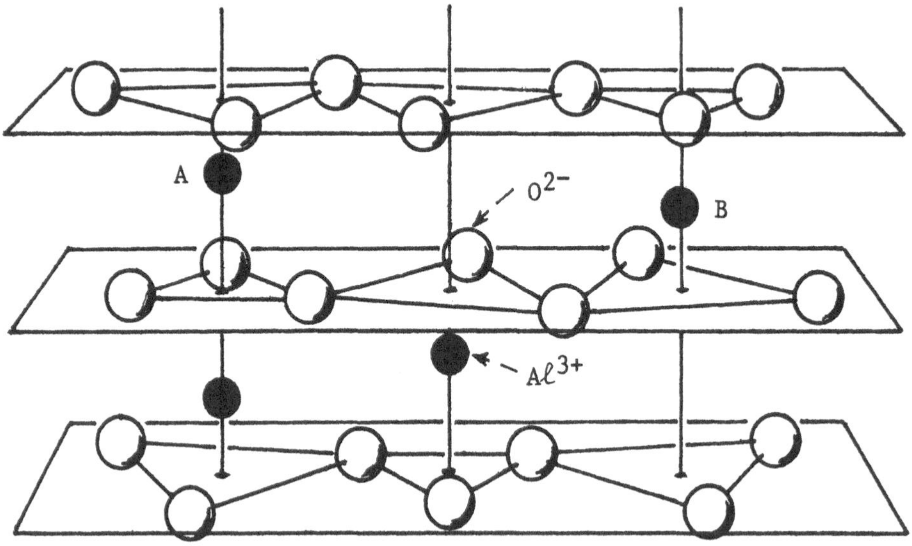

Fig. 11. Position of the $A\ell_2O_3$ crystal structure showing the A and B sites of the $A\ell$ ions. The optic axis is perpendicular to the planes of oxygen ions in the figure.

separate lines because of the ground state splitting. The B ions are unexcited by the laser. (c): The external field is switched off for a specific time, t. This brings the A and B ions back into resonance and allows the excited A ions to transfer energy to the unexcited B ions which are now in resonance with them. If there is a fast resonant transfer between single ions this mechanism should transfer a sizeable portion of the A ion excitation to B ions. (d): The field is switched on again separating the A and B ions, and one looks for a FLN signal from the B ions - indicated by the second narrow shaded line in the figure. The efficiency of the transfer is related to the relative intensities of the FLN signals from the A and B ions. The experiments were conducted at helium temperature.

By varying the time, t, during which the field is switched off, during which time the A → B transfer takes place, one can study the time dependence of the transfer. For the weakly doped crystals the transfer rate varies as $t^{1/2}$, indicative of dipole-dipole coupling between the ions. The conclusion of Chu et al. is that there is nonradiative resonant transfer between single Cr^{3+} ions in ruby, but that it is slow - less than half the excited Cr^{3+} ions making one

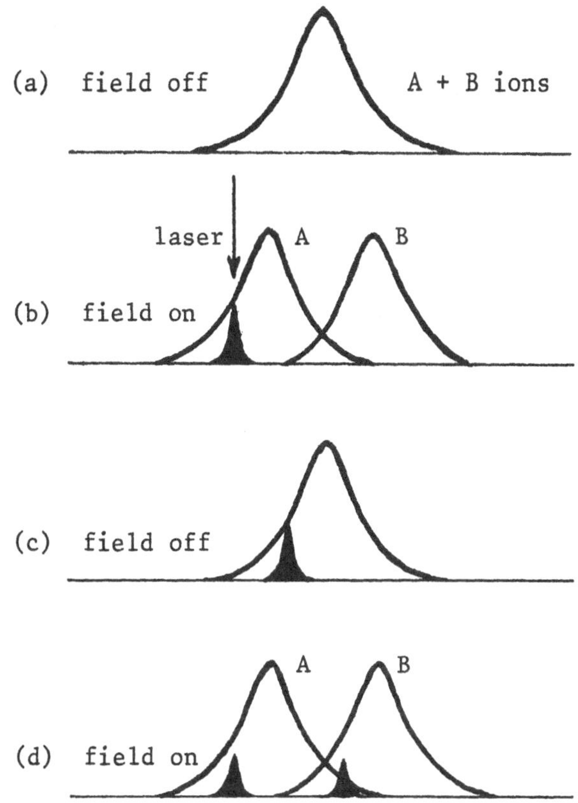

Fig. 12. An external electric field separated the A and B ions (b). A subset of the A ions is excited by the laser and a FLN signal is observed. The A and B ions are brought into resonance (c) and if transfer occurs a separate FLN signal is observed from the B ions as soon as the A and B ions are again separated by the external field (d).

transfer during the fluorescent lifetime in a 0.25 at.% sample. This millisecond transfer time is much slower than the microsecond transfer times initially posulated

Jessop and Szabo [36] conducted very similar experiments on ruby independently of Chu et al. By skillful experimental techniques Jessop and Szabo were able to separate radiative and nonradiative transfer processes, and they, too, concluded that the resonant transfer is slow, of millisecond transfer time in 0.8% ruby, and the transfer could be proceeding by a dipole-dipole interaction.

Gibbs et al. [18] made a careful calculation of the dipole-dipole transfer from A to B ions using the experimentally known dipole matrix elements and using a reasonable value for the homogeneous width of the R_1 line. Their calculated transfer rate is close enough to the rate measured in the electric field experiments of Chu et al. to confirm that a dipole-dipole mechanism can account quantitatively for the transfer observed in the electric field experiments.

Now the evidence that led originally to the idea of a fast resonant nonradiative transfer between single Cr^{3+} ions followed by a slower transfer to pairs was the observation, confirmed by many workers, of a single exponential R_1 decay and of a pair line decay in which the slow component decayed at the R_1 decay rate. If the single ions were not acting collectively, it was argued, the R_1 decay and the slow component of the pair line decay would not be identical. Instead both decays would be non-exponential. Gibbs argues that if one makes a theoretical calculation of the pair decay pattern assuming that single ion → pair transfer occurs by dipole-dipole interaction and that no single ion → single ion decay occurs (the electric field experiments indicate that it is weak) one finds that the deviation of the slow component of the pair decay from a single exponential form is small and difficult to detect. Consequently, he argues, a slow nonradiative resonant transfer by dipole-dipole interaction is not inconsistent with the experimental measurements of Imbusch, Gerlovin, Selzer et al., Chu et al. These electric field experiments would seem to offer convincing arguments against the notion of a fast nonradiative resonant transfer of energy among single ions.

An argument has been made that whereas these electric field experiments would seem to rule out the possibility of fast nonradiative resonant transfer between A and B ions, they do not rule out the possibility that a fast transfer may be occurring among the A ions or among the B ions. Monteil and Duval [37] have carried out uniaxial stress experiments to separate the ions into two sublattices, and these sublattices are distinct from the A and B sublattices which are differentiated by the electric field, and they have examined the transfer between ions in these stress-separated lattices. They find that in contrast to the slow transfer between A and B sublattices, the transfer within each of these sublattices may be fast. In addition, recent magneto-optical studies of ruby luminescence by Viliani [38] have been interpreted by him as supporting the notion of a fast migration of excitation through the single chromium in system before transfer to pairs. The experiments of Monteil and Duval and of Viliani are described elsewhere in this book.

So far in the discussion on Anderson localization attention has focussed on the randomness in the energies of the centres (the diagonal interaction terms) and the localization is attributed to this randomness in the diagonal interaction terms. However, localization may occur because of randomness in the off-diagonal interaction terms which arises because of the random distribution of centres and the consequent random distribution of separations between adjacent centres. This problem has been considered by Huber and Ching [39],[40]. In the case where the interaction between centres is $V_0 \exp(-\alpha r_{ij})$. The numerical studies of Huber and Ching indicate that an Anderson localization occurs below a critical concentration, n_c, where

$$\frac{\alpha}{n_c^{113}} \simeq 2.7 \qquad (11)$$

Applying this to ruby, where an exchange interaction is assumed to act, and taking $\alpha = 1$ Å^{-1} one finds that for ruby with 0.4 at % chromium the disorder parameter α/n^{113} has the value 17, indicating localization at this concentration. These workers predict that an Anderson transition should occur only at a concentration of around 10 at. %.

At present the results of the elective field experiment [34] are regarded as strong evidence that fast migration does not occur between A and B rather, the transfer is slow and proceeds by a dipole-dipole process. Sublattice ions The nature of the transfer within a sublattice would still seem to be uncertain, some experimental results being cited as evidence for a very slow transfer among all Cr^{3+} ions indicating fast transfer. If the transfer is weak then the ruby system does not satisfy the conditions of the Anderson theory and one must look elsewhere for a system which will exhibit unambiguously the elusive transition between localization and delocalization.

REFERENCES

1. T. Förster, Ann. Physik 2, 55 (1948).
2. D. L. Dexter, J. Chem. Phys. 21, 836 (1953).
3. R. K. Watts, in Optical Properties of Ions in Solids, Plenum Press, p.307 (1975).
4. P. W. Anderson, Phys. Rev. 109, 1492 (1958).
5. N. F. Mott, Advan. Phys. 16, 49 (1967).

6. M. H. Cohen, H. Fritzsche, and S. R. Ovshinsky, Phys. Rev. Letters 22, 1065 (1969).
7. E. N. Economu and M. H. Cohen, Phys. Rev. B 5, 2931 (1972).
8. C. Hsu and R. Powell, Phys. Rev. Letters, 35, 734 (1975).
9. S. K. Lyo, Phys. Rev. B 3, 3331 (1971).
10. F. L. Varsanyi, D. L. Wood, and A. L. Schawlow, Phys. Rev. Lett. 3, 544 (1959).
11. A. L. Schawlow, S. L. Wood, and A. M. Clogston, Phys. Rev. Lett. 3, 271 (1959).
12. A. A. Kaplyanskii and A. K. Przheruskii, Sov. Phys. - Solid State 9, 190 (1967).
13. L. F. Mollenauer and A. L. Schawlow, Phys. Rev. 168, 309 (1968).
14. P. Kisliuk, N. C. Chang, P. L. Scott, and M. H. L. Pryce, Phys. Rev. 184, 367 (1969).
15. R. C. Powell, B. DiBartolo, B. Birang, and C. S. Naiman, Phys. Rev. 155, 296 (1967).
16. G. F. Imbusch, Phys. Rev. 153, 326 (1967).
17. P. M. Selzer, D. L. Huber, B. B. Barnett, and W. M. Yen, Phys. Rev. B 17, 4979 (1978).
18. H. M. Gibbs, S. Chu, S. L. McCall, and A. Passner, in NATO Workshop on Coherence Transfer in Glasses, Cambridge England, 1982 (to be published).
19. R. J. Birgeneau, J. Chem. Phys. 50, 4282 (1969).
20. I. Y. Gerlovin, Sov. Phys. Solid State 16, 397 (1974).
21. J. Heber and H. Murmann, Z. Physik B 26, 145 (1977).
22. H. Siebold and J. Heber, J. Lumin. 23, 325 (1981).
23. J. Koo, L. R. Walker, and S. Geschwind, Phys. Rev. Lett. 35, 1669 (1975).
24. S. Chu, H. M. Gibbs, and A. Passner, Phys. Rev. B 24, 7162 (1981)
25. A. Szabo, Phys. Rev. Lett. 25, 924 (1970).
26. A. Szabo, Phys. Rev. Lett. 27, 323 (1971).
27. A. Szabo, Phys. Rev. B 11, 4512 (1975).
28. A. Szabo and M. Kroll, Opt. Lett. 2, 10 (1978).
29. P. M. Selzer, D. S. Hamilton, and W. M. Yen, Phys. Rev. Lett. 38, 858 (1977).
30. T. Holstein, S. K. Lyo, and R. Orbach, Phys. Rev. Lett. 36, 891 (1976).
31. H. J. Eichler, J. Eichler, J. Knof, and C. H. Noak, Phys. Status Solidi 52, 481 (1979).
32. P. F. Liao, L. M. Humphrey, D. M. Bloom, and S. Geschwind, Phys. Rev. B 20, 4145 (1979).
33. D. S. Hamilton, D. Heiman, J. Feinberg, and R. W. Hellwarth, Opt. Lett. 4, 124 (1979).
34. S. Chu, H. M. Gibbs, S. L. McCall, and A. Passner, Phys. Rev. Lett. 45, 1715 (1980).
35. W. Kaiser, S. Sugano, and D. L. Wood, Phys. Rev. Lett. 6, 605 (1961).

36. P. E. Jessop and A. Szabo, Phys. Rev. Lett. 45, 1712 (1980).
37. A. Monteil and B. Duval (to be published).
38. G. Viliani (this book).
39. W. Y. Ching and D. L. Huber, Phys. Rev. B25, 1096 (1982).
40. D. L. Huber and W. Y. Ching, Phys. Rev. B25, 6472 (1982).

ENERGY TRANSFER AND IONIC SOLID STATE LASERS

F. Auzel

Centre National d'Etudes des Télécommunications
196 rue de Paris
92220 BAGNEUX (France)

ABSTRACT

A few years only after the demonstration of the first Neodymium laser, energy transfer was considered as a means to improve the coupling of broad optical pumping sources with narrow absorption lines in active ions. A number of schemes were then proposed.

Somewhat later, however, it was realized that multiple energy transfer or cooperative effects could play either a useful or a detrimental role according to the energy levels positions in a given matrix. In a material research approach, compromises have to be found between useful fast decay and self-quenching.

Both aspects of energy transfers are considered in this article with references to experimental examples and to some of the theoretical aspects developed in companion articles in this book.

I. INTRODUCTION

Since the beginning of the laser era, Cr^{3+} [1] and Nd^{3+} [2] have been the most used ionic systems. This fact can be explained by two special interesting features of their own:

- Cr^{3+} presents broad and strong absorbing bands in the visible region where powerful flash lamp are available though it is a three levels laser system.

- Nd^{3+} presents an ideal four levels system though its absorption bands in the visible are rather narrow.

Then it is no wonder that the first attempt to deliberately apply energy transfer to laser materials has been in a combined Cr^{3+}-Nd^{3+} system [3] with a hope to combine advantages and not drawbacks. Since then, many other pumping schemes with energy transfer have been proposed. However, with a better understanding of energy transfer processes, other positive aspects such as de-activation and cascade schemes have been considered.

On the other hand, negative aspects have also been found to be introduced by energy transfer such as reabsorption by APTE effect (up-conversion by sequential energy transfer) [4] and self-quencying of inverted population levels. Assuming the knowledge of the different energy transfer mechanisms, the aim of this article is to present a picture of the different roles played by energy transfer in ionic laser solid state materials.

II. ENERGY TRANSFER SCHEME FOR PUMPING EFFICIENCY IMPROVEMENT

II.A. Stokes Processes

In this chapter, energy transfer mechanisms which are considered lead to photon emissions at smaller energy than the input pumping photons.

1. Energy Transfer Towards Pumping Levels. One of the most severe limitations on the efficiency of optically pumped ionic solid lasers is the lack of suitable absorption band to match the generally wide spectral output of commercially available pumping lamps.

This is particularly true for trivalent rare-earths lasers which have only relatively weak 4f-4f parity forbidden absorptions. Of course, the use of lasers as pumping sources may partially solve this problem which still exists for the primary laser. The first approach was then to consider sensitizers with stronger absorption transitions expecting the absorbed energy to be transferred to the weaker transitions of the laser ions.

In order to understand what kind of improvements energy transfer can bring to pumping, basic processes developed at large in other articles in this book have to be briefly recalled in Table 1.

As shown, two aspects have to be considered: microscopic and macroscopic. Staying at the microscopic level, one can understand that no spectral improvement can be expected from a resonant transfer alone; whereas a non-resonant transfer allows extension of the absorption range at the sensitizer. However, each sensitizer can usually excite at most one activator at a rate which is the product of transitions probability within a sensitizer and an activator (Förster-Dexter relation); so only a reduction in probability has to be expected at this level.

Then, besides extension in pumping spectral range, improvement by energy transfer is due essentially to the macroscopic process of diffusion among sensitizers: excitation energy on the sensitizer is made available to any activator in the crystal even when activator is at a diluted level. Further, it should be noted that diffusion is obtained through resonant transfer which, because of the self-quenching process to be discussed later, is usually of the non-radiative type.

Figure 1 presents the first example of such improved pumping between 3d to 4f electrons in YAG ($Y_3Al_5O_{12}$) codoped with Cr^{3+} and ND^{3+} [3]. Gain in pumping efficiency due to 1% Cr^{3+} was as high as a factor of 4 when a Hg lamp with discrete spectrum is used as pumping source.

Lanthanides ions (Ln^{3+}) have also been used as sensitizing ions. Sometimes sensitizing ions belong to the crystalling matrix itself [5] or are color centers [6].

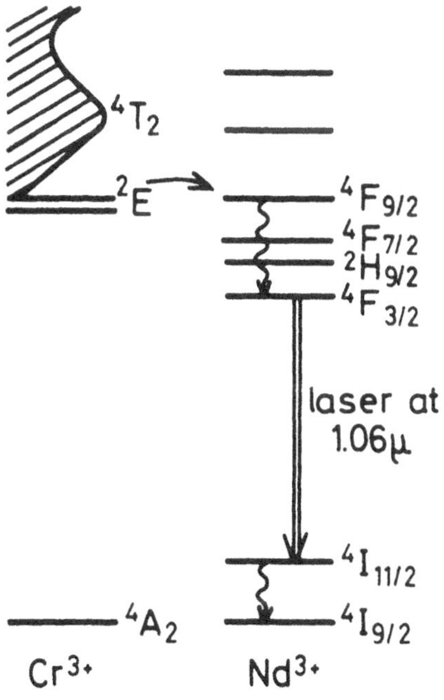

Figure 1. Nd^{3+} laser with energy transfer improved pumping from Cr^{3+} in YAG:Nd^{3+}, Cr^{3+} [3].

Table 1. Summary of Energy Transfer Cases.

A. Microscopic Studies Followed by Macroscopic Statistics

Microscopic Case (theoretical one sensitizer-one activator case).		Macroscopic Case (Experimental all sensitizers-all activator case).
- Resonant - Radiative : $P_{SA}^{(R)} = \frac{C_o^2 \sigma_A}{4\pi R^2 \tau_S} \int g_A(\nu) g_B(\nu) d$ - Non Radiative : $P_{SA}^{(R)} = \frac{(R_o/R)^s}{\tau_s}$; (s = 6,8,10). with : $R_o^s \propto \int g_A(\nu) g_B(\nu) d\nu$	Intermediate Statistics : Average probability for one sensitizer → all acceptors	- Without diffusion among sensitizers : non exponential decay : $I(t) = \exp\left[\frac{-t}{\tau_S} - \Gamma(1-\frac{3}{s}) \frac{C}{C_o} (\frac{t}{\tau_S})^{3/s}\right]$
- Non Resonant : Non Radiative : $P_{SA}^{(R)} = \frac{(R_o/R)^s}{}$; (s = 6,8,10) with : $R_o^s \propto e^{-\beta\Delta E}$	→ $W_{SA} = N_A \int P_{SA}(R) 4\pi R^2 dR$ strictly valid for $g_A(\nu)$ and $g_B(\nu)$ homogeneous	- With diffusion among sensitizers : (long time approximation) exponential decay : $I(t) = \exp\left(\frac{-t}{\tau_S} - \frac{t}{\tau_D}\right)$ "Diffusion limited " : $\tau_D^{-1} = VN_S N_A$ "Fast diffusion" : $\tau_D^{-1} = UN_A$ (rate equation case).

B. Direct Microscopic Quantum Statistics:

- rate equation case (Grant) → $\tau_D^{-1} = UN_A$
- inhomogeneous brodenning case (Anderson's localization)
  - → no diffusion for $\langle H_{SA} \rangle < \Delta E_{inhomogeneous}$ (without thermal effects)
    and $\langle H_{SA} \rangle \propto R^{-s/2}$ for S >6
  - → diffusion if thermal effects (one and two-phonons).

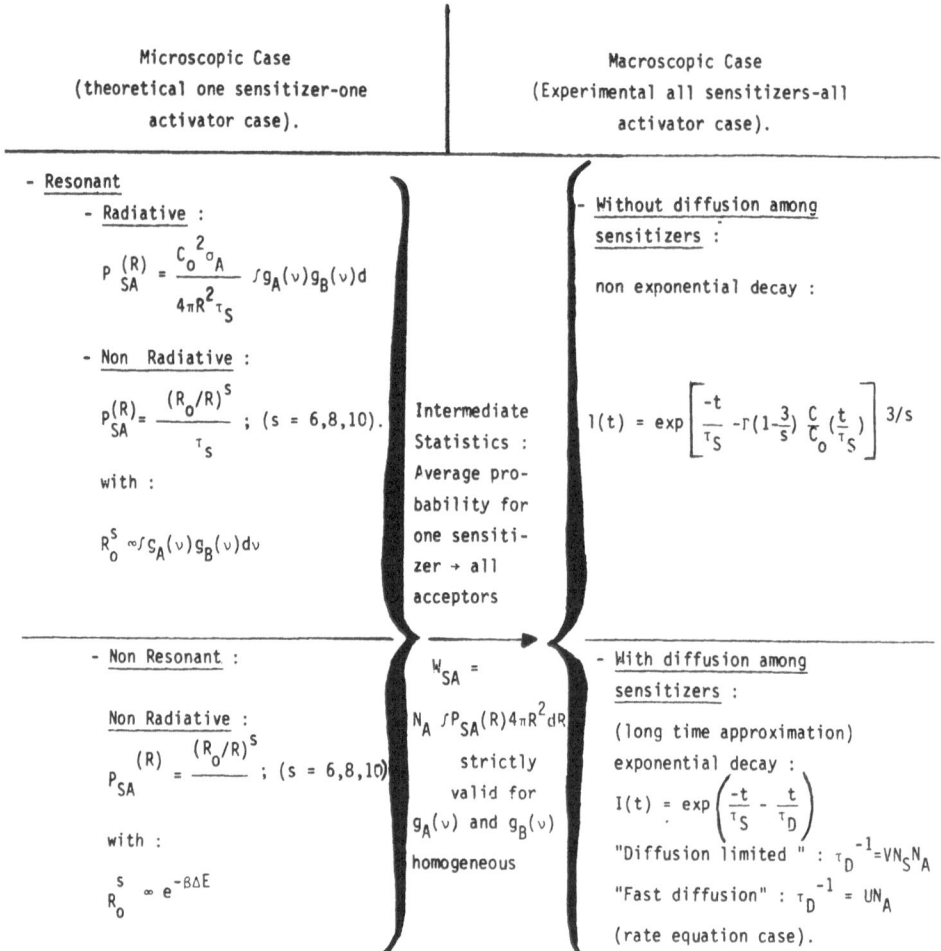

All known energy transfer improved pumping schemes for different hosts have been recently exhaustively reviewed in [7]. Then in Table 2 are shown only the complete sets of ions systems which have been investigated but not all matrices in which they have been included.

2. Deactivation by Energy Transfer of Levels in Self-saturating and Cascade Lasers. When the lifetime of the laser emitting level is shorter than the lifetime of the laser terminating level, then self-saturation may take place: the laser starts in a "four levels" scheme and quickly ends in a "three levels" scheme, because of the bottleneck of population in the terminating level.

Effective lifetime of a given level is usually reduced by multiphonon transitions according to the following equation:

$$\tau = \tau_0/[1 + \tau_0 W_{nR}(0)\exp(-\alpha\Delta E)]$$

where τ_0 is the radiative lifetime of the considered level and $\exp(-\alpha\Delta E)$ the exponential gap law for multiphonon decay [8] from a level with energy separation ΔE to next lower level. Considering the Ln^{3+} energy schemes, one finds that in the heavier half, Ln^{3+} have larger separation for first excited levels than the lighter half. Consequently, self-saturating lasers are rather to be found among the heavier half. Effectively, self-saturating transitions have been found up to now in H_o^{3+} and Er^{3+} with $^5I_7$ and $^4I_{13/2}$, respectively, as terminating levels]7].

Energy transfer was suggested long ago [9] as a means to avoid self-saturation in lasers through reduction of terminating level lifetime. This principle has been effectively used for the Er^{3+} $^4I_{11/2} \rightarrow {}^4I_{13/2}$ transition at 2.69 μm in $Lu_3Al_5O_{12}$:Er^{3+},Tm^{3+}. Here, the lasing ion (Er^{3+}) acts as a sensitizer and Tm^{3+} as a deactivator [10]. See Figure 2.

Closely linked with self-saturating lasers are cascade lasers. The self-saturating level in one ion is emptied either by a second laser effect in the same ion or by energy transfer and laser action in the deactivator as shown in Figure 3 for Er^{3+} - Tm^{3+}, Er^{3+} - Ho^{3+} and Er^{3+} - Tm^{3+} - Ho^{3+}, in doped oxyde and fluoride crystals [11], [12]. Such systems appear to be the most efficient to obtain C.W. lasers in the mid-I.R. region around 3-5 μm. Of course, when laser action is obtained in the second ion (deactivator) self-saturation in the first ion (sensitizer) is precluded which greatly improves the laser efficiency in the first ion. For instance, introduction of Ho^{3+} in an Er^{3+} system gives an efficiency for Er^{3+} in excess of 30% at 2.69 μm [12].

Table 2. Ionic Systems for Improved Pumping of Solid State Lasers (Stokes Processes).

Sensitizers		Activators
Type	Ion	Ion (Laser Transitions)
Transition Metals	Cr^{3+}	$Nd^{3+}(^4F_{3/2} \rightarrow {}^4I_{11/2})$
		$Ho^{3+}(^5I_7 \rightarrow {}^5I_8)$
		$Tm^{3+}(^3F_4 \rightarrow {}^3H_5\ ;\ {}^3H_4 \rightarrow {}^3H_6)$
	Fe^{3+}	$Ho^{3+}(^5I_7 \rightarrow {}^5I_8)$
	Mn^{2+}	$Ni^{2+}(^3T_2 \rightarrow {}^3A_2)$
Lanthanides	$Nd^{3+} + Cr^{3+}$	$Yb^{3+}(^2F_{5/2} \rightarrow {}^2F_{7/2})$
	Nd^{3+}	$Yb^{3+}(^2F_{5/2} \rightarrow {}^2F_{7/2})$
	Ce^{3+}	$Nd^{3+}(^4F_{3/2} \rightarrow {}^4I_{11/2})$
	Gd^{3+}	$Tb^{3+}(^5D_4 \rightarrow {}^7F_5)$
	Er^{3+}	$Dy^{3+}(^6H_{13/2} \rightarrow {}^6H_{15/2})$
		$Ho^{3+}(^5I_7 \rightarrow {}^5I_8)$
		$Tm^{3+}(^3H_4 \rightarrow {}^3H_6)$
	$Er^{3+} + Tm^{3+}$	$Ho^{3+}(^5I_7 \rightarrow {}^5I_8)$
	$Er^{3+} + Tm^{3+} + Yb^{3+}$	$Ho^{3+}(^5I_7 \rightarrow {}^5I_8)$
	$Er^{3+} + Yb^{3+}$	$Tm^{3+}(^3H_4 \rightarrow {}^3H_6)$
	Ho^{3+}	$Er^{3+}(^4S_{3/2} \rightarrow {}^4I_{13/2})$
	Yb^{3+}	$Er^{3+}(^4I_{13/2} \rightarrow {}^4I_{15/2})$
Color Center	Color Center in CaF_2	$Er^{3+}(^4I_{13/2} \rightarrow {}^4I_{15/2})$

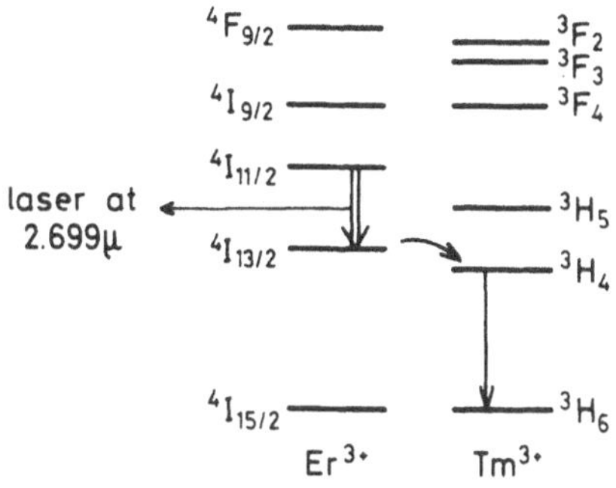

Figure 2. Er^{3+} laser with Tm^{3+} as deactivator in $La_3Al_5O_3{:}Er^{3+},Tm^{3+}$ [10].

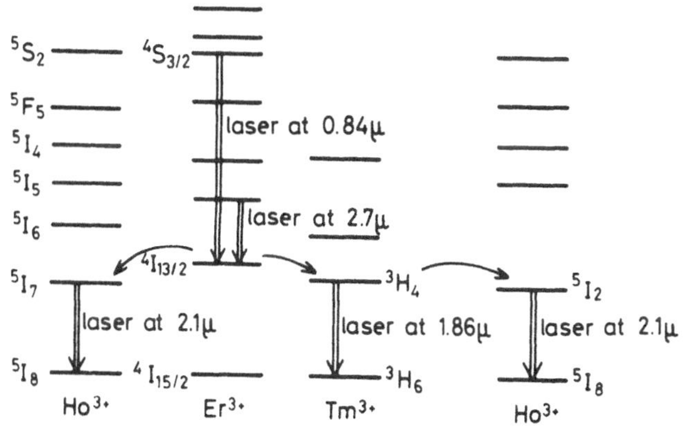

Figure 3. Cascaded laser by non-radiative transfer between $Er^{3+}-Tm^{3+}$, $Er^{3+}-Ho^{3+}$ and $Er^{3+}-Tm^{3+}-Ho^{3+}$ systems [11],[12].

II.B. Anti-Stokes Processes and Up-conversion Pumped Lasers

There exists several up-conversion processes providing one-photon out of the energy of several incoming photons. However, their intrinsic efficiencies are quite different as can be seen from Table 3 which summarizes efficiencies for two-photon processes.

Table 3. Comparison Between Different Multiphonon Up-converting Processes [8].

Mechanisms	Efficiency μ (cm^2/W)	Material
Two successive transfer	$\sim 10^{-3}$	YF_3:Yb:Er
Two-step absorption	$\sim 10^{-5}$	SrE_2:Er
Cooperative sensitization	$\sim 10^{-6}$	YF_3:Yb:Tb
Second harmonic generation	$\sim 10^{-11}$	KPD
Two-photon absorption	$\sim 10^{-13}$	$CaF_2:Eu^{2+}$

The intrinsically most efficient of them is the APTE effect (addition de photons par transferts d'énergie) [13],[14] which, to the difference with all previous transfer mechanisms, involves at least one energy transfer towards an activator already in an excited state [13]. The basic reason for highest efficiency is that real levels are involved at each step which allows a longer interaction time between photons and matter. This effect is so efficient that it has been used [15] to obtain up-conversion pumped lasers in which the pumping photons are at smaller energy than the laser photon. Such lasers were developed from the first $Yb^{3+}-Er^{3+}$ and $Yb^{3+}-Ho^{3+}$ schemes [13],[16] as shown in Figure 4. I would like to point out here that such up-conversion pumping schemes could provide a way to obtain compact visible solid state lsers with the efficient IR semiconductor lasers recently developed as pumps.

III. DRAWBACKS INTRODUCED BY ENERGY TRANSFER

Besides the positive aspects of energy transfer which have been deliberately looked for in laser materials, there are a number of inherent drawbacks which have to be compromised with.

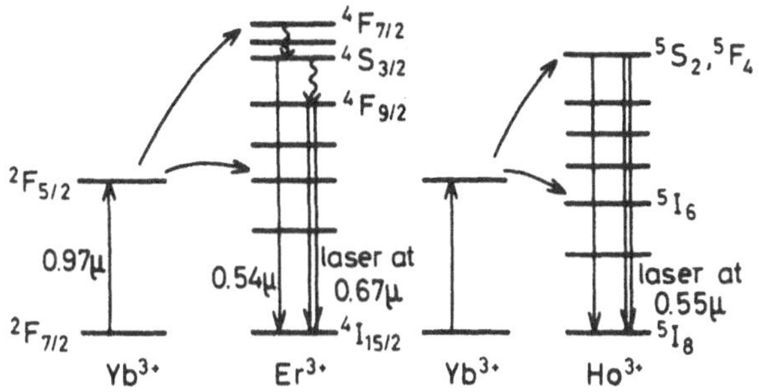

Figure 4. APTE effect in Yb^{3+} - Er^{3+} [13] and in Yb^{3+} - Ho^{3+} [16] couples with up-conversion pumped lasers (double arrows) in $Ba(Y,Yb)_2F_8$ doped with Er^{3+} and Ho^{3+} [15].

III.A. Stokes Processes

Again, unwanted energy transfer can be classified according to the relative energy of photons in and out of the system.

1. Self-quenching by Energy Diffusion and Cross-relaxation. When it is desired to increase the energy density inside a laser material, the active ion density has to be increased. Unfortunately, however, for any luminescent system, a high activator concentration generally leads to poor quantum efficiencies for emissions. This is the pervading effect of self-quenching due to usually the combination of two types of energy transfer between identical activators: resonant diffusion through identical pairs of levels or transfer to unwanted impurities. Figure 5 gives the energy scheme for such processes in Nd^{3+}. In some cases [7] materials with rather small self-quenching may be found such as the so-called stoichiometric laser materials already discussed in a previous publication [8]. NdP_5O_{14} (NdUP) is the archetype of such materials [17].

This effect can be characterized by defining a quenching rate

$$W_0 = 1/\tau - 1/\tau_0$$

where τ_0 is the radiative lifetime and τ the measured lifetime at

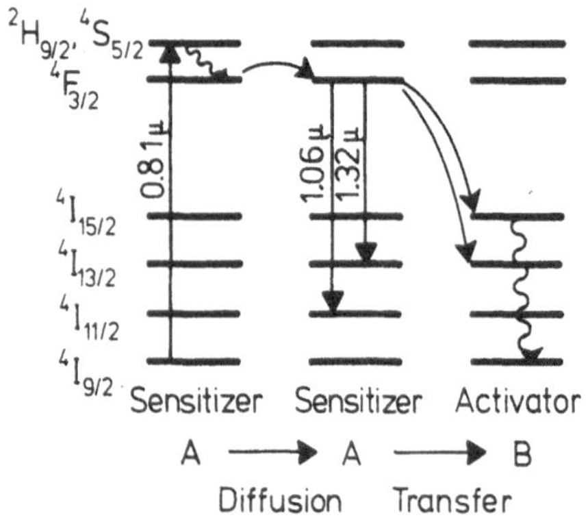

Figure 5. Self-quenching in Nd^{3+} by diffusion and cross-relaxation transfers.

high activator concentration ($>10^{21}$ cm^{-3} for Ln^{3+}).

2. Role of Crystal Field Strength. Comparing for example in Table 4 the behavior of $Nd^{3+}(^4F_{3/2})$ and $Tb^{3+}(^5D_3)$ we have recently demonstrated [18] that self-quenching of a given level and ion cannot be generalized to other ions in a given crystal.

Cross-relaxation, which is the basic loss, is not inhibited by large shortest $Ln^{3+} - Ln^{3+}$ distances, as sometimes believed, but by a poor energy overlap.

For a given ion, crystal field strength is found to mediate the energy overlap [19] (see Figure 6) through the maximum Stark splitting suffered by the free ion in the crystal. In order to study such effects I have defined a scalar crystal field strength parameter:

$$N_{vc} = [\sum_{k,q} (B^k_q)^2 \left(\frac{4M}{2k+1} \right)]^{1/2}$$

where B^k_q are the usual crystal field strength parameter:

Table 4. Comparison of Self-Quenching Rate (W_Q) of $Nd^{3+}(^4F_{3/2})$ and $Tb^{3+}(^5D_3)$ at $N = 4 \times 10^{21}$ cm^{-3} in Two Stoichiometric Materials.

Materials	d_{min} $Ln^{3+}-Ln^{3+}$	$Nd^{3+}(^4F_{3/2})$		$Tb^{3+}(^5D_3)$	
		W_Q	Slope	W_Q	Slope
$NaLn(WO_4)_2$	3.9 Å	$3\times10^5 s^{-1}$	2	$4\times10^4 s^{-1}$	1
LnP_5O_{14}	5.2 Å	$5\times10^3 s^{-1}$	1	$1.3\times10^5 s^{-1}$	2

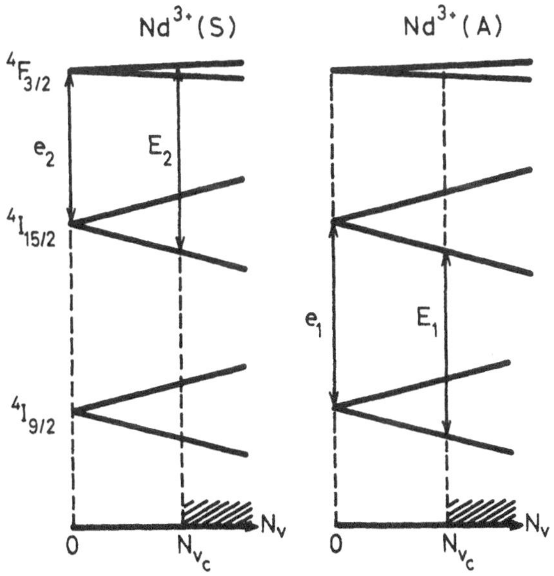

Figure 6. Role of crystal field in resonance condition for S → A cross-relaxation transfer; critical field is given by $E_2 - E_1 \geq 0$.

$$v = \sum_{i,h,q} B_q^k (C_q^k)\ i$$

The maximum Stark splitting (ΔE_J) of a given J-level of Ln^{3+} found to be proportional to N_{vc} [20]

$$\Delta E_J \simeq S[\prod_k |<J||\sum_i C_{(i)}^{(k)}||J>]^{1/3} N_{vc}$$

where S is a statistical coefficient depending on the degeneracy removal of the J-level and $<J||C_{(i)}^{k}||J>$ the reduced matrix elements of the operator $C_{(i)}^{k}$ for spherical harmonics.

This property which was first experimentally verified on a number of Nd^{3+} materials, provides a way to obtain the resonance conditions on field strength for Nd^{3+} cross-relaxation as: $N_{vc} \geq 2800\ cm^{-1}$. Stronger crystal fields material show cross-relaxation, smaller one much less.

III.B. Anti-Stokes Processes Up-conversion and Reabsorption

APTE effect was found while working on Er^{3+} glass laser sensitized with Yb^{3+} [13]. The resulting up-conversion process giving a green photon at 0.54 μm out of two IR ones at 0.97 μm (see Figure 4) was the first proof that energy transfer could efficiently happen between excited states and in that example at the expense of the population inversion in Er^{3+} [4]. This effect shows that it is not sufficient to consider only the intended transfer but also all the undesired ones which shall spontaneously happen provided there exists a populated metastable level involved in a pair of ions in approximate resonance. Multiphonon assisted transfer (see Table 1) with practically energy mismatch up to four-phonons should not be neglected in investigating this type of loss.

Losses in lasers due to single ion reabsorption have been considered since the ruby laser time [21]. When transfer is involved such effect can be unwillingly increased in parallel with pumping spectral range. However, very little consideration has been given to such drawbacks up to now becuase not so many systems simultaneously involve high concentrations and high enough pumping intensity. Yet, in stoichiometric materials Nd^{3+} - Nd^{3+} up-conversion losses accorded to scheme of Figure 7 become noticeable when only 10% of the active ions are pumped in the $^4F_{3/2}$ state [22].

Clearly, in a material research approach, compromises have to be found between desired pumping improvement and undesired up-conversion losses.

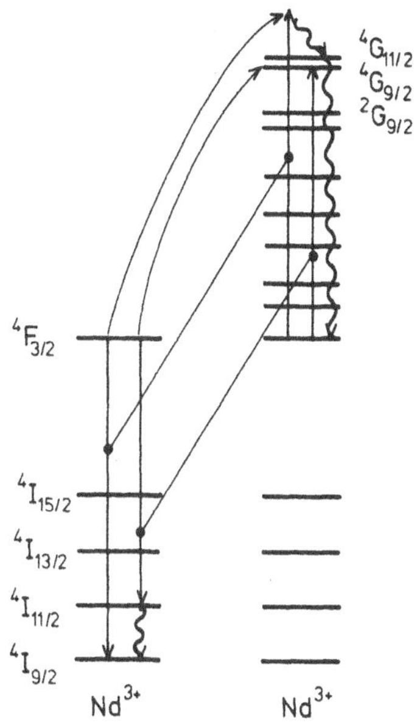

Figure 7. $Nd^{3+} - Nd^{3+}$ self-quenching due to reabsorption by APTE effect (up-conversion).

IV. CONCLUSION

A general picture of the role of energy transfer in ionic solid state lasers has been drawn. The role of diffusion through non-radiative resonant transfer has been stressed in the process of pumping spectral range enlargement for laser active ions. The recent use of deactivators ions decaying spontaneously or by cascade laser has been described. However, up-conversion processes by APTE effects which may have either positive or negative aspects have not yet been given the full interest they deserve. I believe that with the advent of lasers as pumping sources, one of the basic limitations in future ionic state laser shall be through up-conversion enhanced reabsorption leading to gain saturation.

In order to be complete before ending, I have to mention the special role of transfer between identical ions in inhomogeneous sites giving rise or not to spectral diffusion: a problem otherwise well-documented in this school. Their role becomes important when considering finer aspects in the laser functioning such as longitudinal mode structure, spatial and spectral hole burning. However, such aspects were out of our scope in this article.

REFERENCES

1. T.H. Maiman, Nature, 187, 493 (1960)
2. L.F. Johnson, K. Nassau, Proc. IRE 49, 1704 (1961)
3. Z.J. Kiss, R.C. Duncan, Appl. Phys. Lett. 5, 200 (1964)
4. F. Auzel, Ann. Telecom. 24, 363 (1969)
5. J.R. O'Connor, W.A. Hargreaves, Appl. Phys. Lett. 4, 208 (1964)
6. P.A. Forrester, D.A. Sampson, Proc. Phys. Soc. 88, 199 (1966).
7. A.A. Kaminskii, Laser crystals Vol. 14, Springer series in optical sciences. Springer-Verlag, (1981).
8. F. Auzel, Radiationless Processes, (B. Di Bartolo and V. Goldberg, eds), p. 213, Plenum (1980).
9. O. Deustschbein, F. Auzel, Quantum Electronics, (P. Grivet and N. Bloembergen, eds), p. 859, Dunod and Columbia University Press (1964).
10. A.A. Kaminskii, T.I. Butaeva, A.O. Ivanov, I.V. Mochalov, A.G. Petrosyan, G.I. Rogov, V.A. Fedorov, Sov. Tech. Phys. Lett. 2, 308 (1976).
11. A.A. Kaminskii, T.I. Butaeva, A.M. Kevorkov, V.A. Fedorov, A.G. Petrosyan, and M.M. Gritsenko, Izvest. Akad. Nauk SSSR, Neorg. Mat. 12, 1508 (1976).
12. A.A. Kaminskii, Izvest. Akad. Nauk, SSSR, Neorg. Mat. 7, 904 (1971).
13. F. Auzel, C.R. Acad. SC. Paris B 263, 819 (1966).
14. F. Auzel, Proc. IEEE 61, 758 (1973).
15. L.F. Johnson and H.J. Guggenheim, Appl. Phys. Lett. 19 44 (1971)
16. R.A. Hewes and J.F. Sarver, Phys. Rev. 182, 427 (1969).
17. H.G. Danielmeyer and H.P. Weber, IEEE J. Quantum Electronics 8, 805 (1972).
18. F. Auzel, J. Dexpert-Ghys and C. Gautier, J. of Luminescence, 27, 1 (1982).
19. F. Auzel, Mat. Res. Bull. 14, 223 (1979).
20. F. Auzel and O. Malta, J. de Physique, 44, 201 (1983).
21. F. Gires, G. Mayer, Quantum Electronics (P. Grivet, N. Bloembergen eds.) P. 841, Dunod and Columbia University Press, (1964).
22. M. Blätte, H.G. Danielmeyer and R. Verich, Appl. Phys. (Germ.) 1, 275 (1973).

A SCALAR CRYSTAL FIELD STRENGTH PARAMETER FOR RARE-EARTH IONS: MEANING AND APPLICATION TO ENERGY TRANSFER

F. Auzel

Laboratoire de Bagneux
Centre National d'Etudes des Télécommunications
196 rue de Paris
92220 Bagneux, France

ABSTRACT

A scalar crystal field parameter is first introduced experimentally and shown to rule a special energy transfer between Nd^{3+} ions. Its general theoretical meaning is then discussed demonstrating its extension for any ions and j-terms as well as its limitations.

I. INTRODUCTION

Decrease in quantum yield at high concentration (self-quenching) in Nd^{3+} doped materials can be explained by the following crystal field enhanced cross-relaxation [1] :

$$Nd(^4F_{3/2}) + Nd(^4I_{9/2}) \rightarrow 2Nd(^4I_{15/2})$$

As shown on Fig.1, this cross-relaxation can take place only if the energy released at ion $S(Nd^{3+})$ is about equal to the energy which can be accepted at ion $A(Nd^{3+})$. For two near by "free ions" the energy difference (e_2-e_1) being negative, such transfer could not take place. For ions in crystal the Stark effect can provide a resonance condition : $E_{2M} - E_{1m} = 0$ which for phonon assisted transfers can be extended to $E_{2M} - E_{1m} \geqslant 0$.

For a "free" Nd^{3+} ion, (e_2-e_1) is found to be experimentally about a constant (from gravity centers in different materials) :

$$(e_2 - e_1) \simeq - 430 \text{ cm}^{-1}$$

The resonant condition for cross-relaxation is then :

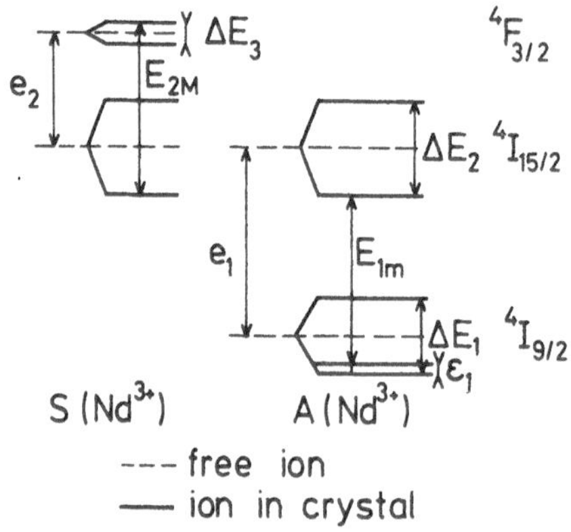

Fig. 1. Energy levels involved in self-quenching for Nd^{3+} in crystals and for "free ions."

$$\Delta E_2 - \frac{\Delta E_1}{2} + \frac{\Delta E_3}{2} = 430 \qquad (1)$$

where ΔE_1, ΔE_2, ΔE_3 are respectively the maximum splitting of $^4I_{9/2}$, $^4I_{15/2}$, $^4I_{3/2}$.

Stark splittings are usually well described by potential V at ion sites :

$$V = \sum_{k,q,i} B_q^k \, (C_q^k)_i$$

where C_q^k are the well known spherical tensorial operators :

$$C_q^k = \frac{4\pi}{2k+1} \, Y_{k.q} \, (\Theta, \phi)$$

J consider V as an Hilbert space function the norm of which is taken as a scalar crystal field parameter 1 :

$$N_v^* = \left[\sum_{k,q} (B_q^k)^2 \left(\frac{4\pi}{2k+1}\right)\right]^{1/2} \simeq N_v = \left[\sum_{k,q} (B_q^k)^2\right]^{1/2} \qquad (2)$$

with $2J \geqslant k$

Good linear relationship are experimentally found for maximum splitting whatever sites symmetry [1] :

$$\Delta E_1 \simeq 0.23\ N_v$$
$$\Delta E_2 \simeq 0.27\ N_v$$
$$\Delta E_3 \simeq 0.026\ N_v$$

The resonance conditions (eq. 1) can simply be written as :

$$\Delta E_1 = 470\ cm^{-1} \quad \text{or} \quad N_v = 1800\ cm^{-1} \qquad (3)$$

ΔE_1 can easily be obtained directly by an absorption spectra of $^2P_{1/2}$ state from $^4I_{9/2}$ at room temperature giving directly the Stark splitting of $^4I_{9/2}$.

Relations (3) indicate simply that weak self-quenching materials will be thoses with

$$\Delta E_1 < 470\ cm^{-1} \text{ or } N_v < 1800\ cm^{-1}$$

Examples of such result is presented on Fig. 2 which in the same time demonstrate experimentally the good linearity between ΔE_1 and N_v. All materials for which simultaneously ΔE_1 and B_q^k are available whatever their sites symmetry are presented by round points.

The purposes of this lecture are to answer the following :
- why N_v describes linearly ΔE_{max} ?
- Is it general ? In other words, can it be used for other ions than Nd^{3+} ? Can it be used for other J-terms than $^4I_{9/2}$, 11/2, 15/2 ?
- What is the theoretical value of the proportionality coefficient ?

II. THEORETICAL INVESTIGATION OF MAXIMUM STARK SPLITTINGS [2].

Assuming small J-mixing we consider the degeneracy removal of a g-fold degenerate J-term in a field potential V (see Fig.3 for meaning of symboles). g_a levels are obtained. g_a equal g if J is integer and equal g/2 if J is half-integer.

Considering the root-mean-square deviation of n Stark levels :

$$(\Delta\varepsilon)^2 = \frac{1}{g_a} \sum_i^n (E_i)^2$$

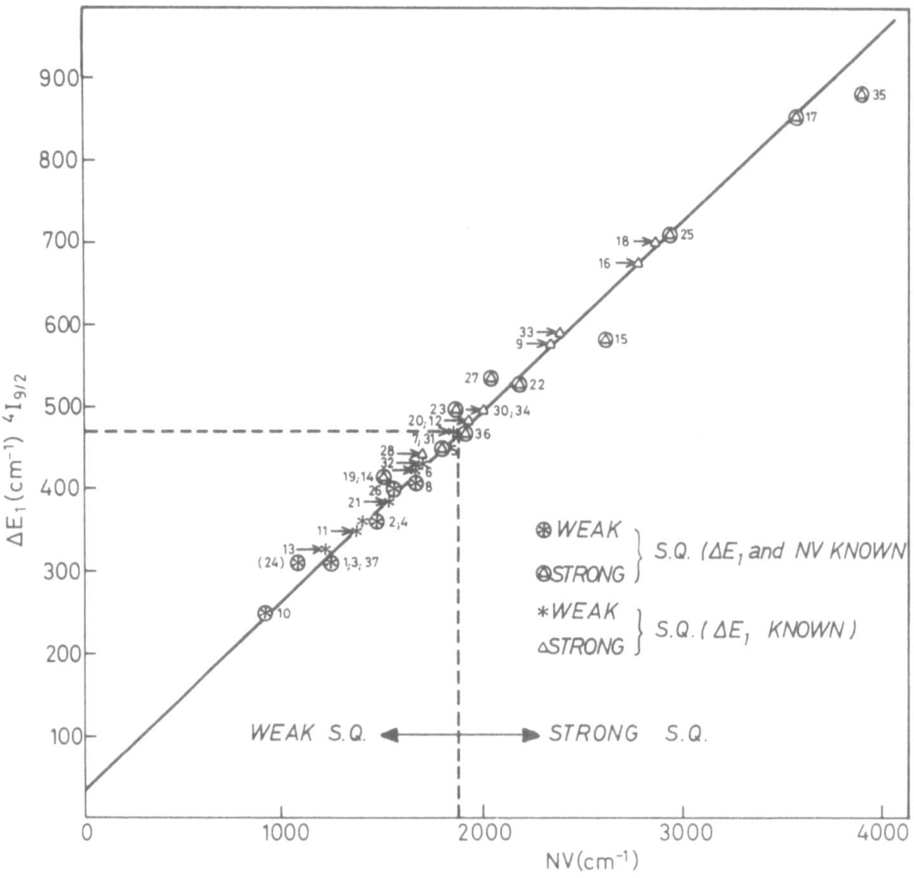

1 NdP_5O_{14}, 2 $LiNd(PO_3)_4$, 3 CLAP, 4 $Nd_5Nd(WO_4)_4$, 5 $Na(La,Nd)(WO_4)_2$,
6 $K_3Nd(WO_4)_4$, 7 $K_5Nd(MoO_4)_4$, 8 (NAB) $NdAl_3(BO_4)_3$, 9 $Na_{18}Nd(BO_3)_7$,
10 $NdCl_3$, 11 $(Nd,La)Ta_7O_{19}$, 12 $LiNdNbO_4$, 13 tellurite glass,
14 $LOS(Nd_2O_2S)$, 15 CaF_2, 16 $YAlO_3(La,Nd)$, 17 $Y_3Al_5O_{12}(La,Nd)$,
18 $Ba_2MgGe_2O_7$, 10 $NdNb_5O_{14}$, 20 Silicate flint glass, 21 $Gd_{17}Nd_{03}(MoO_4)_3$,
22 $Li(Y,Nd)F_4$, 23 $(Nd,La)F_3$, 24 $Nd(C_2H_5ZO_4)_3$, $9H_2O$, weak field quenched by H_2O, 25 $Ca_5(PO_4)_3F$(FAP), 26 KY_3F_{10}, 27 Nd_2S_2, 28 YVO_4
29 $PbMoO_4$, 30 $LiNbO_3$, 31 phosphate glass KLi, 32 $KY(MoO_4)_2$,
33 $CaAl_4O_7$, 34 La_2O_3, 35 $LaOF(Nd^{3+})$, 36 WO_4Ca, 37 Cl_3Nd in ice.

Fig. 2. Maximum Stark splitting of $^4I_{9/2}$ (Nd^{3+}) versus N_v for different crystals with various site symmetry known in literature.

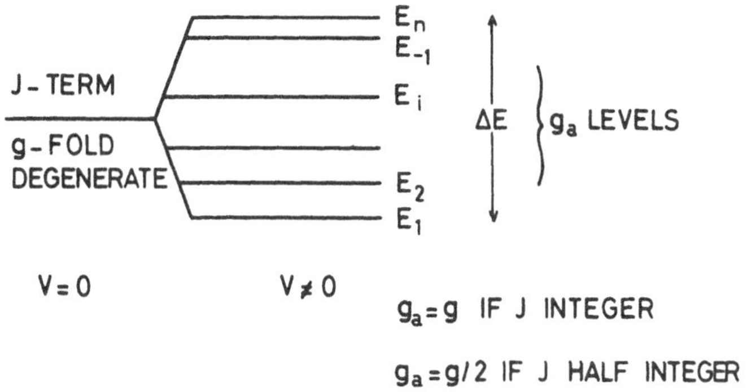

Fig. 3. Crystal field degeneracy removal for a J-term.

from [3] we get:

$$(\Delta\varepsilon)^2 = \frac{1}{g} \text{ tr } (P_o V P_o V)$$

where $P_o = \sum_{M'} |JM'\rangle\langle JM'|$ (projector operator).

$$\begin{aligned} \text{tr}(P_o V P_o V) &= \sum_{M,M'} \langle JM|V|JM'\rangle\langle JM'|V|JM\rangle \\ &= \sum_{\substack{M,M' \\ k,q \\ k',q'}} B^k_q B^{k'}_{q'} \langle JM| \sum_i (C^k_q)_i |JM'\rangle\langle JM'| \sum_i (C^{k'}_{q'})_i |JM\rangle \\ &= \sum_{k,q} \frac{(B^k_q)^2}{2k+1} \, \delta(J,J,k) \, |\langle J|| \sum_i (C^k)_i || J\rangle|^2 \end{aligned}$$

On the other hand :

$$(\Delta\varepsilon)^2 = \frac{1}{g_a} \left(\frac{\Delta\varepsilon}{2}\right)^2 \left[1 + a_2^2 + \dots a_{g_a-2}^2 + 1\right]$$

with $0 < a_i < 1$

Assuming an homogeneous distribution of a large number n of Stark levels :

$$(\Delta\varepsilon)^2 = \frac{2}{g_a}\left(\frac{\Delta E}{2}\right)^2 \sum_{n=1}^{g_a/2} n^2 \left(\frac{2}{g_a}\right)^2$$

$$= \left(\frac{\Delta E}{2}\right)^2 \left(\frac{2}{g_a}\right)^2 \frac{1}{6} \left(\frac{g_a}{2} + 1\right)(g_a+1)$$

Then :

$$(\Delta E)^2 = \frac{12\, g_a^2}{g(g_a+2)(g_a+1)} \sum_{k,q} \frac{(B_q^k)^2}{2k+1} |\langle J\| \sum_i (C^{(k)})_i \| J\rangle^2 \tag{4}$$

Let us look at the sum in (4) as the equation of a quadric surface in the tridimensional k-space. It is an ellipsoid equation. The central approximation we perform at this point is to <u>replace this ellipsoid by a sphere of same volume</u> :

$$\sum_{k,q} \frac{(B_q^k)^2}{(2k+1)} |\langle J\| \sum_i (C^{(k)})_i \| J\rangle|^2 \simeq \sum_{k,q} \frac{(B_q^k)^2}{2k+1} \left[\prod_{k'} |\langle J\| \sum_i (C^{(k)})_i \| J\rangle|^2\right]^{2/3}$$

$$\simeq \frac{1}{4\pi} \left[\prod_k \langle J\| \sum_i (C^{(k)})_i \| J\rangle|\right]^{2/3} N_v^{*2}$$

and :

$$\Delta E \simeq \left[\frac{3g_a}{\pi g(g_a+2)(g_a+1)}\right]^{1/2} \left[\prod_k |\langle J\| C^{(k)} \| J\rangle|\right]^{1/3} N_v^* \tag{5}$$

III. DISCUSSION OF THE APPROXIMATION

The spherical approximation performed in (5) is so much the better that the ellipsoïd is more sphere-like, which is the case when $\langle J\| C^{(k)} \| J\rangle^2$ is stationary in k.

Since $$\langle J\| C^{(k)} \| J\rangle^2 = \langle f\| C^{(k)} \| f\rangle^2 \langle SLJ \| U^{(k)} \| SLJ\rangle^2$$

and $$\langle \ell\| C^{(k)} \| \ell\rangle^2 = (2\ell+1)^2 \begin{pmatrix} \ell & k & \ell \\ 0 & 0 & 0 \end{pmatrix}^2$$

the value of which is almost constant for ℓ=f configuration :

$$\langle f\| C^{(2)} \| f\rangle = -1.36$$

$$\langle f\| C^{(4)} \| f\rangle = 1.13$$

$$\langle f\| C^{(6)} \| f\rangle = -1.27$$

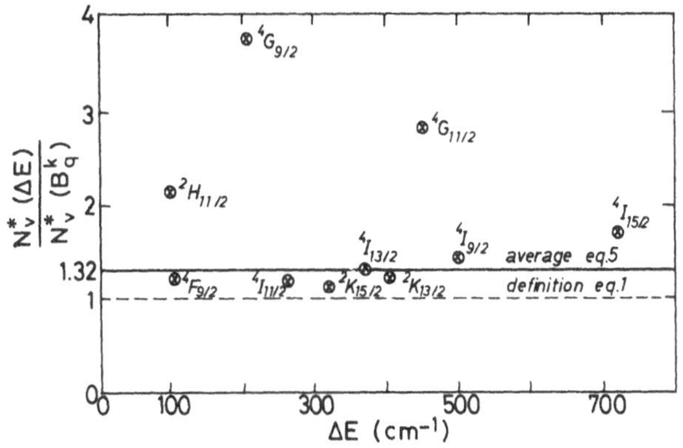

Fig. 4. Ratio of N_v^* parameters obtained from ΔE (eq.5) anf from B_q^k (eq.2) versus maximum splitting of J-terms of Nd^{3+} with 2J ⩾ k.

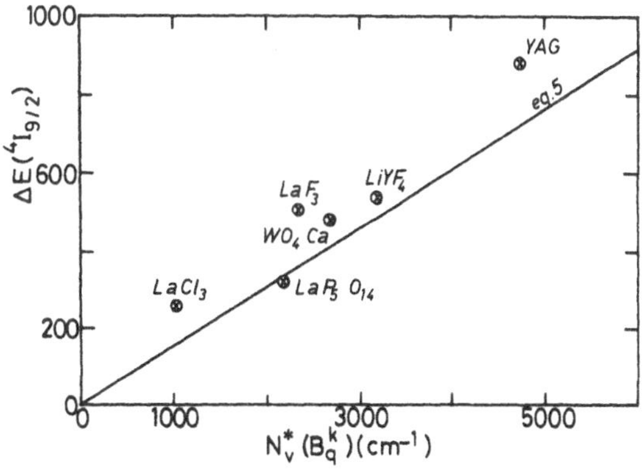

Fig. 5. Maximum Stark splitting of $^4I_{9/2}$ (Nd^{3+}) versus N_v^* as given by eq. 5 compared to experimental points.

Then, variation in k arises essentially from $<\|U(k)\|>^2$. Inspection of Carnall's table [4] plus the 2J > k condition indicates the most "spherical" terms which reciprocally can be used as <u>crystal field probes</u> :

$$Pr^{3+}\ (^3H_4,\ ^3H_6)$$
$$Nd^{3+}(^4I_{9/2},\ ^4I_{11/2},\ ^4I_{15/2})$$
$$Sm^{3+}(^4I_{13/2},\ ^6H_{15/2})$$
$$Eu^{3+}(^7F_5)$$
$$Tb^{3+}(^7F_5)$$
$$Dy^{3+}(^6F_{9/2},\ ^4F_{9/2},\ ^6F_{11/2},\ ^4I_{13/2},\ ^6H_{15/2},\ ^4H_{15/2})$$
$$Ho^{3+}(^5I_4,\ ^5I_6).$$
$$Er^{3+}(^4F_{9/2},\ ^4I_{11/2},\ ^4I_{13/2})$$
$$Tm^{3+}(^3H_4,\ ^3H_6)$$

Of course other J-terms, provided the J-mixing is small, could be used also, the sphere being an average of the ellipsoid considered in (4).

IV. APPLICATION TO MAXIMUM STARK SPLITTING CALCULATIONS ; COMPARISON WITH EXPERIMENTS

IV.A. Given Ion (Nd^{3+}) and Crystal (LaF_3), Comparison of N_v^* from Maximum Splitting to N_v^* from B_q^k's for Different J-Terms

Starting from B_q^k's and maximum splitting given in (4) we can calculate N_v^* from ΔE by eq. (5) and from its definition (eq.2). The ratio of both N_v^* is given on Fig.4 versus ΔE for varous J-terms of Nd^{3+}. Horizontal line shows proportionality of ΔE with N_v^* according to eq.5. Terms such as $^4G_{9/2}$, $^4G_{11/2}$ which are away from the horizontal lines correspond to terms with high J-micing.

IV.B. Given J-Term ($^4I_{9/2}$) of Given Ion (Nd^{3+}), Study of Maximum Splitting for Different Crystals with Different Site Symmetry

In order to compare theory with experiments such as those presented on Fig. 2, we calculate N_v^* from B_q^k's found in literature [4],[5],[6],[7],[8],[9] for crystals of different structures and plot by round points on Fig. 5, the experimental maximum splitting of $^4I_{9/2}$ of Nd^{3+} versus N_v^*. Theory as presented by eq. 5 gives the line obtained without any adjustable parameter. It shows that ΔE is proportional to N_v^* and that the theoretical slope discribes experiments reasonably well.

In order to make comparison between Fig.2 and 5 easier the values of N_v^* (complete) and N_v (approximate) are given for different crystals in Table 1. On the average $N_v^*/N_v \simeq 1.35$.

IV.C Given J-Term ($^4I_{13/2}$) and Crystal (LaF_3), Study for Different Ions

In the following we consider the $^4I_{13/2}$ term which has been found above to be rather "spherical" in a number of ions : Nd^{3+}, Sm^{3+}, Dy^{3+}, Er^{3+}.

For each of these ions we compare the experimental maximum splitting of $^4I_{13/2}$ to the theoretical one as given by eq.5. The result is presented in Table 2 as the ratio $\Delta E_{exp}/\Delta E_{th}$. The different values found arround 1 demonstrate the generality of N_v^* for different ions.

Table 1. N_v^* parameters obtained from B_q^k from literature for different crystals doped with Nd^{3+}.

Crystals	$LaCl_3$	LaF_3	LaP_5O_{14}	$CaWO_4$	$LiYF_4$	YAG
N_v^* (cm^{-1}) complete	1062	2356	2190	2691	3095	4739
N_v (cm^{-1}) approximate 4	928	1841	1406	1860	2194	3575
Ref.	[5]	[4]	[9]	[8]	[7]	[6]

Table 2. N_v^* parameters obtained from B_q^k [4] for different ions in LaF_3 and ratio ($\Delta E_{exp}/\Delta E_{th}$) of experimental to predicted maximum splitting of $^4I_{13/2}$.

Ions	Nd^{3+}	Sm^{3+}	Dy^{3+}	Er^{3+}
N_v^* (cm^{-1})	2356	2098	1905	1638
$\frac{\Delta E_{exp}}{\Delta E_{th}}$	1.32	1.01	1.13	1.11

V. CONCLUSION

It has been shown that the scalar crystal field parameter N_v^* (N_v) describes linearly ΔE_{max}, the maximum Stark splitting of J-terms with small J-mixing and for which $<\| U^{(k)} \|>^2$ is "spherical." This last term meaning that its values do not differ by more than a factor ten for k=2,4,6.

The theoretical proportionality coefficient gives ΔE_{max} by a factor 1 to 1,3 from experiment for different ions, crystals and J-terms, provided the above conditions are respected.

For the first time, strength of crystal fields of different symetry can be compared in a quantitative way giving new interest in B_q^k's tables found in literature.

Another interest in this simple parameter is that some J-terms can be used as simple crystal field probe. For instance as shown on Fig.2 the line given either by theory or experience can be used as a calibration line providing by a simple ΔE_{max} measurement a quantitative value of crystal field. The points on Fig.2, not given by large round points, give such values of crystal field for whole categories of new materials for which crystal field analysis has not yet been performed.

Finally, such N_v^* parameter, as examplified by Fig.2, helps making prediction for energy transfers in Rare-Earth ions doped crsytals.

REFERENCES

1. F. Auzel, Mat. Res. Bull., 14, 223 (1979).
   F. Auzel, in "The Rare-Earths in Modern Science and Technology", Vol.2, ed. G.J. McCarthy, J.J. Rhyne and H.B. Silber, Plenum, 1980, p. 619.
2. F. Auzel and O.L. Malta, J. Physique, 44, 201 (1983).
3. A. Messiah, "Quantum Mechanics", North Holland, 1970, Vol.II, p. 721.
4. W.T. Carnall, H. Crosswhite and H.M. Crosswhite, "Energy level structure and transition probabilities of the trivalent lanthanides in LaF_3 (Argonne National Lab. Argonne, Illinois, 4493).
5. H.M. Crosswhite, H. Crosswhite, F.W. Kaseta and R. Sarup., J. Chem. Phys., 64, 1981 (1976).
6. V. Nekvasil, Phys. Stat. Solidi, B87, 317 (1981).
7. D. Sengupta and J.O. Artman, J. Chem. Phys., 53, 838 (1970).
8. D. Sengupta and J.O. Artman, J. Chem. Phys. 50, 5308 (1969).
9. N. Karayianis, C.A. Morrison and D.E. Wortman, J. Chem. Phys. 64, 3890 (1976).

ENERGY TRANSFER BETWEEN INORGANIC IONS IN GLASSES

R. Reisfeld*

Department of Inorganic and Analytical Chemistry
The Hebrew University of Jerusalem
Jerusalem, Israel

ABSTRACT

Results on energy transfer between Uranyl or first transition metal ions to the rare earths Nd^{3+}, Ho^{3+}, Er^{3+} and Yb^{3+} in inorganic glasses are discussed. The efficiency of the transfer and the transfer probabilities are presented. Suggestions are given for utilization of energy transfer in luminescent solar concentrators (LSC) and Neodymium and Erbium glass lasers.

I. INTRODUCTION

Energy transfer between two inorganic ions can occur either by multipolar interaction or exchange mechanism in dilute systems [1]-[4] or by multi-step migration in concentrated systems [5]. In addition to resonant transfer phonon-assisted transfer is quite a common phenomenon [4].

Usually energy transfer efficiencies and probabilities determined experimentally for macroscopic systems are applied to some average values and not to actual distances between donor and acceptor ions. The theories for resonant energy transfer were developed by Förster [6] and Dexter [7] for multipolar coupling, by Inokuti and Hirayama [8] for exchange coupling, and the theory for nonresonant energy transfer, where the energy mismatch between the energy levels of the donor and the acceptor ions is compensated by the emission or absorption of phonons, was developed by Miyakawa and Dexter [9].

---

* "Enrique Berman" Professor of Solar Energy.

The energy transfer mechanism in dilute systems has been summarized by Watts [10]. At high concentrations of donor and at elevated temperatures the donor-donor transfer may be appreciable. The fluorescence decay curves of the donors behave differently in the two abovementioned cases. If we write the donor-acceptor transfer rate as a/R^s and the donor-donor rate as b/R^s where R is the separation between the interacting ions and s equals 6, 8, 10, for dipole-dipole, dipole-quadrupole and quadrupole-quadrupole interaction then two limiting cases can be considered for $b/a = 0$ where donor-donor interaction is absent and $b/a \gg 1$ where donor-donor interaction is predominant. In the former case the decay curve of the donor fluorescence is non-exponential being the sum of the decay of an isolated donor ion and energy transfer to various accepted ions characterized by the factor $\exp(-At^{3/s})$. In the opposite limit which corresponds to rapid donor-donor transfer the decay is exponential at all times with a rate equal to the total donor-acceptor transfer rate average over all donors. In most cases which fall between these two limits the donor-decay is initially non-exponential but becomes exponential in the long time limit.

The theoretical approach to energy transfer is discussed in detail by Williams, Di Bartolo, Kenkre and Blasse. I shall therefore present mostly experimental data which have recently been obtained in our laboratory dealing with energy transfer between a post-transition or transition metal ion as donor and a rate earth ion as acceptor.

Contrary to energy transfer between a couple of rare earth ions where the transfer probabilities can be obtained from optical transition probabilities using the Judd-Ofelt approach [4], in the present case the transfer probabilities cannot be predicted theoretically as the optical transitions in the donor ions depend strongly on the ligand field and our present theoretical techniques do not allow a priori calculation of these transitions in the condensed phase.

Practical aspect of the energy transfer described here may be found in increasing the pumping efficiency of glass lasers or in recently studied luminescent solar concentrators [11]-[14].

Our discussion here will include energy transfer between UO_2^{2+} and Nd^{3+}, Ho^{3+} and Eu^{3+} in oxide glasses, Bi^{3+} and Nd^{3+} and Eu^{3+} in oxide glasses, Cr^{3+} and Nd^{3+} and Yb^{3+} in lithium lanthanum phosphate glasses (LiLaP) and Mn^{2+} and Er^{3+} in fluoride and oxide glasses.

The formula for the determination of energy transfer probabilities and efficiencies are obtained from experimentally measured lifetimes of the donor ion lifetime with and without addition of the acceptor ion. The decrease of the donor luminescence quantum yield in the presence of the acceptor ion, and the increase of the acceptor emission when excited via the donor ion as compared to the acceptor emission when exciting the acceptor ion directly. This can be summarized in the following Equations:

The efficiency of energy transfer η_{tr} for measured lifetimes is given by Equation (1)

$$\eta_{tr}^{(1)} = 1 - \frac{\tau_d}{\tau_d^o} \qquad (1)$$

where τ_d^o is lifetime of the donor ion in singly doped glasses and τ_d is the lifetime of the donor ion in presence of the acceptor ion.

When the lifetimes are not single exponentials the average lifetimes are used. The average lifetime τ_{av} is defined as

$$\tau_{av} = \frac{\int_0^\infty tI(t)dt}{\int_0^\infty I(t)dt}$$

where $I(t)$ is the emission intensity at a given time t.

The transfer efficiencies η_{tr} for the decrease of the donor luminescent efficiency in presence of the acceptor ion are obtained using Equation (2)

$$\eta_{tr}^{(2)} = 1 - \frac{I_d}{I_d^o} \qquad (2)$$

where I_d is the luminescence quantum efficiency of the donor ion in presence of the acceptor ion and I_d^o is the efficiency of the donor ion in the singly doped glass.

The transfer efficiency can also be calculated from the increase of the acceptor fluorescence using Equation (3)

$$\eta_{tr}^{(3)} = \frac{E_a^d A_a}{E_a^a A_d} \qquad (3)$$

where E_a^d is the emission of the acceptor ion excited at the donor absorption band, E_a^a is the emission of the acceptor ion excited via the acceptor, A_a and A_d are the optical densities of the acceptor and donor at their absorption maxima.

This last Equation provides the net transfer which is the forward minus the backward transfer.

II. URANYL ION AND RARE EARTH IONS

The Uranyl ion has been extensively studied in connection with its photophysics [15] and luminescence. It has been established that in glasses similarly to crystals the Uranyl ion forms a linear molecule with U-O distances of about 1.75 Å which is about 0.6 Å shorter than the distances to 5 - 6 ligating Oxygen atoms from the glass in the equatorial plane. The excited states around 20 000 cm^{-1} arise from electron transfer from the highest M.O. (of odd parity) of the 2p Oxygen orbitals to the empty orbitals of Uranium.

The lowest excited state of the Uranyl ion is the only long-lived state. All the higher levels are nonradiatively de-excited to this level followed by emission in the yellow and green part of the spectrum [16]. The oscillator strengths to the lowest levels are about 10^{-4} for the region between 20 000 - 25 000 cm^{-1} and represent at least 4 electronic transitions [17] from π_u or σ_u orbitals of Oxygen 2p orbitals to the empty 5f shell. There are no parity allowed transitions below 30 000 cm^{-1} hence the relatively weak intensity of the electron transfer band of Uranyl as compared with transition metal complexes.

In phosphate glasses 5 emission maxima peaking at 20 530, 19 710, 18 840, 17 950 and 17 070 cm^{-1} are obtained. The decay curve can be approximated by 2 exponential curves with lifetimes of 115 and 367 μs [16]. The luminescence undergoes concentration quenching and the maximum efficiencies are obtained at about 0.1 molar concentrations. The best efficiencies at room temperature are observed in borosilicate glasses where the efficiencies are about 50% at 0.1 mole % [18].

Efficient energy transfer from the Uranyl group was observed for Eu^{3+} [19], for Sm^{3+} [20], Nd^{3+} and Ho^{3+} [21] and for Er^{3+} [22]. A comparative study in phosphate glasses on energy transfer for the Uranyl group to a number of rare earth ions showed the energy transfer is phonon-assisted and its probability decreases along the series Nd^{3+}, Sm^{3+}, Eu^{3+}, Pr^{3+}, Tm^{3+}, Dy^{3+} [11],[23]. efficiencies of transfer are summarized in Table 1.

A practical consequence of energy transfer between UO_2^{2+} and Nd^{3+} can be applied to LSC's since the existence of the 2 ions in a glass increases the absorption range of the spectrum [21].

III. Bi^{3+}, Eu^{3+} AND Nd^{3+}

The optical characteristics of Bi^{3+} doped oxide glasses are summarized in Reference [24] and References therein. The absorption and excitation spectra of Bi^{3+} peaking at 330 nm arise from the transition $^1S_0 \rightarrow {}^3P_1$ and the fluorescence of Bi^{3+} arises from $^3P_1 \rightarrow {}^1S_0$ transition peaking at about 450 nm. Although this

Table 1. Efficiency of Energy Transfer in Glasses

Glass	Donor ion	Acceptor ion	$\eta_{tr}^{(1)}$	$\eta_{tr}^{(2)}$	$\eta_{tr}^{(3)}$	Ref.
Phosphate	1wt% UO_2^{2+}	1wt% Sm^{3+}	0.22	0.25		[20]
		1wt% Eu^{3+}	0.17	0.36		[20]
		2wt% Nd^{3+}	0.56	0.65		[21]
		2wt% Ho^{3+}	0.29	0.55		[21]
Germanate	1wt% Bi^{3+}	0.05→0.6M% Nd^{3+}		0.2→1.0		[26]
LiLaP	0.308M% Cr^{3+}	0.5M% Nd^{3+}	0.59		0.34	[32]
		1.0M% Nd^{3+}	0.76		0.79	
		2.0M% Nd^{3+}	0.91		1.00	
		3.0M% Nd^{3+}	0.92		-	
		0.5M% Yb^{3+}	0.51			
		1.0M% Yb^{3+}	0.64			
		2.0M% Yb^{3+}	0.78			
		3.0M% Yb^{3+}	0.88			
Fluoride	24M% Mn^{2+}	2M% Er^{3+}	0.77			[34]
Calcium phosphate	2M% Mn^{2+}	2M% Er^{3+}	0.86			[35]

transition is spin-forbidden it has a high oscillator strength of about 0.1 because of strong spin-orbit coupling. The quantum efficiency of Bi^{3+} of its fluorescence in germanate glass is about 2%.

At room temperature energy transfer takes place from the $^3P_1$ level and the transfer probability from 0.1 wt% Bi^{3+} to 1.0 wt% Nd^{3+} is $10^6 s^{-1}$. The same value was obtained for room temperature transfer between Bi^{3+} and Eu^{3+} [25].

Concentration dependence of energy transfer between Bi^{3+} and Nd^{3+} reveals that at a concentration of 1.0 wt% Bi^{3+} and 0.2 wt% Nd^{3+} the pumping efficiency is increased by 175% [26]. The transfer efficiency is given in Table 1.

The low temperature dependence of the above-mentioned bismuth-europium germanate glass and the time-resolved-spectroscopy of these glasses has been studied with the Lyon group [27],[28]. The results of these studies will be discussed by Boulon.

IV. Cr^{3+} AND Nd^{3+} AND Yb^{3+} IN LANTHANUM PHOSPHATE GLASS

Cr^{3+} in glasses has been extensively studied [29] because of the predicted similarity between the celebrated ruby crystal and glasses. References to these studies can be found in [30].

Absorption and emission of Cr^{3+} arises from the parity forbidden electronic transitions in the 3d electronic shell. Crystal field split states of Cr^{3+} in octahedral symmetry are illustrated in the Tanabe-Sugano diagrams. The relative positions of the excited $^4T_2$ and ^2E states depend on the crystal field strength. In cases where $Dq/B < 2.3$ (low field cases) $^4T_2$ is the low state, the emission arises from the $^4T_2$ to $^4A_2$ spin allowed transition. In the case of $Dq/B > 2.3$ (high field cases) the lowest state is ^2E and the luminescence arises from the spin forbidden ^2E to $^4A_2$ transition characteristic for the R-line emission of ruby. The spin allowed transitions are characterized by broad emission spectra and short lifetimes contrary to the spin forbidden emission from the ^2E state (which sometimes is mixed with $^2T_1$ levels) with narrow band and long lifetimes.

All the known emissions of Cr^{3+} in glasses arise from $^4T_2$. Reasons for this may be: 1. $^4T_2$ lies below the ^2E level; 2. The position of two levels may be of equal energy, however the weak ^2E emission may be obscured by the broad emission from $^4T_2$.

The quantum efficiency of the emission arising from ^2E is usually about 100% because of the equal equilibria positions and shape of $^4A_2$ and ^2E levels in the configuration diagrams. The fluorescence efficiency from the ^4T levels varies drastically in various media. Thus, contradictory to the high efficiency of Cr^{3+} in crystalline fluoride materials (elpasolite) and probably crystalline aluminium metaphosphate, efficiency of Cr^{3+} in glasses is low but it may be increased by tailoring the glass composition [31]. The best efficiencies for $^4T_2 \rightarrow {}^4A$ of 23% are obtained for LiLaP glasses. The reason for the nonradiative transitions of Cr^{3+} in

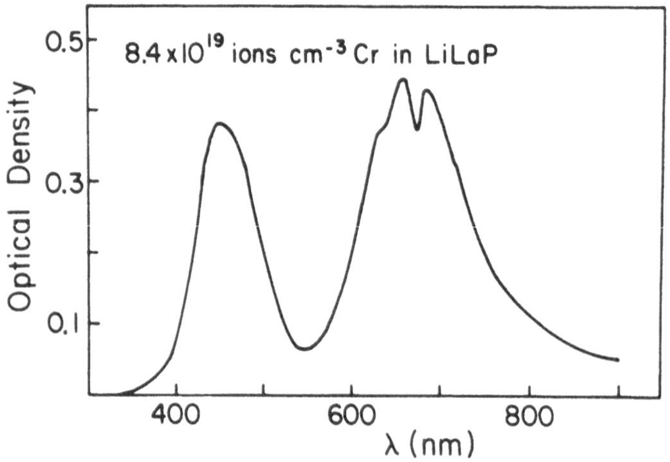

Fig. 1. Absorption spectrum of 0.31M% Cr^{3+}

glasses is still under investigation by the group of Dr. Lempicki of GTE Laboratories and our group. At this point it is absolutely clear that the theory of multiphonon relaxation which holds in the case of rare earth ions cannot be applied to Cr^{3+}.

Our present studies [32] show that in LiLaP glass the probability of energy transfer from Cr^{3+} to Nd^{3+} may be higher than the nonradiative transfer within the Cr^{3+} system. This is probably the reason for good laser performance of such glasses as reported recently by Härig [33].

The integrated lifetime is 24.6 μs at room temperature which is indicative of the proximity of ^4T and ^2E states in the LiLaP glass [31].

Fig. 1 presents the absorption spectrum of 0.31 M% Cr_2O_3 in LiLaP glass. Fig. 2 presents the absorption spectra of 0.15 M% Cr_2O_3, 2 M% Nd_2O_3 and 0.15 M% Cr_2O_3 + 2 M% Nd_2O_3 in LiLaP glass. From this Figure it can be seen that the absorption is additive. The same is true in the case of LiLaP glass with Cr_2O_3 and Yb_2O_3. Fig. 3 shows the absorption spectra of 0.31 M% Cr_2O_3 and 3 M% Yb_2O_3. Fig. 4 presents the emission spectra of Cr^{3+} in LiLaP glass, curve (a) for 0.05 M% Cr_2O_3 and curve (b) for 0.31 M% Cr_2O_3 arising from the $^4T_2 \rightarrow {}^4A$ transition. Fig. 5 presents the emission spectra of Nd^{3+} excited at 585 nm to Nd^{3+} directly and at 647 nm via Cr^{3+}. This curve evidences energy transfer from Cr^{3+} to Nd^{3+}.

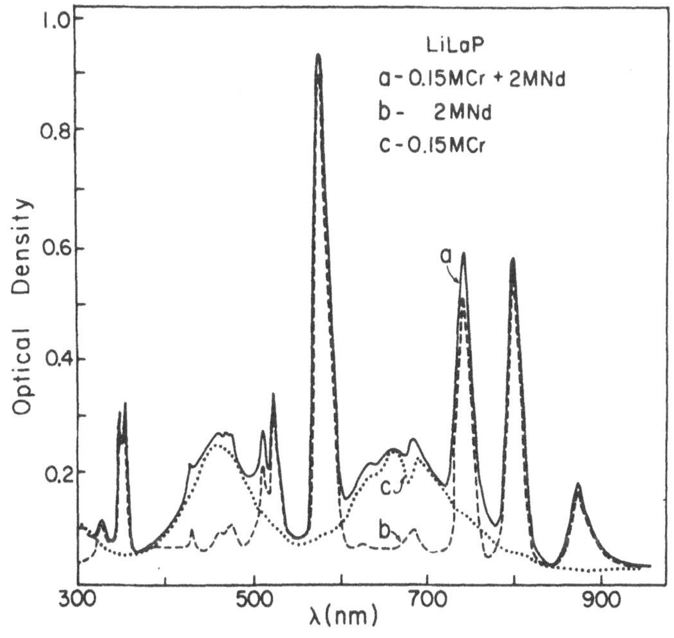

Fig. 2. Absorption spectra of Cr^{3+}, Nd^{3+} and Cr^{3+} + Nd^{3+} (4.2×10^{19} ions cm^{-3} Cr, 5.44×10^{20} ions cm^{-3} Nd)

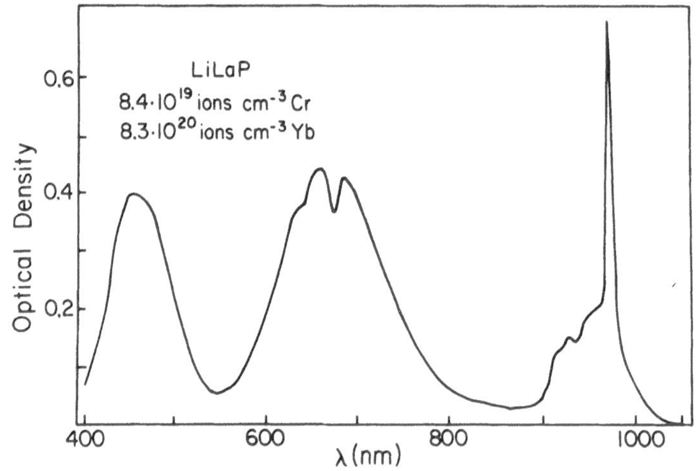

Fig. 3. Absorption spectrum of 0.31 M% Cr^{3+} + 3.0 M% Yb^{3+}

Fig. 6. presents the emission spectra of Yb^{3+} excited directly at 915 nm into the $^7F_{7/2} \rightarrow {}^2F_{5/2}$ transition (left-hand-side of diagram) and excited via Cr^{3+} into $^4T_2$. This right-hand-side part of the diagram again provides evidence of energy transfer between Cr^{3+} and Yb^{3+}.

Examples of energy transfer efficiencies obtained from Equations (1) and (3) between Cr^{3+} and Nd^{3+}, and Cr^{3+} and Yb^{3+} are presented in Table 1. It is not surprising that energy transfer efficiencies between Cr^{3+} and Nd^{3+} are higher than the corresponding transfer in Yb^{3+} because of the much higher spectral overlap between Cr^{3+} and Nd^{3+}.

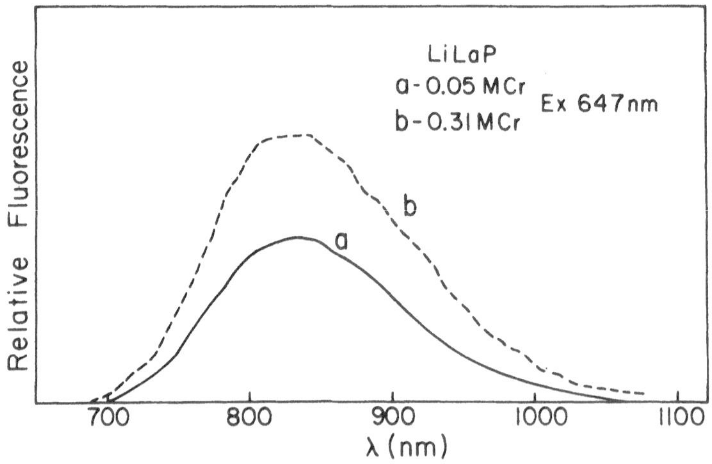

Fig. 4. Emission spectra of Cr^{3+}

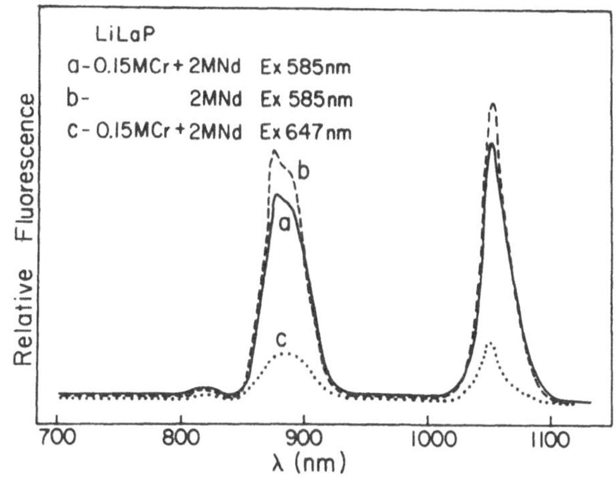

Fig. 5. Emission spectra of Nd^{3+}

V. ENERGY TRANSFER FROM Mn^{2+} TO Er^{3+} IN FLUORIDE GLASSES AND Mn^{2+} TO Nd^{3+}, Ho^{3+} AND Er^{3+} IN OXIDE GLASSES

V.A. Manganese

The absorption spectrum of Mn^{2+} doped fluoride glass [34] of the composition $36PbF_2$, $24MnF_2$, $35GaF_3$, $5A\ell(PO_3)_3$, $2LaF_3$ is

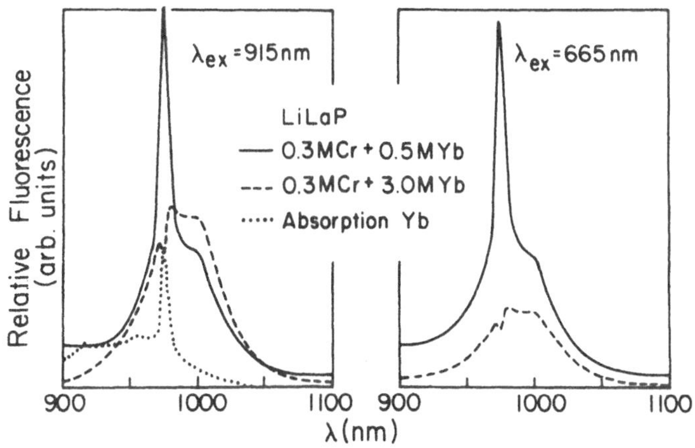

Fig. 6. Emission spectra of Yb^{3+} (0.5M% Yb^{3+} = = 1.36×10^{20} ions cm^{-3}, 3.0M% Yb^{3+} = 8.31×10^{20} ions cm^{-3})

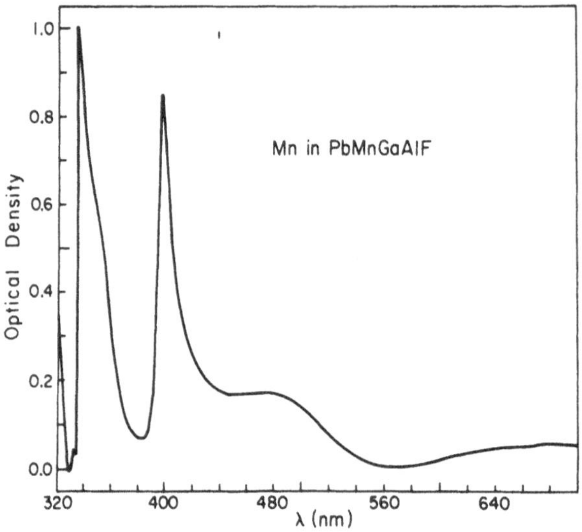

Fig. 7. Absorption spectrum of Mn^{2+} in $36PbF_2$, $24MnF_2$, $35GaF_3$, $5Al(PO_3)_3$, $2LaF_3$

presented in Fig. 7. The two peaks arising from $^6A_1$ to $^4T_1$ at 480 nm and ^6A to $^4A_1$, ^4E at 398 nm are similar to those obtained for Manganese in phosphate glasses [13],[35]. The additional peak 332 nm belonging to the higher ligand field levels coincides probably with the Pb^{2+} absorption. The excitation spectrum of the emission at 626 nm is presented in Fig. 8 where curve (a) belongs to the excitation of a glass containing Mn^{2+} only and curve (b)

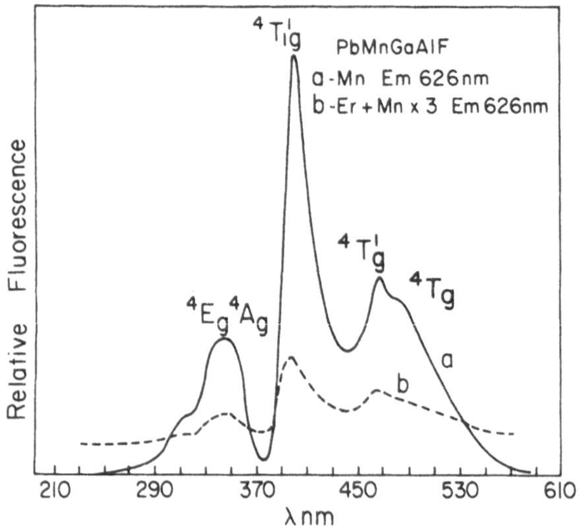

Fig. 8. Excitation spectra of Mn^{2+} in $36PbF_2$, $24MnF_2$, $35GaF_3$, $5Al(PO_3)_3$, $2LaF_3$ and $36PbF_2$, $24MnF_2$, $35GaF_3$, $5Al(PO_3)_3$, $2ErF_3$

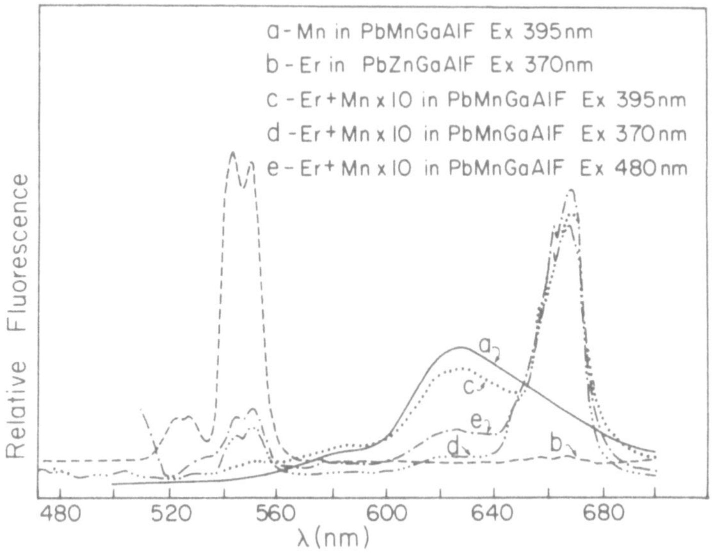

Fig. 9. Emission spectra of Mn^{2+}, Er^{3+} and $Mn^{2+} + Er^{3+}$

is the excitation spectrum of a glass into which 2 mole% Er^{3+} was added. Note that the sensitivity of curve (b) is increased by a factor of 3.3 thus indicating the decrease of the luminescence of Mn^{2+} in presence of Er^{3+} as a result of energy transfer from

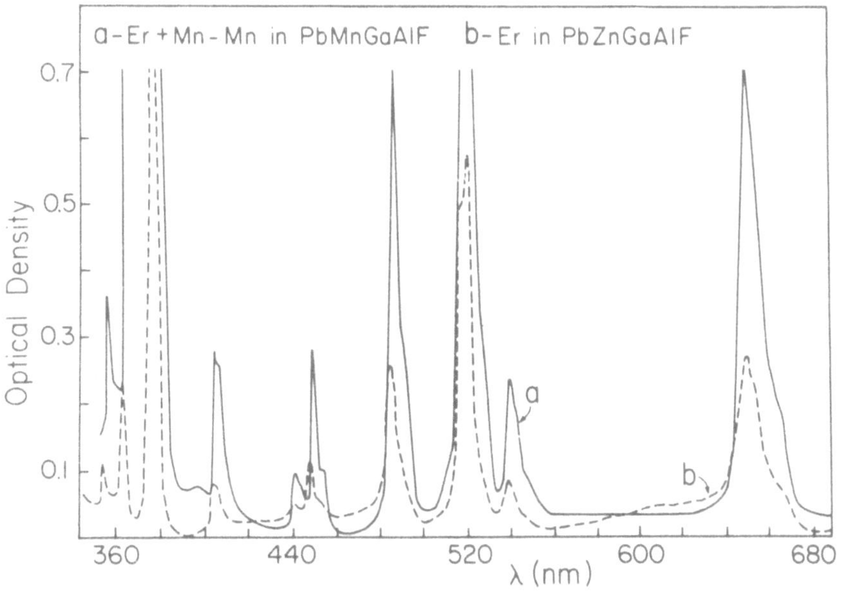

Fig. 10. Absorption spectra of Er^{3+} (two concentrations)

Mn^{2+} to Er^{3+}. The emission spectrum of Mn^{2+} under excitation of 395 nm is presented in Fig. 9 curve (a). This curve consists of a single peak at 626 nm arising from the transition $^4T_1 \rightarrow {}^6A_1$. This emission is similar to the emission of Mn^{2+} in oxide glasses. The lifetime of this emission exhibits a nonexponential behavior which can be roughly described by two lifetimes 1.06 ms and 1.82 ms, and an integrated lifetime of 1.45 ms for glasses consisting of Mn^{2+} only. Glasses containing 2 mole% of Er^{3+} have two decay times, 150 μs and 430 μs, and an integrated lifetime of 340 μs.

V.B. Erbium

The absorption spectrum of Er^{3+} in the glass is shown in Fig. 10 in which curve (a) is due to the absorption in a glass containing Mn^{2+} measured with a similar glass without Er^{3+} and curve (b) is the absorption of Er^{3+} in which Mn^{2+} was replaced by Zn^{2+}. There is no observable difference between the shape (half-band width) and position of bands in the two spectra. Such behavior was observed previously [36] in glasses of composition $46PbF_2$, $22MnF_2$, $20GaF_3$, $2LaF_3$ and $46PbF_2$, $22ZnF_2$, $30GaF_3$, $2LaF_3$.

The excitation spectrum of Er^{3+} for the 543 nm emission due to $^4S_{3/2} \rightarrow {}^4I_{15/2}$ is presented in Fig. 11 curve (a) and of the 666 nm emission due to transition $^4F_{9/2} \rightarrow {}^4I_{15/2}$ in curve (b) in which the latter sensitivity is 20 times higher. This means that the 666 nm emission is negligible compared with the 543 nm emission in glasses doped by Er^{3+} alone.

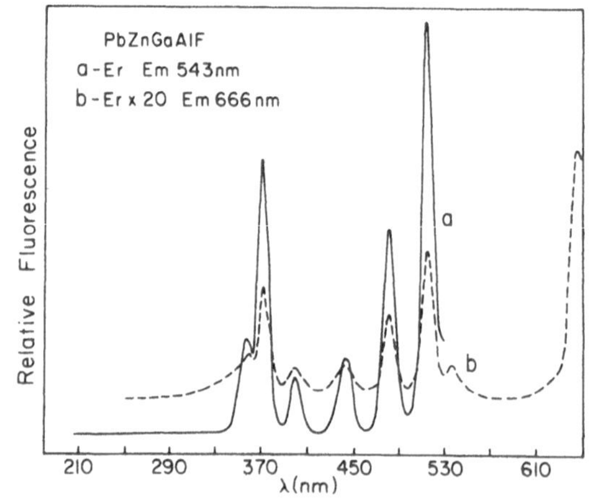

Fig. 11. Excitation spectrum of Er^{3+}

The emission spectrum of Er^{3+} alone excited at 370 nm is presented in Fig. 9 curve (b). Here we see an emission at 525 nm of the transition $^2H_{11/2} \rightarrow {}^4I_{15/2}$, a much stronger emission at 543 nm of transition $^4S_{3/2} \rightarrow {}^4I_{15/2}$ and almost a negligible emission at 666 nm of transition $^4F_{9/2} \rightarrow {}^4I_{15/2}$ corresponding to the very weak excitation shown in Fig. 11.

V.C. Energy Transfer between Manganese and Erbium

The emission spectrum of the composite glass containing 2 mole% Er^{3+} is presented in Fig. 9 curve (c) excited at the Mn^{2+} excitation level 395 nm. This curve consists of an emission band at 626 nm of Mn^{2+} and at 666 nm of Er^{3+}. The emission of Er^{3+} at 543 nm is almost totally absent. Note that this curve (c) is quite different from curve (b) for the 666 nm emission where no Mn^{2+} is present. Curve (d) Fig. 9 shows the emission of the same glass excited at 370 nm (excitation level of Er^{3+}). We observe a much higher emission at 666 nm than at 543 nm and a small peak at 626 nm of Mn^{2+}. This means that we have energy transfer from Er^{3+} to Mn^{2+} as opposed to energy transfer from Mn^{2+} to Er^{3+} in curve (c). The absence of the 543 nm emission of Er^{3+} in curve (c) probably means that when Mn^{2+} is excited the energy is transferred from 626 nm to $^4F_{9/2}$ of Er^{3+}. Energy transfer to the higher level $^4S_{3/2}$ is impossible. On the other hand when Er^{3+} is excited [curve (d)] part of its energy is transferred to Mn^{2+} and additional energy is emitted at 666 nm. The energy transfer probability was calculated from the lifetime by Equation (1). The probability of energy transfer between Mn^{2+} and 2 mole% Er^{3+} is $2.3 \cdot 10^3$ s^{-1}. The back transfer probability from the $^4S_{3/2}$ state of Er^{3+} to Mn^{2+} can only be approximated from the decrease of the lifetime of Er^{3+} (97.8 μs for 0.1 wt% Er^{3+} only to 9.1 μs in 2 mole% Er^{3+} in the Mn-containing glass. Using Equation (1) we obtain $\sim 9 \cdot 10^4 s^{-1}$ for this transfer. The energy transfer efficiencies are presented in Table I.

VI. CONCLUSIONS

It should be noted that the transfer efficiencies as given in Table I are the average values when the averaging includes the inhomogeneous broadening due to a variety of emitting ion sites and the averaging over various transfer mechanisms.

The line-narrowing technique for the donor ions discussed in this paper at low temperatures allows excitation of a smaller number of inhomogeneously broadened sites, however there is no possibility of exciting one type of site only [27]. From the experiments performed so far it appears that the Europium ion is the only ion that can be selectively excited in glasses. For other ions different techniques must be used. We have recently suggested that the donor ion be adsorbed onto porous glasses [37] followed by a successive adsorption of the acceptor ions. In such an arrangement the donor-acceptor distances can be calculated due to the lower dimensionality of the system.

The macroscopic transfer efficiencies are of value in predicting systems with good pumping efficiencies for lasers and for solar concentrators where the high absorption of the donor ion permits high efficiency of trapping of the exciting light.

ACKNOWLEDGMENT

The author would like to express her gratitude to Mrs. E. Greenberg in preparation of the manuscript and to Professor C.K. Jørgensen for helpful discussions.

REFERENCES

1. R. Reisfeld and C.K. Jørgensen, Lasers and Excited States of Rare Earths, Springer-Verlag, Berlin, Heidelberg, New York (1977).
2. R. Reisfeld, Structure and Bonding 13, 53 (1973).
3. R. Reisfeld, Structure and Bonding 22, 123 (1975).
4. R. Reisfeld, Structure and Bonding 30, 65 (1976).
5. R.C. Powell and G. Blasse, Structure and Bonding 42, 43 (1980).
6. Th. Förster, Discuss. Faraday Soc. 27, 7 (1959).
7. D.L. Dexter, J. Chem. Phys. 21, 836 (1953).
8. M. Inokuti and P. Hirayama, J. Chem. Phys. 43, 1978 (1965).
9. T. Miyakawa and D.L. Dexter, Phys. Rev. B1, 2961 (1970).
10. R.K. Watts, Optical Properties of Ions in Solids, Ed. B. DiBartolo, Plenum, New York (1975).
11. R. Reisfeld and C.K. Jørgensen, Structure and Bonding, 49, 1 (1982).
12. R. Reisfeld, E. Greenberg, A. Kisilev and Y. Kalisky, Collection and conversion of solar energy for photoconverting systems. Photochemical Conversion and Storage of Solar Energy, Ed. J.S. Connolly, Academic Press, 364 (1981).
13. A. Kisilev, R. Reisfeld and H. Tzehoval, Book of Abstracts. Fourth Int'l Conf. on Photochemical Conversion and Storage of Solar Energy, 299 (1982).
14. R. Reisfeld, N. Manor and D. Avnir, Solar Energy Materials, in press.
15. C.K. Jørgensen and R. Reisfeld, Structure and Bonding 50, 121 (1982).
16. N. Lieblich-Sofer, R. Reisfeld and C.K. Jørgensen, Inorg. Chim. Acta 30, 259 (1978).
17. C.K. Jørgensen and R. Reisfeld, J. Electrochem. Soc. in press.
18. R. Reisfeld and H. Tzehoval, in preparation.

19. R. Reisfeld, N. Lieblich, L. Boehm and B. Barnett, J. Luminescence 12/13, 749 (1976).
20. R. Reisfeld and N. Lieblich-Sofer, J. Solid State Chem. 28, 391 (1979).
21. R. Reisfeld and Y. Kalisky, Nature 283, 281 (1980).
22. J.C. Joshi, N.C. Pandey, B.C. Joshi and J.Joshi, Ind. J. Pure and Appl. Phys. 15, 519 (1977).
23. R. Reisfeld, Multiphonon Relaxation in Glasses in Radiationless Processes, Eds. B. DiBartolo and V. Goldberg, Plenum Press, New York, 489 (1980).
24. R. Reisfeld and Y. Kalisky, Chem. Phys. Lett. 50, 199 (1977).
25. R. Reisfeld, Y. Kalisky, L. Boehm and B. Blanzat, Proc. Rare Earth Res. Conf. 378 (1976).
26. Y. Kalisky, R. Reisfeld and J.S. Bodenheimer, J. Noncryst. Solids 44, 249 (1981).
27. J.C. Bourcet, B. Moine, G. Boulon, R. Reisfeld and Y. Kalisky, Chem. Phys. Lett. 61, 23 (1979).
28. B. Moine, J.C. Bourcet, G. Boulon, R. Reisfeld and Y. Kalisky, J. Physique 42, 499 (1981).
29. E.J. Sharp, J.E. Miller and M.J. Weber, J. Appl. Phys. 44, 4098 (1973).
30. L.J. Andrews, A. Lempicki and B.C. McCollum, J. Chem. Phys. 74, 5526 (1981).
31. A. Kisilev and R. Reisfeld, Solar Energy, in press.
32. R. Reisfeld, A. Kisilev and Y. Millstein, in preparation.
33. T. Härig, G. Huber and I.A. Shcherbakov, J. Appl. Phys. 52, 4450 (1981).
34. R. Reisfeld, E. Greenberg, C.K. Jørgensen, C. Jacoboni and R. de Pape. To be published in J. Solid State Chem.
35. S. Parke and E. Cole, Phys. Chem. Glasses 12, 125 (1971).
36. R. Reisfeld, G. Katz, N. Spector, C.K. Jørgensen, C. Jacoboni and R. de Pape, J. Solid State Chem. 41, 253 (1982).
37. R. Reisfeld, Chem. Phys. Lett., in press.

NON-EQUILIBRIUM CONCEPTS IN SOLAR ENERGY CONVERSION

P. T. Landsberg*

Department of Electrical Engineering
University of Florida
Gainesville, Florida 32611

ABSTRACT

Among non-equilibrium concepts introduced here are first the uses of fluxes of energy (Φ) and entropy (Ψ) which for black-body radiation leads to $\Phi = (3/4)\Psi T = \sigma T^4$ where σ is Stefan's constant. An application to planetary temperatures is noted. Secondly, diluted black-body radiation, which is a non-equilibrium form of radiation, is introduced and applied to the conversion of black-body radiation. A thermodynamic and statistical mechanical study is made of an infinite stack of solar cells to determine optimum efficiencies. One finds here a chemical potential of non-equilibrium radiation if it is in interaction and in a steady state with electrons and holes, each governed by a quasi-Fermi level.

I. GENERAL CONCEPTS FOR RADIATION

I.A. Introduction

There are many non-equilibrium concepts and ideas one could discuss in these lectures. However, in four lectures one can cover only some of the interesting topics properly, and so I had to omit a great deal of kinetics, problems of photosynthesis and non-equilibrium phase transitions which it was originally planned to include. What can be covered is, however, of importance and has been a focus of recent interests.

---

*Permanent address: University, Southampton S09 5NH, England.

Black-body radiation has been extensively studied and one knows practically all about it; the Planck frequency distribution, the thermodynamic properties such as pressure and entropy as a function of temperature, etc. What is one to do if the radiation incident on a converter is not of the black-body type? It certainly happens: one need only think of the direct solar spectrum at sea level with its many dips due to optical absorption in the atmosphere. How is one to work out the entropy of the radiation in such more general situations? It is to be emphasized that such non-black-body distributions are stable only in an enclosure with perfectly reflecting walls. If normal matter at some temperature T is present these distributions become unstable and go over into black-body radiation corresponding either to that temperature, or to an appropriate equilibrium temperature resulting from averaging of the thermal energy in the matter and in the radiation.

The basic difficulty is that one cannot find a distribution function in this case by maximizing the entropy--that would merely lead to an equilibrium distribution. One wants to amend this usual procedure by using an entropy expression which is valid also away from equilibrium. That such an expression should exist is not surprising, for even maximization of the entropy for equilibrium presumes that the expression used is valid away from equilibrium.

I.B. Photons in Discrete Quantum States

In order to discuss the energy and entropy of a gas of bosons which are not necessarily in equilibrium, let $P(N_1,N_2,\dots)$ be the probability of finding N_1 bosons in single-particle quantum state 1, N_2 in state 2, etc. This distribution enables one to infer the probability, to be denoted by $p_j(N_j)$, of finding N_j bosons in quantum state j, whatever the occupation of the other quantum states. We now assume: (i) the probabilities $p_1(N_1)$, $p_2(N_2)$, ... are independent in the probability sense, so that

$$P(N_1,N_2,\dots) = p_1(N_1)p_2(N_2)\dots \quad (1)$$

(ii) the probability of an additional particle occupying a state j is independent of the number already in this state. Thus $p_j(N_j) \propto q_j^{N_j}$, where q_j is an undetermined positive number. Normalization then gives

$$p_j(N_j) = (1 - q_j)q_j^{N_j}, \ (0 \leq q_j \leq 1). \quad (2)$$

From this apparently flimsy base one can reach all results needed here, without having to assume equilibrium conditions by maximizing the entropy [1].

The parameter q_j, specific to state j, can be related to the mean occupation number n_j of state j, by noting

$$n_j = \sum_{N=0}^{\infty} N q_j^N (1 - q_j) = \frac{q_j}{1 - q_j}. \tag{3}$$

Hence one can express q_j in terms of n_j:

$$q_j = \frac{n_j}{1 + n_j}, \quad 1 - q_j = \frac{1}{1 + n_j} \tag{4}$$

Next, the entropy of the system is

$$S = -k \sum_{N_1=0}^{\infty} \cdots \sum_{N_j=0}^{\infty} \cdots P(N_1, N_2, \ldots) \ln P(N_1, N_2, \ldots)$$

This becomes, with (1) and (2),

$$S = -k \sum_{j=1}^{\infty} \sum_{N_j=0}^{\infty} \left((1 - q_j) q_j^{N_j} \ln(1-q_j) + (1 - q_j) q_j^{N_j} N_j \ln q_j \right)$$

$$= -k \sum_{j=1}^{\infty} \left(\ln(1 - q_j) + \frac{q_j}{(1 - q_j)^2} (1 - q_j) \ln q_j \right).$$

where we have used (2) in the form

$$\sum_{N_j=0}^{\infty} p_j(N_j) = 1,$$

and also

$$\sum_{N_j=0}^{\infty} N_j q_j^{N_j} = q_j \frac{d}{dq} \left(\sum_{N_j=0}^{\infty} q_j^{N_j} \right) = q_j \frac{d}{dq_j} \frac{1}{1 - q_j} = \frac{q_j}{(1 - q_j)^2}.$$

Hence

$$S = k \sum_{j=1} \left(\ln(1 + n_j) - n_j \ln \frac{n_j}{1 + n_j} \right)$$

$$= k\sum_j [(1 + n_j) \ln(1 + n_j) - n_j \ln n_j] \qquad (5)$$

This result, usually obtained from equilibrium stastical mechanics, is therefore of wider significance and represents a non-equilibrium entropy. It is exact for non-equilibrium situations if (1) and (2) hold. The equilibrium Bose distribution can be derived from (5) by entropy maximization. The equilibrium distribution law for ensembles, Stirling's approximation, the grouping together of quantum states, the equiprobability of states in nonequilibrium situations are all unsuitable assumptions for nonequilibrium discussions and have been jettisoned in the present approach. This is its chief merit. An unusual interpretation of q_j will be given in Section II.C.

I.C. Continuous Photon Spectrum

We now pass to a continuous spectrum approximation by replacing the sum over quantum states by an integral. This requires an expression for the number of states of the radiation in a frequency interval $(\nu, \nu + d\nu)$, when it crosses an element of a surface into a solid angle $d\Omega$ in a direction making an angle θ with the normal to the element. For emission or absorption of radiation the surface is part of the surface bounding the volume v of the radiation enclosure, and the normal is drawn outwards. But the surface can also be a fictitious surface drawn inside v. We shall normally consider only isotropic radiation which may be polarized, specified by $\ell = 1$, or unpolarized, specified by $\ell = 2$. In the latter case each translational mode can have one of two directions of polarization, so that the translational density of states is multiplied by ℓ in the general case. The required density of states can be written in various ways:

$$h^{-3} \ell v p^2 \, dp \, d\Omega = c^{-3} \ell v \nu^2 \, d\nu \, d\Omega = h^{-3} c^{-3} \ell v e^2 \, de \, d\Omega, \qquad (6)$$

where p, ν, and e are, respectively, linear momentum, frequency, and energy of a photon in the range. In the isotropic unpolarized case $\ell d\Omega$ can be replaced by 8π after integration over solid angles. The usual density of states used for black-body radiation is then obtained from (6). One finds from (5) and (6)

$$\frac{S}{v} = \iint s_\nu \, d\nu \, d\Omega,$$

$$s_\nu \equiv \frac{\ell k}{c^3} \nu^2 [(1 + n_\nu) \ln(1 + n_\nu) - n_\nu \ln n_\nu], \qquad (7)$$

where n_ν is the mean occupation number of a translational state of frequency ν. The internal energy per unit volume is

$$\frac{U}{V} = \iint u_\nu \, d\nu \, d\Omega, \quad u_\nu \equiv \frac{\ell}{c^3}\nu^2 (h\nu) n_\nu. \tag{8}$$

The quantities u_ν, s_ν are energy and entropy of polarized or unpolarized radiation in the cavity, expressed per unit volume, per unit frequency range, and per unit solid angle. For black-body radiation at temperature T

$$n_\nu = [\exp(h\nu/kT) - 1]^{-1} \tag{9}$$

I.D. Photon Fluxes

For efficiency analyses we need energy and entropy per unit area per unit time which passes through a surface element (or is emitted or absorbed at a surface element), θ being the angle between radiation and the normal to the surface. These energy and entropy flux densities are

$$\Phi = \iint K_\nu \cos\theta \, d\nu \, d\Omega, \quad K_\nu \equiv c u_\nu = \frac{\ell h \nu^3}{c^2} n_\nu, \tag{10}$$

$$\Psi = \iint L_\nu \cos\theta \, d\nu \, d\Omega, \tag{11}$$

$$L_\nu = c s_\nu = \frac{\ell k \nu^2}{c^2} [(1 + n_\nu) \ln(1 + n_\nu) - n_\nu \ln n_\nu].$$

The factor $\cos\theta$ projects unit area of the surface to be normal to the radiation. The extra factor c arises in (10) since the energy passing normally through a surface of area ΔA in time Δt due to radiation in frequency range $d\nu$ and solid angle $d\Omega$ is given by $K_\nu d\nu d\Omega \Delta A \Delta t = u_\nu \, d\nu \, d\Omega \Delta A (c\Delta t)$, because the radiation lies in a cylinder whose axis has length $c\Delta t$. A similar argument holds for the entropy. The quantities K_ν, L_ν are called, respectively, spectral energy radiance and spectral entropy radiance. If K_ν and L_ν are independent of direction within the solid angle considered, then

$$\Phi = B\int K_\nu \, d\nu, \quad \Psi = B\int L_\nu \, d\nu, \tag{12}$$

where

$$B \equiv \int \cos\theta \, d\Omega. \tag{13}$$

Suppose now that we have a source for which the spectral energy and entropy radiance K_ν and L_ν are independent of the angles θ and ϕ

specifying the direction of the emitted radiation. Let the cone subtended by the source at the receiver be of half-angle δ. The angular integral in energy and entropy densities (7) and (8) is then

$$\int d\Omega = \int_0^{2\pi} d\phi \int_0^{\delta} \sin\theta \, d\theta = 2\pi(1 - \cos\delta). \qquad (14)$$

Insertion of this result into (7) and (8) yields formulas already noted by Parrott[2]. For a correction to another part of that paper see Ref. [3].

Similarly the integral which occurs in the energy and entropy flux densities is

$$B = \int_0^{2\pi} d\phi \int_0^{\delta} \cos\theta \, \sin\theta \, d\theta = \pi \sin^2\delta. \qquad (15)$$

For small δ one finds $\pi\delta^2$ in both (14) and (15). This is an approximation which is occasionally used.

I.E. The Case of Black-Body Radiation

First recall the standard black-body expression for the number of photons per unit volume in black-body radiation at temperature T and in frequency range $d\nu$

$$8\pi\nu^2 c^{-3}/[\exp(h\nu/kT) - 1)] \qquad (16)$$

To obtain u_ν, i.e. an energy per unit volume and solid angle, one must multiply by $h\nu/4\pi$ so that

$$u_\nu = 2h\nu^3 c^{-3} \, d\nu/[\exp h\nu/kT - 1]$$

This result will be inserted into the theory of Section I.D. We have here used the fact that the energy of a photon of frequency ν is $e = h\nu$.

For black-body radiation at absolute temperature T one has from (10) and (12) that the energy flux is

$$\Phi = B\int K_\nu(T)\, d\nu = Bc \int u_\nu \, d\nu = B\int \frac{\ell h\nu^3}{c^2} \frac{d\nu}{\exp(h\nu/kT) - 1}$$

$$= \frac{B\ell k^4 T^4}{c^2 h^3} \int \frac{x^3 \, dx}{\exp x - 1} \qquad (17)$$

$$= \frac{\pi^4 B\ell k^4}{15c^2h^3} T^4 \quad \text{(for the integral, see Problem 3, below)} \tag{18}$$

as the integral has the value $\pi^4/15$. For isotropic emission $\delta = \pi/2$ so that $B = \pi$ in (15). Also $\ell = 2$ for unpolarized radiation. Hence finally

$$\Phi = \frac{2\pi^5 k^4}{15h^3c^2} T^4 = \sigma T^4 \tag{19}$$

Stefan's constant $\sigma \sim 5.67 \times 10^{-8}$ Wm^{-2} K^{-4} has thus been obtained from the general formalism, so acting as a check on it. Stefan's law (19) gives the energy emitted by a black-body at temperature T per unit area per unit time. If one wants simply to deduce (19) without the general formalism, a more direct procedure can be used, as shown in Problems 1 and 2.

The entropy flux is by a similar, but slightly longer, argument [see Problem 2]

$$\Psi = (4/3)(\Phi/T) = \frac{4\pi^4 B\ell k^4}{45h^3c^2} T^3, \quad L_\nu(T) = \frac{4\ell h\nu^3}{3c^2T} \frac{1}{e^{h\nu/kT}-1} = \frac{4}{3}\frac{K_\nu(T)}{T} \tag{20}$$

We thus have

$$\Phi = \frac{3}{4}\Psi T = \sigma T^4, \tag{21}$$

a simple result to which we shall return in Section I.G.

I.F. Simple Applications of the Energy Flux Concept [4]

The two examples from the solar energy field to be presented are very simple. Both require only Φ the energy flux and not the entropy flux. Furthermore it is assumed that black bodies are involved. More advanced applications will be given in Sections II.A to II.F.

The first example leads to a value of the solar constant in terms of other data from the solar system. Let a star of radius R_S radiate as a black-body at temperature T_S. Then the energy emission rate can be equated to the reception rate on a larger sphere of radius R_{SP} drawn so that a planet P lies on it. The energy per unit time emitted and received is

$$4\pi R_S^2 \, \sigma T_S^4 = 4\pi R_{SP}^2 G_{SP}$$

Here G_{SP} is the solar constant for planet P, G_{SP} being defined as the energy received per unit area at the planet P from the star S. Thus

$$G_{SP} = \left(\frac{R_S}{R_{SP}}\right)^2 \sigma T_S^4. \tag{22}$$

For the sun-earth system one has

$$G_{SP} = \left(\frac{6.95 \times 10^8 m}{1.5 \times 10^{11} m}\right)^2 (5.67 \times 10^{-8}\ Wm^{-2}k^{-4})\ (5760K)^4$$

$$= 1350\ Wm^{-2}$$

This is a good estimate of the measured solar constant. The calculation assumes merely that the solar surface temperature is known at 5760K and that R_S and R_{SP} are given from astronomical data.

The second example serves to estimate the surface temperature of planets. The solar energy received will heat up the planet until it reaches the solar temperature unless an energy loss mechanism is provided. One can assume that the planet radiates as a black-body at absolute temperature T_P. Since the planet presents a disk to the radiating star S it absorbs energy at the rate $\pi R_P^2 G_{SP}$ and it emits energy at the rate $4\pi R_P^2\ \sigma T_P^4$. Hence, using (22),

$$T_P^4 = \frac{G_{SP}}{4\sigma} = \left(\frac{R_S}{2R_{SP}}\right)^2 T_S^4 \tag{23}$$

This suggests that when a star has several planets, the average surface temperatures of the planets are such that

$$T_P R_{SP}^{\frac{1}{2}} = (R_S/2)^{\frac{1}{2}}\ T_S = \text{same for all planets.} \tag{24}$$

Such a simple result cannot be expected to be either new or accurate. It was obtained by the present writer in the presented in lectures at a summer school on solar energy at ICTP (Trieste) in 1977 [4]. The best reference he has found is the remark made by Arrhenius in 1908 who referred to Christiansen (no initial, no reference) as having made this type of calculation [5]. As to the accuracy of (23), this has been found by D. Wilkins (Oakland University, Rochester, Michigan 48063) [6] to be quite good as Table 1 shows.

Table 1. Some Planetary Temperatures

	R_{SP} (10^{11}m)	T_P Predicted by (23) (K)	Observed [7](K)
Mercury	.580	450	753
Earth	1.50	280	295
Mars	2.28	230	250
Jupiter	7.80	120	134
Saturn	14.31	91	97
Uranus	28.8	64	60
Neptune	45.2	51	57
Pluto	50.1	45	43

The greenhouse effect makes the calculation inapplicable to Venus.

I.G. Fluxes Compared with Equilibrium Quantities

The result (21) is simple and instructive. But is it basically the same as the known relation for energy U in black-body radiation which has entropy S at temperature T, $U = (3/4)TS$?

To answer this question it will be shown that the flux F_Q of an extensive quantity Q, which is carried by particles in a gas, is related to the amount Q of that quantity in volume v by

$$F_Q = \frac{\overline{V}}{4v} Q \tag{25}$$

where $\overline{V}$ is the mean (scalar) speed of the particles in the gas.

Assume that the distribution of molecules in equilibrium is isotropic. Thus all particle velocity vectors of lengths V are uniformly distributed over angles. The probability of such vectors in a solid angle $d\Omega$ is then $d\Omega/4\pi$, since this yields unity when integrated over all distributions (Fig. 1):

$$\int d\Omega = 2\int_0^{\pi/2} \sin\theta \, d\theta \int_0^{2\pi} d\phi = 2[-\cos\theta]_0^{\pi/2} \; 2\pi = 4\pi.$$

The probability of finding a velocity vector of length in the range $(V, V + dV)$ be $f(V)dV$. The probability that a particle chosen at random has a velocity vector in that range as regards magnitude and in the range $(\Omega, \Omega + d\Omega)$ as regards direction is

$$f(V)dV \; d\Omega/4\pi. \tag{26}$$

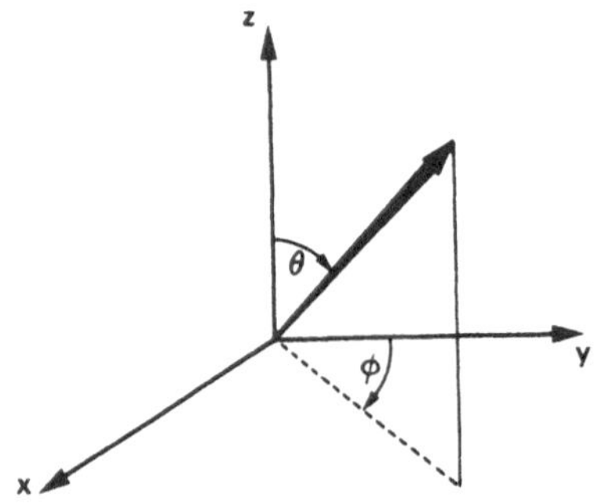

Fig. 1. Integration over a solid angle.

The number of particles dN_h hitting a surface area dA in time dt is the product of three factors: (a) the volume originally occupied by the particles, which is a cylinder of length Vdt and of cross section dA cosθ, i.e. the area projected normally to the beam (Fig. 2); (b) the equilibrium number of particles per unit volume N/v; and (c) the probability (26). This gives

$$dN_h = \int Vdt\ dA\ \cos\theta\ (N/v)\ f(V)\ dV\ d\Omega/4\pi. \qquad (27)$$

The integral is over V and over a hemisphere. This leads to an integral

$$\int_0^{\pi/2} \cos\theta\ \sin\theta\ d\theta \int_0^{2\pi} d\phi = [-1/2\ \cos^2\theta]_0^{\pi/2}\ 2\pi = \pi.$$

Hence the flux of particles is

$$F_N \equiv \frac{dN_h}{dAdt} = \frac{N}{4v}\int_0^\infty Vf(V)\ dV = \frac{\overline{V}}{4v}\ N \qquad (28)$$

This is (25) for Q = N.

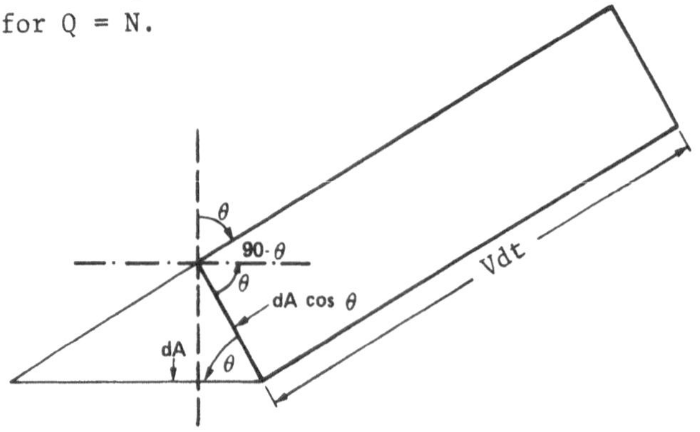

Fig. 2. Projection of area dA normal to beam.

Similarly, the energy received by dA in time dt, dU_h say, is given by (27) with N/v replaced by the equilibrium energy per unit volume U/v. Equation (28) is then replaced by

$$\Phi \equiv F_u \equiv \frac{dU_h}{dAdt} = \frac{\bar{V}}{4v} U$$

Similarly

$$\Psi \equiv F_S \equiv \frac{dS_h}{dAdt} = \frac{\bar{V}}{4v} S$$

On applying the result (25) to (4v/c) x (21) one finds

$$U = \frac{3}{4} TS = \frac{4v}{c} \sigma T^4 \quad (29)$$

which is a correct result for black-body radiation. It is derived independently in Problem 3.

II. DILUTED BLACK-BODY RADIATION

II.A. DBR: Definition and Properties

Consider now the effect of incident black-body radiation being scattered elastically either by atmospheric particles, as in a clear blue sky, or by watér droplets as in a cloud. These processes reduce the occupation number of a mode in the incident radiation. If the incident radiation was black-body, the resulting radiation has been called diluted black-body radiation (DBR) [8a]. Its production may be envisaged as due to the elastic scattering of black-body radiation at temperature T from a smaller solid angle ω_i into a larger solid angle ω_s. Since the scattering is elastic there is no redistribution of energy among the various spectral components of the radiation and so we shall consider for simplicity the effect of scattering on a typical narrow frequency width $\Delta\nu$ of radiation. The incident and scattered energy flux densities are given by (10).

$$\Phi_i = \frac{2h\nu^3}{c^2} \Delta\nu n_{\nu b} \int_{\omega_i} z \, d\Omega = \Phi_s = \frac{2h\nu^3}{c^2} \Delta\nu n_{\nu s} \int_{\omega_s} z \, d\Omega, \quad (30)$$

where $n_{\nu b}(=[e^{h\nu/kT} - 1]^{-1})$ and $n_{\nu s}$ are the mean occupation numbers of a mode in black-body radiation of temperature T and in the scattered radiation, respectively. The quantity z takes the value $\cos\theta$ or unity depending on whether the scatterer is Lambertian or whether it scatters energy evenly into each direction. On the right-hand side of (30) one integrates over a solid angle 2π if the scatterer is of finite size, while a solid angle of 4π can be envisaged for the rather academic case of a point scatterer. The four possible

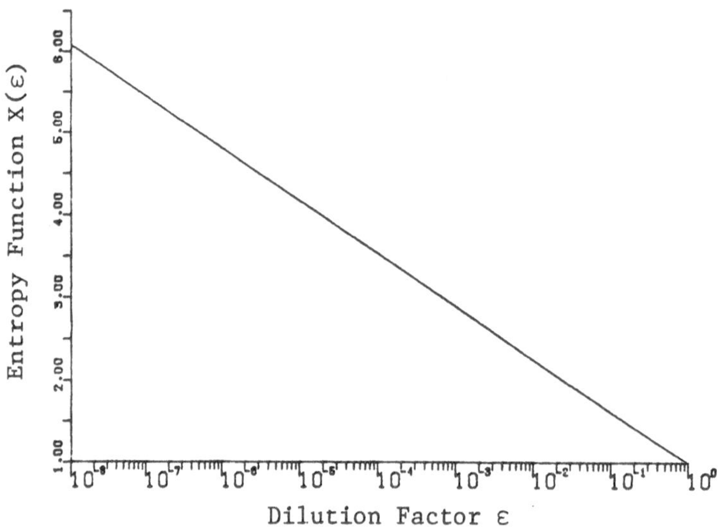

Fig. 3. Increasing dilution (smaller ε) leads to increase in X(ε) and hence fall in the effective temperature T*. The curve is based on (37).

"dilution factors" $\varepsilon = n_{\nu s}/n_{\nu b}$ obtainable by integrating (30) are recorded in Table 2, where δ is the half-angle of the cone of incoming radiation. In applications to solar energy in Section II.D we assume the sun to be directly over the absorber and to subtend a solid angle $\omega_i = \pi \sin^2\delta$ where δ is the half-angle of the cone with the sun as base and vertex at the earth's surface. Approximations for small δ (i.e., also small ω_i) are also given. In the Lambertian case ($z = \cos\theta$) the relation

$$\int_{4\pi} z \, d\Omega = 2\int_{2\pi} z \, d\Omega$$

is used since the total energy flux density leaving the scatter from either side of its surface is required. Diffuse solar radiation can be modeled as DBR by considering the scattering which takes place in the clouds to be nondirectional ($z = 1$) and into a solid angle 4π [8a].

The entries in the Table are obtained in Problem 5. For the

Table 2. The Dilution Factors ϵ of DBR Produced as a Result of Scattering

Type of Scatterer	Dilution Factor of Scattered Radiation Solid Angle = 2π	Solid Angle = 4π
Lambertian ($z = \cos\theta$)	$\sin^2\delta = \omega_i/\pi$	$(\sin^2\delta)/2 = \omega_i/2\pi$
Nondirectional ($z = 1$)	$\omega_i/2\pi$	$\omega_i/4\pi$

mean occupation number of DBR we write

$$n_\nu = \frac{\epsilon}{\exp(h\nu/kT_R) - 1} \tag{31}$$

where T_R is the absolute temperature of the undiluted radiation and ϵ is now a parameter which is characteristic of it. For solar radiation

$$\omega_i = 6.8 \times 10^{-5} \text{ steradians}, \quad \delta = \sqrt{\omega_i/\pi} = 4.65 \times 10^{-3} \tag{32}$$

II.B. Fluxes of DBR

One can now calculate the energy and entropy flux densities of DBR by using (31) and (12). In the case of the energy flux, (10) and (31) show that one must obtain a multiple of the equilibrium black-body result. In fact, since

$$\int_0^\infty \frac{y^3}{e^y - 1}\, dy = \Gamma(4)\zeta(4) = \frac{\pi^4}{15}, \quad y \equiv \frac{h\nu}{kT_R} \tag{33}$$

where ζ is the Riemann zeta function,

$$\Phi = B\frac{\ell h\epsilon}{c^2}\int_0^\infty \frac{\nu^3 d\nu}{e^{h\nu/kT_R} - 1} = \frac{B}{\pi}\frac{\ell}{2}\epsilon\sigma T_R^4 . \tag{34}$$

Here Stefan's constant

$$\sigma \equiv 2\pi^5 k^4/15h^3c^2 \tag{35}$$

has been used. A greater departure from the black body result is found in the case of the entropy flux density

$$\Psi = \frac{4}{3}\frac{\ell}{2}\frac{B}{\pi}\sigma\,\epsilon\, X(\epsilon) T_R^3 , \tag{36}$$

where (see Fig. 3) for n_ν given by (31),

$$X(\epsilon) \equiv \frac{45}{4\pi^4}\frac{1}{\epsilon}\int_0^\infty y^2[(1 + n_\nu)\ln(1 + n_\nu) - n_\nu \ln n_\nu]dy \qquad (37)$$

and n_ν is given by (31). It is immediately obvious that

$$X(1) = 1 \qquad (38)$$

since for $\epsilon = 1$ we must regain (20). One also regains (18) from (34) in that case. However the black-body result (21) $\Phi = (3/4)\Psi T$ is spoiled for DBR by the appearance of $X(\epsilon)$; it can, however, be recovered in some sense, as will now be shown.

The definition of an absolute thermodynamic temperature is

$$\frac{1}{T_A} = \left(\frac{\partial S}{\partial U}\right)_V \qquad (39)$$

In analogy an effective temperature T* for DBR may be defined by

$$\frac{1}{T_R^*} \equiv \frac{d\Psi}{d\Phi} = \frac{X(\epsilon)}{T_R} \qquad (40)$$

As DBR is non-equilibrium radiation, T_R^* is not an absolute thermodynamic temperature. We have from (34), (36) and (40)

$$\Phi = \frac{3}{4} T_R^* \Psi = \frac{B\ell}{2\pi}\sigma^*(T_R^*)^4 \ , \ \sigma^* \equiv \sigma\,\epsilon\,[X(\epsilon)]^4 \qquad (41)$$

This is a result which is analogous to (21).

We consider next the temperature T_{br} of undiluted black-body radiation such that it contains the same energy density as a given DBR with undiluted temperature T_R. From (34) we see that

$$T_{br} = \epsilon^{1/4} T_R.$$

If one regards the entropy flux density of DBR as parameterized by ϵ and T_R then, by (36)

$$\Psi(\epsilon,T_R)/\Psi(1,T_{br}) = \epsilon X(\epsilon) T_R^3/T_{br}^3 = \epsilon^{1/4} X(\epsilon). \qquad (42)$$

On the left we have the ratio of two entropy flux densities corresponding to the same energy flux density, one being for DBR the other being for black-body radiation. The latter must have the greater entropy [9]. Hence the inequality

$$0 \leq \epsilon^{1/4} X(\epsilon) \leq 1 \qquad (43)$$

must hold.

It should be possible to prove (43) directly from (37) as a purely mathematical exercise. This has not been done, and

(43) has the status of a thermodynamically derived inequality which has also been confirmed numerically. The inequality between the arithmetic and geometric mean, as well as related inequalities, can also be derived thermodynamically [10], but in this case purely mathematical arguments are of course available as well.

II.C. DBR as Non-Equilibrium Radiation

It is obvious from (37) that the non-equilibrium entropy expression (5) has been used. It must therefore be confirmed that a quantity q_j, needed to derive this result in section I.B., actually exists for DBR. This is done in this section.

For thermal equilibrium of bosons at temperature T_R the grand canonical ensemble leads to the exact result

$$(n_j)_{eq} = \frac{1}{\exp[(e_j - \mu)/kT_R] - 1}, \tag{44}$$

where e_j is the single-particle energy of state j, and μ is the chemical potential. Assume that there exists constants ϵ_j such that

$$n_j(\epsilon j) = \frac{\epsilon j}{\exp[(e_j - \mu)/kT_R - 1}. \tag{45}$$

There then exists a $q_j(\epsilon_j)$ satisfying (3), namely

$$q_j(\epsilon_j) = \frac{\epsilon_j}{\exp[(e_j - \mu)/kT_R] + \epsilon_j - 1}. \tag{46}$$

From (3) and (44) one finds

$$(q_j)_{eq} = \exp[(\mu - e_j)/kT_R] \tag{47}$$

and therefore we see that

$$n_j(1) = (n_j)_{eq},\ q_j(1) = (q_j)_{eq}. \tag{48}$$

The factors $\epsilon_j < 1$ are the "dilution factors" of the case $\epsilon_j = 1$, $\mu = 0$ which describes black-body radiation. If $\epsilon_j > 1$ they are "intensifying factors". We here consider ϵ independent of both frequency and direction with $\epsilon < 1$ and term the radiation "diluted black-body radiation". It represents a nonequilibrium situation, as it cannot be obtained from (5) by maximization. There will be of course many non-equilibrium situations when either or both of (1) and (2) fail. The temperature T_R which occurs in n_j for $\epsilon < 1$ refers to the <u>undiluted</u> black-body radiation; it is not be be considered as the absolute thermodynamic temperature of the <u>diluted</u> radiation. Thus using the <u>continuous</u> spectrum approximation the mean occupation number n_ν of a translational state of frequency ν is, for DBR, from (45)

$$n_\nu(\epsilon) = \frac{\epsilon}{\exp(h\nu/kT_R) - 1} \cdot \tag{49}$$

A factor ϵ ($\neq 1$) in (49) has two important consequences. While the photon number and the energy of the radiation are scaled down by a constant factor for all frequencies, this does not apply to all thermodynamic functions, as was explained by reference to the entropy in section II.B. Secondly the DBR is no longer equilibrium radiation. This can be seen by noting that the absolute temperature $T_A(\nu)$ of equivalent black-body radiation is different for each near-monochromatic pencil making up the DBR:

$$\frac{1}{\exp[h\nu/kT_A(\nu)] - 1} \equiv \frac{\epsilon}{\exp(h\nu/kT_R) - 1},$$

i.e.,

$$T_A(\nu) = (h\nu/k)(\{h\nu/kT_R + \ln[1 - (1 - \epsilon)\exp(-h\nu/kT_R)] - \ln\epsilon\}^{-1}). \tag{50}$$

Only for the special case $\epsilon = 1$ (undiluted black-body radiation) do we have a frequency-independent absolute temperature and hence equilibrium between radiation of different frequencies.

II.D. Application of DBR to Solar Energy Conversion

It is desirable to derive first an expression for the maximum efficiency of converting incident radiation to work. Imagine we have a converter which absorbs fluxes Φ_p, Ψ_p of energy and entropy from a pump and creates mechanical or chemical work at a rate $\dot{W}$ while delivering heat $\dot{Q}$ to a sink, which will often be the surroundings at temperature T_s. Here s stands for sink or surroundings, not for the sun, which might be the pump. Thus the fluxes to the sink are

$$\Phi_s = \dot{Q},\ \Psi_s = \dot{Q}_s/T_s \tag{51}$$

If $\dot{S}_g$ is the rate of entropy production in the converter, and Φ_c, Ψ_c the fluxes emitted by the converter, then the balance equations for a steady state are, $\dot{S}_g$ being an irreversibly produced rate,

$$\Phi_c = \Phi_p - \dot{W} - \dot{Q},\ \Psi_c = \Psi_p - \dot{Q}/T_s + \dot{S}_g \tag{52}$$

Eliminating $\dot{Q}$,

$$\dot{W} = \Phi_p - T_s\Psi_p - (\Phi_c - T_s\Psi_c) - T_s\dot{S}_g \tag{53}$$

The conversion efficiency is

$$\eta_W \equiv \frac{\dot{W}}{\Phi_p} = 1 - T_s\frac{\Psi_p}{\Phi_p} - \left(1 - T_s\frac{\Psi_c}{\Phi_c}\right)\frac{\Phi_c}{\Phi_p} - \frac{T_s\dot{S}_g}{\Phi_p} \tag{54}$$

Although this is a useful relation for later use (see section II.F), we shall derive an upper limit by neglecting the last two terms. Furthermore we shall assume two distinct pumps, in the form of DBR so that (41) can be used. Hence, writing T_1, T_2 for T_{R1}, T_{R2},

$$\eta_W \leq 1 - \frac{4}{3} T_s \frac{\beta_1 T_1^{*3} + \beta_2 T_2^{*3}}{\beta_1 T_1^{*4} + \beta_2 T_1^{*4}} \equiv \eta_W^{(max)} \tag{55}$$

where

$$\beta_i \equiv \left(\frac{B\ell}{2\pi} \sigma\epsilon [X(\epsilon)]^4 \right)_i \tag{56}$$

and i refers to pump 1 or pump 2. Note that the β_i are independent of temperature.

In applying the inequality (55) four types of dilution are considered [8]:

(i) Dilution due to the transmission of the radiation through the atmosphere. This can be measured by the atmospheric transmission coefficient

$$\epsilon \to t \equiv \frac{\text{Total photon density at absorber } (N_d)}{\text{Photon density of solar radiation just outside the atmosphere } (N_\odot)}$$

For a clear day t is typically 0.65 while t ~ 0.2 for a cloudy day.

(ii) Confinement of attention to direct radiation is equivalent to a dilution factor which can be defined by

$$\epsilon \to d \equiv \frac{\text{Photon density at absorber due to direct part of the solar radiation } (n_d)}{N_d}$$

(iii) Incomplete absorption by the converter is equivalent to a further dilution of the radiation in the converter. This can be measured by the absorption coefficient of the absorber:

$$\epsilon \to \alpha$$

(iv) Dilution due to elastic scattering from a solid angle ω_i to solid angle ω_s

$$\epsilon \to n_{\nu s}/n_{\nu b}$$

The direct part of the solar radiation is regarded as the first pump. It is subject to the dilutions (i) to (iii). For unpolarized radiation (ℓ = 2) and B given by (15) one finds

$$\beta_1 = (\sin^2 \delta)\sigma\alpha td[X(\alpha td)]^4 \quad (57)$$

The second pump is the diffuse part of the solar radiation with $\ell = 2$. The half-angle α subtended by the source and the receiver is in this case $\delta = \pi/2$ as radiation comes from all directions. Hence $B\ell/2\pi = 1$. All four dilutions are effective in this case and

$$\beta_2 = \sigma\alpha t(1 - d)(\omega_i/4\pi)[X(\alpha t(1 - d)(\omega_i/4\pi)]^4 \quad (58)$$

The factor $\omega_i/4\pi$ is the dilution due to scattering from a small solid angle ω_i to the full solid angle 4π during the conversion from direct to diffuse radiation. Both δ and ω_i were introduced formally in section II.A. The factor $1 - d$ is a measure of the proportion of the incident radiation which is not direct, and can be regarded as diffuse.

If one assumes the originating temperature in both cases to be the solar temperature $T_{\odot}$, then (40) tells us that

$$T_1^* = T_{\odot}/X(\alpha td) \quad (59)$$

$$T_2^* = T_{\odot}/X[\alpha t(1 - d)\ \omega_i/4\pi] \quad (60)$$

The lowering of the effective temperature by $\omega_i/4\pi$ is of course an important effect. For a grey converter ($\alpha = 0.9$) and little direct sun ($t = 0.11$) $T_2^* \sim 1200K$ while for a clear day ($t = 1$) and a black converter ($\alpha = 1$), $T_1^* \sim T_{\odot} = 5760K$. These two cases are represented by curves B and A respectively in Fig. 4.

II.E. Discussion

It should be pointed out that by adding the fluxes for the two pumps in (55) the absorption areas presented by the converter to the two pumps have been assumed to be the same. Furthermore, the efficiency is calculated not with respect to the incident flux, but with respect to the flux actually absorbed. Lastly, it has been assumed here that the dilution due to scattering is frequency independent. This is a good approximation for scattering in clouds but not for scattering producing sunlight, where there is a ν^4-dependence (Rayleigh scattering). This is not so important however since the proportion of energy in skylight is always relatively small. Even for sunlight, the DBR model still illustrates qualitatively the correct trends to be expected as the factors considered here are varied. The radiation from the ambient (a possible third pump) to the converter and the emitted radiation (which is thermal radiation from the surface of the conversion device) are here neglected or, alternatively, are assumed to be equal and to cancel each other out.

Typical values for T^* in (55) range from $T_{R1}^* = 5760K$ (the black-body temperature of the solar surface) for an ideal clear day

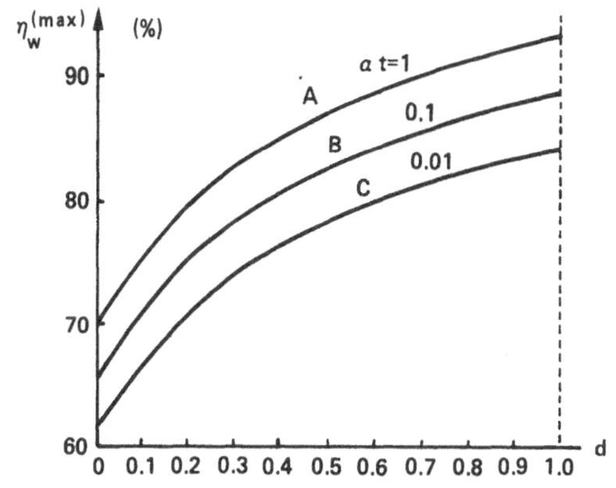

Fig. 4. The maximum conversion efficiency (55) from solar radiation into work is plotted as a function of the proportion d of the radiation which is direct. A. $\alpha t = 1$; B. $\alpha t = 0.1$; C. $\alpha t = 0.01$ ($T_R = 5760K$, $T_s = 300K$, $\delta = 4.65 \times 10^{-3}$).

(t = 1) and a black converter ($\alpha = 1$) to $T^*_{R2} = 1200K$ for a typical cloudy day and a grey converter with absorptivity $\alpha = 0.9$. Figure 4 shows the upper limit of η_W from (55) as a function of the energy proportion d of the solar radiation which is direct (the rest being diffuse). The three curves are for different dilutions due to transmission through the atmosphere and absorption by the converter. Curve A ($\alpha t = 1$) corresponds to the ideal case of a perfectly black converter ($\alpha = 1$) receiving solar radiation which is undiluted by absorption in the atmosphere (t = 1). For a nonblack absorber (e.g. $\alpha = 0.9$) and an illumination intensity corresponding to a typical day with little direct sun (e.g. t = 0.11), curve B represents a fair approximation. We see that in this case the thermodynamic limiting efficiency $\eta_W^{(max)}$ ranges from $\alpha \times 65.5\%$ for completely diffuse sunlight to $\alpha \times 88.8\%$ for completely direct sunlight. Note that the lower effective temperature of the diffuse radiation compared with direct radiation results in lower conversion efficiencies, so that the curves have a positive slope.

A further application of the results is to the conversion of radiation into plant free energy in photosynthesis, which is an equivalent form of energy to that described here as 'work'. One matter, made quantitive by Fig. 4, is that decreasing light intensity (corresponding to decreasing t) reduces the free-energy conversion efficiency, a point made by Duysens [11] and confirmed by Spanner [12].

II.F. A More Rigorous Version of Section II.D

The theory given in section II.D was somewhat simplified in that the last two terms in (54) were neglected. We now include them and suppose that there are n pumps for each of which

$$\Phi_j = \beta_j [T_j^*]^4, \ \Psi_j = \tfrac{4}{3}\beta_j [T_j^*]^3 \tag{61}$$

By virtue of (41) this may be interpreted to mean that each pump is DBR, but it is in fact more general, in that no assumption will be made about the β_j's except that they are temperature-independent constants. Indeed, for n = 3 pumps, the third pump will be interpreted as the energy incident on the converter from the surroundings in so far as it is not solar direct and diffuse radiation. Thus the third pump can cover thermal radiation of various terrestial objects and it may not be possible to model this effect by DBR. In any case the main equation (53) now reads

$$\dot{W} = \sum_{i=1}^{n} \beta_i T_i^{*4} [1 - 4T_s/3T_i^*] - \beta_c T_c^{*4} [1 - 4T_s/3T_c^*] - T_s \dot{S}_g \tag{62}$$

Although non-black-body radiations cannot be in equilibrium one can get close to this situation in effective equilibrium defined by

$$T_1^* = T_2^* = \ldots = T_n^* = T_c^* , \ \dot{Q} = \dot{S}_g = 0. \tag{63}$$

This is a kind of equilibrium in the sense that no work can be extracted from the system. To prove that, note that if energy and momentum flux balance, equations (52) are

$$\left(\sum_{i=1}^{n} \beta_i \right) T_c^{*4} = \beta_c T_c^{*4} + \dot{W} \tag{64}$$

$$\frac{4}{3}\left(\sum_{i=1}^{n} \beta_i \right) T_c^{*3} = \frac{4}{3} \beta_c T_c^{*4} \tag{65}$$

Hence from (65) and (64), used in that order,

$$\sum_{i=1}^{n} \beta_i = \beta_c, \ \dot{W} = 0, \tag{66),(67}$$

as was to be proved. Note that for effective equilibrium to be possible fluxes of greater dilution must correspond to higher absolute temperatures.

Equation (66) is independent of temperature and is therefore valid also away from effective equilibrium. One can use (66) to eliminate β_n from (62) to find for n = 3

$$\dot{W} = \sum_{i=1}^{2} \beta_i \left[T_i^{*4} \left(1 - \frac{4}{3}\frac{T_s}{T_i^*}\right) - T_3^{*4}\left(1 - \frac{4}{3}\frac{T_s}{T_3^*}\right)\right]$$

$$- \beta_c T_c^{*4}\left[\left(1 - \frac{4}{3}\frac{T_s}{T_c^*}\right) - T_3^{*4}\left(1 - \frac{4}{3}\frac{T_s}{T_3^*}\right)\right] - T_s\dot{S}_g \qquad (68)$$

To simplify this result it will be assumed that there is effective equilibrium between the ambient and the converter

$$T_3^* = T_c^* \qquad (69)$$

Hence

$$\dot{W} = \sum_{i=1}^{2} \beta_i T_i^{*4}(1 - 4T_s/3T_i^*) - \left(\sum_{i=1}^{2} \beta_i\right) T_c^{*4}(1 - 4T_s/3T_c^*).$$

The efficiency is $\dot{W}$ divided by $(\Phi_{p1} + \Phi_{p2})$. Hence, omitting $\dot{S}_g$,

$$\eta_W \leq 1 - \frac{4}{3}T_s \frac{\sum_{i=1}^{2} \beta_i T_i^{*3}}{\sum_{i=1}^{2} \beta_i T_i^{*4}} - \frac{\left(\sum_{i=1}^{2} \beta_i\right) T_c^{*4}}{\sum_{i=1}^{2} \beta_i T_i^{*4}}\left(1 - \frac{4T_s}{3T_c^*}\right) \qquad (70)$$

This gives the correction term (to the earlier formula (55), but the correction is less than 1/4% in the case of solar radiation [8b].

The result (70) is symmetrical in the two pumps. If only one pump is present it simplifies to

$$\eta_{Wi} \leq \lambda\left(\frac{T_s}{T_i^*}\right) - \left(\frac{T_c^*}{T_i^*}\right)^4 \lambda\left(\frac{T_s}{T_c^*}\right) \equiv \eta_{Wi}^{(max)} \qquad (i = 1,2) \qquad (71)$$

where the so-called Landsberg efficiency is

$$\lambda(x) = 1 - \frac{4}{3}x + \frac{1}{3}x^4 \qquad (72)$$

As $T_s \sim T_c^*$ and $\lambda(1) = 0$, the approximation

$$\eta_{Wi}^{(max)} \sim \lambda(T_s/T_i^*) \qquad (i = 1,2)$$

is adequate where T_1^* is given by (59) with d = 1 and T_2^* by (60) with d = 0.

The λ-function (see Fig. 7) lies below the corresponding Carnot efficiency $\eta_C(x) = 1 - x$, except for the common points at x = 0 and x = 1:

$$\eta_c(x) - \lambda(x) = (x/3)(1 - x^3)$$

It was first emphasized in [13], [14], [15], [9]. A fuller account of earlier work is given in [18].

II.G. An Argument from Availability

The λ-function involved in the efficiency as given in (72) can also be obtained by considering the availability (or 'exergy')

$$A = U - T_s S - p_s v$$

of a system with energy U, temperature T and volume v which is in surroundings at temperature T_s and pressure p_s. Then the reduction of the initial system availability A_i to that of the surroundings, A_s, is the work which can maximally be done.

Consider a system with $U = avT^n$ where n and a are constants. Then, as shown in Problem 6,

$$U = (n - 1)pv = \frac{n - 1}{n}TS = avT^n.$$

It follows that if the volume is not changed

$$A_i - A_s = U_i - U_s - T_s S_s + T_s S_i$$

$$= av_s(T_i^n - T_i^s) - \frac{n}{n - 1} av_s(T_i^{n-1}T_s - T_s^n)$$

$$= av_s T_i^n \left[1 - \frac{n}{n - 1}\frac{T_s}{T_i} + \frac{1}{n - 1}\left(\frac{T_s}{T_i}\right)^n\right]$$

If the initial energy U_i is taken as the denominator of the efficiency one does indeed find

$$\frac{A_i - A_s}{U_i} = 1 - \frac{n}{n - 1}\frac{T_s}{T_i} + \frac{1}{n - 1}\left(\frac{T_s}{T_i}\right)^n \xrightarrow{n = 4} \lambda\left(\frac{T_s}{T_i}\right)$$

This derivation has been noted before [8a], but for an efficiency calculation it must be understood that other denominators may be more appropriate, for example the initial energy excess of the system over the surroundings at the same volume:

$$U_i - U_s = av_s(T_i^n - T_s^n).$$

In that case the λ-function would not be obtained, and some reservations must therefore be retained regarding the above argument.

III. STATISTICAL THERMODYNAMICS OF CASCADE CONVERTERS

III.A. Some Thermodynamic Results

In section II.D. the thermodynamic efficiency of a converter was obtained in terms of such parameters as energy and entropy fluxes emitted by both a pump and a converter, and also in terms of the temperature T_s of the surroundings. The last two terms of equation (54) were discarded and the results were discussed in sections II.D and II.E. In the fuller theory of section II.F only the converter entropy generation rate $\dot{S}_g$, i.e. the last term of equation (54) was neglected and an improved theory led to the more accurate efficiency formula (70) and it also led to the occurrence of the λ-function (72) in the upper limit to the efficiency. A further advance in accuracy would be to identify $\dot{S}_g$ for a particular model, and this will be done in the present section III.

A schematic representation of the general model underlying (54) is given in Fig. 5 and one has from (54)

$$\eta_W = 1 - \frac{T_s}{T_{Fp}} - \left(1 - \frac{T_s}{T_{Fc}}\right)\frac{\Phi_c}{\Phi_p} - \frac{T_s\dot{S}_g}{\Phi_p} \tag{73}$$

where the flux temperature is by (41)

$$T_{Fi} = \frac{\Phi_i}{\Psi_i} \left[= \frac{3T_i}{4X(\epsilon)} \right] \qquad (i = p,c) \tag{74}$$

where the identification in square brackets refers to DBR, based on black-body radiation at temperature T_i. However, attention will be confined to ordinary black-body radiation ($\epsilon = 1$, $X(\epsilon) = 1$) when

$$\eta_W = 1 - \frac{4}{3}a - (1 - \frac{4}{3}\frac{a}{b})b^4 - \frac{T_s\dot{S}_g}{\Phi_p} \tag{75}$$

where

$$a \equiv T_s/T_p \;, \; b \equiv T_c/T_p \tag{76}$$

The pump temperature which occurs here is conceptually clear. The first appearance of T_c, however, calls for a remark concerning its relation to the temperature T_s of the surroundings. It must be possible in the case of a solid state device for the lattice to be in approximate equilibrium with the surroundings even if the current carriers which are excited by the pump are not. The interpretation of T_c must then be based on its defining expression Φ_c/Ψ_c. It is then seen that the T_c governs the emission of black-body radiation by the converter. The model used thus assigns two temperatures to the converter: T_c for emission and T_s for its lattice.

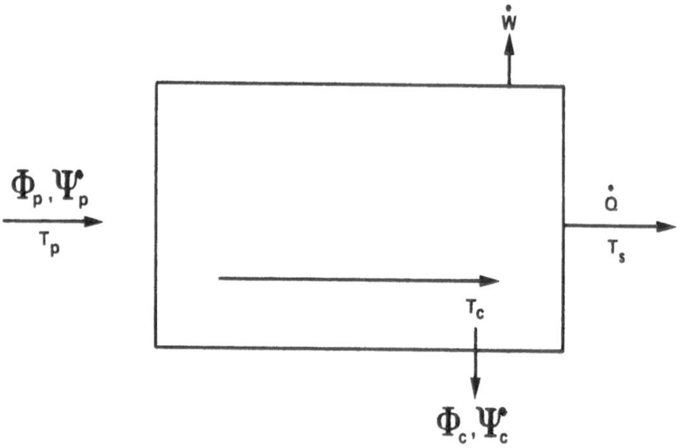

Fig. 5. A general model for energy conversion. It is treated by the use of balance equations and leads to the efficiency η_W of (75) which is subject to inequalities (79).

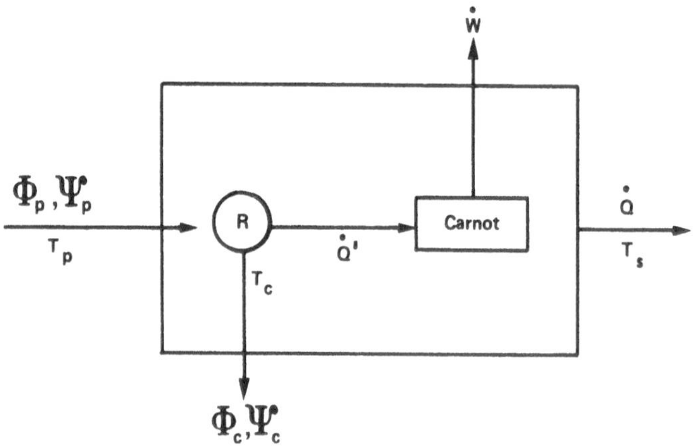

Fig. 6. A more specific interpretation of the converter of Fig. 5. It leads to the efficiency η_2 of (80).

If one discards the $\dot{S}_g$ term then, since the entropy generation rate cannot be negative, one obtains an upper limit of η_W:

$$\eta(a,b) \equiv 1 - b^4 - \frac{4}{3}a(1 - b^3) \qquad (0 \leq a \leq b \leq 1) \tag{77}$$

The inequality $a \leq b$ is needed since the surroundings have a temperature which cannot exceed that of the converter emission temperature T_c. If one puts $a = b$ in (77), one finds again the λ-function (72):

$$\eta(a,a) \equiv \lambda(a) = 1 - \frac{4}{3}a + \frac{1}{3}a^4 \qquad (78)$$

In fact

$$\lambda(a) - \eta(a,b) = b^4 + \frac{1}{3}a^4 - \frac{4}{3}ab^3 = (b-a)^2(b^2 + \frac{2}{3}ab + \frac{1}{3}a^2)$$

The last form of writing shows that $\lambda(a) \geq \eta(a,b)$, so that

$$\eta_W \leq \eta(a,b) \leq \lambda(a). \qquad (79)$$

Leaving now the general "black box" kind of results appropriate to Fig. 5, let us give a special interpretation of the model based on Fig. 6. The entropy generation occurs now in a radiator R which absorbs the incident black-body radiation at absolute temperature T_p and re-radiates it at absolute temperature T_c. Heat flux $\dot{Q}'$ at the converter emission temperature T_c falls on the Carnot engine which works at the rate $\dot{W}$ and rejects a heat flux $\dot{Q}$ at T_s. The externally visible quantities Φ_p, Ψ_p, T_p, $\dot{W}$, $\dot{Q}$, T_p are exactly the same her (fig. 6) as they were for the "black box" of Fig. 5. The new efficiency can be obtained exactly on the basis of this model as

$$\eta_2 = \frac{\dot{W}}{\Phi_p} = \frac{\dot{Q}'}{\Phi_p}\frac{\dot{W}}{\dot{Q}'} = \frac{\Phi_p - \Phi_c}{\Phi_p}\left(1 - \frac{T_s}{T_c}\right),$$

so that

$$\eta_2 = (1 - b^4)(1 - a/b) \qquad (80)$$

We have here used the fact that Φ_p and Φ_c are both of the black-body type and that the Carnot efficiency can be used. In the special case of Fig. 6 the model of Fig. 5 must still be valid so that (74) for Fig. 5 can be equated to (80) for Fig. 6. This yields [16]

$$\frac{a\dot{S}_g}{\sigma T_p^3} = 1 - b^4 - \frac{4}{3}a(1 - b^3) - [1 - \frac{a}{b} - b^4 + ab^3] = [\frac{1}{3}b^3 + \frac{1}{b} - \frac{4}{3}]a$$

This is an essentially positive quantity, as may be seen by writing it as

$$\frac{3b\dot{S}_g}{\sigma T_p^3} = (1 - b)^2(b^2 + 2b + 3) \qquad (81)$$

If converter and pump emission balance, one has b = 1 and both η_2 and $\dot{S}_g$ vanish: neither work nor entropy is generated in this case. These thermodynamic results will be discussed in the next section.

The flux temperature $T_F = \Phi/\Psi$ of (74) is convenient for balance equations involving fluxes, and in the case of quasistatic conduction it can coincide with the absolute temperature (see section III.B). The effective temperature (40), on the other hand, is

$$T^* \equiv d\Phi/d\Psi [= T/X(\epsilon)],$$

where the term in square brackets gives the evaluation for DBR based on black-body radiation at temperature T. The effective temperature is somewhat more fundamental as it lends itself to a description of effective equilibrium (63). Furthermore in the case of black-body radiation of temperature T, one has $T^* = T$, but $T_F = (3/4)T$.

III.B. Discussion

One obtains the Carnot efficiency from (73) if (i) the processes involved are reversible ($\dot{S}_g = 0$) and (ii) if no fluxes are emitted by the converter ($\Phi_c = \Psi_c = 0$); this means that the bracketed term disappears in this equation. In the normal deduction of Carnot's efficiency the working fluid is certainly not in a steady state; instead it is acting periodically and produces the same work in each cycle. This can be covered by the approach via fluxes, if one agrees to look at the working fluid only periodically, namely when it is some standard state. This "stroboscopic" method leads one to $\Phi_c = \Psi_c = 0$, as required. The hot reservoir has temperature T_{Fp} which is its actual thermodynamic temperature T_p, since for reversible or quasistatic, i.e. very slow, withdrawals of heat Q, the pump entropy changes in time t by

$$A\Psi_p t = Q/T_p = A\Phi_p t/T_p = AtT_{Fp}\Psi_p/T_p.$$

Hence $T_{Fp} = T_p$. Here A is the area over which the flux is measured. Fig. 7 shows η_{Carnot} as a function of $a \equiv T_s/T_p$.

The efficiencies $\eta(a,b)$ and η_2 are also shown in Fig. 7. For given $b = T_c/T_p$ both are falling straight lines which start to fall from $\eta = 1 - b^4$ at a = 0. The downward slopes are

$$-\frac{1 - b^4}{b} \text{ for } \eta_2 \qquad -\frac{4}{3}(1 - b^3) \text{ for } \eta(a,b)$$

so that the slopes are steeper for η_2, provided $0 \leq b < 1$. Thus η_2 gives a more severe constraint on permitted efficiencies provided the model of Fig. 6 is applicable.

The $\eta(a,b)$ curves are valid for $a \leq b$ and they therefore terminate at points a = b. These points generate the curve $\lambda(a)$ which

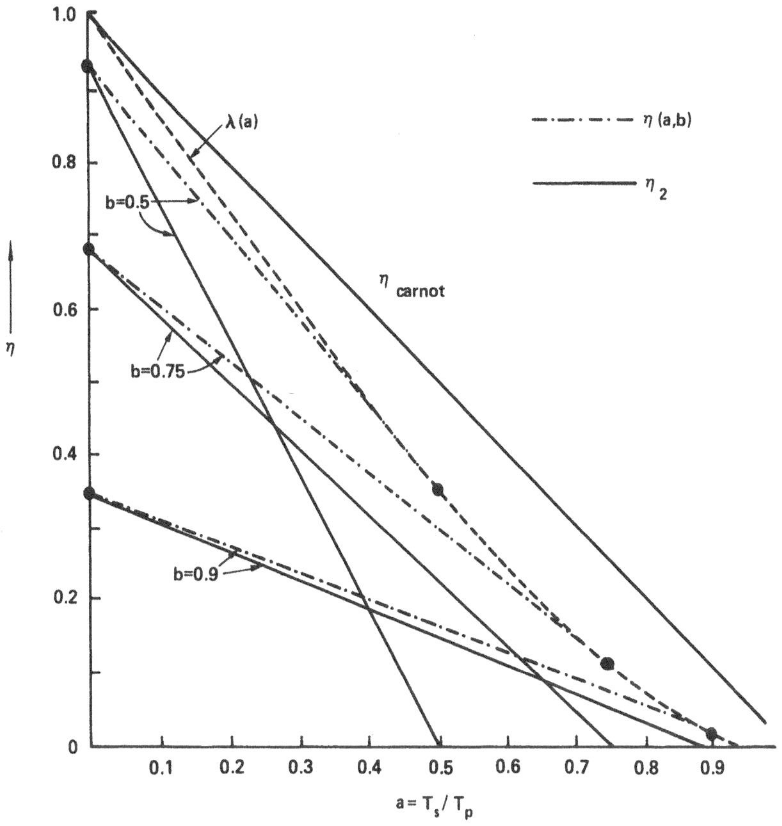

Fig. 6. The six lines emerging in pairs from the points on the $a = 0$ axis represent $\eta(a,b)$ and η_2 in each case for different values of $b = T_c/T_p$. The $\eta(a,b)$-lines end at the marked points when $a = b$. The curve $\lambda(a)$ is the locus of all such points.

lies above the other efficiency curves. Thus we have

$$\eta(a,b) \leq \lambda(a) \leq \eta_{Carnot}$$

The efficiency $\lambda(a)$ is pulled down below the Carnot efficiency because of the black-body emission from the converter which does not contribute to the useful work output. Of all the efficiencies which take this effect into account $\lambda(a)$ is the highest. The reason is that it assumes converter emission at the ambient temperature T_s. As T_c floats above T_s the emission loss mechanism increases and the efficiency is pulled down.

III.C. The Absorption Coefficient; the Photon Chemical Potential

A quantum statistical interpretation of the above thermodynamic scheme is provided by solar cells which convert radiation into electricity. To study it one needs the statistics of radiation processes. These consist of spontaneous emission, stimulated emission and absorption. In the first two atoms or electrons or other systems lose energy by the emission of radiation. The spontaneous emission rate u^{sp} does not involve the photon occupation number n_ν, whereas the stimulated emission rate u^{st} is proportional to it if the transition is between some quantum states I and J whose energies satisfy $E_I - E_J = h\nu$. In absorption the system is excited from a state J of lower energy to one of higher energy, the energy being supplied by a photon. In addition the expressions for the three rates involve an appropriate transition probability per unit time per unit volume whose form is given here though it is not needed later. If δ denotes an energy-conserving delta function for the emitted photon, it is

$$B_{IJ} = \frac{2\pi e^2 h}{V^2 m^2 \kappa^2 E}\,\delta(E_I^o - E_J^o \pm \hbar\omega)\,\left|M_{IJ}\right|^2_{av}$$

The last term is an average of a square of a momentum matrix element and κ is the refractive index of the medium.

Let the states I and J belong respectively to groups i and j of quantum states, such as conduction and valence bands, which have quasi-Fermi levels $kT_s\gamma_i$ and $kT_s\gamma_j$, where T_s is the lattice temperature of the converter. Further let p_I be the probability that state I is in a condition in which it can give up an electron and let q_I be the probability that the system is in a specific quantum state which can be converted into state I by the addition of an electron. These probabilities are easy to formulate for states in bands, but somewhat more complicated for trapping states [17]. The recombination rates per unit volume are

$$u_{IJ}^{sp} = B_{IJ} p_I q_J \tag{82}$$

$$u_{IJ}'^{abs} = u_{IJ}^{abs} - u_{IJ}^{st} = B_{IJ}(p_J q_I - p_I q_J) n_\nu \tag{83}$$

where n_ν is the mean occupation number of a photon mode of given polarization and of frequency ν, as in (9). The absorption coefficient is the net number rate of photon absorption divided by the photon flux. This is

$$\alpha_{IJ} \equiv \frac{u_{IJ}'^{abs}}{(c/\kappa v) n_\nu} = \frac{\kappa v}{c}\left(\frac{p_J q_I}{p_I q_J} - 1\right) u_{IJ}^{sp}$$

$$= \frac{\kappa v}{c}\left(\exp\frac{h\nu - qV}{kT_s} - 1\right) u_{IJ}^{sp} \tag{84}$$

Here κ is the refractive index of the material. See Problem for the photon flux expression. The last step depends in the simplified case of non-localized electronic states on

$$\frac{q_I}{p_I} = \frac{\exp(\eta_I - \gamma_i)/[\exp(\eta_I - \gamma_i) + 1]}{1/[\exp(\eta_I - \gamma_i) + 1]} = \exp(\eta_I - \gamma_i)$$

where

$$\eta_I \equiv (1/kT_s) \times \text{energy of state I}$$

Hence

$$X \equiv \frac{q_I}{p_I}\frac{p_J}{q_J} = \exp\ (\eta_I - \gamma_i - \eta_J + \gamma_j) \qquad (85)$$

This gives the required result if

$$\eta_I - \eta_J \equiv h\nu/kT_s,\ \gamma_i - \gamma_j \equiv qV/kT_s \qquad (86)$$

T_s is the temperature of the surroundings. The Fermi levels will be assumed to be flat, so that the difference qV is independent of position. Thus V is a voltage which represents the influence which causes the separation of the quasi-Fermi levels. This may be an applied voltage, incident radiation, injection of current carriers or other influences which bring about a departure from equilibrium.

Write (84) as

$$\alpha_{IJ}(x) = \alpha_0 g(x),\ (g \equiv \frac{\kappa\nu}{c}\ u_{IJ}^{sp}\ ,\ \alpha_0 \equiv \exp\frac{h\nu - qV}{kT_s} - 1) \qquad (87)$$

One sees that negative α means that stimulated emission dominates absorption and one has a laser amplifer. One finds as an incidental result the threshold condition for a laser amplifier

$$qV \geq h\nu \qquad (88)$$

This is also the condition for the more energetic state to have the higher occupation probability:

$$\frac{1}{\exp(\eta_I - \gamma_i) + 1} > \frac{1}{\exp(\eta_J - \gamma_j) + 1}$$

implies (88). This is the so-called condition for population inversion.

If one has two sets of states i, j, each at a given energy, with transition between them and the occupation of each set specified by its quasi-Fermi level, then the photons have a simple distribution in a steady state. For the upward and downward transitions must then occur at an equal rate:

$$u_{IJ}^{sp} + u_{IJ}^{st} = u_{IJ}^{abs}. \tag{89}$$

Eq. (89) applied to (84) yields

$$n_\nu = \left(\exp \frac{h\nu - qV}{kT_s} - 1\right)^{-1} \qquad (h\nu > qV) \tag{90}$$

In true thermal equilibrium V = 0 by (86) and the Planck distribution results.

The thermodynamic relation

$$dU = TdS - pdv + \sum_i \mu_i dN_i$$

shows that at constant entropy and volume a minimum energy (and hence chemical equilibrium) holds if

$$\sum \mu_i dN_i = 0 \tag{91}$$

Suppose electron-hole pairs are generated by photons and in turn generate photons by recombination. We treat the system as an intrinsic semiconductor with three variable concentrations: electrons in the conduction band (chemical potential μ_e), holes in the valence band (chemical potential μ_h. We shall use μ_v as the chemical potential of electrons in the valence band) and photons (chemical potential μ_r). The conservation condition is that an electron and a hole are created when a photon disappears. Hence we have

$$\mu_e dN_e + \mu_h dN_h + \mu_r dN_r = 0$$

$$dN_e = -dN_r,\ dN_h = -dN_r$$

Since for two groups of levels in the electron description $\mu_e - \mu_v = qV$ it follows that

$$\mu_r = \mu_e + \mu_h = \mu_e - \mu_v = qV \quad (\text{i.e. } \mu_h = -\mu_v) \tag{92}$$

thus furnishing an interpretation of qV in (90) as a chemical potential of radiation. This holds for thermal equilibrium. If, in addition, chemical equilibrium holds with a system for which $\mu = 0$ such as black-body radiation or phonons, then $\mu_e = \mu_u$ and $\mu_r = 0$.

III.D. Solar Cell Equation in Terms of Photon (Number) Fluxes

The dependence of the photon mean occupation number $n_\nu(x)$ on position will be determined next. In an increment of length dx (the photon generation rate) x (time needed to cross dx) is $dn_\nu(x)$.

$$\text{i.e. } dn_\nu = v(u^{sp} + u^{st} - u^{abs}) \times (\kappa/c)\, dx \tag{93}$$

$$= (1 - u'^{abs}/u^{sp})u^{sp}(\kappa v/c)dx$$

$$= (1 - \alpha_0 n_\nu) g(x)dx \tag{94}$$

The solution is

$$n_\nu(x) = \alpha_0^{-1} - [\alpha_0^{-1} - n_\nu(0)] \exp[-\alpha_0 G(x)] \tag{95}$$

where

$$G(x) \equiv \int_0^x g(x')dx'$$

It will be assumed that at the right-hand side of the converter (a semiconductor solar cell typically) the following approximation is adequate

$$n_\nu(d) = \alpha_0^{-1} - [\alpha_0^{-1} - n_\nu(0)] \exp[-\alpha_0 G(d)] \sim \alpha_0^{-1}$$

It follows that we must again use formula (90), but this time as an approximation:

$$n_\nu(d) = [\exp \frac{h\nu - qV}{kT_s} - 1]^{-1} \tag{96}$$

It is the x-dependence of the quantities involved and the finite thickness of the cell which forces one to make the approximation. The photon flux emitted by the cell is, using Problem 4,

$$\Lambda_e(V) \equiv \int_{\nu_G}^{\infty} \Lambda_e(V,\nu)\, d\nu \tag{97}$$

where

$$\Lambda_e(V,\nu) = \frac{2\pi\nu^2}{c^2} n_\nu(d) = \frac{2\pi\nu^2/c^2}{\exp[(h\nu - qV)/kT_s] - 1}$$

For a two-sided cell the emitted flux per unit area is $2\Lambda_e(V)$. If the incident photon flux is Λ_i, and all photons are converted to carriers or even carrier pairs, the current density voltage characteristic of the cell is [20]

$$I(V) = -q[\Lambda_i - 2\Lambda_e(V)], \tag{98}$$

where q is the charge on the current carriers, and may be positive or negative. This assumes that all the incident photons are absorbed for frequencies $\nu \geq \nu_G$ and that non-radiative recombination can be neglected. For incident black-body radiation at absolute temperature T_p (the pump temperature)

$$\Lambda_i = \int_{\nu_g}^{\infty} \frac{2\pi\nu^2}{c^2} \frac{d\nu}{\exp(h\nu/kT_p) - 1} \qquad \text{(BBR)} \tag{99}$$

Note for later use that any increment of current density due to an increment of frequency is $-qd(\Lambda_i - \Lambda_e)$ for black-body radiation incident on one side of the cell:

$$dI_\nu(V) = \frac{2\pi\nu^2 q}{c^2} \left[\frac{1}{\exp[(h\nu-qV)/kT_s] - 1} - \frac{1}{\exp(h\nu/kT_p) - 1}\right] d\nu \tag{100}$$

where emission from only one side of the cell has been assumed. It is written in an alternative notation below [see (108) and (114)].

Treating (98) as a solar cell equation, let the open-circuit voltage be denoted by V_{oc}. Then

$$I(V_{oc}) = 0 \text{ and } \Lambda_i = 2\Lambda_e(V_{oc}). \tag{101}$$

The short-circuit current density is $I_{sc} = I(0)$. Then

$$I_{sc} = -q\Lambda_i + 2q\Lambda_e(0) = 2q[\Lambda_e(0) - \Lambda_e(V_{oc})] \tag{102}$$

Dividing (98) by (102)

$$\frac{I(V)}{I_{sc}} = \frac{\Lambda_e(V_{oc})/\Lambda_e(0) - \Lambda_e(V)/\Lambda_e(0)}{\Lambda_e(V_{oc})/\Lambda_e(0) - 1} \tag{103}$$

This shows clearly that $I(0) = I_{sc}$, and that $I = 0$ implies $\Lambda_e(V) = \Lambda_e(V_{oc})$. We shall return to this result in section III.G.

III.E. The Maximum Efficiency of an Infinite Stack of Solar Cells

Consider now a stack of solar cells of the simple type discussed in section III.D, whose energy gaps vary from zero at the back side of the stack to infinity at the illuminated side. This is the limiting case of an infinity of cells. Under these conditions the incident radiation loses progressively the more energetic photons, photons of each frequency range $(h\nu, h\nu + d(h\nu))$ being absorbed by that component solar cell whose energy gap is correctly adapted to it: $E_G = h\nu$. Thus E_G is now to be considered as variable and integrations over ν are now integrations over contributions from all the solar cells in the stack. Such converters are said to be arranged in tandem. They are also said to form a cascade, the word tandem being restricted to two cells.

The relation between T_c and T_s may be understood from the following consideration. The absorption of a photon by the lattice of the device, i.e. by the working medium, increases its entropy by $(h\nu - qV)/T_s$. The promotion of an electron to a higher level as part of the absorption process represents stored and recoverable energy, which yields work. Only the remaining energy is to be regarded as heat and so contributes to entropy. On the other hand that part of the cell which represents the emission of radiation at temperature T_c loses entropy $h\nu/T_c$ upon emission of a photon. In a steady state the entropy of the whole system is constant and so is not changed by these processes. Hence V adjusts itself for a stack so that T_c is independent of ν and V. Thus V is proportional to ν as one passes from cell to cell:

$$\frac{h\nu - qV}{T_s} = \frac{h\nu}{T_c} \quad \text{or} \quad \frac{qV}{h\nu} = 1 - \frac{T_s}{T_c} \qquad (104),(105)$$

The efficiency of the stack of solar cells will now be estimated. The denominator of such an expression is the total incoming photon energy flux. This is by (13)

$$\Phi = B\ell c^{-2}\int_0^\infty h\nu^3 n_\nu \, d\nu \longrightarrow 2\pi c^{-2}\int_0^\infty h\nu^3 n_\nu d\nu \longrightarrow$$

$$2\pi c^{-2}\int_0^\infty h\nu^3(\exp h\nu/kT_p - 1)^{-1} \, d\nu \qquad (106)$$

The first arrow means that the usual assumption of isotropic $(B = \pi)$ and nonpolarized $(\ell = 2)$ radiation is made. The second arrow means that the assumption of incident black-body radiation at absolute temperature T_p is added, so that one has a definite expression for n_ν. This yields as in (19)

$$\Phi_i = \sigma T_p^4 \ . \qquad (107)$$

If the stack is fully surrounded by the pumping radiation a factor 2 must be added since flux would be incident on front and back faces. We suppose here that the photosensitive surface lies in the bounding disc of a hemisphere at temperature T_p.

The numerator of the efficiency gives the useful power per unit area which is in this case

$$-\int_{\nu=0}^{\nu=\infty} V dI_\nu(V) = \int_0^\infty (1 - \frac{T_s}{T_c})\frac{h\nu}{q}\,\frac{2\pi\nu^2 q}{c^2}\,[n_\nu(T_p) - n_\nu(T_c)]d\nu \qquad (108)$$

The integral is over all the characteristic frequencies ν_g of the individual cells, and (104) and (100) have been used, assuming again black-body radiation. The integration yields

$$(1 - T_s/T_c)\sigma(T_p^4 - T_c^4) \qquad (109)$$

On dividing (109) by (107) one recovers precisely the result (80) already found on thermodynamic grounds (see also [23]):

$$\eta_2 = (1 - T_s/T_c)(1 - T_c^4/T_p^4). \tag{110}$$

The stack of solar cells thus furnishes a quantum statistical realization of the thermodynamic model. The entropy generation rate (81) is therefore applicable also to the present problem.

Using (108) the contribution of a typical cell of the stack to the total power can be maximized by evaluating

$$\frac{d}{dV}[VdI_\nu(V)] = 0. \tag{111}$$

Thus

$$V\frac{dI_\nu}{dV} + dI_\nu = 0$$

i.e.

$$\frac{qV}{kT_s}\exp\left(\frac{h\nu - qV}{kT_s}\right)\left(\exp\frac{h\nu - qV}{kT_s} - 1\right)^{-2} + \left(\exp\frac{h\nu - qV}{kT_s} - 1\right)^{-1} - \left(\exp\frac{h\nu}{kT_p} - 1\right)^{-1} = 0$$

This yields a transcendental equation for V in terms of ν, T_p and T_s [24]:

$$\left(1 + \frac{qV}{kT_s}\right)\exp\frac{h\nu - qV}{kT_s} - 1 = \frac{[\exp\{(h\nu - qV)/kT_s\} - 1]^2}{\exp h\nu/kT_p - 1}$$

This formula gives the voltage across each cell specified by $\nu \equiv \nu_G$ in the stack for optimum power conversion. Writing

$$x \equiv qV/kT_s,\quad u \equiv h\nu/kT_s,\quad a = T_s/T_p \quad \text{as in (76)} \tag{112}$$

the equation is

$$\frac{(1 + x)\exp(u - x) - 1}{[\exp(u - x) - 1]^2} = \frac{1}{\exp au - 1} \tag{113}$$

Figure 8 shows that there are two solutions $x_s < u$ and $x_\ell > u$ for x. The smaller one gives a maximum for η_2.

The output power is with $B_\nu \equiv 2\pi\nu^2c^{-2}qA$ when A is the area of the cell, and using (100)

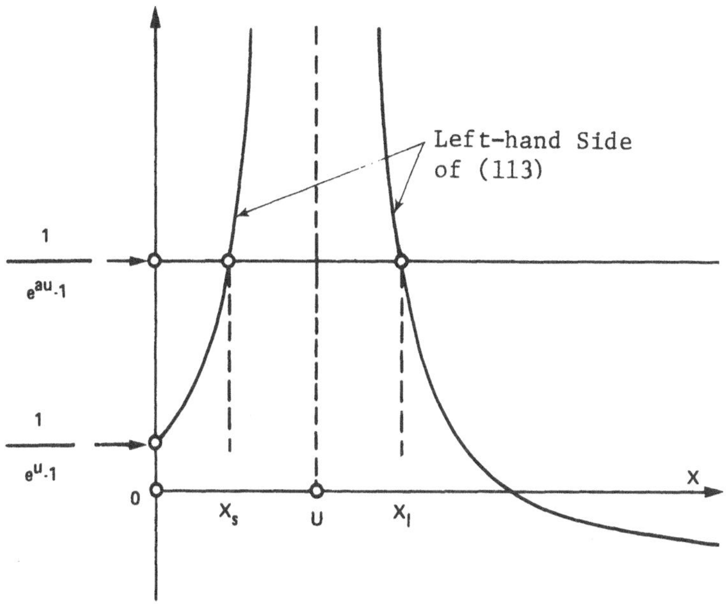

Fig. 8. Solutions of equation (113).

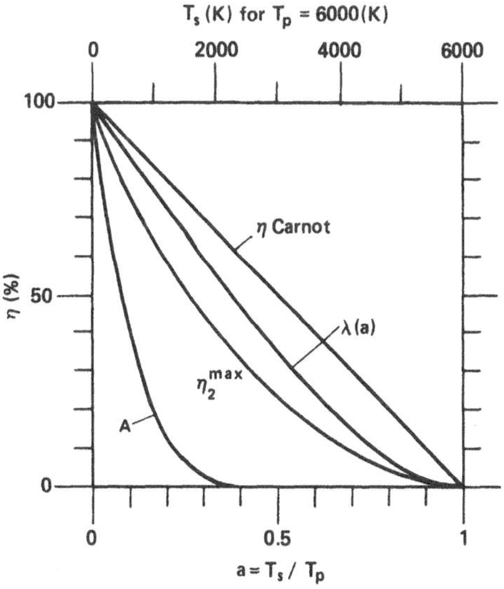

Fig. 9. Comparison of various thermodynamic efficiencies. Curve A refers to one-sun illumination ($f = 2.18 \times 10^{-5}$) as obtained in Eq. (147), whereas η_2^{max} refers to maximum concentration ($f = 1$). For one sun a lower maximum efficiency is expected from Fig. 4, since there is a greater dilution of solar radiation in this case.

$$P = -\int V dI_\nu(V) = \int B_\nu \left\{ [\exp au - 1]^{-1} - [\exp (u - x) - 1]^{-1} \right\} V d\nu . \tag{114}$$

Eliminating the term with a by (113)

$$P = \int_0^\infty \frac{(1 + x) \exp(u - x) - 1 - \exp(u - x) + 1}{[\exp(u - x) - 1]^2} V B_\nu d\nu$$

$$= \int_0^\infty \frac{x \exp(u - x)}{[\exp(u - x) - 1]^2} V B_\nu d\nu$$

$$= \frac{2\pi A}{c^2} \frac{(kT_s)^4}{h^3} \int_0^\infty \frac{u^2 x^2 \exp(u - x)}{[\exp(u - x) - 1]^2} du \tag{115}$$

For black-body radiation the input power is the area A multiplied by the energy flux (19)

$$P_{in} = A \sigma T_p^4 \tag{116}$$

The maximum efficiency is therefore [24]

$$\eta_2^{max} = \frac{P}{P_{in}} = \frac{15}{\pi^4} a^4 \int_0^\infty \frac{u^2 x^2 \exp(u - x) du}{[\exp(u - x) - 1]^2} , ^* \tag{117}$$

x being the smallersoultion of (113) for each u. This dependence of x on u reflects the fact that for given T_s the optimum bias voltage of a cell depends on ν, i.e. on its energy gap.

Fig. 9 gives a comparison between η_2^{max} of de Vos and Pauwels [20], (117), $\lambda(a)$, (78) of [14] and the Carnot efficiency

$$\eta_{Carnot} = 1 - a .$$

It shows that the lowest upper efficiency bound η_2^{max} applies to the most specialized model. The thermodynamic efficiencies increase with the generality of the model. It would be expected that lack of specificity prevents thermodynamics from yielding the lowest upper bounds for specific models.

The occurrence of exponents $u - x = (h\nu - qV)/kT_s$ is of interest and can be traced back to α_0 (87) and $n_\nu(d)$ (96). This shows that we have here non-equilibrium radiation which appears to have a non-zero chemical potential [25],[26] as already noted in section III.C.

---

* Since the appropriate value of x from (113) must be inserted for each u, (117) does not imply $\eta_2^{max} \propto a^4 \propto T_p^{-4}$.

III.F. Additional Comments: Independent Derivation of $\dot{S}_g$ for the Stack

That the entropy generation rate is (81) in a stack of solar cells is due to the fact that (110) is identical with (80).

$$\frac{\dot{S}_g}{\sigma T_p^3} = \frac{1}{3}b^3 + \frac{1}{b} - \frac{4}{3} \qquad \left(b \equiv \frac{T_c}{T_p}\right) \tag{118}$$

It is desirable to obtain this result directly by calculations based on the solar cell model [20].

The converter receives by (12) and (17) an energy flux $BK_\nu(Tp)$ from the pump and emits a flux $BK_\nu(T_c)$, using (104). This leads to an entropy generation rate per unit area for the whole system of $B[K_\nu(T_p) - K_\nu(T_c)]/T_c$. We equate it to the entropy gain rate $B[L_\nu(T_p) - L_\nu(T_c)] + \dot{S}_{g\nu}$, where $\dot{S}_{g\nu}$ is the irreversible entropy generation rate in the converter. Hence this latter rate, per unit frequency interval per unit area, is

$$\dot{S}_{g\nu} \equiv (B/T_c)[K_\nu(T_p) - K_\nu(T_c)] + B[L_\nu(T_c) - L_\nu(T_p)]. \tag{119}$$

Integration over ν as in (17) and (21) leads to the entropy generation rate for the stack (per unit area):

$$\dot{S}_g \equiv \int \Psi_{c\nu} d\nu = \frac{\sigma}{T_c}(T_p^4 - T_c^4) + \frac{4}{3}\sigma(T_c^3 - T_p^3),$$

whence one does indeed find (81):

$$\frac{\dot{S}_g}{\sigma T_p^3} = \frac{1}{3}\left(\frac{T_c}{T_p}\right)^3 + \frac{T_p}{T_c} - \frac{4}{3}.$$

The corresponding result for the pump is

$$\dot{S}_{pg\nu} = (B/T_p)[K_\nu(T_c) - K_\nu(T_p)] + B[L_\nu(T_p) - L_\nu(T_c)]. \tag{120}$$

The frequency-integrated entropy generation rate for the pump is given per unit area by

$$\frac{\dot{S}_{pg}}{\sigma T_p^3} = \left(\frac{T_c}{T_p}\right)^4 - \frac{4}{3}\left(\frac{T_c}{T_p}\right)^3 + \frac{1}{3} \tag{121}$$

The total entropy generation rates are therefore (per unit frequency range and integrated over frequency respectively, both per unit area)

$$\dot{S}_{Tg\nu} \equiv \dot{S}_{cg\nu} + \dot{S}_{pg\nu} = B[K_\nu(T_p) - K_\nu(T_c)]\left(\frac{1}{T_c} - \frac{1}{T_p}\right) \tag{122}$$

$$\frac{\dot{S}_{Tg}}{\sigma T_p^3} \equiv \frac{\dot{S}_{pg} + \dot{S}_{cg}}{\sigma T_p^3} = \left(\frac{T_p}{T_c} - 1\right)\left(1 - \left(\frac{T_c}{T_p}\right)^4\right) \tag{123}$$

The result (123) can be studied graphically together with $dI_\nu(V)/d_\nu$ of (100) and $T_c(\nu)$ of (105), where

$$\frac{dI_\nu(V)}{d\nu} = \frac{2\pi\nu^2 q}{c^2}\left([\exp \frac{h\nu}{kT_c} - 1]^{-1} - [\exp \frac{h\nu}{kT_p} - 1]^{-1}\right)$$

and

$$\frac{qV}{h\nu} = 1 - \frac{T_s}{T_c}$$

This is done in Table 3 and Fig. 10. One sees that there is zero entropy generation rate in a particular cell when it is at open-circuit, as one would expect since no current passes. Also on the present model one needs $qV < h\nu$ as one would otherwise have divergencies.

Figure 10a shows a schematic current-voltage characteristic of a solar cell which is found also for cells which are not part of a stack, when the vertical axis is simply the current density. It is clear that the output power is zero at both short-circuit, because the voltage across the cell vanishes, and also at open-circuit, because the current vanishes. Between these two limits there is always a maximum power output P^{max}, which is also shown. The integral $-\int V dI(V)$ for maximum power never sweeps the entire shaded rectangle of area $I_{sc}V_{oc}$, and so the term "fill-factor" (F) is used for

$$F \equiv P^{max}/I_{sc}V_{oc} \tag{124}$$

A simple theory of it is given in section III.G. Typical values range from 0.7 to 0.8.

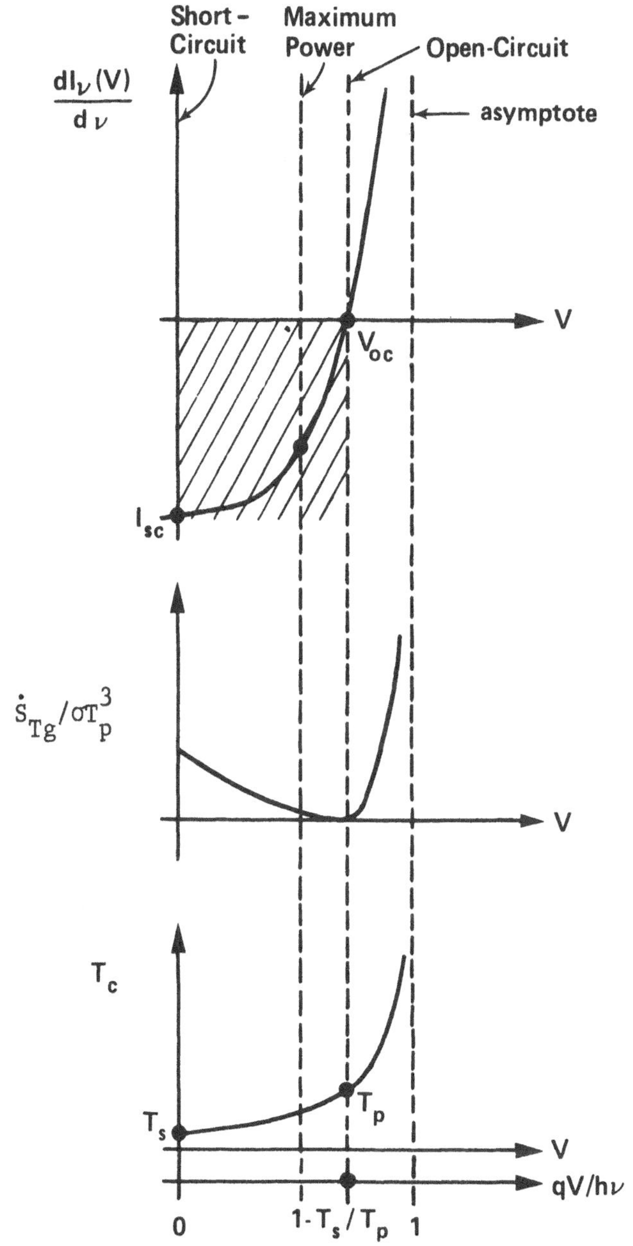

Fig. 10. Schematic diagrams illustrating for given incident radiation the voltage dependence for a given cell, i.e. for given ν, of the current, total entropy generation and cell temperature. Since the figure refers to maximum concentration C = 45900 suns, open-circuit conditions correspond to the stack radiating like the pump, i.e. it behaves like a mirror. The region $V > V_{OC}$ is not normally accessible for solar cell operation.

Table 3. The Effect of Three Special Values of $qV/h\nu$ on Converter Emission Temperature, Entropy Generation Rate per Unit Frequency Interval per Unit Area, and Current Density per Unit Frequency Range

$qV/h\nu$	0	$1 - T_s/T_p$	1
T_c	T_s	T_p	∞
$\dot{S}_{Tg\nu}$	$B[K_\nu(T_p) - K_\nu(T_s)]\left(\frac{1}{T_s} - \frac{1}{T_p}\right)$	0	∞
$dI_\nu(V)/d\nu$	$-2\pi\nu^2 qc^{-2}[n_\nu(T_p) - n_\nu(T_s)]$	0	∞

III.G. Additional Comment: The Solar Cell Equation and Standard Approximations

Returning to the solar cell equation (103) for an individual cell in the stack in terms of emitted photon fluxes Λ_e and incident photon fluxes $\Lambda_i = 2\Lambda_e(V_{oc})$, note that it can be written

$$\frac{I(V)}{I_{sc}} = \frac{w_{oc} - w}{w_{oc} - 1}, \tag{125}$$

where with $x \equiv qV/kT_s$, $u \equiv h\nu/kT_s$

$$w \equiv \int_{x_G}^{\infty} \frac{u^2 du}{\exp(u - x) - 1} \Bigg/ \int_{x_G}^{\infty} \frac{u^2 du}{\exp u - 1} . \tag{126}$$

For w_{oc} one simply replaces x by $x_{oc} \equiv qV_{oc}/kT_s$. (125) shows at once, even without the interpretation (126), that

$$I(V_{oc}) = 0 \text{ and } I(0) = I_{sc}.$$

This general form of (125) was noted in [27].

If one regards the denominators in (126) as resulting from sums of geometrical progressions,

$$\frac{1}{b^{-1} - 1} = \frac{b}{1 - b} = b[1 + b + b^2 + \dots], \tag{127}$$

then, putting $v \equiv \exp u$, $y \equiv \exp x$, and

$$J_r = \int_{x_G}^{\infty} u^2 v^{-r}\, du$$

one sees that

$$w = \frac{\int u^2 [\frac{y}{v} + \left(\frac{y}{v}\right)^2 + \ldots]\, du}{\int u^2 [\frac{1}{v} + \left(\frac{1}{v}\right)^2 + \ldots]\, du} = \frac{yJ_1 + y^2 J_2 + \ldots}{J_1 + J_2 + \ldots} \tag{128}$$

This assumes of course that $y/v = \exp(x - u) = \exp[(qV - h\nu)/kT_s)] < 1$. This is just the reasonable proposition $qV < h\nu$ which is also obvious from Fig. 10. If $yJ_1 \gg y^2J_2$, $J_1 \gg J_2$ etc., one finds

$$w \sim yJ_1/J_1 = \exp x = \exp qV/kT_s. \tag{129}$$

which was the interpretation of w adopted in [22], and other papers. Notably, this choice of w was implicit in the pioneer paper [28].

The approximation (129) is consistent with other approximations to equations derived earlier, and they will be noted below because of their frequent occurrence in the literature.

We shall work in analogy with sections III.D and E. However, instead of using (96), we shall argue within more standard p-n junction theory. This leads to an increment of current density due to an increment of incoming frequencies

$$dI_\nu = \frac{2\pi\nu^2 q}{c^2}\left(\frac{\exp x}{\exp u - 1} - \frac{1}{\exp au - 1}\right) d\nu \tag{130}$$

Incident black-body radiation has been assumed and this is represented by the second term. This is the same pumping term $q\Lambda_i$ as in (100). However the term to account for the loss of photons due to radiative emission at bias voltage V is changed. The non-equilibrium photon occupation number is now taken to be $\exp qV/kT_s$ times its zero-voltage value. We interrupt our discussion of (130) to explain this point [29].

Let $Q(E_G,V,T_p)$ be the number of electron-hole pairs created by the incident radiation per unit time and multiplied by the electron charge divided by the area to yield a current density when the voltage across the cell is V. Although V is itself related to the incident radiation, this is a convenient notation. Q depends also on the surrounding temperature T_s, but this is not shown explicitly. If the cell is in thermal equilibrium with the surroundings at the same temperature $T_p = T_s$, then one speaks of the cell "in the dark" and one has V = 0. Therefore by detailed balance

$$\begin{Bmatrix}\text{generation current}\\ \text{density}\end{Bmatrix}_0 = \begin{Bmatrix}\text{recombination}\\ \text{current density}\end{Bmatrix}_0 = Q(E_G,0,T_s).$$

The suffix 0 refers to thermal equilibrium. In this case recombination and generation are in balance with each other and with black-body radiation at temperature T_s.

The recombination current density in the dark (i.e. at temperature T_s) at voltage V across the cell, can be obtained by assuming the recombination rate at a plane x' in the active region to be proportional to the product of the carrier concentrations. For non-degenerate semiconductors assuming quasi-Fermi levels $kT_s\gamma_n(x)$, $kT_s\gamma_p(x)$ to exist, and denoting the intrinsic carrier density by n_i, one has an integral over the region considered:

$$\text{(i)}\quad Q(E_G,0,T_s) = C\int [n(x')p(x')]_0 dx' = C\int n_i^2(x')\,dx',$$

where C is a recombination coefficient, and

$$\text{(ii)}\quad \frac{n(x')p(x')}{[n(x')p(x')]_0} = \exp[\gamma_n(x') - \gamma_p(x')] \quad \left(= \exp\left(\frac{eV}{kT_s}\right)\right) \qquad (131)$$

If the Fermi levels have a constant separation eV in the transition region in the usual way, one finds the last expression in (131) provided recombination through traps or surface effects do not upset it. If the separation is not quite constant the contributions to $Q(E_G,V,T_s)$ which arise from different slabs of thickness dx are different, and an ideality factor $\alpha \neq 1$ may be expected. Hence

$$Q(E_G,V,T_s) = C\int n(x')p(x')\,dx' = C\exp(eV/kT)\int [n(x')p(x')]_0 dx'$$

i.e.

$$\frac{Q(E_G,V,T_s)}{Q(E_G,0,T_s)} = \exp\frac{eV}{\alpha kT_s}$$

Now the current density in the dark is

$$\begin{Bmatrix}\text{recombination current density}\\ \text{leading to radiative emission}\end{Bmatrix} - \begin{Bmatrix}\text{generation current density}\\ \text{weakly dependent on V}\end{Bmatrix}.$$

As the carriers have to overcome the field on recombination [27,29], the first term is expected to depend strongly on V. It is in fact

$$C\int n(x')p(x')\,dx' = Q(E_G,V,T_s) = Q(E_G,0,T_s)\exp x,$$

as was explained in connection with (130).

We continue the interuption by obtaining some basic results we now have at hand. The generation current is normally field-assisted so that changes in V do not greatly affect it [27,29]. Hence we have obtained the standard diode equation

$$I(V) = I_r(\exp x - 1) \quad (x \equiv qV/kT_s)$$

with the interpretation $I_r = Q(E_G,0,T_s)$. The ideality factor will be omitted. For large reverse bias, $x << -1$, $I(V) \sim I_r$ and this is therefore called the reverse saturation current density.

If illumination, specified by the parameters T_p, is applied, the generation current density is changed to $Q(E_G,V,T_p)$. Here T_p can be regarded as a set of parameters which specifies the spectrum of the pumping radiation and which is the absolute temperature in the case of black-body radiation. One finds for the current

$$I(E_G,V,T_s,T_p) = Q(E_G,0,T_s)\exp\frac{eV}{kT_s} - Q(E_G,V,T_p)$$

$$= Q(E_G,0,T_s)\left[\exp\frac{eV}{kT_s} - 1\right] - [Q(E_G,V,T_p) - Q(E_G,0,T_s)]$$

Thus the solar cell equation

$$I = I_r(\exp x - 1) - I_L \tag{132}$$

has been obtained with the interpretation

$$I_r(E_G,T_s) \equiv Q(E_G,0,T_s) = \left\{\begin{matrix}\text{reverse saturation current}\\ \text{density in the dark}\end{matrix}\right\} \tag{133}$$

$$I_L(E_G,T_p) \equiv Q(E_G,V,T_p) - Q(E_G,0,T_s) = \left\{\begin{matrix}\text{light induced}\\ \text{current density}\end{matrix}\right\} \tag{134}$$

A dependence of $I_L(E_G,T_p)$ on voltage V, which (134) suggests, implies a correction to (132): one cannot just take the dark characteristic of I versus V and shift it by a voltage-independent quantity J_L, though this so-called "shift theorem" is often a good approximation [30].

Return now to the question of an optimum efficiency. Maximizing the contribution of an individual cell to the output power implies again $d[VI_\nu(V)]/dV = 0$ (111), this time applied to (130). This gives for a stack of cells since (130) has been used

$$\frac{(1 + x)\exp x}{\exp u - 1} = \frac{1}{\exp au - 1} \tag{135}$$

in contrast with (113). The values of x determined by this equation as u is varied have to be inserted into the power output expression

$$P = -\int V dI_\nu = \frac{2\pi qA}{c^2}\int \frac{kT_s x}{q}\left(\frac{kT_s}{h}\right)^3 u^2 \left(\frac{1}{\exp au - 1} - \frac{\exp x}{\exp u - 1}\right) du$$

to give

$$P^{max} = \frac{2\pi (kT_s)^4 A}{c^2 h^3}\int \frac{x^2 u^2 \exp x}{\exp u - 1}\, du \tag{136}$$

The efficiency is P^{max} divided by the power received from the pump. Regarding the pump as a black body, this is $A\sigma T_p^4$, whence

$$\eta^{max} = \frac{15}{\pi^4} a^4 \int_0^\infty x^2 \frac{u^2 \exp x}{\exp u - 1}\, du \tag{137}$$

This result is graphically indistinguishable from the curve of η_2^{max} given in Fig. 9 [24]. For example for T_p = 6000 K, T_s = 300 K, (117) and (137) yield respectively 0.8682 and 0.8686. Additional consideration of maximum efficiencies of tandem cells are given in [31]-[34].

In conclusion, turn to the above "interruption" in order to obtain from it some other standard results for individual solar cells. From (132)

$$I_{sc} = -I_L \text{ and } I_r \exp x_{oc} = I_r + I_L \tag{138}$$

give the short-circuit current density and the open-circuit voltage. If one puts $w \equiv \exp x$, $w_{oc} \equiv \exp x_{oc}$,

$$\frac{I(V)}{I_{sc}} = \frac{I_r(w - 1) - I_L}{-I_L} = \frac{I_r w_{oc} - I_r w}{I_r w_{oc} - I_r} = \frac{w_{oc} - w}{w_{oc} - 1}$$

This shows that the present model is via (132) in agreement with (125).

Similarly a maximum power condition can be obtained from (132). Denoting x for maximum power by x_m, $d(VI)/dV = 0$ gives

$$(1 + x_m) \exp x_m = 1 + I_L/I_r = \exp x_{oc} \tag{139}$$

Hence

$$\frac{q}{kT_s} P^{max} = x_m I_r (\exp x_m - \exp x_{oc}) \tag{140}$$

The optimal fill factor (124) is

$$F^{max} = \frac{x_m}{x_{oc}} \frac{\exp x_{oc} - \exp x_m}{\exp x_{oc} - 1} \tag{141}$$

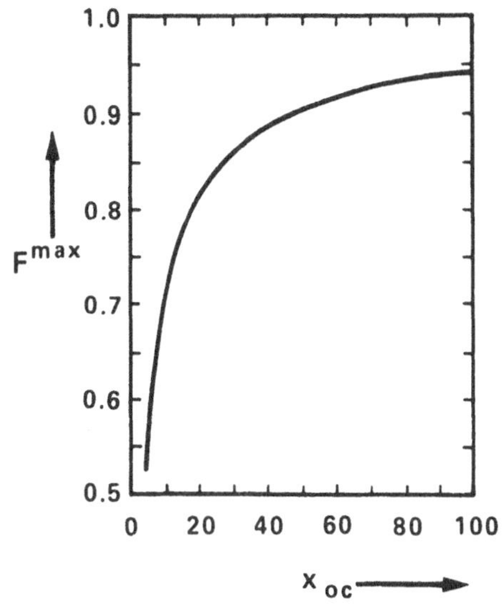

Fig. 11. The optimal fill factor F_m as a function of eV_{oc}/kT for a solar cell satisfying equation (132).

and is shown in Fig. 11. It is a universal function of x_{oc} since x_m is determined by (139) [35].

III.H. The Finite Stack [36]

Imagine a stack of cells arranged in a linear series so that the incident radiation encounters first the cell with the largest energy gap $E_1 = h\nu_1$, and finally the last, nth, cell with the smallest gap $E_n = h\nu_n$, so that

$$\nu_1 \geq \nu_2 \geq \cdots \geq \nu_n.$$

The photons with the smallest energy will survive unabsorbed longest, those with the higher energies are absorbed first, provided of course one has ideal semiconductors whose absorption thresholds lie at E_j ($j = 1,2,\ldots,n$). The simple p-n junction model (132) will be adopted:

$$I_j(V_j) = I_{rj}(e^{x_j} - 1) - I_{Lj}, \quad x_j \equiv qV_j/kT_s \tag{142}$$

Here $I_j(V_j)$ is the current-voltage characteristic of the jth cell, I_{rj} being the reverse saturation current density. This is determined by the radiative recombination on the assumptions (not normally correct) that non-radiative transitions can be neglected. Making this assumption following [28] and later workers,

$$I_{rj} = 2\int_{\nu_j}^{\infty} \frac{2\pi\nu^2 q}{c^2} n_{j\nu} d\nu \tag{143}$$

The factor 2 arises from the double-sided emission from the cell, and $n_{j\nu}d\nu$ is the mean photon number of frequencies in the range $(\nu, \nu + d\nu)$ in cell i. The light-generated current density I_{Lj} for the jth cell is

$$I_{Lj} = I'_{Lj} - I_{rj}, \tag{144}$$

Radiative emission due to radiative recombination in the cell subtracts carriers which would otherwise contribute to the current, and this accounts for the subtracted term. The term I'_{Lj} is due to radiation absorbed by cell j from the pump and the electroluminescent radiation from the neighboring cells. Thus

$$I'_{Lj} = q\left(f\int_{\nu_j}^{\nu_{j-1}} \frac{2\pi\nu^2}{c^2} n_{p\nu}\, d\nu + e^{x_{j-1}} \int_{\nu_{j-1}}^{\infty} \frac{2\pi\nu^2}{c^2} n_{j-1\ \nu}\, d\nu + e^{x_{j+1}} \int_{\nu_j}^{\infty} \frac{2\pi\nu^2}{c^2} n_{j+1\ \nu} d\nu \right) \tag{145}$$

Here f is a geometrical factor which takes account of the limited solid angle under which the pumping source may be seen. It is $\delta^2 = \omega_i/\pi$ of (32) and has the approximate value 2.18×10^{-5}. Also $n_{p\nu}d\nu$ is the photon occupation number of modes lying in the range $(\nu, \nu + d\nu)$, and $n_{j\nu}d\nu$ is the analogous photon number in cell j. For black-body radiation $n_{p\nu}$ is the Bose occupation number for temperature T_p, while $n_{j\nu}$ is the same expression with temperature T_s.

The electric power output P of the stack is given by (142) and (144):

$$\frac{q}{kT_s} P = -\frac{q}{kT_s} \sum_{j=1}^{n} V_j I_j(V_j) = \sum_{j=1}^{n} \left(x_j I'_{Lj} - x_j I_{rj} - x_j I_{rj} (\exp x_j - 1) \right)$$

where $x_j \equiv qV_j/kT_s$. The optimal working point $I_j(V_j)$ of an individual cell i is obtained from $\partial P/\partial x_i = 0$. This gives

$$I'_{Li} + x_{i-1} \frac{\partial}{\partial x_i} I'_{Li-1} + x_{i+1} \frac{\partial}{\partial x_i} I'_{Li+1} - I_{ri} \exp x_i -$$

$$x_i I_{ri} \exp x_i = 0$$

Assume that all cells are at the same temperature so that $n_{j\nu}$ is independent of j. The condition is for i = 1,2,...,n

$$I'_{Li} + qx_{i-1}\, e^{x_i} \int_{\nu_{i-1}}^{\infty} 2\pi\nu^2 c^{-2} n_\nu \, d\nu + qx_{i+1}\, e^{x_i} \int_{\nu_i}^{\infty} 2\pi\nu^2 c^{-2} n_\nu \, d\nu =$$

$$(1 + x_i)\, e^{x_i} I_{ri}$$

The obvious modifications for i = 1 (when the term in i - 1 does not arise) and for i = n (when the term in i + 1 does not arise) will not be made explicitly. Thus

$$(1 + x_i)e^{x_i} = \frac{I'_{Li}}{I_{ri}} + \frac{1}{2}\, x_{i-1}\, e^{x_i} \frac{\int_{\nu_{i-1}}^{\infty} \nu^2 n_\nu \, d\nu}{\int_{\nu_i}^{\infty} \nu^2 n_\nu \, d\nu} + \frac{1}{2}\, x_{i+1}\, e^{x_i} \quad (146)$$

The set of equations can be solved for the x_i given f and the incident spectrum $n_{p\nu}$. Since n_ν depends on $h\nu/kT_s$, T_s must also be given. From this information the optimal energy gaps and working point $I_i(V_i)$ of the various cells and the efficiency of the stack is obtainable. The results [36] are given in Tables 4 and 5 for T_s = 300 K and incident black-body radiation at 6000 K, as an approximation to the solar spectrum. The first case of unconcentrated sunlight is for n = ∞ represented by curve A in Fig. 9. The second case of maximum concentration corresponds to the stack being hemispherically surrounded by solar radiation.

Although the treatment here follows [36], related work should be noted [37], as well as references cited in these papers.

Consider now the limit n → ∞. We shall see that the above theory leads back to section III.G. The somewhat tedious limiting process is described below.

The three terms in (145) have the following limits in an obvious notation:

$$\text{(i)} \longrightarrow qf\, 2\pi\nu^2 c^{-2} n_{p\nu}\, \Delta\nu$$

Table 4. The Optimal Set of Bandgaps E_{Gi} for n stacked cells in Unconcentrated Sunlight ($f = 2.18 \times 10^{-5}$) [36]

n	η(%)	E_{G1}(eV)	E_{G2}(eV)	E_{G3}(eV)	E_{G4}(eV)
1	30	1.3	--	--	--
2	42	1.9	1.0	--	--
3	49	2.3	1.4	0.8	--
4	53	2.6	1.8	1.2	0.8
∞	68	$E_{Gn} = 0.25$ eV			

Table 5. The Optimal Set of Bandgaps E_{Gi} for n stacked cells in Sunlight Concentrated in the Ratio 45 900:1 (f=1) [36]

n	η(%)	E_{G1}(eV)	E_{G2}(eV)	E_{G3}(eV)	E_{G4}(eV)
1	40	1.1	--	--	--
2	55	1.7	0.8	--	--
3	63	2.1	1.2	0.6	--
4	68	2.5	1.6	1.0	0.5
∞	87	$E_{Gn} = 0$			

$$\text{(ii)} \longrightarrow q e^{x_i} e^{-\Delta x_i} \left[\int_{\nu_i}^{\infty} 2\pi\nu^2 c^{-2}\, n_\nu\, d\nu - \int_{\nu_i}^{\nu_{i-1}} 2\pi\nu^2 c^{-2}\, n_\nu\, d\nu \right]$$

$$\longrightarrow q e^{x}(1 + \Delta x) \int_{\nu}^{\infty} 2\pi\nu^2 c^{-2}\, n_\nu\, d\nu - 2\pi\nu^2 c^{-2}\, q e^{x}\, n_\nu\, \Delta\nu$$

$$\text{(iii)} \longrightarrow e^{x} q(1 - \Delta x) \int_{\nu}^{\infty} 2\pi\nu^2 c^{-2}\, n_\nu\, d\nu$$

Hence

$$I'_{Lj} \longrightarrow 2\pi c^{-2} q \left(f\nu^2 n_{p\nu} \Delta\nu + 2e^x \int_\nu^\infty \nu^2 n_\nu d\nu - \nu^2 e^x n_\nu \Delta\nu \right)$$

Also

$$I_{rj} \longrightarrow 4\pi q c^{-2} \int_\nu^\infty \nu^2 n_\nu d\nu$$

Hence

$$\frac{I'_{Lj}}{I_{rj}} \longrightarrow \frac{f}{2} \frac{\nu^2 n_{p\nu} \Delta\nu}{\int_\nu^\infty \nu^2 n_\nu d\nu} + e^x - \frac{e^x \nu^2 n_\nu \Delta\nu}{2\int_\nu^\infty \nu^2 n_\nu d\nu}$$

The other two terms in (146) become

$$\frac{1}{2}(x + \Delta x)e^x \left(1 - \frac{\nu^2 n_\nu \Delta\nu}{\int_\nu^\infty \nu^2 n_\nu d\nu} \right) + \frac{1}{2}(x - \Delta x)e^x$$

$$\longrightarrow xe^x - \frac{1}{2} xe^x \frac{\nu^2 n_\nu \Delta\nu}{\int_\nu^\infty \nu^2 n_\nu d\nu}$$

Collecting terms, $(1 + x)e^x$ is seen to cancel in (146), and so does the integral in the denominator of the remaining terms. Hence the condition (146) becomes simply

$$(1 + x)e^x = f\, n_{p\nu}/n_\nu = f(e^u - 1)/(e^{au} - 1) \quad (147)$$

This is just (135) as section III.G presumed $f = 1$. The smallest energy gap must be chosen so that $x = 0$, i.e.

$$f \exp \frac{E_{Gn}}{kT_s} - 1 = \exp \frac{E_{Gn}}{kT_p} - 1$$

so that for $f = 1$ one has $E_{Gn} = 0$. This consistency with section III.G is of course essential, although consistency with section III.E would be even better, but, as has been observed, the two approaches differ only slightly in the results to which they lead.

The maximum electrical power output is in the limiting case given by (136), with limits extending from E_{Gn}/kT_s to infinity.

Experimental studies on systems of two cells approaching the desirable energy gaps E_{G1} = 1.9 eV, E_{G2} = 1.0 eV have been carried out on the Al Ga As Sb—Ga As Sb cell (1.8/1.2 eV) [38]. The theoretical maximum efficiency of Table 4 is reduced from 42% to about 33% for AMI and 1 sun conditions by more detailed computer modeling. Al Ga As—Ga As (1.9/1.4 eV) cells are also being studied and computer modeling suggests that they should attain 24% efficiency under AMI and 1 sun conditions [39]. For recent work, including cascade cells based on amorphous silicon, see [40].

The finite-cascade has here been worked out for the "standard" or "conventional" theory, based equivalently on (132) or (142). In the limit of an infinite number of cells, this leads therefore to the conventional efficiency (137). Furthermore, the theory for sections III.E,F,G has been given for a converter which is completely surrounded by the sun, i.e. the geometrical factor f of (145) has been put at unity. For this reason there is also no allowance for absorption from the ambient in these sections.

The author is grateful to Professor J. Parrot, Dr. A. de Vos and Dr. P. Würfel for comments on the manuscript.

IV. PROBLEMS

Problem 1 Using the result of (26) and equation (16), give a simple direct derivation of Stefan's law (19).

Problem 2 Find the entropy flux for (20) of black-body radiation from the general relation (12).

Problem 3 The Riemann zeta function is here defined for convenience as an integral

$$\zeta(s+1) \equiv \frac{1}{\Gamma(s+1)} \int_0^\infty \frac{x^s\,dx}{e^x - 1} \qquad (s \geq 0)$$

s	0	1	2	3	10
$\xi(s+1)$	∞	$\frac{\pi^2}{6} = 1.645...$	1.202...	$\frac{\pi^4}{90} = 1.082...$	1.001

$\Gamma(r)$ is the gamma function whose values are $\Gamma(0) = \infty$, $\Gamma(1) = \Gamma(2) = 1$. Assuming the result (16), establish that for black-body radiation in a volume v at temperature T

$$U = \frac{3}{4}TS = \frac{4v}{c}\sigma T^4 = 3\ pv$$

which incorporates the result (29), now obtained by standard statistical mechanics.

Problem 4 If the occupation number of photon modes in the frequency range $(\nu, \nu + d\nu)$ is n_ν at the surface of a body, show that the photon number flux emitted in this frequency range is for a Lambertian surface

$$2\pi\nu^2 c^{-2} n_\nu d\nu$$

What is the expression for n_ν if the emitter is black?

Problem 5 Obtain the entries in Table 2.

Problem 6 Establish the result for u in section II.G

Problem 7 Equation (10) suggests the formula

$$\ell c^{-2}\nu^2\ n_\nu\ d\nu\ \cos\theta\ d\Omega$$

for the photon number flux for increments of frequency $d\nu$ and of solid angle $d\Omega$. Derive from this the formula $(c/\kappa\nu)n_\nu d\nu$ used in (84) for the case of discrete states.

Problem (8) Obtain (90) and related results in the context of a continuous density of states function, rather than assuming discrete states as in the text.

V. MAIN SYMBOLS USED AND REFERENCES

A area
a T_s/T_p (76)
B Einstein coefficient (III.C), $\int \cos\theta\ d\Omega$ (13)
b T_c/T_p (76)
C a recombination coefficient (III.G)
c velocity of light
d, n_d/N_d, a dilution factor (II.D)
e energy of a photon

E_G band gap energy
F fill factor
F_Q flux for a quantity Q
f a geometrical factor (III.H)
G solar constant
g see (87)
h Planck's constant

I, J quantum states
I current density
I_ν current density per unit frequency interval
I_{sc} short-circuit current density
I_r reverse saturation current density
I_L light-induced current density

i, j groups of quantum states

K_ν spectral energy radiance
k Boltzmann's constant
L_ν spectral entropy radiance
ℓ 1 or 2 for polarization (6)

N_d photon density at absorber
N_j occupation number of state j
n_j mean occupation number of state j
n_d photon density at absorber due to direct radiation
P output power
P, p probabilities
p momentum

Q heat
$Q(E_G, V, T_p)$ current density expressing the number of electron hole pairs created by incident radiation
q a parameter (2)
q a probability (82)
R distance
S entropy
$\dot{S}_g$ entropy generation rate
s_ν entropy per unit frequency range

T, T_A absolute temperature
T^* effective temperature $d\dot{\Phi}/d\dot{\Psi}$
T_c converter emission temperature
T_F flux temperature $\dot{\Phi}/\dot{\Psi}$
T_p temperature of a pump
T_R temperature of radiation
T_s temperature of surroundings or a sink
$T_\odot$ sun's temperature

t atmospheric transmission coefficient
t a dilution factor (II.D)
U mean internal energy
u_ν internal energy per unit frequency range
$u = h\nu/kT_s$ (112)
u recombination rate per unit volume
u' (absorption rate minus stimulated emission rate) per unit volume

V a voltage
V velocity
V_{oc} open-circuit voltage
v volume
W work
w see (126)
X product of probabilities (85)
$X(\epsilon)$ see (37)
$x = qV/kT_s$ (112)

α absorption coefficient
α_0 see (87)
$\frac{4}{3}\beta_i = \Psi_i/\Phi_i$ see (55)

γ Fermi level $/kT_s$ or quasi-Fermi level$/kT_s$

δ, θ, ϕ angles

ϵ dilution factor
Λ_e photon flux emitted by a solar cell
Λ_i photon flux incident on a solar cell
λ see (72)
μ chemical potential
ν frequency

η efficiency
η_I (energy of quantum state I)$/kT_s$

κ refractive index of a material

σ Stefan's constant

VI. APPENDIX

We give here a more formal derivation of (105) for the cell temperature T_c, by using the balance equations. We chose the latter in the form (53):

$$\dot{W} = \phi_p - \phi_c - (\psi_p - \psi_c)\, T_s - T_s\, \dot{S}_g \tag{148}$$

In order to extract work qV, suppose that we require a net energy input f h ν at temperature T_c to the carrier system and g h ν at temperature T_s to the lattice system. Then

$$\begin{aligned} \phi_p - \phi_c &= (f + g)\, h\, \nu \\ \psi_p - \psi_c &= f\, h\, \nu/T_c + g\, h\, \nu/T_s \end{aligned} \tag{149}$$

Substituting (149) into (148)

$$qV = (f + g)\, h\, \nu - [f\, h\, \nu\, (T_s/T_c) + g\, h\, \nu] - T_s\, \dot{S}$$

i.e.

$$\frac{qV}{fh\nu} = 1 - \frac{T_s}{T_c} - \frac{T_s\, \dot{S}_g}{f\, h\, \nu}$$

This result is independent of the net energy input to the lattice and for f = 1 and $\dot{S}_s = 0$ it is precisely (105).

REFERENCES

1. P. T. Landsberg, Proc. Phys. Soc. 74, 486-488 (1959).
2. J. E. Parrott, Solar Energy 21, 227-229 (1978).
3. J. E. Parrott, Solar Energy 22, 572-573 (1979).
4. P. T. Landsberg, Lecture on the mathematical aspects of solar energy at the British Association for the Advancement of Science, Edinburgh, 1979, and lectures on solar energy conversion, International Center of Theoretical Physics, Trieste, 1977.
5. A. Arrhenius, Das Werden der Welten (Translated from the Swedish, Leipzig: Akademische Verlagsgesellschaft, 1908), p. 42.
6. D. Wilkins, personal communication.
7. J. A. Wood, The Solar System (Englewood Cliffs, New Jersey: Prentice Hall, 1979).
8. P. T. Landsberg and G. Tonge, (a) J. Phys. A 12, 551 (1979); (b) J. App. Phys. 51 R1 (1980).
9. P. T. Landsberg, Thermodynamics and Statistical Mechanics Oxford University Press), 1978.
10. E. D. Cashwell, C. J. Everett, Am. Math. Monthly 74, 271 (1967); P. T. Landsberg. J. Math. Anal. and Applic. 76,209 (1980).
11. L. M. N. Duysens, Brookhaven Symp. Biol. 11, 18 (1958).

12. D. C. Spanner, Nature 198, 934 (1963).
13. P. T. Landsberg and J. R. Mallinson, in Int. Colloquium on Solar Electricity (CNES, Toulouse, 1976), p. 27.
14. P. T. Landsberg. Photochemistry and Photobiology 26, 313 (1977).
15. W. H. Press, Nature 264, 734 (1976).
16. P. T. Landsberg, J. App. Phys. 54, 2841 (1983).
17a. G. J. Lasher and F. Stern, Phys. Rev. 133, A 553 (1964).
17b. P. T. Landsberg, in Festkörperprobleme Vol. VI (Braunschweig: Vieweg, 1967)(Ed. O. Madelung).
17c. M. J. Adams and P. T. Landsberg, Theory of the Injection Laser in Gallium Arsenide Lasers (Ed. C. H. Gooch; London: Wiley 1969), p. 5-79.
18. M. G. A. Bernard and B. Durraffourg, Phys. Stat. Sol. 1, 699 (1961).
19. P. T. Landsberg, Phys. Stat. Sol. 19, 777 (1967).
20. A. De Vos and H. Pauwels, Appl. Phys. 25, 119 (1981).
21. A. F. Haught 3rd, Intern. Conf. on Photochem. Conversion and Storage of Solar Energy (August 1980), Boulder, Colorado.
22. W. Ruppel and P. Würfel, IEEE Trans. ED 27, 877 (1980).
23. S. M. Jeter, Solar Energy 26, 231 (1981).
24. A. de Vos, C. C. Grosjean and H. Pauwels, J. Phys. D15, 2003 (1982).
25. P. T. Landsberg, J. Phys. C 14, L 1025 (1981).
26. P. Würfel, J. Phys. C 15, 3967 (1982).
27. P. T. Landsberg, Solid-State Electronics 18, 1043 (1975).
28. W. Shockley and H. J. Queisser, J. Appl. Phys. 32, 510 (1961).
29. P. T. Landsberg, in Photovoltaic and Photoelectrochemical Solar Energy Conversion (Eds. F. Cardon, W. P. Gomes, and W. Dekeyser)(New York: Plenum Press, 1981).
30. F. A. Lindholm, J. G. Fossum, and E. L. Burgess, Proc. 12th IEEE Photovoltaic Specialist Conference, Baton Rouge, p. 33 (1976).
31. J. Parrott, J. Phys. D 12, 441 (1979).
32. H. Pauwels and A. de Vos, Solid-State Electronics 24, 835 (1981).
33. J. Parrott and A. Baird, Proc. of 15th IEEE Photovoltaic Specialist Conference, p. 383 (1981).
34. C. H. Henry, J. Appl. Phys. 51, 4494 (1980).
35. J. Landmayer, COMSAT Tech. Rev. 2, 105 (1972).
36. A. de Vos, J. Phys. D 13, 839 (1980).
37. N. Gokcen and J. Loferski, Solar Energy Materials 1, 271 (1979).
38. M. L. Timmons and S. M. Bedair, 15th PVSC, Orlando, p. 1289 (1981).
39. S. M. Bedair, J. A. Hutchby, J. Chiang, M. Simons and J. R. Hauser, 15th PVSC, Orlando, p. 21 (1981).
40. Proceedings of the 16th Photovoltaic Specialist Conference,

San Diego, USA, September 1982. The following are authors of relevant papers given there:

J. Loferski (Brown University, Providence)
H. B. Curtis and M. P. Godlewski (NASA-Lewis, Ohio)
M. W. Wanlass and A. E. Balskeslee (SERI, Colorado)
B. Beaumont, F. Raymond and C. Verié (CNRS, Meudon, France)
C. M. Fraas, D. E. Sawyer, J. A. Cape and B. Shin (Chevron Research Company, Richmond, California)
R. J. Markunas, M. L. Timmons, J. A. Hutchby and M. Simons (Research Triangle Institute, North Carolina)
G. Nakamura, K. Sato and Y. Yukimoto (Mitsubishi Electric Corporation, Itami, Japan)
Y. Kuwano et. al (Sanyo Electric Co., Osaka, Japan)
C. Flores (Milano, Italy)

MAGNETO-OPTICAL STUDY OF ENERGY TRANSFER IN RUBY

M. Ferrari, L. Gonzo, M. Montagna, O. Pilla,
and G. Viliani

Eepartimento di Fisica
Università di Trento
38050 Povo, Trento, Italy

ABSTRACT

The ion-pair transfer mechanisms were studied under magnetic field. Both resonant and non-resonant contributions are found from the magnetic field dependence of the transferred intensities. For transfer to third neighbors, non-resonant, magnetic field independent transfer is found to dominate, while the contrary is observed for transfer to fourth neighbors. The effects of magnetic field were investigated at several temperatures and at different emission wavelengths.

I. INTRODUCTION

Since the pioneering work of Imbusch [1], the problem of energy transfer in ruby attracted much experimental and theoretical work. Several transfer mechanisms are conceivable, namely resonant and non-resonant exchange-induced transfer, radiative transfer, dipole-dipole interaction, etc. The experimental data lead to somewhat conflicting models, especially as regards the transfer rate among single ions (S-S), and from single ions to pairs (S-N1 and S-N2 respectively) [2],[3].

Application of an external magnetic field will produce important effects on the transfer rate by bringing out of resonance degenerate energy levels, or by producing level-crossings; such experiments could thus help in understanding whether the S-N transfer is direct or it proceeds via some intermediate centers. At low magnetic fields (H < 6 KG) it has been found [4] that the S-S transfer rate is increased when there is a crossing among the levels of the coupled "isolated" ions; this, in turn, is seen as an increase of the S-N rate. The observed resonance effect is only a few percent [4].

We have studied the magnetic field effect up to 60 KG; the results are reported and discussed in the present work.

II. EXPERIMENTAL

The sample was placed in the bore of a superconducting magnet with the field parallel to the C_3 axis; emission and excitation spectra, and lifetime measurements were carried out at various temperatures as a function of the applied field. The lifetime measurements were performed by focussing an Ar-ion laser beam on a mechanical chopper and then by refocussing it on the sample; R1-line and N2-line emission intensities were detected by a photon counting apparatus and recorded on a multichannel analyzer. The excitation spectra were taken by illuminating the sample by a tungsten lamp and a monochromator, plus photon counting detection, for the R1, N1, and N2 emissions.

The effect of the magnetic field on the S-N transfer was directly observed by monitoring the N1 and N2 emission intensities as a function of the field. Since the N emissions show two different lifetimes, the faster one (less than 1 msec) deriving from direct excitation and the slower one (4-10 msec, depending on reabsorption) deriving from transfer, in order to discriminate between the two components phase-sensitive detection was employed. The phase of the lock-in amplifier was adjusted by zeroing the slow component of the N signals; in this way, in the counterphase output the (slow) transferred component was maximized, while in the in-phase output only the (fast) directly excited component was observed.

Finally, excitation spectra of the N lines were taken by the phase-sensitive technique also, in order to discriminate "fast" from "slow" peaks in the spectra.

It should be noted that since the magnetic field makes S-S radiative reabsorption weaker, the phases were slightly field-dependent. We adjusted the phase at high field; in this way the spectra needed to be corrected only in the first few kilogauss.

III. EXPERIMENTAL RESULTS

III.A. Lifetime Measurements

The results for N2 show that the transfer rate decreases

markedly (by a factor of about 2) in passing from zero-field to 10 KG; after this field value it is roughly constant up to about 35 KG and shows a marked and sharp peak centered at about 40 KG. This behavior shows the importance of the magnetic field for the transfer process, and we decided to study such effect in a more systematic way by directly monitoring both the N1 and N2 emission intensities as a function of the applied field.

III.B. Transferred Intensities vs. Magnetic Field

The counterphase emission intensities of N2 and N1 lines, after excitation at 19600 cm^{-1} in the high-energy tail of $^4T_2$, as a function of the magnetic field are reported in Figures 1 and 2 respectively for three different temperatures; the band-pass of the detection monochromator was about 3 cm^{-1}, larger than the full width of the observed emissions. The corresponding in-phase spectra are flat and are not reported here. In Figure 1 we note an initial decrease of the transferred signal to which some structures at about 10, 15, 25, and 30 KG are superimposed, and a sharp and intense peak at 40 KG. The effect of increasing temperature is that of washing out these features, and of increasing the continuous background, while new structures appear at higher fields (about 48 and 55 KG). At even higher temperatures ($T > 170$ K) the whole spectrum becomes flat, except for a slight continuous descent.

A rather different behavior is observed in Figure 2 for the N1 emission: the spectrum is flat up to 30 KG, where an intense peak appears, followed by another one at 50 KG; here also, new weaker peaks appear at higher temperatures and all structures smooth out for $T > 70$ K.

The same spectra as in Figures 1 and 2 were taken also under two different experimental conditions; (a) Excitation with a monochromator (band-pass 3-10 cm^{-1}) at different wavelengths ranging from the low-energy tail of $^4T_1$ to the R1-line; (b) Excitation with the laser at 5145 Å and revealing at higher resolution (0.5 cm^{-1}) in different parts of the emission lines.

As regards N2, the experiments of type (a) showed the following effects: (1) By exciting in the R2 and (especially) R1 lines the 40 KG peak looks weaker and broader; (2) The initial descent is slower by exciting in R1 and R2; (3) The weaker resonances at 10 and 15 KG do not seem to be very affected by varying the excitation energy, except for a broadening by exciting in R1 and R2.

Still regarding N2, the experiments of type (b) show the following effects: (1) By revealing at the center of the line, all peaks are intense and well-defined, comparable to Figure 1(a), while in the tails they become much weaker and broader; (2) The initial descent is much steeper in the tails and the zero-field

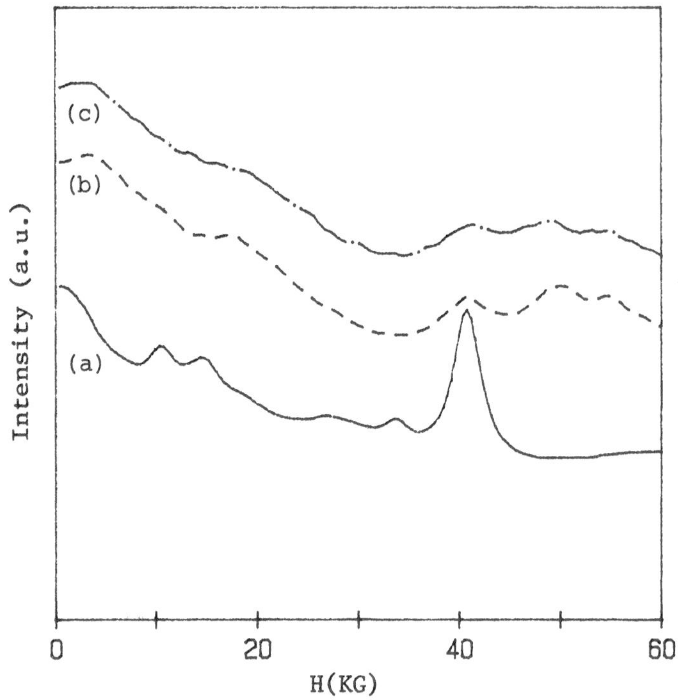

Figure 1. N2-line emission intensity as a function of the applied magnetic field. (A) T = 7 K; (b) T = 65 K; (c) T = 90 K.

intensity is in this case several times as high as the one of the prominent 40 KG peak.

Let is consider now N1. Type (b) measurements show the following features: (1) Here also, the intensity of the two main peaks decreases in passing from the line-center to the tails; (2) New structures appear (at low temperature) at approximately the same field values as the weaker structures of Figures 2(b) and 2(c), but not when revealing in the low energy tail of the emission line.

Type (a) measurements do not show any significant change with respect to Figure 2(a), with the exception of excitation in the R2 line; in this case: (1) There is a marked initial increase up to about 6 KG; after this field value there is a decreasing background upon which the two structures of Figure 2(a) are superimposed; (2) The high-field peak becomes weaker than the low-field one. Note that by exciting in R2 there is no direct excitation of third neighbor pairs and the observed intensity is due only to transfer.

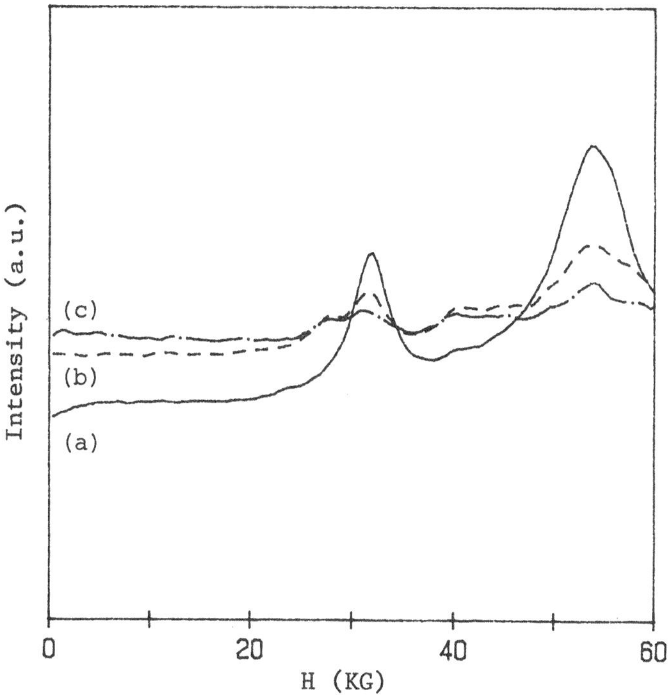

Figure 2. N1-line emission intensity as a function of the applied magnetic field. (a) T = 7 K; (b) T = 25 K; (c) T = 45 K.

III.C. Excitation Spectra

Since the spectra of Figures 1 and 2 are quite different from each other, we have investigated the possibility that the peaks arise from level-crossings between Zeeman sublevels belonging to the $\bar{E}$ multiplet (single ions), and excited states relative to pairs. We have recorded excitation spectra in the energy region near the R1 and R2 lines for the three emissions at different fields. In the pair spectra several peaks appear besides the R1 and R2 lines, as shown in Figure 3 at zero-field.

The origin of the peaks A, B, and C in Figure 3(b) is not completely known; the multiplet (A,B) was tentatively assigned to R' centers (single-ions weakly coupled to the pairs) [2], but its behavior under magnetic field as observed by us suggests that it corresponds to an S = 2 excited state of the fourth neighbor pairs. The peaks in the low-energy side of the R1-line, which also were attributed to R' centers [2], behave under magnetic field like the $^4G_3$ and $^4H_3$ [5] excited states of fourth neighbor pairs. As regards third neighbor pairs, the origin of peaks D and E of Figure 3(a) is not completely understood either, although they could correspond to yet unobserved excited states of the pair;

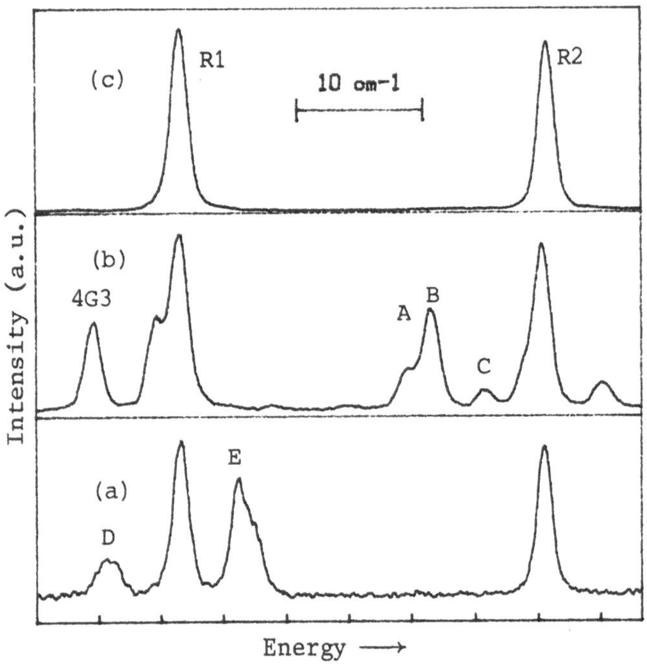

Figure 3. Excitation spectra of single-ion and pair lines at zero magnetic field and T = 7 K. (a) N1-line; (b) N2-line; (c) Sideband of R1-line. Cr-concentration 1800 ppm.

Selzer et al [2] showed that peak E is a rapid channel for excitation of N1 luminescence.

We would like to remark that the same excitation spectra taken by phase-sensitive detection show that all the features of Figures 3(a) and (b), other than R1 and R2 lines, are rapid channels for the excitation of pair emission.

IV. DISCUSSION

Since the longer lifetime of the N lines is the same, or very similar to the R-radiative lifetime, both in presence and in absence of reabsorption among single ions, it is reasonable to think that all of the single ions contribute to the S-N transfer, and not only those ions which are strongly coupled to the pairs (if any). This is further supported by our observation that the center of the R-line contributes to the S-N transfer more than the tails, in agreement with the measurements of Chu et al [6]. On the other hand, the difference between Figures 1 and 2 shows that the observed magnetic field effects are not due to the S-S transfer. Moreover, it can be shown that by assuming the single ions to be coupled by exchange interaction, the magnetic field splits initially degenerate levels, and this results in a marked decrease of the resonant transfer rate; a level crossing at 17 KG should give a peak at this field value [7]. These considerations seem to exclude a transfer mechanism consisting of fast S-S diffusion followed by excitation of the pair with unity-efficiency when the excitation is found within some "capture radius" of the pair. In fact, in this case the S-N transfer rate would be proportional to the diffusion coefficient [8], which in turn is proportional to the exchange interaction between single ions. In any case, whichever be the S-S interaction, if the above diffusion model were true the N1 and N2 spectra (Figures 1 and 2) should have some common features.

From transient electric field measurements [9] it results that an upper limit for the non-radiative S-S transfer time is of the order of a fraction of a millisecond at our concentrations (see, however, [3]); the transfer time as deduced from our data is of the order of 1 second which implies that the last step in the S-N transfer is the slowest. Our data indicate that both resonant and non-resonant mechanisms are active in the S-N transfer process. In fact, non-resonant, phonon-assisted transfer is not expected to be affected by a magnetic field because the required phonon energies (roughly, 150 cm^{-1} for S-N2 and 220 cm^{-1} for S-N1) are much greater than any Zeeman splitting. This seems to be the predominant mechanism for S-N1 transfer at any temperature, while it seems to be less important for S-N2 at low temperature. In the latter case, in fact, the residual transferred intensity at high field is of the order of 10% of the initial one in our samples. In any case, at high temperature the non-resonant contribution predominates and the field-dependence vanishes.

The peaks observed for both N1 and N2 are by us assigned to resonant transfer due to level crossings between excited states of single ions and pairs. By taking excitation spectra at various magnetic fields, we have verified that the resonances of Figure 1(a) correspond to the S-N2 cross relaxations indicated in Table I. These resonances could correspond either to radiative reabsorption or to non-radiative transfer. The fact that the most intense resonances correspond to σ^+-σ^+ transitions would favor the former hypothesis, while the total m_s is not conserved. On the other hand, no appreciable change is observed in the spectra by changing the excitation geometry.

Let us now consider the prominent resonance at 40 KG; by using the g-values appropriate to the relative levels the width of the resonance ($\sim$3 KG) corresponds to a convoluted width of the two resonant levels of $\sim$0.5 cm^{-1}. The widths of the involved lines are ΔR1(hom) $\sim$ 0, ΔR1(inh) $\sim$ 1 cm^{-1}, while as regards the $^4G_3$ transition it can be assumed that its inhomogeneous width is comparable to that

Table I. Observed resonances for N2-line (Figure 1(a)) and proposed cross-relaxation transitions between single ions and pairs.

Resonance (KG)	Single Ion Trans. $m_s(E) \rightarrow m_s(^4A_2)$	Polar.	GHI Pair Trans. $m_s(S=3) \rightarrow m_s(S=2)$	Polar.
10	$1/2 \rightarrow 3/2$	σ^+	$-2 \rightarrow -1$	σ^+
15	$-1/2 \rightarrow -1/2$	σ^+	$-2 \rightarrow -1$	σ^+
32	$-1/2 \rightarrow 1/2$	π	$-3 \rightarrow -2$	σ^+
40	$1/2 \rightarrow 3/2$	σ^+	$-3 \rightarrow -2$	σ^+

of N2, that is ~ 1 cm^{-1}, but its homogeneous width can be appreciable because of the limited lifetime of the excited G level. In a radiative reabsorption the full (homogeneous + inhomogeneous) width should contribute to the observed width, producing a much broader resonance than observed. On the contrary, if the inhomogeneous width is mainly due to macroscopic strain and if the cross relaxation is non-radiative, the homogeneous width will give the main contribution. It is therefore possible that the observed width of 0.5 cm^{-1} corresponds mainly to the homogeneous broadening of the G state.

Considering now the initial descent of N2 (Figure 1), it is more important by observing the tails of N2; no such effect is observed for N1. Such different behavior can be understood by noting that the fourth neighbor pairs have excited levels (H,I) which are nearly degenerate (within 2 cm^{-1}) with the $\bar{E}$ single ion level, while this is not the case for third neighbor pairs. From the excitation spectra of N1 (Figure 3(a)) it appears that the nearest peaks resonate with the R1 line at magnetic fields corresponding to the resonances of Figure 2. This suggests that the zero-field transfer in N2 is mainly due to a resonant mechanism. If we assume that the excited I and/or H states also have a homogeneous width of the order of 0.5 cm^{-1}, and that they are nearly resonant with $\bar{E}$, it can be understood why the tails of N2 are more favored for resonant transfer than the line center. In fact, emission from the tails of N2 derives from slightly distorted pairs whose excited I and/or H levels may be nearer in energy to the $\bar{E}$ levels of the single ions of their domains, if macroscopic strain is mainly active. Application of a magnetic field, by splitting both single-ion and pair levels, will reduce the number of resonant states, thereby decreasing the transferred intensity. In this way one can understand why the resonances, when monitored on the tails of N2, are broader and less intense.

In conclusion, our measurements show that different transfer mechanisms exist in our samples. The relative importance of such mechanisms may be concentration-dependent; such measurements are in progress.

ACKNOWLEDGEMENT

This work was partially supported by the Consiglio Nazionale delle Ricerche.

REFERENCES

1. G. F. Imbusch, Phys. Rev. 153, 326 (1967).
2. P. M. Selzer, D. L. Huber, B. B. Barnett, and W. M. Yen, Phys. Rev. B17, 4979 (1978), and references therein.
3. E. Duval and A. Monteil, these proceedings give an account of more recent controversies; see also A. Monteil, Thèse d'Etat, Lyon (1982), unpublished.
4. Y. Fukuda, T. Muramoto, and T. Hashi, J. Phys. Soc. Japan 50, 2369 (1981).
5. P. Kisliuk, N. C. Chang, P. L. Scott, and M. H. L. Pryce, Phys. Rev. 184, 367 (1969).
6. S. Chu, H. M. Gibbs, and A. Passner, Phys. Rev. B24, 7162 (1981).
7. M. Ferrari, unpublished.
8. H. C. Chow and R. C. Powell, Phys. Rev. B21, 3785 (1980).
9. S. Chu, H. M. Gibbs, S. L. McCall, and A. Passner, Phys. Rev. Lett. 45, 1715 (1980); P. E. Jessop and A. Szabo, Phys. Rev. Lett. 45, 1712 (1980).

SPECTROSCOPIC STUDIES OF ENERGY TRANSFER IN SOLIDS

G. Boulon

Laboratoire de Physico-Chimie des Matériaux Luminescents
E.R.A. 1003 du CNRS - Université Lyon I
43 Bd du 11/11/1918 - 69621 Villeurbanne Cédex - France

ABSTRACT

We present some experimental results obtained in Lyon's Group about the energy transfer process between active ions in doped or stoichiometric fluorescent materials. We have chosen various examples illustrating the different types of energy transfer for which we give the useful data about their theoretical approaches. Fluorescent ions concerned are divalent (Eu^{2+}) and trivalent (Eu^{3+}, Gd^{3+}, Tb^{3+}) rare-earth ions, Mn^{2+} transition ion and also Bi^{3+} heavy ion incorporated in some oxyde, fluoride and chloride crystals and also germanate glass.

I. INTRODUCTION

Energy transfer processes are widely studied in our laboratory since many years as well in doped crystals and glasses as stoichiometric materials. We would like to contribute to the understanding of the basic processes involving the dynamical properties of the condensed matter under optical pumping. In the first, part, we give the main theoretical results used on the energy transfer and migration processes field. In the second part, we give some examples for each type of energy transfer studied in our materials. These examples illustrate the resonant radiative energy transfer, the resonant non-radiative energy transfer either by direct energy transfer or by fast-diffusion and also the up-conversion mechanism like two-successive transfers, for example, giving rise to additonal pumping to higher excited states above the absorption energy level.

II. MATERIALS AND EXPERIMENTAL EQUIPMENT

The phosphors selected in this presentation are TbF_3, $GdCl_3$ $(6H_2O)$, $LaCl_3(Gd^{3+})$, $KY_3F_{10}(Eu^{2+})$, $Lu_2Si_2O_7(Bi^{3+} - Eu^{3+})$ and germanate glass doped by Bi^{3+} or Eu^{3+} or codoped by Bi^{3+} and Eu^{3+} and manganese compounds ($CsMnF_3$, $RbMnF_3$, $RbMnCl_3$). The most experiments are done by using pulsed tunable dye-laser and lamp techniques in the uv and visible range like site selection spectroscopy and time resolved spectroscopy. The pump laser is either a nitrogen laser (1 mJ per pulse - τ = 4ns) or a YAG-Nd powerful laser with a frequency doubler or tripler (300 mJ per pulse at λ = 5320 Å; 130 mJ per pulse at λ = 3550 Å) followed by a three stage amplifier dye laser and others frequency doublers. So we may selectively excite from 2100 Å to 8000 Å (linewidth 0.1 cm^{-1} - τ = 15 ns). The fluorescence data are recording and processing over 256 channels with a maximum resolution of 2 µs per channel by using an Intertechnique Model IN 90 multichannel analyser. The fluorescence decays at shorter times are recorded with a PAR boxcar integrator and the profile are fitted by using a Tektronix 4051 computer.

III. USEFUL DATA ABOUT THEORETICAL APPROACHES OF THE ENERGY TRANSFER

In this paper we distinguish between the resonant radiative energy transfer and the resonant non-radiative energy transfer. It is important to mention that the microscopic theoretical approache uses the interaction between two ions labelled S (sensitizer or donor ion) and A (activator or acceptor ion) whereas the experimental results reflect the average behaviour of a lot of sites in the host. So we have to choose some statistical analysis over all the macroscopic scale of the sample. The different cases of energy transfer processes have been treated by Auzel [1] , Reisfeld [2] Riseberg and Weber [3] , Watts [4] . In this part, we shall give both the transfer probabilities and physical parameters derived from the microscopic model and the intensity decay of the emission of S and A ions.

III. A. Resonant Radiative Energy Transfer

This mechanism depends on the geometric configuration of the crystal. In this case the normalized emission spectrum of S ion $g_S(\nu)$ and the absorption spectrum of A ion $g_A(\nu)$ overlaps and, then, the structure of the S emission spectrum is A ion concentration dependent. More, the S decay time (τ_S) does not vary with A ion concentration because the spontaneous fluorescence of S ion is emitted in all directions.

The probability of such transfer between two S and A ions at distance R is:

$$W_{SA}(R) = \frac{\sigma_A}{4\pi R^2} \frac{1}{\tau_S} \int g_S(\nu)\ g_A(\nu)\ d\nu \tag{1}$$

σ_A is the absorption integrated cross-section of A ion. We will see an example with Bi^{3+} - Eu^{3+} co-doped germanate glass.

III.B. Resonant Non-radiative Energy Transfer

The coupling of S and A ions can arise via exchange interaction if the wave functions overlap or via electric or magnetic multipolar interactions. So the optical excitation absorbed by S ion may be transferred to A ion before S ion will be able to emit the fluorescence.

The first process involving resonant non-radiative energy transfer between S and A ions has been treated by Förster and Dexter [5],[6] for a multipolar or an exchange interaction. Inokuti and Hirayama [7] have applied the statistical analysis in order to get the profile I(t) of the fluorescence decay rate of S ions.

1. Without Diffusion Among S Ions. The S excitation density $\Phi(t)$ satisfies the equation:

$$\frac{\partial\Phi(R,t)}{\partial t} = -\frac{1}{\tau_{SO}}\Phi(R,t) - \sum_A W_{SA}(R_i) \cdot \Phi(R,t) \cdot \tag{2}$$

τ_{SO}: the intrinsic decay rate of S in absence of the A ion; $W_{SA}(R_i)$ is the probability for energy transfer between S and A ions at position R_i.

Inokuti and Hirayama [7] find the relation

$$\Phi_S(t) = \Phi_S(o)\ \exp\{-[\frac{1}{\tau_{SO}} + \gamma(t)] \cdot t \tag{3}$$

$\Phi_S(o)$: the intensity of S ions at t = o; $\gamma(t)$: function which depends on the interaction between S and I ions.

- Exchange interaction

Dexter [6] has derived the following expression for $W_{SA}(R_i)$:

$$W_{SA}(R)\ \alpha\ \exp(-\frac{2R}{L}) \cdot \int g_S(\nu) \cdot g_A(\nu)\ d\nu \tag{4}$$

and $\gamma(t)$ is predicted by Inokuti and Hirayama [7]:

$$\gamma(t) = \frac{4}{3}\Pi C_A \cdot R_0^3 \cdot \gamma^{-3} g\left(\frac{e^\gamma \cdot t}{\tau_S}\right) \quad (5)$$

C_A: A ions concentration; R_0: transfer critical distance at which the probability for radiative and non-radiative transfers are equal; $W_{SA}(R_0) = 1/\tau_{SO}$; $C_0 = 3/4\Pi R_0^3$: critical transfer concentration; $\gamma = 2R_0/L$; L: the effective average Bohr radius; and

$$g(Z) = 6Z \sum_{m=0}^{\infty} \frac{(-Z)^m}{m!(m+1)^4} \quad (6)$$

The non-exponential decay may be written:

$$\Phi(o) = \Phi_0 \exp -\left[\frac{t}{\tau_{SO}} + \gamma^{-3} \cdot \frac{c}{c_0} \cdot g\left(\frac{e^\gamma t}{\tau_S}\right)\right] \quad (7)$$

- Electric multipolar interaction

The dependence of the transfer probability W_{SA} on the S-A distance may be obtained from Dexter [6] on the following form:

$$W_{SA} = \frac{1}{\tau_S} \cdot \left(\frac{R_0}{R}\right)^s \quad (8)$$

s = 6, 8 or 10 for dipole-dipole, dipole-quadrupole and quadrupole-quadrupole interactions respectively. The transfer critical distance is, for instance, for dipole-dipole interaction:

$$R_0^6 = \frac{3}{64\Pi^5} \frac{1}{n^2} \frac{\sigma_A \cdot \eta_{SO}}{\bar{\nu}^4} \int \underset{\text{emission}}{g_S(\nu)} \cdot \underset{\text{absorption}}{g_A(\nu)} \, d\nu \quad (9)$$

η_{SO}: the quantum efficiency in the absence of A ions; n: the refractive index; $\bar{\nu}$: the average wavenumber of the transition.

Now we have a $\gamma(t)$ function written as:

$$\gamma(t) = \frac{4}{3}\Pi C_A R_0^3 \, \Gamma\left(1 - \frac{3}{s}\right)\left(\frac{t}{\tau_{SO}}\right)^{3/s} \quad (10)$$

and the intensity of the S fluorescence decay is as follows:

$\Gamma(1 - \frac{3}{s})$: Euler's function (s = 6, 8 or 10).

The decay is characterized by an initial non-exponential portion followed by an exponential decay with a time-constant τ_{SO}. The analysis of decay curves thus enables the various energy transfer processes to be distinguished. Such process has been seen with the Bi - Eu^{3+} codoped glass (and silicate crystal) and also with $KY_3F_{10}(Eu^{2+})$.

2. <u>With Diffusion Among S Ions</u>. In the previous part we dealt with the direct transfer between S and A neighboring ions but in many doped or fully concentrated materials the transfer between S ions occurs. We distinguish two cases depending on the coupling strength between S-S and S-A ions respectively:

- diffusion limited energy transfer for which both the radiative emission probability of S ions, the diffusion processes probability among S ions and the direct energy transfer probability between S and A ions are comparable.

Yokota and Tanimoto [8] showed that the S excitation density $\Phi(R,t)$ satisfies the diffusion equation by using the same notations:

$$\frac{\partial\Phi(R,t)}{\partial t} = -\frac{1}{\tau_{SO}}\Phi(R,t) - \sum_{A} W_{SA}(R_i)\cdot\Sigma(R,t) + D\nabla^2\Phi(R,t) \qquad (12)$$

D: the diffusion constant.

The main solution of the diffusion-limited energy transfer given for a dipole-dipole interaction takes a complicated form. But, under assumption of uniform distribution of S ions, at long time, $\Phi(t)$ decays exponentially with a time constant τ:

$$\tau^{-1} = \tau_{SO}^{-1} + \tau_D^{-1} \text{ with } \tau_D^{-1} \propto D^{3/4} \qquad (13)$$

- Fast-diffusion energy transfer, for which the diffusion probability among S ions is higher than any other process. This case may be dominant when the S concentration is increasing as in the fully concentrated materials. At long time, the intensity $\Phi(t)$ decays exponentially such as:

$$\frac{1}{\tau_D} \propto C_A \qquad (14)$$

The decays are exponential as when S ions are well-localized without A ions in the host. We shall see the evidence of such mechanism with TbF_3 for which we have also a cross-relaxation process.

III.C. Up-conversion Processes by Energy Transfer

S and A ions are both in their excited states prior to energy transfer. These kinds of processes (two-successive transfers with two or three ions, the cooperative sensitization of luminescence, the cooperative energy transfer, the exciton-exciton annihilation) give rise to anti-Stokes spontaneous emission. They have been reviewed by Auzel [1] and Watts [4]. The main physical criteria used to distinguish between multiple successive transfers are 1) the differences between the energy levels; 2) the pump intensity dependence of the anti-Stokes fluorescence (for instance, a quadratically variation when 2 photons are involving); 3) the S and A ions concentration dependence of the anti-Stokes fluorescence (for instance, a quadratically variation with C_S if 2 S ions are involved and a linear variation with C_A if only one A ion is concerned); 4) the presence of the rise-time of the decay intensity of the anti-Stokes fluorescence and the numerical value of the time constant τ. If the 2 S ions or the 2 S and A ions are identical it is easy to show that τ is equal to half-value of the time-constant of the pumped excited state. More, the absence of the initial rise in the time scale usually experimentated above few nanoseconds characterizes a two-photon absorption on one ion. We shall meet these processes with TbF_3, $LaCl_3(Gd^{3+})$ and $GdCl_3$-$6H_2O$.

III.D. Influence of the Traps in the Materials

We shall point out that some spectroscopic properties mainly in the concentrated materials due to the presence of activator ions perturbated by impurities inside the host. They are assigned to defects or traps or still irregular or perturbated ions by opposite to regular or unperturbated or still intrinsic ions. So we can both of the intrinsic and trap fluorescence but it is clear that energy transfer may occur between the intrinsic and trap energy levels respectively. The model used is presented schematically in Figure 1 with two kinds of traps. The majority of the activator ions forms the exciton band (level 2) and the trap energy level 1 is fed by energy transfer from the intrinsic exciton band. The transfer of excitation between the various levels is described in terms of transition rate W_{ij} which are shown in Figure 1. W_r is the radiative transition probability practically the same for the perturbated and unperturbated ions because the perturbation remains weak. W_q is the transition rate from level 2 to other radiative and non-radiative traps deeper than level 0. The inverse transitions from level 1 to level 2 and from level 0 to level 2 require thermal phonons and their rates are proportional to exp $(-\Delta_{12}/kT)$ and $\exp(-\Delta_{02}/kT)$ respectively. The transition probability W_{02} or W_{12} are temperature dependent so that they may be negligible

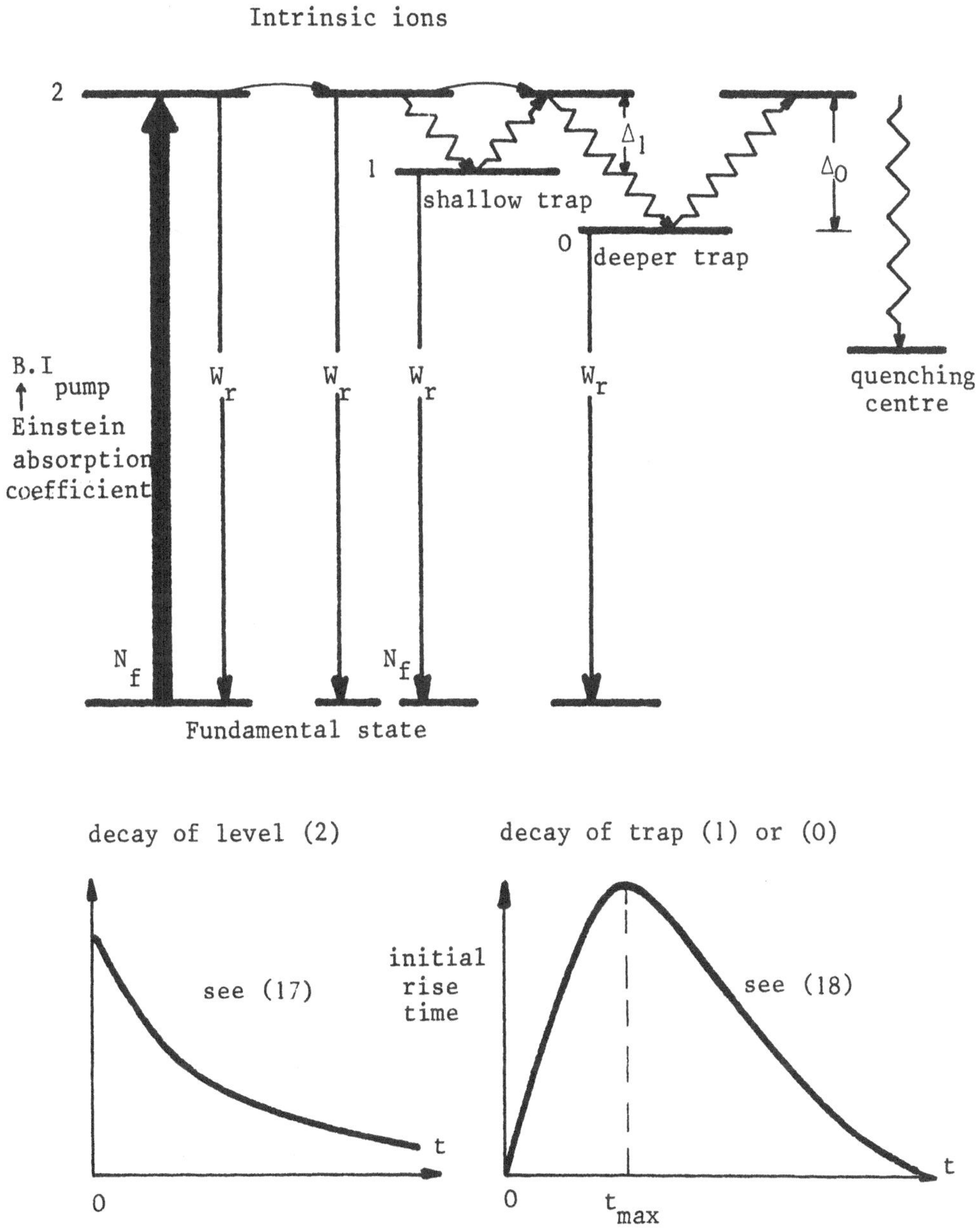

Figure 1. Model for the energy transfer between two kinds of traps.

for a deep trap at low temperature (if $T = 4K$, $kT = 2.8\ cm^{-1}$). We assume also that the excitation trapped by a quenching trap is irretrievably lost for the system.

The temperature dependence of the relative intensity of the trap line emission I_1 with respect to the intensity of the intrinsic line emission I_2:

$$I_2/I_1 = \exp(-\Delta_{12}/kT) \tag{15}$$

From the total probability of deexcitation for trap level 1, we have:

$$\tau^{-1} = \tau^{-1}_{(o)} + K \exp(-\Delta_{12}/kT) \tag{16}$$

K is a constant.

Consequently, the energy gap Δ_{12} (and Δ_{02}) is deduced from the temperature dependences both of the relative intensity and the inverse lifetime τ^{-1} of the trap fluorescence. It is also possible to establish the rate equations of each level and to assume that at $t = 0$ the initial populations are $N_1(0) = 0$, $N_o(0) = 0$ and $N_2(0)$ has a finite value just after the excitation pulse. We get the following expressions of the populations: (where A, B, C, D depend on W_{ij} rate)

$$N_2(t) = \frac{N_2(0)}{m_o}\left\{ A \exp(m_+t) + B \exp(m_-t)\right\} \tag{17}$$

$$N_1(t) = \frac{N_2(0)}{m_o}\left\{ W_{21} \exp(m_+t) - \exp(m_-t)\right\} \tag{18}$$

$$N_o(t) = \frac{N_2(0)}{m_o}\left\{ \begin{array}{l} C(\exp(m_+t) - \exp(-W_rt) + \\ D(\exp(m_-t) - \exp(-W_rt) \end{array}\right\} \tag{19}$$

$$m_o = W_{21} + W_{12} + W_{20} + W_q + 2W_r \tag{20}$$

$$m_+ = -(W_{21} + W_{12} + W_{20} + W_q + W_r) \tag{21}$$

$$m_- = -(W_r + \frac{W_{12}(W_{20} + W_q)}{m_o} - \frac{W_r}{m_o}) \tag{22}$$

At short time, $N_1(t)$ is increasing as $\{1 - \exp(m_-t)\}$ whereas $N_2(t)$ is decreasing as $\exp(m_-t)$ so that we observe an initial rise time on the time dependence of the trap intensity.

IV. ENERGY TRANSFER IN DOPED MATERIALS

We present some examples with doped materials where although the weak concentration the energy transfer may be seen. The diffusion between donor ions can be neglected and the exchange interaction is very weak so that the interaction has probably a multipolar nature.

IV.A. Bi^{3+}-Eu^{3+} Codoped Germanate Glass or $Lu_2Si_2O_7$ Crystal

The excitation spectra of 0.8% Bi^{3+} germanate glass present two broad bands with maxima at 2780 Å and 3190 Å at T = 4.2 K and 2860 Å and 3250 Å at room temperature. By pumping in the first band $^1S_0 \rightarrow {}^3P_1$ with a nitrogen laser, it is easy to observe the broad blue fluorescence band: 4490 Å at 295 K due to $^3P_1 \rightarrow {}^1S_0$ transition and 4650 Å at 4 K due to $^3P_0 \rightarrow {}^1S_0$ transition. At 4.2 K we observe only the $^3P_0 \rightarrow {}^1S_0$ band unsplit by crystal field having a time constant long enough for accurate measurements (τ_0 = 700 μs) [9]. We have selected a subset of ions in the quasi-continuum of multisites in glass, so the fluorescence spectra and the decay curves were obtained for various concentrations of Eu^{3+} acceptor ions up to 5%.

Transfer occurs by radiative mode because the dip in the maximum of the $^3P_0 \rightarrow {}^1S_0$ broad band corresponding to the $^7F_0 \rightarrow {}^5D_2$ forced electric dipolar transition (Figure 2). The radiative transfer efficiency η_R can be approximately calculated by comparing the area of the dip to the area of the emission band included in the spectral range of the $^5D_2$ absorption: η_R = 0.05 with 0.5% Eu^{3+} and η_R = 0.3 with 5% Eu^{3+} in agreement with the Formula (1) of the previous section III.A.1 and η_R is temperature independent. Transfer occurs, too, by non-radiative mode: the integrated fluorescence of Bi^{3+} ion is decreasing when the Eu^{3+} concentration is increasing and, the other hand, when Eu^{3+} ions are added, the Bi^{3+} decays initially deviates from a single exponential dependence as can be seen in the Figure 3 where the Bi^{3+} decay intensity for different Eu^{3+} concentrations are plotted in a logarithmic scale. The final portion of all the decay curves tends to the same slope. These results have been interpreted by a resonant energy transfer process between Bi^{3+} and Eu^{3+} ions in the multipolar case treated by Inokuti and Hirayama (11) without diffusion among donor ions because τ_D^{-1} (13) is negligible at any concentration. The absence of diffusion within the Bi^{3+} ions may be a result of the Stokes shift in the Bi^{3+} ion which means there is no resonance matching between the donor ions. The best fittings of the experimental decays by the relation (11) indicate a dipole-dipole and dipole-quadrupole couplings for high concentrations (3 to 5 % Eu^{3+}) whereas at lower concentration the distance between Bi^{3+} and Eu^{3+} are larger and the transfer is due to dipole-dipole coupling. The critical transfer distance R_0 of the Inokuti and Hirayama's formula is R_0 = 9 A. This parameter has also been evaluated from the Dexter's formula (9) by taking account of n^4 = 6, of the overlap integral and the oscillator strength of the $^7F_0 \rightarrow {}^5D_2$ ($2.76 \cdot 10^{-7}$): we find R_0 = 5,3 Å. This result disagrees with the previous one but it seems not valuable because the presence of the random distribution of the multisites which does not participate to the energy transfer. So, the overlap integral will take a higher numerical value and R_0 will increase. R_0 = 9 Å seem to be a valuable result [10].

Some analogous results were obtained with Bi^{3+} - Eu^{3+} codoped $Lu_2Si_2O_7$ crystal [11]. In this material the $^3P_0 \rightarrow {}^1S_0$ fluorescence band maximum is 3690 Å in uv range and the excitation in the $^3P_1$ level is 2880 Å. The energy levels of the Eu^{3+} ion involving in this energy transfer are $^7F_0 \rightarrow {}^5L_6$, $^5G_{2,3,4,5,6}$, $^5L_7$. We found a dipole-dipole coupling for Eu^{3+} low weight concentration like the glassy phase with R_0 = 9 Å. Unfortunetely it was not possible to growth silicate crystals doped by concentrations higher than 2 % Eu^{3+} to test other interactions.

IV.B. $KY_3F_{10}(Eu^{2+})$

Another interesting case of energy transfer was studied with KY_3F_{10} doped from 0.1 to 3 % Eu^{2+}. Despite a selective doping on K^+ sites, Eu^{2+} ions enter both K^+ (weak field) and Y^{3+} sites (strong field) in the fluoride. The fluorescence consists both of some narrow uv lines around 3583 Å due to $^6P_7 \rightarrow {}^8S_{7/2}$ transition of the K^+ site and a weak violet band ascribed to the $4f^6$ - 5d $^8S_{7/2}$ transition of the Y^{3+} site. As the absorption spectrum of the emission band overlaps totally with these lines, energy transfer between the two kinds of sites has a chance to occur. We find it, under nitrogen laser 3371 Å excitation above T = 150 K because the $4f^6$ - 5d absorption band shifts to higher energy if temperature decreases (Figures 4-a and 4-b). It appears that the decay curve of the principal Stark component of the uv line is non exponential. Concerning the fluorescence, the decay profile is exponential at 4.2 K (τ = 0.55 µs) because we excite only $4f^6$ - 5d energy levels of the Y^{3+} site but above 150 K, the decay becomes non-exponential: a fast one is observed as in the low temperature range and a longer one (a few ms) which seems surprising for such a Laporte rule allowed transition. This long time tail points out definitively the energy transfer between Eu^{2+} in K^+ site (donor ion) and Y^{3+} site (acceptor ion).

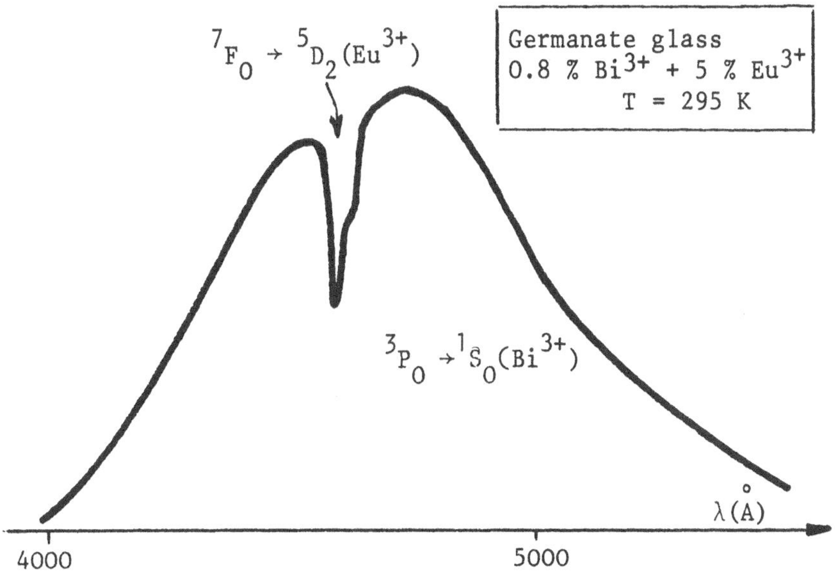

Figure 2. Fluorescence spectrum at 295 K of the germanate glass.

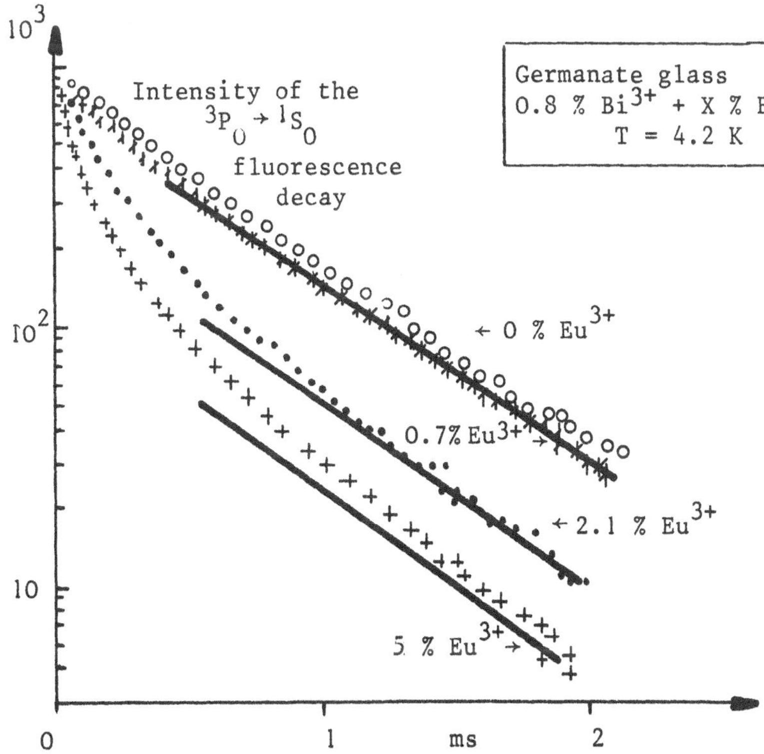

Figure 3. Fluorescence decay at 4.2 K of the germanate glass.

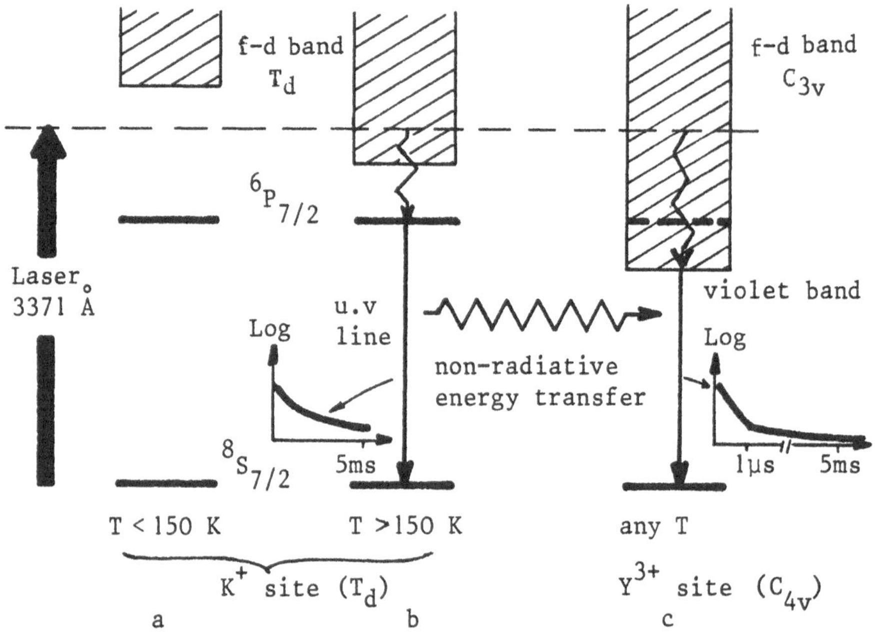

Figure 4. Energy transfer model between Y^{3+} and Eu^{2+} in KY_3F_{10} (Eu^{2+}).

We have used Inokuti and Hirayama's formula (11) to fit the non-exponential decay of uv line by assuming a dipole-dipole coupling because the Eu^{2+} ions are very diluted inside the host. The best fitting is reached with R_0 = 11 Å, τ_{so} = 2.03 ms and C_A = 0.3 C where C is the Eu^{2+} initial concentration in good agreement, with experimental τ at 4.2 K equal to 1.98 ms, with the reasonable assumption of the absence of a selective doping only in K^+ site and also with the nature of the involved transitions [12].

IV.C. $LaCl_3$ - Gd^{3+}

The spectroscopy study of Gd^{3+} ions doped materials as $LaCl_3$ is particularly interesting due to the characteristics of this rare-earth activator: i) a high energy of Stark components of the first excited $^6P_{7/2}$ level (3.114,4 Å; 3112,8 Å; 3111,6 Å; 3111,3 Å in $LaCl_3$) excluding any non-radiative transitions to the $^8S_{7/2}$ ground state, contrary to spectra of other rare-earths which exhibit numerous multiplets in the visible range. ii) the relatively long time constant of $^6P_{7/2}$ (a few ms) which may be used to probe the energy transfer by an accurate analysis. iii) the very weak splitting of the $^8S_{7/2}$ ground state (<0.2 cm^{-1}). So the observed fluorescence lines are related to the excited level structure and at low temperature the splitting of $^6P_{7/2}$ level gives information about the environment of Gd^{3+} ions. In $LaCl_3$, we find three unequivalent sites. They could correspond to perturbation on at least one of the Cl^- ion of the three types of Cl^- ions around Gd^{3+} [13]. Site selection spectroscopy and time resolved spectroscopy show clearly an energy transfer between these sites (site I: 3114,4 Å; site II: 3114,6 Å; site III: 3114,9 Å) with very near time constant values (site I: τ = 5.4 ms; site II: τ = 6.4 ms; site III: τ = 4 ms). For instance, under the excitation in the site I (32100 cm^{-1}) the decay intensity of the site II exhibits an initial rise time maximum at 0.4 ms which may be analyzed with a three-level scheme like f, 1 and 2 levels of Figure 1, where the 1 and 2 levels belong to different Gd^{3+} sites. Another fascinating experimental result was recorded: under pumping in the $^6P_{7/2}$ level (site I), an intense anti-Stokes emission $^6I_{7/2} \rightarrow {}^8S_{7/2}$ has been pointed out at T = 4.2 K. The integrated fluorescence intensity depends quadratically on the excitation power of the laser beam confirming a two-photon mechanism either on the same ion or on two different ions (see section III.C) involving a high excited level at 2 x 32100 cm^{-1} of the f-d configuration giving rise to a Laporte's rule allowed transition ($^6P_{7/2} \rightarrow$ f-d). However, we can distinguish them by the time constant and the initial rise of the anti-Stokes emission from $^6I_{7/2}$ level populated by the non-radiative relaxations (see section III.C). In our experiments, the decay of $^6I_{7/2}$ was measured in different parts of the fluorescence line under excitation on the site I. We observe an initial rise time (t_{max} = 0.55 ms) different from those found previously except for the higher energy wing and we add that the time constants are equal to 4.5 ms, thus higher than the half-time constant of $^6P_{7/2}$. Accordingly, it is difficult to choose between the two models more specially as the intensity at t = 0 is far from zero. Probably, the two mechanisms act simultaneously giving a resulting decay and its complexity can also be increased by energy transfer processes between the three types of sites.

V. ENERGY TRANSFER IN STOICHIOMETRIC MATERIALS

Substantial efforts have been dedicated in our laboratory to the investigation of the luminescence properties of the fully concentrated solids. One generally observes a diminution of the radiative quantum yield when the active ion concentration increases, a very annoying effect which severely limits the number of materials which can be efficiently used for lighting or laser applications. From a fundamental point of view, various processes seem to be responsible for concentration quenching, most of them being related to the existence of very efficient energy transfer between the active ions themselves and between these active ions and uncontrolled centers acting as non-radiative sinks (crystallographic defects non-luminescent chemical impurities noted traps in the previous section III.D). The experience shows, however, that reality needs more information and the necessity appears more crucial when one deals with some fully concentrated materials in which strong luminescence is observed.

Our group started this investigation with a heavy ion concentrated crystal $Bi_4Ge_3O_{12}$ (BGO) [14]. This now well-known crystal (new scintillator) presented some very particular optical properties difficult to interpret. The absorption bands of the crystal were at higher energies than a usual Bi^{3+} doped compound and the fluorescence time constant increased asymptotically when one lowered the temperature below 10 K instead of remaining constant due to the influence of the $^3P_0$ metastable level. The absorbing centers could not be interpreted any longer in terms of isolated ions. We have to introduce bismuth germanate clusters in which the excitation is self-trapped at low temperature and delocalized in the crystal at the benefit of fluorescing traps when the temperature is high enough. Because we dealt with broad bands we could not distinguish clearly between the various spectral features; therefore no further proofs could have been added to our qualitative interpretations. That is why we turned our attention to the fluorescence of fully concentrated solids made with transition or rare-earth ions such as Mn^{2+}, Tb^{3+} or Gd^{3+} and we could work with well-defined sharp spectra. A full lecture has been presented by one of us at the last Erice's school about manganese compounds as $CsMnF_3$, $BaMnF_4$ and $RbMnCl_3$ [15] and we cannot present them again but we would like to comment.

V.A. Manganese Compounds

In antiferromagnetic manganese systems the energy transfer processes are exclusively allowed by short range magnetic interactions, that is to say superexchange interactions between ferromagnetically coupled near neighbor Mn^{2+} ions. This has been proved in various cases by looking at the fluorescence properties of these systems when submitted or not to an external magnetic field. For example, in $CsMnF_3$, a three-dimensional system where only third near neighbors are coupled by ferromagnetic exchange, the applied magnetic field changes the nearest neighbor interactions from antiferromagnetic to slightly ferromagnetic which removes the magnetic selection rule and extends the delocalization of the optical excitations to all these ions. As expected, this effect manifests in fluorescence by a reduction of the emission intensity and decay time of the donor system formed by the intrinsic Mn^{2+} ions at the benefit of the acceptor traps; quenching traps as well as fluorescing traps (see section III.D). These shallow traps are Mn^{2+} ions sites perturbated by first, second and third nearest neighbor impurities such as Mg^{2+}, Zn^{2+}, Ca^{2+} and the deeper traps are formed by impurities such as Ni^{2+} or Fe^{3+} ions in concentration of a few ppm. With $RbMnCl_3$, it has been shown that the entire fluorescence is now seen to be essentially intrinsic without trap emission and it was possible to conclude to exchange character of the dominant interactions responsible for the energy transfers in concentrated Mn^{2+} systems. Moreover, the results indicate a complete inefficacy of the electric multipolar mechanisms which could compete when exchange type energy transfers are forbidden by the selection rules [16].

V.B. Rare-earth Compounds

First, we have studied the energy transfer mechanisms in TbF_3. The continuous (CW) and time-resolved spectra (TRS) of the blue fluorescence $^5D_4 \rightarrow {}^7F_6$ (20604 cm^{-1}; 4852.1 Å) under selective dye laser excitation in the lowest Stark component of the $^5D_4$ manifold, show the presence of both intrinsic and impurity induced trap emission lines (see Figure 1 of section III.D). At 4.2 K most of the emission originates from traps while at 10 K the intrinsic fluorescence is stronger. The thermal behavior indicates the presence of an efficient transfer between unperturbed ions and perturbed Tb^{3+} ions. On the basis of the experimental results a fast diffusion model (see section III.D) including shallow traps (N_1 of Figure 1) and deeper traps (N_0 of Figure 1) is proposed. The analysis of temperature dependence of the exponential tail of the deeper trap fluorescence decay, from 4.4 to 7 K, we have deduced an energy gap Δ_{02} = 35 cm^{-1} (see formula (16)) that may be compared favorably with the spectral energy separation of 40 cm^{-1}. In the same way, we get $\Delta_{12} \simeq 10$ cm^{-1}. The whole fitting of the intrinsic fluorescence decay from two level with the formula (17), we obtain $m_- = -5586$ s^{-1} and $m_+ = -250$ s^{-1}. As a proof of the model, these values m_+ and m_- were used to describe the decay curve of the trap with the aid of Eq. (19). The initial rise time as well as the decay curve were very well-fitted indicating a quite satisfactory agreement. In the fast diffusion regime the donor exciton decay should be essentially exponential with a rate constant τ given by formula (13) which applies here, τ = 179 μs. This allows us to estimate an isotropic diffusion coefficient $D = 8.6.10^{-10}$ cm^2 s^{-1} at 4.4 K [17].

By exciting in the $^5D_4$ manifold, we observe a uv anti-Stokes fluorescence from the upper $^5D_3$ level (26386 cm^{-1}; 3788.7 Å). The whole excitation spectrum of the most intense uv fluorescence line compares very well with the excitation spectrum of the blue fluorescence originating from the $^5D_4$ level and its integrated emission intensity depends first quadratically on the excitation pump power for low $^5D_4$ exciton densities, then, linearly. The model which is proposed to interpret the anti-Stokes fluorescence is based on both the exciton-exciton annihilation and the cross-relaxation process between $^5D_3 \rightarrow {}^5D_4$ (5725 $\pm$ 60 cm^{-1}) and $^7F_6 \rightarrow {}^7F_1$ (5703 $\pm$ 97 cm^{-1}) due to good energy matching. So, when the intrinsic Tb^{3+} ions are excited into the lower $^5D_4$ Stark component, the $^5D_4$ exciton density may be sufficient to allow exciton-exciton interaction. The coupled $^5D_4$ excitons then annihilate to create a biexciton having twice their energy (41208 cm^{-1}) which corresponds to the absorption tail of the excited configuration $4f^7 - 5d$. After a rapid relaxation to the $^5D_3$ level the $^5D_3$ level population is depleted both by cross-relaxation process (W_{cr}) mentioned above and by radiative process to ground state. Taking into account the deexcitation probabilities W_2 and W_3 of the levels $^5D_4$ and $^5D_3$, one can write the following rate equations (see Figure 5) [18],[19].

$$\frac{dN_2}{dt} = N_f \cdot B \cdot I - W_2 N_2 - W_T N_2^2 + W_{cr} N_3 N_f$$

$$\frac{dN_3}{dt} = \frac{1}{2} W_T \cdot N_2^2 - W_{cr} \cdot N^3 \cdot N_f - W_3 N_3$$

In the low pumping regime and assuming $W_3 << W_{cr}$ the $^5D_3$ population is:

$$N_3(t) = \frac{W_T N_2^2(o)}{W_{cr} N_f - 2W_2} \left\{ \exp^{-2W_2 t} - \exp^{-W_{cr} N_f \cdot t} \right\}$$

This predicts at long times, an exponential decay with a time constant half that of the intrinsic blue fluorescence. We find τ_{uv} = 89 μs which agrees fairly well with the half value of the $^5D_4$ exciton τ_{blue} = 179 μs as predicted above. In addition, the time t_{max} = 0.5 μs at which the uv fluorescence intensity reaches its maximum nicely correlates with:

$$t_{max} = \frac{1}{W_{cr} N_f - 2W_2} \cdot \mathrm{Log} \frac{W_{cr} N_f}{2W_2}$$

Here, the short rise time behavior is related to the fast cross-relaxation process between $^5D_3$ and $^5D_4$ and the observed decay time of the $^5D_3$ level is τ = 70 ns.

$GdCl_3 - 6H_2O$ is another concentrated material analyzed by the group. We have encountered some difficulties because the hygroscopic properties of this crystal but we were able to observe similar results with those presented previously. There are a lot of Gd^{3+} perturbed ions playing an efficient trapping effect below 20 K. We see the perturbed ions sharp lines in the lower energy wing from 3118 Å to 3130 Å of the $^6P_{7/2} \rightarrow {}^8S_{7/2}$ intrinsic emission located at 3117.5 Å. Their decays show different initial rise times with a gap energy dependence (see Figure 1). An anti-Stokes fluorescence has also been pointed out from $^6I_{7/2}$ levels under pumping in $^6P_{7/2}$ levels of perturbed ions as unperturbed ions illustrating the easy way of the two-photon absorption mechanism. At last, we must add that the diffusion process seems weaker than

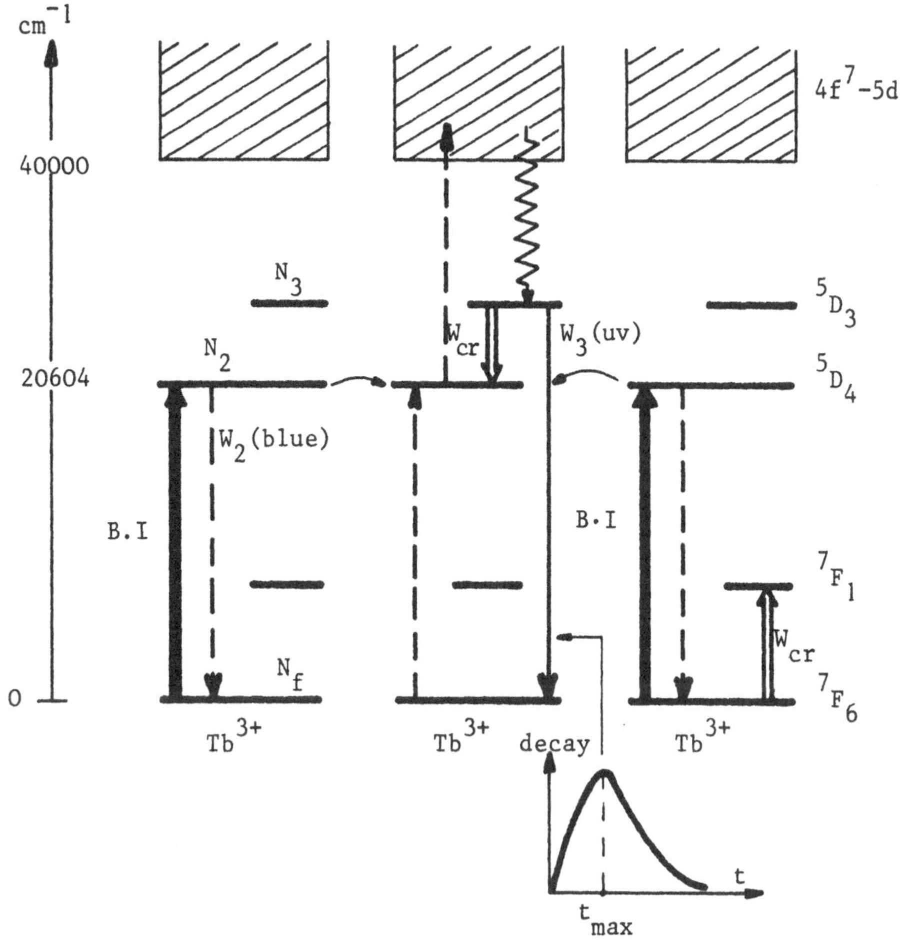

Figure 5. Energy transfer mechanism in TbF_3.

in TbF_3 host. Actually, the studies of the spectroscopic properties of $GdCl_3$ fully concentrated material are in progress in order to determine the nature of the interactions between the Gd^{3+} perturbed and unperturbed ions.

VI. SUMMARY

We have presented various results obtained in our laboratory on the energy transfer processes in doped and stoichiometric materials using the new laser spectroscopy techniques. Now, the mechanisms may be studied as well for the Stokes fluorescence as anti-Stokes fluorescence which is still not very well understood. Further works are needed to know the exact electronic interactions between active ions, then, to give a better description of the dynamical properties as the energy transfer in the hosts.

ACKNOWLEDGEMENTS

I want to dedicate this article to Madame F. Gaume, Director of our group Physico-Chimie des Matériaux Luminescents in Lyon. I wish to acknowledge with thanks all members of this group working on the energy transfer research for several years (B. Jacquier, R. Moncorgé, B. Moine, M. F. Joubert, J. C. Bourcet). I am indebted to C. Madej, M. Blanchard and A. Lagriffoul for their helpful technical assistance.

REFERENCES

1. F. Auzel, Proc. IEEE 61, 758 (1973); Radiationless Processes (ed. B. Di Bartolo), Plenum Press (1980).
2. R. Reisfeld, Structure and Bonding 13, 53 (1973); 30, 65 (1976).
3. L. A. Riseberg and M. J. Weber, Progress in Optics (ed. E. Wolf) 14 (1976).
4. R. K. Watts, Optical Properties of Solids (ed. B. Di Bartolo), Plenum Press (1975).
5. T. Förster, Ann. der Physik 2, 55 (1948).
6. D. L. Dexter, J. Chem. Phys. 21, 836 (1953).
7. M. Inokuti and F. Hirayama, J. Chem. Phys. 43, 1978 (1965).
8. M. Yokota and O. Tanimoto, J. Phys. Soc. Japan 22, 779 (1967).
9. G. Boulon, B. Moine, J. C. Bourcet, R. Reisfeld, Y. Kalisky, J. Luminescence 18-19, 924 (1979).
10. B. Moine, J. C. Bourcet, G. Boulon, R. Reisfeld, Y. Kalisky, J. de Physique 42, 499 (1981).
11. F. Bretheau-Raynal, Thèse de Doctorat ès-Sciences, Université Paris VI, 22 mai 1981.
12. M. F. Joubert, G. Boulon, F. Gaume, Chemical Physics Letters 80, 367 (1981).
13. C. Linarès, Ph. Jung, G. Boulon, F. Gaume, J. of Less Common Metals (to be published).
14. R. Moncorgé, G. Boulon and B. Jacquier, J. of Luminescence 12-13, 467 (1976) and 14, 337 (1976).
15. R. Moncorgé and B. Jacquier, Collective Excitations in Solids (ed. B. Di Bartolo), Plenum Press (1982).
16. R. Moncorgé, B. Jacquier, C. Madej, M. Blanchard, J. de Physique 43, 1267 (1982).
17. M. F. Joubert, B. Jacquier, R. Moncorgé, G. Boulon, J. de Physique 43, 893 (1982).
18. M. F. Joubert, B. Jacquier, R. Moncorgé, G. Boulon, Solid State Chemistry 1982, Proc. 2nd European Conference, 6-7 June 1982, Veldhoven (Netherlands).
19. M. F. Joubert, B. Jacquier, R. Moncorgé, Phys. Rev. B (to be published).

DYNAMICAL MODELS OF ENERGY TRANSFER IN CONDENSED MATTER

J. Klafter

Corporate Research Science Laboratories
Exxon Research and Engineering Company
Clinton Township
Annandale, N.J. 08801, U.S.A.

and

A. Blumen

Lehstuhl für Theoretische Chemie
Technische Universität München, Lichtenbergstr 4
D-8046 Garching, West Germany

ABSTRACT

We present a study of direct and of indirect energy transfer in disordered materials. In the direct case we focus on the ensemble averaged survival probability of an excited donor. The indirect transfer is treated in the framework of the continuous time random walk (CTRW) for which we derive the distribution of stepping times from the survival probabilities of the direct transfer. This distribution of stepping times is used to mimic the random walk on a disordered system. Based on the CTRW, we obtain the diffusion coeffficient D(t) for a trap-free random system; furthermore we introduce trapping centers and establish the decay laws due to trapping. We find that trapping shows a rich time dependence and that the onset of the asymptotic behavior for diffusion and for trapping may occur on widely different time scales.

I. INTRODUCTION

Time dependent energy transfer among impurity molecules in

condensed media has recently been of considerable interest both experimentally and theoretically [1]-[6], since time dependent studies provide valuable insight in understanding the various mechanisms of excitation transfer among the impurities. Although we refer throughout this paper only to energy transfer, the theory accounts well also for problems of charge transport and recombination in amorphous materials [7]-[12]. Important examples of energy transfer are the electronic [1]-[6], the vibrational energy transfer[13] and the spin migration[14].

The energy transfer exhibits two limiting cases: on the one hand the direct, and on the other hand the indirect, multistep, mechanism. In the direct transfer case an excited donor transfers its energy directly to randomly distributed acceptors. One is then interested in the ensemble averaged decay law of the excited donor [3],[15],[16],[17]. This decay law can be expressed exactly for acceptors randomly distributed on a lattice or in a liquid, if the back-transfer to the donor is unimportant[17]. The indirect transfer problem where the excitation randomly hops in many steps over active sites in a trap free system, (eventually until encountering a trap), is more complicated [3],[4],[8],[9],[18].

In this paper we will consider a random environment which arises from the substitutional occupancy by active centers of sites on an ordered lattice. The random walk will then take place over the active centers only. When treating trapping by impurities, these will be taken to replace active centers. The trapping problem will thus include two types of disorder: the randomness of the carrier medium (active centers) and the random distribution of traps. In the diffusion problem one, however, faces only the first type of disorder, namely that due to the random distribution of the active centers. This can be viewed as a problem of a random walk on random lattices. In this case the situation is considerably more complex than for regular lattices[18] and one is forced to make use of approximating schemes.

An approach which has been shown to be powerful for the general problems of the multistep transport in disordered systems is the continuous time random walk (CTRW). The CTRW ideas were suggested by Montroll and Weiss[19] who introduced the distribution function $\psi(t)$ into the usual random walk formalism to discuss a random walk with random stepping times. This approach was elaborated later by Scher and Lax [8],[9] who applied the CTRW to transport in disordered systems by generalizing the functions $\psi(t)$ to forms $\psi(\underset{\sim}{r},t)$ which couple the spatial and temporal disorder aspects. Applications of a decoupled scheme (i.e. $\psi(\underset{\sim}{r},t) = f(\underset{\sim}{r})\psi(t)$) was also proposed by Scher and Montroll[10]. The connection of the CTRW to the master equation, which is the basic equation to describe the transport of a particle in a random medium was discussed by Bedeaux, Lakatos-Lindenberg and Shuler [20] and the connection to the generalized master equation was pointed out by Kenkre, Montroll and Shlesinger[21]. This later connection was studied only in a decoupled scheme, which corresponds to the above mentioned Scher-Montroll approximation [10]. Recently Klafter and Silbey [22],[23] proved that the original random walk problem on a random lattice can be rigorously reduced to a CTRW formulation of the Scher-Lax types. The derivation makes use of projection operator techniques, and $\psi(\underset{\sim}{r},t)$ is related to a self-energy (or a memory) function. However, the calculation of the self-energy is difficult and one has to look for the best possible approximations. Our purpose here is to provide concise framework for the problems of energy diffusion and trapping in random sysems in terms of the CTRW by starting from a single site type of approximation. We focus on the ensemble averaged probability $\Psi(t)$ that no transfer from an initially occupied active center occurs during the time t[17]. As we will show in Sec. II this quantity can be used to obtain forms for the stepping time distributions $\psi(t)$ and $\psi(\underset{\sim}{r},t)$. These approximate forms were used quite successfully to mimic the random walk on a random system [8],[9] for which the exact $\psi(\underset{\sim}{r},t)$ is not known. In Sec. III we will treat the transport in the absence of traps and will establish the diffusion coefficient D(t) which follows from $\psi(\underset{\sim}{r},t)$ in the CTRW framework. In Sec. IV we introduce the traps and calculate the CTRW decay law due to trapping showing that different behaviors for the decay obtain for different classes of stepping time distributions $\psi(t)$.

II. DIRECT TRANSFER

In this section we present the ensemble averaged transfer from a donor molecule to randomly distributed acceptor molecules. We start our considerations by assuming that the positions of the donor and of the acceptors are fixed, so that the acceptors form a certain configuration $\underset{\sim}{K}$ around the donor. The probability $(\underset{\sim}{K};t)$ that the donor did not transfer its energy during time t to any of the acceptors is exponential

$$\Psi(\underset{\sim}{K};t) = \exp[-t \sum_{\underset{\sim}{r}\varepsilon \underset{\sim}{K}} w(\underset{\sim}{r})] = \prod_{\underset{\sim}{r}\varepsilon \underset{\sim}{K}} \exp[-tw(\underset{\sim}{r})] \tag{1}$$

In Eq. (1) $w(\underset{\sim}{r})$ denotes the transfer rate to the acceptor at position $\underset{\sim}{r}$ and the sum and product extend over all acceptors. Often encountered forms for $w(\underset{\sim}{r})$ are

$$w(\underset{\sim}{r}) = \frac{1}{\tau}\frac{d^s}{r^s} \tag{2}$$

for multipolar interactions and

$$w(\underset{\sim}{r}) = \frac{1}{\tau} \exp\,[\gamma(d - r)] \tag{3}$$

for exchange interactions. The parameter s in Eq. (2) equals 6 for dipole-dipole, 8 for dipole-quadrupole and 10 for quadrupole-quadrupole interactions. The quantity γd in Eq. (3) is a measure of the range of the exchange interaction and is larger for shorter range interactions. We have chosen to scale (2) and (3) in terms of the transfer rate τ^{-1} of the donor-acceptor pair at nearest distance d. Forms (2) and (3) are only approximations to realistic transfer rates which may have a considerably more complex structure.

The quantity of interest experimentally is not $\Psi(\underset{\sim}{K},t)$, which is due to a particular acceptor configuration, but the ensemble average of Ψ over all acceptor configurations

$$\Psi(t) = \sum_{\underset{\sim}{K}} p(\underset{\sim}{K})\Psi(\underset{\sim}{K},t) \tag{4}$$

where $p(\underset{\sim}{K})$ denotes the probability with which configuration $\underset{\sim}{K}$ occurs.

We now consider a system in which active molecules ocupy the lattice sites in a random, non-correlated way with probability p. Eq. (1) thus reads with the donor located at the origin:

$$\Psi(\underset{\sim}{K};t) = \exp[-t \sum \zeta(\underset{\sim}{r})w(\underset{\sim}{r})] \tag{5}$$

where $\zeta(\underset{\sim}{r})$ are random variables which take only the values 1 and 0 with probability p and 1-p respectively, depending on whether the site at distance $\underset{\sim}{r}$ is occupied or not. The configurational average of Eq. (5) is (see Ref. (24) for other derivations):

$$\begin{aligned}\Psi(t) \equiv \langle\Psi(\underset{\sim}{K};t)\rangle_{\{\zeta(\underset{\sim}{r})\}} &= \langle\Pi'\exp[-t\zeta(\underset{\sim}{r})w(\underset{\sim}{r})]\rangle_{\{\zeta(\underset{\sim}{r})\}} \\ &= \prod_{\underset{\sim}{r}}{}'\langle\exp(-t\zeta(\underset{\sim}{r})w(\underset{\sim}{r})]\rangle_{\zeta(\underset{\sim}{r})} \\ &= \prod_{\underset{\sim}{r}}{}'\ [1-p + pe^{-tw(\underset{\sim}{r})}]\end{aligned} \tag{6}$$

Here use was made of the fact that the $\zeta(\underset{\sim}{r})$ are uncorrelated. In

Eq. (6) the product extends over all lattice sites (excluding the origin) and the final form depends only on p and not on the configurations $\underset{\sim}{K}$. Eq. (6) is exact[17]. It is valid for all times and all concentratins of active molecules and it explicitly includes the structure of the underlying lattice. It also holds regardless of the particular form of $w(\underset{\sim}{r})$ and of the detailed structure of the transfer law for each donor-acceptor pair, $\exp[-tw(\underset{\sim}{r})]$.

Eq. (6) gives the time dependent survival probability. There are however situations where one is interested in the ensemble averaged radiative efficiency η[16]. For example, much interest has recently arisen in the problem of luminescence quenching due to non-radiative tunneling in amorphous semiconductors[25,26]. In these luminescence quenching studies the excitation is initially localized at a radiative center from which it can either radiatively decay with rate τ_0^{-1} (τ_0 is the radiative lifetime) or tunnel to a non-radiative center. The problem is a direct transfer problem, but we have to add the radiative lifetime to our earlier considerations. using Eq. (1) the modified decay N(t) for a particular configuration $\underset{\sim}{K}$ of acceptors (or non-radiative centers) is

$$N(t) = \exp(-t/\tau_0)\Psi(\underset{\sim}{K};t) \tag{7}$$

The radiative efficiency η is the branching ratio

$$\eta = \frac{\tau_0^{-1}}{\tau_0^{-1} + \sum_{\underset{\sim}{r}\in\underset{\sim}{k}} w(\underset{\sim}{r})} \tag{8}$$

We are now interested in the value of η averaged of all possible acceptor configurations. We notice that η and $\Psi(\underset{\sim}{K};t)$ are related through a Laplace transform. The Laplace transform of $\Psi(\underset{\sim}{K};t)$ is

$$\Psi(\underset{\sim}{K};u) = \int_0^\infty dt \exp(-ut)\, \Psi(\underset{\sim}{K};t) = \left[u + \sum_{\underset{\sim}{r}\in\underset{\sim}{K}} w(\underset{\sim}{r})\right]^{-1} \tag{9}$$

and, therefore

$$\eta = \tau_0^{-1} \left[\Psi(\underset{\sim}{K};u)\right]_{u=\tau_0^{-1}} \tag{10}$$

Since both the Laplace transform and the ensemble averaging are linear operators, it follows that

$$\langle n\rangle = \tau_o^{-1}\,[\langle L\Psi(\underset{\sim}{K};t)\rangle]_{u=\tau_0^{-1}} = \tau_0^{-1}[L\Psi(t)]_{u=\tau_0^{-1}} \tag{11}$$

where L represents the Laplace transform. Eq. (11) directly relates the averaged quantities $\langle n\rangle$ and $\Psi(t)$. Thus, if $\Psi(t)$ is known, $\langle n\rangle$ follows by a simple integration[27].

Under the assumption of a low acceptor concentration a number of approximate forms for the donor transfer law have been derived using different procedures [15],[16],[28],[29]. We retrieve now these forms from the exact equations by taking the logarithm of both sides of Eq. (6) and by expanding with respect to p. From Eq. (6) [24],[30]:

$$\begin{aligned} \ln \Psi(t) &= -\sum_{\underset{\sim}{r}} \ln\,(1-p\,\{1-\exp\,[-tw(\underset{\sim}{r})]\}\,) = \\ &= -\sum_{\underset{\sim}{r}} \sum_{k=1}^{\infty} \frac{p^k}{k}\,\{1-\exp\,[-tw(\underset{\sim}{r})]\}^k \end{aligned} \tag{12}$$

where the last line is valid for small p ($p < 1$). For a densely packed medium the sum in Eq. (12) may be replaced by an integral (continuum approximation), thus obtaining for very small p, $p \ll 1$

$$\Psi(t) = \exp\,(-p\rho\int\{1-\exp\,[-tw(\underset{\sim}{r})]\}\,d\underset{\sim}{r}) \tag{13}$$

where ρ is the density of molecules in the medium. Thus $p\rho$ is the concentration of active molecules. Interestingly, the approximate expression (13) occurs frequently in many fields [9],[10],[15]-[17], [28]-[30].

For isotropic interactions $w(\underset{\sim}{r})$ we have, therefore, [17]

$$\Psi(t) = \exp\{-\Delta V_\Delta \rho p\int[1-\exp(-tw(r))]r^{\Delta-1}dr\} \tag{14}$$

where Δ is the dimensionality of the lattice and V_Δ the volume of a unit sphere in Δ-dimensions. We insert now the forms (2) and (3) into Eq. (14) and obtain for isotropic multipolar interactions[24]

$$\Psi(t) = \exp[-V_\Delta p\rho d^\Delta \Gamma(1-\Delta/s)\ (t/\tau)^{\Delta/s}] \tag{15}$$

where $\Gamma(x)$ is Euler's gamma function, and for isotropic exchange interactions[31]

$$\Psi(t) = \exp\left[-V_\Delta p\rho\gamma^{-\Delta} g_\Delta\left(\frac{t}{\tau}\exp(\gamma d)\right)\right] \tag{16}$$

where the function $g_\Delta(z)$ is defined through

$$g_\Delta(z) = \Delta\int\{1-\exp[-z\exp(-y)]\}y^{\Delta-1}dy \tag{17}$$

A list of the properties of the function $g_\Delta(z)$ is given in the appendix of Ref. (31). Formula (15) for $\Delta=3$ and s=6 was derived in the energy transfer field by Förster[15], for $\Delta=3$ and s arbitrary the result was found in Inokuti and Hirayama[16].

An interesting extension of the results in Eqs. (14) - (17) obtains for structures which are not regular but have a fractal behavior[32]. Fractal structures are characterized by non-integer dimensionalities. It has recently been suggested that many disordered materials appear to be fractals; for instance linear and branched polymers [33],[34], and epoxy resins [35]. On fractal structures the density of sites is not constant but length dependent[32]. This means that we have to replace the constant density in Eq. (13) by $\rho(r)$, where [32]-[34]

$$\rho(r) \sim r^{\bar{\Delta}-\Delta} \tag{18}$$

$\bar{\Delta}$ is the fractal dimension[32], and Δ is the space in which the fractal structure is embedded $\bar{\Delta} \leq \Delta$. The survival probabilities for direct transfer to acceptors randomly distributed on fractals are then:

$$\Psi(t) = \exp(-pAt^{\bar{\Delta}/s}) \tag{19}$$

for multiple interactions, and

$$\Psi(t) = \exp\left(-pBg_{\bar{\Delta}}(Ct)\right) \tag{20}$$

for exchange interaction. A, B, and C are independent of time[36].

Another useful extension of Eq. (14)ff is the decay law of the donor due to transfer to moving acceptors. For acceptors moving along given paths $\underset{\sim}{r}(t)$ the decay due to a particular acceptor configuration is[37]

$$\Psi(\underset{\sim}{K};t) = \prod_{\underset{\sim}{r}\in\underset{\sim}{K}} \exp\left[-\int_0^t w[\underset{\sim}{r}(t)]dt\right] \tag{21}$$

The average over the different molecular distributions follows as above[37]:

$$\Psi(t) = \prod_{\underset{\sim}{r}}{}' \left[(1-p) + p\exp\left[-\int_0^t w[\underset{\sim}{r}(t)]dt\right]\right] \tag{22}$$

Eq. (6) is a special case of Eq. (22), obtained by keeping all $\underset{\sim}{r}(t)$ fixed. Through $\underset{\sim}{r}(t)$ the exact decay law (22) is explicitly motion dependent. For acceptors diffusing in a liquid the decay law is given by Eq. (22) averaged over all paths $\underset{\sim}{r}(t)$ accessible to the liquid molecules. If the energy transfer is rapid on the time scale of the particle motion, the decay law for small p (p << 1) is given through[37]

$$\Psi(t) = \exp\left[-p\rho\int[1-\bar{E}(t,\underset{\sim}{r})]d\underset{\sim}{r}\right] \tag{23}$$

where $\bar{E}(t,\underset{\sim}{r})$ is the solution of the differential equation

$$\frac{\partial\bar{E}}{\partial t} = \left[D\nabla^2_{\underset{\sim}{r}} - w(\underset{\sim}{r})\right]\bar{E} \tag{24}$$

with initial condition $\bar{E}(o,\underset{\sim}{r}) = 1$. In Eq. (24) D is the sum of the diffusion coefficients of the donor and the acceptor and $\nabla^2_{\underset{\sim}{r}}$ is the Δ-dimensional Laplace operator. In the case of isotropic multipolar interactions, Eq. (2), the decay law turns out to be[37]

$$\Psi(t) = \exp\left[-V_\Delta p\rho d^\Delta\Gamma\left(1-\frac{\Delta}{s}\right)\left(\frac{t}{\tau}\right)^{\Delta/s} + V_\Delta p\rho\sum_{n=1}^{\infty} C_n D^n t^{n-2n/s+\Delta/s}\right] \tag{25}$$

The first term of Eq. (25) reproduces Eq. (15); the first three coefficients C_1, C_2, C_3 are given explicitly in Ref. [37] as functions of Δ and s and reduce for $\Delta=3$ and s=6 to the expressions of Yokota and Tanimoto[38].

III. DIFFUSION IN THE FRAMEWORK OF THE CTRW

We now consider the indirect multistep transfer over a disordered system in the CTRW framework. We follow the Scher-Lax model[8],[9], namely the transport occurs on an ordered lattice, while the disorder is contained in the distribution of stepping times $\psi(t)$. We will use the expression calculated in Sec. II for $\Psi(t)$ in order to derive the distribution of stepping times $\psi(t)$. We will follow the approximation suggested by Scher and Lax [8],[9] to relate $\psi(t)$ to $\Psi(t)$:

$$\psi(t) = -\frac{d}{dt}\Psi(t) \tag{26}$$

From Eq. (6) follows,

$$\psi(t) = \sum_{\underset{\sim}{r}} pw(\underset{\sim}{r})e^{-tw(\underset{\sim}{r})}\{\prod_{\substack{\underset{\sim}{r}'\neq\underset{\sim}{r}\\ \underset{\sim}{r}'\neq\underset{\sim}{0}}}[1-p+pe^{-tw(\underset{\sim}{r}')}]\} \tag{27}$$

$$\equiv \sum_{\underset{\sim}{r}} \psi(\underset{\sim}{r},t)$$

In Eq. (27) $\psi(\underset{\sim}{r},t)$ is the probability of an excitation jumping a distance $\underset{\sim}{r}$ with stepping time t. In the continuum approximation and for low p, $p \ll 1$,

$$\psi(\underset{\sim}{r},t) \simeq pw(\underset{\sim}{r})\exp[-tw(\underset{\sim}{r})]\,\Psi(t) \tag{28}$$

where we have replaced the product in Eq. (27) by its form (6). Interestingly, Eq. (28) represents an example for a coupled stepping time distribution [8],[9],[39],[40].

Within this model the probability $P(\underset{\sim}{r},t)$ of finding the walker at site $\underset{\sim}{r}$ at time t is given in terms of its Fourier and Laplace transforms [8],[22]:

$$P(\underset{\sim}{k},u) = \frac{1-\psi(u)}{u}\cdot\frac{1}{1-\psi(\underset{\sim}{k},u)} \tag{29}$$

where $\psi(\underset{\sim}{k},u)$ is the Fourier-Laplace transform of $\psi(\underset{\sim}{r},t)$ defined in Eq. (28). The Laplace transform of the mean squared displacement $\langle r^2(t)\rangle$ of the walker is given by (L represents the Laplace transform)

$$\langle r^2(u)\rangle = L\,\langle r^2(t)\rangle = \sum_{\underset{\sim}{r}} r^2 P(\underset{\sim}{r},u) = -\nabla^2_{\underset{\sim}{k}} P(\underset{\sim}{k},u)|_{\underset{\sim}{k}=o} \tag{30}$$

from which with (29) it follows

$$\langle r^2(u)\rangle = \frac{1}{u[1-\psi(u)]} \sum_{\underset{\sim}{r}} r^2 \psi(\underset{\sim}{r},u) \tag{31}$$

The time dependent diffusion coefficient D(t) is defined as

$$D(t) \equiv \frac{1}{2\Delta}\frac{d}{dt}\langle r^2(t)\rangle \tag{32}$$

So that

$$D(u) = \frac{1}{2\Delta[1-\psi(u)]} \sum_{\underset{\sim}{r}} r^2 \psi(\underset{\sim}{r},u) \tag{33}$$

Thus, $\psi(\underset{\sim}{r},t)$ is sufficient to determine in the CTRW approach the diffusion coefficient D(t) uniquely. In the continuum approximation we obtain by using (28)

$$\sum r^2 \psi(\underset{\sim}{r},t) = L\,[\Psi(t)\Omega(t)] \tag{34}$$

with

$$\Omega(t) = p\rho \int d\underset{\sim}{r}\; r^2 w(\underset{\sim}{r}) e^{-tw(\underset{\sim}{r})} \tag{35}$$

from which Eq. (33) can be evaluated as:

$$D(u) = L\,[\Psi(t)\Omega(t)] / [2\Delta u \Psi(u)] \tag{36}$$

$\Psi(t)$ was extensively studied in Sec. II. The form (36) for D(u) is the main result of this section and we now consider for example multipolar interactions in 3 dimensions. With Eq. (2) one has:

$$\Omega(t) = C_1 t^{5/s-1} \tag{37}$$

with

$$C_1 = \Gamma(1-5/s)\,\frac{4\pi p\rho}{s}\left(\frac{d^s}{\tau}\right)^{5/s} \tag{38}$$

From Eq. (15) one has:

$$\Psi(t) = \exp[-C_2 t^{3/s}] \tag{39}$$

with

$$C_2 = \Gamma(1-3/s)\,\frac{4\pi}{3}\,p\rho\,\left(\frac{d^s}{\tau}\right)^{3/s} \tag{40}$$

In order to obtain the diffusion coefficient one generally has to evaluate the inverse Laplace transform of Eq. (36). In our case for short and long times the behavior of D(t) can be found directly from $\Omega(t)$ and $\Psi(t)$ using Tauberian theorems[41].

For short times $\Psi(t)$ is a slowly varying function of t, so that

$$\Psi(u) \sim \frac{1}{u} \tag{41}$$

and, using Eq. (37)

$$L\,[\Psi(t)\Omega(t)] \sim \frac{C_1\Gamma(5/s)}{u^{5/s}} \tag{42}$$

From Eq. (36) it now follows:

$$D(t) \sim L^{-1}\left[\frac{C_1\Gamma(5/s)}{6u^{5/s}}\right] \sim \frac{C_1}{6}\,t^{5/s-1} = \Gamma(1-5/s)\,\frac{2\pi\,p\rho}{3s}\left(\frac{d^s}{\tau}\right)^{5/s} t^{5/s-1} \tag{43}$$

This result concurs with the expression obtained by expanding the functions in powers of $\frac{1}{u}$ by Godzik and Jortner[39].

For long times (small u) one has

$$L[\Psi(t)] = \sum_{j=0}^{\infty} \frac{(-u)^j}{j!}\,\tau_{j+1} \tag{44}$$

and

$$L[\ (t)\ (t)] = C_1 \sum_{j=0}^{\infty} \frac{(-u)^j}{j!}\,\tau_{j+5/s} \tag{45}$$

with

$$\tau_k \equiv \int dt\, t^{k-1}\, \Psi(t) = \frac{s}{3}\,\Gamma(ks/3)\; C_2^{-ks/3} \tag{46}$$

Since the moments τ_j of Eqs. (44) and (45) exist, one has from the Tauberian theorems for longer times[41]

$$D(t) \sim \frac{C_1\tau_{5/s}}{6\tau_1} + O(1/t) \tag{47}$$

typical for time independent diffusion.

We now consider the case of the distribution of stepping times $\psi(\underset{\sim}{r},t)$ being decoupled. For nearest neighbors one then gets from Eq. (33)

$$D(u) = \ell^2/2\Delta\; \frac{\psi(u)}{1-\psi(u)} \tag{48}$$

where ℓ is the nearest neighbor distance on the regular lattice. Eq. (48) may also be derived from simple random walk arguments such as will be used in the next section for the trapping case. We let the walk take place on a regular lattice of unit length ℓ. Thus the mean squared distance $\langle r_n^2\rangle$ travelled in n steps and averaged over all random-walk realizations is $\langle r_n^2\rangle = n\ell^2$. We set $\phi_n(t)$ for the probability of having performed exactly n steps in time t; then:

$$\langle r^2(t)\rangle = \sum_{n=0}^{\infty} \langle r_n^2\rangle \phi_n(t) = \ell^2 \sum_{n=0}^{\infty} n\phi_n(t) \tag{49}$$

The generalized diffusion coefficient in Δ-dimensions is given through Eq. (32).

For simplicity, we restrict ourselves to the three-dimensional case, and have from Eq. (49):

$$D(t) = (\ell^2/6) \sum_{n=0}^{\infty} n\dot{\phi}_n(t) \tag{50}$$

Furthermore, in our model, all steps occur with the common stepping-time distribution $\psi(t)$. Thus $\phi_n(t)$ obtains from $\psi(t)$. Setting f(u) for the Laplace transform of f(t), one has for

$\phi_n(u)$[42-45]:

$$\phi_n(u) = [\psi(u)]^n\,[1-\psi(u)]/u\ , \tag{51}$$

the result of a renewal process. The Laplace transform of $\phi(t)$ is, therefore:

$$L[\dot{\phi}_n(t)] = [\psi(u)]^n\,[1-\psi(u)] - \delta_{n,0} \tag{52}$$

since $\phi_n(t=0) = \delta_{n,0}$. Inserting Eq. (50) it follows that:

$$\begin{aligned} D(u) &= (\ell^2/6)\sum_{n=1}^{\infty} n\,\{[\psi(u)]^n - [\psi(u)]^{n+1}\} \\ &= (\ell^2/6)\sum_{n=1}^{\infty} [\psi(u)]^n \\ &= (\ell^2/6)\,\psi(u)/[1-\psi(u)] \end{aligned} \tag{53}$$

IV. TRAPPING IN THE CTRW MODEL

In this section we consider the trapping problem of an excitation which randomly hops over in a disordered system. Here we follow closely the procedures developed with Zumofen in Refs. [46]-[49]. The traps are taken to be randomly distributed on the lattice, with probability p, and the excitation is to be trapped instantaneously at the first encounter of a trap. Denoting by R_n the number of distinct sites visited in n steps one has for the probability Φ_n that the walker survived the first n steps [46]-[48]:

$$\Phi_n = \langle (1-p)^{R_n - 1} \rangle \tag{54}$$

where the average extends over all realizations of the walk. In Refs. [46] to [48] there is a discussion of the forms ϕ_n for different lattice types and dimensions, by using cumulant expansions of the distribution of R_n. For simplicity only the first cumulant S_n, the mean number of sites visited in n steps, $S_n \equiv \langle R_n \rangle$ will be considered here. Furthermore, we approximate S_n by $S_n = an$, an expression which holds well in three dimensions for larger n[19],[48]. Then for a low trap concentration, $p \ll 1$, one obtains the decay law $\Phi_n \approx \exp(-pan)$. This simplified form does not have any explicit random-walk dependence. Following now the same development as in Eq. (49), one obtains:

$$\Phi(t) = \sum_{n=0}^{\infty} \Phi_n \phi_n(t) = \sum_{n=0}^{\infty} e^{-pan} \phi_n(t) \tag{55}$$

Using Eq. (52), the Laplace transform of $\Phi(t)$ is readily summed [42]-[45]:

$$\Phi(u) = \frac{1-\phi(u)}{u} \sum_{n=0}^{\infty} [e^{-pa}\psi(u)]^n$$

$$= [1-\psi(u)] / \{u[1-e^{-pa}\psi(u)]\} \tag{56}$$

Equation (56) relates the decay law to $\psi(u)$.

Concentrating on the diffusion coefficient and on the decay law, one is led to consider the role played by $\psi(u)$ in Eqs. (53) and (56). The simplest case obtains for an exponential stepping time distribution, $\psi(t) = \tau_1^{-1} \exp(-t/\tau_1)$. Then $\psi(u) = (1+u\tau_1)^{-1}$ and thus $D(u) = \ell^2/(6u\tau_1)$ and

$$\Phi(u) = [(1-e^{-pa})\tau_1^{-1} + u]^{-1}$$

The inverse Laplace-transform is here straightforward: $D(t) = \ell^2/(6\tau_1)$ (a constant) and

$$\Phi(t) = \exp[(e^{-pa}-1)t/\tau_1]$$

i.e an exponential decay for all times[42], less straightforward is the situation for general forms of $\psi(t)$. As discussed in Ref.[42],two cases of particular interest are encountered: In the first case, all moments $\tau_j = \int_0^\infty t^j\psi(t)dt$ are finite; then $\psi(u)$ admits an expansion in powers of u:

$$\psi(u) = \sum_{j=0}^{\infty} (-1)^j \tau_j u^j/j! = 1-u\tau_1 + \ldots \tag{57}$$

In the second case, $\psi(t)$ has a long-time tail, $\psi(t) \sim t^{-1-\beta}$ $(0<\beta<1)$ so that already $\int_0^\infty t\psi(t)dt$ diverges. Then

$$\psi(u) = 1 - \Gamma(1-\beta)u^\beta/\beta + \ldots \tag{58}$$

Using $\psi(u)$ given by Eqs. (57) and (58) one has, respectively, $D(u) = u^{-1}F(u)$ and $D(u) = u^{-\beta}F(u)$. Here we let F(u) denote a slowly varying function of u, for $u \to 0$ [49],[50]. From Tauberain theorems we may now obtain some information about D(t), but the procedure required additional conditions. On physical grounds D(t) should be a positive, continuous and monotonous function of time. Then, from Theorem 3 on page 551 of Ref. [50], one concludes that [49], for long times:

$$D(t) \sim \ell^2/(6\tau_1) \tag{59}$$

if $\psi(u)$ is given by Eq. (57), whereas:

$$D(t) \sim \ell^2 \beta t^{\beta-1}/[6\Gamma(\beta)\Gamma(1-\beta)] \tag{60}$$

if $\psi(u)$ is given by Eq. (58).

Turning now to the case of the decay law, we remark that from Eq. (56) one has for $\psi(t) \sim t^{-1-\beta}$ [43]:

$$\Phi(u) \sim [\Gamma(1-\beta)/\beta]\, u^{\beta-1}F(u)$$

where again F(u) denotes a slowly varying function of u. Again arguing that $\Phi(t)$ is positive, continuous and monotonous we obtain as above (see also Refs. [43] and [49]):

$$\Phi(t) \sim t^{-\beta}/(pa\beta) \tag{61}$$

for long time, and recalling that p is small, $p \ll 1$. For $\psi(u)$ given by Eq. (57) one has:

$$\Phi(u) \sim \tau_1[1-e^{-pa}+\tau_1 u e^{-pa}]^{-1} \sim [u + pa/\tau_1]^{-1} \tag{62}$$

Under the same physical conditions, one obtains from the same Tauberian theorem the rather disappointing result that for $t \to \infty$ the integral $\int_0^t \Phi(\tilde{t})d\tilde{t}$ is of the order of $[\frac{1}{t} + \frac{pa}{\tau_1}]^{-1}$ i.e. it has an upper bound of the order τ_1/pa. As in Ref. [43], a form obviously compatible with Eq. (62) is the exponential decay:

$$\Phi(t) = e^{-pat/\tau_1} \tag{63}$$

This form is, however, by far not the only one that leads to $\Phi(u) \sim [u + pa/\tau_1]^{-1}$. In Ref. [42] it was shown (for some stepping times in current use in the theory of energy transfer) that, after the short onset period, Eq. (63) provides a reasonable description of the decay for moderately long times. Also in Ref. 49, by Zumofen, Klafter and Blumen the inverse Laplace-transform of Eq. (57) was analyzed numerically for several classes of $\psi(t)$, and it was demonstrated that in general, for very long times, behaviors very different from Eq. (63) may develop.

ACKNOWLEDGEMENTS

The authors are very thankful to Dr. G. Zumofen for his collaboration in many of the works whose results were presented here. The support to A. Blumen from the Deutsche Forschungsmemeinschaft and from the Fonds der Chemischen Industrie is gratefully acknowledged.

REFERENCES

1. R. Kopelman, In Radiationless Processes in Molecules and Condensed Phases, edited by F. K. Fong (Springer, Berlin, 1976), p. 297.
2. J. C. Wright, in Radiationless Processes in Molecules and Condensed Phases, edited by F. K. Fong (Springer, Berlin, 1976), p. 239.
3. D. L. Huber, in Laser Spectroscopy of Solids, edited by W. M. Yen and P. M. Selzer (Springer, New York, 1981), p. 83.
4. S. Alexander, J. Bernasconi, W. R. Schneider and R. Orback, Rev. Modern Phys. 53, 175 (1981).
5. R. C. Powell and Z. G. Soos, J. Lumin. 11, 1 1975).
6. J. Klafter and A. Blumen in Random Walks in the Physical and Biological Sciences, edited by M. F. Shlesinger and B. J. West (AIP Conference Proceedngs, 1983).
7. G. Pfister and H. Scher, Adv. Phys. 27, 747 (1978).
8. H. Scher and M. Lax, Phys. Rev. B7, 4491 (1973).
9. H. Scher and M. Lax, Phys. Rev. B7, 4502 (1973).
10. H. Scher and E. W. Montroll, Phys. Rev. B12, 2455 (1975).
11. J. M. Hvam and M. H. Brodsky, Phys. Rev. Lett. 46, 371 (1981).
12. Z. Vardeny, P. O'Connor, S. Ray and J. Tauc, Phys. Rev. Lett. 44, 1267 (1980).
13. A. Blumen, J. Manz and G. Zumofen, Nuovo Cimento 63B, 59 (1981).
14. B. E. Vugmeister, Fiz. Tverd, Tela 18, 819 (1976), [English translation: Sov. Phys. Solid State 18, 469 (1976)].

15. T. Förster, Z. Naturforsch, Teil A4, 321 (1949).
16. M. Inokuti and F. Hirayama, J. Chem. Phys. 43, 1978 (1965).
17. A. Blumen, Nnovo Cimento 63B, 50 (1981).
18. G. H. Weiss and R. J. Rubin, Adv. Chem. Phys. 52, 363 (1983).
19. E. W. Montroll and G. H. Weiss, J. Math. Phys. 6, 167 (1965).
20. D. Bedaux, K. Lakatos-Lindenberg and K. E. Shuler. J. Math. Phys. 12, 2116 (1971).
21. V. M. Kenkre, E. W. Montroll and M. F. Shlesinger, J. Stat. Phys. 9, 45 (1973).
22. J. Klafter and R. Silbey, J. Chem. Phys. 72, 843 (1980).
23. J. Klafter and R. Silbey, Phys. Rev. Lett. 44, 55 (1980).
24. A. Blumen and J. Manz, J. Chem. Phys. 71, 4694 (1979).
25. C. Tsang and R. A. Street, Phil Mag. B37, 601 (1978).
26. T. M. Searle, Phil Mag. B46, 163 (1982).
27. A. Blumen and J. Klafter, Phil Mag. B47, L5 (1983).
28. M. D. Galanin, Zh. Eksp. Teor. Fiz. 28, 485 (1955) [English translation: Sov Phys. JETP 1, 317 (1955)].
29. A. Blumen and R. Silbey, J. Chem. Phys. 70, 3707 (1979).
30. V. P. Sakun, Fiz. Tverd. Tela 14, 2199 (1972) [Sov. Phys. Solid State 14, 1906 (1973)].
31. A. Blumen, J. Chem. Phys. 72, 2632 (1980).
32. B. Mandelbrot, The Fractal Geometry in Nature (W. H. Freeman, San Francisco, 1982).
33. S. Alexander and R. Orbach, J. Phys. Lett. 43, 625 (1982).
34. R. Rammal and G. Toulouse, J. Phys. Lett. 44, 13 (1983).
35. S. Alexander. C. Laermans, R. Orbach and H. M. Rosenberg (submitted).
36. J. Klafter and A. Blumen, J. Chem. Phys. (submitted).
37. K. Allinger and A. Blumen, J. Chem. Phys. 72, 4608 (1980), J. Chem. Phys. 75, 2762 (1981).
38. M. Yokota and O. Tanimoto, J. Phys. Soc. Jpn. 22, 779 (1967).
39. K. Godzik and J. Jortner, Chem. Phys. Lett. 63, 428 (1979). K. Godzik and J. Jortner, J. Chem. Phys. 72, 4471 (1980).
40. B. E. Vugmeister, Phys. Stat. Sol. (b) 76, 161 (1976); 90, 711 (1978).
41. W. Feller, An Introduction to Probability Theory and Its Applications (Wiley, New York, 1971), Vol. II, p. 445.
42. A. Blumen and G. Zumofen, J. Chem. Phys. 77, 5127 (1982).
43. A. Blumen, J. Klafter and G. Zumofen, Phys. Rev. B27, 3429 (1983).
44. W. P. Helman and K. Funabashi, J. Chem. Phys. 71, 2458 (1979).
45. M. Tachiya, Radiat. Phys. Chem. 17, 447 (1981).
46. G. Zumofen and A. Blumen, J. Chem. Phys. 76, 3713 (1982).
47. G. Zumofen and A. Blumen, Chem. Phys. Lett. 88, 63 (1982).
48. A. Blumen and G. Zumofen, J. Stat. Phys. 30, 487 (1983).
49. G. Zumofen, J. Klafter and A. Blumen, J. Chem. Phys. (in press).

50. G. Doetsch, Handbuch der Laplace-Transformation, Bd. 1 (Birkauser, Basel, 1971).

ENERGY TRANSFER AND ELECTRON TRANSFER IN PHOTOBIOLOGICAL, PHOTOCHEMICAL, AND PHOTOELECTROCHEMICAL PROCESSES
(Abstract)

A. Nozik

S.E.R.I.
1617 Cole Boulevard
Golden, Colorado 80401, USA

ABSTRACT

The conversion of radiant energy into stored chemical potential is an extremely important process. It occurs in biological photosynthesis, as well as in the nonbiological approaches to the conversion of solar energy into fuels and chemicals. All of these biological and nonbiological approaches to the photoconversion of light into energètic chemical bonds can be classified as either photobiological, photochemical, or photoelectrochemical processes. The former process depends upon natural photosynthesis, while the latter two processes depend on inorganic and/or organic photoactive materials, such as semiconductors or melecular chromophores. All of the processes rely on the efficient transfer of excitation energy from the light-absorbing site to another region of the system that consequently exhibits enhanced electrochemical potentials for oxidation and reduction reactions. The importance and role of energy transfer vis à vis charge transfer processes are widely different in photobiology, photochemistry, and photoelectrochemistry. These roles were described and compared for the three photoconvereion approaches to solar energy utilization.

AN EXAMPLE OF IDENTIFYING THE SPECIFIC MECHANISM OF RESONANT ENERGY TRANSFER: $Sb^{3+} \rightarrow Mn^{2+}$ IN FLUOROPHOSPHATE PHOSPHORS (Abstract)

R.L. Bateman

General Electric Company
Nela Park, Comp. 1364
Cleveland, Ohio

ABSTRACT

The kinetics of energy transfer from antimony donor to manganese acceptor in fluorophosphate phosphors has been studied for the purpose of identifying the specific mechanism of resonant energy transfer (1). By comparison of the manganese concentration dependence of the experimental donor quantum yield and of the emission decay curves with the theoretical calculations for dipole-dipole, dipole-quadrupole, and exchange mechanisms, the energy transfer was found to take place through the exchange mechanism. The probability per unit time for the energy transfer by exchange is given by $P=KR^{16}e^{-2R/L} \sin^2 \Theta \cdot \cos^2 \phi$ where the empirical parameters were found to be K-48.7A $\text{Å}^{-16} \mu$ sec^{-1} and L=0.55Å. In the absence of manganese acceptors add for the low donor concentrations investigated, the antinomy emission decay curve was found to be exponential over more than two orders of Magnitude with a lifetime of 7.65 + 0.05μ sec.

REFERENCE

(1) T.F. Soulves, R.L. Bateman, R.A. Hewes, and E.R. Kreidler, Phys. Rev. B 7, 1657 (1973).

ENERGY TRANSFER AND ANDERSON LOCALIZATION IN RUBY ELECTRIC FIELD AND UNIAXIAL STRESS EFFECTS

E. Duval and A. Monteil

Spectroscopie des Solides - E.R.A. 1003 du CNRS

69622 Villeurbanne - France

ABSTRACT

Measurements of the effects of an electric field and a σ_{xz} uniaxial stress on the single to pair energy transfer, would suggest a new model for energy diffusion among single ions. The non radiative resonant energy transfer would be rapid inside E-sublattices made inequivalent by electric fields and slow between these E-sublattices. It is shown that earlier experimental results do not conflict with this model.

I. INTRODUCTION

Since the work of Imbusch [1] on energy transfer between single ions and pairs it has been generally believed that excited Cr^{3+} ions in pink and red ruby resonantly transfer their energy to neighbouring ions rapidly. The exponential decay of R_1 line [2] would indicate that the diffusion among the single ions is rapid in comparison with the direct transfer from single ions to pairs. This situation favoured the existence of the Anderson transition from delocalization to localization of the optical energy. Effectively Koo, Walker and Geschwind [3] revealed the Anderson localization and mobility edges in ruby by measuring the single to pair transfer rate.

Since 1980 most researchers think that the resonant energy transfer among Cr^{3+} ion is slow. The conclusion is based on the electric field experiments performed by Jessop and Szabo [4], and by Chu et al [5] . The results of these experiments show that the resonant energy transfer is slow (close to 1 m sec^{-1} for 0.8 Cr % ruby from Jessop and Szabo) between Cr^{3+} ions which can be differentiated by an electric field directed along the optical axis. They apparently agree with grating experiments via degenerate four - wave mixing. Liao et al [8] found that the diffusion length for the optical energy is less than 20 nm. Recently Jessop and Szabo [9] did not observe Anderson transition by measuring the coherence time T_2' by photon echoes technics : the transition from a localized to an extended state would be followed by a change in T_2'. Finally, theoretical study of the influence of off-diagonal disorder on the localization would confirm the absence of Anderson transition for chromium concentration less than 10 % [10] .

In spite of these results the evidence is not conclusive with regard to a slow resonant energy transfer among Cr^{3+} ions and to the absence of an Anderson transition in ruby. First of all if a slow resonant energy diffusion exists the energy transfer from an excited chromium to a pair must be direct, which conflicts with the exponential decay of the R_1 line. On the other hand a direct single to pair transfer without diffusion cannot explain the results of Koo et al.

The experiments which are considered as decisive by most researchers and which agree with a slow resonant energy transfer are the Jessop and Szabo [4] and Chu et al [5] transient electric field experiments. It is shown that the resonant energy transfer is slow between the two sublattices (E-sublattices) differentiated by an electric field directed along the optical axis. But these experiments do not exclude the possibility of a rapid energy diffusion in each E-sublattice. A uniaxial stress directed in a specific direction differentiates two sublattices (σ - sublattices) which are different from the E-sublattices. We have measured the effects of such a uniaxial stress and an electric field separately on the single to pair transfer. We found that the uniaxial stress effect is strong and that of an electric field negligible.

In section II the two types of sublattices, E-sublattices and σ -sublattices, are compared. In section III the experimental results are described. In section IV our experimental results are discussed in comparison with the numerous experimental results obtained by other researchers. It is shown that the hypothesis of a resonant energy transfer among Cr^{3+} ions, which is rapid inside an E-sublattice and slow between the E-sublattices, can interpret most experimental results.

II. E-SUBLATTICES AND σ-SUBLATTICES

In Figure 1, the projection of the corrundum lattice on a (2 1 1 0) plane is shown. The cation sites are not centrosymmetric, but the sites I and II are symmetric by inversion (i). Then an electric field orientated along the trigonal symmetry axis differentiates the sites of type II from sites of type I or type II. It is easily seen that the cations of type I are arranged along

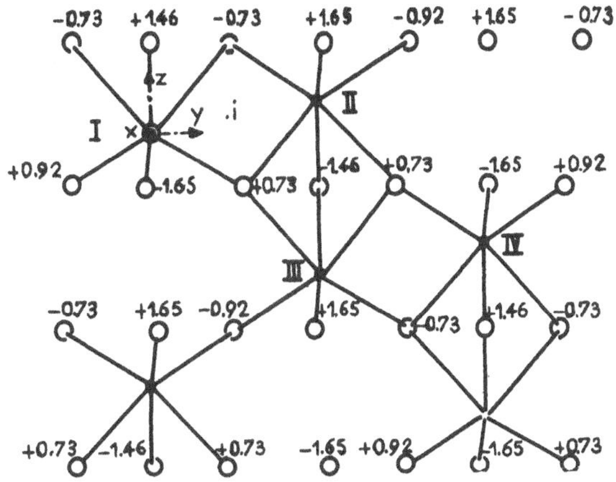

Fig. 1. Projection of α .Al_2O_3 on (2 $\bar{1}$ $\bar{1}$ 0). The vertical distances (in Å) from the oxygens to the plane of projection are given in parentheses. The metal ions lie at height zero [14]. • Aluminum ions, o oxygen ions.

different lines where they are respectively situated in 4th N.N. positions like the cations I and III or II and IV. Due to the trigonal symmetry there are three equivalent lines which go through a same cation. Therefore in an E-sublattice of corundum the cations are situated in 4th N.N. positions and there are two different E-sublattice.

Sites II and III are symmetric by a π-rotation around an axis parallel to the x-axis (Figure 1), and, as has already been noticed, sites II and III are symmetric by inversion. From the symmetry, if we apply a σ_{xy} uniaxial stress, sites I and II are equivalent, sites II and III are not equivalent and therefore I and III are not equivalent. A σ_{xz} uniaxial stress creates two σ-sublattices different from the two E-sublattices : it lifts the degeneracy of two Cr^{3+} ions in a same E-sublattice. The existance of the two different σ-sublattices has been clearly indicated by the effect of a σ_{xz} stress on the ${}^4A_1 \rightarrow {}^4T_2$, ${}^3A_1$ (${}^3T_2$) $\rightarrow {}^3T_2$ and ${}^1A_1 \rightarrow {}^1T_1$ zero phonon lines of Cr^{3+} [11], V^{3+} [12], Co^{3+} [13] ions respectively.

III. EXPERIMENTAL

The aim of the experimental work was to compare the effect of an electric field and the effect of a σ_{xz} uniaxial stress effect on the single (R_1 line) to 4th N.N. pair (N_2 line) transfer. From the Jessop and Szabo [4] and Chu et al [5] transient electric field experiments, we expect a negligible effect with an electric field, and a greater effect with a σ_{xz} uniaxial stress, if the resonant transfer among single ions is rapid.

The experimental results were presented by Monteil in his thesis [14] and by Monteil and Duval [15] . In Figure 2, the

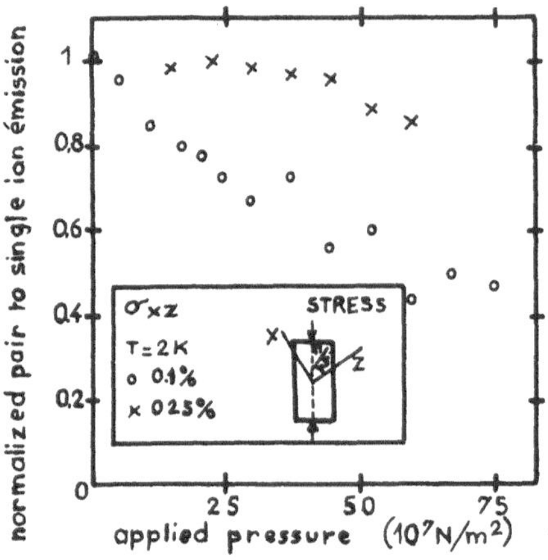

Fig. 2. Normalized pair (N_2line) to single ion emission (R_1 line) in ruby as a function of a pressure applied along an axis in the (xz) plane orientated at $\pi/3$ with respect to the z trigonal axis.

I_{N_2}/I_{R_1} ratio is represented as the function of a pressure applied along an axis in the xz plane orientated at $\pi/3$ with respect to the z trigonal axis : for this orientation the crystal is submitted to a σ_{xz} stress. The measured intensity I_{N_2} corresponds to the N_2 fluorescence excited by transfer from single ions. Figure 2 shows that, at very low temperature (T = 2K), I_{N_2}/I_{R_1} decreases rapidly with the applied pressure in a 0.1 Cr % crystal, and remains practically constant, at least with the lowest pressures, in more concentrated crystal (0.25 Cr %). The σ_{xz} stress effect decreases with the temperature and it is equal to zero for temperatures higher than 77 K. As is expected, the electric field effect is negligible in agreement with the works of Jessop and Szabo [4] and Chu et al [5]. However we must point out that pressure applied along the y axis or along the z-axis has an effect on the single to pair transfer, but particularly at the lowest pressures this effect is not so great as at a pressure which induces a σ_{xz} stress. This last result is surprising and must be explained.

Unfortunately it is not possible to observe directly the splitting of the R_1 line due to the σ_{xz} uniaxial stress which lifts the degeneracy of the two σ-sublattices, because it is smaller than the inhomogeneous broadening. From the corresponding splitting measured on the ${}^4A_2 \rightarrow {}^4T_2$ zero-phonon line [11] it is possible to estimate that the splitting of R_1 line is close to 0. 1 cm^{-1} for the highest applied stress.

IV. DISCUSSION

IV.A. Rapid or Slow Nonradiative Resonant Transfer?

At present the experimental data obtained indicate two opposing viewpoints regarding the energy transfer. (1) The non radiative resonant energy transfer among the single Cr^{3+} ions is slow. This assumption involves a dipole-dipole interaction responsible for this transfer ; a microscopic inhomogeneous broadening ; the non-existence of the Anderson transition from a localized state to an extended state ; and a direct single to 4th N.N. pair energy transfer at very low temperature. (2) The resonant energy transfer among the single Cr^{3+} ions is rapid inside an E-sublattice and slow between E-sublattices : this rapid resonant energy transfer in relatively concentrated crystals proceeds from the super-exchange interaction ; a macroscopic inhomogeneous broadening is crucial to this rapid transfer ; the Anderson transition can exist in relatively small domains ; the single to pair transfer goes through the non-radiative resonant energy diffusion at very low temperature.

In a recent publication entitled "Resonant energy transfer in ruby and Anderson localization" Gibbs et al [16] summarize the experimental results which lead to the first viewpoint ; (1) Grating experiment via degenerate fourwave mixing [6]-[8], (2) Transient electric fields [4][5], (3) Fluorescence line narrowing [2],[17],[18]. We have already noticed that the grating experiments are not decisive for the first viewpoint. As is suggested by Liao et al [8] a rapid single ion diffusion can occur on distances smaller than the smallest distances(∿ 20 nm) which it is possible to determine for diffusion by the transient grating experiments. On the other hand, the transient electric field experiments show only that the single-single transfer is slow between E-sublattice but not inside an E-sublattice.

The FLN experiments are interpreted by a microscopic inhomogeneous broadening if we assume a slow resonant non radiative transfer. On the other hand, they can be interpreted by a macroscopic inhomogeneous broadening if we assume a rapid transfer. But they do not prove either point of view. Due to the $\Delta(=0.38\ cm^{-1})$ splitting of the ground level and the inhomogeneous broadening, when the R_1 line is excited at the ν energy, three lines at $\nu - \Delta$, ν and $\nu + \Delta$ are observed. If rapid resonant transfer exists between two Cr^{3+} ions submitted to different crystal fields so that the energy difference is Δ , we would observe a fourth line at $\nu + 2\Delta$. Selzer and Yen [17] found that this fourth line arises slowly by radiative resonant transfer but not by non-radiative resonant transfer. Assuming rapid non radiative resonant transfer between Cr^{3+} ions submitted to a same crystal field they conclude that there is a macroscopic inhomogeneous broadening. On the contrary Gibbs et al [16], generalizing the slow resonant transfer observed by a transient electric field experiment, conclude that the experiment of Selzer and Yen [17] confirms that the resonant non radiative transfer is slow. Furthermore they claim that the non radiative spectral energy transfer which is independent of the energy mismatch as observed by Selzer et al, is consistent with a microscopic inhomogeneous broadening. Their reason is that a gradual spreading of the narrow components until they fill the full inhomogeneous line is not observed as would be expected. But, it is easy to imagine that resonant Cr^{3+} ions in a same domain or in distinct domains are in contact with non-resonant Cr^{3+} ions in different domains. We will propose a model for such a situation.

To sum up, these three types of experiments are not decisive for the non-existence of a rapid resonant non radiative transfer inside domains. The effect of the uniaxial σ_{xz} stress and of the electric field on the single to pair transfer agrees with non radiative resonant single-single transfer which is rapid inside an E-sublattice and slow between E-sublattices. Furthermore these experiments are interpreted more easily by a superexchange interaction responsible for the rapid transfer than by a dipole-dipole interaction. When a σ_{xz} is applied the concentration of resonant Cr^{3+} ions is halved. For a 0.1 Cr % ruby it is possible to find that the mean transfer probability is divided by a factor three, and by a factor two for a 0.2 Cr % ruby, if the superexchange interaction is effective. If the dipole-dipole interaction is responsible for the transfer, the transfer probability would be divided by a factor two at any concentration. Experimentally, at the strongest applied pressure ($7.5 \cdot 10^7$ N/m^2) and for a 0.1 Cr % ruby the transfer probability would be divided by a factor greater than two. Therefore it is not possible to interpret these experimental results by a dipole-dipole interaction responsible for the rapid resonant transfer.

The reduction of the σ_{xz} effect when the concentration is increased from 0.1 Cr % to 0.25 % is surprising and interesting. At first we should point out that for a superexchange interaction the effect of a σ_{xz} uniaxial stress on the transfer probability decreases with the Cr^{3+} concentration, as described above. But this decrease with the concentration is not sufficient to explain the experimental data. Another explanation is the following. The splitting induced by a σ_{xz} uniaxial stress which lifts the degeneracy of the two E-sublattices is smaller than the inhomogeneous broadening. Then the effect of such a σ_{xz} uniaxial stress is possible in our model if the broadening is macroscopic. However, the broadening increases with the concentration : it is multiplied by a factor 1.5 when the concentration is changed from 0.1 % to 0.25 %. One could argue that the microscopic broadening increases with the concentration with respect to the macroscopic broadening, which would explain the reduction of the σ_{xz} uniaxial stress effect with the Cr concentration. Another explanation will be discussed in the sequel : the splitting of the two E-sublattice would be smaller than the mobility edge E_c of Koo et al [3] in 0.25 Cr % crystal.

It is also observed that the effect of the σ_{xz} stress decreases with the temperature. This experimental result can be interpreted easily. When the temperature increases the phonon assisted transfer becomes stronger than the resonant transfer.

This model does not explain the effect of a σ_{yy} uniaxial stress. A trivial explanation could be an imperfect orientation of the crystal or an inhomogeneous applied uniaxial stress. We have verified that the orientation error is less than 3 degrees. Another explanation would be the following. If the cation sites are equivalent with respect to the σ_{yy} stress in a pure corrundum crystal it is not necessarely true in a ruby crystal in which the summetry is broken by the presence of Cr^{3+} ions or other impurities. In an E-sublattice two Cr^{3+} ions separated by an even number of Al-O bonds are non-equivalent and the effect of a σ_{yy} stress can be different on the one and the other. Therefore the σ_{yy} stress would have a qualitatively similar effect as the σ_{xz} stress on the resonant diffusion.

At the present time it is not possible to claim with certainty that the single to pair transfer goes through a rapid non radiative resonant transfer among single ions in E-sublattices. If the direct resonant transfer from singles ions to 4th N.N. pairs is accepted the effect of uniaxial stress effect can be interpreted by a shift of the R_1 line with respect to the 4th N.N. transitions which are very near to the R_1 lines, like 4K1 and 4H3 and perhaps 4I3 transitions [19] . Such a shift is observed with respect to 4K1 and 4H3. It is approximately equal to 0.1 cm^{-1} at 10^8 N/m^2 [14] . It would decrease the resonance between R_1 line and pair lines.

Several objections to this last interpretation can be raised : (1) we do not observe any difference in the shift of R_1 line with respect to pair lines when the applied pressure is along the y-axis or along an axis at $\pi/3$ from z-axis in the (xz) plane. Now the effect for the second orientation when a σ_{xz} uniaxial stress is present is larger than for the y-orientation (Fig. 3 and 5). (2) The maximum of the single to pair transfer corresponds to the maximum of the R_1 line and it is not shifted towards the pair lines [14] .(3) The decrease of the uniaxial stress effect observed when the concentration is increased, is not explained easily by the direct transfer. When the concentration of chromium is changed from 0.1 % to 0.25 % the width of the R_1 line is multiplied by 1.5. We would expect a reduction of the uniaxial stress effect by about the same factor. Experimentally this factor is stronger since the effect becomes negligible for a 0.25 Cr % ruby (Fig. 2).

IV.B. Anderson Transition

The aim of the research on the energy transfer in ruby is the possible existence of an Anderson transition between localized and extended states. The experiment of Koo et al [3] which showed the presence of mobility edges was very exciting. Unfortunately new experiments performed to find the mobility edges again were unsuccessful [9]-[20]. Certainly the observation of these mobility edges depends strongly on the nature on the samples and more precisely on the inhomogeneous broadening.

According to Anderson [21] a critical concentration exists below which the excitation is localized if the variation as a function of the distance between the Cr^{3+} ions is faster than $1/r^3$. The exchange or superexchange interaction which varies exponentially with r can give rise to an Anderson transition. Measurements of energy transfer time τ_2 from single ion to 4th N.N. pairs performed by Monteil and Duval [14],[22] as a function of the chromium concentration are well interpreted by a diffusion among single ions due to a superexchange interaction. If C is the chromium concentration the experimental data at very low temperatures are fitted by the following equation :

$$\frac{1}{\tau_{2(0)}} = 2.3 \times 10^5 \; C \exp(-\,0.69\; C^{-1/3}) \; s^{-1} \qquad (1)$$

At temperatures lower than 50 K, $1/\tau_2$ is linear with the temperature like the spectral diffusion among single ions [2] . It is found [14],[23]:

$$\frac{1}{\tau_2} = \frac{1}{\tau_{2(0)}} + A_2 T \qquad (2)$$

Similar equations are determined for the transfer towards the 3rd N.N. pairs. This last result is an argument for a single to pair transfer through the single-single phonon assisted transfer. Furthermore it is confirmed that the superexchange interaction is responsible of this transfer because the variation of A_2 as a function of the chromium concentration C is identical with the variation of $1/\tau_2(0)$:

$$A_2 = 2.3 \times 10^3 \; C \exp(-\,0.65\; C^{-1/3}) \; s^{-1} \; K^{-1} \qquad (3)$$

These experimental results would show that the Anderson transition is possible in ruby. However to observe mobility edges on the inhomogeneous broadened R_1 line the microscopic broadening must be larger than the macroscopic one. If the macroscopic inhomogeneous broadening is preponderant the Anderson transition can be effective in domains. From the transient grating experiments the dimensions of these domains would be smaller than 20 nm [8] . We can consider that in the sample used by Koo et al [3] the inhomogeneous broadening was rather microscopic while it was macroscopic in samples used by Chu et al [20] , and Jessop and Szabo [9], who did not observe mobility edges.

In our experiments which show the effect of a σ_{xz} uniaxial stress on the single to pair transfer, the quenching of this effect, which appears when the chromium concentration increases, could be interpreted by the presence of mobility edges. In fact the splitting induced by σ_{xz} stress between two σ-sublattices in one E-sublattice where the transfer can be rapid, is a perfectly microscopic energy mismatch. Now in the 0.1 Cr % ruby this splitting would be larger than the mobility edge Ec, and smaller in the 0.25 Cr % ruby.

In a recent paper Huber and Ching [10] found a mobility edge when the Cr concentration is equal to 10 at %. By their calculation the localization occurs at this high concentration because of the off-diagonal interaction. However, from Elyutin's [24] theoretical result concerning the influence of the off-diagonal disorder on the localization, a much weaker concentration would be found. Furthermore the result of Huber and Ching conflicts with the theoretical works of Antoniou and Economou [25] , and Klafter and Jortner [26] which show that a negligible effect of the off-diagonal disorder on the localization regarding the diagonal disorder exists.

IV.C. Internal Electric Fields and Energy Transfer

Several reasons can be found to explain the more rapid resonant transfer in a E-sublattice : the angles between Al-O bands are great, the lines of Al-O bonds are not broken. Therefore the superexchange would be strongest in these sublattices. However, a simple reason would be that the ionic impurities with a charge different from the ones of Al^{3+} and O^{2-} respectively, or defects, create random electric fields in the crystal which separate, like the applied electric fields, the Cr^{3+} ions into two E-sublattices. This assumption is very likely. For a concentration of an impurity such Mg^{2+} equal to 10 ppm a Cr^{3+} ion would be submitted to a mean electric field equal to approximately 15 kV/cm. If this electric field is orientated parallel to the trigonal axes the corresponding splitting of the R_1 line would be equal to 0.1 cm^{-1}. Furthermore Jessop and Szabo [27] suggested that the broadening of the R_1 line in their dilute ruby crystals is characterized by a Holtsmark shape specific to random electric fields.

With these internal electric fields the inhomogeneous broadening between E-sublattices would be microscopic, and macroscopic in an E-sublattice where fast resonant transfer can occur. It is an additional fact which prevents the observation of mobility edges:

the inhomogeneous broadening between E-sublattice hides the inhomogeneous broadening in a E-sublattice.

On the other hand, the inhomogeneous broadening due to internal electric fields would explain the two surprising characteristics of the spectral energy diffusion [2] : (1) independence of the energy mismatch, (2) one-phonon assisted transfer. The energy mismatch can be relatively large even between two next neighbour Cr^{3+} ions in two different E-sublattices : the spectral diffusion would be governed by the microscopic inhomogeneous broadening between E-sublattices, and the resonant spatial energy diffusion by the macroscopic inhomogeneous broadening (see section 1 of this discussion and the paper of Gibbs et al [16]).

The one-photon assisted energy transfer among single ions is surprising, principally because the wavelength of the phonons which have an energy equal to the energy mismatch is much longer than the distance between two Cr ions which transfer their energy [28],[29]. If in the ruby crystal there are two E-sublattices split by internal electric fields, we can assume that the spectral transfer between the two sublattices is induced by the interaction with optical phonons which are odd with respect to the cation site. The component of the vibration parallel to the optical axis has an opposite effect on the two different sublattices, like the electric field. The modulation of the energy of two next neighbour Cr ions in two different E-sublattices is out of phase, and then the energy transfer can occur between these two non-resonant next neighbour Cr ions.

V. CONCLUSION

The measurements of the effects of a σ_{xz} uniaxial stress and of an electric field on the single to pair energy transfer are well interpreted if we assume a non radiative resonant transfer which is rapid inside the E-sublattices differentiated by an electric field, and slow between these two sublattices. Without an applied electric field these two E-sublattices would be still split by internal electric fields induced by impurities or defects : which explains the slow resonant transfer between E-sublattices. The rapid resonant transfer takes place in domains, the size of which is smaller than 20 nm and is determined by the distribution of impurities inducing electric fields.

The inhomogeneous broadening due to internal electric fields is macroscopic inside one or other of the E-sublattices, and microscopic between the two E-sublattices. The presence of internal electric fields which lifts the degeneracy of the two E-sublattices would explain the one phonon assisted spectral transfer which is independent of the energy mismatch.

It is possible that Anderson transition exists in ruby inside domains. However mobility edges cannot be observed by sweeping the inhomogeneous R_1 line when the inhomogeneous broadening due to internal electric fields is dominant or when the macroscopic inhomogeneous broadening is larger than the microscopic broadening. However, the quenching of the σ_{xz} uniaxial effect on the single to pair transfer with the increased concentration could be due to the presence of mobility edges.

The model which has been presented for the non radiative transfer agrees with all the crucial experiments. However it must be confirmed by experiments which enable direct observation of the rapid resonant transfer inside E-sublattices. Such experiments are in progress.

REFERENCES

1. G. F. Imbusch, Phys. Rev. 153, 326 (1967).
2. P. M. Selzer, D. S. Hamilton and W. M. Yen, Phys. Rev. Lett. 38, 858 (1977).
3. J. Koo, L. R. Walker and S. Geschwind, Phys. Rev. Lett. 35, 1669 (1975).
4. P. E. Jessop and A. Szabo, Phys. Rev. Lett. 45, 1712 (1980).
5. S. Chu, H. M. Gibbs, S. L. McCall and A. Passner, Phys. Rev. Lett. 45, 1715 (1980).
6. H. J. Eichler, Opt. Acta 24, 631 (1977).
   H. J. Eichler, J. Eichler, J. Knof and C. H. Noak, Phys. Status Solidi 52, 481 (1979).
7. D. S. Hamilton, D. Heiman, J. Feinberg and R. W. Hellwarth, Opt. Lett. 4, 124 (1979).
8. P. L. Liao, D. M. Bloom, L. M. Humphrey and S. Geschwind, Bull. am. Phys. Soc. 24, 586 (1979).
   P. F. Liao, L. H. Humphrey, M. Bloom and S. Geschwind, Phys. Rev. B20, 4145 (1979).
9. P. E. Jessop and A. Szabo, Phys. Rev. B26, 420 (1982).
10. D. L. Huber and N. Y. Ching, Phys. Rev. B25, 6472 (1982).
11. R. Louat, R. Lacroix, E. Duval and B. Champagnon, Phys. Status Solidi B69, 33 (1975).
12. B. Champagnon, E. Duval and B. Champagnon, Phys. Status Solidi B69, 339 (1975).
13. E. Duval, R. Louat, B. Champagnon, R. Lacroix and J. Weber, J. de Phys. (Paris) 36, 559 (1975).
14. A. Monteil, Thesis Lyon (1982).
15. A. Monteil and E. Duval, to be published.
16. H. M. Gibbs, S. Chu, S. L. McCall and A. Passner, Nato Workshop on Coherence and Energy Transfer in glasses, Cambridge Sept. 13-17 (1982).
17. P. M. Selzer and N. M. Yen, Opt. Lett. 1, 90 (1977).

18. P. M. Selzer, D. L. Huber, B. B. Barnett and W. M. Yen, Phys. Rev. B17, 4979 (1978).
19. P. Kisliuk, N.C. Chang, P. L. Scott, M. H. L. Pryce, Phys. Rev. 184, 367 (1979).
20. S. Chu, H. M. Gibbs and A. Passner, Phys. Rev. 324, 7162 (1981).
21. P. W. Anderson, Phys. Rev. 109, 1492 (1958).
22. A. Monteil and E. Duval, J. Phys. C13, 4565 (1980).
23. A. Monteil and E. Duval, J. Phys. C12, L415 (1979).
24. P. V. Elyudin, J. Phys. C14, 1435 (1981).
25. P. D. Antoniou and E. N. Economou, Phys. Rev. B16, 3768 (1977).
26. J. Klafter and J. Jortner, J. Chem. Phys. 71, 2210 (1979).
27. P. E. Jessop and A. Zsabo, Appl. Phys. Letters 37, 510 (1980).
28. R. Orbach, in Optical Properties of Ions in crystals, ed. by H. M. Crosswhite, H. W. Moos (Interscience, New-York 1967) p. 445.
29. R. Holstein, S. K. Lyo and R. Orbach, in Laser Spectroscopy of Solids ed. by W. M. Yen and P. M. Selzer. Topics in Applied Physics vol. 49 (Spring Verlag 1981).

TIME-RESOLVED STUDIES OF ENERGY TRANSFER

R.C. Powell

Department of Physics
Oklahoma State University
Stillwater, OK 74078, USA

ABSTRACT

The dynamics involved in the transfer of excitation energy among impurity ions in solids is directly reflected in the time dependence of the energy transfer rate. In this lecture experimental techniques are discussed which allow the determination of this time dependence. Special emphasis is given to the techniques of site-selection spectroscopy for studying spectral energy transfer and four-wave mixing spectroscopy for studying spatial energy transfer. The fundamental physical properties of the system which can contribute to the observed time dependence of the energy transfer are summarized and the methods used to connect the experimental results to these physical properties through theoretical interpretation of the data are discussed. Several examples are given of different types of projects involving time-resolved studies of energy transfer.

I. INTRODUCTION

In order to characterize the properties of energy transfer between ions (or molecules) in solids, the fluorescence spectra is monitored under specific static or dynamic conditions. The information of interest is the mechanism of the ion-ion interaction responsible for the energy transfer, the strength of the interaction, whether it is a single or multi-step process, the role played by phonons in the transfer, and the influence of the statistical distribution of the ions. The most important experimental parameters that can be varied in energy transfer experiments are the concentration of active ions, temperature, and time. The greatest amount of work has been done by measuring the changes in fluorescence intensities and lifetimes as a function of concentration. Although useful information concerning the efficiency of energy transfer can be obtained by this method, the results are generally not sensitive enough to distinguish between different mechanisms of energy transfer and the type of transfer taking place may be quite different at high and low concentrations. Temperature dependent studies are useful in distinguishing between phonon-assisted and resonant processes and in elucidating the effects of phonon scattering in multi-step processes. The development of pulsed laser experimental techniques have made time-resolved spectroscopy the most powerful method of characterizing energy transfer properties. The time dependence of the energy transfer rate is very sensitive to the mechanism and the dynamics of the transfer processes. In this lecture we will focus our attention on the physical origins of different time dependences which can be observed and the experimental techniques which have been developed for obtaining time dependent information. Examples are given of experimental results for several different cases.

II. ORIGIN OF THE TIME DEPENDENCE

A very general expression for the evolution of energy with time away from an initially excited ion is given by

$$dP_i(t)/dt=-(\beta+\sum_j W_{ij} + \sum_{n\neq i} w_{in})P_i(t)+\sum_j W_{ji}P_j(t)+\sum_{n\neq i} w_{ni}P_n(t) \quad (1)$$

where $P_i(t)$ is the probability of finding the excitation on sensiizer ion i at time t, β is the intrinsic fluorescence decay rate, W_{ij} describes the transfer of energy from sensitizer i to activator ion j and W_{ji}, is the rate of back transfer, and w_{in} describes the migration of energy among sensitizer ions before transfer to an activator. The solution to this equation and its relationship to an experimental observable such as fluorescence intensity requires performing a configuration average over the distribution of all possible ion-ion interactions and inclusion of the initial excitation condition.

The time evolution of the observed energy transfer is determined by four factors: the ion-ion interaction mechanism causing the transfer; the spatial distribution of ions involved in the transfer; the dynamics of the processes (i.e., single-step vs multistep, coherent vs incoherent, etc.); and the initial excitation conditions. Figure 1 shows some examples of several common situations which occur. In part (A) a random distrubution of excited sensitizers is created and each one can transfer energy to an activator ion located at a specific distance away. This pairing of sensitizer and activator ions can occur due to problems

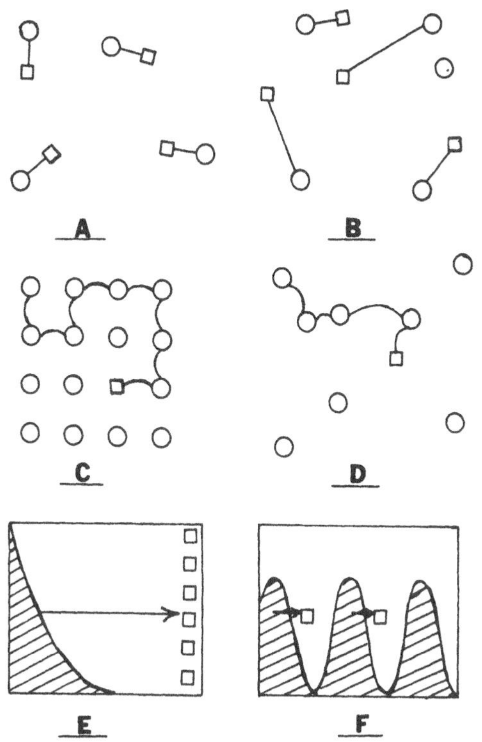

Fig. 1. Different types of distributions resulting in different time dependences for energy transfer.

with charge compensation or lattice distortion. The consequence is a time independent energy transfer rate. Part (B) depicts a random distribution of initially excited sensitizers transferring to randomly distributed activators. This is the most common physical case. The energy transfer rate for the system decreases as a power law with time with the specific dependence determined by the ion-ion interaction mechanism.

Parts (C) and (D) both involve multistep energy migration among the sensitizer ions before trapping at an activator site occurs. It is important to note the observed energy tranfer rate in this case is made up of contributions from two distinct physical processes: the energy migraton among the sensitizers and the transfer of the energy to the activators. These may take place through different ion-ion interaction mechanisms. The time dependence of the measured energy transfer depends on the types of both of these mechanisms and the relative importance of the two interactions[1]. Another important aspect of the problem is whether the energy migration takes place on a uniform lattice of sensitizers as shown in Part (C) or a random distribution of sensitizers as shown in Part (D)[2]. Also the dimensionaligy of the migration (ie. 1-dimension, 2-dimension, or 3-dimension) is important. In addition the distortion of the lattice by the activator impurity ions can bias the migration process[3],[4]. If back transfer from activator traps is present, trap modulated migration occurs[5]. All of these factors can contribute to the time dependence of the energy transfer.

The final two parts of Fig. 1 show two important cases of special initial distributions. In Part (E) the sample consists of one region containing sensitizer ions with a specific part of the region being excited and another region containing the activator ions. In this type of experiment energy transfer occurs over controlled spatial distances. The observed time dependence of the transfer is determined by both the type of interaction mechanism and the transfer distance[4]. In Part (F) the sample contains a ramdom mixture of sensitizer and activator ions but the region of initially excited sensitizer ions has a geometric pattern, in this case shown as a sine wave. Experimental measurements of the decay of this grating pattern indicate how efficiently energy transfers from sensitizers in the peak regons of the grating to activators in the valley regions.[6] The observed time dependence is related to the mechanism of energy transfer as well as the size and shape of the geometric pattern initially produced.

III. METHODS OF THEORETICAL ANALYSIS

The ideal way to theoretically interpret experimental results of energy transfer studies would be to find the exact solution to

Eq. (1) for the most general case which includes all possible physical situations such as those discussed in the previous paragraphs. So far such a general analytical solution has proven to be impossible and all solutions which have been obtained starting from this master equation approach have involved highly restrictive assumptions. The most comprehensive theoretical treatment of this type is the work of Huber and coworkers[7]. They have employed the average T-matrix approximation (ATA) formalism to develop the solution of the master equation up to a certain point where assumptions specializing the solution to specific cases are required. The advantage of his approach is that one formalism can be used to treat a variety of special cases.

A second way to approach the theoretical analysis of energy transfer data is using phonomenological rate equation models[5]. This appraoch is the inverse of the master equation approach described above which has as its starting point ion-ion interactons on a microscopic level and involves complex mathematical manipulations to relate theoretical predictions to macroscopic observable parameters. On the other hand, the rate equation approach involves formulating a phenomenological model which describes the physical situation under investigation and solving the model equations for the experimentally observed parameter. Then by fitting the theoretical prediction to the experimental results the properties of the energy transfer rate parameter for the system are obtained. This energy transfer rate is the primary parameter obtained from this approach. Secondary parameters describing the microscopic physical details of the processes which are taking place (such as the ion-ion interaction strength, diffusion coefficient, etc.) are then determined from the properties of the transfer rate. This approach has the advantage of mathematical simplicity and is useful in cases where the general characteristics of energy transfer in the system are considered to be the most important result. The disadvantage with using this approach is that it is sometimes difficult to unambiguously identify the details of the physical processes giving rise to the energy transfer.

One of the most difficult physical situations to treat theoretically is the case involving high concentrations of sensitizers so there is multistep migration amoung the sensitizers in addition to the transfer to activators. Several different phenomenological models have been developed to treat different limiting situations in this case, and it is important to test the limiting criterea to determine the validity of a given model for a specific case[2]. One important problem is how to describe the dynamics of the energy migration among the sensitizers. Usually this is done with a random walk or diffusion mathemtical formalism. For many situations this approach is valid and a variety of special cases have been treated in the initial work of Montroll et al., [8].

and the recent work of Blumen[9]. There are two situations where different theoretical approaches are required. The first is when the exciton moves with a long mean free path so that "coherence" becomes important. Although the quantum mechanical formalism for treating purely coherent motion exists, intermediate cases are still difficult to handle theoretically[10]. The second situation where standard random walk or diffusion formalism can not be simply applied is when the migration occurs on a random distribution of sunsitizers. So far the only methods developed for treating the effects of this randomness are computer simulation techniques such as Monte Carlo [2], [11] or percolation theory approaches [1]. The former is most appropriate for cases where the distribution is completely random and the most important Monte Carlo treatments have been developed by Lyo et al., [11]. Percolation theory is most appropriate at concentrations that are so high that the sensitizers are distributed in clumps with the transfer within a clump being much more efficient than transfer between clumps. This formalism has been developed by Kopelman and et al., [12].

IV. EXPERIMENTAL TECHNIQUES

The most important empitus for studying the time dependence of energy transfer has been the development of pulsed laser sources and boxcar integrators. Figure 2 shows the schematic of a standard time-resolved spectrosocpy experimental setup. The window on the boxcar integrator is triggered to look at the spectrum at a specific time after the laser pulse. By monitoring the time evolution of fluorescence spectrum the transfer of energy from sensitizers to activators can be followed in real time. With more elaborate detection techniques such as streak cameras or optical delay lines this type of investigation can be extended to the picosecond time regime.

An additional important aspect of laser excitation is the high degree of spectral resolution which can be obtained. This has lead to the development of site-selection spectrosocpy in which the same type of ions located in nonequivalent crystal field sites play the role of sensitizers and activators. The narow laser line selectively excites ions in one type of site and energy transfer ocurs to ions in other types of sites. This type of investigation has the advantage of not having to use a second type of activator ion which may introduce lattice distortions in the host. The case where the laser excitation width is much less than the inhomogeneous width of the spectral transition under investigation is referred to as fluorescence line narrowing (FLN). These types of studies have been especially useful with disordered materials and glasses. It is important to note that both SSS and FLN techniques measure spectral transfer of energy.

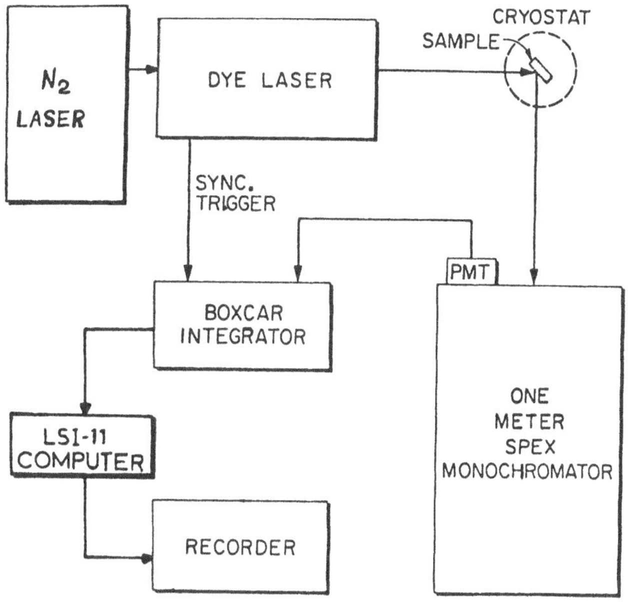

Fig. 2. Experimental setup for time-resolved spectroscopy.

Four-wave mixing spectroscopy is a powerful new technique that has been developed to measure spatial transfer of energy with or without spectral transfer. The schematic diagram of this experimental technique is shown in Fig. 3. The two pump beams are incident on the sample from one side and interfere to write a sine wave hologram. The weak probe beam enters the sample from the opposite side conjugate to one of the pump beams and Bragg diffracts off the sine wave grating. The diffracted signal beam is detected by a photomultiplier tube. By chopping the pump beams on and off the decay of the signal beam can be detected and this represents the decay of the grating which is related to the diffusion length of the excitons. The limitaton of this technique is that it can only detect energy transfer over distances at least the size of the peak to valley distance of the grating which is formed. This is approximately half of the wavelength of the light being used. Dispite this severe limitation, FWM has been useful in studying long range exciton migration in both organic and inorganic materials[6],[13].

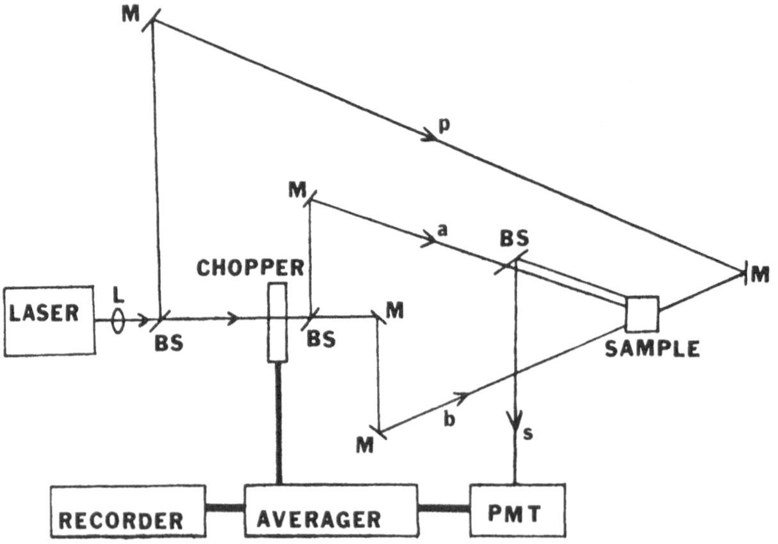

Fig. 3. Experimental setup for four-wave mixing spectroscopy.

There are numerous other types of experimental techniques which can give information on energy transfer rates. Two examples are the dephasing times measured by coherent transient spectrosocpy techniques[14] and the slope of the hysteresis curves in optical bystability measurements[15]. Although these methods are not commonly used by energy transfer studies at the present time, they have the potential for being powerful tools in future studies.

V. EXAMPLES OF TIME-RESOLVED ENERGY TRANSFER STUDIES

In this section we review examples of time resolved energy transfer studies which illustrate different important aspects of the problem.

V.A. Energy Transfer Between Eu^{3+} Ions in $Eu_xY_{1-x}P_5O_{14}$ Crystals

Time-resolved site-selection spectroscopy was used to characterize the transfer of energy between Eu^{3+} ions in non-equivalent crystal field sites in $Eu_xY_{1-x}P_5O_{14}$ crystals[16]. One of the $^5D_0$ - $^7F_2$ transitions was chosen for monitoring the energy transfer because it is "hypersensitive" to the local environment of the ion. The spectral resolution of the experimental setup was less than 0.5 Å and the temporal resolution was better than 10 ns. The laser excitation was in the $^5D_1$ level.

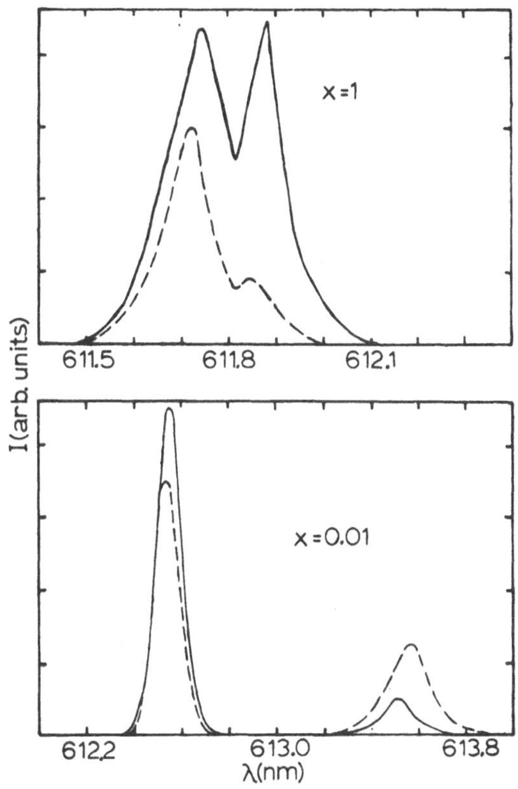

Fig. 4. Fluorescence spectra of $Eu_xY_{1-x}P_5O_{14}$ at two times after the laser pulse at 12 K. The solid line is for 0.05 ms after the excitation pulse and the broken line 1.6 ms delay.

Figure 4 shows typical spectral results obtained at 12 K for samples containing 100% and 1% Eu^{3+}. For both samples two different lines appear for this specific transition and their relative intensities vary with laser excitation wavelength. The fluorescence lifetimes associated with both lines are the same. These lines represent transitions from Eu^{3+} ions in two types of nonequivalent crystal field sites. The positions and relative splitting of these transitions are significantly different for the two samples. In the 100% sample the transition energy difference is $\Delta E_{sa} \simeq 5$ cm^{-1} and 0.5% of the Eu^{3+} ions are in activator sites while for the 1% sample $\Delta E_{sa} \simeq 26$ cm^{-1} and about 36.3% of the Eu^{3+} ions are in activator sites.

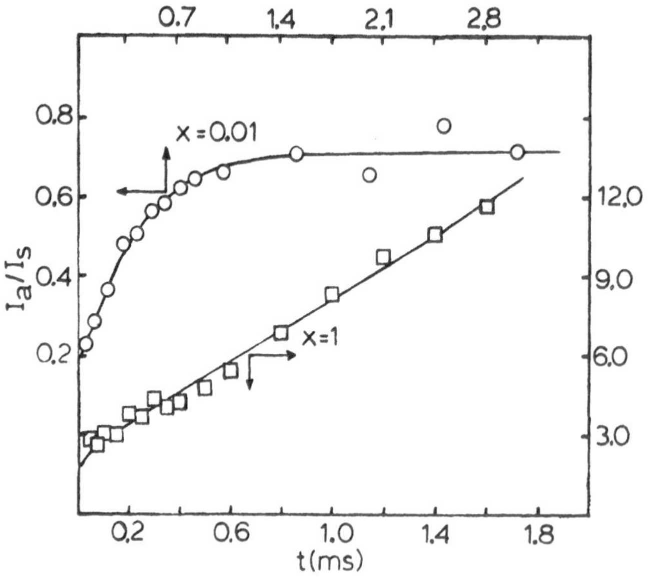

Fig. 5. Time evolution of the ratios of the fluorescence intensities of sensitizer and activator transitions for $Eu_xY_{1-x}P_5O_{14}$ at 12K.

The time evoluton of the spectra can be used to ascertain the properties of energy transfer between ions in these two different types of sites. As seen from the figure, these properties are quite different for the two samples. In the case of the 100% sample the specific initial excited state distribution created by the laser excitation leads to energy transfer from ions in the sites giving rise to the lower energy transition to ions in sites associated with the higher energy transitions. In the 1% sample the energy transfer goes from the high energy to the low energy sites. The time dependence of the integrated fluorescence intensity ratios are shown in Fig. 5. Rate equation models were developed to obtain the best fits to these data. For the 100% sample the model giving the best fit to the data assumes single-step transfer, electric dipole-dipole interaction, no back transfer, and a random distribution of sensitizers and activators. This gives the solid line shown in the figure. For the 1% sample the model giving the solid line fit to the data assumes electric dipole-dipole interaction between sensitizer-activator pairs at fixed nearest neighbor distances with similar values for transfer and back transfer rates. These results demonstrate how different the energy transfer characteristics can be for the same ions in similar hosts.

V.B. Energy Transfer Among Nd^{3+} Ions in Lightly Doped Solids

Energy transfer among Nd^{3+} ions in various crystalline and solid hosts has been a problem of great interest because of the importance of these materials in laser applications[17]. In many cases the transfer has been found to be a multitep migration process and time-resolved site-selection spectroscopy has been useful in characterizing the transfer of energy between Nd^{3+} ions in nonequivalent types of crystal field sites. The type of interactons causing the transfer has been identified a a two-phonon assisted processes involving a real intermediate state[18]. The most important aspect of these results is that the energy transfer has been shown to play an important role in lowering the quantum efficiency of laser materials such as Nd-YAG[19] and this understanding has helped point the way for the development of new methods of heat treatment and crystal growth to enhance the quantum efficiency of this materials[20].

One of the basic points raised in these studies is how different are the energy transfer characteristics if the sensitizer excitation migrates on a random lattice instead of a uniform lattice? One way to answer this question is to compare the results of analyzing time-resolved spectroscopy data using the usual methods which assume a uniform lattice with the results obtained using a Monte Carlo approach[2]. In the latter approach, sensitizer excitons are generated, allowed to hop from site to site, and disappear when they hop onto an activator. The observed fluorescence intensity is related to the number of excitons alive at a given time after the excitation pulse. The random nature of the lattice is accounted for by using the configuration-averaged distribution of hoping times. This is constructed by first generating a uniformly distributed set of random numbers and then weighting this according to the Hertzian distrubution which describes the occupancy of nearest neighbors in a random distribution of available sites in three dimensions. These numbers are then converted to a set of hopping times by a second weighting factor which reflects the variation of the energy transfer rate with distance r. For electric dipole-dipole interaction this is $(R_0/r)^6$ where R_0 is the critical interactions distance. At each step in the random walk of each exciton a number from this double weighted set is randomly chosen as the hopping time for the next step. This method accounts for the possibility of transferring to any other sensitizer site in the lattice. After the hopping time is generated for a given step in the random walk, a test is performed to check to see if the exciton has decayed by fluorescence emission during that time. This is done by comparing a number from uniformly distributed random set with the probability of fluorescence decay over the time interval given by the hopping time. Also after each step in the random walk, a test is performed to determine if the exciton has hopped onto an activator. This is done by calling a number from a uniformly generated random set and comparing it with the fractional occupancy of lattice sites by activators.

Figure 6 shows experimental data obtained from time-resolved spectroscopy measurements on Nd^{3+} doped garnet and vanadate crystals. In both cases there is no back transfer and the fluoresence decay rates are the same for ions in the sensitizer and activator sites. In the Monte Carlo model these results can be described by

$$I_a/I_s=\{[n_a(0)/n_s(0)]+1\}\exp(-\beta_a)/N(t)-1 \tag{2}$$

where N(t) is the normalized number of excitons alive at time t. The good fits in the data shown in the figure were obtained when 5000 excitons where generated. These fits are extremely sensitive

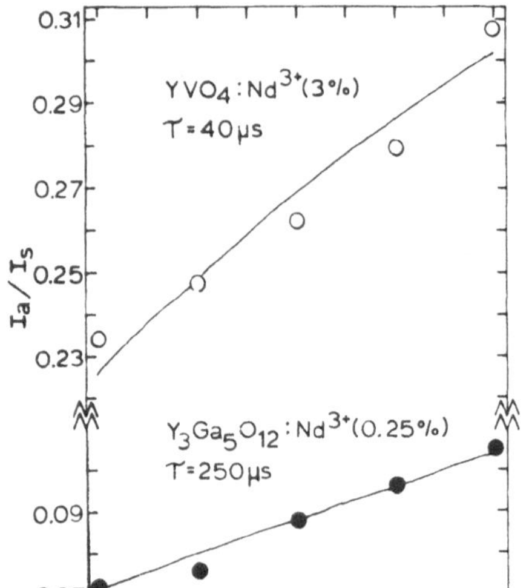

Fig. 6. Time dependence of the ratios of the fluorescence intensities of sensitizer and activator transitions at 100 K. Solid lines represent the data from Ref.[18] and the circles represent the Monte Carlo fits to the data.

to the choice of R_0. The fitting results are given in Table 1. Similar fits to the data are obtained with the standard model of a random walk on a unform lattice[21]. In this case the results are described by the expression

$$I_a/I_s=\{[I_a(0)/I_s(0)-1]\exp(-\beta_a t)/I_s(t)\}-1 \tag{3}$$

where

$$I_s(t)=I_s(0)\bar{n}_s(t)\exp(-t/t_0) - t_0^{-1}\int_0^t I_s(t')\bar{n}_s(t-t')\exp[-(t-t')/t_0]dt'. \tag{4}$$

Here t_0 is the average hopping time and

$$\bar{n}_s(t)=\exp[-\beta_s t-(4/3)\pi R_0{}^3 n_a(\pi\beta_s t)^{1/2}]. \tag{5}$$

This equation was solved by numerical iterative methods. The fitting results are again given in Table 1.

Comparing the results of the two approaches to fitting the time-resolved spectrosocpy data described above shows that it is possible to obtain a good description of the results by either assuming a uniform lattice of sensitizers or by explicitly acounting for the random nature of the sensitizer distribution. However, the values obtained for the critical interaction distance in the former approach are only half as large as those obtained using the latter approach. The theoretically calculated values for R_0 are very close to those obtained from the Monte Carlo fitting routine. This indicates that uniform lattice models significantly underestimate the values of the critical interaction distance needed to describe exciton migration on a random lattice. This may be due to the fact that on a random lattice an exciton can become "bottlenecked" in regions of well separated sensitizers and this never occurs on a uniform lattice.

V.C. Exciton Diffusion in $Nd_xLa_{1-x}P_5O_{14}$ Crystals

The two examples of time-resolved spectroscopy research discussed above involve site-selection techniques which elucidate the properties of spectral migration of energy. Applying similar methods to samples of $Nd_xLa_{1-x}P_5O_{14}$ have shown little or no spectral energy migration[22]. Howevr long range spatial migration of energy can be observed directly through four-wave mixing transient grating techniques[6].

The equation describing the exciton population in a transient grating experiment is

$$\partial n(x,t)/dt=D\partial^2 n(x,t)/\partial x^2-N(x,t)/\tau. \tag{6}$$

Solving this for the initial condition of a sine wave distribution gives

$$n(x,t)=1/2e^{-t/\tau}\{1+e^{-16\pi^2 Dt\sin^2(\theta/2)/\lambda}\cos[4\pi\sin(\theta/2)x/\lambda]\}. \quad (7)$$

Table 1. Monte Carlo Versus Uniform Lattice Fitting Parameters

Sample	Parameter	Monte Carlo	Uniform Lattice	Theoretical Prediction
$Y_3Ga_5O_{12}$ Nd(0.25%)	R_0(Å)	21	11	20
	t_0(s)	1.6×10^{-4}	7.76×10^{-3}	
	$I_a(0)/I_s(0)$	0.059	0.05	
YVO_4 Nd(3.0%)	R_0(Å)	14	7	12
	t_0(s)	2.1×10^{-6}	1.33×10^{-4}	
	$I_a(0)/I_s(0)$	0.202	0.16	

The observed four-wave mixing signal intensity is proportional to the depth of the grating

$$I_s(t)=I_p(I_e\Delta n)^2 = I_p I_e{}^2 e^{-Kt} \quad (8)$$

where

$$K=(32\pi^2 D/\lambda^2)\sin^2(\theta/2)= 2/\tau. \quad (9)$$

Note that this pure exponential decay holds only for diffusive exciton motion. For nondiffusive motion the expression is more complicated[10].

Measurements of this type have been made on samples of $Nd_xLa_{1-x}P_5O_{14}$ as a function of concentration, temperature, laser power and excitation wave length[23]. Pure exponential decays of the signal beam intensity were observed showing that the excition migraton is diffusive. The grating decay constant was measured versus the pump beam crossing angle. The results are shown in Fig. 7 for two of the samples at room temperature. The data is shown to vary linearly with $\sin^2(\theta/2)$ and to extrapolate to $2/\tau$ at $\theta=0^o$ as predicted by Eq. (9). The values of the exciton diffusion coefficients are found from the slopes of the curves in Fig. 7. These are 5.1×10^{-6} cm^2sec^{-1} for the sample with x=1 and $5.2 \times 10^{-7} cm^2sec^{-1}$ for the sample with x=0.2. These values of D imply diffusion lengths of the order of tenths of microns.

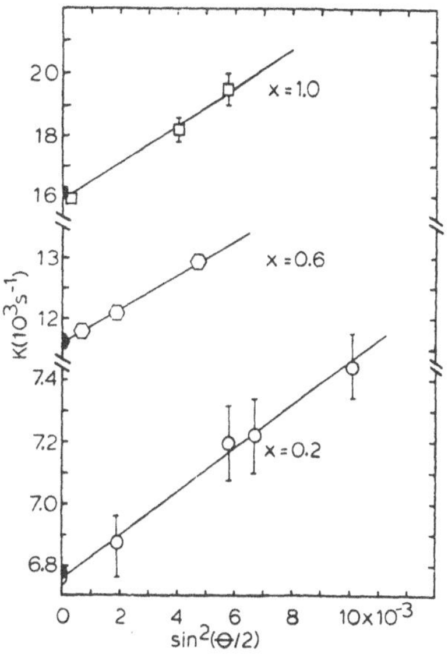

Fig. 7. Variation of the grating decay constant with the pump-beam crossing angle at room temperature in $Nd_xLa_{1-x}P_5O_{14}$. The shaded points are twice the fluorescence decay rates.

VI. CONCLUSIONS

Time-resolved studies have developed into the most powerful techniques for characterizing the properties of energy transfer among ions and molecules in solids. Some of the most important areas to be developed in this field can be catagorized in terms of experimental techniques, materials, theory, and applications.

Although there is still much information that can be gained from site-selection experiments, especially with regard to the effects of disorder and phonon-assistance, the two most exciting experimental possabilities are techniques to observe coherent exciton motion and techniques for ultrafast measurements. At the present time it appears that four-wave mixing transient grating spectrosocpy is the best method for observing coherent exciton migration since the difference in signal shape for diffusive and nondiffusive motion is possible to detect quite easily. Ultrafast techniques rely on the use of streak camers for detection and the development of femptosecond lasers as sources. Pushing measurement capabilty into this time regime allows the possability of studying aspects of energy transfer which could not previously be monitored. Two examples are following in real time the transfer of energy between strongly coupled pairs of ions or molecules and following the process of energy transfer before and during excited state relaxation.

A number of very interesting studies of energy tranfer in specially tailored materials have been performed recently. One example is the use of radiation induced defects to increase the oscillator strengths of the impurity ions and thus greatly enhance energy transfer efficiency[24]. Another example is monitoring the changes in energy transfer characteristics in doped alkali halides when the impurity ions form aggregates and defect phases[25]. In this case the energy transfer can be used as a direct measurement of the defect substructure of the material.

As greater detailed information becomes available from more sophisticated experimental techniques, there is an increased demand for better theoretical descriptions of energy transfer. The major theoretical problem which has still not been treated satisfactorally is performing the configuration average over the ensamble of sensitizer and activator ions in a way that accounts for spatial and spectral disorder. So far the models developed for doing this strictly apply only to special physical cases and not in general. Another interesting theoretical problem is explicitly accounting for the discrete structure of the lattice on the properties of energy transfer[16].

Finally, the field of energy transfer in solids has matured to the point where it should now be possible to start designing materials for solid state device applications which utilize energy transfer processes. There are already important examples of both gas lasers and liquid dye lasers which utalize energy transfer between pumping and emitting species to optimize laser device properties. Similar possabilities exhist for improving solid state lasers especially those being developed for broad band tunable applications. Energy transfer has been used in some commercial solid state scintillators for wave length shifting from the ultra-violate to the visible spectral region[26]. It should also be possible to develope materials which work in the opposite direction to give infrared to visible unconversion through energy transfer[27]. Demonstrating the relavance of energy transfer by developing important applications is crucial to justifying further research in this field.

REFERENCES

1. R. C. Powell and G. Blasse, in Structure and Bonding Vol. 42: Luminescence and Energy Transfer. (Springer-Verlag, Berlin, 1980), p. 43.
2. H. C. Chow and R. C. Powell, Phys. Rev. B 21, 3785 (1980) and C.M. Lawson, E. E. Freed, and R. C. Powell, J. Chem. Phys. 76, 4171 (1982).
3. Z. G. Soos and R. C. Powell, Phys. Rev. B 6, 4035 (1972).
4. R. C. Powell and Z. G. Soos, J. Lumin. 11, 1 (1975).

5. R. C. Powell and Z. G. Soos, Phys. Rev. B 5, 1547 (1972).
6. C. M. Lawson, R. C. Powell, and W. K. Zwicker, Phys. Rev. B 25, 4836 (1982); Phys. Rev. Lett. 46, 1020 (1981).
7. D. L. Huber in Laser Spectroscopy of Solids, ed. W. M. Yen and P. M. Selzer, (Springer-Verlag, Berlin, 1981), p. 83 and references therein.
8. E. W. Montroll and G. H. Weiss, J. Math, Phys. 2, 167 (1965) and A.A. Maradudin, E. W. Montroll, G. H. Weiss, R. Herman, and H. W. Mines, Acad. Rov. Belg. Classe Sci. Mem. Collection in 4° 14, no. 7 (1960).
9. A. Blumen and G. Zumofen, Chem. Phys. Lett. 70, 387 (1980); G. Zumofen and A. Blumen, Chem. Phys. Lett. 78, 131 (1981); A. Blumen and G. Zumofen, J. Chem. Phys. 75, 892 (1981); and G. Zumofen and A. Blumen, Chem. Phys. Lett. 88, 63 (1982).
10. See for example M. D. Fayer and C. B. Harris, Phys. Rev. B 9, 748 (1974).
11. S. K. Lyo, Phys. Rev. B 20, 1297 (1979); ibid 22, 3616 (1980).
12. R. Kopelman in Radiationless Processes in Molecules and Condensed Phases, ed. F. K. Fong (Springer-Verlag, Berlin, 1976) p. 297.
13. J. R. Salcedo, A. E. Siegman, D. D. Dlott, and M. D. Fayer, PHys. Rev. Lett. 41, 131 (1978).
14. R. M. Shelby and R. M. Macfarlane, Phys. Rev. Lett. 45, 1098 (1980).
15. D. A. B. Miller, Laser Focus 18, 79 (1982).
16. J. K. Tyminski, C. M. Lawson, and R. C. Powell, J. Chem. Phys. 77, 4318 (1982).
17. R. C. Powell in Nd-YAG Lasers, ed. L.G. DeShazer (Springer-Verlag, Berlin, 1983).
18. L. D. Merkle and R. C. Powell, Phys. Rev. B 20, 75 (1979); M. Zokai, R. C. Powell, G. F. Imbusch, and B. DiBartolo, J. Appl. Phys. 50, 5930 (1979); and D. K. Sardar and R. C. Powell, J. Appl. Phys. 51, 2829 (1980).
19. R. C. Powell, D. P. Neikirk, and D. K. Sardar, J. Opt. Soc. Am. 70, 486 (1980).
20. D. P. Devor, L. G. DeShazer and R. C. Pastor, Phys. Rev. B. (1983) to be published; D. P. Devor, R. C. Paster, and L. G. DeShazer, J. Chem. Phys. (1983), to be published.
21. A. I. Burshtein, Sov. Phys. JETP 35, 882 (1972).
22. J. M. Flaherty and R. C. Powell, Phys. Rev. B 19, 32 (1979).
23. J. K. Tyminski, R. C. Powell, and W. K. Zwicker, Phys. Rev. B (1983) to be published.
24. K. H. Lee and W. A. Sibley, Phys. Rev. B. 12, 3392 (1975).
25. J. Garcia-Sole, M. Aguilar G., F. Aguillo-Lopez, H. Murrieta S. and J. Rubio O., Phys. Rev. B. 26, 3320 (1982).
26. R. C. Powell, J. Chem. Phys. 55, 1871 (1971); R. C. Powell and L. A. Harrah, J. Chem. Phys. 55, 1878 (1971).
27. F. E. Auzel, Proc. IEEE 61, 758 (1973).

TRENDS IN SCIENTIFIC COMPUTING

C.K. Landraitis

Department of Mathematics
Boston College
Chestnut Hill, MA 02167, U.S.A.

ABSTRACT

The computing environment of the scientific user is changing rapidly. This paper surveys some of the coming changes and some underlying trends in computer hardware and software.

I. THE PROBLEM-SOLVING CYCLE

Figure 1 is a rough model of the steps involved in solving a problem by computer. The algorithm is the scientist's formulation of the method by which the problem is to be solved, while the program is the expression of that method in FORTRAN or another programming language. The broken lines lead to actions taken a second time when results indicate a need for changes in the algorithm or program. The remaining steps of the cycle are then repeated.

So-called batch operating systems have predominated until recently. Under a batch system, user's jobs are queued and do not begin running until previously submitted jobs are completed, sometimes many hours after submission. The user cannot communicate with the computer while entering or running his job. As a result, even the most trivial errors necessitate resubmission and another extensive delay in receiving results. The consequences of such delays are so undesirable that much effort is typically expended in preventing errors from reaching the machine. While this does promote careful preparation, it makes computer use a slow and frequently frustrating experience. One could say that men and computers are only able to communicate with great difficulty in a bach operating system environment.

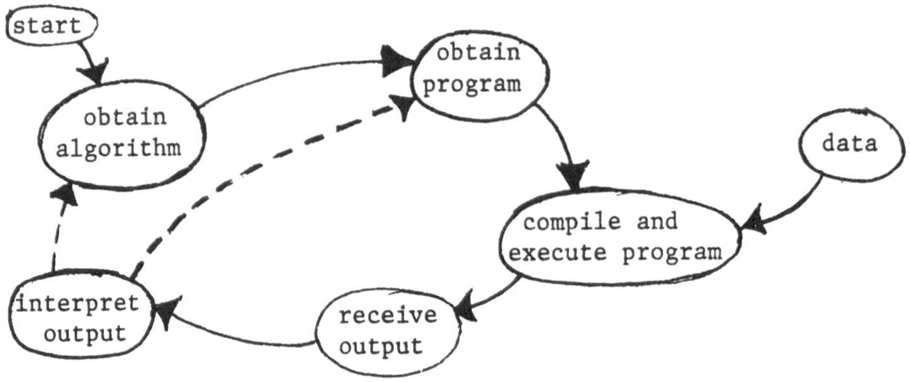

Figure 1 The Problem-Solving Cycle

Today's interactive time-sharing operating systems permit man-machine communication during job preparation and execution and virtually immediate return of results. Facilities for error detection and correction are much improved. The overall result is that of a real man-machine dialogue, faster return of output and greater user satisfaction. Even with these advances, the scientist can hope for an expect further improvements, as the next section indicates.

II. PROVIDING A BETTER ENVIRONMENT FOR SCIENTIFIC COMPUTING

In spite of the progress alluded to in the previous section, there is much room for improvement in fulfilling the potential of the computer as an aid to the scientist. The following subsections address some of the areas where progress is both possible and likely to yield significant benefits.

II.A. Computer-Based Local and Long-Distance Networks

Some networks specifically for the use of the scientific community are in existence. They are expanding as others are being created. Along with the physical capability to communicate with remote scientific installations, there is a need for cooperation between scientists to make known and provide the means for use of locally-written software which may be useful at other sites.

II.B. Hardware and Software Tools

1. Video Display Technology. A drawback of the conventional video display terminal is that the user can view material related to only one task at a time. Display technology now permits the division of the screen into seversl "windows" whose size and location are controlled by the user. Each window acts in effect as an independent

terminal, displaying the status of a different file or computation. For example, one window can be used for data entry while a second follows a running computation and a third displays an electronic mail message just received. Parts of the display can be overlaid temporarily by a newly created window, then reappear.

The capability to generate specialized scientific symbols and represent complex geometric objects has been available only on specialized graphics terminals, but is now being incorporated into high-volume products by manufacturers.

2. Direct Input. Direct means of computer input such as voice and hand movements promise significance gains in convenience. Computer recognition of speech, although progress has been relatively slow, is now feasible on a limited basis. It requires significant expenditure of processing power and memory. Hand movements, by contrast, can be processed at very limited expense. One popular approach is to control the screen cursor via a hand-held "mouse". Moving the mouse on a flat surface induces corresponding movement of the cursor on the screen.

3. Software. Programming languages close to standard scientific prose could ease the programmer's task, but these may be long in coming, since progress in language development proceeds more slowly than in computer hardware. FORTRAN, a language of the 1950's still predominates in scientific applications in the U.S.A. One contributing factor is surely the cost of converting programmers to another language and making new programs in a new language compatible with older software. But, it may be argued that no newer language offers significant enough benefits to induce scientific users to make a change.

Many computer scientists view the current state of computer languages as one of the several most significant obstacles to further progress in computer applications. Some seek a link between languages and long-established patterns of computer design. Section IV of this paper returns to this matter.

Operating environments, the system software which provides the user with access to files, editors, programming languages and the other resources of his computer, have improved notably in recent years. Both productivity and ease of use have been enhanced.

III. ACHIEVING FASTER COMPUTATION

III.A. The Need for Faster Computation

While progress in computer hardware since the advent of the digital computer has been dramatic, scientific and engineering users have an insatiable appetite for computing resources, especially instruction-processing speed and high-speed memory. Signal-processing (including voice recognition), explosives research, computer-assisted design and weather forecasting are among the applications demanding more computing power. Some applications, or their widespread use, are only feasible when more speed at less expense becomes available.

III.B. Historical Trends

A standard measure of processor speed is the number of floating point arithemetic operations performed per second (FLOPS). Large computers of the early 1960's could perform about 10,000 FLOPS, while today's faster "super computers" perform in the range of 100 million FLOPS. Although the improvement is dramatic, it has been achieved primarily by means which do not promise indefinite future improvements of the same magnitude: decreasing the switching speeds of circuit components and reducing the length of connections between them. Consequently, computer scientists are now seeking to complement progress at the micro-level with advances at the level of system design and programming languages.

III.C Conventional Processor Designs

Current digital computer designs are, for the most part descended from designs of the 1940's for which John von Neumann, a mathematician of genius and broad interests, is given a major share of the credit. They feature common storage of instructions and data in memory, which is organized into strings of binary digits of uniform length, usually a power of 2:8, 16, 32 or 64. Each memory location has an address that can be included as part of an instruction that reads from or writes the contents of the memory location. Program instructions are stored in succeeding memory locations and fetched, one by one, by the CPU (central processing unit) for execution. This one-by-one execution of instructions is termed "serial" to distinguish it from simultaneous execution of two or more instructions, which is "parallel" or "concurrent".

In the conventional von Neumann computer design or architecture a data pathway runs between the high-speed memory and the CPU, where instructions are interpreted and executed. This pathway carries heavy traffic. Instructions travel from memory to the CPU. The instructions typically operate on data which must also be fetched from memory to the CPU. Frequently, the instruction contains not the address of the data, but the address where the address of the data may be obtained (indirect addressing). Finally, the result of execution of an instruction must often be returned to memory. All parts of the computer system which are not involved in executing the current instruction remain idle long enough for its several data transfers to take place. This undesirable situation has led to efforts to find computer architectures permitting more at one time than the execution of a single instruction in a single central processing unit. The next several sections survey the characteristics of some of these architectures.

III.D. Concurrent Operation of Computer Subsystems

One approach to improving upon the strictly sequential execution of instructions is to allow a variety of devices to operate concurrently by freeing the CPU from the necessity of continuous supervision of them. For example, specialized communications computers often handle the routine aspects of communication between a large host computer and remote terminals. Another example is provided by "direct memory access", under which high-speed data transfer can take place between memory and peripheral devices. The CPU initiates the transfer and is notified of its completion. In between, it is free for other tasks.

III.E. Pipelining

In this approach to speeding computation, the execution of consecutive instructions is overlapping. A second, Third, or even fourth instruction may be initiated before the first is completely executed. Some large computers now utilize pipelining.

III.F. The Potential for Exploiting Parallelism

A still largely unexploited avenue for increasing processing speed is based on the observation that parts of many computations may proceed independently, of, and simultaneously with other parts. A simple example is that of the addition of two n by n arrays $[a_{ij}]$ and $[b_{ij}]$ of floating point numbers. Suppose that the result is $[c_{ij}]$, with each element c_{ij} equal to $a_{ij} + b_{ij}$. In principle, all of these n^2 additions could be carried out simultaneously in an elapsed time independent of the value n. To accomplish this, n^2 computing elements capable of addition would suffice, since no one of the additions requires the results of any other. In ordinary scientific discourse, we say "Let $[c_{ij}] = [a_{ij}] + [b_{ij}]$." to indicate this operation, thus making no explicit commitments to serial or parallel execution. However, most existing computers compute $[c_{ij}]$ in serial fashion. Conventional programming languages, with some exceptions, typically reflect this operational characteristic by requiring the programmer to specify the serial order of the additions, as in

```
for I=1 to n do
    for J=1 to n do
        C(I,J):=A(I,J)+B(I,J)
    end
end
```

The difficulty with this approach is not only the undersirable complexity of the programmer's code as it compares with "Let $[c_{ij}] = [a_{ij}] + [b_{ij}]$." but also the slowness of serial as compared with parallel execution. There is now a trend toward the design of specialized "array processors or "vector processors" with multiple pathways for simultaneous execution of arithmetic instructions. The CRAY-1 supercomputer is of this type. Some other array processors are smaller, modestly priced add-ons designed to enhance the performance of conventional computers.

Array processors represent a relatively conservative departure from the conventional von Neumann design and fail to exploit the full potential of parallelism, which involves the possible concurrent execution of many instructions of different types. It remains to be seen what performance benefits can be realized in practice from this. If n instructions can be concurrently executed, then a list of size $m < n$ can be sorted in time proportional to m. This compares with a theoretical best of $m \log_2 m$ for serial execution of a sorting algorithm. The parallel algorithm improves on the serial one by a factor proportional to $\log_2 m$, a significant difference if it could be realized in practice.

The next section gives a sketchy account of some approaches to building machines capable of fully exploiting the potential for parallelism.

III.G. Radical Innovations in Processor Design

The conventional computer employs an instruction counter, which contains a memory address, to determine the location of the next instruction to be executed. One innovative processor design, the "data flow computer"[1], determines which instructions will next be executed in a different way. As soon as the data for an instruction is made available, that instruction with its data is put in line for execution by any of the computer's network of processing elements which happens to be idle. The result is then routed to wherever it is needed. By this means, many processing elements can potentially be in simultaneous operation.

Another innovative approach, known as the "reduction computer," executes an instruction only when its result is called for by another waiting to execute. This architecture may be well suited to the "functional" or "applicative" languages which may represent the next wave of progress in programming languages [2]. These languages are based on the twin concepts of function and recursion. They seem not to execute efficiently in conventional computers, even though related languages, such as LISP, are in widespread use.

IV. COMPUTATIONAL MODELLING AND SIMULATION

The methods of classical analysis were developed long before the appearance of computers. They provide many methods for extracting approximate results without extensive computation. However, the need to avoid large scale computation is less and less compelling. Computation - intensive models for physical processes with more precision than conventional models based on classical methods are now feasible in many applications.

REFERENCES

1. P.C. Treleaven et al., "Data Driven and Demand Driven Computer Architectures," ACM Computing Surveys, Vol. 14, No. 1, (1982).

2. J. Backus, "Can Programming be Liberated from the von Neumann Style? A Functional Style and its Algebra of Programs," Comm. ACM, Vol. 21, No. 8, pp. 613-641 (1978).

PHOTOCONDUCTIVITY OF INDIUM IN SILICON

R. Lindner

Universität Erlangen-Nürnberg
Institut für Angewandte Physik
Gluckstrasse 9
D-8250 Erlangen, FRG

ABSTRACT

R. Lindner reported on the photoconductivity spectra obtained for the system Si:In in the near IR-region. The role of In as a shallow acceptor was investigated at a given concentration ($10^{17}/cm^3$) and at various temperatures. The experimental technique was explained and a comparison was done between the photoconductivity-spectrum and the absorption of the Si:In.

NONLINEAR ENERGY TRANSFER IN SEMICONDUCTORS YIELDING OPTICAL BISTABILITY

K. Bohnert

Physikalisches Institut der
Johann Wolfgang Goethe-Universität
Robert-Mayer Strasse 2-4
6000 Frankfurt am Main, FRG

ABSTRACT

Nonlinear energy transfer from a high intensity photon field to the electronic system of a semiconductor resulting in an absorptive optical bistability was investigated. The high intensity photon field creates a high density of free carriers forming an electron hole plasma. In the plasma density-dependent renormalisations occur: the band gap is strongly shifted to lower energies giving rise to excitation dependent indicies of refraction and absorption. K. Bohnert explained how these effects were applied for a realization of absorptive optical bistability. For this purpose high intensity nanosecond laser pulses in the spectral range below the excitonic resonances were transmitted through a thin CdS sample. During the temporal development of the pulse the band gap shifts from its original position to energies below the laser energy and back again to its original value. When it passes the laser energy there are transitions from two to one photon absorption and from one to two photon absorption when the gap goes downward and upward, respectively. These transitions show up as jumps of the transmitted intensity. Because they occur at different intensities of the incident laser light, they give rise to an absorptive optical bistability. The switching times are in the subnanosecond range. The dependence of the bistability on photon energy was investigated and discussed.

LUMINESCENCE AND ENERGY TRANSFER IN YAlG:Nd,Ce

J. Mares

Institute of Physics
Czechoslovak Academy of Sciences
Na Slovence 2
180 40 Prague 8, Czechoslovakia

ABSTRACT

Selective laser excitation was used by J. Mares for studying the visible and near IR luminescence of YAlG:Nd, Ce. The Nd^{3+} luminescence at 1.02 μ can be excited by pumping selectively either into the Nd^{3+} absorption bands or into the broad blue Ce^{3+} absorption band where no Nd^{3+} absorption occurs. In addition to this, Nd^{3+} absorption structures were observed in the Ce^{3+} luminescence spectrum in the region 480 to 700 nm. These observations can be interpreted as evidence of a radiative energy transfer $Ce^{3+} \rightarrow Nd^{3+}$. Another possible energy transfer mechanism could be through the Ce^{3+} excited states: $Ce^{3+} \rightarrow Ce^{3+} \rightarrow Nd^{3+}$.

ENERGY TRANSFER EFFECTS IN $NaEuTiO_4$

P.A.M. Berdowski

Rijksuniversiteit Utrecht
Fysisch Laboratorium
Princetonplein 5, Postbus 80.000
3508 TA Utrecht, The Netherlands

ABSTRACT

Energy migration among Eu^{3+} ions has been observed in a lot of concentrated Eu systems. Transfer at low temperatures (~1.2K) has never been observed for non-glasses. This has always been explained by the fact that at low temperatures resonant multipole-multipole transfer could only occur via $^7F_0 \rightarrow {}^5D_0$ transitions. In most compounds that $^5D_0 \leftrightarrows {}^7F_0$ transition is strictly forbidden, so that energy migration does not occur. P. Berdowski investigated the energy transfer properties of a compound, $NaEuTiO_4$, in which this $^5D_0 \rightarrow {}^7F_0$ transition is no longer strictly forbidden, due to the influence of a linear crystal field at the Eu^{3+} site. Both the concentration dependence of the intensity and decay-time measurements show that in this system transfer occurs, even at the lowest temperatures. Best decay-fits can be made by assuming a 2-dimensional diffusion process. This can be explained by the plane structure of $NaEuTiO_4$. Diffusion increases rapidly in the temperature region between 1.2 and 40 K. This is probably due to the phonon assistance which is required for the migration.

ENERGY TRANSFER IN ANTIFERROMAGNETIC ALKALI MANGANESE HALIDE CRYSTALS

U. Kambli

Institut für Anorganische Chemie, Universität Bern
Freiestrasse 3
CH-3000 Bern 9, Switzerland

ABSTRACT

Energy transfer has been investigated in pure and Er^{3+} and Nd^{3+} doped crystals of $CsMnBr_3$ Rb_2MnCl_4, $RbMnCl_3$ and $CsMnCl_3$. These compounds adopt different structures, in which the manganese ions are coupled one-, two-or three-dimensionally. Evidence of efficient energy transfer is given by the intense Lanthanide luminescence on excitation into the manganese absorption bands, the time dependence of the Lanthanide emission as well as the shorter decay times of the manganese emission in the doped samples. Quantitative information on transfer rates is obtained from the temperature and concentration dependence of the decay times. In all compounds the energy transfer within the manganese hosts is a thermally activated process. The activation energies however differ markedly and they are related to the splitting of the manganese $^4T_{1g}$ state due to an axial crystal field. It has not been possible to establish the low dimensional character of the energy transfer in both compounds with strongly anisotropic crystal structures, $CsMnBr_3$ and Rb_2MnCl_4.

AUGER EFFECT DUE TO SHALLOW DONORS IN CdF_2:Mn LUMINESCENCE

A. Suchocki

Institute of Physics, Polish Academy of Sciences
Al. Lotnokow 32/46, Pl-02-668, Warsaw, Poland

ABSTRACT

In conducting crystals doped with emitting localized impurities the Auger effect is a dominant luminescence quenching mechanism. At low temperatures, only the Auger effect due to the weakly bound electrons is expected to be effective. This communication reports for the first time the analysis of a mechanism of such a process observed in CdF_2:Mn, Y crystals. Due to a statistical distribution of distances between Mn emitters and occupied Y donors acting as quenchers, the luminescence kinetics is strongly nonexponential. The degree of nonexponentiality, as well as the relative luminescence yield depend on the donor concentration at the constant Mn doping level. Among the various mechanisms of energy transfer involved only the dipole-dipole /DD/ or exchange mechanisms /EX/ were found to be efficient. For the DD mechanism the characteristic parameter is the critical radius R_o. Its value estimated from the experiment was about 32Å, while its upper bound estimated value from the absorption spectra of Mn^{2+} and shallow Y donors was only about 25 Å yielding approximately an order of magnitude small quenching efficiency as compared to the observed one. The alternative mechanism is the exchange interaction between the excited Mn^{2+} ions and shallow donors. Here the effect is exponentially dependent on the distance between Mn and donor impurities scaled by the effective Bohr radius L of the more extended wave-function in the interacting pair. Its value estimated from the experiment was between 6 and 10Å in accordance with the known value of the shallow donor Bohr radius in CdF_2 equal to about 7Å. This allows us to conclude that the dominant mechanism of the Auger effect due to shallow donors in CdF_2 is the exchange mechanism.

ENERGY TRANSFER PROCESSES IN ZnSe:Ni, Fe

A. Karipidou

Department of Physics
Boston College
Chestnut Hill, Massachusetts 02167

ABSTRACT

A photoluminescence band exhibiting fine structure with no-phonon line at 11178 cm^{-1} was recorded in ZnSe:Ni crystals. In view of the coincidence of this no-phonon line with the corresponding no-phonon line of the absorption, this luminescence was interpreted as a $^3T_1$ (P) → $^3T_1$ (F) transition of the Ni^{2+} ions. Coincidence between absorption and emission exists also for the $^3T_1$ → $^3T_1$ (F) transition, indicating that T_2 is the lowest level of the $^3T_2$ (F) term. Iron doped ZnSe shows a luminescence band with two no-phonon lines at 11154 and 11128 cm^{-1}. In analogy to similar results in ZnS:Ni (1) this band was interpreted as due to the $^3T_1$ (H) → ^5E(D) transition of the Fe^{2+} ions. In the case of the Fe doped ZnSe a quenching of the Ni^{2+} luminescence was observed even at very low Fe concentrations (Fe:6 ppm; Ni:25 ppm). Because of the frequency overlapping of the two transitions [$^3T_1$ (F) → $^3T_1$ (P) in Ni^{2+}: 11178 cm^{-1}; $^3T_1$ (H) → ^5E(D) in Fe^{2+}: 11154 cm^{-1}] at energy transfer mechanism was suggested as a possible explanation for the quenching. Further investigations concerning selective excitation and lifetime measurements are goind to follow.

REFERENCE

(1) M. Skowronski and Z. Liro, J. Luminescence 24/25, 253 (1981).

ON THE ROLE OF NONLOCALIZED EXCITATION MECHANISMS IN THE GENERATION OF RED Er^{3+} EMISSION IN CdF_2:Er,Yb

A. Stapor

Institute of Physics
Polish Academy of Sciences
Al. Lotnikow 32/46
PL-02-668 Warsaw, Poland

ABSTRACT

Several models have been proposed to explain the origin of the red Er^{3+} luminescence in the infrared to visible upconverting crystals doped with Er and Yb ions. All but one postulate that population of the red emitting $^4F_{9/2}$ Er^{3+} state is due to different nonradiative relaxation mechanisms of the green emitting $^4S_{3/2}$ Er^{3+} state. To clarify this problem, the red to green emission intensity ratio and temporal evolutions of the $^4F_{9/2}$ and $^4S_{3/2}$ states were investigated in CdF_2:Er,Yb crystals at 77 K. The green 514.5 nm argon laser line was used as an excitation source. The rate equations were examined for each model and their solutions were compared with the experimental results. It was shown that localized processes, such as multiphonon relaxation and cross-relaxation. play an important role only for low excitation density or low Yb^{3+} concentration. In the other cases the spatial energy diffusion among Yb^{3+} ions should be considered and nonlocalized mechanisms dominate, giving up to 98% of red Er^{3+} emission intensity under certain conditions. Their relative probabilities were discussed.

THE GENERAL THREE-DIMENSIONAL HAKEN-STROBL MODEL

I. Rips

Department of Chemistry
Tel-Aviv University
Ramat-Aviv, Tel-Aviv, Israel

ABSTRACT

The general three-dimensional Haken Strobl model was presented by I. Rips. The second and fourth order moments of the displacement were calculated as well as the memory functions in the generalized master equation. It was shown that in the case of localized initial conditions the average energy of the particle is both time- and momentum independent. A solution of the model under arbitrary initial conditions was derived. It was also shown that the system possesses the equilibrium state corresponding to completely localized particle. The time period during which the system reaches the equilibrium state was shown to coincide with the period of time after which the diffusion sets in. This time period is given by inverse of the phenomenological parameter 1' describing the correlative properties of the model. Finally, the basic physical assumptions of the model and the range of its applicability were discussed.

THE LUMINESCENCE SPECTRUM OF UO_2MoO_4

W.M.A. Smit

Rijksuniversiteit Utrecht
Fysisch Laboratorium
Princetonplein 5, Postbus 80.000
3508 TA Utrecht, The Netherlands

ABSTRACT

The luminescence spectrum of powdered UO_2MoO_4 has been studied from 4.2 to 150 K. Using pulse excitation the intrinsic luminescence spectrum can be obtained at 4.2K at a very short time after the pulse. At longer time after the pulse trap emission is showing up, indicating energy transfer among the intrinsic uranium centers. By raising the temperature trap emission disappear one after the other. Above ~ 70K only intrinsic emission can be observed. Decay times of the traps range from 60 -90 μs at 4.2 K, whereas the intrinsic decay shows a complicated behaviour, starting with decay times from 0.5-3 μs and possibly ending with a linear part of 72 μs. The trap depth obtained from the τ vs T curves of the two most prominent traps is in full accordance with the trap depth obtained from the luminescence spectrum. The excitation spectrum at 4.2 K shows a fourfold splitting of the zero-phonon line due to correlation field splitting. The intrinsic luminescence spectrum shows strong vibronic lines mainly representing progressions associated with the internal vibrations of the uranyl group. Also the different traps show vibronic lines: for the traps the ratio of the vibronic intensity to the zero phonon line intensity is smaller than for the intrinsic centers revealing the lower symmetry of the trap sites.

CONTRIBUTORS

F. Auzel, Centre National d'Etudes des Télécommunications, Bagneux, France
R.L. Bateman, General Electric Company, Cleveland, Ohio, U.S.A.
P.A.M. Berdowski, Rijksuniversiteit Utrecht, Utrecht, The Netherlands
J.E. Bernard, University of Delaware, Newark, DE, U.S.A.
D.E. Berry, University of Delaware, Newark, DE, U.S.A.
G. Blasse, Rijksuniversiteit Utrecht, Utrecht, The Netherlands
A. Blumen, Technische Universität München, Garching, F.R. of Germany
K. Bohnert, J.W. Goethe Universität, Frankfurt am Main, F.R. of Germany
G. Boulon, Université de Lyon I, Villeurbanne, France
B. Di Bartolo, Boston College, Chestnut Hill, MA, U.S.A.
E. Duval, Université de Lyon I, Villeurbanne, France
M. Ferrari, Università degli Studi di Trento, Povo (Trento), Italy
L. Gonzo, Università degli Studi di Trento, Povo (Trento), Italy
G.F. Imbusch, University College, Galway, Ireland
J. Jortner, Tel Aviv University, Tel Aviv, Israel
U. Kambli, Universität Bern, Bern, Switzerland
A. Karipidou, Boston College, Chestnut Hill, MA, U.S.A.
V.M. Kenkre, University of Rochester, Rochester, NY, U.S.A.
J. Klafter, Exxon Research and Eng. Co., Annandale, N.J., U.S.A.
C. Klingshirn, J.W. Goethe Universität, Frankfurt am Main, F.R. of Germany
E.E. Koch, Deutsches Elektronen Synchrotron, Hamburg, F.F. of Germany
C. Landraitis, Boston College, Chestnut Hill, MA, U.S.A.
P. Landsberg, University of Southampton, Southampton, U.K.
R. Lindner, Universität Erlangen-Nürnberg, Erlangen, F.R. of Germany
J. Mares, Czechoslovak Academy of Sciences, Prague, Czechoslovakia
M. Montagna, Università degli Studi di Trento, Povo (Trento), Italy
A. Monteil, Université de Lyon I, Villeurbanne, France
A. Nozik, S.E.R.I., Golden, Colorado, U.S.A.
O. Pilla, Universitá degli Studi di Trento, Povo (Trento), Italy

R.C. Powell, Oklahoma State University, Stillwater, OK, U.S.A.
R. Reisfeld, The Hebrew University of Jerusalem, Jerusalem, Israel
E. Rips, Tel-Aviv University, Ramat-Aviv, Tel-Aviv, Israel
N. Schwentner, Freie Universität Berlin, Berlin, F.R. of Germany
W.M.A. Smit, Rijksuniversiteit Utrecht, Utrecht, The Netherlands
A. Stapor, Polish Academy of Sciences, Warsaw, Poland
A. Suchocki, Polish Academy of Sciences, Warsaw, Poland
G. Viliani, Universitá degli Studi di Trento, Povo (Trento), Italy
F. Williams, University of Delaware, Newark, DE, U.S.A.

INDEX